T0332036

Eaches or Pieces Order Fulfillment, Design, and Operations Handbook

Series on Resource Management

Eaches or Pieces Order Fulfillment, Design, and Operations Handbook

David E. Mulcahy

Series on Resource Management

Auerbach Publications
Taylor & Francis Group
Boca Raton New York

Auerbach Publications is an imprint of the
Taylor & Francis Group, an informa business

Auerbach Publications
Taylor & Francis Group
6000 Broken Sound Parkway NW, Suite 300
Boca Raton, FL 33487-2742

© 2007 by Taylor & Francis Group, LLC
Auerbach is an imprint of Taylor & Francis Group, an Informa business

International Standard Book Number-10: 0-8493-3522-1 (Hardcover)
International Standard Book Number-13: 978-0-8493-3522-8 (Hardcover)

Library of Congress Cataloging-in-Publication Data

Mulcahy, David E.
    Eaches or pieces order fulfillment, design, and operations handbook / David E. Mulcahy.
        p. cm.
    Includes index.
    ISBN 0-8493-3522-1
    1. Physical distribution of goods--Management--Handbooks, manuals, etc. 2. Warehouses--Management--Handbooks, manuals, etc. 3. Order picking systems--Handbooks, manuals, etc. 4. Business logistics--Management--Handbooks, manuals, etc. I. Title.

HF5415.7.M846 2006
658.7'88--dc22   2006040670

Visit the Taylor & Francis Web site at
http://www.taylorandfrancis.com

and the Auerbach Web site at
http://www.auerbach-publications.com

Preface

The objective in writing this practitioner (practical how-to-do-it) book is to provide a book that contains insights and tips for warehouse, distribution or logistics professionals to make their eaches or pieces order-fulfillment operation more efficient and cost effective and to provide on-time and accurate customer service to their e-mail, catalog, direct market or TV market customers. Each chapter focuses on a particular order-fulfillment operation activity or concept and gives the reader a quick and easy reference. These chapters cover equipment applications, concepts and implementation practices whether your operation is large, medium or small. The book contains illustrations, forms and tables that assist in developing your order-fulfillment operation to (1) reduce unit of product (UOP) damage, (2) enhance UOP and customer order (CO) flows, (3) increase employee productivity, (4) improve customer service, (5) reduce operating costs and improve profits, (6) maintain on-schedule deliveries and (7) assure asset protection.

The purpose of this book is to help develop the skill and knowledge of its readers to design, organize and operate an order-fulfillment operation or concept.

To assist in this objective, line art illustrations and sketches are used to visually present a piece of equipment or material handling concept. Because the profession of order-fulfillment operations management, material handling concept design and logistics is constantly changing, this book may not include the latest changes in state-of-the-art references to new technologies, equipment applications or material handling concepts.

It is also necessary to recognize that this book cannot cover all the available equipment applications, technologies and material handling concepts in the field of eaches and pieces warehouse distribution and order-fulfillment operations, but this book assists in training and will help the reader obtain practical experience, which has no substitute.

It is important for the reader to use the collection of data, concepts and forms as a guide. Prior to the purchase and installation of your new order-fulfillment concept or equipment, it is essential that you develop and project correct, accurate and adequate facility, inventory, stock-keeping units (SKU), transaction data and design factors. Because these are the design bases for your proposed order-fulfillment equipment application or facility, it is prudent for you to gather and review vendor literature and to visit existing facilities that utilize the order-fulfillment concept or equipment application. These activities will familiarize you with the operational characteristics of the order-fulfillment concept or equipment application to be implemented in your facility. The concept and performance specifications and physical design and installation characteristics are subject to redesign, improvement and modification. They are also required to meet vendor and local governmental standards and specifications.

Each chapter of this book deals with key aspects and issues of planning and managing an order-fulfillment operation. Some of these issues are: how your facility layout and UOP location affect your employee productivity; when to use the 80/20 rule and where to locate your power SKUs; how to route your order pickers and organize their work for the best productivity; how to determine what is the best small item, hanging garment, carton and pallet load pick concept; how to control the batch release; what is required for an across-the-dock operation; what are the most efficient and cost effective small item, carton and pallet load pick concepts for the business; how to the identify the pick position; how to control pick position replenishment and decide what is the best sortation concept for your business.

Most logistics professionals have learned from experiences with a preplanned and organized order-fulfillment operation—they have increased accurate and on-time deliveries, reduced costs and improved profits. By creating and maintaining an order-fulfillment concept as outlined in this book, your existing order-fulfillment operation will be improved and you will learn useful strategies for your next warehouse or plant facility.

The author would like to express his thanks to all material handling, warehouse and distribution and logistics professionals with whom he has had an association, whether at various companies, as fellow managers, as a client, as a speaker at seminars or as publishers.

About the Author

David E. Mulcahy was a project manager with the QVC Corporation International Logistics Group. He received his MBA from the University of Dallas, Texas. With the AMWAY Corporation, Mr. Mulcahy participated as a project manager for the design, building, installation and start-up for order-fulfillment operations in Japan, Korea, Taiwan, New Zealand, Australia, China, the UK and Italy that included pick-to-light and wire-guided VNA storage areas and order-fulfillment operation remodeling in warehouses in Germany, Spain, the Netherlands, Mexico and Central Europe.

As a QVC project manager he was involved with the QVC Germany existing order-fulfillment operation remodel with a remote reserve storage and a new facility and order-fulfillment concept installation. The new operation equipment included pick-to-light, pallet and carton AS/RS, warehouse management system and an extensive conveyor network. He retrofitted the QVC Japan existing operation with remote storage and is involved with a new facility design that includes pick-to-light, sort to light, small carton and filler material insert into a CO pick container, pallet and carton AS/RS, WMS and extensive conveyor network that has direct pallet and carton AS/RS transfer to pick areas on elevated floors.

Mr. Mulcahy has been a speaker at various conferences, and is a contributing author to many journals. He is the author of *Warehouse Management Handbook* and *Materials Handling Management*.

In 1981 he designed the multilayer case selection concept that won the 1981 Materials Handling Institute award at the Material Handling Show.

Contents

Chapter 1

Order-Fulfillment Objectives and Their Impact on Company Profit and Customer Service

Introduction

This chapter defines order-fulfillment objectives for a company and its customers, whether the company is an e-commerce, catalog, or direct-mail company, or a retail store or manufacturing plant. The chapter objectives are to:

- Review order-fulfillment activities
- Provide insights, tips and techniques for the development, design and control of a storage operation and warehouse management system (WMS), including facility construction and equipment installation
- Review methods for projecting future volumes
- Review methods for controlling annual operations expenses

Overview

This chapter overview serves to provide the reader with a review of the major areas of focus for Chapter 1:

- Reviewing order-fulfillment activities
- Improving a distribution operation
- Explaining the details operating budget

- Discussing capital investments
- Describing the impact of productivity on your operation and productivity standards

Design of an Order-Fulfillment Operation

In the design of a small-item or flatwear apparel order-fulfillment operation, the order-fulfillment concept and building design are based on specific design parameters:

- Number of customer orders (COs), lines, and pieces per peak day
- CO mix and cube
- CO/delivery cycle time
- Radio frequency (RF) tag reader or bar code label location and line of sight for scanner
- Vendor-delivered master carton and pieces or units of product (UOPs) that have company SKU identification

If a change is made to one or more of the design parameters, it impacts each order-fulfillment operation activity. These activities include:

- Yard and vendor delivery truck control
- Receiving and quality assurance (QA) activities
- In-house transport
- Moving product into, and withdrawing product from, storage
- Pick and sort operations
- Check and pack operations
- Value-added activities
- Manifest, load, and ship operations
- Security, maintenance, sanitation, and customer returns and rework operations

Order-Fulfillment Activities

If a WMS includes a storage operation, whether manual, mechanized, or automatic, order-fulfillment activities for any size supply chain can be categorized as pre-storage, storage, or post-storage activities.

The major activity areas of a small-item or flatwear apparel order-fulfillment operation are receiving, in-house transport, storage, and pick, pack, and ship.

The order-fulfillment operation objective is to provide accurate and on-time CO service at the lowest possible cost per unit. Most order-fulfillment operations are defined as a customer or demand pull UOP flow. To satisfy this objective, the most important order-fulfillment activities are the pick, pack, and ship activities.

In a CO demand pull UOP flow, the major order-fulfillment operation activity characteristics are:

- A pre-determined time span for completing operations
- Large employee or stock-keeping unit (SKU) numbers
- Facility space and order-fulfillment and storage equipment

After a CO is received at an order-fulfillment operation, the order-fulfillment activity time span characteristics are:

- Longest time in the receiving and in-house transport and storage activities
- Shortest time span for the pick, pack, and ship activities

Order-fulfillment employee and SKU number characteristics are:

- Fewest employees and SKU numbers in the receiving activity
- Few employees but a large SKU number in the storage area
- Large employee and SKU numbers in the pick, pack, and ship activities

Order-fulfillment operation facility activity area size characteristics are:

- Greatest square feet area for the in-house transport and storage activities
- Medium square feet area for the pick, pack, and ship activities
- Smallest square feet area for the receiving area

Unit of Product Handling Characteristics

As UOPs flow through your small-item or flatwear apparel order-fulfillment operation, there is a good possibility that there is a change to the product characteristics. For a small-item order-fulfillment operation and the associated UOP flow patterns, include product changes receiving SKUs as master cartons or pallets and sending individual pieces and master cartons to a company customer.

Small-item or flatwear flow is defined as a store-and-hold UOP flow concept. With a store-and-hold UOP flow concept, WMS-identified small-item or flatwear SKUs in master cartons or totes are transferred through WMS-identified storage and pick positions. Per a CO, a WMS-identified small-item or flatwear master carton or garment on hanger (GOH), is withdrawn from a WMS storage position or pick position, sent to a pack station and the completed CO is sent to a CO dock staging area or is loaded directly onto a CO delivery truck.

In an order-fulfillment operation, WMS-identified small-items in master cartons are:

- Moved from the receiving area to the storage area
- Deposited in a WMS-identified storage position
- Withdrawn for a pick line set-up or replenishment to a WMS-identified pick position
- Moved from a WMS pick position to a pack station
- Moved from a pack station as a CO package through a manifest area and into a CO staging area or directly onto a CO delivery truck

Order-Fulfillment Operation Objective

The small-item or flatwear order-fulfillment operation objectives are to ensure that each SKU meets your company quality standards, and that the correct SKU is transferred from a storage position and in sufficient quantity at the appropriate time. A CO must be withdrawn in the right quantity, in the correct condition, on schedule, and then packaged in a labeled protective shipping container that is properly manifested and delivered to a CO delivery location. A successful order-fulfillment activity accomplishes this within the CO order and delivery cycle time and at the lowest possible operating cost per unit.

Order-Fulfillment Operation Activities

To achieve these order-fulfillment objectives, the facility and order-fulfillment equipment layout, as well as vendor delivery UOP and CO piece and information flows are designed so as to minimize piece handlings, ensure an efficient and cost-effective operation, assure accurate and on-time piece and information flow through the order-fulfillment supply chain, and operational transactions are completed to satisfy COs.

Order-fulfillment operation activities are categorized as:

- Pre-order pick activities, which include:
  - Vendor or customer delivery truck yard control
  - Vendor delivered piece unloading and palletizing
  - Receiving and SKU QA or quality and quantity check and as-required customer rework
  - Ensuring that a company identification is placed on each SKU

- For some small-items or flatwear to package, price tickets are labeled and placed on each SKU
  - In-house transport
  - Deposit in a storage position
  - Inventory control
  - Pick line set-up or replenishment
- Order pick activities, which include:
  - Information technology department CO entry and, for paper pick concepts, download to print pick documents or pick labels; for paper-less pick concepts, download to a pick area or automatic pick machine microcomputer
  - Tote, carton make-up, label, and optional pack list attachment or insertion into a pick container
  - Manual order pick or computer impulse to release a SKU from a pick position and sort
  - Replenishment to a depleted pick position
  - Transport to a pack station
  - Trash removal
- Post-order pick activities, which include:
  - Checking order for quality and quantity
  - Packing CO pack slip, SKUs, and fill voids into a shipping container
  - Sealing and labeling shipping container
  - Manifesting shipping container
  - Loading and shipping
  - Handling customer returns, such as out of season inventory and transfers
  - Maintenance, sanitation, and security

Yard Control

Yard control activity ensures that the appropriate vendor delivery truck or ocean-going container is spotted at the correct receiving dock and that a CO delivery truck is spotted at the correct shipping dock. Whenever possible, a receiving dock assigned to a vendor delivery truck will minimize your in-house UOP transport distance from the dock to a storage position or from a CO manifest area to a CO shipping dock door.

With some store-and-hold operations, yard control activity includes spotting vendor rail cars on your rail spur. A rail car spot on a rail spur is assigned to ensure the shortest internal transport distance from a rail car receiving dock to a storage position.

Unloading

With an order-fulfillment operation, the vendor delivery unloading activity involves unloading small-item master cartons, master cartons, or pallets from a vendor delivery vehicle onto your receiving dock staging area. After a quality and quantity check, the UOPs are transferred according to your internal transport procedure from the receiving area to the storage area. After a WMS-identified SKU travels on the in-house transport and arrives in the storage area, the SKU is placed into a WMS-identified storage position.

Vendor-Delivered Piece Quality and Quantity Check

In an order-fulfillment operation, your QA department assures that the vendor-delivered piece quality matches your company purchase order specifications. Per your QA department, the vendor-delivered UOPs are either unloaded and received or held for vendor rework or disposition.

Receiving

In the receiving activity, a company employee physically transfers a SKU from a vendor delivery vehicle, enters the SKU quantity into your WMS, the WMS identifies each UOP and transports the SKU from a receiving area to the storage area. Per your company procedures, ≠piece samples are taken to the QA department for inspection. During the QA activity, the vendor-delivered UOPs are WMS identified and moved from a receiving area to a WMS-identified storage position. In the storage position, these WMS-identified SKUs are placed on QA hold status until the QA department approves or rejects the samples. If the QA department approves the UOP quality, WMS-identified SKUs are available for an order-fulfillment pick activity. If the QA department rejects the piece quality, pieces are held in a storage area for disposition and are not available for order-fulfillment pick activity.

Ticketing

In some retail store order-fulfillment operations, a subactivity to a SKU identification activity is the individual SKU price ticket activity. An employee or machine places a price ticket onto each individual SKU or sale piece. This price ticket activity is very common in flatwear, small-item, or master carton (ready for retail sale) distribution operations.

In a price-ticketing activity, a mechanical printer prints the price tickets, which are labels that are glued, clipped, stitched, or hooked to the SKU or placed on the SKU exterior surface.

Packaging

A SKU packaging activity is an activity that has an employee place an individual SKU or a SKU group into a material handling or shipping container. The containers may be plastic bags, paper bags, chipboard boxes, or cardboard boxes such as for jewelry SKU placed into a display case. In an operation that receives flat-pack garments, transferring clothing from a vendor box to hangers is considered a packaging activity. Another packaging activity is to place an individual GOH piece into a plastic or paper bag.

In-House Horizontal or Vertical Transport

The in-house horizontal or vertical transport activity moves SKUs from the receiving area to a storage or pick area. Manual or mechanized transport concepts are used in this activity.

Storage

In an order-fulfillment operation, the UOP storage activity provides a distribution operation with a physical position to store a SKU. Each position has a WMS identification. When required for a CO, a SKU is transferred from a WMS-identified storage position and transferred to a WMS-identified pick position.

Deposit

In a deposit activity, an employee or computer-controlled machine places a WMS-identified UOP into a WMS-identified storage position. An accurate and on-schedule deposit activity completion ensures that the right SKU is in the proper place, in the proper quantity, in the correct condition, and at the correct time. The deposit activity permits a pick line set-up or replenishment activity to occur.

Inventory Control

The inventory control activity ensures that UOPs are transferred to the correct WMS storage position or pick position and in the correct quantity. Other WMS inventory control concerns are: (1) proper piece rotation, (2) accurate SKU counts, (3) minimal "stock outs" at a pick position, (4) tracking a UOP flow through an operation, and (5) verifying each storage and pick transaction completion.

Information Technology Customer Order Download

The IT department activity involves vendor UOP purchase order and CO receipt, CO entry processing, and sending customer-ordered piece quantity to a pick line printer, unloading dock printer, microcomputer, or automatic pick machine. The IT CO download activity ensures that the proper receiving documents, pick documents, self-adhesive labels, individual CO cube, suggested CO shipping carton, and SKUs are ready for the order pick activity, and that COs are properly sequenced and available for a paperless pick line start station or for traveling through an automatic pick machine.

Carton or Shipping Container or Captive Tote Make-Up

For a small-item, flatwear, or GOH pick fulfillment operation, a CO carton, container make-up, or captive tote make-up and entry to a pick line activity starts a CO order-fulfillment process. In a pick/pass or pick/sort pick fulfillment operation, an employee picks customer-ordered SKUs into a captive tote or shipping carton. In a batched CO pick, transport, sort, and pack operation, CO shipping carton make-up activity occurs at a pack station.

An employee or machine transfers a formed carton or CO shipping container onto a pick line, pick vehicle, or pack table, according to CD delivery requirements, company practices, SKU quantity, SKU cube characteristics, container size, and weight. A small-item one-piece corrugated carton has sealed bottom flaps for easy handling and transport over a conveyor travel path. If a two-piece carton is used on a pick line, a carton top piece is placed under the carton bottom piece or the top piece is added to the carton at the pick line end.

Captive pick container types are plastic, wood, or metal. Small-item pick containers have optional open tops, handles or hand grips, meshed or solid bottoms or sides, stacking or nesting features, and three sides with one open side.

Customer Order Pick

In a small-item or flatwear operation, a CO pick activity obtains a CO pick instruction from a printer or microcomputer. Per each CO or batched CO pick instruction, an employee or automatic pick machine transfers the proper SKU and SKU quantity from a pick position to a transport conveyor, into a container, onto a vehicle load-carrying surface, or into an employee's hands or apron.

Sort

In an order-fulfillment operation with batched COs, a customer-ordered and CO-picked SKU sort activity is the key to accurate and on-time pick and pack activities. The sort activity allows an order-fulfillment operation to handle a high CO and CO piece volume. In a batched order-fulfillment operation, the sort activity is the first post-pick activity. The batched CO concept means that a computer has grouped eight to ten customer-ordered SKUs in one pick wave. For each pick wave, a computer prints an exact pick label number for each SKU on any CO. A pick label has a CO discreet identification. When a single item CO pick activity is in the batched CO mode, the sort activity per a CO label or SKU identification separates each customer-ordered and CO-picked SKU from a CO-ordered, -picked, and CO-labeled SKU mix. The sort activity verifies that a SKU was withdrawn from a pick position and was transported to the sort and pack area. The sort activity components are:

- A human, machine, or human/machine readable bar code or symbology on each SKU exterior surface
- A SKU transport and sort travel path with a constant travel speed
- A communication network that involves a barcode scanner or symbology reader, tracking device, microcomputer, and divert device
- A divert lane with queue space

CO-ordered, -picked, and -labeled SKUs are singulated on a sort travel path and, with a proper gap between two SKUs, a barcode scanner or RF symbology reader reads the bar code or symbology. The barcode scanner or RF reader sends the SKU data over a communication network to a microcomputer. While maintaining a constant travel speed, a microcomputer and track device activates a divert device at the appropriate time to transfer a SKU from a sort travel path to a CO pack station or shipping container. Depending on whether the sort application is a CO piece sort concept or CO package sort concept, the sort location is a bin, container, carton, chute, or conveyor.

The sort options are:

- Manual
- Mechanized concepts that use one of the following
  - An active sorter
  - A passive sorter
  - An active-passive sorter
- Each SKU has a bar code, manual code, or RF tag and the SKU induction options are:
  - Manual
  - Automatic discrete induction

Pick Line Set-Up or Replenishment

In an order-fulfillment operation involving single-item or flatwear SKUs that have fixed pick positions or have new SKUs each day, SKU pick line set-up is a pre-pick activity and replenishment is a post-pick activity. SKU pick line set-up or replenishment ensures that the correct SKU is removed from a WMS-identified storage position on schedule and in the proper quantity, and is placed into a WMS-identified SKU pick position. In many operations with a WMS program, after a SKU quantity is scanned and physically transferred into a WMS-identified pick position, the microcomputer releases the CO pick instruction to the pick line.

SKU pick line set-up or replenishment steps are:

- Listing SKU or WMS pick positions that require replenishment
- Prioritizing SKU quantity withdrawal from a WMS-identified storage position
- Transfering a SKU quantity to the SKU pick position
- Verifing a replenishment transaction to a WMS-identified pick position
- Transferring a replenishment transaction to the inventory system for storage and pick position update

Pack

In an order-fulfillment operation, CO pack activity is a post-order pick activity. The pack activity objective is to ensure that customer-ordered, CO-picked, and -packed SKUs are protected from damage during delivery from the facility to a CO delivery location and that the SKU is received by the customer in satisfactory condition. In the distribution business, the CO package exterior and SKU condition inside the package is your customer's first impression of your company's order-fulfillment service. This is especially true for catalog, e-commerce, and direct mail businesses.

The pack activity functions are to verify a customer-ordered and CO-picked SKU quality and quantity, with filler material used to fill the voids in a CO shipping carton, add a CO pack slip and sales literature in the package, seal or close a shipping container bottom and top flaps or bag mouth, place the CO delivery address label in a shipping carton assigned location, and transfer a CO package onto take-away in-house transport equipment.

Customer Order Container Seal

At a small-item or GOH order-fulfillment operation, the CO pack activity ensures that the CO shipping carton or bag does not open during transport from the operation to the CO delivery address and that when the package is delivered to the CO delivery address the SKUs and CO pack slip are in the package.

The package seal concept is for plant delivery to a production line to pack multiple SKUs loosely in one large container or to pack an individual SKU or SKUs into an employee- or computer-suggested container, carton, or bag. A CO shipping container is sealed with a self-seal method, tape, plastic bands, or plastic wrap.

Customer Order Package Weight and Manifest

CO shipping carton or bag weight and manifest activity objectives are to ensure each CO package receives the correct transport fee or postage, to list each CO discreet identification number and weight, to send the package by the most cost-effective transport method, to have proper documentation in a CO shipping carton, to obtain the exact weight and manifest per a CO discreet identification, and to verify that a CO was completed by your operation.

CO shipping carton weight and manifest activities use a scale that shows the exact CO carton weight to verify that the actual carton weight and computer-projected weight match within a variance, and to ensure that a CO discreet identification is listed on a transport document. With a computer-projected method, a scale interface is direct to a host computer or computer. With a computer, the computer receives a CO discreet identification and projected weight from a host computer.

Customer Order Load and Ship

At an order-fulfillment operation, CO package loading and shipping activities ensure that your customers' shipping packages are transferred from a sortation travel path or pack station onto the CO delivery truck. Loading and shipping activity is a direct load activity or units are placed on carts or pallets for later loading onto the CO delivery truck.

Customer Order Returns, Out-of-Season, Transfers, and Vendor Rework

The CO return activity occurs in all industries. It is most evident (varying from 5 to 38 percent of the shipped volume) in the catalog, e-commerce, and direct mail industries. Your CO return activity ensures that your CO returned quantity was received at your facility, your customer received the appropriate credit, and the returned SKUs physically flow through your facility and are returned and entered into inventory, placed in a SKU pick position, sent to an outlet store, donated to charity, disposed into trash, or returned to the vendor.

Out-of-season SKU and transfer activities occur in the retail industry. The out-of-season activity is a distribution activity that holds, temporarily in a storage position, SKUs that did not sell at a retail store. With your company management approval, a retail outlet places a SKU into a package and returns the out-of-season

SKUs to the distribution facility for temporary storage. At a later date, when all stores have returned the out-of-season SKUs to the distribution facility, the company management decides how to handle them.

SKU transfers are overstock units that move from one retail store location with low sales, through the distribution or transportation network system, to another retail store location that has high SKU sales. With your company management approval and with proper documentation, the SKUs become a transfer package that flows from one retail store to another through your company distribution and transportation operations.

Vendor rework occurs after a vendor-delivered SKU is rejected by your QA department for a minor off-standard or specification situation. Vendor-delivered rework activity has approval from the vendor and merchandise department. After the rework activity, the vendor-reworked and -delivered SKUs are sent through the QA department and per the QA approval are received into inventory.

Maintenance, Sanitation, and Security

The remaining key order-fulfillment operational activities are the maintenance, sanitation, and security activities. The objectives of these activities are to protect your company assets and ensure that the inventory, building, and equipment are available to satisfy your customer orders and operate at the lowest possible cost per unit.

On-Schedule and Accurate Order-Fulfillment Operation Performance and Customer Order Delivery Activities Means Profits and Satisfied Customers

The effective and efficient order-fulfillment operation, from completion of pre-pick, pick, and pack activities, to CO delivery, ensures that a company's customers are satisfied with the best service. When activities are completed on schedule and at the lowest cost per unit, the SKU package, SKU, and documentation make a positive and lasting impression on a customer. This impression ensures that an order-fulfillment operation is profitable and has satisfied customers.

Why an Accurate, Efficient, and Cost-Effective Order-Fulfillment Operation Is Important

An accurate, efficient, and cost-effective order-fulfillment operation is important to a company because customer service quality is determined by a CO that is delivered on-time, accurate, and with no damaged SKUs and because order-fulfillment activities have a large employee number and occupy a large area.

Comparing order-fulfillment concepts indicate that:

- A manual operation has the largest employee number and uses a large area
- A mechanized operation has a medium employee number and uses a medium-sized area
- An automatic pick machine operation has the smallest employee number and uses a small area

The Standard for a Good Order-Fulfillment Operation

The standard for a good order-fulfillment operation is your company basic performance standard. This basic performance standard is common to a company of any size in any industry, whether it serves a retail, industrial, e-mail, direct mail, catalog, or personal consumer customer group.

The order-fulfillment operation basic performance standard is to provide the best customer service at the lowest possible cost per unit. Stated in the broadest terms, the order-fulfillment operation standard is to ensure that the right SKU is in the correct pick position, in sufficient quantity at the appropriate time, is withdrawn in the right quantity, in the correct condition, on schedule, as required with a packing list, and is properly manifested and delivered to the CO delivery location at the lowest possible operating cost per unit.

How to Improve Your Order-Fulfillment Operation

Whether your order-fulfillment operation has a manual, mechanized, or automated concept, potential ways to improve vendor delivery, customer service, and operating cost per unit include the following:

- Look at employee numbers for each activity
- Develop a vendor and customer service standard
- Develop employee manning or productivity schedules that are based on actual or best-estimated vendor delivery and CO pieces, lines, and order packages
- Consider pallet handling with dual-cycle activities
- Reduce travel time and distance
- Improve the pick aisle SKU hit activity concentration and hit density
- Use the ABC or power SKU allocation method
- Kit or family group SKUs
- Use clear, simple pick and storage transaction instructions with a WMS position and WMS UOP identification
- Use an employee routing sequence and clear transaction instruction
- Provide queue space

- Level the work volume over a week's work days
- Use part-time employees
- Change from a paper document concept to a self-adhesive label, RF paperless, or automatic concept
- Develop a good equipment layout, product flow, and facility layout
- Provide clean and clear aisles
- Assure well-lighted aisles
- Change from a human-paced to a machine-paced concept
- Use non-powered or powered equipment
- Use automatic identification in the operation
- Provide the employee with sufficient work each work day
- Assure an arithmetic progression in a pick aisle
- Keep a picker in a pick aisle the majority of the time

Combined Family Group and ABC

If a small-item order-fulfillment operation services a large number of customers or a retail shops with small sales aisles layout, a combined family group with an ABC arrangement should be considered for the pick area layout. To have a family group with an ABC layout means that the order-fulfillment operation aisles or pick zones mirror a retail store sales aisles. In this concept, CO-picked SKUs from the order-fulfillment operation are in a CO container or containers that are located in one retail store aisle.

The concept benefits are:

- At an order-fulfillment operation, high-volume or A-moving SKUs are located at each aisle entrance.
- At the retail store, there is high restock productivity and it is easy to control over stock.

Look at the Employee Numbers

To look at employee numbers means to identify order-fulfillment operation activities that have the highest employee number, labor cost plus fringe benefits, or operational cost factor. Other operational factors include overtime, piece, building or equipment damage, overs, shorts or errors, off-schedule CO deliveries, and employee injuries.

Employee productivity improvement in an order-fulfillment operation with a high employee number has greatest impact on your order-fulfillment operation in total employee productivity, annual operating expenses, and the ability to provide on-schedule and accurate customer service.

Develop a Standard

An order-fulfillment employee productivity standard is a figure that states order-fulfillment productivity as number of customer-ordered pieces, lines, or order packages completed vs. total employee work hours or costs. Order-fulfillment employee productivity is the key factor in projecting the labor expense line for the company's annual expense budget.

Pick/Pass or Batched Customer Orders and Dual-Cycle Activities

In an order-fulfillment operation, a pick/pass or batched CO pick activity and storage area dual-cycle activity is most effective in improving your order-fulfillment operation. The pick/pass concept employs a CO-ordered SKU quantity within a pick zone for an order picker to transfer a SKU from a pick position directly into a CO container or shipping carton and pass the CO carton to the next pick zone. In a pick-and-sort concept, batched COs mean that a computer groups together a specific CO number and prints an order pick instruction for these CO-ordered SKUs. The CO-initiated SKU print sequence is determined by the SKU pick position number on self-adhesive pick labels or sequenced by pick position lines on a paper pick document or downloaded to an RF pick device. This SKU pick position sequential print for picker instructions permits an order picker to pick the ordered SKU quantity for all COs with the batch or wave. Batched CO pick activity options are:

- With a paper pick document or RF device, to place a picked SKU immediately into a CO holding position.
- After a label pick activity, the picked and labeled SKU is placed into a discreet CO sorting location.
- After a label pick activity, picked and labeled SKUs are sent as a group to a sorting area. Here, the SKUs are separated to the assigned CO sort/pack/stage location.

The pick/pass and batched CO pick methods increase order picker productivity due to reduction in order picker travel distance and repetitive order pick activities. For best results, a pick/pass concept requires a pick to light concept with a conveyor pick line. When picked into a CO shipping carton, both pickers and packers have improved productivity. A batched CO concept requires a computer program, on-time orders, and a sorting concept that has space, time, and labor. The batched CO pick concept provides picker productivity improvements that substantially offset the sort costs.

In a carton or pallet storage area, a dual-cycle activity has the forklift truck or AS/RS crane:

- Enter a storage aisle, travel down the aisle to the microcomputer-assigned or employee-directed storage position, and perform the storage deposit transaction
- Continue traveling in the same storage aisle to another storage position, complete a withdrawal transaction, exit the storage aisle, and transfer the product onto a pick-up and delivery P/D station

A dual-cycle activity improves forklift truck or AS/RS crane productivity by approximately 20 percent over that of a single cycle activity. To achieve this productivity increase, storage areas have:

- Storage aisle pallet P/D stations. A P/D station provides an opportunity to balance the in and out pallet storage transactions per aisle.
- Forklift truck or AS/RS crane storage transactions are preprinted on a document, an RF device, or on-line.

Reduce Travel Time and Distance

Reducing your order picker travel time and distance between two pick positions improves order picker productivity. Small-item or flatwear order-fulfillment operations have a large SKU number that varies in size, shape, weight, and pick volume, and the order-fulfillment operation pick area has many aisles and pick positions. A picker walking through a pick aisle or along a pick line between two pick positions represents unproductive travel time and distance. In a small-item or flatwear order-fulfillment operation a dramatic order picker productivity increase results from decreasing travel time and distance between two pick positions.

Improve SKU Hit Concentration and Hit Density

Allocating a WMS-identified master carton or pallet SKUs according to Pareto's law (80/20), which states that 80% of the volume is derived from 20% of the SKUs, should improve SKU hit concentration and SKU hit density.

The SKU hit concentration means the (SKU) number that is ordered by your COs or CO lines (stops or hits) within a particular aisle or the WMS position number within one aisle that has SKUs for a CO. A SKU hit density is the SKU number (quantity) that a CO has for a particular SKU (one pick position) or hit concentration is the number (hit quantity) for one SKU to complete a CO line.

A high SKU hit concentration and hit density dramatically improve order picker productivity due to a reduction in travel time and distance between two positions. For best results in a single-item pick onto a conveyor or into a carton or tote order-fulfillment operation, each order picker is assigned to a pick line zone or specific pick aisles. Grouping slow-moving (low-hit) SKUs improves hit concentration and hit density, which increases employee productivity.

Use ABC or Power Allocation

When an order-fulfillment operation area layout is based on SKU popularity, it is based on Pareto's law (after Vilfredo Pareto, 1848–1923; Italian economist). This law states that 80 percent of the wealth is held by 20 percent of the people. In a small-item or flatwear order-fulfillment industry, this law indicates that 80 percent of the volume shipped to the customers is derived from 20 percent of the SKUs. Many studies have indicated that another 10 percent of the volume shipped to customers results from another 30 percent of the SKUs and that an additional 5 percent of the volume shipped to customers is attributed to 55 percent of the SKUs. In the catalog or direct-mail business with two to four catalogs introduced within a year, 90 percent to 95 percent of the business is from 5 percent of the SKUs. This is due to the fact that each catalog has a different inventory of SKUs. Recent studies show that 95 percent of the volume shipped to customers is obtained from 55 percent of the SKUs. This is referred to as "Pareto's law revisited."

The ABC Theory

When order-fulfillment professionals refer to the three zones of Pareto's law, their reference is to the ABC theory. The ABC theory states that an order-fulfillment operation has three zones:

- The A zone, which is allocated to the fast moving SKUs that are few in numbers and have a large inventory per SKU
- The B zone, which is allocated to slower moving SKUs that are medium in number but have a medium inventory per SKU
- The C zone, which is allocated to the slow moving SKUs that are large in number and have a small inventory per SKU

If an order-fulfillment operation layout has receiving and shipping docks located on a facility front side and a SKU pick position location that is based on the ABC theory, it locates the fast moving SKUs at a pick aisle front.

If an order-fulfillment operation has receiving and shipping docks located on opposite sides of a building the fast moving SKUs are located by its unloading and loading ratio. The unloading and loading ratio compares the trip number that unloading employees require for a delivery truck. When employee unloading trip number equals an employee loading trip number, a SKU pick position location is near the shipping docks or any location in a row. When employee unloading trips are more numerous than employee loading trips, a SKU pick position is located near the receiving docks. This feature reduces your in-house transport employee total travel distance and time.

The Golden Highway, or Power or Fast-Moving SKUs in One Pick Area

To have the power or fast-moving SKUs located in one pick aisle or zone is considered like a Golden Zone but due to several high-volume SKUs in one aisle, we refer to a pick aisle as a Golden Highway. This concept has an inventory (SKU) allocation program that locates all fast moving SKUs to pick positions that are within one pick aisle and pick positions that are adjacent to one another. This philosophy has all promotional, seasonal, special sale, and fast moving SKUs in a pick aisle or zone. This SKU arrangement increases your picker hit concentration (pick number per aisle or zone) and hit density (hits per SKU number). A high hit concentration and hit density means high order picker productivity due to a short travel distance between two pick positions. Key to an accurate, efficient, and on-time small-item or flatwear pick operation with power SKUs in a pick area concept is to have storage pieces available to replenish a pick position and to have a completed CO or picked SKU take-away transport concept.

The Kit or Family Group

A kit or family group pick layout is dictated by a company requirement that SKUs sent to a customer are by family group. With this guideline, SKUs are assigned to specific pick positions that are either one or multiple pick aisles or one or multiple pick zones within a pick aisle.

This pick aisle layout philosophy requires that the pick positions are designed to accommodate SKUs that have similar characteristics, such as similar dimensions, weight, and SKU material components; components; for the same final product, located in the same aisle in a retail store, require normal, refrigerated, or freezer conditions; contain high security, toxic or nontoxic material; one style or all sizes such as shoes. Other characteristics could be edible or non-edible, flammable or non-flammable, or stackable or non-stackable products.

Keep It Simple and Clear

The best picker instruction method is to keep pick employee instruction as simple as possible. This allows a picker to read the pick instruction, clearly understand an instruction, and complete a pick transaction. Pick instruction concepts include a:

- Paper document
- Self-adhesive label
- Bar code/RF tag or lighted display panel that is on a RF device and micro-computer control

Each pick instruction concept presents a pick instruction as alphabetic characters, number digits, or a combination and a bar code. Each pick position has a WMS identification. Each WMS-identified pick position identification is attached to a pick position structural support member. During the pick activity and when a picker matches the pick instruction to a WMS position identification this match serves as a signal to an order picker that the picker has arrived at the correct pick position. At this pick position a picker physically completes a pick transaction.

Sequential Pick Aisle Routing Patterns

In an order-fulfillment operation, a sequential picker routing pattern is the preferred pattern for improving order picker productivity. A sequential routing pattern has an arithmetic progression to the pick position numbers through a pick aisle. This means that the lowest SKU pick position number is 1 or 0 and is located at the entrance to a pick aisle. The highest pick position number is 99 or 100 (or greater) and is located at the exit to a pick aisle.

In a sequential routing pattern, a picker starts at the first required pick position in a pick aisle. As a picker travels down a pick aisle the next required pick position is as close as possible to the previous pick position. In an order-fulfillment operation, any sequential routing pattern provides your order-fulfillment operation with an efficient and cost-effective picker group.

Advantages include reduced non-productive travel time, reducing trips per aisle by two or more, which increases productivity and minimizes employee confusion.

Cube Your Pick Activity or Automate

If an order-fulfillment operation pick instruction method "cubes out" or divides each order picker or pick machine activity, the pick area computer program determines:

- Small-item, flatwear, or master carton number for each CO-ordered SKU
- Picker number per pick aisle
- Shipping carton size

- The SKU number that is released by an automatic pick machine for a ship-ping carton cube. This means that the cube program ensures that a picker is capable of handling the picked SKU number without additional trips in a pick aisle or along a pick line.

The other factors that determine the order pick cube or SKU quantity per CO carton are carton, human, or vehicle load-carrying surface and weight and ship-ping carton fill utilization, and delivery company maximum weight per shipping carton.

If an order-fulfillment operation has no match for the SKU to a human carry or carton capacity, there is a high probability that an order picker section for a CO exceeds the human or carton capacity. To complete a CO portion, this situation has a high potential to have an order picker make unnecessary, unproductive, or less than optimum travel trips between a pick aisle and a CO pack station, to cause additional order picker hand movements into a carton to rearrange SKUs inside the carton, or to cause an automatic pick machine to overfill a carton and create piece damage and pick errors.

Smooth or Level the Work Volume

In an order-fulfillment operation, leveling the work volume to be evenly allocated over a workday increases employee productivity. Level the week's work volume keeps the daily CO handling volume at relatively the same quantity for each work-day. This averaged daily work volume assures that your order-fulfillment operation employees have sufficient and productive work quantity for each workday. This means that there is no non-productive employee time to change jobs within the operation.

Use Part-Time Employees

Using part-time employees can improve employee productivity or cost per piece. If a small-item or flatwear order-fulfillment operation business experiences peak vol-ume periods, hire full-time employees for the average business volume. During the peak business volume conditions, instead of paying overtime for regular full-time employees, use part-time and lower-cost employees to complete the above-average work volume. This part-time employee approach enables you to:

- Maintain your labor cost per UOP due to the fact that part-time employees do not have the same fringe benefits as regular or full-time employees
- Not pay overtime premiums

■ Permit experienced full-time employees to perform key order-fulfillment operation activities and have part-time employees perform non-productive or less skilled order-fulfillment operation activities.

Reduce Customer Order Pick and Pack Spikes

To have a low cost per CO order-fulfillment operation, a small-item order- fulfillment operation minimizes the CO spike frequency number and magnitude. With a high demand pull or CO number, SKU mix and company customer service standard (customer order/delivery cycle time) a small-item order-fulfillment operation develops a game plan to reduce CO spikes. The opportunity factors are:

■ Having a sales program or merchandizing department that identifies promotional SKUs
■ Using past sales records to identify SKU piece volume sold as single COs and multi-line COs
■ SKU know inventory quantity
■ Suggested SKU single CO shipping carton size
■ Increase the SKU number as slapper label cartons

Within the operations opportunity area, the operation manager options are:

■ Complete-pack a SKU that requires a slapper label
■ Semi-complete-pack a carton with a SKU and bottom fill that requires top fill, CO pack slip, tape, and label

Sales Program Identified Promotional SKUs

A company sales program and merchandizing department identifies the SKUs that are expected to have high sales or are promotional SKUs for each work day. The merchandizing department and WMS program identify the promotional SKU piece or WMS-identified UOPs. The estimated SKU sales volume allows the order-fulfillment operations to determine the SKUs for each workday that are candidates for pre-pack activity. From the receiving dock schedule, ASN, purchase order, or WMS inventory files, the operations manager identifies the candidate SKUs that have UOPs in the inventory or the day that the UOPs are scheduled for vendor delivery.

The sales program identifies each candidate SKU product classification.

Past Sales Program SKU Piece Volume as Single and Multi-Line Customer Orders

Historical sales program records indicate a SKU piece volume that was sold as single-line COs and multi-line COs as a percentage of a SKU inventory piece quantity and SKU product classification. From SKU historical sales data, an operations manager has a PC or staff member project or forecast each candidate SKU single CO and multi-line CO volume. With projected pre-pack SKU volumes, an operations manager estimates labor hours that are required to pre-pack a candidate SKU quantity. Pre-pack labor hours are based on the budgeted productivity rates.

SKU WMS Identification UOP Inventory Quantity

A SKU WMS-identified UOP inventory quantity is the SKU piece quantity that is available for sales (quality department approved) and is in a WMS-identified storage position. For each candidate SKU, the operations manager estimates (based on historical SKU sales volume, such as 50 percent to 95 percent), a SKU piece quantity for a pre-pack activity. For each proposed pre-pack SKU, the operations manager determines the labor hours for each pre-pack activity.

Suggested SKU Single Customer Order Shipping Carton Size

To prepare a pre-pack activity, the manager verifies for a candidate SKU, each SKU-suggested shipping carton size. With the projected pre-pack volume, the operations manager reviews the suggested shipping carton inventory quantity. After verifying the required SKU shipping carton size, the manager is ready to set up a fast pack line.

How to Pre-Pack a Promotional SKU

After the operations manager has finalized the SKU and shipping carton inventory quantities, the pre-pack options are to pre-pack a SKU as a slapper label or semi-complete a CO that requires top fill, CO pack slip, tape, and label.

Pre-Pack Activity

With a slapper label used for a pre-pack promotional SKU, the operations manager can then:

- Select a fast pack line
- Schedule the WMS-identified UOP and shipping carton to be transferred from the storage area to an assigned fast pack line

- Decide to apply a new WMS identification to each pre-pack pallet
- Decide the WMS identification and SKU identification procedures
- Identify a WMS-identified UOP storage location

Select a Fast Pack Line and UOP and Shipping Carton Transfer

Selecting the fast pack line and UOP and shipping carton transfer are straightforward activities.

WMS Pre-Packed UOP Identification Concept

Prior to a pre-pack activity, an operations manager has options for WMS UOP identification:

- Complete a WMS-identified UOP transfer or move transaction and piece quantity depletion for the existing SKU WMS identification. This option has an employee deplete or throw a UOP WMS identification into the trash. An employee completes a piece count, applies a new WMS UOP identification to a pallet and enters a new SKU and WMS identification with the piece quantity into inventory. The option assures an accurate inventory transaction and employee transfer transaction instructions are simple.
- Reuse an existing WMS UOP identification that is applied to a pre-packed SKU pallet. This option has an employee transfer a WMS UOP identification from a pallet to a pre-pack pallet. This option has an employee verify that the UOP piece quantity on a pallet is equal to the old WMS UOP identification quantity. This option has the potential for low employee productivity, additional employee work, and potential SKU inventory imbalance.

SKU Identification Concepts

Available SKU identification options are:

- Keep the same SKU identification number for the pre-pack SKU. This is a simple employee activity, but requires the WMS program to allocate first the WMS-identified pre-pack UOPs for a single CO pack activity and after all single COs have been completed to allocate the non-pre-pack WMS-identified UOPs to multi-line COs.
- Change the SKU number, which requires a complete WMS UOP flow activity.

Pre-Pack SKU as a Slapper Label SKU

Pre-packing a promotional SKU as a slapper label package concept has a fast pack line pack a SKU with as-required filler material in a shipping carton. Each shipping carton is sealed and stacked onto a pallet. Each full pallet receives a new WMS UOP identification and an employee counts the SKU piece quantity.

The WMS UOP identification scan transaction and SKU piece quantity are sent to the WMS computer. The WMS computer associates the WMS UOP identification with the piece quantity, which permits the SKU pre-pack pallet to be placed into a WMS identification storage position that is specific for a single-line CO.

When required to complete a single-line CO, the WMS computer allocates a WMS-identified UOP from the specific WMS-identified storage position. The features are:

- One day or pre-pack day additional labor cost to pre-pack SKUs, which increases a day's cost per unit
- Increases management discipline, procedures, and controls
- WMS to properly track and allocate WMS-identified UOPs
- Some additional WMS-identified UOP handling
- Low cost per unit on a CO day due to fewer activities to complete a CO
- High-throughput volume
- Evens the pick and pack volume over a two day period

Semi-Completed or Packed Pre-Packed SKU

Pre-packing a promotional SKU as a semi-packed or completed SKU requires that a fast pack line pack a SKU with as-required bottom filler material in a shipping carton. Each shipping carton is stacked onto a pallet. With the potential for unstable cartons on a pallet, each pallet is wrapped in plastic and receives a new WMS UOP identification and SKU piece quantity per pallet. A WMS UOP identification scan transaction and piece quantity are sent to the WMS computer. The WMS computer associates the WMS UOP identification with the piece quantity that permits the SKU pre-pack pallet to placed into a WMS-identified storage position. The WMS-identified pre-packed UOPs are allocated to single-line COs. The next day, the WMS program allocates these WMS-identified UOPs to single COs.

Proper Elevation and Physical Location of the Golden Zone or Pick Position

A SKU pick position elevation and physical location on a pick shelf or level has an impact on picker productivity. Many order-fulfillment professionals refer to the best pick position elevation and location as a Golden Zone. A SKU physical location in a hand stack rack, shelf, or carton flow rack pick position concept is a very important factor that contributes to picker and replenishment employee productivity. To achieve maximum picker or replenishment employee productivity, a SKU Golden Zone pick position elevation reduces picker or replenishment employee reach or bend to complete a pick or replenishment transaction. In a hand stack, shelf, or carton flow rack pick concept, this means that the top and bottom (pigeonhole) pick position levels are least desirable pick positions. Golden Zone pick position levels are located between 20 in. elevation above the floor and 5 ft. 6 in. elevation above the floor. Golden Zone levels are the preferred pick positions. Elevations for a Golden Zone vary slightly according to your average employee height.

Change from a Paper Document to a Paperless or Automated Concept

Improvement in small-item, flatwear, or GOH picker productivity is realized from a change in pick instruction format, from a paper document to a paperless pick instruction format or to an automatic concept. When a manual or mechanized order-fulfillment concept uses a paper document, the paper document has lines and columns. Most paper pick documents use standard style black ink characters and digits on white paper. To obtain the instruction, a paper document concept has a picker read the proper line for a SKU pick position and SKU quantity. After a pick position transaction completion, the a picker places a mark adjacent to the appropriate line or location on the paper document. This mark verifies the SKU pick activity completion.

A self-adhesive label placed onto a SKU exterior surface, a paperless (pick to light or RF device) pick instruction or an automatic pick concept reduces several order picker non-productive activities, including reading the document and locating the specific line for a pick transaction; and placing a mark on a paper document that verifies transaction completion. Eliminating these activities reduces employee non-productive clerical activities in a pick activity.

Develop a Good Equipment and Facility Layout and Flow Pattern

A good pick area, equipment and facility layout, and small-item or flatwear UOP and CO flow pattern permits pickers, UOPs and COs to easily move between two facility locations, through pick aisles to pick positions, through a sort concept and from a pack station to a shipping dock. Pick and pack activities require equipment and aisles that occupy space. A good equipment, aisle and facility layout and UOP and CO flow pattern assures:

- Minimal picker travel time and distance between two facility locations and a continuous UOP and CO flow
- Good facility space utilization
- Accurate and on-time pick line set-up and replenishment transactions completed with minimal damage
- Sufficient aisle width to assure an on-time transaction completion with minimal damage. These factors improve employee productivity at the lowest operating cost.

Maintain Clear Aisles and Good Housekeeping

In a pick operation, to maintain clear pick aisles (clean and without obstacles) permits a picker and replenishment employee to:

- Travel at maximum allowable travel speed through an aisle
- Assure proper travel speed without unintentional stops
- Minimize accidents

Many order-fulfillment professionals have stated that good housekeeping in an order-fulfillment operation enhances employee productivity by 5 percent.

Maintain Good Lighting in a Pick Aisle or Pick Line

Good lighting in a forklift truck storage aisle improves pick transaction productivity. The light fixture options for your pick aisles are to have the light fixtures hung either directly above the pick aisle center or perpendicular to the pick aisle. This arrangement has the light fixtures above the pick positions. Both light fixture arrangements require you to specify the desired lighting (lumen) level at 30 in. above the floor surface.

With light fixtures hung directly above the pick aisle center, light fixtures hang from the ceiling joists or cross support members that are attached to pick position structural support members. This light fixture arrangement has fixtures that illuminate the entire length and width of the pick aisle. When required to replace the light fixture, maintenance employees have easy access to the fixture. Order pickers are generally more familiar with this light fixture arrangement.

When a light fixture arrangement is perpendicular to a pick aisle and above the pick positions, it has light fixtures that are hung from a ceiling joist or pick position structural support members. In this arrangement a light fixture row begins above the first pick row, continues above the other pick rows and aisles and ends above the last pick row. The center spacing between the light fixture rows ensures that the light (lumen) level is as specified by your written functional specifications. To replace light fixtures, maintenance employees have a more difficult task with this arrangement due to the fact that light fixtures are above the pick positions, thus making them less accessible. Pickers are unlikely to have this light fixture arrangement in their homes or retail shops and are less familiar with it.

Change Your Employee Work from a Human-Paced Concept to a Machine-Paced Concept

To change an order-fulfillment activity from a human-paced to a machine-paced concept means that you identify and modify the order-fulfillment operational activities that have large employee number, highest over time, UOP, building or equipment damage, highest employee injuries, high off-scheduled activities, and high errors.

To improve an order-fulfillment activity, match the activity with the equipment. Consider using:

- Non-powered or powered queue conveyor, AGVs, or another transport concept that moves multiple pieces, master cartons, or pallets between two work stations
- Mechanized or automated master carton, or pallet delivery and pick-up at a work station
- A WMS bar code/RF tag attached to a master carton or pallet, a WMS bar code/RF tag attached to each storage and pick position and using barcode scanners or RF readers to complete storage or replenishment transaction scans
- Guided aisles for mobile vehicle travel

Add Material Handling Transport Equipment to Your Employee Work

In an order-fulfillment operation, dramatic employee efficiency improvements are realized from using equipment to maximize the number of master cartons or pallets transported per trip between two work stations, with the fewest handlings. The GOH, master cartons or pallet in-house transport concepts are:

- Manually operated carts or pallet trucks
- Internal combustion or electric battery-powered forklift trucks, pallet trucks, AGVs, or towed cart train
- Non-powered or electric, powered conveyors
- Guided-powered vehicles
- Computer to schedule labor, equipment, and vendor deliveries
- Mechanized master cartons or pallet transport concepts

- Identify each GOH, master carton, or pallet at a vendor facility and at your receiving dock
- Unitize and secure the largest quantity of master cartons or pallets
- Provide sufficient master cartons or pallet queues ahead of each activity station

Use Automatic Identification

An automatic identification method is a WMS concept that has a human- or machine-readable discreet code on a label or paper that is attached to a master carton or pallet and attached to each storage and pick position. A labeled master carton or pallet is placed onto a transport concept device. This bar code/RF tag is read by a forklift truck driver with a hand-held or fixed position barcode scanner or RF tag reader that transmits the code data accurately and on-line to a microcomputer. The microcomputer controls another machine that responds to the code information or stores information in memory. In a WMS program that is supported by an automatic identification concept, the system handles high volume and a wide piece mix, has accurate information and on-line information flow, uses few employees, reduces data transfer errors and minimizes UOP or CO transfer errors.

Provide Sufficient Work

Providing an employee with sufficient work for an employee work day improves productivity. Using the projected order-fulfillment operation, small-item or flat-wear vendor delivery UOPs or CO SKU volume, along with your expected employee productivity rates for unloading, receiving, in-house transport, storage deposit, storage withdrawal, and pick line set-up and replenishment, pick, sort and pack, in-house transport, manifest and loading, you forecast the employee number needed to complete vendor delivered UOP and CO piece volume or transactions. If the actual employee number exceeds the projected required employee number, you allocate the extra employees to order-fulfillment operation value-added activities. This preplanning for an employee to relocate to another activity minimizes non-productive time for an employee.

Consider Off-Line Pick and Pack Activity for High-Volume, Single-Item SKUs

An order-fulfillment productivity guideline is that if your order-fulfillment operation has a high volume for a few single-item SKUs, you should consider whether to pick and pack SKUs and COs off-line. The off-line pick and pack activity is a single-item activity that occurs when your order-fulfillment operation has a high single-line CO volume that has the same SKU or pair (two SKUs) on your COs. The pick and pack activity has one or two additional steps in the normal pick activity. When compared to the normal or individual pick activity, there is a slight decrease in the order pick productivity because a picker picks, packs, seals, and labels a CO shipping container. This total pick and pack activity performed by one employee or employee group dramatically improves your total order- fulfillment operation and piece flow because the merchandise is picked *en masse* and the picked SKUs and COs bypass the normal pack and check station activities. If a SKU is picked *en masse* and transferred to a specific pick and pack station, order picker productivity is very high. With an off-line pick and pack activity, one picker productivity is very high. With an off-line pick and pack activity, one picker picks all SKUs, packs a CO pack slip and SKU into a CO shipping container, seals and labels a CO carton, and transfers a completed CO carton onto a transport concept.

An off-line order-fulfillment employee productivity improvement guideline is to have multiple pick lines that handle the top 20 percent of high-volume SKUs. Each off-line pick line has a transport concept for completed CO transport to the next pick line station and a second transport method for all partially completed COs that have additional picks for SKUs in a slow-moving pick area.

On-Time and Accurate Pick Line Set-Up and Replenishment Transactions

On-time and accurate pick line set-up and replenishment transactions assure that the correct SKU is in the correct pick position in sufficient quantity to satisfy a CO quantity. During an order-fulfillment operation, a stock-out occurs as an order picker arrives at an assigned small-item or flatwear pick position to complete a pick transaction and the pick position is depleted or has an insufficient SKU quantity to complete a pick transaction.

When a stock-out occurs with an order, the picker has completed the travel distance and time and not completed a pick transaction. A "no-pick" means low-order picker productivity and a potentially dissatisfied customer. On-time and accurate pick line set-up and replenishment to a pick position are supported by a WMS concept that has:

- WMS identification on each UOP.
- WMS identification at each pick and storage position.
- For each storage or replenishment transaction, an employee scans a WMS product identification and a WMS storage or pick position identification and the scans are transmitted on-line or delayed to a WMS inventory files.

Where to Start Customer Orders

In a pick/pass order-fulfillment operation, a major productivity consideration is to have a CO start at a pick line front. Pick/pass and pick aisle layouts locate fast- moving, heavyweight and high-cube SKUs in one pick area and slow-moving, small-cube and lightweight SKUs are located in a separate pick line or zone. To obtain high-order picker productivity in a pick/pass or batched CO pick concept, your question is, in what section do you start a CO or locate an automatic pick machine? In a pick/pass or batched CO fulfillment operation, fast moving SKUs (or 20 percent) account for 80 percent of the pick volume and the slow moving SKUs (or 80 percent) account for 20 percent of the pick volume.

On a pick/pass order pick line, CO start locations are:

- To start a CO in the fast-moving, heavyweight and high-cube pick line section as the next pick line section to have the medium to slow-moving, lightweight and small-cube SKU or automatic pick machine
- To start a CO in the slow- to medium-moving, lightweight and small-cube SKU pick line section or automatic pick machine and the fast-moving, heavyweight and high-cube pick line section as the next pick line section

If a CO starts in the fast-moving, heavyweight, and high-cube SKU pick line section, the results are improved pick position replenishment, high volume handled by the pick line and a more continuous CO flow from a pick line to a check area due to faster CO completion. With slow-moving, lightweight, and small-size SKUs at a pick line end, it is easier for a picker to place SKUs into a CO shipping carton. There is a decrease in CO handling number and this method ensures high-order picker productivity due to the fact that fast-moving SKUs with high hit density and concentration and high-cube and heavyweight SKUs are easily read on a picker instruction form (at the top) and are the first SKUs transferred from pick positions into a CO carton.

If a CO starts in the slow-moving, lightweight, and small-cube SKU pick section or with an automatic pick machine section, the results are lower picker productivity due to the fact that a picker in the fast-moving, heavyweight, and high-cube SKU pick line area handles two cartons for a CO. This is because one CO container is partially full with slow-moving SKUs and a second CO carton is required for the remaining fast-moving, heavyweight, and high-cube SKUs. With a paper pick instruction it is more difficult to read a pick instruction document because fast-moving SKUs are located on a pick instruction document bottom. On a conveyor travel path, a slow-moving SKU partially full carton (due to light weight) is more difficult to transport over a conveyor travel path. It is difficult to keep replenishment activities coordinated with the order pick activities due to the fact that slow-moving SKU replenishment is handled prior to fast-moving SKU replenishment. There is lower picker productivity due to the fact that for a picker to transfer SKUs into a pick carton or to make space in a CO carton for the fast-moving, heavyweight, and high-cube

SKUs in a partial dull carton, a picker rearranges the SKUs in a CO carton. The SKU rearrangement time in a pick carton is unproductive picker time. In a pick/pass CO pick concept your pick line transport concept has a full carton or completed CO take-away conveyor.

Arithmetic Progression through a Pick Aisle or on a Pick Line

An arithmetic position progression through a pick aisle means that there is a sequential arrangement for pick position numbers in a pick aisle. An arithmetic progression method routes a picker to complete a pick transaction as a picker walks or rides through a pick storage aisle. An arithmetic progression has the first pick position at an aisle entrance as the lowest pick position number and the last pick position has the highest pick position number. In a small-item or flatwear order-fulfillment operation, picker productivity improvement is realized from the arithmetic progression through a pick aisle. Arithmetic pick position progressions through pick aisle routing patterns may have pick position numbers that end with even numbers on an aisle right side and pick position numbers that end with odd numbers on an aisle left side. To maximize picker productivity, whenever possible all picker truck routing patterns have pick position numbers that end with even digits on an aisle right side and pick position numbers that end with odd digits on an aisle left side. This numbering pattern reduces confusion because employee daily activities have a similar numerical arrangement and when a picker reads a pick transaction instruction format, the even and odd number pick position scheme automatically directs the picker to the proper aisle side. With pick positions on both sides of the aisle, as a picker enters a pick aisle left side pick positions are identified with a number such as aisle 1 and are aisle right side pick positions are identified with another number, such as 2.

Other factors that influence a pick position identification scheme are:

- Small-item or flatwear pick aisles that have a serpentine picker routing pattern through aisles in a pick area
- In a decked rack, shelf, or carton flow rack pick line application, the preferred picker routing pattern is a horizontal scheme rather than a vertical routing pattern. A horizontal picker routing pattern has a picker start at the highest right-hand or left-hand corner pick position. From this pick position, the next possible pick position is on the same level and is adjacent to the first pick position. After all picks on the top level, a picker routing pattern is continued on the next lower bay pick level. This pattern is repeated for all pick levels.

A major WMS concept component is that each WMS small-item or flatwear storage or pick position has a unique identification and that a pick or storage position identification progression through a pick or storage aisle assures an efficient and cost-effective picker, forklift truck driver, or AS/RS routing pattern.

Advantages are that it is easy to read and follow, high-volume and -cube items occupy multiple locations per the storage philosophy and it is easy to upgrade from a paper to a paperless transaction instruction concept.

How to Control Your Order-Fulfillment Operation

To have an effective and cost-efficient order-fulfillment operation, some requirements are:

- Accurate projected small-item or flatwear, master carton, or pallet vendor delivery, or CO volume
- An accurate annual operating expense budget
- A short interval schedule concept

Design Vendor-Delivered and Customer Order Piece Volumes

To design a pick fulfillment operation concept for a new facility or to consider new equipment for an existing operation, design key functional areas to handle a specific small-item, flatwear, master carton, or pallet volume. To project small-item, flatwear, master carton, or pallet storage operational volume, you determine your vendor delivery and CO volume and associated order-fulfillment transactions. Your system should be design to handle UOP or CO volume levels that are:

- Average volume
- Some volume between the average and peak volume
- Peak or spike volume

If an order-fulfillment operation concept is designed to handle an average volume and an actual volume exceeds the average volume, the consequences are:

- Off-schedule activities or UOP vendor or CO deliveries
- Employee overtime
- Increased building, equipment and product damage, and employee injuries
- High customer complaints

Features of an average volume level are small land and building size, low capital investment, and low fixed material handling cost per piece.

If an order-fulfillment operation concept is designed to handle a volume that is somewhere between average and peak volume levels, when actual volume exceeds design volume, the consequences are:

- Employee over-time
- In-house transport vehicle congestion
- Increased product, equipment, and building damage
- Off-schedule activities and vendor UOP or CO deliveries
- Some customer complaints

Advantages of a volume design level somewhere between an average and peak volume levels are medium land site and building size, capital investment, and fixed cost per unit.

If a pick fulfillment operation concept is designed to handle a peak volume, when an actual volume arrives at peak volume, an order-fulfillment operation does not experience off-schedule activities or vendor UOP or CO deliveries and customer complaints.

The peak design volume level features are large site and building size and additional investment, high fixed cost per unit, and excellent customer service.

Employee Productivity Must Be Tied to an Order-Fulfillment Operation

The true value of an order-fulfillment operation's employee productivity program is that it provides an order-fulfillment operation manager with an employee productivity rate that is an accurate forecast tool. This forecast tool is used to project and control order-fulfillment labor expense dollars and provide on-schedule service to vendor UOP deliveries and COs.

The order-fulfillment employee productivity rate is the basis for calculating:

- Annual operational budget labor expense dollars
- Budget dollar justification for a capital expenditure
- The labor hours, expense, pieces, vehicle deliveries, and equipment and labor schedules or employee number per shift

After an employee productivity program is implemented in an order-fulfillment operation, this employee productivity rate is tied to the operating labor expense budget and is related to a capital expenditure justification. If order-fulfillment employees do not achieve this projected employee dock productivity rate, performance is below par, which increases the overall company operational cost per piece or customer order and lowers the company profits.

Short Interval Scheduling

Short interval scheduling (SIS; see Figure 1.1) is a method that tracks productivity of an order-fulfillment operation employee in-house transport, picker, or other

SHORT INTERVAL SCHEDULE

	PICK RATE	PICK AREA	PICK PIECES	700	COFFEE BREAK			LUNCH			COFFEE BREAK			1700
					800	900	1000	1100	1200	1300	1400	1500	1600	
PICKER NAME OR ID														
ESTIMATED TIME														
ACTUAL TIME														

Figure 1.1 Short interval schedule.

activity. The SIS method can be implemented in a small-item or flatwear order-fulfillment operation that is manual, mechanized, or automated, in a small, medium, or large company. This SIS method is used to track individual employee handling productivity for an entire shift, or work activities that are required to complete a specific transaction. The SIS concept is designed as a manual or personal computer-based concept.

Steps in the SIS method are to:

- Identify an order picker or in-house or forklift truck.
- Give an employee pick instructions or self-adhesive labels.
- Project picker activity time to complete a task.
- Identify a picker travel path.
- Complete the pick operation. Per instruction, a picker stops at a pick position and completes a pick transaction. After completing all picks, a picker drops a CO at a pack station or transfer location and returns to the dispatch station to receive other pick transaction instructions.

Short Interval Scheduling for Your Order-Fulfillment Activities

The SIS method is designed for a manual pick, in-house transport, or other manual-based concept activities in an order-fulfillment operation. The SIS method has a clerk who is located at a control desk, where pick area transactions or portions are divided by a goal-set employee productivity rate per hour. The goal-set employee productivity rate is predetermined for each CO section or pick aisle, in-house transport travel path, or other activity. This feature does recognize the difference in travel minutes that a picker or in-house transport employee travels from a control station to a pick aisle and all activities to assure a pick transaction completion. The result is the time that is projected for a picker or in-house transport vehicle to travel from a control station, in an aisle, complete a transaction, and return to a control station.

How Short Interval Scheduling Works

An employee or computer performs short interval scheduling using a bar chart with columns across the top of the page. Columns are divided into 15-minute or other predetermined time intervals. Horizontal rows under each column represent individual employee activity. Each employee has a bar chart page.

On the first horizontal bar chart line is written the employee name. The next two horizontal lines are used for each storage area transaction. On the first horizontal row, a line is drawn from the issue time to the projected return time. After the employee returns to the control station, on a second row a second line is drawn from the issue time to the actual return time. The difference between the two horizontal lines is the difference between the projected employee productivity time and the actual employee productivity time. This difference shows whether the employee is on-schedule or off-schedule. Each subsequent employee activity is listed below on the two horizontal row sets.

At the start of each workday and at a control station, each employee is given a pick transaction group. Upon completion of the first pick transaction group the employee returns to the control desk, where the employee receives another pick transaction group. The actual employee return time is indicated on the employee corresponding event horizontal bar under the appropriate time column. Comparison of projected return time to the actual employee return time indicates the employee's actual productivity and provides the on-time and exact review for each employee's productivity performance. The employee's actual performance is compared to your employee productivity standard.

If your employee actual time is consistently before the projected return time, your order-fulfillment operation performance is better than your annual expense budgeted labor dollar amount. This excellent order-fulfillment operation performance means that the actual labor dollar expense is below the budgeted labor dollar expense.

If the employee actual return time consistently exceeds the projected return time, your order-fulfillment operation performance is lower than the annual expense dollar budget amount. A poor order-fulfillment operational performance means that the actual labor expenses exceed the annual dollar budget expense amount.

How to Project Your Order-Fulfillment Operational Budget

Top management teams are requesting the area operation manager to prepare an annual order-fulfillment operating budget. Reasons for an increase in the use of annual order-fulfillment operating expense budget requests are to:

■ Plan for labor and equipment adjustments that are due to business volume fluctuations

- Anticipate and control company cash flow
- Review actual labor expense and budget labor expense variance
- Provide a basis for a capital investment request for a new order-fulfillment concept

With a large labor dollar expense amount, an order-fulfillment operation budget is prepared in a sophisticated and realistic manner.

Annual Order-Fulfillment Operating Expense Budget Methods

Order-fulfillment operation annual expense budget preparation methods are calculated by a staff member or by computer and use a simple percentage increase method or a detailed line item and man-hour (SIS)-based budget method.

Simple Percentage Increase Annual Operating Expense Method

A simple percentage increase annual order-fulfillment operating expense budget uses the last fiscal year annual operating expense dollar amount and a percentage increase for the next fiscal year business volume. After the two values have been determined, multiply the last fiscal year annual operating expense amount by the next fiscal year business volume percentage increase. This provides an annual operating expense budget for the next fiscal year.

Detailed Expense Item and Man-Hour (SIS) Budget Method

A detailed expense item and man-hour (SIS) budget method projects your order-fulfillment operation's annual operating expense. To be an effective management tool, a projected order-fulfillment operational expense is shown for each period and fiscal year quarter. The company fiscal year is for a 52-week period. This 52-week period agrees with the calendar or starts on a day other than January 1.

A detailed line item and man-hour (SIS) budget method recognizes that an order-fulfillment concept has:

- Operational expenses that fluctuate with business volume
- Expenses that are:
  - Controllable by management staff and are related to fluctuations in your business volume
  - Non-controllable, which are expenses that are not controlled by management staff and generally do not fluctuate with business volume

A detailed expense item and man-hour (SIS) budget method requires that you project annual controllable and non-controllable operating expenses for your facility.

Controllable Expenses

Controllable order-fulfillment operation expenses are based on the anticipated employee productivity rates and the projected business volume and hourly wage rates.

Some controllable expenses are labor, labor fringe benefits, UOP, equipment and building damage, and employee injuries, supplies, and repairs.

Non-Controllable Expenses

Non-controllable annual operating expenses are based on your financial department accounting methods and company non-labor contracts. Non-controllable expenses include labor wage rate and fringe benefits, depreciation, taxes, rent or lease rates, and third-party logistics contracts.

How to Work with a Detailed Expense Item and Man-Hour (SIS) Budget Method

After an order-fulfillment operation establishes annual operating objectives for the next fiscal year, as well as employee productivity rates and business volume, you perform the calculations for a dollar expense budget. A budget expense is calculated for each period or fiscal year quarter. A budget process has each expense item on an order-fulfillment operation's next fiscal year budget compared to expense item actual dollar expense amount for the last fiscal year. If management requires a justification for increase or decrease in a budget expense item for the next fiscal year, a detailed expense item and man-hour budget method provides a complete analysis and permits a statement to be made to substantiate a dollar increase or decrease in the expense item. This expense fluctuation is related to the projected business volume for the next fiscal year or is substantiated by your justification.

A detailed expense item and man-hour budget separates an order-fulfillment operation's annual operating expense into groups including wages, salaries and fringe benefits as a subtotal, other controllable expenses as a second subtotal, and non-controllable expenses as a third subtotal.

Calculations in a detailed expense item and man-hour annual operating expense budget are to:

- Identify, qualify, and quantify for a fiscal year each operating expense item as a controllable or non-controllable expense item or income
- Obtain next fiscal year's period, quarter, and annual operating calendar end dates
- Determine a dollar value for the company's next fiscal year period, quarter, and annual non-controllable expenses

- Project period completion for a capital expenditure and its impact on next fiscal year annual operating expenses, such as depreciation and labor productivity improvements
- Determine next fiscal year period projected employee productivity rate for each storage operation activity
- Obtain next fiscal year annual business volume
- Obtain the projected hourly wage rate increase and period that an increase is to occur for each employee job classification, as well as fringe benefits, and payroll tax rates
- Forecast all expected and unexpected activities and their impact on annual operating expenses
- Determine the next fiscal year item and man-hour annual operating expenses
- Identify, qualify, and quantify all annual income items for a fiscal year
- Review with your financial manager each item of projected expenses

The first step in the detailed line item and man-hour budget method is to identify, qualify, and quantify each major expense item for the last fiscal year as a controllable or non-controllable expense item. The most important expense classification is the controllable expense due to the fact that dollar expenses fluctuate with an increase or decrease in the volume. An expense dollar value increase or decrease is controlled by the order-fulfillment manager's decisions. Operations controllable expenses are wages for all labor groups, fringe benefits, trash disposal, pallet purchases, product, building and equipment damage and employee injury, operating supplies, guard service, sundry items, and telephone and fax expenses.

During a review meeting with the financial manager, obtain the dollar value for each controllable expense item for the past fiscal year. If figures do not cover the 12 or 13 months, an order-fulfillment manager uses the required periods from the previous fiscal year to obtain or estimate a total for the present fiscal year periods. Also obtain the next fiscal year calendar or schedule for each month and quarter end dates. The schedule of the calendar end dates indicates each of the 13 period calendar end dates. Each period has four weeks. Three-quarters of the calendar year has three periods or 12 weeks. One quarter has four periods or 16 weeks. The financial manager determines the 16-week quarter position on the fiscal calendar.

Non-Controllable Expense Projection

Non-controllable expense items for an order-fulfillment operation do not fluctuate with an increase or decrease in business volume. They include building and equipment rent or leases, property and other taxes, insurance, utilities, building and equipment depreciation, and office administration. Non-controllable expense items are not within the operations manager's control. They are determined by the financial manager, and the operations manager has an opportunity to review the budgeted dollar amount.

After the financial manager and the operations manager agree on each expense item for the next fiscal year's operating budget, the operations manager enters each annual expense item budgeted figure onto an individual budget expense account sheet, which shows a budget expense for each line item. The manager enters onto the individual account sheet all information that relates to the budgeted dollar value calculations.

After all individual non-controllable expense account sheets are completed, the annual figure for each expense item is transferred from the individual expense account sheet to the appropriate line and column in the operating plan sheet, and the manager calculates a total which gives the operating plan subtotal for the fixed expense line subtotal.

Controllable Expense Projection

The next step in the budget process is that the operation manager determines the next fiscal year's controllable expense dollar budget value for each expense item. The financial and operations managers determine a weighted average wage rate for each order-fulfillment concept labor job classification. The next step is a meeting with the sales manager. At this meeting the manager receives the business product volume projections for the year.

After this meeting, the operations manager enters each business volume projection onto an appropriate line of the direct labor wage work sheet. On this work sheet, the operations manager enters each direct labor activity per shift that is required to move a vendor-delivered UOP or CO volume through an operation.

At a review meeting with the operations management staff, the operations manager projects direct labor productivity rate for each shift for the next fiscal year. Direct labor employees are considered employees who physically move UOPs or COs or control storage or pick equipment that moves UOP or CO. With the next year's employee productivity rates and business volume, the operations manager calculates each shift direct labor gross available hours. The gross available hours or total hours are separated into subtotals for each full-time or part-time employee's straight (regular) hours and over-time (OT) hours. With the gross available hour figures, the operations manager enters a corresponding average hourly wage rate for each job classification type. This hourly wage rate includes night-time premium per hour. If an employee contract has an hourly wage increase, the operations manager indicates the hourly wage increase in a period that the hourly wage increase occurs and states this fact on a work sheet section. The hourly wage rate is a weighted average wage rate that was calculated by the financial and operations managers.

The next required direct labor budget calculation is to determine the dollar expense values that are associated with hourly direct labor. This calculation has gross available hours for each labor group multiplied by the corresponding hourly wage rate. After all calculations are totaled for a fiscal year, the operations manager enters the figures onto a direct labor line of the quarter storage direct labor wage work sheet and operating plan sheet.

The remaining labor budget line items include indirect labor. Indirect labor is labor that is required to support an operation and the employees do not touch a product or control storage or pick equipment that moves UOPs or COs. The indirect labor classification includes supervision, clerical, cleaning or sanitation, general, and security and maintenance labor.

Each indirect labor classification has its individual work sheet. As required, the expense increase or decrease justification is explained at the bottom of the sheet.

Management Salaries

An indirect labor budget expense is the salaries for an order-fulfillment operation management staff. To budget this expense item, an operations manager enters each supervisor name or anticipated position on the operation facility indirect salary labor work sheet. After each manager staff name or position, an X is placed that indicates a salary expense for the fiscal year. On a total line, the operations manager enters a total management dollar value expense for the fiscal year. When management salary increases are granted to the management staff, it is indicated in the appropriate period. After the management team work sheet is completed, the operations manager transfers the bottom line dollar expense total for each period to the appropriate line of the operating planning sheet.

To determine other hourly and indirect labor job wage expenses, the operations manager determines indirect labor job number that has regular pay, over-time pay and average hourly wage for the fiscal year. After the dollar calculations are completed for the fiscal year, the operations manager transfers the dollar value from the operation facility indirect hourly wage sheet to the operating planning sheet.

Non-Work Hour Expenses

The next calculation is the non-working hour calculation, that is the dollar expense value for non-working and employee paid jobs. Non-working hours include sick days, sick leave, and jury duty.

The non-working hours are additions to the net available straight or regular employee hours. On a non-working hour sheet, each expense is indicated and in an explanation section the operations manager makes a statement as to the non-working hour type and average hour wage rate. The dollar expense is calculated for each hour type and is entered in the appropriate period. After the non-working hour sheet is totaled, figures for the fiscal year are transferred to the appropriate line of the operating planning sheet.

On the operating planning sheet, the operations manager adds the year's wage lines to obtain a wage bottom-line figure which is the operation wage expense for a fiscal year.

Fringe Benefits and Taxes

The next calculation for the detailed expense item and man-hour budget method is the employee fringe benefits and payroll tax line item. The operations manager refers to the employee contract and payroll tax laws. This information is obtained from the financial manager and determines fringe benefits and payroll tax budget expense.

Fringe benefits include holidays, pensions, thrift plan, group insurance, vacations, workmans compensation, and cafeteria or baby care. Payroll taxes are FICA and federal and local taxes.

Each fringe benefit and payroll tax has an individual account work sheet. After each fringe benefit and payroll tax expense is calculated for each period, the operations manager transfers each period expense from an individual work sheet to an appropriate line on the fringe benefit and payroll tax summary sheet. For vacations, historical employee payroll records indicate a period that the fringe benefits were taken by an employee. If this information is not available, the storage operations manager projects the period for these expenses.

After all fringe benefits and payroll tax budget figures are transferred to the fringe benefit and payroll tax summary sheet, the operations manager adds the expense items to obtain a fringe benefit and payroll tax bottom-line figure. This dollar figure is the operation fringe benefit and payroll tax budget expense for the fiscal year. This bottom-line figure is transferred to the appropriate line on the operating planning sheet.

After fringe benefit and payroll tax transfer, the operations manager totals direct labor, indirect non-working, fringe benefits, and payroll tax lines. The sum is a bottom-line figure for the operating planning sheet wage and salary controllable expense line.

Other Controllable Expenses

The last detailed expense and man-hour budget expense group is the other controllable expense group, which includes miscellaneous purchases, repairs, trash disposal, damage, operating supplies, guard service, and telephone.

On a controllable individual budget expense account work sheet, the operations manager determines the fiscal year budget expense. Explanation statements are made on this sheet, which substantiates an expense item budget expense increase or decrease. The statements are contract agreements and anticipated special events.

After completing each controllable individual expense account work sheet, the operations manager transfers the fiscal year budget expense amount to the appropriate line on the operating planning sheet and totals the line items. The sum is the other controllable bottom line expense.

To obtain the gross total budget expense for the operation, the operations manager adds the three major groups of the budget fiscal subtotal lines.

The disadvantages of this method are that it is an involved task, requiring exact past records, and projected storage operation productivity rates and calculations can require several days. A computer program simplifies a budget process or task and reduces the time that is required to make calculations or revisions. The advantages are that it provides top, middle, and lower management with the ability to plan, control, and review an operation and gain an understanding of actual and budget dollar expense variance, because each dollar expense is related to the next fiscal year's projected volume.

Why Use Manning or Employee Productivity Schedules?

The objective of using manning or employee scheduling is to fine tune labor allocation in your order-fulfillment operation. It permits your operations manager to meet the budgeted labor expense and achieve projected employee activity productivity rates, and it assures controlled vendor-delivered UOP and CO order flows through your operation, and on time CO delivery.

A manning or employee scheduling projection requires several pieces of operational information, including:

- Estimated or actual
  - Vendor delivery UOPs that are based on your receiving dock schedule program
  - CO pieces, lines, or COs that are based on the IT download
- Your budgeted employee activity productivity rate or pieces per hour for each activity
- A PC or staff member with a calculator to complete the calculations

Using a PC, for each order-fulfillment operation activity you have pre-entered into the appropriate column cell the budgeted employee activity productivity rate or pieces per hour (Figure 1.2). In the associated column cell, per the order-fulfillment activity you enter estimated vendor delivery pieces, based on your receiving dock schedule, and CO pieces that were downloaded from the IT department.

The PC performs the necessary calculations and indicates:

- Each order-fulfillment activity for the total employee hours
- Full time employee number, if full-time employees are used

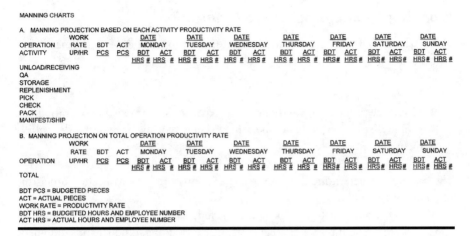

Figure 1.2 Manning charts: production per activity and production per total operation.

- Part-time employee number, if part-time employees are used
- Extra labor that is available for reallocation to another activity or additional part-time labor that is required for the order-fulfillment operation to maintain your customer service standard

From a manning or productivity schedule, the operations manager finalizes the full-time and part-time employee number for each activity. Prior to an operation start-up, the operation management team has adjusted employees to transfer from one activity to another activity or have fewer part-time employees at the operation.

If a manning or employee productivity schedule is completed for a total operation productivity rate (see Figure 1.2), the need for manager to be able to visualize each order-fulfillment operation activity and allocate full-time and part-time employees is reduced and the management team functions to fine tune the operation.

Why Have Capital Investments?

In an order-fulfillment operation, a major capital expenditure is related to maintaining business growth, improving manual concepts, and replacing manual order-fulfillment equipment with a mechanized or automated concept.

Objectives of an order-fulfillment concept are to:

- Increase the vendor-delivered UOPs and CO ordered SKU handling number
- Lower an operation's annual expenses
- Maintain or improve on-schedule and accurate customer service

What Is a Capital Expenditure Justification?

The major part of a financial justification for the proposed order-fulfillment concept is based on:

- Projected business volume
- Employee productivity and operating expenses for the existing operation
- Employee productivity and operating expenses for the proposed order-fulfillment concept

The cost figures determine the labor expense savings and the associated order-fulfillment concept savings.

Other economic justification factors include facility construction costs; land costs, and site preparation costs; other costs such as building, product and equipment damage and employee injuries, and investment and related depreciation expenses. Savings between an existing and proposed concept provide a return on investment that is a payback for a capital investment.

It is also understood that other costs and non-economic factors exist, including customer service standard and business growth and SKU expansion. Whenever possible, the non-economic factors are used to justify an operation's capital expenditure.

How Depreciation Expense Affects Your Income Statement and Balance Sheet

Capital expenditure on a new order-fulfillment concept has a tremendous impact on the operation budget and actual operational expenses. Investment in a new concept is an asset that appears on the company's balance sheet. Good accounting practice has a new concept listed as an asset that is depreciated over a period of years. This depreciation is in the form of an annual depreciation expense. Depreciation expense is calculated by the accounting department and is an asset write down that occurs over time due to wear and tear. This annual depreciation expense is charged against the company income.

How Your Employee Productivity Affects Your Income Statement

Employee productivity is measured as UOPs or number of COs that one employee handles within a given time period, typically one hour. Employee productivity determines total employee hours or the number of employees who work in an operation. It is a component of the financial justification for a new concept. Because the company pays an hourly wage to each employee, it is an expense item that appears on the company's income statement. An expense item represents a reduction in the company's gross income. After all expense items are subtracted from gross income, the result is the company's net income.

What Good Employee Productivity Means to a Company's Income Statement

In an order-fulfillment operation, depreciation and employee wage expense together have tremendous impact on income. If an increase in an expense item is above the annual budgeted dollar amount, it results in a decrease in the company's net income. (Decrease in net income is not the operation's objective!) On the other hand, a decrease in an expense item below the annual budgeted dollar amount results in an increase in the company's net income. This is considered a company objective!

When a company invests in a new concept and the manager achieves or exceeds the company-projected employee productivity rates that were used to justify the investment, there is no increase in the cost to the company. This means that there is no change in the company's projected profit.

If company-projected employee productivity rates are not used to justify a capital expenditure, there is an increase in operation expenses. Additional labor expense raises cost per unit or cost as a percentage of sales above company projections. This situation increases company cost and means that there is lower company net profit.

Reasons for Economic Justification and Justification Factors

Top management teams are requesting order-fulfillment operations managers to provide comparative financial and non-financial justification for order-fulfillment concept capital expenditures. This is necessary because of the high dollar value for a modern order-fulfillment concept, high interest rates for borrowed money and limited company cash flow or capital expenditure funds.

A capital expenditure investment is a project that has a life of more than one year or extends an asset life. A capital expenditure investment is allocated as a depreciation expense over a company's future income. A low dollar equipment replacement project is considered a company expense and is an expense against the current year's income.

With a high capital investment dollar amount that is associated with a proposed modern order-fulfillment concept, a financial evaluation analysis compares alternatives in a realistic manner. A company accounting manager is a project team member who provides expertise to perform economic justification calculations and assures that cash inflow and outflow classifications are within company and Internal Revenue Service (IRS) or federal government tax on income policies.

Why High Employee Productivity Is Important

Employee productivity is defined as the output (tons, piece quantity) that is handled or moved by an employee as a ratio of total number of work hours or cost. To obtain a measure of employee productivity, total vendor UOPs or COs are multiplied by the total work hours or cost.

In order to have an efficient and productive operation, your management staff should ensure:

- Proper order-fulfillment equipment
- Proper operational vendor UOP or CO travel path layout
- Simple and clear operational instructions
- Motivated employees
- Scheduled activities that include vendor UOP deliveries, labor, and equipment
- Level flow of the daily piece and CO volume and vendor UOP or CO through the facility
- Realistic operational budget and objectives

Employee productivity improvements are one means to off-set other operational cost increases in salaries, wages, and operational controllable and non-controllable expenses.

Business Factors that Affect Productivity

Order-fulfillment business factors that affect employee productivity are:

- Vendor UOP or CO mix and volume
- Pick positions or delivery locations that are located over a large area
- Variations in size of vendor UOP or CO
- Time required to complete a CO or workstation order and delivery cycle
- Travel path for a vendor UOP or CO delivery

Areas for Improvements in Employee Productivity

To ensure that installing new equipment or enhancing existing equipment results in budgeted employee productivity:

- Employees should use the specified methods to perform their work
- There should be adequate management and employee training
- There should be written rules and procedures

- There should be properly posted signs in the work area
- Employee work activity should be reviewed regularly

Ways to improve employee productivity include the following:

- Rearranging work methods
- Changing how work is performed
- Moving the largest number of vendor UOPs or COs with fewest handlings and handle the largest volume per trip
- Scheduling and utilizing the equipment and labor at the maximum rate
- Improving physical surroundings and vendor UOP or CO travel paths
- Providing employee and management incentives
- Promoting pride in work
- Changing from a human-paced concept to a mechanized or automated order-fulfillment concept

Change Employee Work Methods

Employee productivity can be improved by making changes to employee work methods. In conjunction with this approach, look to implement:

- Sequences for vendor UOP or CO travel path and load pick-up and deliveries to reduce empty load-carrying surfaces, vehicles, or frequency of employees dispatch
- Provide adequate queue area, such as queue conveyor or P/D stations before workstations
- Apply the ABC theory of SKU movement from order-fulfillment positions to pick positions
- Kit or family group SKU allocation to order-fulfillment pick area
- Complete the CO order and delivery cycle with the fewest vendor UOP or CO handlings

Add Equipment to Employee Work

In an order-fulfillment operation, the most dramatic employee efficiency improvements are realized from introducing mechanized equipment to handle the maximum number of vendor UOPs or COs per trip with fewest handlings. Examples are:

- Manually operated pick carts or pallet trucks

- Internal combustion or electric battery-powered forklift trucks, pallet trucks, or AGVs
- Gravity- or electric-powered conveyors
- Guided powered order pick vehicles
- Computer-scheduled labor, equipment, and vendor UOP or CO deliveries
- Mechanized or automated vendor UOP or CO pick or transport concepts
- Barcode scanners and human/machine-readable codes
- At the vendor facility receiving dock, unitize and identify vendor UOPs
- Unitize or palletize and secure vendor UOPs
- Provide sufficient vendor UOP or CO queue prior to each storage area or pick line activity

Enhance Use of Employees and Equipment

Another way to improve employee productivity is to enhance employee available work hours or equipment scheduling, and vehicle utilization. Some methods do not require investment, whereas others will require an investment. They include:

- Using part-time employees
- Purchasing or leasing equipment that is flexible and can be used in several other operational areas or is able to perform a specific task with an attachment
- Smoothing the work week peaks and valleys to improve equipment and labor utilization
- Ensuring controllable vendor UOP or CO queues prior to each workstation
- Ensuring that there is sufficient direct or indirect work for each full-time employee work day

Improve the Work Area

Ways to improve the design, appearance, and physical surroundings include:

- Keeping the aisles and vendor UOP or CO travel paths clear
- Improving housekeeping, sealing floors, and painting equipment and facility interior walls
- Painting the facility ceiling and walls with a light color
- Providing adequate lighting in pick aisles or on pick lines
- Eliminating damaged or salvage pieces and obsolete equipment from the facility
- Providing adequate storage and pick aisle widths and recommended equipment clearances

- Painting lines on floor surfaces in queue areas, mobile equipment travel paths, and personnel travel paths
- Providing good order pick or vehicle routing and sufficient aisles
- Eliminating bottlenecks and jams
- Providing good and clear vendor UOP or CO and vehicle dispatch and order pick instructions
- Eliminating work steps from vendor UOP or CO flow or travel path

Provide a Work Incentive Program

One can also increase employee productivity by motivating employees with a work incentive program. A work incentive program gives an employee additional money or something of value for extra effort or achievement. In some companies, an incentive program has increased employee productivity by 10 to 15 percent.

A good employee incentive program is well researched and presented, and is:

- Measurable
- Achievable
- Understandable
- Clear
- Fair
- Administered equitably

If the work changes, the employees' incentive program should change too.

Increase Employee Pride

Motivating employees to take pride in their company, their department, and their work improves productivity. Employee pride can be promoted by participation in such company-sponsored activities as a forklift truck rodeo or family outings. Other ways to enhance employee pride include recognition for accurate, high productivity, injury-free work days and work completed on schedule; providing uniforms; giving awards for being on time and for good attendance records and for team work; and providing a clean and properly lit work area and safe equipment.

Guidelines for a Successful Employee Productivity Program

Guidelines for implementing an employee productivity program include the following:

- Categorize all employee jobs or tasks in an operation.

- Identify activities, functions, or jobs that have incurred highest costs over time; also identify the greatest vendor UOP, CO, or equipment and building damage, frequent errors, off-schedule deliveries, and highest employee injuries.
- Total the number of employees and the cost of wages and benefits that are associated with each activity.
- Select a unit of measure for each employee activity. This figure is the productivity measurement for the present and proposed activities.
- After implementation of a new concept, compare your actual employee productivity figures to the budgeted employee productivity figures.

Use an industry average or another company's employee productivity figure for comparison. Understand your company's operational policies, procedures, operations, business, vendor UOP or CO characteristics, and employee productivity figure calculations.

Identify Labor Activity That Has the Highest Cost

Identify employee activity that uses the highest number of employees or has the highest associated operating expense. Improvement in employee productivity for these activities will have the greatest impact on company operational expenses and service to customers.

A good employee productivity program is simple, complete, and clearly understood by employees and supervisors.

It Must Be Cost Effective

The method used to obtain information on employee work activity should not require the employee, supervisor, or clerk to spend a large amount of time. If the record- keeping process requires a large amount of time, employee time spent to record work activity offsets employee productivity improvements.

Results Must Be Timely

Timely information permits your management team to review performance of an individual employee or shift. This information, with projected UOP or CO volume, permits the management staff to project the next day's labor, delivery vehicle, and equipment schedules.

How to Measure Employee Productivity

One characteristic of a good employee productivity program is that it is based on a unit of measure. The best unit of measure in an order-fulfillment operation is the number of vendor UOPs or COs handled by an order-fulfillment concept. The unit of measure is a single-item, flatwear or GOH customer order. If an operation handles vendor UOPs or COs, the method of measurement depends on the UOP or CO mix. If there is no change in vendor UOP or CO mix, a good measure of productivity is employee activity. If there is wide vendor UOP or CO mix, it is better to use an average conversion factor that relates vendor UOPs or COs to productivity.

Vendor UOP and CO units of measure that are used in an order-fulfillment operation are:

- Tons or weight
- Accident-free days
- Expenses as a percent of sales
- Overs, shortages, and damaged pieces
- Errors
- Returns
- Off-schedule CO deliveries
- Sanitation rating
- Machine downtime, number of vehicles or machines in maintenance

When tons or expenses as a percent of sales are used as the unit of measure, daily employee productivity varies due to the piece weight or dollar value. Vendor UOP or CO order volume variation occurs by work day, month, or season of the year and market conditions.

Alternative Techniques to Measure an Operation

Storage activities move vendor UOPs or COs from the receiving dock area, through an order-fulfillment operation, through a value added or pick operation and to the shipping dock. Within each work area, the vendor UOP or CO is transformed or picked from large piece quantities (pallets or master cartons) into individual pallets, master cartons, GOH, or loose single items per a CO and assembled for shipment to satisfy customer orders.

Resources used in the receiving of vendor UOPs or COs include labor, mechanical equipment, and the work area that is used as a travel path for employees or vehicles to move UOPs or COs. Each of these resources has a cost associated with its use. Measuring productivity includes taking a detailed look at the cost of each of these components to provide economic justification for improvements in these areas. This cost justification approach affords the operations manager an opportunity to show how proposed changes will reduce operating costs by:

- Improvements in labor productivity
- Increased capacity to handle a greater vendor UOP number or a larger CO volume by adding mechanical equipment
- Reducing vendor UOP, CO, or vehicle travel path length or width, which is building space
- Reducing vendor UOP or CO, equipment or building damage and reducing personnel injuries
- Assuring on-time and accurate vendor UOP or CO deliveries

To implement a change to improve an operation, the manager should be aware of how well the operation measures up to company expectations; and if there is an opportunity to improve a company operation, how to project future employee productivity.

However, the cost of vendor UOPs or COs handled does not provide the total picture of productivity of an order-fulfillment operation. A more practical approach to measuring productivity is to use ratios that provide deeper insights into the cost components. There are seven ratios, which we describe below in the section, "Seven Order-Fulfillment Operation Ratios." Before we discuss those, we outline the cost components that affect productivity in an order-fulfillment operation. These include:

- Dollar cost measurement
- Concept expenses as a percentage of sales
- Weight handled per employee hour
- Vendor UOPs or COs handled per employee hour
- Labor ratio
- Direct labor handling loss ratio
- Vendor UOP or CO movement and operation ratio
- CO cycle efficiency ratio
- Space utilization efficiency
- Equipment utilization ratio
- Aisle space potential ratio

Dollar Cost Measurement

This cost measurement technique is used because:

- It is common to all order-fulfillment concepts
- It impacts the company's financial statements and ratios
- It can be related to the management bonus program
- It is a key factor in providing economic justification for a new concept
- It is directly related to the operation's annual budget

To calculate the dollar cost or operating expense measurement for an operation, the financial department provides the actual expenses or budget dollars.

The disadvantages are accounting dollar figures that show past performance of an order-fulfillment concept, possible inconsistency between two accounting periods and what is considered a cost factor, and costs that are considered non-controllable operational factors. The advantages are that it is easy to calculate, tied to many important management activities, used in many companies, and can be tracked and compared to the historical company data.

Measurement of Cost per Piece or Customer Orders Handled

Components of a formula for measuring cost per vendor UOP or CO include the costs of order-fulfillment equipment divided by the total vendor UOPs or COs that were handled by an operation.

The disadvantages and advantages are similar to the dollar measurement technique, except that the cost per piece or customer order handled measurement varies by vendor UOP or CO volume.

Measurement of Cost (Expenses) as a Percent of Sales

Measurement of cost as a percent of sales uses the total order-fulfillment concept costs and total vendor UOPs or COs that were moved or stored. To calculate the cost as a percent of sales measurement, the total concept operational costs is divided by company total sales.

The disadvantages are (1) that there are fluctuations by cost per vendor UOP or CO and (2) costs are affected by factors that are non-controllable to an order-fulfillment concept manager. The non-controllable costs are the operational costs that are not directly controlled by the concept manager.

Weight for Measurement of Pieces or Customer Orders Handled

Weight for vendor UOPs or COs handled is measured by total weight for total vendor UOPs or COs that were handled by an operation and total number of employee hours that were required to move the UOPs or to complete COs.

To calculate weight for vendor UOPs or COs handled:

- Multiply vendor UOPs or COs by the associated weight for each UOPs or COs to obtain the total weight handled
- Divide the total weight by total employee hours

This measures the total weight of vendor UOPs or COs that were handled and it shows employee hours that were required to move or store vendor UOPs or to complete COs. When a UOP or CO mix has a consistent shape and weight, the advantages of this measurement are that it is relatively easy to calculate and one period can be compared to another. If the UOP or CO order mix has a wide variety in shape and weight, the total weight is calculated using an average weight for vendor UOPs or COs or by summing the individual vendor UOP or CO weights.

Measurement of Vendor Pieces or Customer Orders Handled

Measurement of vendor UOPs or COs handled is similar to measuring weight for vendor UOPs or COs completed. The similarity is that both measurement methods are based on vendor UOP or CO number completed by an operation and the total number of employee hours.

To calculate vendor UOPs or COs completed, the total number of UOPs or COs that were completed by an operation is divided by the total number of employee hours. If a concept handles a wide piece or customer order mix with a wide weight variation, measurement of vendor UOPs or COs completed allows comparison of an operation's productivity over several periods.

Employee Hours Must Be Consistent

The value to an organization of consistent measurement of productivity in the order-fulfillment operation is that it allows comparisons between two or more time periods and permits tracking; it can be used as an annual operational expense budget tool; and it can be used to schedule employees, equipment, and vendor delivery trucks.

In order to have consistent measurement, number of employee hours must also be consistent for each period.

Management may wish to compare your company's productivity measures to those of another company. The difficulty in doing this is that the employee activities that are considered operational activities at one company may differ from activities considered operational at another company. However, industry professionals agree that employees who have direct contact with a vendor UOP or CO or employees who control equipment that touches or moves vendor UOPs or COs are considered operational.

Disagreement among industry professionals and managers concerns employee activities (hours) that are associated with indirect or support employee activities (hours). Indirect or support employee activities include supervision, clerical, maintenance, and sanitation hours. These employee activities or employee controlled equipment do not directly touch or move vendor UOPs or COs. However, without their support activity an operation has a difficult task to maintain its budgeted operational costs, projected employee productivity, and on-schedule customer service. Therefore, some industry professionals consider some or all of these employee support activity hours as operational hours.

In my view, components of productivity measurement for an order-fulfillment operation are:

- One subtotal for direct employee hours.
- A second subtotal for indirect employee hours.
- The sum of all direct and indirect employee hours. It is noted that these are major employee classifications and that within direct and indirect labor classifications there are subclasses.

This order-fulfillment concept measurement method for separate labor hour classification provides a complete analysis for an order-fulfillment concept operation and a tool for your employee work scheduling and preparation of your annual expense budget. Separating employee hours shows how employee hours, vendor UOP or CO volume variance affect into direct and indirect hours employee classifications and hours, productivity, and costs. To justify an actual dollar cost to budget dollar cost variance, this employee hour classification provides the data to explain the variance.

Seven Order-Fulfillment Operation Ratios

The seven order-fulfillment concept ratios appeared in the *Bases of Material Handling*, a publication that was prepared by the Material Handling Institute. As stated in this publication, each ratio represents a way to establish quantitative figures for a storage concept. The advantages of using these ratios are:

- They are useful in determining effectiveness of an order-fulfillment operation
- They make use of historical operational data
- The ratios show improvement in the operation by allowing comparison of past, present, and projected figures
- They affect short- and long-term investment strategies for an order-fulfillment concept

The seven order-fulfillment concept ratios are:

- Employee (labor) ratio
- Direct employee handling loss ratio
- Piece or customer order movement operation ratio
- Customer order-fulfillment cycle efficiency
- Space utilization efficiency ratio
- Equipment utilization ratio
- Aisle space potential ratio

Employee (Labor) Ratio

The employee (labor) ratio indicates the number of employee hours that were assigned to operational activities as a percentage of the company's total operational employee number. Employees who were assigned to a task that involved moving, or controlling equipment that moved or stored, a UOP or completed CO make up the numerator in this formula. Employees who are assigned to an operation but do not perform a direct function, including clerical, supervisory, maintenance and sanitation employees, are included in the total operational employee number.

The advantages of this ratio are that it is easy to calculate, and provides an indication of the proportion of direct employee or labor hours to total employees. Thus, it can be used to demonstrate improvements by showing labor savings or by allowing historical ratio data to be tracked graphically.

Direct Employee Handling Loss Ratio

The direct employee handling loss ratio measures direct operational employee time for employees who are allocated to non-order operational activities but performed an operational activity. The employees completed or in the course of their work performed an operational activity. This represents additional time that is allocated to the operational activity time and reduces the direct employee operational activity time. The direct employee handling loss ratio is calculated by the total non-operational employee time divided into the operational time lost by non-operational direct employee time.

In addition to the advantages listed above for the direct employee ratio, advantages are that it is easy to calculate, to consider a new order-fulfillment concept, or a change to an existing order-fulfillment concept, it identifies all order-fulfillment concept activities that are directly affected by an order-fulfillment concept and this lost employee time can be used as part of a new concept economic justification, due to the fact that employee hours are converted into a cost reduction.

The disadvantage is that the non-operational direct employee loss time is accurate. Steps to verify the non-operational direct employee loss time accuracy is to have an outside consultant evaluate the operation or perform a time study for the specific operational activity.

Piece or Customer Order Movement Operation Ratio

The vendor UOP or CO movement or operation ratio is considered the number of times that an order-fulfillment concept is required to move vendor UOPs or COs that are delivered to a pick line to complete a CO. It is expressed as vendor UOP or CO movement activity station or pick line operation ratio. In an order-fulfillment operation, the ratio is calculated by dividing the number of pick line stations into a CO move number that includes receiving, storage replenishment, pick, pack, manifest, and load.

The advantages are that it determines the efficiency of your order-fulfillment on a pick line operation, identifies slow-moving SKUs that are grouped together, and reduces the number of pick position replenishment trips.

Customer Order-Fulfillment Cycle Efficiency

This ratio shows the actual operational time that is spent to pick/pack and ship a customer order out of the shipping door. This ratio indicates the time that is required for a vendor UOP or CO to move through the storage operation. It represents the time it takes for a UOP or CO to move from the receiving door to the shipping door.

The customer order-fulfillment cycle efficiency is calculated by dividing the total time that is required to complete (pick, pack, and ship) a CO into the total operation time. It measures the total time required to move UOPs or COs from the receiving area, to the storage area, from the storage area to pick line, from the first pick position to the pack and manifest position and through the shipping door. With this door-to-door analysis, this ratio shows the length of time that a UOP or CO is in the storage area, in a storage concept and order-fulfillment area or the time it takes for a CO to travel over your storage and order-fulfillment concept transport and sortation concept. The value of this ratio is that it shows the value of a just-in-time replenishment to a pick position or program for a company supply chain logistics strategy.

Space Utilization Efficiency Ratio

The space utilization efficiency ratio measures use of the space for an order-fulfillment concept such as racks, pick positions, or the travel path used in transporting UOPs or COs in relationship to total space for the concept. Examples are:

- Occupied storage positions to number of total storage positions in a storage area
- Occupied pick positions to total number of pick positions in a pick aisle, on a pick line, or in an automatic pick machine
- Receiving and shipping dock staging area to total area
- Space for back-up vehicle or personnel aisles along a powered vehicle travel path to the combined total space for all aisles in a facility

The space utilization ratio determines the occupied storage/pick positions as a percentage of total storage/pick positions in the storage/pick area concept. To calculate the space utilization ratio, determine the total storage/pick positions or cubic feet occupied and divide by the total number of storage/pick positions or cubic feet in a storage/pick concept area. When the cubic feet or ¼-, ½-, or ¾-full pallet analysis is made, space utilization indicates a more exact picture. For example, in a single-item, GOH, master carton, or pallet storage area, space occupied is based on storage positions with single-item, GOH, cartons or pallets divided by total number of storage positions in a storage area.

The benfit of the space utilization efficiency ratio is that it identifies functional space that is not effectively used. This represents space for expansion, for additional pallet storage positions, or for adjustments to the rack position height.

Equipment Utilization Ratio

The ratio of occupied storage/pick positions to the total number of storage/pick positions shows the effective use of available storage/pick positions. To determine the storage/pick equipment utilization ratio, the total number of storage/pick positions is divided into the actual occupied storage/pick positions. To determine a pick equipment utilization ratio, occupied pick positions to total pick position number in a pick aisle, along a pick line, or in automatic pick equipment is divided into actual occupied pick positions in a pick aisle, along a pick line, or in the automatic pick equipment.

The advantages are that it shows the available equipment capacity. Completed on a daily or weekly basis, it indicates the slowest time period which is available for maintenance, equipment modification, or new equipment installation. Calculated for each workstation on a value-added process line, in a pick aisle, along a pick line, or in an automatic pick machine, it shows potential low pick productivity or vendor UOP or CO queue areas.

Aisle Space Potential Ratio

The aisle space potential ratio shows the effective aisle area use and is very similar to the space utilization efficiency ratio. The aisle space potential ratio shows recommended replenishment and pick aisle space and the actual replenishment and pick aisle space.

Recommended aisle space is by the minimum aisle width pick or storage equipment manufacturer. In the storage area, the minimum pick or storage vehicle aisle width is the clear distance between two storage or pick racks (product) and it is the dimension that is required for a vehicle to make a right angle (stacking) turn or two order pickers to travel in an aisle. The right angle turn allows the storage vehicle to complete a storage transaction.

In other facilities, minimum aisle widths allow a vehicle to travel from one aisle to an adjacent aisle, or allow two-way vehicle traffic through an aisle.

In a conveyor operation, the minimum travel path is the actual space that permits vendor UOPs or COs to move on the conveying surface, or the actual space that is required for the structure to support the conveying surface and vendor UOP or CO clearances.

Actual aisle space is the actual distance between two storage or pick racks, pieces, two pieces of equipment, or from equipment to a building wall or obstacle. The recommended aisle floor space is calculated by the equipment manufacturer recommended aisle width multiplied by the aisle length. The actual aisle floor space is calculated by the actual aisle width and actual aisle length.

It identifies available space that is occupied by a traffic aisle or personnel or mobile equipment travel path. When remodelling or a new order-fulfillment concept that requires a narrower aisle is being considered, this ratio helps provide the economic justification for the alternative order-fulfillment concept.

Various Measurement Standards

For an employee productivity program to be useful, employee productivity is tracked and compared against a company standard or budgeted employee productivity measurement. Measurement standards for comparison may be an agreed-upon standard, industry standard, own company standard, time study standard, or regression analysis.

Agreed-Upon or Budgeted Standard

The first step in deciding on an agreed-upon or budgeted employee measurement standard is to predetermine the number of times that an employee performs the activity. This activity frequency is agreed upon by your management staff and employee group. The performance standard becomes an employee productivity rate to perform the activity. This employee productivity rate is the hour rate that is used to project the annual labor expense for the annual operational budget or to economically justify the concept expenditure.

Industry Standard

When an order-fulfillment operation uses another company's employee productivity measurement as its employee productivity measurement standard, the potential problems are:

- Differences in vendor UOP or CO mix and characteristics, nature of the business, and size.
- Confidence in the accuracy of the other company's employee productivity figure. It is important to know how the productivity figure was calculated and what factors was the other company used in calculating the productivity figure.

- Differences in employee productivity figure components.
- Accounting uniformity for the employee productivity.

Company Standard

Company historical or goal-set employee productivity figures as a measurement uses employee productivity figures obtained from your company's past storage concept records.

Time Study Standard

The time study productivity measurement is based on observations over time of the employees who perform the activity. This is usually done by an industrial engineer or a member of the management staff. The observed times are averaged and form the basis of an employee productivity standard for future employees to perform the activity.

Regression Analysis

The regression analysis method is a complex mathematical calculation that involves multiple variables.

Chapter 2

Receiving, Manifesting, Loading, and Shipping Concepts

Introduction

Topics covered in Chapter 2 include:

- Dock locations
- Dock designs
- How to bridge a dock edge and delivery truck gap
- Dock staging concepts
- Dock seals and shelters
- Delivery vehicle load and unload concepts: (a) mobile equipment, (b) conveyor types, (c) automatic types
- Each or piece SKU identification concepts and master carton pallet label options
- Shipping carton sort concepts: (a) manual or mechanized, (b) travel path layouts and options
- Shipping lane concepts, design, direct load, unitize onto pallets or BMCs
- In-house transport concepts: (a) manual and (b) powered

Design of Receiving and Shipping Docks

Your small-item, flatwear, or GOH order-fulfillment operation starts at your receiving docks and ends at your shipping docks. Your receiving and shipping dock areas should be designed to ensure a cost-effective, efficient, smooth, and continuous movement of UOPs through the order-fulfillment operation. A receiving activity ensures the vendor delivery matches your company's purchase order and that there is a UOP moved from the dock area to the storage area. A shipping activity ensures flow of CO ordered, picked, packed, and manifested packages from your process area into a CO delivery vehicle. Cost-effective and efficient receiving and shipping dock activities require a vendor UOP delivery or CO vehicle spotted at the correct dock and at the appropriate time. An additional dock incremental cost is a small investment when compared with the order-fulfillment operational costs that are associated with the increase of an in-house transport concept cost, low employee productivity, poor vendor-delivered piece flow, late customer order (CO) delivery, or inability to unload or load critical vendor delivered pieces (UOP) on-time.

Several parameters and operational factors assure an accurate and on-time UOP and CO delivery truck movement within your truck yard. Elements of a cost-effective and efficient unloading activity of a vendor UOP delivery truck and loading activity of a CO delivery truck are:

- Accurately projecting the number of vendor UOP and CO delivery trucks and associated unloading and loading times
- Good design of the receiving and shipping UOP and CO dock staging areas
- Easy vendor UOP and CO delivery truck access to a truck yard, best truck traffic flow pattern within a truck yard, and sufficient truck yard space in front of each dock location
- Adequately designed delivery vehicle holding area in a truck yard
- Properly designed and equiped vendor UOP receiving and CO delivery truck docks

Delivery Company Load Time

Objectives of the order-fulfillment operation are to achieve a fast CO delivery cycle time and maintain the company's delivery standard. The two major factors that affect on-time CO delivery are (1) order-fulfillment pick and pack productivity and (2) the time it takes for a delivery company to deliver a CO package to the delivery address. The time it takes for a CO package to arrive at the customer's delivery address depends on truck departure time from your facility's dock and arrival at the delivery company's terminal; time to process the CO package through the terminal and load it onto a local delivery truck; and for a local delivery truck to travel from the terminal to the delivery address. If the delivery truck does not depart on-time from your shipping dock, it arrives off-schedule or late at the delivery company's freight terminal. Late arrival means your CO package is not unloaded and sorted onto that day's local delivery truck, but waits in the terminal for sorting and loading onto the next day's delivery truck.

Factors that affect on-time delivery truck departure from your shipping dock include number or cube (volume), full-load delivery trucks, travel time from your shipping dock to your freight company terminal, your freight company sort start time and length of sort time, and your freight company local delivery truck capacity (package number and cube).

To assure the maximum number of freight company delivery truck departures for each work day your order-fulfillment pick and pack activities should start as early as possible to assure that the maximum delivery truck number arrives at the freight terminal. To establish the start time, you should determine the time that is required for each of the factors listed above. The start factor is your delivery company truck unload and sort time.

Receiving and Shipping Dock Locations

Receiving and shipping dock locations directly affect small-item, flatwear, GOH, master carton, or pallet flow and employee productivity. Location of receiving and shipping docks should reduce in-house transport distance between the receiving or shipping dock area and the storage area (see Figure 2.1). The best receiving dock location is determined by the SKU storage position in the storage area. The incremental cost for an additional receiving dock is a small investment when compared to in-house transport investment increase, low employee productivity and poor product flow. The basic delivery truck receiving and shipping dock concepts are combination docks, separated docks, and scattered docks.

Combination Docks

In a combination (shipping and receiving) dock concept, the receiving and shipping activities are performed in one building area, which means the operation has few dock positions. Combined receiving and shipping activities use the same building area, equipment, and employees. For best results, a combination concept has a delivery truck dock schedule with (inbound) vendor-delivered UOPs delivered in the morning and CO packages (outbound) loaded in the afternoon. This concept is best for small facilities that handle low UOP and CO volumes and each delivery truck has a small master carton or pallet quantity. With receiving and shipping docks on the same wall, UOP flow is a horseshoe or U-pattern.

The disadvantages are an increase in in-house transport, the need for exact inbound and outbound delivery truck schedules and difficulty compensating for pallet delivery problems and business fluctuations. The advantages are maximum facility utilization, equipment and employee flexibility, increased supervision, easy-to-assign delivery trucks, and improved security.

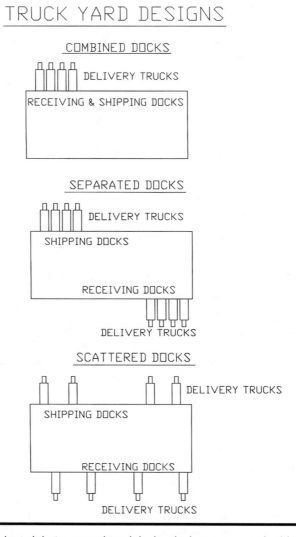

Figure 2.1 Truck yard designs: combined docks, docks on separate building sides, and separate docks on building sides in remote locations.

Separated Docks

A separated dock location concept has the receiving activities and shipping activities performed in separate building areas and with separate equipment, employees, and supervisors. This concept reduces in-house transport distances and activities. Separated dock locations work best for a large operation that has a high volume, large SKU number, and large SKU mix. With receiving and shipping docks in separate locations, the dock operations handle any UOP or CO flow pattern.

The disadvantages are increased building and dock equipment investment and an increase in employee number. The advantages are flexibility in scheduling trucks, increased capacity to handle business fluctuations, improved UOP flow, and reduced in-house transport costs.

Scattered Docks

A scattered dock concept is variation of the separated dock concept that has the receiving docks located along one entire building wall. One wall or area is assigned as the inbound building side, with each dock located directly across from a SKU storage area. This dock arrangement permits master cartons or pallets to flow directly (shortest distance) in a straight line from the delivery truck dock area to the assigned storage area. The shipping docks are located along the opposite building wall, thus allowing UOPs to flow from a storage area, through a pick area, and from a pack/manifest area as a CO to the shipping dock area. A scattered dock concept dictates a one-way straight UOP flow pattern.

The disadvantages are increased design and planning activity and management discipline and control. The advantages are continuous and uninterrupted UOP flow and shortest in-house travel distance.

Projecting Required Number of Truck Docks

Your facility design team determines the number of receiving and shipping delivery truck docks that are required, based on time to unload a vendor UOP or to load a CO delivery truck. Slightly different factors are used to determine requirements for these.

Determining Number of Vendor Truck Docks Required

The number of receiving truck docks that is required at your facility is determined by manual calculation, manual dock simulation, or computer dock simulation.

Manual Dock Calculation Method

Manually calculating the required number of vendor delivery truck docks is easily done by your receiving department, as follows:

- Tabulate for a time period, either is a week or month, the number of daily truck deliveries for the peak and average days plus growth
- Identify the time of day and frequency of vendor delivery truck arrivals
- Determine the delivery truck load type, that is floor stack or pallets

- Estimate the time that is needed to unload and stage the vendor-delivered UOPs on a receiving dock or unload UOPs onto a powered conveyor transport concept
- Identify and determine the vendor delivery vehicle type, such as ocean-going container, back-haul trailer, common carrier, Federal Express, United Parcel Service, United States Postal Service, etc.
- In the dock area, estimate the number of unloading conveyors
- Include the time for coffee and lunch breaks, vendor truck maneuvering in the truck yard, vendor truck driver time to sign the delivery document, and time to break the seal; also determine a reasonable safety factor such as 20–25 percent

A simple formula to determine the number of receiving truck docks required is to multiply the projected number of delivery trucks for the year by the number of hours it takes to unload each delivery truck, and divide that by the total number of work hours per year. You should provide an adequate number of docks to unload the maximum number of delivery trucks that are scheduled to arrive at your facility. You should also add at least one dock space for a ramp or dock lift to permit forklift travel between the dock and ground level, a maintenance dock with a high door, as well as docks for trash, paper bales, and delivery services such as UPS and USPS.

Manual Dock Simulation

A manual dock simulation method projects the number of vendor delivery truck docks required for the vendor delivery vehicle types. A manual delivery truck dock simulation provides data on:

- Number of delivery trucks that arrive at your facility during an average day. This includes floor stacked or palletized loads.
- Vendor delivery truck or container rear bed heights.
- Vendor delivery truck arrival time at the facility.
- Number of hours remains that each vendor delivery truck at a dock. This time includes truck spot, driver document signing the delivery document, and departure times.
- Number of standard and special dock.
- Your company's anticipated growth rate and peak activity factor.

In addition, you will need a delivery truck dock simulation form that shows the number of docks along the top and hours of the work day in 15-minute intervals down the side of the form. This form is used to track the number of delivery trucks that arrived at the facility for each time period. The form should show number and lengths of coffee and lunch breaks.

With this receiving dock data, you have the necessary information to perform a manual dock simulation for your existing or proposed receiving dock area. A manual dock simulation shows:

- Dock number utilized per hour
- Total available dock hours for one weekday
- Number of receiving docks that is required to handle your projected vendor delivery trucks

This is based on your company's experience and policy. Options are to provide a dock for every delivery truck as it arrives at your facility, or to have a fixed number of dock positions at your facility. With the latter arrangement, if the number of delivery trucks arriving exceeds the number of docks at your facility, there should be a temporary holding area to which the trucks are sent.

A manual vendor delivery truck dock simulation shows:

- How to improve your present vendor delivery dock operation
- Requirement to develop a vendor delivery truck arrival schedule for better vendor delivery truck dock utilization
- The number of vendor docks required for your present or proposed facility

A truck dock simulation is carried out as follows:

- On the simulation form described above, enter the time that each delivery truck is assigned to a receiving dock.
- Fill in the squares under the appropriate receiving dock that represents the delivery truck unloading time. Remember that coffee breaks and lunch breaks represent delivery truck dock occupancy time even though delivery truck unloading activity does not actually occur.
- As other delivery trucks arrive at the facility, the other docks become occupied with delivery trucks.
- If a new delivery truck arrives at the facility and all existing docks are occupied by delivery trucks, the new delivery truck is assigned to the temporary holding area until a dock becomes available. If a delivery truck dock becomes available, the oldest delivery truck from the holding area is assigned to the available dock and you repeat the steps listed above.

Computer Dock Simulation

When you use a computer dock simulation to design your vendor delivery dock receiving operation, you provide the same receiving delivery truck activity, vendor delivery truck volume, employee productivity, dock number, and other dock area design and operational information that was mentioned in the sections on manual dock calculation and manual dock simulation. After this receiving delivery truck data is entered into the computer, the computer program projects each receiving delivery truck dock utilization, dock number, and other dock area design information.

Determining Number of Shipping or Customer Order Delivery Truck Docks Required

The number of CO packages per shift is a key factor in determining the number of docks required for CO shipping. If your operation has a direct-load CO package concept, the number of CO delivery or shipping truck docks required is equal to the number of CO packages that are scheduled for the shift. If your operation has CO packages that are placed in a staging area for later loading onto the delivery truck, you would use the manual calculation, manual dock simulation, or computer simulation methods described above to determine the number of shipping docks that are required.

Staging Area or Conveyor Network

In the dock receiving and shipping areas should be provided sufficient floor space for a dock staging area or queue conveyor. The dock staging area is based on the vendor-delivered UOP and CO package flow concept. If a dock operation has vendor-delivered UOP or CO packages held for a period of time, the dock operation has a requirement for a dock staging area. Factors to consider in designing a dock staging area are anticipated piece quantity, ability to utilize storage racks, portable stacking racks or frames, and employees and vendor delivery schedule or CO delivery truck availability.

If a receiving dock operation has a WMS identification attached to each vendor-delivered UOP for deposit into a WMS-identified storage position, the vendor-delivered and -received UOP flow is more dynamic and requires a smaller dock staging area. If completed CO flow is direct from a pack area, past a manifest scanner and into a CO delivery vehicle, a shipping dock area requires less space. If the completed CO flow is from the pack area, past a manifest scanner, to be unitized onto a pallet or four-wheel cart, the shipping dock area has a large dock staging area.

Delivery Truck Access to Docks

Vendor UOP and CO delivery vehicle access and exit to your facility truck yard is a factor that determines receiving and shipping dock locations. At a large facility that is located on a site with sufficient land, vendor delivery truck access is designed to improve delivery truck flow through the truck yard and reduces the risk of accidents.

A well-planned truck access road allows vendor or customer delivery trucks to be driven forward from a public highway into a truck yard safely, rapidly and with minimal of delivery truck maneuvering. An exit road permits a vendor or CO delivery truck to be driven forward from a truck yard onto a public highway. To permit good vendor or CO delivery truck flow in the truck yard, the access–exit road is at least twice the longest tractor and trailer vehicle length. Between the public road and truck yard, a vendor or CO delivery truck passes through a security gate and past a guardhouse. A security guard ensures that only authorized vehicles are allowed onto company property, the delivery truck has arrived at the correct address and contains UOPs that are assigned to the company facility, the seal is not broken and delivery truck arrival is on time.

The security guard notifies the receiving department of the delivery truck arrival. A receiving employee advises the security guard to direct the vendor delivery truck to a preassigned receiving dock or delivery truck-yard holding spot.

Because the delivery truck drivers are on the left side of the tractor cab in the United States, vendor delivery truck traffic flow around a facility is a counterclockwise pattern. A counterclockwise pattern provides the truck driver with clear visibility while driving the truck through the truck yard or backing the truck up to the receiving dock position. When a truck driver backs up to a receiving dock position from a clockwise traffic flow pattern and sits on the left side of a tractor, there is low productivity because of a "blind side" or difficulty to spot the truck at the truck dock. A straight truck (cab and bed on one frame) has a different turn dimension than a tractor and trailer combination. A tractor and trailer turns on the rear wheel and requires additional space. In accordance with the site configuration, local codes, or architectural requirements, the delivery truck-receiving docks are located in the rear or along side of a rectangle-shaped building. A vendor delivery truck is driven to the rear or appropriate building side. With a delivery vehicle of at least 40,000 lb. traveling on a service road and truck yard, a vehicle travel path requires a minimum of 9 in. of crushed gravel, 5 in. of asphalt, proper lighting, highway lines, and proper location or building door signage.

Truck Yard Traffic Flow Patterns

In a truck yard, delivery truck flow patterns are either one-way or two-way.

A One-Way Truck Yard Pattern

In a one-way truck yard traffic pattern, the facility is encircled by the truck road and yard. The yard should be designed to allow at least a 13-ft. wide space from the building that allows for the longest parked tractor-trailer combination plus 13 ft. and maneuvering area that extends outward from the delivery truck dock. For an employee walkway adjacent to the truck traffic lane, add 4 ft. to the above calculated width. The disadvantages of a one-way pattern are the necessity for increased service road construction and additional land investment. The advantages are better delivery truck flow and improved truck yard safety.

A Two-Way Truck Traffic Pattern

A two-way truck yard traffic and road pattern allows the vendor delivery trucks to travel in both directions between the guard station and the delivery truck dock. The concept does not require the road to encircle the facility but has at least the longest tractor and trailer combination length, maneuvering area, and a 26-ft. wide truck lane. An employee walkway adds an additional 4 ft. to the width of the delivery truck lane.

The disadvantages are decreased truck yard safety and lower vehicle control. One advantage is improved vendor delivery vehicle receiving dock spotting time.

Delivery Truck Temporary Holding Area

A temporary holding area for delivery trucks is an important truck yard feature. A holding area is located in a fenced area. When a vendor delivery truck arrives ahead of its scheduled receiving time, it is allowed to exit the public highway and is assigned to a temporary parking area. Figure 2.2 shows options for design of the holding area.

The Block or Square Pattern

In a block or square parking pattern, the delivery trucks are parked along the building area perimeter, in a straight line against the security fence. Alternatively, delivery trucks are parked in the interior of the holding area, in a single row, back to back. For parking along the security fence, an option is to place wheel bumpers on the pavement. These stop the truck's rear wheels and prevent the rear of the trailer from hitting the fence. A block pattern provides a holding area with the maximum number of trailer parking positions with the widest maneuvering area.

Note that delivery trucks with rear doors that face the perimeter fence pose a potential security problem. This is reduced if the trucks are parked on the interior, because they are back to back, with one truck rear door against another.

The Angled Parking Pattern

In a 45-degree angled parking pattern, the delivery trucks are parked along the holding area perimeter at a 45-degree angle to the security fence and to the middle

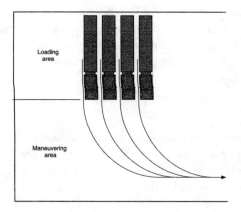

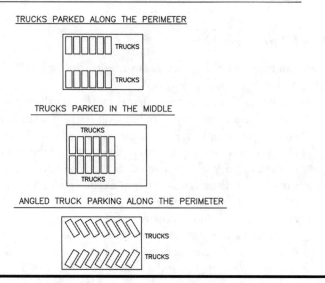

Figure 2.2 Truck temporary holding locations: maneuvering area, along the perimeter, in the middle, angled along the perimeter.

of the truck yard. When compared to the block delivery truck parking pattern, the 45-degree angle delivery truck pattern provides fewer delivery truck parking positions and a narrower delivery truck maneuvering area.

Back-to-Back and Side-to-Side Parking

In a back-to-back and side-to-side delivery truck parking pattern, trucks are parked in a block or 45-degree parking pattern on the interior of the holding area. With this back-to-back and side-to-side pattern, delivery truck is backed up to the rear door of another truck and its side is adjacent to the next delivery truck. This pattern reduces delivery truck rear and side door security problems, with most delivery truck rear doors blocked by another delivery truck rear door.

Landing Gear Pad

At dock operation, without being hooked to a tractor, delivery trailers are parked or dropped in a temporary holding area or in a truck yard in front of the receiving dock. The delivery trailer landing gear or legs stand on a landing gear pad, which is a reinforced concrete strip 48 in. wide. The length is determined by the number of docks. The pad provides a solid base for the landing gear. Without a landing gear pad, during hot or warm weather, a full delivery trailer has the potential to sink below the delivery tractor fifth wheel proper elevation and make it difficult to move the trailer.

Truck Yard Security

The main features of adequate truck yard security are:

- A security fence and berm that is located on the truck yard perimeter. The security fence and berm combination restricts unauthorized personnel from entering the property, reduces the outside view of the truck yard and reduces the noise traveling from a truck yard to the neighbors in the area.
- Cameras that view the parked delivery vehicles and dock locations. Each camera view is transmitted to screens in a guard house.
- Adequate lighting in the truck yard and along the building perimeter.
- A guard station and a security guard on a mobile vehicle that travels through the truck yard. A guard station maintains your truck yard policies and procedures, stops all vehicles or personnel as they enter a truck yard to verify that the vehicle has arrived at the correct location, with direction from the receiving department assigns a delivery vehicle to a truck yard spot, and assures that as a delivery vehicle exits the truck yard there are no unauthorized pieces on the vehicle and that the delivery vehicle is properly sealed with an appropriate discreet numbered seal.

Factors that render the truck yard a more efficient and safe area include:

- Clear and easy-to-read dock identification and striping of truck yard parking spots, traffic lines and arrows, speed and stop signs, maps and facility identification signs, truck dock guides, and guardrails and bumpers.

Delivery Truck Canopy

When a delivery vehicle is backed up to a receiving or shipping dock position, dock facilities use a truck canopy to prevent rain water or melting snow water from falling onto the dock leveler. A truck canopy is a solid wood or metal member that is attached to a facility wall and extends outward from the facility wall. The extension is slightly angled downward and extends outward by 6 or 12 ft. The truck canopy deflects rain water falling through any gap that exists between the delivery truck top and building wall onto the dock lever, thus preventing water from accumulating on the surface. This improves dock safety, because the powered pallet trucks travel over a surface that is free of standing water.

The Delivery Truck Unloading and Maneuvering Area

The delivery truck unloading and maneuvering area is located in front of the facility docks. A truck yard unloading area is designed for the overall length of the longest tractor-trailer combination. A minimum unloading area length is 65 ft. from the dock edge.

The width of the unloading area is designed for the overall widest delivery trailer plus 3 ft. on each truck side. This overall width dimension compensates for a delivery tractor's two side mirrors. If the unloading area is paved with asphalt and delivery trailers are spotted on the landing gear, a land gear pad that is 4 ft. wide and made of reinforced concrete is installed at the required distance in front of and parallel to the facility docks. A normal location for the landing gear on a vendor delivery trailer is 10 ft. behind the delivery truck front. A reinforced concrete landing gear pad has sufficient strength to prevent the delivery landing gear from sinking below the surface. The landing gear pad supports a 40,000 lb. load on 6 ft. centers.

Truck yard maneuvering space is required to allow the longest overall tractor-trailer combination to back up to the receiving dock. With a counterclockwise delivery truck traffic pattern, the truck yard maneuvering area is a minimum of 70 ft. extending outward from the truck yard loading area. For the normal delivery tractor-trailer length of 65 ft., total truck yard loading and truck yard maneuvering area that extends outward from the dock edge is 135 ft. If short delivery truck combinations are used in your vendor or CO delivery activity, the truck yard maneuvering area maybe decreased in length. For a clockwise delivery vehicle traffic pattern, the truck yard maneuvering area is doubled in length and the truck yard maneuvering length is 140 ft. with a total truck yard loading and truck yard maneuvering area of 205 ft. from the dock edge.

In most truck yards, the loading area of the delivery truck yard has a slight grade or slope from the maneuvering area to the dock edge. The degree of slope is a minimum of 3 percent to a maximum of 10 percent. The preferred slope is 6 percent in geographic areas with snow and ice conditions. Your delivery truck length, local building codes, and company policy determine the exact slope. To assist with water drainage, the truck yard loading area that is adjacent to the facility is slightly sloped from the dock edge toward the truck yard maneuvering area to a drain. This outward slope is not extended beyond 3 ft. from the dock edge. If the extension is greater than 3 ft., the slope meets the truck rear wheels. These wheel and slope characteristics have an effect on a delivery truck rear edge elevation to a delivery dock edge and top of the delivery truck to the dock door overhead curtain.

Since truck yard maneuvering, unloading, and dock areas have the sole purpose to facilitate delivery truck unloading and loading, the areas are designed to accommodate the widest delivery truck range and sizes. At some companies, separate receiving docks are designed to handle oversized delivery trailers.

Delivery Truck Dock Dimensions

In designing a delivery truck dock height, the truck's rear bed height is the most important dimension. This truck dimension determines your receiving truck dock edge height above the ground. If your dock operation owns or leases its delivery trucks, to obtain data on delivery vehicle height and other delivery trailer and tractor dimensions is an easy task. If your dock operation uses a common carrier service, an employee is assigned the task to obtain actual delivery truck data or contact a truck company for the delivery trailer and tractor data. Important information to have when designing truck dock dimensions include the overall length and overall height (including dimension from side mirror to side mirror and additional height for a wind deflector) of tractor and trailer or delivery truck.

Other Important Truck Yard Features

Other important truck yard features that improve truck movement, safety and employee productivity are a maintenance building, truck wash rack, fuel island, forklift truck ramp or dock lift, and, in cold climates, delivery truck engine heaters.

Delivery Truck Dock Concepts

Factors to consider in designing truck dock type and location include climate, weather conditions and prevailing winds, land availability, security, delivery vehicle type, delivery load type, delivery truck traffic flow and control, employee comfort and safety, and available funds for investment.

Figures 2.3 and 2.4 show various truck dock concepts:

- A flush dock design that is either a (a) cantilever flush dock or (b) vestibule flush dock

- An open dock design that is either (a) open dock or (b) open dock with curtains or sliding panels
- An enclosed dock design which can be either (a) a straight-in-entrance enclosed dock, (b) a side-loading or finger dock, (c) or a drive through or in-facility dock
- A staggered or saw-tooth dock
- A pier dock
- A freestanding dock or dock house

The Flush Dock Design

A flush dock design is one of the most common receiving dock designs. A dock facility has an unloading or loading conveyor or pallet truck concept and powered pallet truck or conveyor transport and sort concept. Flush dock designs are cantilever and vestibule.

The Cantilever Flush Dock

A cantilever flush dock is basically a hole in a building wall with a dock door and dock seal along the outside of the doorframe. This concept provides excellent weather protection and security. For unloading, a delivery truck is backed up to the building wall and the truck rear door is secured against the dock seal. The delivery truck remains outside the facility and the truck driver enters the building through a personnel entrance. Inside the building, the dock staging area is open to the storage area. A cantilever flush dock has a dock leveler with a truck ICC bar-restricting device dock plate or chock block. To prevent unexpected delivery truck yard movement during the unloading activity, it also requires:

- Excellent communication between the truck driver and dock employee.
- A stop/go light system. With a cantilever flush dock design, the building wall is set back to provide at least 8 in. of clearance between a delivery truck rear and a building wall at 6 ft. elevation above a dock. Your building architect approves a minimum 6 in. clearance between a delivery trailer top and a building

Dock bumpers that extend outward from the building wall achieve this clearance. Solid rubber strip bumpers are attached to the building wall and absorb the impact from a truck backing up to a dock. The bumper dimensions are determined by the opening size, the loading area shape, the distance between a delivery truck rear bed and rear wheel center. A bumper concept is used with yard jockey tractor, where a trailer slope is increased as the yard tractor lifts the delivery trailer front end higher

ENCLOSED DOCK DESIGNS

ENCLOSED DOCK ONE ENTRANCE & EXIT

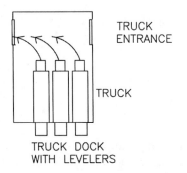

ENCLOSED DOCK ONE DOOR ONE TRUCK OR ONE DOOR TWO TRUCKS

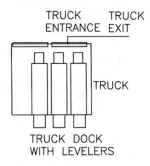

ENCLOSED DOCK FINGER

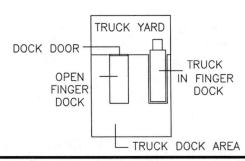

Figure 2.3 Enclosed dock designs.

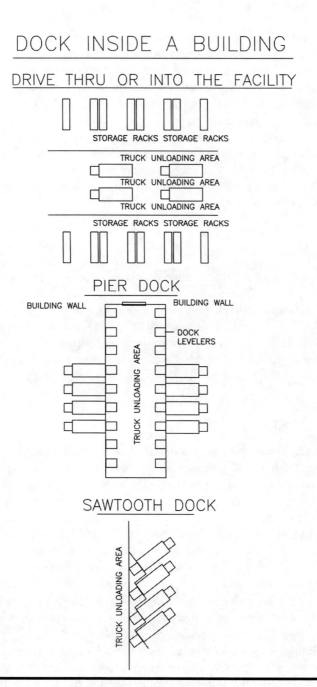

Figure 2.4   Truck dock designs: enclosed dock, one entrance and exit; enclosed dock straight-in; enclosed finger dock; drive through; pier dock; and saw-tooth.

than the normal delivery tractor. Without the bumper and with this delivery trailer angle and height, the tractor-trailer has the potential to damage the building wall. With a flush dock concept, to prevent the risk of rain or snow condition to destroy a dock seal top, a small canopy extends over the seal top and delivery truck rear section. When your dock facility ground floor is on ground level, a flush dock with a depressed driveway is designed to provide a delivery truck dock position. A driveway (loading area) is depressed to create the proper height between the dock edge and the truck bed rear. Special consideration is given to the dock bumpers, water or snow drainage, hand railings, and a window in each dock door. The window in the door allows a dock employee, without opening a door, to observe the truck yard status in front of the dock.

The Vestibule Flush Dock Design

A vestibule design is a variation of the flush dock type. The difference and a vestibule dock is inside the building. With a vestibule dock, there is an open area between the building exterior wall and a second interior wall. The design features several forklift truck aisles, wall penetrations for GOH on a powered trolley conveyor unload, transport, sort and load concept, and with GOH on a cart or pallet dock operation, this dock design has space for a staging area. This design requires building with a large area.

Open Dock Design

The open cock design is a low-cost design that features a concrete platform that extends outward from the exterior building wall. An open dock depth between a dock plate building side and a building wall provides sufficient open space for pallet truck travel onto a dock plate and pallet truck travel on a dock. An open dock width and length provides access to all building doors. Since an open dock concept is exposed to bad weather, employee and power pallet travel is a safety concern. When the open dock design is considered for a dock operation, a canopy is extended at least 4 ft. to 12 ft. beyond a truck dock edge. The canopy prevents rain and snow from falling onto the open dock surface. For powered pallet trucks, water makes a dock leveler and dock area travel path surface area slippery. A wet dock leveler surface or travel path area is considered a hazardous condition for operating powered forklift trucks. Additional open dock concept safety and security options are (1) to install sliding panels or dock curtains on the open dock perimeter. This feature reduces the negative impact of rainy or snowy weather and improves security for an open dock operation. (2) Yellow-coated concrete posts and chains on the dock edge perimeter increase safety.

Dock Curtains or Sliding Panels

To convert an open dock into a semi-enclosed dock facility, sliding panels or dock curtains are installed on the open dock edge perimeter. This feature creates a solid barrier between the dock area and truck yard. When required to perform unloading dock activities and after a truck is backed up to a dock, sliding panels or dock curtains are opened a sufficient width that creates a passageway for powered pallet truck travel between the dock area and the delivery truck. Dock panels and curtains maintain a solid barrier on the dock edge and delivery vehicle door. When not required to perform unloading dock activities, the sliding dock curtains are closed to create a solid barrier between the dock area and truck yard.

The Enclosed Truck Dock Design

With an enclosed delivery dock design, a truck (combined tractor and trailer length) unloading area is inside the building area. At the building exterior is a set of building doors that control the delivery truck entry to the truck unloading and dock area. The interior enclosed dock area has an open or flush dock design. With an internal, open enclosed dock design, a facility design has building dock edge doors that permit unloading onto the internal open dock. When external building doors are opened for delivery truck entry into a truck well, doors help keep the cold air or dust in the truck well from entering the building. To improve enclosed dock area safety and truck stopping efficiency, the following measures can be taken:

- Exterior metal truck entry guides
- Painted guidelines along the full length of the truck well, which should a finished surface
- Two yellow-painted strips along the rear wall of the truck well
- Adequate lighting on the truck driver side
- If located in a cold climate, sprinkler pipe protection from freezing in the cold
- Adequate ventilation is required to prevent truck engine fumes from entering and queueing in the building area
- Adequate water drainage

When considering an enclosed truck dock design, options are:

- A side entrance enclosed dock
- A straight-in-entrance enclosed dock
- A side-unloading or finger-enclosed dock
- A drive through or in-facility dock
- A staggered or saw-tooth enclosed dock design
- A pier dock design
- A freestanding dock or dock house design

The Side Entrance Enclosed Dock Design

With a side entrance enclosed dock design, building construction costs are high. In this design, there are two sets of doors (one on each side of the building) and truck unloading and maneuvering areas are inside the building's four walls. This feature dictates one-way delivery truck traffic flow through the building (in one door on the building's right side and out the second door on the building's left side). The side entrance enclosed dock design is best combined with the staggered (sawtooth) dock design described below. When a side entrance enclosed dock concept is designed for a dock operation that has a manual or mechanized unload, stage area, transport, and sort concept for single items or flatwear. A side entrance enclosed dock, with its high cost, is not the preferred dock concept.

The Straight-in-Entrance Enclosed Dock Design

A straight-in-enclosed dock design has: one exterior building door per truck dock or one extra-large wide exterior building door for every two truck docks. With all straight-in enclosed docks, a delivery truck maneuvering area is outside the building in the truck yard. Since the delivery truck unloading area is inside the building, the delivery truck backs straight up into the unloading area, to the dock edge.

There are two design options: (1) to have a short truck unloading lane for a delivery trailer dropped inside the enclosed area onto a land gear pad while the tractor travels to the exterior truck yard, or (2) to have a long truck unloading lane to accommodate both tractor and trailer in the enclosed area. A straight-in-enclosed dock concept is designed for an across the dock operation that has a manual or mechanized unload, stage area, transport, and sortation concept for single items or flatwear. When a side entrance enclosed dock is compared to the straight-in enclosed dock concept, the is preferred because of lower investment and operational costs.

The Side-Unloading or Finger Dock Design

A side-unloading dock or finger dock design is used primarily for flat-bed delivery trucks, open-sided vans or side-opening vans with side doors that open for unloading pieces. This dock design is a cutout in a building dock area floor. A finger dock cutout is directed inward from the building exterior wall into the dock staging area. A finger dock in a dock area floor has sufficient length and width to permit a flat-bed delivery trailer or open-sided van to back full length into an opening. Concrete platforms on both sides of the cutout permit forklift right angle turns to unload pallets. To improve dock safety, yellow-coated posts and chains surround the cutout on the platform edge. A finger enclosed dock concept is designed for a dock operation that has a forklift truck pallet unload, stage, transport, and sort concept.

The Drive-Through or In-Facility Dock Design

A drive-through delivery truck dock or in-facility dock design is used for flat-bed or open-sided delivery vans. A flat-bed or open-sided delivery truck drives into the building and across the floor. At the proper location, forklift trucks side-unload pallets. This delivery truck dock design has one-way truck travel through the facility or delivery trucks are backed into the facility. The drive-through delivery truck dock design is not considered for a small-item or flatwear dock operation.

The Staggered or Saw-Tooth Dock Design

When there is limited truck yard maneuvering area on a small site, a staggered or saw-tooth dock design is a good option. The staggered dock design has one-way traffic flow in the truck yard. This delivery truck flow matches the approach to a saw-tooth dock angle. The saw-tooth dock design requires a counterclockwise or clockwise delivery truck flow pattern. For an efficient unloading activity, the delivery truck rear bed height is equal to the dock edge height. Disadvantages are approximately twice the building dock space for fewer docks and increased building construction costs. A saw-tooth enclosed dock concept is designed for a dock operation that has a manual or mechanized unload, stage, transport, and sort concept for single items, master cartons, or pallets.

The Pier Dock Design

If the building delivery dock side does not have sufficient wall space for the required number of delivery docks or the building interior layout does not permit construction of a dock area, a pier dock design an option. In this design a wide enclosed section or very wide aisle extends outward from the building wall. Docks are located on each side of the extension. This means one entrance side is the receiving truck dock area and the other side is the shipping dock area. The width of the extension permits a forklift truck turning aisle and a travel path. With the pier dock design there is a truck yard maneuvering area on both sides of the pier dock. A pier enclosed dock concept is designed for a dock operation that has a manual or mechanized unload, transport, and sort concept for small-items, flatwear, master cartons, or pallets.

Dock House and Free-Standing Dock Designs

A free-standing or dock house design is placed in front of a building door. A dock house is a shell that has two walls and a roof. A shell and facility door permits forklift truck travel between a delivery truck and the dock area. A dock house dock is not preferred for an across the dock operation. A free-standing dock design is a hardened metal mobile ramp with a flat top section. A ramp slope and travel path surface permits a forklift to travel up the slope and turn into the delivery truck.

Delivery Truck Dock Door Design Considerations

A delivery truck dock door is a building device. When a dock door is open, it permits pallet truck or cart passage between a dock space and delivery trailer and when it is closed it provides security and energy conservation. Common dock door types are: vertical lift, vertical and horizontal movement, and roll-up. The type of door selected is determined by the building floor surface to ceiling design considerations and available funds. Important dock door options are:

- Windows in a door panel. A small window reduces the need for an employee to open the dock door and verify that a delivery truck is at the dock.
- Lockable. A lockable dock door is designed with a device to lock the door with an industrial paddle lock from the building interior. This feature as well as the dock door window improves security at the facility.
- Weather stripping. In cold and windy climates, weather stripping is placed along the dock door, frame, and bottom to reduce energy loss.
- Folding gate. A folding gate covers the entire dock door passageway. In warm climates a folding gate is pulled open and locked in position across the door opening. The gate prevents personnel from passing through the door a cool breeze to enter the building, while maintaining security.
- Short dock house. In a warm climate, a flush dock option to reduce energy loss is a short dock house that extends outward from the exterior of the building. The short dock house surrounds the delivery truck rear with a passageway through the building wall and reduces movement of warm air into the building. Door dimensions are determined by the delivery truck rear door dimensions, dock seal, or shelter type and required internal environmental conditions. A dock door is considered an energy loss and security problem. In a conventional store and hold operation, the number of docks is kept as small as possible. With a dock operation, maximum dock number is preferred to assure a constant piece flow between the receiving and shipping areas.

For the average dock facility, dock door widths range from 8 to 9 ft. to match an 8 to 8 ft. 6 in. delivery truck rear dimension. When a delivery truck is backed up to a dock, this design permits the delivery truck to have a slight variation from an exact center position. The trend of 102-in. wide delivery trucks permits an increase in the pallet capacity, the 8 ft. 6 in.- or 9 ft.-wide door is preferred for a new across the dock facility. With a 4-ft. high edge of dock, dock door heights range from 8 to 10 ft. The 9-ft. high dock door permits a 7-ft. to 7 ft. 6-in. high pallet on a pallet truck to travel across the dock leveler. A 9 ft. high dock door with 3 in. permits an 8 ft. 9-in. high pallet on a pallet truck to travel across the dock leveler. The 10-ft.

high dock door handles a high cube delivery truck. A dock door height principle is that at least one dock door has a 13 ft. 6 in. to 14 ft. clear height to accommodate the height of the tallest pallet truck with a collapsed mast.

Delivery Truck Dock Height

The best delivery truck dock height provides the smallest variance between the dock edge and the rear of the delivery trailer bed. Designing the dock height to accommodate variations in truck height creates a height differential between the delivery truck bed and the dock edge as well as a gap between the delivery truck bed and dock edge. As the height differential between the dock edge and delivery truck bed increases, the bridge device incline and decline slope increases. If this slope increases to a high degree, as powered pallet trucks or extendible conveyor wheels travel over the dock leveler, a pallet truck or cart hang-up problem occurs on the dock leveler. A hang-up problem decreases dock employee productivity, increases pallet truck and dock equipment maintenance, and decreases dock safety. An edge of dock height for most delivery trucks range between 46 and 52 in. If there is a wide variety of delivery trailer bed heights, dock design options are:

- To design separate delivery truck docks for specific delivery truck bed heights
- To use an extra long dock leveler (up to 12 ft. long) with an 18 in. up-down (vertical) travel from the dock edge
- To install a delivery truck leveler to raise or lower the entire delivery truck in the yard area
- To use rear wheel risers to elevate the delivery truck rear bed height

How to Bridge the Gap between the Dock Edge and Delivery Truck Bed

A dock plate or leveler is a device that permits the gap to be bridged between the dock edge and the delivery truck bed. It allows pallet trucks to enter and exit a delivery truck to unload pallets. The dock plate or leveler extends from the facility into the truck rear bed. The type of leveler selected depends on the frequency of deliveries, loading and unloading equipment, the dock type, piece type, available dock space, edge of dock and truck height variance, and available capital.

Figure 2.5 shows examples of equipment used:

- Portable walk ramp
- Portable dock plate with a T-bar on the underside
- Portable dock board

- Mobile dock
- Recessed, pit, manual, or hydraulic dock leveler
- Manual or hydraulic vertical-stored dock leveler
- Front of dock
- Edge of dock
- Hydraulic jack in the floor
- Scissors lift
- Pit-installed scissors lift
- Dock lift leveler
- Truck trailer gate
- Wheel lifts

A Portable Walk Ramp

A portable walk ramp is an aluminum ramp that is designed with a skid-resistant deck and permits an employee with a two-wheel cart to load or unload a delivery vehicle. A portable walk ramp is not preferred for a dock operation.

A Portable Dock Plate

A portable dock plate is made out of aluminum, and is 2 to 5 ft. in length, 5 to 6 ft. wide, with an equalizing bend. It allows a dock edge to truck bed height differential of from 3 to 10 in. The shorter the length of the plate, the less the difference in height that it can accommodate. To prevent pallet truck or cart skids on the plate, the surface of the travel path has a diamond pattern instead of a smooth surface. The dock plate has a locking leg (T-bar) attached to the underside, which extends downward and fits into the gap between the truck bed and the dock edge. The plate has handles on both sides to allow it to be moved easily between docks by an employee.

The portable dock plate is a light-weight, low-cost device, which has sufficient structural strength to handle a pallet truck. Safety yellow strips along both sides of the travel path minimize the possibility of pallet trucks running off the sides. This kind of device should not be used with a mobile extendible belt carton conveyor or with electric-powered pallet trucks.

A Portable Dock Board

A portable truck dock board with an equalizing bend is an aluminum device that permits an electric pallet truck or forklift to unload or load a delivery truck. To reduce pallet truck skids on a dock board, its travel path surfaces have a diamond pattern. To secure the dock board in the truck and building gap, locking legs (T-

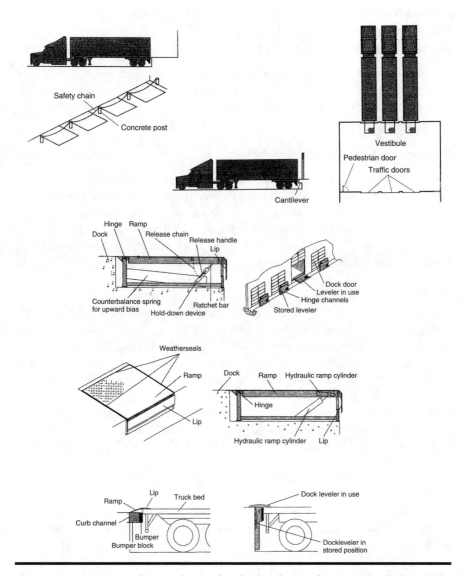

Figure 2.5 Equipment that can be used to bridge the gap between the dock and the truck bed: cantilever dock; vestibule dock; hydraulic dock leveler; manual dock leveler; edge of dock leveler; front of dock leveler.

bars) are attached to the underside, extend downward, and are inserted into the gap between the truck and building. Yellow-painted safety curbs help eliminate pallet truck run-offs. The dock board requires a powered forklift truck to move and position it in the required dock. When compared to a portable dock plate, a portable dock board has the same restrictions against its use with master carton conveyor but can be used for forklift trucks and electric- powered pallet trucks.

Mobile Yard Ramp, Scissors Lift, Bascule Bridge Dock, In-Floor Hydraulic Lift, Tailgate Trailer, Manual Wheel Lift, and Dock Leveler Lift

Although these devices can bridge the gap between the dock edge and delivery truck bed, they have limited use in dock operations. The reasons are low productivity, higher investment, and limited flexibility.

Recessed Dock Leveler

For operations that have a dock edge 4 ft. above ground level, a recessed dock leveler is the most commonly used device to bridge the gap. The recessed dock leveler is installed in a pit or floor cut-out. It has a flexible lip that extends outward and is lowered to rest on the delivery truck bed (Figure 2.6). The standard length of the lip is 16 in., permitting a 12-in. extension into the truck bed. For safety the travel path surface is diamond-patterned. The length of the lip projection is especially critical for a refrigerated delivery truck, where the pallet truck wheels could become caught in drains in the truck bed.

The leveler is either hydraulic and automatic or mechanical and manually operated. The structural strength of the recessed leveler permits forklifts or pallet trucks and master conveyors to travel between the dock and the delivery truck bed. Dimensions depend on the type of loading and unloading equipment utilized at the facility (see Figure 2.7). A common length is 8 ft., which can accommodate a 10 in. height differential. Typically, an electric pallet truck requires a 6-ft. long leveler that bridges a 7- to 8-in. height differential. A 12-ft. long leveler can accommodate a height differential between 13 and 18 in. and is used for an electric pallet jack or gas-powered forklift truck. The width of the recessed dock leveler is between 6 and 7 ft., the optimum width for an 8-ft. wide delivery truck. The extensible lip is tapered on the sides to compensate for a truck that is parked off-center or for a narrower delivery truck.

A hydraulic recessed dock leveler has a hydraulic pump and motor in the pit that moves the travel path surface to the desired elevation. The dock leveler is controlled by push buttons that are located on the interior building wall adjacent to the dock. When a delivery truck arrives at the dock, an employee presses a button that causes the dock leveler to automatically rise and lower to a flat level position with the lip inside on the delivery truck bed. With the lip on the delivery truck rear bed, the travel path surface is relatively level as a bridge between the dock edge and delivery truck bed. At this level position, pallet trucks or master carton conveyors can enter and exit the delivery truck. With few mechanical parts and structural members in the pit, the hydraulic recessed dock leveler is very reliable, requires little maintenance, and permits pit cleaning. When compared to a dock plate or board, a recessed dock leveler has a higher capital investment. A hydraulic recessed

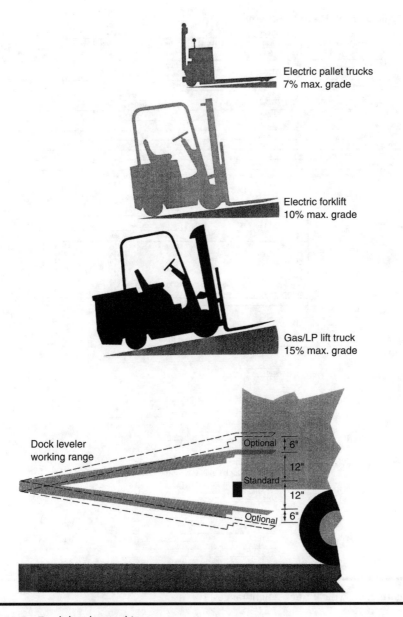

Figure 2.6 Dock leveler working range.

dock leveler is used for a dock operation that has a manual or mechanized unload, transport and sort concept for single items, master cartons, or pallets.

A mechanical recessed dock leveler has a diamond-patterned travel path surface and an upwardly biased ramp with a spring mechanism that is held down by a releasable ratchet device. To operate, an employee walks onto the ramp and pulls

Typical Truck Heights Survey

Type of Truck	Truck Bed Height				Overall Height				Rear Ledge Projec-tion	Percent Of All Trucks Serviced
	Total Range		Actual Range		Total Range		Actual Range			
	Min.	Max.	Min.	Max.	Min.	Max.	Min.	Max.		
Container	56"	62"			12'2"	13'6"				
Reefer	50"	60"			12'6"	13'6"				
Double Axle Semi-Trailer	44"	52"			12'	13'6"				
City Delivery	44"	48"			11'	12'6"				
High Cube Van	36"	42"			13'	13'6"				
Furniture Van	24"	36"			13'	13'6"				
Step Van	20"	30"			8'6"	10'				
Panel Truck	20"	24"			8'	9'				
Straight Truck	36"	48"			10'6"	12'				
Flatbeds	48"	60"			—	—				
Special										

*NOTE: Some states permit greater maximum heights than shown; consult your state's highway commission.

	Dock height, in	Truck overall dimensions, ft		
Truck type		Length	Width	Height
Flatbed	52	55–70	8–8½	12–14
Over-the-road	48	55–70	8–8½	12–14
City trailer	46	55–70	8–8½	12–14
Container	55	40–55	8–8½	12–14
Refrigerated	52	30–35	8–8½	12–14
High-cube	40	55–70	8–8½	12–14
Straight truck	44	15–35	8–8½	12–14

Dock Heights and Leveler Lengths

Truck Combinations	Recommended Dock Height	Recommended Leveler Length
Semi only	50"	6' or 8'
Semi and Reefer	51-52"	8'
Semi and Straight	48-50"	6' or 8'
Semi, Straight & Reefer	50-51"	8'
Semi, Straight & City	48"	8'
Semi & High Cube	43-44"	12'
High Cube & Reefer	44-45"	10' or 12'
Flatbed and Semi	52"	8'
Container and Semi	52"	8'
Supervan	Consult Rite-Hite Sales Representative	

Minimum Dock Leveler Lengths

Height Differential Truck to Dock	Pallet Jack Operation	Electric Fork Lift Operation	Gas Fork Lift Operation
1-2"	5'	2'-4"	2'-4"
3-4"	6'	5'	5'
5-6"	8'	6'	5'
7-8"	10'	8'	6'
9-10"	10'	8'	8'
11-12"	12'	10'	8'
13-14"	12'	10'	10'
15-16"	N/A	12'	10'
17-18"	N/A	N/A	12'

Figure 2.7 Typical truck height ranges.

on the release chain that is connected to the ratchet mechanism. This releases the ramp, which rises to its uppermost position. During the upward movement, the ramp lip is automatically extended forward. An employee walks forward onto the extended ramp, which is forced down toward the delivery truck bed. With the lip on the delivery truck bed, the dock leveler is ready for the unloading or loading activity. The advantage is that it costs less than the hydraulic model. A mechanical recessed dock leveler is used in a dock operation that has a manual or mechanized unload, transport and sortation concept for single items, master cartons, or pallets.

Dock safety, energy conservation, and sanitation can be improved by the following features:

- A full range of tow guards on the dock leveler sides
- An ICC bar or truck restraints to prevent delivery truck rollaway
- A dock restraint lip to prevent forklift trucks from running off an open dock
- Weather seals between the dock leveler edge and floor curb channel to improve sanitation and conserve energy
- Dock occupancy lights or stop/go lights to improve safety

The disadvantages of the recessed dock leveler are that it requires a higher capital investment and there needs to be a leveler for each dock. Advantages include that it has the capability to compensate for a wide range of height differences between the dock edge and delivery truck beds; it can accommodate powered forklift trucks, pallet trucks, and master carton conveyors with a low-grade range; and can handle heavier loads than other types of dock levelers. The recessed dock leveler also has the flexibility to service a wide range of delivery truck dimensions. It offers higher employee productivity, greater dock area safety and makes a variety of loading and unloading methods possible

Vertically Storing Dock Leveler

A vertically storing dock leveler has a ramp that is lowered from a vertical position inside the building. The ramp pivots on hinges that are installed in a step-down installed on the interior wall parallel to the dock. A step-down is a continuous pit that runs on a steel channel along the wall. Because there is no recessed pit, this type of leveler has reduced sanitation problems. A standard vertically storing ramp is 5 ft. in length, with a 6- to 7-ft. wide travel path surface. The standard lip projection is 16 to 18 in. long and the bend has a grade break of 12.5 percent. These dimensions accommodate a 6-in. height differential between the dock edge and delivery truck bed.

The vertically storing leveler is either hydraulic and automatic or mechanical and manually operated. If the leveler is hydraulic, an employee controls the lip and ramp with push buttons on a control panel located on the interior building wall. A manually operated ramp is lowered by an employee onto the delivery truck bed. Counterbalanced springs assist in the lowering effort. A vertically storing ramp can be moved from one dock to another as needed. When not in use it is locked in the vertical position. A vertically storing dock leveler is used in an operation that has a manual or mechanized direct or temporary stage unload, transport and sort concept for single items, flatwear, master cartons, or pallets.

Advantages are lower maintenance costs, energy control improvements, and lower total dock investment due to increased ramp mobility. Disadvantages are higher cost per dock leveler and the step-down collects debris.

Edge of Dock Leveler

An edge of dock leveler is a short ramp with a flexible lip. It is mounted on a steel channel that is embedded in the front of the dock. This type of ramp is available in 27- and 30-in. lengths and 66- and 72-in. widths. The standard lip is 15 in. long, with a projection of 11.5 in., or 17 in. long, with a 13.5-in. projection. Edge of dock levelers are available in both standard and low-profile types. The low-profile leveler is best used for a forklift or pallet truck with a low clearance. It should not be used with high-speed forklift or pallet trucks with low under-clearance.

The edge of dock leveler can be either hydraulic or manually operated. The hydraulic model is controlled by push buttons located inside the building and, when activated, the ramp is pushed upward with the lip in the extended position. After reaching full extension, the lip is lowered onto the delivery truck rear bed. After the delivery truck leaves the dock the ramp is returned to its stored position. This type of leveler requires a low capital investment but is only able to handle a height differential of 5 in.

Front of Dock Leveler

A front of dock leveler is bolted to the concrete dock wall, and has built-in bumpers. This type of leveler has a standard ramp length of 30 in. and width of 72 in., with a flexible lip projection of 11 in. The front of a dock leveler can be either high- or low-profile. The high-profile type accommodates a difference of 6 in. above and 4 in. below the dock; the low-profile type a difference of 4 in. above and 1.5 in. below the dock. The ramp is operated by an employee who manually lifts the ramp with the help of counterweights, flips it into the horizontal position and lowers the leveler onto the truck bed. When the truck leaves the dock, the leveler is automatically lowered to the stored position. The front of dock leveler can reduce the total investment if powered conveyors are used in loading and unloading activities. This type of leveler is not recommended for an operation that uses high-speed forklift or pallet trucks with low under-clearances. The disadvantages of this type of leveler are that it accommodates only a limited height differential and does not service all types of loading and unloading equipment. The advantages are that it requires a low capital investment and is easy to relocate.

Hydraulic Wheel Lift

A hydraulic wheel lift is located in a pit installed directly below the dock. The surface width is 10 ft. wide by 14 ft. long and the travel path is diamond-pattoned. A 5 in. high wheel locator is an elevated middle section of the platform. When the delivery truck is backed up to the dock, the wheel locator is between the delivery truck's two rear wheels. Hydraulic controls are located on the building wall inside and are used to raise and lower the delivery truck rear end to the required dock edge and truck height (minimum height differential) for easy pallet truck or cart travel over a dock board or leveler. When using a wheel lift device attention is given to the new slope of the delivery truck and the delivery truck and dock height. The delivery truck bed is sloped higher at the dock edge and lower at a vendor or customer delivery truck nose, which creates an upgrade or pull for an employee unloading with a pallet truck. During a loading activity, the new slope creates a downgrade. To reduce delivery truck or product damage or employee injury, the unloading or loading employee needs to be extremely careful due to the decline or incline that is created inside the delivery truck.

Important Dock Area Facility Dimensions

For design of a dock facility, important dimensions to consider are:

- The space between building columns
- The space between the floor surface and the bottom of the lowest ceiling obstruction
- Complete pallet and bottom support device dimensions
- Clearances or open spaces between two pallets and between a pallet and a rack structural member
- Clearance between two pallet rack upright posts
- Aisle space between a dock leveler and pallet staging area
- Aisle space between the pallet staging area and storage area

Clear Space between Two Building Columns

The clear space between two building columns is the open space between the column bases. This dimension indicates the floor space between the four building walls that is available for a dock operation. Important building column dimensions are the center line to center line or the dimension between two building column midpoints, and the open or clear span between two building column stands or base plates.

With a dock staging are inside a building, the open space between two building columns is an important dimension. It determines the number of floor stack pallets or rack bay base plates that fit between two building columns.

The Need for Sufficient Clear Floor Space and Aisles

For pallet receiving/shipping activities to be performed on schedule and with as little damage as possible, dock areas should have adequate clear floor space. There should be adequate aisle space between pallets, between the dock leveler and pallet staging area, and between the staging and storage areas. To ensure an efficient and safe dock operation, there should be a minimum aisle width of 6 ft. between the dock leveler and pallet staging area. The dock staging area should be between the dock door and the main traffic aisle, and is designed to handle pallets from a delivery truck, ocean-going container, or railroad car. It should have sufficient space for an employee to count master cartons on each pallet and to place a WMS identifier on each pallet. The aisle width between pallet storage positions should be adequate for a forklift or pallet truck to make a right-angled turn, or for a guided vehicle to perform a receiving pallet pick-up transaction.

There should be an aisle between the shipping or receiving dock and the storage area and this aisle should be wide enough to facilitate efficient vehicle movement to the storage area. It should accommodate two-way transport vehicle traffic, and the recommended width is specified by the vehicle manufacturer. The aisle permits efficient pallet transport from the staging area to the storage area and uninterrupted vehicle traffic in the main traffic aisle.

Rack and Facility Dimensions

Important dimensions to consider are:

- The space between the floor surface and the lowest ceiling obstruction
- Dimensions of the pallet and bottom support device
- Clearances or open spaces between two pallets and between a pallet and the rack structural member

In an operation with pallet racks as the staging area, there should be clearance between the upright posts of the pallet racks.

The space between the floor surface and the lowest ceiling obstruction or elevated floor support members determines the number of pallets in a floor (block) stack, number of load beams in a rack section, and the forklift truck pallet lift requirement to complete a receiving transaction. This open or clear space includes pallet dimensions, required forklift truck operational dimension, fire protection or sprinkler space, and rack structural members such as load beams or stacking frames.

The Receiving and Shipping Dock Staging Area

The dock operation requires queue floor space, racks, a staging area, and conveyor. Design of the staging area is based on the type and volume of master cartons and pallets, the product, as well as the flow concept and in-house transport equipment. On a receiving dock, master cartons or pallets are unloaded from a vendor delivery truck to the staging area, where they are identified according to WMS, master carton or pallet identification procedures. A sample is sent for QA checking and the master cartons or pallets are transported to a storage location. Space considerations in the design of a receiving dock staging area are:

- Peak quantities for pallets and master cartons
- Potential vendor UOP or CO delivery truck no-show
- Potential conveyor travel path or scanner mechanical or electric power failure
- Insufficient number of employees to handle delivery truck unloading volume

If a receiving dock operation involves vendor-delivered cartons that are unloaded, counted, identified for the WMS, with the piece quantity entered in the WMS files, and released for transport to the storage area, the requirements for a staging area are minimal. However, if product is held in the staging area for a period of time at the receiving dock, there is a higher requirement for an adequate staging area. Factors to consider in planning the staging area are master carton or pallet quantity, availability of storage racks, portable stacking racks or frames, as well as number of employees, the availability of CO delivery trucks, and the capacity of in-house transport equipment to handle the volume.

Figure 2.8 shows some options for the staging area. They include the use of floor stacks, stacking frames, pallet cages or portable racks, and standard, push back, or gravity flow racks.

Floor Stack or Block Concepts

In floor stack or block staging concept pallets are placed directly onto the dock floor, in two lanes that are located 6 to 8 ft. directly behind a dock door aisle. It has the capacity to hold the number of pallets carried by one delivery vehicle. Each lane accommodates pallets six to ten deep, stacked one, two, or three high. The height of the pallet stack is determined by the clear space between the floor and ceiling, strength of the master carton side wall to support the stacked load weight, and receiving procedure to attach a WMS identification and count the master carton or piece quantity load onto each pallet. With a very high pallet stack (above employee reach) a forklift truck lowers the pallet for a receiving clerk to complete the receiving activity. The lane length is designed with pallet fork openings to face the main traffic or take-away aisle and to provide an open space for a receiving clerk to inspect the condition of the pallet or master carton and product, and to apply WMS identification. The open space between two pallets is 30 to 36 in. The lane width is designed to handle a pallet width.

Each receiving door has two lanes, with white lines 6 in. wide markng lane locations. Space lines at predetermined locations along the lane borders ensure accurate forklift truck placement. After WMS identification is applied to the pallet in an approved location, it serves as a signal for a forklift truck driver to move the pallet from the receiving area to the in-house transport concept.

To effect a pallet unloading transaction, a forklift truck unloads the pallet from the delivery truck, enters a staging lane, travels along the lane to the pallet position in the main traffic aisle, deposits the pallet and backs out from the lane to the dock aisle. After the WMS identification is applied to the pallet, a forklift truck enters from the main traffic aisle, picks up the pallet, backs out onto the main traffic aisle, and travels to the appropriate storage location. This floor stack staging concept handles a wide SKU mix, any number of pallets per SKU, and any pallet volume.

Stacking Frames, Pallet Cages, or Portable Racks

A pallet cage, tier rack, or pallet-stacking frame is used when a master carton that is placed into a floor stacked position concept is crushable, extends beyond the pallet edges, or is not self-supporting. This concept is used to make a stackable and uniform unit load. The devices are stackable, which optimizes the receiving area cube or air space. They are handled in the same way as floor stack pallets.

A pallet cage is a wire-meshed, four-walled container that is attached to a pallet. At the corner of each wire-meshed container is an upright support member with a pad or section. This pad or section is designed to support another pallet cage. To assure access to cartons inside the pallet cage, it is designed with a hinge or does not extend the full height of the pallet.

A tier rack has four legs at each corner. Each leg base is attached to a pallet. The four legs are connected or intersect in the middle at the top. Nails are driven through the tier rack base plate, which makes the tier rack a more secure attachment to the wood pallet. When tier racks are not in use, they are disassembled and stacked to reduce storage space, but this requires employee assembly and disassembly time.

A stacking frame is a hardened welded and coated metal structure that has four upright legs that are permanently attached to a base. It has the structural strength to support three full stacking frames. The tops of the legs have horizontal structural support members and devices to support another stacking frame. The base has two

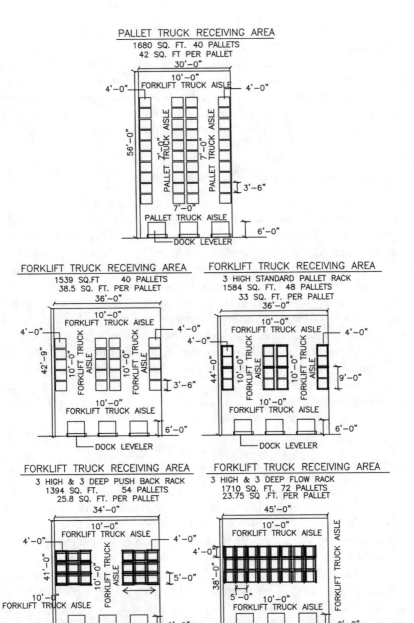

Figure 2.8 Truck dock staging concepts: pallet truck and floor stack; forklift truck and two high floor stack; forklift truck and standard pallet racks; forklift truck and push back racks; forklift truck and flow racks.

forklift truck openings. The clearance between the four legs and the base handles a full pallet or hand-stacked cartons. Stacking frame designs feature four legs that are permanently attached to the base and extend straight upward from the base. At the top, there are four horizontal cross members that tie the four legs together. A nestable stacking frame has three horizontal structural upright members that connect the four legs at the top and creates one open side. Frames that are not being used can be stacked eight to ten high to reduce storage space.

Fork Sleeves or Stir-Ups and Other Options

During a receiving transaction or transport activity, to prevent a metal stacking frame from sliding on the metal forks of the forklift truck, fork sleeves or stir-ups are added to the base. The base of the sleeve is slightly above the floor surface. The dimensions (downward extension, width, and opening permits the forks of the truck to easily enter the sleeve. The reduce the possibility of cartons accidentally falling from a stacking frame or tier rack, rubber bands, stretch wrap, or netting are wrapped around the four sides of the frame or rack. If a requirement is to have no indents on the exterior surfaces of the master cartons, they are hand stacked. Local fire codes should be reviewed prior to implementing a pallet cage, stacking frame, or tier rack concept. The disadvantages of these methods are the additional capital investment and low forklift truck driver productivity. In addition, for some models, when the racks or frames are not being used, they either occupy storage space or require employee time to assemble and disassemble. One advantage of these models is that they accommodate odd-shaped SKUs.

Standard Pallet Racks

A standard pallet rack staging concept has a single rack row or back-to-back rows. Single rack rows are tied overhead to the other racks or to a wall; back-to-back rows are tied together in the back. If the back of a single rack row faces an aisle, safety netting is attached to the back. Rack rows start 6 to 8 ft. behind the dock leveler, or sufficient distance for a forklift truck turn, and extend to the main traffic aisle.

Dock doors are 8 to 9 ft. wide and there should be 3 to 4 ft. between doors. With a single 44-in. deep pallet rack in a 4 ft. long staging area, the rack row is located between the two dock doors. One rack row side accesses an aisle that is between the dock door and main traffic aisle. The other side has safety netting and all racks are tied overhead to each other. The rack row layout is repeated for each dock door.

A back-to-back row layout has the two end rack rows a single layer deep against the wall or aisle. The other rows are back to back. In this layout the space between the dock leveler and the start of the rack row is very important. To accommodate a forklift truck, a right-angled stacking aisle should have at least 9 to 10 ft. between two rack faces.

The standard pallet racks are upright frames and load beams that are connected together to form pallet positions. The pallet positions are floor level high and elevated on a load beam pair. The pallet positions are two or three levels high. Each floor level and elevated load beam pair holds ballets in one, two, or three rows. With all pallets on the floor or load beams, the pallet rack concept handles both stackable and non-stackable merchandise. Some stacking options are pallets two high set on the floor or one pallet on the floor, one pallet on an elevated level and two pallets high on the highest load beam level. Rack layouts are specified by the manufacturer and the racks are manufactured with additional strength for the increased load beam vertical height and weight. The rack installation design is approved by the rack manufacturer. Rack post protectors are placed on the aisle transaction-side posts and dock leveler and main traffic upright frame to reduce forklift truck damage. For the floor-level pallets, the receiving clerk performs the count and places a WMS pallet identification onto each pallet, then enters the associated piece quantity and WMS pallet or master carton identification into the WMS inventory files. The pallets in elevated positions are transferred onto the floor for the receiving clerk to complete these activities.

Advantages of pallet racks are increased space or cube utilization and the need for smaller dock area. The disadvantages are increased cost, employee training, and pallets on the elevated levels require double handling.

Push Back Racks

A push back rack concept is a stand-alone or back-to-back rack concept with one aisle that permits a forklift truck to complete pallet transactions. A push back rack concept has the same structural components and design characteristics as gravity flow racks (described below) with some exceptions. The concept is designed to hold three to four pallets deep and three to four pallets high. Pallet weight and height determines the slope and pitch to flow rack lanes. A push back rack concept is designed as a single row that is installed along a building wall, in a building location that permits one aisle for a forklift truck to perform all transactions or as back-to-back rack rows. A single push back rack is 3 ft. 6 in. wide and the back-to-back racks have a 7-ft. width. Back-to-back rack rows have the same aisle distances between the dock leveler and the start of the rows, and between the end row and the main traffic aisle. The width of the aisle between the dock leveler and the push back rack row is very important. The row should be located 6 to 8 ft. behind the dock leveler, or sufficient distance for a forklift truck turn. An open space between two back-to-back rack rows allows a forklift truck to complete a right angle turn. With a counterbalance forklift truck, a right angle stacking aisle is at least 10 ft. between two racks, which means some aisles are not directly behind the dock doors. To complete a pallet transaction, a forklift truck places a pallet against an existing pallet that is in a lane aisle position. When an existing pallet is sufficiently pushed back into a lane, it creates a new pallet space. To withdraw a pallet, the forklift truck slowly raises the pallet 2 to 3 in. high above the lane conveyor and backs out with the pallet from the lane. As the pallet is removed from the lane, gravity moves the next pallet into the lane exit position.

Push back racks can be either standard or telescoping. The standard conveyor push back rack does not require brakes but has end stops on both flow lane ends. A telescoping push back rack has three pallet carriages that ride on a set of tracks. Carriage design permits the carriage that travels into the interior pallet position when not in use (without a pallet) to nest over the empty carriage that is adjacent to the aisle.

For all interior and elevated pallets, for a receiving clerk to complete receiving, including WMS pallet identification and WMS pallet scan transactions, a forklift truck completes a letdown transaction. The push back receiving staging concept handles a limited SKU mix, medium to high pallet number per SKU, and medium to high volume.

The advantages are increased space or cube utilization and shorter dock area. The disadvantages are increased cost, employee training, and double pallet handling.

Gravity Flow Racks

A gravity or flow through rack staging rack concept is designed as a single or stand alone rack concept that has one aisle for pallet in-feed and another aisle for pallet out-feed. The pallet gravity flow rack has upright frames, upright posts, braces, brakes, end stops, and skate wheel or roller conveyors that make individual pallet flow lanes. In a conventional forklift truck operation, pallet flow lanes are three or four levels high and three to eight pallet lanes deep. Pallet weight and height determines the slope and the flow rack lane pitch. The pallet unit height/length ratio is 3 to 1. If pallets exceed this ratio, there is a potential for uneven flow through or hang-ups in a flow lane. In most pallet flow rack systems, to ensure smooth flow through the lane, a pallet is placed onto a captive or slave pallet. In some flow rack concepts, conveyor rollers have flanged wheels that act as guides for a pallet as it flows through the rack. To prevent rack damage, there are lane entry guides, upright post protectors, forklift truck stops, and sufficient forklift truck turning aisle widths at the entry and exit positions. The pallet flow rack design has the pallet flow rack end face the receiving dock aisle. The receiving dock aisle width is 9 to 10 ft. from the dock leveler to permit a forklift truck to complete a pallet transaction. The pallet flow rack lanes are sloped toward the storage area. The discharge end is 9 to 10 ft. wide to permit a WA forklift truck to complete a pallet withdrawal transaction.

After a WA forklift truck unloads a pallet from a delivery truck, a receiving clerk verifies the SKU identification pallet master carton and piece quantity. The receiving clerk places a WMS identification on the pallet side that does not face the forklift truck, has the forklift truck transfer a WMS-identified pallet into a pallet flow lane and enters the WMS identification and SKU piece quantity into the WMS files. With a WMS identification on the pallet lead end in a flow lane, it assures the discharge end has the WMS pallet identification facing the withdrawal aisle. After a forklift truck in aisle 'A' or receiving dock aisle places a pallet onto a gravity conveyor lane, gravity and the pallet weight on the rollers allow the pallet to flow through the flow lane to the gravity flow lane exit end. At the gravity flow exit position, the pallet has the WMS identification facing the forklift truck driver. The pallet is removed by a forklift truck in aisle 'B' or the storage area aisle from the lane exit end. This activity permits the next pallet in the lane with the WMS identification in proper orientation to index forward to the withdrawal position. In a long concept, to reduce line pressure and product damage, pallet brakes and a pallet separator at the exit position are installed in the flow lane. After a pallet is moved from the pallet flow lane, the WMS identification on the pallet faces the forklift truck driver and is ready for transport from the receiving area to the storage area. Pallet gravity flow concept indexes pallets from the deposit (entry) position to the withdrawal position that allows each flow lane to accommodate one SKU per lane.

Product Pattern in a Delivery Vehicle

The main receiving and shipping objective is to transfer master cartons or pallets between the delivery vehicle and the dock area. The pallet unloading activity moves pallets from a vendor delivery truck, ocean-going container, or rail car onto the receiving dock. Each vendor delivery vehicle concept has a different unloading location at a facility, but each vendor delivery has a unique method to load pallets onto a delivery truck. A storage operation has the potential to receive product as:

- Master cartons stacked on the floor of a delivery vehicle, rail car, or ocean-going container
- Standard or non-standard pallets
- Slip sheets

The method of arranging the pallets or slip sheets in a delivery vehicle has a direct influence on unloading employee productivity.

Floor-Stacked Master Carton Concept

In the floor-stacked master carton method, vendor master cartons are set directly onto the floor of the delivery vehicle. To unload master cartons from a delivery vehicle, employees lift and transfer master cartons onto a cart, platform truck, semi-live skid, pallet, or conveyor travel path. Disadvantages are employee injury and low employee productivity. This method requires the largest number of employees and more docks. It takes four to five hours to unload a delivery truck or ocean-going container and five to six hours to unload a rail car. The advantages are excellent delivery vehicle cube utilization and no extra weight on the delivery vehicle.

Pallet Concept

A pallet is a master carton support device with top and bottom deck boards and support stringers or blocks. A pallet stringer or block has the surface dimension to accept a WMS identification, the location for a WMS identification line of sight and the pallet material assures that a WMS identification remains on a pallet stinger or block. A forklift truck or pallet truck with a set of forks is inserted into a pallet fork opening and lifts the pallet. This feature permits a pallet truck or forklift truck to transport a pallet from a delivery vehicle onto the receiving dock. The pallet stringer end or block has a surface for a front face or wrap-around WMS pallet identification. If a vendor pallet delivery is not a standard pallet, receiving options are manually or with a powered clamp device to transfer the master cartons from a non-standard pallet onto a standard pallet; or to place a non-standard pallet onto a standard pallet.

In a storage area, this feature requires additional storage position height. A normal delivery truck or ocean-going container has an 18- to 20-pallet capacity, a truck is unloaded in one to two hours and it takes two to four hours to unload a rail car. To minimize UOP damage and assure high employee productivity, cartons are secured onto a pallet. Disadvantages are empty pallet storage area, pallets are an additional expense and add weight to a delivery vehicle; there is some cube loss in the delivery vehicle, and it requires a pallet truck or forklift truck. The advantages are high employee productivity, and the need for fewer employees and fewer dock positions due to quicker delivery vehicle turns; and low employee injury.

The Slip Sheet Concept

With a slip sheet concept, prior to slip sheet transfer onto a delivery vehicle, the master cartons are unitized onto a corrugated, plastic, or fiberboard sheet that has a lip. The lip extends forward from the unit load and permits a pallet truck or forklift truck with a slip sheet attachment to clamp onto the lip and lift the unit load. After the unit load is secured onto a forklift truck slip sheet attachment, it is transferred from the delivery vehicle onto a pallet in the receiving dock staging area. WMS identification options are either on the pallet stringer end or block surface or the master carton on the lowest layer.

In a carton order-fulfillment operation that uses this WMS identification procedure, the pick employees are instructed to assure that the WMS-identified carton remains on the unit load and is the last master carton removed from the unit load. To handle a delivery truck or ocean-going container with 18 to 20 slip sheets, the slip sheet method takes two to two and a half hours. A slip sheet in a delivery truck or ocean-going container has a single lip that faces the rear door or truck side. In a railcar, a multiple lip or four lip slip sheet permits maximum cube utilization and unload activity takes three to four hours. The disadvantage is slip sheet attachment cost and the advantages are less weight and maximum cube utilization.

Delivery Vehicle UOP Loading Pattern

Pallets, slip sheets, or containers are placed inside a delivery vehicle in a manner that maximizes the product load capacity, assures good employee load and unload productivity, and permits dunnage to be placed in void spaces between two unit loads or between the load and the vehicle side wall. With a delivery truck or ocean-going container, pallets or slip sheets are placed inside the delivery vehicle to have the fork openings or slip sheet lip face the delivery vehicle doors. With a rail car, possible pallet arrangements are:

- A two-way pallet that has fork openings facing the rail car door
- A four-way pallet in a pin-wheel pattern
- A single-lip slip sheet facing the rail car door
- A multi-lip slip sheet in a pin-wheel pattern

Options to Improve Dock Area Safety and Efficiency

For an efficient and safe dock operation, additional devices may be installed in front of and behind the dock doors. These include wheel chocks, exterior guard posts or metal truck guides, dock bumpers, interior door guard posts, dock leveler lip, truck canopy, dock ladders and personnel entrance, fifth wheel lock jack, guidelines, dock or door identification, and truck delivery stripes or guide rails.

Dock Seals and Shelters

Dock seals and dock shelters (see Figure 2.9) are exterior dock door frames and doorjambs that extend outward from the building wall. The permit the rear door, sides, and top of the delivery truck to rest flush against this extension. The devices ensure that there is no gap between the delivery truck and building dock door frames and doorjambs. This improves dock security and energy conservation.

Dock Seals

A dock seal has two foam-filled pads or two air pads on each dock door frame and one head curtain along the doorjamb top. Some dock seals have moveable head curtains that are manually adjusted vertically to match a delivery truck height. Yellow strips or pads with yellow strips are options on the exterior sides of a dock seal. Yellow strips help a delivery truck driver to properly locate a delivery truck at the dock. Pads extend the life of the dock seal by reducing wear from delivery truck rear movement up and down as it is unloaded or loaded by dock employees.

Dock Shelters

A dock shelter conserves energy and improves dock area security. The dock shelter has two fixed-frame structures or flexible-frame structures with a head curtain that extends outward from the building wall. The two sides of the dock shelter frames hang onto the delivery truck's rear sides and roof. These three sections conform to the delivery truck rear door. Both dock shelter types have steel channel protectors to reduce damage to the dock shelters as the delivery truck is backed up to the receiving or shipping dock. Yellow strips on the pads or plain pads serve the same purpose as the yellow seal pads.

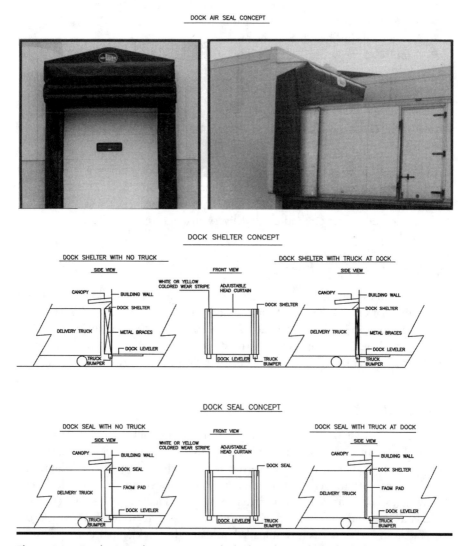

Figure 2.9 Dock air seal concepts: truck dock seals; truck dock shelters; and air padded dock seals.

Dock Lights

Dock lights are adjustable and moveable devices that provide light for the delivery truck interior. A dock light is a single lamp that has a flexible arm. Dock lights are located between two delivery truck dock doors and service both dock doors. A dock light arm provides light to a single dock door. Lockable screens protect the lamp from damage and pilferage. A dock light option is a wire prong that extends in a dock door path. As an employee opens a dock door, the wire prong automatically turns a dock light on as the dock door is moved into the open position. As an employee closes the dock door, the wire prong comes in contact with the dock door and turns off the dock light.

ICC Bar Lock or Hook Device

The ICC bar lock or hook device is a dock leveler option that improves forklift truck safety in the dock area. The ICC lock or hook is a device in the dock leveler pit area and is controlled from the dock area to extend forward and upward. When a delivery truck is spotted at a dock, the hook motion and new position are in front of the delivery trailer or truck ICC bar. The hook in this position restricts the forward movement of the delivery trailer, thus preventing delivery truck rollaway.

Stop-and-Go Lights

Stop-and-go lights are red and green lights that are located inside and outside each dock door. When a delivery truck is at a dock with the dock door open, the light is red and signals to both forklift truck operator and the delivery truck driver that the delivery truck is at the dock and is not ready for departure. When the light is green, it signals to the forklift truck driver and delivery truck driver that the delivery truck is ready to depart from the dock and it is not safe to enter the delivery truck.

Receiving Office

A typical receiving office has a receiving area supervisor and clerk for document control. Office sections are restrooms for employees and separate toilets for common carrier truck drivers. The common carrier or vendor delivery truck driver area has pay phones, vending machines, and a sliding window on an office wall for document transfer.

Dock Ramp

A dock ramp is a permanent building structure or portable metal structure with side guards and handrails that permit powered forklift trucks to travel between the facility floor surface and the ground. The ramp slope or degree of pitch depends on the height differential between the facility floor surface and the ground, forklift or pallet truck gradability and grade clearance, and space for the travel path.

Dock Lift

A dock lift is designed to replace the dock ramp as the means for a forklift truck or pallet truck to move between the facility dock floor surface and the delivery truck. It serves as a dock leveler that bridges the gap between the facility dock edge and delivery truck rear bed, as well as a vertical movable and adjustable dock leveler that bridges the gap between a facility dock edge and a lower or higher than normal delivery truck rear bed height. Adjustability permits a non-powered or powered vehicle to transfer pallets between the dock area and delivery truck.

The advantages are less facility space and land, permits electric battery- powered and non-powered equipment to travel between the ground level, delivery truck, and the floor dock area, safe operation, and it handles a wide facility dock edge and delivery truck rear bed height differential range.

Unloading and Loading Concepts

When you consider receiving and shipping dock activity, you are considering the physical movement of the vendor-delivered pieces from a delivery truck onto your receiving dock and completed CO packages from your manifest area onto a CO delivery truck. The concepts to move vendor-delivered pieces from a vendor delivery truck or CO packages onto a CO delivery truck are manual, mechanical, or automated.

Manual Unloading and Loading Concepts

With a manual unloading or loading concept (see Figures 2.10 and 2.11), an employee carries pieces or pulls/pushes a wheeled carrier with pieces between a delivery truck and dock. The carrier may be either a manual pallet truck for master cartons or pallets, or a small-item four-wheel cart.

Employee Walk Concept

An employee walk unloading or loading concept is the simplest and most basic unloading or loading delivery truck concept. To improve employee productivity, manual concepts have an employee use a two-wheel hand truck, roller dolly or pallet roller, four-wheel platform truck, freight pry bar, or a four-wheel cart with shelves or a hang bar. Manual concepts have an employee carry UOPs or push/pull a wheeled carrier between a delivery truck and dock. Due to the slow activity, a manual concept has a large dock staging area and requires large amounts of time to unload a delivery truck. Other disadvantages include risk of employee injury, greatest employee number, greatest dock number, low employee productivity and slow turning of delivery trucks. The advantages are low capital investment, handles all SKU types, and when other concepts are down, the concept can be used.

Manual Low-Lift Pallet Truck

A manual low-lift pallet truck unloading or loading concept has an employee push or pull a manual-powered low-lift pallet truck. The manual pallet truck has an operator handle, two load-carrying wheels, two load-carrying forks, two steering wheels and a hydraulic mechanism to elevate and lower the set of forks. Two forks extend forward from the wheel base and the length matches the pallet length. To lift a pallet, the employee pushes the set of forks into the pallet opening and operates a hydraulic mechanism to lift a pallet above the finished floor surface. To deposit a pallet in a delivery truck, the employee releases a hydraulic mechanism and pulls the forks from the pallet opening. To load or unload a delivery truck, the employee pushes or pulls the pallet truck into the truck, picks up a pallet, and pushes or pulls the pallet from the delivery truck to the dock area for deposit. With a manual pallet truck concept, the unloading of a delivery truck takes from two to three hours.

Disadvantages are that this method handles a medium volume, requires a pallet, and travel up grades is difficult. Advantages are less employee physical effort and risk of injury, low cost, improved employee productivity, improved vendor delivery truck turning time, the back rest reduces piece damage, fewer docks are needed, the pallet is deposited in the dock area, and it can be used in other distribution operation activities.

Mechanical Unloading or Loading Concepts

Equipment for mechanical loading or unloading uses gravity, electrical, or fuel-powered vehicles or conveying surfaces to move master cartons or pallets between the delivery truck and dock. They include:

- Electric-powered pallet trucks with forks
- Electric-powered pallet trucks with slip attachments
- Powered forklift trucks with forks
- Powered forklift trucks with slip sheet attachments
- Non-powered or powered extendible and retractable skate wheel, roller, and belt conveyors

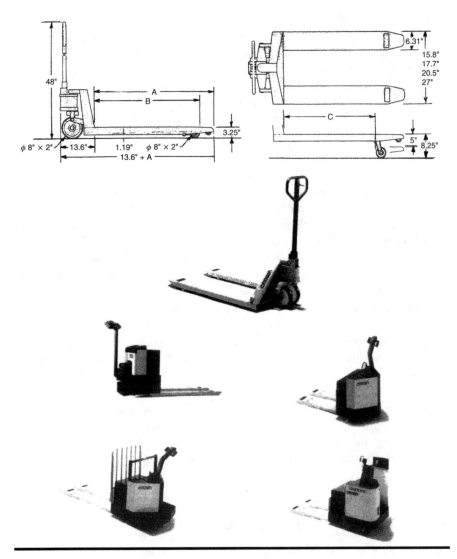

Figure 2.10 Battery-powered and manual-powered pallet trucks.

Electric-Powered Pallet Truck with Forks

An electric-powered pallet truck is a three-wheeled vehicle with one wheel as a rubber-covered drive and steer wheel. The other two wheels are hardened plastic and are the load-carrying wheels under each fork. The wheels and forks elevate or lower a pallet onto the floor. Optional full height load backrests and fork entry wheels improve pallet-handling productivity and reduce product and equipment damage. Vehicles can be walkie or walk behind, walkie and rider, or rider. To unload a delivery truck, an employee steers or drives a pallet truck the a delivery truck, picks up

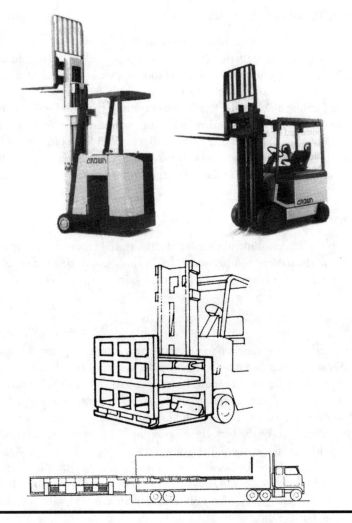

Figure 2.11 Fork lifts: powered forklift truck; lift truck with slip sheet device; and extendible conveyor.

a pallet, and drives the pallet truck with the pallet from the delivery truck onto the dock staging area. A palletized delivery truck is unloaded in one to two hours.

Disadvantages are greater need for employee training, need for a battery-charging area, it is difficult to handle double-stacked pallets, and does not transfer the pallet to an elevated position and moderate grade clearance. The advantages are moderate capital investment, capacity to handle a high volume, travels up grades, reduces employee injury, requires fewer docks, and can be used in other activities.

Optional equipment for the pallet truck improves pallet-handling productivity and reduces UOP and equipment damage. Options include fork entry wheels, pallet stop forktips, and backrests

Fork Entry Wheels

Each fork entry wheel has a hardened metal bracket, axle and bearings, and a hardened plastic wheel. Each pallet truck fork has an entry wheel at the end or tip. As the pallet truck forks enter the pallet fork openings the fork entry wheels assist the load-carrying wheels to travel over the pallet bottom deck board exterior side. This feature permits quick pallet truck fork entry into the pallet fork openings and reduces standard pallet truck load wheels pushing against the pallet deck board, which can cause the pallet truck to push a pallet across a floor surface until the pallet hits an object. This situation can cause damage or if a nail protrudes beyond a bottom deck board, it can cause floor damage.

Pallet Stop

A pallet stop is a hardened metal member that is attached to the base of the fork and is full width of the battery compartment. In this location, during the pallet pick-up activity, the pallet stop ensures that a pallet is square on the pallet forks.

Fork Tips

Pallet truck fork tips are designed to guide the pallet truck forks into the pallet fork openings. Fork tips may be either standard or extended. A standard fork tip is 9 in. wide. The space between two standard forks tip is 4 to 9 in. wide. The internal space between the two forks permits the pallet truck forks to enter the pallet fork openings and to have the pallet stringer or middle block between two forks. In a pallet fork opening, the ends of the fork tip are positioned under the deck board. In this position, the forks support the pallet and do not extend beyond it. Extended fork tips are 9 in. wide. The space between the tips is 4 to 8 in. The narrower fork tips make entry into the fork openings easier. The tips extend beyond the pallet deck board and support the pallet. However, employees must take care not to have the fork tip cause damage or become hung up on an object

Backrests

A backrest has several members made of hardened metal that extend from the fork base to the height of the pallet. They are evenly spaced across two fork widths to ensure that master cartons on a pallet do not fall through the open space between two members. The members have the structural strength to resist pallet movement.

Powered Forklift Trucks

Powered forklift trucks may use an electric battery or have an internal combustion engine. Electric battery-powered trucks are preferred because they produce less fumes, but in an open dock area a diesel or liquid propane (LP) gas forklift truck can be used. Powered forklift trucks have three or four wheels, an overhead guard, a set of forks that extend outward from a mast and move pallets vertically on the mast, and an operator's area, which can be either "sit down" or "stand up." Many dock forklift trucks have masts with full free lift, which allows the forks to be elevated approximately 4 ft. without having to extend the mast upward. The mast collapses to permit the forklift truck to enter a delivery vehicle.

The free lift feature permits a forklift truck inside a delivery vehicle to de-stack the top pallet from a double-stacked pallet. For maximum unloading employee productivity and to reduce piece damage, optional features are forksthat are capable of side-shift and tilt, and spotlights on the mast.

There are several types of forklift truck, depending on the power source, ability to travel outside a storage area, and aisle width required to complete a transaction. They can be identified as:

- ▪ Wide aisle (WA) forklift trucks
- ▪ Narrow aisle (NA) forklift trucks
- ▪ Very narrow aisle (VNA) forklift trucks, which are very heavy, have very tall masts, require an aisle guidance system, and are only able to travel up very slight grades, which limits their use in a dock operation

Wide Aisle Forklift Trucks

Wide aisle (WA) forklift trucks operate in a 10- to 13-ft. wide aisle. There are two types: sit-down counterbalanced and stand-up counterbalanced.

A sit-down counterbalanced forklift truck has a long wheel base, good under-clearance, low mast, and overhead guard. Some models have a short wheel base. These features make this type of forklift truck very maneuverable and versatile and can be used with pallets or slip-sheets in the receiving dock, transport, storage, and shipping dock areas, and is able to travel over a normal floor surface. It is capable of traveling up a 15-degree grade or ramp, to enter and exit vendor or customer delivery trucks. The forklift truck is designed with a set of load-carrying forks that elevate and lower on a telescopic mast and has an operator area with a seat. From the seat, the operator has access to all controls that move the vehicle and the forks. Some models have see-through masts that provide the operator with a full view of the vehicle's travel path. A counterweight is located in the rear chassis. The vehicle has a short or long wheel base and high undercarriage clearance. These features provide the forklift truck with the capability to offset a pallet or slip-sheet weight, to travel over grades or ramps, and to complete a pallet transfer transaction to a height of 12 to 15 ft. In a receiving operation with rack staging, this allows the forklift truck to reach pallets two to three levels high. After the pallet is transferred onto the receiving dock area, a receiving clerk places a WMS identification onto the pallet block or stringer or master carton on the bottom layer.

A rechargeable electric battery powers most common indoor counterbalanced forklift trucks. When a forklift truck is used as an outdoor vehicle in an open dock area, the power source may be an internal combustion engine that uses diesel, LP gas, or gasoline fuel. This type of vehicle has either three or four wheels, with one wheel as a drive wheel, and a steering wheel. The wheels are fitted with pneumatic tires for outdoor use and cushion, solid polyurethane, or rubber tires for indoor use.

Counterbalanced forklift trucks are available with one-, two-, three-, or four-stage masts that provide the forklift truck with up to an 18-ft. stacking height. Counterbalanced forklift trucks are used to handle a standard pallet or slip-sheet load that has a weight range from 2000 to 4000 lb. Some forklift trucks have additional counterweight and wheel-base length to handle heavier loads. Optional devices that increase operator productivity and reduce product and equipment damage include fork side shift and mast tilt devices and dual mast lights. An experienced forklift truck operator makes 20 to 25 transactions per hour. A four-wheel counterbalanced forklift truck, with its wheel base and chassis weight, can be fitted with various attachments, such as a slip-sheet push/pull device or double-pallet-length forks. With side-loaded delivery trucks, double-pallet-length forks permit a forklift truck to handle two pallets, thereby increasing employee productivity. For any given forklift truck, verify with the manufacturer that the forklift truck is compatible with the attachment, load weight, new load center, and can perform an elevated transaction without the forklift truck moving forward. If the forklift truck moves forward under these conditions, the manufacturers may add chassis weight to the rear or you may need to purchase a new forklift truck.

Most three-wheel counterbalanced forklift trucks are used in receiving and shipping dock areas. These trucks are designed to handle the pallet weight with free-lift masts, stacking height, and short right-angle turn requirements.

Forklift truck masts determine the lift height of the forks, and should allow a storage transaction at the highest position in the operation. Theheight of the collapsed mast allows a forklift truck to enter and exit a delivery truck. The overall clearance height is also affected by the height of the overhead guard height which may be a possible restriction to access to a delivery truck. There are different types of mast with which a forklift truck may be fitted:

■ A single mast, which has limited application in a dock operation.
■ A two-stage, nontelescopic mast, with has little or no fork free lift, which. means that as the set of forks rises, the mast also rises. This type of mast has limited application in the dock area.
■ A standard two-stage mast.
■ A two-staged mast with forks that have high free lift, which makes the forks able to rise upward for a distance above the floor surface without upward movement of the mast. This allows unloading of a delivery vehicle with double-stacked pallets.
■ A three-staged mast, which is similar to the two-stage mast with full free lift.

- A four-stage mast.
- A rigid mast. The height of four-stage and rigid masts prevents entry into a delivery truck.

The load capacity and stability of a counterbalanced forklift truck is derived from the front wheels' centerline, which act as a fulcrum (balance point). The pallet weight plus the weight of any attachments is counterbalanced by the counterbalance weight, which includes the battery weight and counterbalance weight, and the wheel base, which is the distance between the truck's front and rear wheel centerlines. To increase a forklift truck's counterweight, the manufacturer adds weight to the forklift truck chassis rear.

The forklift truck capacity is the amount of weight that the forklift truck safely lifts. This capacity that is stated by the manufacturer. Capacity decreases the longer the load that is handled by the forklift truck. A forklift truck handles a normal pallet up to the weight limit that is specified by the manufacturer. As the pallet length increases outward beyond the standard 24-in. center dimension, the forklift truck is able to handle less weight. If the pallet is heavier than a standard weight or over the standard length, a bigger, more expensive, and heavier forklift truck is needed to handle the pallet or the manufacturer adds weight to the forklift truck chassis rear.

Forklift truck maneuverability refers to the ability of the forklift truck to make a right angle turn or to turn into an intersecting aisle. A right angle turn is the distance that is required for the truck to turn and make a transaction from a floor position or rack position. This distance is the aisle width. An easy way to determine the aisle width is to take a conventional pallet length or stringer length plus the distance from a set of forks to face the forklift truck drive wheel centerline plus the outside turning radius. For excellent forklift truck operator productivity and reduced product and equipment damage, 6 to 12 in. is added to the above calculation. If product overhangs a pallet, the overhang dimension is added to the pallet or stringer length. In planning dock area aisles with a forklift truck manufacturer's aisle width recommendation, for good lift truck operator productivity and low product and equipment damage, the dock area aisle width is the forklift truck manufacturer product-to-product aisle width plus 6 to 12 in.

The wheel-base on a counterbalanced forklift truck is the distance between the front and rear wheel centerline. The wheel-base determines the aisle width. A forklift truck with a short wheel-base can operate in a narrow right angle turning aisle, which makes the truck very maneuverable. A forklift truck with a long wheel-base requires a wider right angle turning aisle, but has improved ride, steering traction, and stability. A second factor that affects maneuverability is the ability of the forklift truck to climb a ramp or travel over a dock leveler, referred to as gradeability and underclearance, respectively.

Forklift truck gradeability is the steepest percent grade (ramp or dock leveler slope) that a forklift truck can climb with a pallet on the forks. The steepest grade for an electric forklift truck is 15 percent. The grade for an internal combustion engine forklift truck is 20 to 25 percent.

The forklift truck underclearance measures the steepest grade that the truck can travel at a ramp or dock leveler without the truck undercarriage or the forks hitting the floor. Important underclearance locations are midway between the wheel-base and the bottom of the mast because of the mechanical operating parts in these areas.

As a forklift truck climbs or travels over a ramp or dock leveler to enter and exit delivery trucks and perform outdoor operations, gradeability and underclearance are two important factors in avoiding future repair expenses and forklift truck downtime. A forklift truck with a short wheel base and high underclearance easily travels over ramps and dock levelers, and can perform outdoor work.

A stand-up rider counterbalanced forklift truck is included in the wide aisle (WA) group. The forklift truck has similar design features and operational capabilities as a sit-down rider counterbalanced forklift truck. The major difference is that the operator stands on a platform, which allows a shorter wheel-base. With a short wheel-base, a forklift truck is able to make a right angle turn in a 9- to 10-ft. wide aisle and place a 2000- to 3000-lb. pallet into a pallet position 18 ft. above the floor. A 24-volt electric-powered rechargeable battery-powered mobile aisle forklift truck is very maneuverable and versatile, performs all activities on a normal floor and with a level dock leveler enters and exits all delivery vehicles. Models have an option for a retractable overhead guard that permits easy entry and exit, but cannot handle a slip sheet device.

Narrow Aisle Forklift Trucks

The narrow aisle or NA forklift truck group includes straddle, straddle reach, and double deep-reach forklift trucks. These trucks operate in 7- to 9-ft. wide aisles. A straddle reach forklift truck is designed with two load-carrying wheels and one drive-and-steer wheel. Other forklift truck features are that the forks elevate and lower on a telescopic mast; there is an overhead guard; two stabilizing straddles or outriggers that extend outward from the truck mast; and an operator platform. Some models have a hardened metal member to protect the operator platform area. The operator platform area has both horizontal truck and vertical fork movement controls. To elevate or lower and transport a pallet, the outriggers contain the majority of the pallet weight and the pallet length. This type of forklift truck requires the delivery truck to be in the correct spot at the shipping dock, and no dock leveler deflection or bow. If there is dock leveler deflection or bow, a straddle truck has the potential for hang up. The forklift truck cannot handle attachments or a slip sheet device.

To operate a forklift truck-unloading concept, an employee drives the forklift truck into a delivery truck, pick ups a pallet, and drives with the pallet from the truck onto the dock area. In the dock area, the driver deposits the pallet in an assigned area or delivers the pallet to a customer delivery truck.

Disadvantages of NA trucks are high cost and the need for additional employee training. Advantages are the free-lift mast handles stacked pallets; this type of truck has good underclearance and grability, and can travel over a higher grade clearance. With proper chassis weight and wheel base an NA forklift truck handles a heavier and longer load, can be used in other facility functions, and some counterbalanced models are equipped to handle slip sheet loads. There are also fewer employee injuries.

Non-Powered and Powered Extendible or Retractable Skatewheel, Roller, or Belt Conveyors

The extendible and retractable conveyor group has non-powered and powered conveyor concepts (see Figure 2.12). These conveyors have the ability to extend or retract along a travel path between the dock area and a delivery truck. In most carton-across-the-dock operations, the carton conveyor is used to unload vendor delivery trucks and load customer delivery trucks. The extendible or retractable conveyors are located in the receiving or shipping dock areas and each one is the connecting link between a delivery truck and the in-house dock carton transport or sort concept.

When considering an extendible and retractable conveyor concept for a dock receiving operation, the factors to take into account include:

- Number of docks serviced by a conveyor, frequency of use, available electrical power, investment funds, how to bridge the gap between the delivery truck bed and the dock edge, and to assure that a conveyor travel transition is smooth. Various methods to bridge the gap were reviewed previously in this chapter.
- Delivery truck bed length. For maximum unloading and loading employee productivity, an extendible conveyor length extends to within 3 ft. from the longest delivery truck bed nose.
- Master carton characteristics: conveyable cartons are moved over a non-powered conveyor travel path. This feature assures a constant carton flow between the dock and a vendor or customer delivery vehicle and good employee unloading/loading productivity.
- Off-set an extendible or retractable conveyor travel path to a dock door side. To load or unload non-conveyable cartons from a delivery truck, this feature allows easy and quick pallet truck travel between the delivery truck and receiving dock area and minimizes UOP or CO damage and employee injury.
- Ability for employees to move non-powered conveyor sections.
- A conveyor with locked rear wheels or a rear section that is anchored to the facility floor improves forward or reverse movement of the conveyor.
- A guide track under the rear section improves retraction.

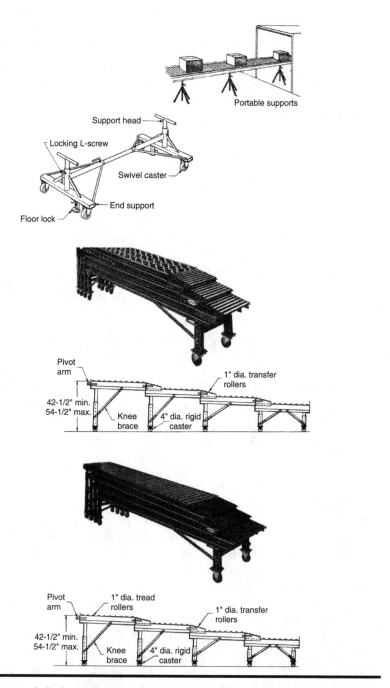

Figure 2.12 Truck dock conveyors: manual set-up conveyors; manual extendible skate wheel conveyor; hytrol conveyor; and manual extendible roller conveyor.

- Merge concept for an unloading extendible conveyor travel path with an in-house powered transport and sort conveyor.

An extendible/retractable unloading conveyor that merges with the in-house carton transport and sort powered conveyor travel path assures that the cartons continuously move between a vendor or CO delivery truck to an in-house carton transport and sort powered conveyor concept. Possible designs are to attach an incline/decline conveyor section to the extendible/retractable conveyor section, or locate an incline/decline conveyor section at each merge location.

A powered incline/decline belt conveyor extends upward from the extendible/retractable conveyor bed to merge with the in-house carton conveyor travel path. When compared to the other extendible/retractable conveyor merge designs, this feature reduces total conveyor cost and requires a wider clear path for extendible/retractable conveyor movement between two docks. This feature means a larger receiving or shipping dock area, but a lower material-handling equipment cost.

The second option is to have an incline/decline conveyor section located at each in-house conveyor merge location. This incline/decline conveyor section extends downward from the in-house carton conveyor travel path to a mobile extendible or retractable unloading conveyor travel path. This design feature increases the total carton conveyor cost but requires a narrower path for the extendible/retractable conveyor to move between two docks. This feature means a smaller receiving or shipping dock area and additional conveyor cost.

There are various types of extendible and retractable carton conveyor types:

- Non-powered, manual set-up skate wheel straight conveyor travel path
- Non-powered, manual, nestable (retractable) and extendible skate wheel straight or flexible conveyor travel path
- Non-powered, manual, nestable (retractable) and extendible roller straight or flexible conveyor travel path
- Nestable (retractable) and extendible powered skate wheel conveyor travel path
- Nestable (retractable) and extendible powered roller conveyor travel path
- Mobile position nestable (retractable) and extendible powered belt conveyor travel path
- Fixed position nestable (retractable) and extendible powered belt conveyor travel path
- Fixed position, fixed bed mobile forward and reverse powered belt conveyor travel path

Non-Powered Manual Set-Up Skate Wheel or Roller Straight Unloading Conveyor

A non-powered, manual set-up skate wheel or roller straight conveyor travel path conveyor concept requires dock employees to set up a conveyor travel path on a dock and in the delivery truck. For the unloading and loading activity, as the delivery truck carton loaded/unloaded quantity changes, the employees are required to add or remove conveyor travel path sections.

A non-powered skate wheel conveyor concept has small wheels that are attached to and turn on an axle. Skate wheels have a plastic, steel, aluminum, or rubber travel path surface that comes in contact with the bottom exterior surface of the master carton. With most straight travel path conveyor models each axle is attached to a conveyor frame member. A flexible model has each skate wheel attached to a bracket and the bracket base is connected to a large wheel on a swivel caster. To have optimum master carton travel over a skate wheel conveyor travel path, your shortest length carton has at least three skate wheels and two axles under the carton. These axles and skate wheels have the structural strength to support the maximum carton weight.

The non-powered roller conveyor concept has a series of rollers, each of which is attached to and turns on an axle. Rollers have a plastic, steel, aluminum, or rubber surface that comes in contact with the bottom surface of the master carton. With a straight conveyor travel path model each axle end is attached to a conveyor frame member. The flexible travel path conveyor has each roller attached to a bracket, with the bracket base connected to a large wheel on a swivel caster. To have optimum carton travel over the roller conveyor travel path, your shortest length master carton has at least three rollers under the carton. The rollers and axles have the structural strength to support maximum carton weight. When a steel conveyor travel path and conveyor frame is compared to an aluminum conveyor travel path and conveyor frame, a steel travel path and conveyor frame is heavier and the aluminum is lighter.

A comparison of the non-powered skate wheel conveyor to the roller conveyor shows that:

■ The skate wheel conveyor is lighter in weight
■ The skate wheel conveyor has faster carton travel speed and provides carton tracking over a conveyor travel path
■ The roller conveyor is heavier
■ The roller conveyor handles a wider carton mix but a meshed skate wheel conveyor handles smaller-sized cartons
■ The roller conveyor allows slower carton travel speed over a conveyor surface
■ The roller conveyor has a slightly higher cost
■ The roller conveyor is more durable to withstand carton impact from being transferred onto a conveyor travel path

Conveyors are supported along the travel path in one of two ways: (1) a fixed conveyor travel path that is permanently attached to a delivery truck bed, and (2) two or three tripod conveyor stands under each conveyor frame. A fixed conveyor travel path in the delivery truck is a master carton conveyor unloading or loading concept that is used with a captive company delivery truck fleet. The delivery trucks have non-powered conveyor sections that are permanently mounted or attached on the floor of the delivery truck. The conveyor travel path is located to one side of the delivery truck; the preferred side the left side because soft shoulders of public highways are on the right side of the public highway and public roads are lower on the right side.

The disadvantages are increased weight and loss of some space in the delivery truck; it is used only for floor-stacked loads; and requires additional investment in the delivery truck. The advantages are reduced unloading conveyor set-up time, lower investment in extendible conveyors, no facility space is required to store conveyor sections, and it can be used at all locations in your supply chain logistics strategy.

With the tripod conveyor concept, employees set-up portable tripods (support stands) on the dock floor surface and on the delivery truck floor. With this arrangement, the carton conveyor travel path is extended from the dock area to the delivery truck nose. In comparison to a fixed conveyor travel path in a delivery truck, the portable tripod conveyor travel path method features are longer set-up time, carton conveyor travel path has aluminum rollers or skate, and lower conveyor investment.

Non-Powered Nestable, Retractable, and Extendible Skate Wheel Straight Conveyor

In a dock operation, this type of conveyor can be extended into a delivery vehicle or retracted onto the dock area. The travel path can be either straight or flexible.

The retracted (nested) nestable and extendible skate wheel conveyor has all skate wheel conveyor frame sections pushed (retracted) under the rear conveyor frame section. In this retracted stage, a nestable extendible conveyor is in the shortest (fully retracted) length. A nestable extendible skate wheel conveyor is 11 ft. to 11 ft. 6 in. when it is fully retracted. In the fully extended (active) stage, the conveyor rests inside a delivery vehicle. The number of conveyor sections determines its fully extended length. With two conveyor sections, a full conveyor travel path extension is 20 ft. With three conveyor sections, a full conveyor travel path extension is 30 ft. With four conveyor sections, a full conveyor travel path extension is 40 ft.

Non-powered nestable, retractable, and extendible skate wheel conveyor components are support members and casters/wheels, straight conveyor frame sections and skate wheel, transfer rollers, or carton travel path.

The manual nestable extendible skate wheel conveyor support members are sleeve legs that adjust a conveyor travel path surface elevation above the dock area floor or delivery truck bed. The leg adjustment feature permits a conveyor travel path charge end to have a 10-in. elevation change, a conveyor travel path discharge end to have a 2-in. elevation change, and to adjust a conveyor travel path slope or pitch. This conveyor travel path elevation change assures the best carton travel speed and carton travel over the conveyor travel path. To improve conveyor travel path rigidity and stability, knee braces are attached at right angles to support the legs and conveyor frame. Since a straight conveyor is extended and retracted in one direction, each of the conveyor section legs has a 4-in. diameter wheel on a rigid caster. The dimensions from rail to rail, or carton travel path are either12 in., 15 in,. 18 in., or 24 in. wide.

The straight skate wheel conveyor design serves as a guide as the various conveyor sections are nested (retracted) under the rear conveyor section.

The non-powered, manual, nestable extendible skate wheel conveyor travel path has skate wheels and transfer rollers. Various skate wheel types include:

- Steel or zinc plated at 1 ⅝ in. diameter with a ⅝ in. face that provides a 30 lb. capacity
- Aluminum at 1 ⁵/₁₆ in. diameter with a ⅝ in. face that provides a 15 lb. capacity
- Plastic or nylon at 1 ⁵/₁₆ in. diameter with a ⅝ in. face that provides a 30 lb. capacity

Most skate wheels are placed on two ½ in. centers on a ¼ in. thread axle that has both ends locked to the conveyor frame. For good carton conveyance, there are at least three skate wheels under the shortest carton length.

The hinged transfer or transition roller section is located on each conveyor section discharge end, except at a conveyor travel path discharge end. These transfer roller sections have 1-in. diameter 18 gauge rollers on 1¼ in. centers. As a hinged component, a roller transfer device bridges the gap between two extendible conveyor sections at a downward angle. The conveyor travel path discharge end has a pull handle and a carton stop device.

As an unloading conveyor, the conveyor section inside a delivery truck is at a higher elevation and declines onto a conveyor section on the dock.

Non-Powered Nestable, Retractable, and Extendible Roller Straight Conveyor

A non-powered nestable, retractable, and extendible roller conveyor concept has the same stages, conveyor travel configurations, and carton travel path features as the skate wheel concept described above. The non-powered manual nestable and extendible roller conveyor travel path has rollers and transfer rollers. The roller types are:

- Steel galvanized roller that has ⅜ in. diameter of 11 gauge steel
- 1.83 in. diameter roller of 16 gauge steel
- 1.83 in. diameter roller that is .83 in. thick aluminum roller

All roller types have free-floating ball bearings and spring-loaded axles and most rollers are placed on 1 in. or 1 ½ in. centers along the conveyor frame. For good carton conveyance, there are at least three rollers under the shortest carton length.

A roller hinged transfer section has the same design and function as a skate wheel hinged transfer section. As an unloading conveyor, the conveyor section inside the delivery truck is at a higher elevation and declines onto the conveyor section on the dock.

The skate wheel conveyor is lighter in weight than the roller conveyor. A 40-ft. long skate wheel straight conveyor travel path has the following range of width and weight:

- 12-in. wide travel path that weighs 583 lb.
- 15-in. wide conveyor travel path that weighs 626 lb.
- 18-in. wide conveyor travel path that weighs 681 lb.
- 24-in. wide conveyor travel path that weighs 789 lb.

A 40-ft. long roller straight straight travel path has the following width and weight range:

- 12-in. wide conveyor travel path that weighs 694 lb.
- 15-in. wide conveyor travel path that weighs 790 lb.
- 18-in. wide conveyor travel path that weighs 863 lb.
- 24-in. wide conveyor travel path that weighs 1000 lb.

Non-Powered, Nestable, Retractable, and Extendible Skate Wheel and Roller Conveyor

This type of conveyor is flexible, expandable and nestable. A flexible conveyor travel path means that there can be curves or bends in the conveyor travel path. The flexible carton unloading conveyor concept has a support leg with a 5-in. or 6-in. diameter swivel caster. The flexible conveyor has aluminum scissor components that expand and contract as it is moved outward or retracted by an employee. At the top of the scissor is a skate wheel or roller axle. A flexible skate wheel or roller conveyor expansion and contraction ratio is 4 to 1, which means a 3-ft. retracted conveyor can be extended to 12 ft.

A skate wheel conveyor is lighter in weight than a roller conveyor. A 30-ft. long skate wheel conveyor with a 30-in. wide travel path weighs 406 lb., whereas a 30-ft. long roller conveyor h with a 30-in. wide travel path weighs 546 lb.

Extendible and Retractable Electric-Powered Skate Wheel Flexible Conveyor

An extendible and retractable electric-powered skate wheel flexible conveyor concept is similar to an extendible and retractable non-powered skate wheel carton unloading conveyor concept.

An electric-powered skate wheel flexible conveyor has a motorized lead end that is activated and manually controlled to move forward on its wheels from the dock area into the delivery truck. A unique feature of this type of conveyor is that the conveyor travel path has grooved sheaves and is powered by one or more small electric drive motors. The drive motor is a ¼ hp motor. A drivetrain is located every 12 ft. Each conveyor drive motor is controlled by an off–on switch at the drivetrain charge and discharge end. The conveyor travel path speed is adjustable from 0 ft. to 100 ft. per minute.

The drivetrain has two number 20 roller chain strands. Roller chain travel paths are in the middle of the travel path width. Each roller chain is laced over the first skate wheel axle top and under the second skate wheel axle which is a zig-zag route pattern. As the roller chain is routed over an axle top, the chain travel path is over a tooth-grooved sheave. The roller chain travel path under the axle is over a smooth plain face sheave. All sheaves have sealed ball bearings. The drivetrain chain travel path is designed with sufficient flexibility to prevent any employee limb injury but has sufficient force to turn a carton conveyor load carrying surface that propels a carton forward over a carton conveyor travel path.

Extendible and Retractable Electric-Powered Flexible Roller Conveyor

An electric-powered roller flexible conveyor concept is similar to the non-powered roller conveyor concept. They have similar travel path surface type and roller center spacings/axles along the conveyor travel path. The conveyor expansion and retraction ratio is 3 to 1.

The electric-powered roller flexible conveyor concept has the same features as the powered skate wheel concept described above. The unique feature of the electric-powered roller conveyor concept is that in addition to a chain travel path, the sheave has two grooves for polyurethane O-ring drive belts. Each O-ring drive belt is looped over a sheave groove and over a carton-carrying roller groove. This feature means that the conveyor travel path rollers are driven carrier rollers. The second top sheave is a smooth sheave which is the chain return travel path to the drive motor and sprocket.

Mobile Powered Extendible and Retractable Belt Conveyor

A mobile powered extendible and retractable powered belt conveyor concept (see Figure 2.13) improves employee productivity to unload a delivery truck, reduces employee injuries, and, compared to a fixed powered belt conveyor concept, decreases the total dock area investment. The mobile powered belt conveyor concept has booms with a powered belt conveyor travel path surface that extends into the delivery truck and retracts onto the dock area and traverses the dock area floor between two docks.

A mobile powered conveyor has structural support members, a boom and boom drive unit, traversing rails, wheels and traversing drive, belt carrying travel path

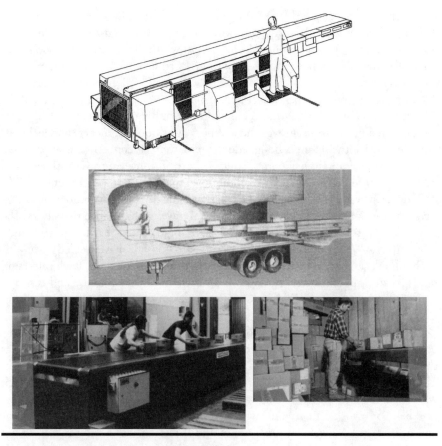

Figure 2.13 Powered extendible belt conveyor.

surface, conveyor belt drive motor, drivetrain, end pulley, and take-up device, guards, and controls. The support section is made of ¼-in. thick steel plates and $^3/_{16}$-in. thick steel slide bed plates. Steel plates are welded structural members and coated per the manufacturer's standard color or per your company specifications. The components provide enclosed housing for the electrical and mechanical components and drivetrain, structural support for the boom and belt conveyor travel path, and means and locations for counterweight and wheel attachment.

The extendible and retractable boom, boom drive motor, and boom drive components provide support for a carton travel path between a delivery truck and dock area and power to move a carton conveyor travel path between the in-land delivery vehicle and dock area.

The boom has $^3/_{16}$-in. thick steel slider bed plates, structural angle members, and side plates. The steel slider bed plates are welded structural members and coated per the manufacturer's standard color or per your company specifications. The extendible and retractable boom types are two boom sections per unit which have a 26 ft. 4 in. retracted length, with a 37 ft. long maximum boom extension. This overall conveyor belt travel path, which includes base and boom extension, is 36 ft. 45 in. long with a nominal 67 in. overall width. There are three boom sections per unit each with a retracted length of 18 ft. 6 in., and a maximum boom 37 ft. long extension.

This overall conveyor belt travel path includes base and boom extension 55 ft. 6 in. long and a nominal 72 in. overall width. Both boom types have a 1 hp drive motor with a flexidyne, a sheave, and a V belt. With tooth sprockets and a closed-loop link chain, the boom is designed to extend the belt conveyor travel path in a cantilevered position. The maximum outward carton conveyor travel path length is 37 ft. The boom extension and retraction travel speed is 40 ft. per minute.

Traversing rails, traversing wheels, and traversing drive motor components allow a conveyor unit base to travel over a fixed metal travel path between two docks. The conveyor unit base traversing rails are two 3 in. high "crane" hardened steel guide rails with a 1 ¾ in. top travel path or face. Traversing rails have welded sides with anchor hooks and are recessed and anchored into a pit in the floor. Each traversing rail is located approximately 32 ⅝ in. from each conveyor unit base and runs the entire traversing distance. End stops at each traversing rail end are provided to stop a conveyor unit base travel.

A mobile powered belt conveyor has four traversing wheels. The four wheels permit a mobile belt conveyor unit base to travel over the two rails and to arrive at the appropriate dock. The traversing wheels have 8 ⅝ in. diameter with a flange. The wheels are made from crane hardened steel and have sealed ball bearings. Each wheel face matches the traversing rail iron face. Power to traverse or move a powered belt conveyor unit between two docks is provided by a traversing drive motor and drivetrain. A traversing drive motor and drivetrain has a 1-hp motor with a flexidyne drive sheave, V belt, and sprocket and chain. The conveyor unit traversing travel speed is 30 ft. per minute.

The mobile powered extendible and retractable conveyor is the carton travel path between the delivery truck and dock area. A standard conveyor belt is an 18-in. wide three-ply bare duck belt that is installed up or with the smooth rubber side facing down or onto the boom plate surface. A conveyor belt is routed from the drive pulley, over the top surface of the boom, over the various end pulleys, through the snubber pulley, and through the take-up device.

The extendible and retractable powered belt conveyor unit application as a receiving or unloading conveyor determines a conveyor belt direction of travel. The receiving or unloading conveyor application has the belt conveyor direction of travel from the delivery truck onto the dock.

Belt drive motor, drivetrain, end pulley, and take-up device components provide a belt conveyor travel speed and assure that the conveyor belt travels at the appropriate speed over the conveyor travel path. The conveyor belt drive motor is a 3-hp motor that is located at the conveyor unit charge end. A drive motor tooth sprocket, associated open-link closed-loop chain and drive pulley tooth sprocket turns the 12-

in. diameter drive pulley with a laggard and crown surface. The various end pulleys are located at strategic locations to assure that the conveyor belt completes the belt travel path around the end of the extendible boom and through the base. The next belt drivetrain component is the snubber. The snubber is a 4-in. diameter pulley that helps with the belt take-up and belt tracking through the belt travel path. Per the conveyor belt travel direction, on the receiving and unloading conveyor application, the snubber is located past the belt travel around the drive pulley.

A belt take-up device assures that the belt conveyor surface moves at the desired travel speed, so thatthe conveyor belt is at the proper tension and has a constant and required conveyor travel speed. As required by the conveyor belt condition, the take-up pulley or pulleys are moved forward from the drive pulley to compensate (increase the conveyor belt travel path) for a slack conveyor belt and are moved toward the drive pulley to compensate (decrease the conveyor belt travel path) for a tense conveyor belt. During boom retraction from the extended position the belt take-up device provides the available space inside the housing to store the unused conveyor belt length and during the boom extension it allows the required conveyor belt transfer from the housing to the boom carton conveyor travel path surface.

On a receiving or unloading conveyor application, the take-up pulleys are located past the conveyor belt travel around the drive pulley. With a two-boom belt conveyor unit, the take-up pulley arrangement has one fixed pulley and two counterweight pulleys. With a three-boom belt conveyor unit, the take-up pulley arrangement has two fixed pulleys and three counterweight pulleys. A belt conveyor drive motor and drivetrain assure that the conveyor belt travel speed is 85 ft. per minute.

The conveyor has guards to protect the operator. During the conveyor unit travel, there are safety tow plates with E stop movement devices along the bottom of both housing bases. The conveyor unit stops moving when these traversing E stop plates strike an object or an employee's limb. Other guards include an extendible boom drive guard or hinged safety plate at the front of the boom,, which stops boom extension if the plate strikes an employee or object. Side guards or guard rails are considered options that are attached to the top of the mobile conveyor unit base. The guard rails assure carton flow across an extendible conveyor travel path surface.

The mobile powered belt conveyor has controls located at the operator's station. A conveyor belt start and stop button and a unit E stop safety switch are located at the front of the boom.

Stationary Powered Extendible and Retractable Belt Conveyor

A stationary powered extendible and retractable belt conveyor can improve employee productivity to unload or load delivery trailers and reduce employee unloading injuries. The stationary powered belt conveyor concept increases the total receiving dock investment but has a lower cost per extendible powered belt conveyor unit.

The stationary powered extendible and retractable belt conveyor has booms with a travel path surface that extends into the delivery truck bed and retracts onto a fixed dock location. Components of a stationary powered conveyor are base structural members, boom drive motor and boom drivetrain unit, support legs and foot plate anchors, belt carton carrying travel path surface, conveyor belt drive motor, drivetrain, end pulley and take-up device, guards, and controls.

The stationary conveyor has a base permanently anchored to the floor of the receiving dock area. This feature minimizes the counterweight amount in the belt conveyor housing rear section. With no conveyor travel path or boom traversing capability, the side guard plates, switches, and wiring are eliminated from a stationary conveyor belt unit. Since a stationary belt conveyor does not traverse, the control devices are the various push button stations and limit switches for a conveyor belt travel and boom extension and retraction. The controls are located at the operator's station and at the boom front end is located a conveyor belt start and stop button and a boom extension E stop safety switch.

Powered Extendible and Retractable Conveyor Unit with Fixed Support

A stationary powered extendible and retractable belt conveyor with a fixed support has structural members that provide support for the full length of the conveyor travel path, which is permanently fully extended. On the dock, this carton conveyor travel path has a fixed length and has space on the dock that is directly behind the dock door. This type of conveyor improves employee productivity and reduces employee unloading injuries. Compared to the mobile powered conveyor, it requires a higher dock investment but has a lower cost per conveyor unit.

The stationary powered extendible and retractable conveyor with fixed support has four wheels, drive motors and drivetrains, base structural members and a guide path, belt carton carrying travel path surface, conveyor belt drive motor, drivetrain, end pulley and take-up device, guards, and controls.

The four wheels are pneumatic wheels with a rubber tread cover and a rigid caster. The wheels are located at the corners of the rectangular conveyor unit. Fixed casters and large pneumatic wheels permit the conveyor unit to enter and exit a delivery truck over the device that is used to bridge the gap between the delivery truck bed and the dock edge and to travel on the dock floor. To assist with the forward and reverse conveyor unit travel, guide rails are attached to the finished floor surface. The two guide rails are located on the inside of the conveyor wheel's travel. A 1-hp drive motor and drivetrain provides the force to move the conveyor unit across the dock floor, over the dock ramp, and into the delivery truck. The drive motor and drive train provide the power to move the conveyor unit forward and to reverse. With a fixed conveyor belt conveyor travel path, the powered belt take-up device is a manually adjusted screw device or an automatic counterweight take-up device.

Since this belt conveyor is moved forward and reversed this fixed carton conveyor travel path is supported on four wheels. With one wheel under each corner of the rectangle-shaped conveyor travel path, the counterweight is minimal in a conveyor base rear section. With this fixed carton conveyor travel path feature and when not in a delivery trailer, the conveyor unit requires dock space. With no traversing capability and no required side guard plates, switches and wiring are eliminated from a conveyor belt unit. Since the conveyor belt concept does not traverse, the control devices are the various stop and start push button stations and limit switches for conveyor belt travel and conveyor fixed travel forward and reverse movement. These controls are located at the operator's station and at the boom front end is located the conveyor belt start/stop buttons and E stop safety switch.

Pallet Handling Device

A pallet handling device (PHD) is a stationary or mobile fully automatic device that is used to unload pallets from a delivery truck. The PHD has a base on the receiving dockand a powered roller pallet conveyor travel path on the top of the base, which provides a staging area for pallets. In the unloading process, a delivery truck pallet pattern is pre-loaded into the PHD's microcomputer and the PHD sensor device determines the first pallet location.

With the PHD forks extended forward, fork sensors determine the pallet openings on the delivery truck floor. With the proper analysis, the PHD concept extends the set of forks forward into the pallet fork openings and elevates the pallet onto the set of forks. With the pallet on the set of forks, the PHD concept returns from the delivery truck to the receiving dock position and raises the pallet for placement onto the roller conveyor travel path that is on the base top.

With a single set of forks, the PHD concept device handles a 3500 lb. pallet and a full delivery truck of 18 pallets in 30 minutes. With a double set of forks, a PHD concept device handles a total of 3500 lb., the combined total for two pallets. When two pallets are handled per trip, the unloading activity time is reduced by 20 minutes to 25 minutes for each delivery truck. When used with the fork sensors and the PHD concept handles a pin wheel-loading pattern, which requires four-way pallets or containers, the PHD concept is capable of handling stacked pallets or containers. To assure dock area safety and minimize damage, the PHD concept has a safety stop device and a safety stop light curtain extends on both sides of the receiving dock area from the dock door to the PHD concept base. A PHD concept has the capability to handle all delivery truck lengths.

Disadvantages are the high cost, it requires a 36- to 40-ft. dock area, and to achieve the objective unloading pallet rate, the pallets are handled in a fluid manner without interruption to the pallet flow. Advantages are minimal labor expense, no dock leveler investment or security guarding for the dock doors and walls, minimal requirement for dock area and delivery truck dock lights, reduced building dock area and mobile warehouse unloading equipment investment, and reduced UOP, building and equipment damage, and maintenance.

Pallet Flow Device

A pallet flow device concept has air pressure, powered strands of chain, mobile rails, and conveyors that move the pallets between the delivery truck and facility dock. The delivery trucks have specially designed floors and the facility dock floor has a specially designed push and pull device and conveyors. To operate, after the delivery truck is located at the dock, an employee activates the dock area section to push the pallets from the dock area onto the conveyorized delivery truck floor. The entire load is pushed into the delivery truck. The delivery truck driver ensures that the load is secured in the delivery truck and the building power electrical cable is disconnected from the dock.

The disadvantages are higher capital investment in the delivery truck, high facility dock area investment, increased maintenance, and both the receiving and shipping locations need to be designed to handle the unique load. The advantages are that it requires less dock time, fewer docks, and fewer employees and equipment.

Customer Order Carton Dock Sort Concepts

The CO carton dock sort concepts assure a CO carton is accurately identified and transferred on time from cartons with mixed CO identifiers for travel to a shipping dock staging area or directly into a CO delivery truck. Carton dock sort concepts can be manual or mechanized.

GOH, Pallet, or Master Carton Identification

The purpose of a WMS-identified master carton or pallet is to allow tracking of the UOP inventory that is placed into an assigned WMS storage position. Each UOP or SKU identification discreetly identifies each master carton or pallet and distinguishes it from the other master cartons or pallets in inventory.

A properly located WMS identification on a pallet or master carton minimizes label damage, is easily recognized by employees, and provides line of sight to an employee eye, RF tag reader or barcode scanner.

In a pallet or master carton operation, a WMS identification label is placed in a consistent location, which could be on the lower right-hand corner of the pallet, attached to a pallet right-hand stringer or block, or in the right-hand corner of the master carton. These locations ensure line of sight for easy barcode scanning or RF tag reading of the WMS identification.

Single-Item, Flatwear Sequence of Operations

In a single-item or flatwear operation, the receiving department separates each SKU delivered into its style, color, or size, which ensures that ticketing and SKU identification activities are cost effective and efficient and each storage master carton or tote receives a WMS identification.

Flat Pack or Master Carton Open Activity

After master cartons containing flatwear SKUs are unloaded from a delivery truck and inspected and approved, each master carton receives a WMS identification and flows directly to a storage area. Merchandise that has not been identified flows through a product SKU identification process. Master cartons with no SKU identification are placed onto the assigned opening lanes. If there is no space available on one opening lane, the extra master cartons are placed into temporary storage positions that are floor-stacked pallets, stacking frames, or portable racks, storage racks, or carton conveyor lanes that are directly behind the opening lane. When an opening lane or workstation becomes available, the master cartons are transferred from the temporary storage position to the assigned vacant opening lane or workstation. The opening lane may be a table top and gravity or non-powered carton conveyor flow lane.

SKU and WMS Identification

Before a SKU arrives at the storage area, it receives a SKU identification. Each master carton receives a WMS identification. In many operations, the receiving department SKU identification activity is considered vendor rework and per the purchasing department and vendor agreement is charged to a vendor or absorbed by the purchasing department.

Table Top SKU and Master Carton Label Concepts

A table top SKU and master carton WMS identification label concept has a rectangular table 4 to 6 ft. by 30 in., with a full pallet or non-powered conveyor located on the right end. An employee transfers one master carton from the pallet or conveyor onto the table top and removes SKUs from the master carton. During this process, the employee verifies the SKU quantity and attaches identification to each SKU. Identified SKUs are returned to the empty master carton or placed into a new master carton, plastic container, or captive tote. Each full master carton or tote receives a SKU and WMS identification with the SKU quantity entered into the WMS files. is the master cartons are then placed onto an empty pallet or a second non-powered conveyor located on the left side of the table. All WMS-identified master cartons or pallets are transferred by a pallet truck or powered conveyor from the identification area to the storage area. Alternative transport concepts are four-wheel carts or non-powered roller or skate wheel conveyors.

The disadvantages of this method are that it handles a low volume, low employee productivity, and increased merchandise handlings. The advantages are low investment, it is simple to operate, and requires little employee training.

Gravity or Non-Powered Conveyor SKU Identification Concept

This concept has gravity conveyors with a powered zero-pressure queue roller and powered belt conveyors, conveyable cartons, totes, and work shelves and platforms. SKU and WMS identifications are placed on each master carton and quantity is entered into the WMS files.

Master cartons are placed onto opening lanes in an arrangement that has one SKU (size, color, and style) per lane. A lane is a carton gravity flow conveyor travel path that permits a master carton to flow from the receiving area to an opening position. On the lane opening side is a fixed end stop. If there is more than one SKU per vendor delivery, the SKUs are arranged by color and style, with the largest SKU size on the right-hand side of the flow lane (to the right of a receiving person facing the gravity conveyor lanes).

As a master carton arrives at the opening station, an employee removes the SKU pieces from the master carton, counts them and places the pieces in an empty master carton or tote. Empty master cartons or plastic totes are supplied by a non-powered or powered conveyor travel path. SKU pieces are placed in the tote in an arrangement that has the identification label facing the direction of travel of the tote. When possible the SKU pieces stand vertically in a tote. The employee enters all SKU piece counts onto a tally sheet. Empty vendor master cartons and other trash are thrown into the trash take-away system. The SKU identification labels are printed.

An employee attaches a product label to each piece. After each SKU in a master carton or tote has been identified, the tote or master carton receives a WMS identification label. The master carton or tote piece quantity is entered into a WMS inventory file and transferred and queued in another gravity flow lane. Totes are released from the SKU labeling lane onto a take-away gravity flow conveyor lane for travel to a distribution gravity flow conveyor lane. When SKU styles, sizes, and colors are labeled and each master carton or tote has a WMS identification label, a lane release employee releases all WMS-identified totes from the product label lanes to SKU distribution lanes. As totes arrive at a distribution gravity flow conveyor lane, a lane control employee directs the totes into a vacant gravity flow lane. Adjacent gravity flow conveyor lanes are used for the same SKU or the next lead tote with new SKU or new WMS identification master cartons or totes arriving at the transfer station. At the distribution gravity flow lane discharge end, the SKU pieces are allocated to COs.

Disadvantages are cost, employee training, and management control and discipline. Advantages are that it handles a high volume and wide UOP mix, work positions are flexible, labor is allocated by skills, and queues prior to workstations.

Master Carton or Pallet Identification

Master carton or pallet identification at the receiving dock can be either permanent or disposable. Permanent identification is associated with a fixed asset that does not change. In a storage operation, each pallet rack storage position has a permanent identification. A permanent identification is listed in computer files and remains an asset. A permanently identified pallet storage position has pallet-identified deposit or withdrawal transactions to an identified pallet position. Other assets with permanent identifications are totes, trolleys, stacking frames, portable racks, Automatic guide vehicles (AGVs), carts, or a transport device load-carrying surface and pallet storage/pick position.

A disposable identification is associated with an asset that has a limited life. In a storage operation, a pallet or master carton has a disposable identification. After all master cartons have been removed from a pallet, the pallet identification has served its purpose. The pallet identification is removed from the pallet and thrown into the trash. In a master carton or small-item operation, the master carton or small-item identification serves the same purpose as a pallet identification.

There are a number of options for location of a pallet identification:

- Ladder orientation
- Picket fence orientation
- Hybrid
- Facing one direction
- Wrapped around a master carton or pallet stringer or block
- On a pallet component or bottom layer master carton

The identification can be attached to the pallet stringer or block tape, staple, self-adhesive label, or sleeve insert.

Ladder Identification Orientation

A ladder orientation on a pallet storage position has the white spaces and black bars on the bar code oriented to resemble steps on a ladder. When a person looks at an identification in a ladder orientation, a bar code's black bars and white spaces are vertical and quiet zones are at the bottom and top. In a conventional forklift truck operation, the light beam on the hand-held scanner has to be vertical, so that the forklift truck driver does not have to turn the scanner to read the identification. In an AS/RS operation, a fixed-position scanner device is easily turned to read a pallet identification in the ladder orientation. The pallet identification is wrapped around a pallet stringer or block or master carton; the ladder identification has the same bar code printed on a wrap around label both sides.

Picket Fence Identification Orientation

In a picket fence orientation, the white spaces and black bars on the bar code resemble the slats in a fence. When a person looks at identification in this orientation, the black bars and white spaces are horizontal and quiet zones are at the right and left. If the pallet identification is wrapped around a pallet stringer or block or master carton, the bar code in the picket fence orientation is printed on the face of the wraparound label.

Hybrid Identification

A hybrid identification has a combination of a ladder and picket fence bar codes on a UOP (master carton or pallet). A hybrid label permits maximum flexibility in an operation that uses manual forklift trucks or AS/RS crane storage vehicles. A hybrid identification means the same bar code (black bars and white spaces) is printed in both orientations on one white surface. With a hybrid label on a UOP, a fixed-beam scanner easily reads the ladder label and a forklift truck driver with a hand-held scanner easily reads the picket fence label. A hybrid label ensures maximum good read number with minimal scanner or label cost.

Identification Facing One Direction

In this method the identification faces one direction of travel, with the pallet fork opening in the lead. The identification label has one face in a ladder or picket fence orientation that is attached to a pallet stringer or block or master carton to ensure line of sight for an employee hand-held scanner or a fixed-position scanner.

Identification Wrapped around the Stringer or Block or Master Carton

A pallet identification that is a wrap-around label has the bar code facing two directions of travel. This option has two bar codes or one long barcode on a label that is attached to the pallet exterior stringer or block or master carton. In this position the pallet identification is in either the ladder or picket fence orientation and faces a pallet fork opening and an exterior stringer or master carton on the pallet's two exterior sides. The location ensures employee hand-held scanner or a fixed-position scanner line of sight.

Customer Order Carton Transport and Sort Concept

If your CO carton dock operation handles a large carton volume for large numbers of COs, to have an on-time, effective, and cost-efficient shipping operation, you use a CO carton transport and sort concept. A discreetly identified CO carton travels over a transport and sort conveyor travel path, a sort concept sorts each discreetly identified CO carton onto an assigned CO divert travel path.

Design parameters for a CO transport and sort concept are:

- A discreet identification label on each carton that is at least human or human/machine readable
- A conveyor transport and sort travel path concept with the ability to queue cartons prior to arriving at the induction station and before each unitizing station or direct load station
- The capacity to handle CO volume
- Capacity to handle the daily vendor deliveries
- Number of CO delivery trucks, with the associated divert transactions

Before designing and implementing a CO carton transport and sort concept, you should have the following data:

- Dimensions for the cartons the operation will handle, including minimum average, and maximum lengths, widths, heights, and weight, as well as top, side, and bottom exterior surfaces
- Cartons per average and peak weekday and per customer order
- Work hours for each work day and customer order number
- Barcode type and label size, and whether the operation will use RF tag and tag reader
- Barcode scanner and sort device, the gap between two cartons
- Carton crushability and fragility
- Ensure that all cartons are sealed and have a conveyable bottom exterior surface
- Provide sufficient queue prior to an induction station or at the divert lane end

Implementing a CO transport and sort conveyor concept with insufficient carton queue capability creates an out-of-balance situation between the scanning, sorting, and loading functions, which prevents loading employees from achieving the budgeted productivity rates.

After these design parameters have been determined for a CO carton sort concept, you design a building to house the CO sort concept or design the carton sort concept to fit within an existing building. CO transport and sort dock concepts may be either manual or mechanized.

Manual Dock Sort Concepts

A manual dock sort concept (see Figure 2.14) is a basic concept with a simple conveyor transport travel path and human readable label. An employee places the label onto the carton exterior surface in a location that is easy to read.

The CO cartons are labeled and transported over a conveyor travel path to a manual sort area that is located in the shipping dock area. As cartons travel on a conveyor travel path, a sort employee reads the carton labels and matches the identification number to a CO human-readable identification on the employee-assigned sort position. When there is a match, the employee transfers the carton onto the

MANUAL SORTATION

MANUAL SORTATION SINGLE CONVEYOR TRAVEL PATH

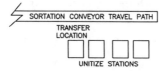

MANUAL SORTATION WITH ELEVATED DID NOT SORT CONVEYOR

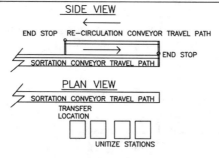

MANUAL SORTATION WITH RE—CIRCULATION CONVEYOR

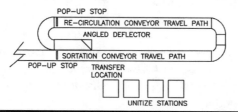

Figure 2.14   Manual sort concepts.

assigned CO shipping cart or pallet. Full pallets or carts are transferred to a staging area or loaded onto a CO delivery truck.

Few and large characters or digits on the identification label means better employee productivity and fewer sort errors. In a manual carton sort concept, a sort employee handles 10 to 15 cartons per minute with little damage impact on carton contents. For best employee carton sort productivity, maximum carton weight is 40 to 50 lb. and the average weight is 20 to 30 lb. per carton. To ensure efficient carton handling, a sort conveyor travel path maintains a 3- to 6-in. gap between cartons and the carton sort conveyor is a low or zero pressure queue conveyor.

The manual dock carton sort concept can be one conveyor, double-stacked conveyors, a recirculation conveyor, or apron conveyor.

One Conveyor Travel Path

In this manual dock carton sort concept, the conveyor travels along all the CO work stations. A low- or zero-pressure queue conveyor has a 1-in. high guard rail on the far side (from the work station) that runs the full length of the conveyor, and the conveyor has an end stop. This guard rail ensure that the cartons are retained on the conveyor travel path. A low–zero pressure queue conveyor travel path allows a sort employee:

- With no new cartons on a sort conveyor travel path, to move a "did not" sort carton in a reverse direction on top of the conveyor travel path to the appropriate CO sort location
- To easily create a gap between two cartons
- To stop a carton on the sort conveyor travel path without forward pressure from the cartons behind

The disadvantages of using one conveyor are possible sort errors, the need for a large sort label, it can only handle a low volume and few customers, it is difficult to handle "did not" sort cartons, requires a large number of employees and has a high potential for employee injury. The advantages are low impact on cartons, requires minimal capital investment, is simple to operate and control and as required with the guard rail is a bilateral sort concept.

Double-Stacked Conveyor Travel Paths

A double-stacked conveyor design has two conveyor travel path levels that are located along sort fronts. The lower conveyor travel path is a low- or zero-pressure queue conveyor that is a sort conveyor. The upper conveyor travel path level is a "did not" sort carton conveyor travel path. This upper conveyor travel path is a low- or zero-pressure queue or non-powered roller conveyor. If a sort employee at the first CO sort station "did not" sort a CO carton, an employee at the next sort station removes the carton from the sort conveyor travel path and transfers it onto the upper carton conveyor travel path. The direction of travel of the elevated conveyor returns the "did not" sort carton to the first CO sort station.

This concept has the same disadvantages as the one-sort conveyor travel path concept, and in addition requires additional investment and uses more employees. In addition to having the same advantages as the one-sort conveyor travel path concept, another advantage is that it minimizes the employee's effort to transfer "did not" sort cartons.

Recirculation or Loop Conveyor

Components of this type of concept are:

- One conveyor travel path that has two straight conveyor travel path sections. One section is an in-feed conveyor travel path and the other a recirculation conveyor travel path.
- Two 180-degree curves or two right angle transfer sections.
- Separate merge locations for new cartons and "did not" sort cartons.

From the in-feed conveyor travel path, cartons travel past CO sort stations. When a "did not" sort carton passes a CO sort station, the carton continues to travel on the sort conveyor and is reintroduced to the sort conveyor travel path, either manually or using a photo-eye to detect "did not" sort cartons.

A manually controlled loop conveyor has:

- One straight conveyor travel path that is the sort conveyor travel path and a second straight conveyor travel path that is a re-circulation conveyor travel path
- An end stop on each sort conveyor
- Guard rails between the sort and recirculation conveyor travel paths or a gap plate the full length of the conveyor travel path

A gap plate is a solid piece of metal or hardened plastic with a slight bow in the middle and is attached to each conveyor bed top. A gap plate serves as a transition plate for a sort employee to push a "did not" sort carton from the sort conveyor onto the recirculation conveyor and a guard rail to direct carton flow over the sort and recirculation travel paths. If a "did not" sort carton appears on a sort conveyor travel path, a sort employee at the next sort station gently pushes the carton across the gap plate onto the recirculation conveyor travel path. The recirculation conveyor moves the carton to a location ahead of the first sort station. At this location, a sort employee pulls the carton from the recirculation travel path across the gap plate onto the sort conveyor and the "did not" sort carton is reintroduced to the sort conveyor travel path.

A powered 180-degree curve carton sort concept has:

- Two powered 180-degree curves.
- An in-feed conveyor with a photo-eye-controlled stop device.
- One straight conveyor travel path from the conveyor which is the sort conveyor travel path and a second straight conveyor travel path as the recirculation loop. Each straight conveyor travel path has a photo-eye. The photo-eye on the sort conveyor controls the in-feed conveyor travel path. The photo-eye on the recirculation conveyor controls the "did not" sort recirculation conveyor section.
- Two merge conveyor sections.
- As-required guard rails on all conveyor travel paths.

Disadvantages of the loop conveyor are the same as for a doubled-stacked conveyor. An additional disadvantage is higher cost. The additional advantages are that employees do not lift "'did not" sort cartons and can handle a higher volume.

Circular Apron Conveyor

The circular apron conveyor sort concept has a photo-eye and conveyor network that allows cartons to enter the sort conveyor travel path and "did not" sort cartons automatically recirculate onto the sort conveyor travel path.

The advantages and disadvantages are the same as the recirculation loop conveyor concept, except that there is an additional investment in the apron conveyor and photo-eye controls.

Mechanized Carton Sort

A mechanized carton sort operation has all CO discreetly identified master cartons transferred from a transport and sort conveyor travel path. A shipping dock transport and sort conveyor travel path has each carton travel through an induction station, over a sort conveyor travel path to the assigned divert station for transfer from a sort conveyor travel path onto a CO-assigned sort unitizing station or shipping lane. From the shipping lane the cartons are either loaded directly onto a CO delivery truck or unitized onto a pallet or cart for transfer at a later time from the CO-assigned shipping dock staging area onto a CO delivery truck.

For a mechanized sort concept to operate efficiently, each carton has a human- or human/machine-readable discreet identification label that contains the assigned CO sort location. After a human, RF tag reader, or barcode scanning device enters a carton discreet identification into a microcomputer, the microcomputer and tracking device controls the sort activity. This sort activity transfers a carton from the sort conveyor travel path onto a CO shipping staging lane for transfer onto a pallet or loading directly into a CO delivery truck.

Components of a mechanized carton dock sort concept are a transport conveyor travel path, induction station area with a brake, meter and induction conveyors, a "no-read" conveyor, a sort conveyor travel path with divert device, customer shipping lane, or unitizing station, clean-out station, and recirculation conveyor travel path.

When you consider a carton dock sort concept implementation, you ensure that there is adequate and clean electrical power in the area and that there is a drain for the air compressor condensation.

Transport Conveyors

A transport conveyor travel path ensures that there is sufficient carton queue and that cartons travel from the pack area to the induction station area. Prior to the induction station, cartons are lined up in a single file. On a transport conveyor travel path a gap is created between two cartons by brake and metering belt conveyors, conveyor travel speed, or a stop device. This gap between two cartons ensures that the induction employee, RF tag reader, or the barcode scanning device reads or enters the correct carton identification label data into the microcomputer and that the divert device has sufficient space to complete a carton divert from a sort conveyor travel path onto a CO-assigned shipping lane.

Induction Conveyor

An induction or in-feed station is the location at which the identification is entered into the sort conveyor microcomputer. The conveyor travel path design allows sufficient conveyor queuing to provide a constant carton flow to the induction area and onto the sort conveyor travel path. The induction equipment can be manual or keypad, a semiautomated or hand-held scanner, or automated with a fixed-position scanning device.

Manual Induction

In a manual induction concept, each carton has a large human-readable label on its top or side. Adjacent to the carton travel path is a keypad that is connected to a microcomputer and tracking device. As a carton passes the induction station, an induction employee reads the identification on each carton and keys it into a keyboard. The keyboard transmits the CO identification to the microcomputer and tracking device. Most keypads have ten digit (0 to 9) buttons, as well as a repeat button, scan off–on button, error signal, cancel key, five-digit LED (light-emitting diode) display, emergency stop button, and a send key. To achieve high manual induction productivity, the human-readable discreet code has large, bold print and contains the lowest number of digits possible. Most keypad applications prefer two digits.

The disadvantages are increased need for employee training, the concept handles a low volume, potential induction errors, it requires a larger number of employees, and at an induction station, all customer discreet codes need to face one direction of travel. Advantages are low capital cost and it serves as a back-up concept for other induction methods.

Semiautomatic or Hand-Held Scanning/Reading Induction

In a semiautomatic induction concept each carton has a human- and human/machine-readable CO label on its top or side. As a carton travels on a conveyor travel path past an induction station, a brake and metering belt conveyor or speed controls create an open space (gap) between two cartons. At an induction station, an employee takes a hand-held scanning device (finger, wand, or gun) and scans the carton label. The unique carton identification is sent to the microcomputer. The carton passes the induction photo-eye and is released onto a sort conveyor travel path.

Cartons traveling on the sort conveyor travel path are under the control of a computer and tracking device. At an assigned location on the sort conveyor travel path, the devices trigger an assigned divert device to transfer the carton from the sort conveyor onto a CO-assigned shipping lane. The computer sends a CO discreet identification to a manifesting system.

Disadvantages are the additional investment and additional employee training required. Advantages are reduced induction errors, the concept handles a higher volume, and information transfer is accurate and on-line.

Automatic Scanning/Reading Induction

An automatic induction or fixed-position barcode scanning concept requires a barcode label or RF tag with the CO identifier to appear on the front, side, top, bottom, or rear of each carton. After the carton is properly spaced on a transport conveyor travel path, it travels to the scanning station, where the CO identification is read by a fixed-position, fixed-beam scanning device, waving beam, or moving-beam scanning device.

As a carton passes an induction photo-eye, it is transferred to a sort conveyor travel path. The CO discreet identification is transmitted to a sort conveyor computer and tracking device to ensure an accurate and complete sort. A CO discreet identification is held in the computer for manifest preparation. This automated induction concept is used on a smooth top belt conveyor, roller conveyor, tilt tray, tilt slat, SBIR, or gull-wing transport, and sort conveyor travel path.

The carton travels on a sort conveyor travel path past the induction photo-eye and is transferred onto a sort conveyor travel path. A CO discreet identification, RF tag, or barcode location is sent to the microcomputer. The sort conveyor travel path constant travel speed ensures that a carton arrives at an assigned divert location and the computer and tracking device activates the assigned divert device to transfer the carton from a sort conveyor travel path onto a CO-assigned conveyor lane.

The disadvantages are capital investment and human/machine readable label. Advantages are reduced induction errors, few employees, handles a high volume, and information is accurate and on-line.

"No Read" Conveyor

In all carton dock sort concepts and especially in an automatic induction concept, a second activity is the reintroduction of "no read" cartons. A "no read" carton is a carton that had a CO discreet identification entry problem at an induction station. After the carton passes the automatic barcode label scanning station, a "no read" divert device and short conveyor returns the "did not" read cartons to the induction station. A "no read" divert device and conveyor spur on a sort conveyor travel path is located a nominal 12 to 15 ft. from the barcode scanner or RF tag reader station and is the first divert location past the barcode scanning/RF tag reading station. When the carton is returned to the induction station, it is automatically read by the barcode scanner a second time or is manually or semiautomatically inducted onto a sort conveyor travel path.

Various Sort Surfaces or Conveyor Travel Paths

The carton sort conveyor is a major component of the carton dock sort concept. The conveyor travel paths can be a roller conveyor, smooth-top belt, slat tray, tilt tray, SBIR, or moving belt conveyor and gull wing. The conveyor surface or travel path assures that the carton travels at a constant speed and provides a path from the induction station to all sort locations.

The sort conveyor travel path can be designed as a single straight-line conveyor travel path or an endless loop conveyor travel path.

A Single Straight-Line Sort Conveyor Travel Path

A straight-line dock sort conveyor travel path follows a straight line (see Figure 2.15) from the induction station to the last sort station. After the last sort station, the sort conveyor travel path ends and the conveyor travel path declines to the floor surface. Alternative straight-line sort conveyor travel path designs are L-shaped or straight-line layout and horseshoe- or U-shaped layout.

Unique characteristics of a straight-line conveyor travel path are that there is no "did not" sort carton automatic recirculation to an induction station. In this scheme, "did not" sort cartons are created by a full sort lane or a divert device malfunction. Options for handling these cartons are to manually return them to an

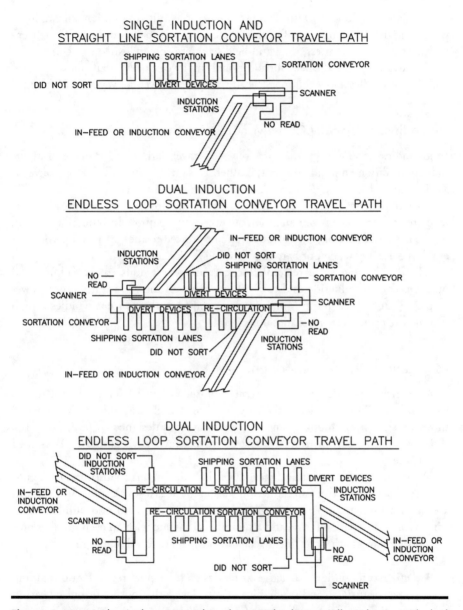

Figure 2.15 Mechanical sort travel path: straight line, endless loop, and dual induction.

induction station or manually transfer them to a unitizing station. The "did not" sort cartons are either removed from the travel path and diverted to the last divert location, or, after the last divert station, the conveyor travel path declines to the floor on a run-out conveyor lane to form a divert lane end, where an employee unitizes cartons onto a cart or pallet. Cartons are manually transported to the induction station or a proper CO unitizing station. At the unitizing station, an employee with a hand-held scanner verifies that the carton was loaded onto a CO delivery truck. To create a manifest, the hand-held scanner is down loaded to a computer.

Disadvantages are the need for additional employees to physically move cartons, and that it handles a low volume. The advantage is a lower conveyor investment.

An Endless Loop Sort Conveyor Travel Path

In an endless loop design, the sort conveyor travel path starts at one end of the induction station end and ends at the other. Designs for an endless loop conveyor are L-shaped, U-shaped, O-shaped, elliptical, and rectangular.

In each endless loop conveyor travel path all "did not" sort cartons are automatically recirculated via a recirculation conveyor travel path to an induction station. At the induction station, the cartons are reintroduced to the sort travel path by an employee, RF tag reader, or barcode scanning device.

The disadvantages are recirculation carton volume adds to the volume and increases carton induction and sort volume. The advantages are it requires fewer employees and there are minimal employee injuries; the concept handles a high volume with automatic recirculation.

Mechanical Divert Component

A mechanical divert device on a sort conveyor (see Figure 2.16) provides the means to transfer a CO carton from the sort conveyor travel path onto a CO-assigned unitizing station or shipping lane. A divert device is designed to pull, push, tip, slide, or pass the carton from the sort conveyor travel path onto a CO-assigned divert lane.

A carton sort conveyor travel path divert lane design has an arithmetic progression for a sort station from an induction station to the last sort station. Sort stations are in an arithmetic progression with numbers that end with odd numbers on a carton sort travel path left side and all numbers that end with an even number on a carton sort travel path right side. Mechanical carton divert designs are:

- Active sorter design. An active sorter design has a powered induction station, conveyor surface, and mechanical divert device that pushes or pulls a carton from the sort travel path onto a CO assigned sort lane.
- Active–passive sorter design. An active–passive sorter design has a manual induction station and a powered conveyor surface that tips or plows a carton from the sort travel path onto the assigned CO sort lane.
- Passive sorter design. A passive sorter design has a manual induction station and a powered conveyor with gravity force to remove a carton from the sort travel path to a CO-assigned station.

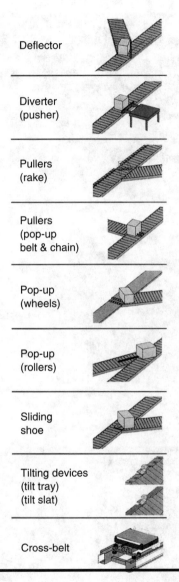

Figure 2.16 Mechanical sort devices.

Figure 2.17 shows various divert devices, which include solid metal deflector, pusher diverter, powered belt diverter, plow diverter, SBIR or moving belt, tilt tray, Nova Sort, tilt slat, gull wing, sliding shoe, pop-up diverts that include (a) pop-up wheel, (b) pop-up chain, and (c) pop-up chain with a blade, rotating paddle, and flap sorter.

Figure 2.17 Mechanical sort travel path: cross belt, tilt tray, and sliding shoe.

Solid Metal Deflector

A solid metal deflector is a passive sort concept that provides a low-cost means of diverting cartons from a mechanized carton sort travel path. This basic deflector is a solid metal bar or arm that is pneumatically or hydraulically activated. To divert a carton the deflector is extended at an angle across the sort conveyor travel path. At the extended angle, the deflector blocks the carton travel path. The conveyor travel path power, carton forward movement, travel speed, and the deflector angle guide the carton from the smooth top belt or roller sort conveyor travel path onto an assigned customer divert lane. When not required to divert a carton from a sort conveyor travel path, the deflector is retracted to the side of the sort conveyor. A solid metal deflector does not change the direction of travel of a carton as it is transferred from the sort conveyor travel path to a shipping lane. This device handles a low volume, 20 to 40 cartons per minute. To achieve a high sort rate, a carton sort conveyor concept handles a carton slug for one divert location. Carton weight is 10 to 50 lb. and a 3 to 5 ft. distance between two divert locations and the carton impact is medium.

Pusher Divert

A pusher carton divert concept is an active sorter device that is either side mounted or an overhead paddle. The device can be an attachment on a sort conveyor travel path or a device that moves across the surface of the sort conveyor.

The pusher is located on the far side of the travel path. After a carton is inducted onto the sort conveyor and as it carton arrives at a CO discreetly assigned sort location, a pusher divert device is activated by the sort concept microcomputer and an impulse from the tracking device. The divert device pushes a divert blade across the sort conveyor travel path, and engages the carton side. As the pusher blade extends forward, the carton is pushed from the sort travel path onto a customer- assigned divert lane. A divert device moves the carton onto a shipping lane. A carton traveling on a shipping lane has a different direction of travel than a carton on a sort conveyor. The average pusher sort number per minute is 25 to 30 cartons, maximum carton weight is 125 lb., spacing between two divert locations is a maximum carton length plus 6 to 12 in. and the diverter impact on a carton is medium.

A puller divert device is similar to a pusher sort concept except that the carton is pulled across the sort conveyor travel path onto a CO assigned sort location. The sort rate is 35 to 40 cartons per minute, carton weight capacity is 10 to 100 lb. and diverter impact on a carton is medium to rough.

An overhead mounted pusher diverter concept handles approximately 30 diverts per minute. An overhead pusher diverter with CO divert locations on both sides of the sort conveyor travel path has the ability to push/divert a carton from the carton sort conveyor travel path on the outgoing stroke and to push another carton on the return stroke onto a second divert station. With a properly identified carton sequence on the sort conveyor travel path, the overhead mounted pusher diverter concept completes 35 to 40 carton sorts per minute but requires increased carton control in-feed and carton spacing on the sort conveyor travel path.

In most dock sort applications, a carton discreet identification label sequence on the sort conveyor travel path is a random occurrence. This reduces the maximum carton sort rate to a normal cart sort rate. The carton sort conveyor travel paths are roller conveyor or smooth top belt conveyor.

Powered Belt Diverter

A powered belt diverter is an active sort divert concept, with arms located along the sort conveyor travel path. The sort conveyor is a powered roller conveyor. Each divert arm has a small powered belt. After a carton is inducted onto the sort conveyor and prior to its arrival at a CO divert location, a divert arm swings across the conveyor travel path and the divert arm belt starts to move. The divert arm grabs the carton side and moves the carton from the sort conveyor onto a CO-assigned sort station. The direction of travel of the carton is the same as on the sort conveyor. The typical carton weight is 1 to 50 lb., divert locations are 3 to 5 ft. apart, and the diverter impact on the carton is medium to rough.

A powered belt divert concept handles 20 to 30 cartons per minute. When there is a carton slug for one CO location, the divert rate is increased to 30 to 40 cartons per minute due to the arm not being required to move for the next carton divert. In a dock sort application, a carton or CO discreet identification label sequence on a sort conveyor travel path is a random occurrence. This means that the sort concept has a normal sort rate.

Plow Diverter

A plow diverter is an active–passive dock sort device that is a curved metal divert arm or plate. Divert arms are programmed to swing across the carton sort conveyor travel path. The sort conveyor is a roller conveyor. A plow diverter is designed with one plow end attached to the far side of the sort conveyor travel path while the other end moves toward the near side of the sort conveyor travel path. In an extended position, the plow diverter moves across the sort conveyor travel path in the direction of travel of the cartons. The curved divert arm position causes the carton to slowly follow the curve of the divert arm, directing carton from the sort travel path onto the appropriate shipping lane. A plow diverter is best used on a roller conveyor that sorts a carton slug that is diverted onto a shipping conveyor lane. There should be 2 to 5 ft. between divert locations. A plow diverter does not change the direction of travel of the carton on the sort travel path. With a relatively slow divert plow movement, a plow diverter concept handles 15 to 20 cartons per minute. Cartons have a weight range of 1 to 50 lb. and impact on the carton is medium and gentle.

In a dock sort application, a carton or discreet identification label sequence on a sort conveyor travel path is considered a random occurrence. This feature has the plow divert concept achieve a normal sort rate.

Moving Conveyor Belt

A moving conveyor belt carton sort concept is an active sort concept that has individual carton-carrying surfaces. Each carton-carrying surface is an individual belt conveyor that is mounted on four wheels. The individual belt conveyors are designed to travel in a closed-loop travel path. Each cross belt carton-carrying surface travels pass an induction station where the CO identification on each carton is read by a barcode label scanner or RF tag reader device and transmitted to a microcomputer. The computer, constant travel speed, and tracking device assure that the individual belt-carrying surface is activated as the carton arrives at the CO-assigned sort location. The CO divert location is on the right or let side of the load-carrying surface. At the CO-assigned sort location, the load-carrying belt surface turns forward and discharges the carton onto a customer shipping lane.

Carton orientation or direction of travel on a CO sort lane is different from carton direction of travel on the carton sort conveyor travel path. A carton sort concept has a gentle impact or no impact on a carton and handles a carton weighing up to 50 lb. The divert device performs a high sort number per minute, a two-way sort concept, and has a capital investment.

Tilt Tray

A tilt tray concept is an active–passive carton sort concept. The tilt tray concept is made up of slightly concave tilt trays with four wheels that constitute a carton-carrying surface (platform). Trays ride on a closed-loop travel path and are pulled by a motor-driven chain or electric buss bar and aluminum interior. Tilt tray concepts have the capability to handle one long carton on two trays. Tilt tray chain driven concepts are:

- Bull wheel and drive, which is old technology.
- Caterpillar drive that drives most existing tilt tray sort systems.
- New tilt tray technology that has an electric bus bar propel a tilt tray forward over a travel path. The forward force is created by the reaction from an electric wave pushing forward an aluminum tilt tray foot.

After a carton with a bar code facing upward or RF tag is manually or automatically inducted onto a tilt tray sort travel path, the tilt tray travels under a barcode scanner or RF tag reader that sends the bar code or RF tag discreet identification to a microcomputer. As a tilt tray with a CO discreet label travels on the conveyor travel path, the tilt tray is under the control of the microcomputer and a tracking device. The tilt tray is electrically or pneumatically tilted at the CO-assigned sort location. This CO sort location is on the left or right side of the tilt tray chain travel path. For best results, odd-numbered divert locations are on the chain travel path left and even-numbered divert locations are on the chain travel path right side. Most tilt tray chain applications have the lowest numbered sort location as the first sort location past the induction station.

The tilt tray tilting action creates a gravity force and the tilt tray forward travel speed causes a carton to slide from the tilt tray onto a CO-assigned shipping lane. After the tray is tilted and prior to the tilt tray conveyor travel path induction station, all trays pass a clean-out lane that tips each tray. This tray tilting action removes all unsorted or "did not" sort cartons from the tilt tray sort travel path. These cartons are reintroduced to the tilt tray induction station or are manually delivered to the assigned CO shipping lane discharge end. If the cartons are manually delivered to the shipping lane, a carton is manually scanned by a hand-held or finger scanner and the carton is placed onto a tray and is downloaded to the computer. After this clean out or "did not" sort station, a tilt tray travel path device relatches or levels all the tilt trays. At the induction station, each tilt tray receives another carton.

Most carton tilt tray sort concepts have the trays set on 27-in. centers and sort locations are on 3- to 4-ft. centers. The tilt tray and divert centers vary according to the maximum carton size, carton weight, and travel speed of the tilt tray chain. Nominal tilt tray conveyor tilt rates are 180 to 241 trays per minute and the handling capacity is 25 lb. maximum per one tilt tray. The carton direction of travel is either changed or not changed on a CO sort lane and the impact on a carton is medium.

An employee or an automatic induction concept on a tilt tray that travels over a fixed closed-loop sort conveyor travel path inducts a carton discreet identification label. A sort conveyor travel path has CO sort locations on a tilt tray conveyor travel path one or both sides. In many applications, the tilt tray sort concept is designed with dual induction stations. Dual induction stations have two tilt tray clean-out chutes, two tilt tray relatch stations, and two induction stations. A dual-induction station has increased tilt tray sort capacity.

Nova Sort. A Nova Sort is a modular concept that has a closed-loop fixed-sort travel path. A Nova Sort has three to four motor-driven trays. Each wheeled tilt tray has a 500-lb. load-carrying capacity and tilts to the left or right of the travel path. Each modular tilt tray train has a computer, a fixed travel path with an electric-powered bus bar, and sensing devices to detect an object or employee limb in the Nova Sort tilt tray travel path.

A Nova Sort vehicle travel path is ceiling hung or floor supported, travels vertically over a slight grade, and travels over 90 and 180 degree curves. A sort travel path directs a Nova Sort tilt tray vehicle train past an induction station and CO sort stations. At an induction station, an induction employee or automatic transfer belt places a discreet identified carton onto an empty tray and a barcode scanner or RF tag reader reads a bar code or RF tag. A barcode scanner or RF tag reader sends the information to a computer. The discreet information assigns a carton to a specific CO sort location. As a tilt tray train travels over a travel path, the computer and tracking device assure that the appropriate vehicle tilt tray tilts and discharges a carton from a tilt tray onto the assigned CO sort location.

With the on-board microcomputer, at a sort location a tilt tray train is unloaded and loaded by a sort location employee. After all tilt trays are tilted at the required CO sort locations, a tilt tray train travels on the fixed travel path and queues prior to the induction station or continues traveling to the maintenance spur. On the maintenance spur, a tilt tray train waits or queues for the next assignment. As an option, each tilt tray train has a fixed load-carrying surface and has the capacity to handle one carton on one tray or one long carton on two trays. Expansion for additional CO sort locations is provided when additional sort travel path tracks are added to the existing travel path.

The Nova Sort design parameters and operational characteristics are similar to the tilt tray sort concept.

The disadvantages are: handles a medium volume, capital investment, single tray travel path could have queues, and induction platform requires employee egress and access paths. Advantages are as required cartons are transported over the travel path, concept is ceiling hung and floor supported, both loading and unloading is performed at a sort station, and easy expansion.

Tilt slat. A tilt slat concept is an active–passive sort concept, which has wheeled slats that ride on a closed-loop travel path. A motor-driven chain pulls the slats. A predetermined number of slats serve as a carton load-carrying surface. After a carton is inducted by an employee or automatic barcode scanning or RF tag reader, the carton is transferred by an employee or automatic transfer device (several short belt conveyors) onto the tilt slat concept. The CO discreet identification is transferred to the microcomputer and tracking device. Constant travel speed and these devices assure that a CO-assigned carton arrives at a CO-assigned sort station. At this station, slats are tilted and the carton slides onto a CO-assigned shipping lane. CO sort stations are located on the right or left side of the closed-loop travel path.

The tilt slat operational characteristics and disadvantages and advantages are similar to the tilt tray operational characteristics and disadvantages and advantages. After the carton sort onto a CO shipping lane, the carton has a new direction of travel. With the capability of the tilt tray concept to handle a long carton on two trays, the tilt slat technology is considered old technology.

Gull wing. A gull wing concept is an active–passive carton sort concept, so named because it is shaped like a bird's wings, or a V. The gull wing carton sort concept has four-wheeled carton load-carrying surfaces that ride on a closed-loop travel path. CO sort locations are on both sides of the sort conveyor travel path. A motor-driven chain pulls the gull wing-shaped devices over the travel path.

After a carton is placed onto a gull wing carrier, it is inducted onto a conveyor sort concept. As the gull wing travels past a barcode label scanning/RF tag-reading station, each carton discreet identification is entered into a sort computer. When a carton arrives at it assigned CO sort location, the computer activates the appropriate tipper device along the sort travel path. The gull wing tipping action uses gravity force and the sort conveyor forward travel speed to slide the carton from the gull wing carrier onto a CO shipping lane.

The gull wing operational characteristics and disadvantages and advantages are similar to the tilt trays. The gull wing sort concept handles a wide carton variety, low capacity, and on a CO shipping lane the carton direction of travel is changed from that on a sort travel path.

Sliding shoe. A sliding shoe concept is an active–passive carton sort concept, with many powered slat surfaces with shoes located between two slats. The open space allows a shoe to slide across the carton travel path. A sliding shoe is designed either as a single sliding shoe or a dual sliding shoe sort concept. Sliding shoes are located and travel along opposite sides of a sort location. This feature has a slat sort conveyor travel path width that accommodates the widest carton plus the sliding shoes or shoe dimensions.

At an induction station, a carton has its discreet identified barcode label or RF tag CO sort location read and sent to the sort computer. A computer, constant travel speed and tracking device assures that a carton travels on a sort conveyor and arrives at an assigned sort station. Prior to a carton arrival at the assigned sort location, the sliding shoes start to move across the carton direction of travel path, from the far side of the sort window to the near side. The position of the sliding shoes on the sort conveyor travel path forms an angle that causes a carton to move from the sort travel path onto a CO shipping lane. The carton direction of travel or orientation on a CO shipping lane is the same as the carton direction of travel on the sort conveyor. A sliding shoe carton sort concept handles 130 to 150 cartons or sorts per minute or a 400 fpm sort conveyor travel speed with a gentle impact on a carton.

Pop-up diverter concepts. Pop-up diverters are a group of active sorters: pop-up wheel, pop-up chain, pop-up slat, and pop-up roller concepts. These carton conveyor sort concepts have similar characteristics and are installed slightly below the live roller conveyor travel path surface. Another similarity is these carton sort concepts handle the carton bottom exterior surface.

A pop-up wheel and roller diverter concept has one wheel, two strings of wheels, or a predetermined number of short rollers that are located on the sort conveyor travel path. The top surfaces of a pop-up device are set at an elevation that is slightly below the wheels or roller tops. At the start end of the carton sort conveyor, these pop-up wheels and rollers are angled (skewed) to the left or right of the conveyor travel path, toward a CO sort location.

After an inducted carton leaves the induction station, it travels along the sort conveyor under control of a microcomputer and tracking device. As the carton arrives at its CO sort location, the computer activates the pop-wheels or rollers to rise slightly above the conveyor travel path surface or the wheel or roller tops. The wheel or roller angle and turning action grabs the carton bottom exterior surface and directs the carton from the sort conveyor travel path onto a CO shipping lane.

After the carton is moved onto the CO shipping lane, the pop-up wheels or rollers return to the lowered elevation that is slightly below a carton sort conveyor travel path surface. This wheel or roller position allows other cartons to travel across a sortation location onto a customer-assigned sort location. The pop-up wheel carton sort concept handles 65 to 150 sorts per minute with a gentle impact on a carton, carton interface is from a carton bottom, sort location spacing is 4 to 5 ft. and handling capacity is a carton weight of 30 lb. The pop-up roller carton sort concept handles 15 to 20 cartons per minute with sort locations on closer centers and handles a carton weight of 200 lb.

The pop-up wheel and pop-up roller carton sort concepts maintain the same direction of travel as the carton sort conveyor travel path.

A pop-up chain concept is an active carton divert concept that transfers a carton from a roller conveyor sort travel path onto a right angle CO shipping conveyor travel path. This carton divert concept changes the direction of travel of the carton on a CO shipping conveyor.

After a carton is inducted onto the sort conveyor travel path and the carton arrives at the assigned sort location, two chain strands sandwiched between the conveyor rollers are raised slightly above the roller conveyor travel path surface. Rollers engage the bottom of the carton bottom. With the carton on top of the chain, the carton is elevated above the sort conveyor surface and is transferred from the sort conveyor onto a CO shipping lane.

The pop-up chain diverter has an attached blade or bar. As a carton is diverted from the sort conveyor travel path, the chain turns and a blade engages the bottom of the carton. As the chain moves forward, the carton moves over the smooth roller conveyor travel path onto a CO shipping lane. A pop-up chain carton sort concept handles approximately 10 to 20 cartons of sorts per minute and a carton weight up to 75 lb. Pop-up chain divert locations is on 4 to 5 ft. centers, has a gentle impact on a carton, and interfaces with a carton bottom.

Rotating paddle. A rotating paddle pusher is an active carton sort concept. A rotating paddle is a spherical sort device with three to four paddles or three to four flat surfaces that extend outward toward the sort travel path when activated.

As a carton arrives at an assigned CO sort location, a paddle rotates and makes contact with a carton side. This paddle contact and carton forward travel speed over the smooth roller conveyor surface forces the carton from the sort conveyor travel path onto a CO shipping lane. A rotating paddle concept handles a 1 to 75 lb. carton with a medium to rough impact on the carton and performs 50 to 70 sorts per minute. On a carton sort conveyor travel path, CO sort locations are on 9 ft. centers.

Flap sorter. A flap sorter concept is an active–passive sort concept. It has fixed belt conveyor sections that make up the sort conveyor travel path. The belt sections are between CO sort locations.

The flap belt sort conveyor section is designed as a unidirectional conveyor section that moves to angle downward at the end. This downward-angle serves to divert a carton from the sort conveyor travel path onto a shipping conveyor travel path.

The shipping sort location is located below the sort location. After a carton sort or after a predetermined period, the sort belt conveyor end returns to the horizontal level. The horizontal conveyor travel path allows other CO cartons to travel pass the sort location. The flap sorter concept is design with a fixed "did not" divert location or a closed-loop sort conveyor travel path with a recirculation conveyor section.

After a discreet coded carton or RF tag is read by a scanning or reader device and communicated to a computer, the computer and tracking device ensure that the carton is sorted at a CO sort location. This sort location is directly past and under the flap belt conveyor section. As a carton on a sort conveyor travel path arrives at a CO sort location by traveling over fixed belt conveyor and flap belt conveyor sections, the computer and tracking device trigger the assigned flap belt conveyor section to swing down. This downward action combined with the carton forward travel movement and carton acceleration due to gravity sorts the carton from the sort conveyor travel path into a CO sort location.

The disadvantages are capital investment and one-way sort concept. Advantages include that it handles batched customer cartons and it handles a high volume.

Customer Shipping Lane or Conveyor

A dock carton sort concept customer shipping lane or conveyor (see Figure 2.18) is a carton travel path from a sort conveyor travel path to a CO palletizing station or direct load conveyor. To minimize carton damage, along each CO shipping lane photo-eyes are located in the middle of the shipping lane and near the sort conveyor travel path. When a carton blocks the photo-eye located in the middle of the shipping lane, the microcomputer activates the partial full shipping lane alarm. This is a signal to management and employees that there is a problem at a shipping lane loading end that is causing cartons to queue on a shipping lane. When a queued carton blocks the photo-eye near the sort conveyor path, the computer deactivates the CO sort device on that sort conveyor, which causes additional assigned COs to recirculate on a closed-loop sort conveyor travel path or, on a one-way sort conveyor, to be discharged at the "did not" sort location.

Full-length side guards are attached to the carton shipping travel path. The side guards ensure that the cartons are retained on the travel path. A CO shipping lane conveyor travel path can be:

■ Gravity metal, strand metal, fiber-glass, or plastic-coated wood slide or chute
■ Gravity skate wheel conveyor or roller conveyor
■ Powered belt conveyor
■ Roller-controlled low-pressure live-roller conveyor

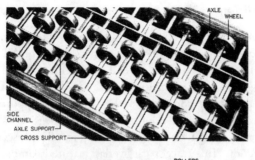

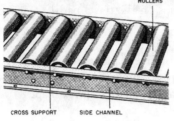

SHIPPING CONVEYOR TRAVEL PATH OPTIONS

STRAIGHT DIVERT & LOADING CONVEYOR CONCEPT

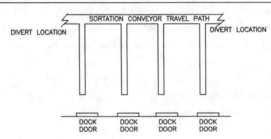

SERPENTINE OR LOOP DIVERT & LOADING CONVEYOR CONCEPT

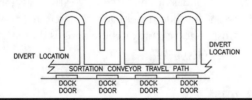

Figure 2.18 Shipping sort lanes: straight, loop, and roller or skatewheel.

Solid Gravity Slide or Chute

A solid gravity slide is basically galvanized sheet metal, fiberglass, or coated wood that has a flat or concave surface for the carton travel path. All gravity slides or chutes have a travel path that has a charge end, a higher elevation at each CO divert location, and a lower elevation at the discharge end. The length of the gravity slide determines the length of the side runners, the number of supports under the conveyor, cross support members, and floor surface support legs or ceiling hangers. If a slide travel path has more than one section, the top travel path section overlaps the next travel path section. This travel path overlap feature is similar to roof shingles and reduces carton hang-ups on the travel path. To ensure smooth carton travel over a slide travel path:

- The slide charge transition end has a crown shape
- Slope and material is determined from carton flow tests
- The discharge transition end has a convex shape

A concave bottom slide configuration provides some carton flow or travel control over a travel path. This controlled carton flow minimizes carton jams. A slide travel path relies on gravity to convey a carton from a sort conveyor travel path to a palletize or direct load station.

To achieve some carton queue on a slide travel path, a step is added in the slide travel path bottom and an adjustable end stop is added to the slide discharge travel path end. With a closed adjustable end stop, cartons queue in the slide travel path and step permits cartons to ride on the queued cartons and queue against an end stop.

A gravity slide can be designed either with astraight travel path, which requires a larger building area but has a lower cost, or a spiral travel path, which requires a smaller building area but has a higher cost.

A slide or chute interfaces with an active or passive carton sort conveyor concept and provides limited carton queue. A slide or chute handles a wide variety of carton sizes and shapes. Due to uncontrolled line pressure, a slide only handles noncrushable or nonfragile cartons. Slides or chutes convey cartons a short travel distance. To reduce carton jams on the slide travel path, the preferred travel path design is the concave design. Most concave slide or chute designs do not change the carton direction of travel on the travel path.

Gravity Strand Slide

A gravity declining strand slide concept has coated metal strands that are the full length of the carton travel path. Each slide has a minimum of three strands that are evenly spaced across the slide width and fit under the carton. Additional strands can be added to the travel path to accommodate a mix of carton widths and weights.

The gravity strand slide concept has the same design parameters and operational characteristics as the solid bottom gravity slide travel path.

Gravity Skate Wheel or Roller Conveyor

A gravity skate wheel or roller conveyor travel path has a carton sort conveyor travel path customer divert charge end that has a higher elevation than a CO divert location or discharge end. The floor support legs or ceiling hangers are determined by conveyor frame sections and the queued weight on the conveyor travel path.

A gravity skate wheel conveyor has several skate wheel conveyor sections connected together. A skate wheel conveyor section has skate wheels on an axle span. An axle is attached onto two frame sections. Conveyor frames are floor supported or ceiling hung. As a carton flows over a skate wheel conveyor travel path, the skate wheels turn on the axles. To assure good carton flow, there are at least three skate wheels under the narrowest carton bottom surface.

A gravity roller conveyor has roller conveyor sections connected together. A gravity roller conveyor section has rollers that span the width of the conveyor travel path. Each roller axle end is attached onto two frames. Conveyor frames are floor supported or ceiling hung. As a carton flows over the roller conveyor travel path, each roller turns on its axles. To assure good carton flow, there are at least three rollers under the narrowest carton bottom exterior surface.

The gravity skate wheel conveyor or roller conveyor is an extension of an active sort device and the carton has the same direction of travel or orientation on a CO shipping lane as on the sort conveyor travel path. To reduce carton hang-ups, carton damage, and smooth carton travel:

- Carton sizes and shapes are standard
- Cartons are noncrushable or nonfragile
- The charge transition end has a non-powered nose-over and the discharge transition end is designed with a convex plate
- Side guards are full length
- To determine the proper conveyor travel path slope, carton flow tests are conducted by a conveyor manufacturer

If the carton travel speed is excessive on a gravity conveyor CO shipping lane, the cartons can be slowed in one of the following ways:

- The conveyor travel path is tilted slightly to one side, which causes one carton side to ride or rub against a guard rail
- The skate wheel or roller is restricted with a washer
- Several plastic stripes extend downward from the overhead structure into the carton travel path and rub on the tops of the cartons
- A brake device is added at a location on the conveyor travel path

Powered Belt Conveyor

A CO shipping lane concept with a powered belt or low-pressure roller conveyor with brake rollers has a sort conveyor travel path set at a slightly higher elevation than the charge CO sort location, or it has a powered conveyor transfer section. A powered belt and roller concept interfaces with an active sort concept. Powered conveyor CO shipping lane concepts are similar to gravity conveyor CO shipping lane concepts, except that a powered conveyor shipping lane concept requires a higher investment, handles a wider carton mix, and the belt conveyor has controlled carton travel.

Carton Conveyor Customer Shipping Lane Design

A carton conveyor CO shipping lane can be designed either as a straight travel path or a loop. A straight travel path has a conveyor that runs along the opposite wall from the shipping docks and slopes from a divert device at a high elevation to a lower elevation on the shipping dock floor. A straight shipping conveyor lane travel path design requires a large building area but has a lower cost. A serpentine or loop travel path has a sort conveyor that runs along a wall above the shipping docks. As the conveyor travel path approaches the wall opposite the shipping docks, it makes a 180 degree turn back towards the shipping dock then slopes downward to a lower elevation on the shipping dock floor. A serpentine shipping conveyor uses a smaller building area but has a slightly higher cost.

Factors that determine effectiveness of a dock conveyor concept include carton travel speed and carton queueing prior to each activity station along a sort and transport conveyor travel path.

To obtain proper carton travel speed across a conveyor as the travel path progresses from one activity area to the next, the conveyor travel speed should increase progressively. In most conveyor transport applications, a conveyor activity area travel speed is 5 fpm faster than the previous conveyor activity area travel speed. This is enough of a difference to ensure that carton movement is controlled over the entire dock conveyor travel path.

To ensure that there is a constant carton flow to each activity station on a conveyor sort or transport travel path (such as a merge location at a loading station, check weigh, barcode scanner, or RF tag reader, fill carton, or seal machine), most conveyors are designed as low- or zero-queue travel paths prior to each activity station. The queue conveyor length and type is determined by your conveyor manufacturer and company transport concept design parameters.

A carton queue conveyor can be non-powered or powered, with rollers or skate wheels, depending on the conveyor travel path slope or decline. If a conveyor travel path section prior to the activity station is a horizontal travel path, the conveyor travel path queue section is powered live roller or skate wheel low- or zero-pressure accumulation conveyor. In most across the dock conveyor sort and transport concepts, these activity stations occur prior to a merge station, prior to the check weigh station, prior to a carton fill station, prior to a carton seal station, and prior to a barcode scan or RF tag reader station.

Powered live roller queue conveyors are either low pressure or zero pressure. When a carton is halted on a conveyor travel path a zero-pressure live roller conveyor does not exert forward-movement pressure on the carton. A low-pressure live roller conveyor exerts a minimal forward-movement pressure on each stopped carton.

A conveyor travel path section that is an incline is a powered belt conveyor, spiral belt, or vertical lift. An incline is used to transport cartons from a pack area to an elevated transport or sort conveyor travel path or to make a controlled carton elevation change between two conveyor travel paths.

A powered tail and nose over with a slope between 15 to 20 degrees ensures that the cartons travel up an incline. A belt conveyor can be designed in one of the following ways:

- Slider bed. A slider bed conveyor concept has a solid sheet-metal full-length or under a belt conveyor travel path.
- Roller bed. A slider roller bed belt conveyor has skate wheels that are evenly spaced and protrude through a sheet-metal surface to provide support for the belt conveyor travel path. This feature provides a solid conveyor travel path surface with a low coefficient of friction or drag on a conveyor belt.
- Slider roller bed. A roller bed belt conveyor concept has several rollers that are evenly spaced under the conveyor travel path to provide support. A roller bed belt conveyor provides a low coefficient of friction.

A spiral belt or hard plastic conveyor is another carton conveyor concept that transports cartons over an inclined travel path. The spiral conveyor travel path is considered a circular travel path inclined around a center post. The conveyor travel slope increases as the vertical travel completes the circle. A vertical conveyor travel surface provides control of the carton travel, resulting in evenly spaced cartons that ensure minimal line pressure. This concept requires a small area.

A vertical lift concept has several carton-carrying slats that move over a fixed vertical travel path between two horizontal travel paths on different elevations. The space between two slats is determined by the tallest carton. The carton-carrying slats are in constant movement and from a low-level transport path pick up a carton, move the carton up to the elevated level, and discharge the carton onto a transport path. With a vertical lift conveyor, there is no line pressure, requires a small area, and has controlled carton movement.

If a conveyor travel path section is a declining travel path, options are to use a powered belt conveyor, powered low-pressure live roller conveyor with selected brake rollers, vertical lift, or a non-powered queue skate wheel or roller conveyor.

Important considerations are the slope and line pressure, type of carton bottom exterior surface, and conveyor design.

The powered belt conveyor provides a high coefficient of friction to ensure maximum carton control, minimal line pressure, handles flat surface cartons, and functions with a 15 to 20 degree slope. With the cartons evenly spaced on the belt conveyor travel there is minimal carton queue.

A live roller conveyor has brake rollers on the declining travel path section. The distance between the brake rollers is determined by the carton length and weight. As a carton starts the decline, the brake roller is activated to control or momentarily stop the carton travel to ensure minimal line pressure and minimal carton queue.

The vertical lift conveyor has the same design parameters and operational characteristics as the incline vertical lift except that the carton travel path has a different direction of travel.

The declining non-powered or gravity carton conveyor travel path is designed to take into account required elevation change, available building space, carton physical characteristics including bottom exterior surface, desired controlled travel speed, and line pressure. A non-powered conveyor is a skate wheel or roller conveyor. A skate wheel conveyor consists of a series of evenly spaced skate wheels each of which is attached to an axle and spans and to each conveyor frame. Enclosed ball bearings allow the skate wheel to turn as the carton flows over the conveyor travel path.

The roller conveyor is a round-shaped object that has an axle and with encased bearings on each end. The ball bearings allow the roller to turn on the axle as the carton flows over the conveyor travel path. The roller spans the open space between the two frames and roller axles are connected to the frames. These frames are ceiling hung or floor supported to maintain the conveyor travel path proper slope.

With either non-powered carton conveyor type, there are three skate wheels or rollers under the narrowest carton bottom exterior surface. To assure smooth and continuous carton travel over the conveyor travel path, the carton must have a smooth, hard bottom surface and the carton must be placed onto the conveyor travel path in the proper direction of travel, with the bottom flaps or carton length in the direction of travel over the conveyor travel path.

When a carton travels over a non-powered decline travel path, its travel speed is uncontrolled. Side guards and photo-eyes improve safety, minimize carton damage, and assure a constant carton flow.

There are various methods of controlling carton travel speed over a declining non-powered conveyor travel path, including:

- A low degree of slope.
- A conveyor travel path that is slightly tilted to one side so that the carton side rides or rubs along a side guard.
- Retarders can be placed on rollers or skate wheels along the conveyor travel path.
- Plastic strips can be positioned directly above the conveyor travel path so they hang down into the travel path and ride over the tops of the cartons.

Side guards or guard rails along the length of the conveyor ensure that cartons are retained on the travel path. Side guards can be solid sheet metal, one- or multilevel C channels, skate wheel, pipe side guards, and angle iron.

The photo-eyes are carton control flow devices along the conveyor travel path that communicate with a microcomputer. The photo eye has a light beam, reflector, receiver, communications network to a computer, and an alarm. When the photo light beam is blocked by a stopped carton, a message is sent to a computer that activates an alarm or deactivates a divert device. The types of photo-eye are:

- Partial full line. The partial full line photo-eye is located in the middle of the travel path. When the photo-eye is blocked by a stationary carton on the decline conveyor travel path, a computer receives the signal and activates a visual or audible alarm to indicate the conveyor travel path condition. Employees or a sort travel path divert device continue to divert cartons onto a shipping lane while management takes corrective action.
- Full line. The full line photo-eye is located near the travel path charge end. When a stationary carton on the conveyor travel path blocks the photo-eye, the microcomputer receives a signal to deactivate the appropriate divert device. Cartons on the sort travel path that are assigned to the deactivated divert station are recirculated or diverted onto a clean-out lane until employees or management take corrective action.

Direct or Fluid Load Cartons

In a direct load operation, CO cartons travel from a sort travel path onto an extendible conveyor that extends into a CO delivery truck. In the delivery truck, cartons are transferred from the conveyor onto the truck floor. Then extendible conveyor is a powered belt, powered roller, or skate wheel or non-powered roller or skate wheel conveyor.

The disadvantages of a direct load operation are large investment in the conveyor, the need for a recirculation conveyor travel path, or increased employee handling of "did not" sort cartons. The advantages are that it requires less dock space and fewer dock doors, less holding area requirement, no carton double handling, and the CO delivery trucks at the dock.

Shipping Carton Unitizing

In a carton unitizing concept, CO cartons travel from a sort conveyor to a unitizing station. At the unitizing station, cartons are transferred onto a cart or pallet. After a cart or pallet is unitized to a predetermined height, the full cart or pallet is identified with a CO discreet code, and transferred from the unitizing station to an assigned outbound staging area or is placed onto a CO delivery truck.

The disadvantages are the need for a cart pallet unitizing area, some double handling, additional shipping dock space, and pallet handling device. The advantages are permits easy loading, increases unloading conveyable and non-conveyable CO delivery load, CO cartons handled without a CO delivery truck at the shipping dock, and preferred for a CO with a cart delivery concept.

Unitizing onto Carts or Direct Load

In a small-item or flatwear operation, CO package loading is an activity that moves a CO carton from a distribution operation onto a CO delivery vehicle. If an operation services an in-house or plant customer, the CO package transport concept interfaces with the plant in-house transport and the carton leaves the pick/pack operational area for delivery to a company process or manufacturing operation.

In the retail, catalog, direct marketing, or direct mail industry, a CO package delivery location has a company-owned or -contracted delivery company perform CO package delivery service. CO packages are unitized on carts or pallets—a loading employee transfers CO packages from a conveyor onto a pallet or four-wheel cart, which is referred to as a BMC or bulk mail carrier. Full carts or pallets are loaded onto a delivery vehicle.

Direct loading requires additional dock space and more employees to handle carts or pallets. In some retail store operations, carts are used to move product to the sales floor, and with a sort concept permits zone skipping with a CO small (pallet or cart) quantity that is assigned to a freight company terminal.

Zone skipping occurs due to the fact that a full cart or pallet for a regional terminal is not handled at the home terminal as individual CO packages but as a full cart or pallet instead that is moved onto a regional delivery truck. This activity can save one day on a CO delivery. Unitizing station are laid out with one divert line to service stations on both sides of the divert lane (see Figure 2.19). This provides the maximum number of unitizing stations.

If a full delivery truck has presorted cartons for a regional freight terminal, the delivery truck bypasses the home terminal sort activity and travels from the home terminal to the regional terminal. A mixed delivery truck goes to the home base terminal for sort to regional terminal truck. The concept features are additional conveyor cost, less dock space and docks, requires a large volume to zone skip, and one to two less employees.

Receiving and Shipping Dock Interface with an In-House Transport Concept

The in-house transport concepts move pallets from the receiving dock area to the storage area. The in-house transport concept is determined by type of UOP or CO,

PALLET OR BMC UNITIZING STATIONS

BMCS ON DIVERT CONVEYOR BOTH SIDES

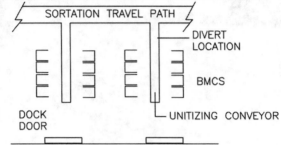

BMCS ON DIVERT CONVEYOR ONE SIDE

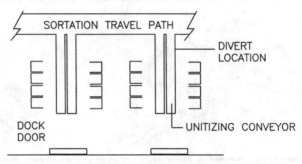

ANGLED DIVERT CONVEYOR BMCS ON ONE SIDE

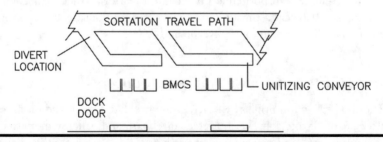

Figure 2.19 BMC or pallet unitizing concept: both conveyor sides, on one conveyor side, angled conveyor.

power source, and WMS pallet identification. Master carton in-house transport concepts include (see Figure 2.20) a four-wheel multiple shelf cart, two-wheel hand truck and platform truck, carton or tote conveyor powered travel path, baskets on a powered trolley travel path, and pallets on a pallet truck or forklift truck.

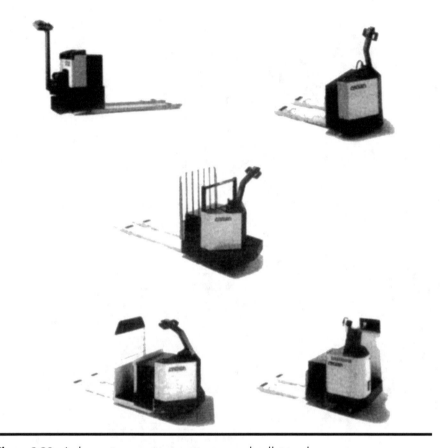

Figure 2.20 In-house transport concepts powered pallet trucks.

Pallet in-house transport concepts include a platform truck, pallets on a pallet truck or forklift truck, a tugger with a cart train, in-floor towline, powered conveyor, monorail, overhead tow conveyor, and AGVs.

Master Carton In-House Transport Concepts

The four-wheel and multiple shelf cart, two-wheel hand truck and platform truck, master carton or tote conveyor powered travel path and pallet with master cartons on a non-powered and powered pallet truck or forklift truck are the same vehicles that are used to load and unload master cartons or pallets between a delivery vehicle and dock.

Pallet Transport Concepts

Pallet transport concepts are manual control and variable travel path concepts or electric-powered fixed closed-loop travel path concepts with code readers and diverts.

Two-Wheel Hand Truck

A two-wheel hand truck is a manual-powered variable horizontal transport concept that moves four to five cartons or totes between two locations. A two-wheel hand truck is considered a good transport concept to move a few cartons (nominally 12 in. high) a short travel distance. A two-wheel hand truck has a carton- or tote-carrying surface that is a lip or nose, two fixed wheels with an axle or axles, bearings and axle bracket back rest or frame or legs, and two handles.

Two-wheel trucks are either eastern style, with a tapered frame and two wheels outside the frame, curved or flat cross members and curved or straight handles; western style, with two frame members parallel to the handles and a lip with a curve or flat cross members with curved handles; or a specially designed truck. The two-wheel hand truck is used for replenishment to pick positions.

Semi-Live Skid

A semi-live skid is a variable travel path manual-powered vehicle. The skid has a forward hardened metal plate or extension with a slot or eye hole or stud/pin, two rigid fixed wheels, two front rigid legs or stands, a skid deck that is 1 in. thick, and a skid handle jack (pull handle) with a stud/pin or slot/eye hole. Size ranges are 24 in. width by 48 in. length; 36 in. width by 48 in. length; 30 in. width by 60 in. length; 36 in. width by 60 in. length; and 36 in. width by 72 in. length.

The semi-live skid is similar to a four-wheel cart except at a storage or pick position the cartons or totes remain on the skid deck. In a dynamic order-fulfillment operation, a semi-live concept is not a preferred transport.

Dolly

A dolly is a manual-powered variable path vehicle that transports a carton or tote stack between two locations. A dolly component has three swivel wheels, three or four frames, one of which has a rope or tow line hole. Standard deck sizes range from 18 to 24 in. wide and from 24 to 36 in. long.

In a storage or pick position, the cartons or totes are hand-stacked into a position or remain on the deck. In a dynamic order-fulfillment operation, a dolly concept is not a preferred transport.

Platform Truck

A platform truck is a manual-powered variable path vehicle that has a carton or tote load-carrying deck. It has two rigid casters/wheels, two swivel casters/wheels, a removable push handle, and a solid wood deck. A platform truck can be tilt or non-tilt. The tilt-type truck has greater maneuverability with four corner wheels that have the same diameter and two wheels with a larger diameter, whereas the non-tilt type has four wheels that have the same diameter.

Standard deck sizes are 20 in. width by 42 in. length, 20 in. width by 48 in. length, 20 in. width by 60 in. length, 30 in. width by 48 in. length, 30 in. width by 54 in. length, 30 in. width by 60 in. length, 30 in. width by 72 in. length, 36 in. width by 60 in. length, 36 in. width by 72 in. length, and 36 in. width by 96 in. length.

In a storage or pick position, the cartons or totes are hand-stacked into a position or remain on the deck. In a dynamic order-fulfillment operation, a platform truck concept is not a preferred transport.

Non-Powered or Powered Pallet Truck or Forklift Truck

Non-powered or electric battery-powered pallet or forklift trucks are manually controlled and variable path concepts that have the same design and operational characteristics as pallet trucks that are used to unload and load delivery vehicles. A non-powered pallet truck is used for short travel distances. A powered double-pallet truck is used to transport two pallets over long travel distances.

Tugger with a Cart Train

A powered and manually controlled tugger is an electric battery-powered vehicle with a steering handle and operator's platform. From the operator's platform an operator has access to all vehicle movement and speed controls. A tugger with a cart train (see Figure 2.21) has the capacity to pull four to five carts with pallets over a variable travel path. With a cart train a travel path requires wider turning aisles. The tugger and each cart has a manually activated coupler and hitch to make a cart train. Each cart surface handles one or two pallets. The cart wheel/caster type and location under the cart load-carrying surface permits good cart trailing action. At the dispatch or delivery location, a forklift truck transfers a pallet between a cart and location. At each dispatch or delivery station, a set-down area permits a forklift truck to complete a pallet transfer between the cart and location and allows other cart trains to move through the aisle. If multiple pallets are required at one delivery location with a long travel distance from a dispatch station, a powered tugger with a cart train is an option, but to transport pallets between a cart and dock/storage location, it has a forklift truck at each delivery and dispatch station.

Powered Pallet Conveyor

A powered pallet conveyor (see Figure 2.22) is a fixed travel path between a dock area and a storage area. A powered conveyor has an electric-powered drive motor and drive chain that turn rollers or pull a drag chain. After a forklift truck places

TRAILING INFORMATION

Chart I gives the aisle widths (A) required to make a 90° turn with a number of trailers in specific sizes; Chart II, the space (B and C) required to make a 180° turn for a number of trailers in specific sizes. Conversely, both charts can be used to determine the number of trailers that can be used in existing situations. Please contact us for any additional information required . . . . particularly regarding unequal aisles and/or unusual situations.

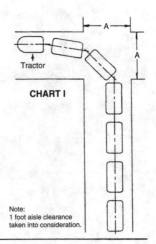

CHART I

"A" = aisle width required for 90° turn

Deck size	Number of trailers					
	1	2	3	4	5	6
	Dimension "A" (caster steer with automatic couplers)					
30 × 60"	6'-6"	7'-6"	8'-3"	9'-0"	9'-6"	10'-0"
36 × 60"	7'-2"	8'-0"	8'-10"	9'-5"	10'-0"	10'-6"
36 × 72"	7'-8"	8'-6"	9'-3"	9'-10"	10'-5"	10'-10"
36 × 92"	8'-6"	9'-6"	10'-5"	11'-2"	11'-10"	12'-5"
Deck size	Dimension "A" (4 wheel knuckle steer)					
36 × 72"	6'-9"	7'-9"	8'-6"	9'-0"	9'-6"	10'-0"
36 × 96"	7'-5"	8'-6"	9'-5"	10'-2"	10'-10"	11'-4"
48 × 96"	8'-5"	9'-6"	10'-5"	11'-2"	11'-10"	12'-4"

Note:
1 foot aisle clearance taken into consideration.

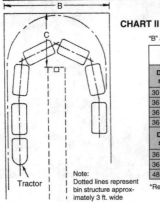

CHART II

"B" and "C" = space required for 180° turn

Deck size	*Dim. "B"	Number of trailers					
		1	2	3	4	5	6
		Dimension "C" (caster steer with automatic couplers)					
30 × 60"	17'-6"	6'-10"	7'-8"	8'-5"	9'-1"	9'-8"	10'-2"
36 × 60"	18'-0"	7'-4"	8'-2"	8'-11"	9'-8"	10'-3"	10'-10"
36 × 72"	19'-9"	8'-0"	8'-10"	9'-9"	10'-7"	11'-1"	11'-6"
36 × 92"	24'-6"	8'-10"	10'-2"	11'-6"	12'-7"	13'-7"	14'-6"
Deck size	*Dim. "B"	Dimension "C" (4 wheel knuckle steer)					
36 × 72"	17'-6"	6'-6"	7'-3"	7'-11"	8'-6"	9'-1"	9'-8"
36 × 96"	21'-0"	7'-0"	7'-9"	8'-7"	9'-4"	10'-1"	10'-10"
48 × 96"	22'-3"	7'-9"	8'-6"	9'-2"	9'-10"	10'-6"	11'-1"

Note:
Dotted lines represent bin structure approximately 3 ft. wide

*Regardless of number of trailers

Figure 2.21 Powered tugger.

Figure 2.22 In-house transport concepts: overhead conveyor, powered forklift truck, pallet roller conveyor, and in-floor tow cart.

a pallet with WMS identification onto a conveyor in-feed station, a powered roller or drag chain pulls the pallet over the travel path. Photo-eyes and sensing devices along the travel path ensure that the pallet is diverted or arrives at the assigned location, where a forklift truck driver with a hand-held scanner or an AS/RS crane-fixed position fixed-beam scanner scans the WMS identification, a forklift truck or divert device moves the pallet from the conveyor travel path to a storage location. The WMS pallet transaction is updated in the WMS inventory files.

In-Floor Towline

The in-floor towline concept has a towline cart move at 60 to 90 ft./min. over a fixed closed-loop travel path. Along the travel path automatic divert and in-feed locations have spurs or sidings that permit pallet loading and unloading as other towline carts move over the travel path. The towline has an electric motor-driven sprocket that pulls a link chain through a C-channel. To minimize wear, the C-channel has wear bars. As the towline chain is pulled over the travel path chain links (dogs) with a cavity are arranged at predetermined intervals. As the towline cart tow pin is inserted into the C-channel, the tow pin slides over the links, the tow pin slides into the cavity and the chain pulls the cart forward. To ensure smooth travel through a curve, each cart has four underside wheels with two front wheels set inside.

At a dispatch station, an employee sets the dispatch pin into the cart pin holder or applies a bar code or RF tag to the cart front. On the travel path as the cart travels past a symbology or code reader, a divert device is activated outward into the cart tow pin travel path and moves the cart from the travel path to a spur or siding. To prevent cart jams, there is a slight slope on the spur floor and employees assure the diverted carts are moved from the travel path. Unloading and loading pallets between a tow cart and station is assisted by a forklift truck. With a long delivery distance between two locations, and frequent, on–schedule deliveries, the tow-line concept is a potential transport concept.

Overhead Monorail

The self-powered monorail is a fixed path transport concept. The monorail has an electric bar or self-powered carrier that rides over an overhead track. The monorail carrier has 220 ft./min. travel speed on a hardened metal beam travel path. The monorail carriers have delivery code, with sensing devices along the travel path to read the code and activate a divert device, loading and unloading sidings or spur, anticollision devices, and are forklift truck-assisted to load and unload pallets. The monorail transport concept is very similar to the AGV concept except the travel path is elevated above the floor.

Overhead Tow Conveyor

The overhead tow-conveyor concept is a fixed travel path concept that has an overhead motor-driven chain and trolleys. Each trolley has a downward extending loop that has an employee engage the loop onto a four-wheel cart rigid mast or manual pallet truck handle. The overhead tow conveyor has similar operational and design characteristics as the in-floor tow line.

Automatic Guide Vehicle

An AGV is an electric battery-powered driverless vehicle that has a load-carrying surface or cart-towing ability. An AGV follows a fixed closed-loop travel path between a dispatch station and delivery location. An AGV has three or four wheels that match the floor, with two drive wheels and an undercarriage guide path-sensing device or on-board laser devices. A wire, tape, paint, or laser beam-guided travel path has a separate path or runs parallel to the main travel path. The travel path has loading and unloading spurs that permit (1) other AGVs to move over the travel path and (2) a forklift truck-assisted or pallet self-handling AGV to transfer a pallet between the AGV and P/D station. If the AGV is a pallet self-handling vehicle, at each load and unload station, a pick-up and delivery (P/D) station interfaces with the AGV pallet-carrying surface. AGVs are equipped with safety bumpers, safety flashing light, off-travel path stop system, undercarriage, or on-board guide path sensors and an on-board sensing device that communicates to a microcomputer anticollision or zone-blocking system. AGVs are available as single-pallet, two-pallet or cart-towing types that communicate to a microcomputer that controls AGV dispatch, travel over the guide path, and battery charging. With a long delivery distance between two locations, and deliveries that are frequent and on schedule, the AGV concept is a potential transport concept with a fixed position line of sight to an AGV WMS identification. The WMS identification scan transactions are similar to a WMS identification on a pallet.

Chapter 3

Master Carton and Pallet Storage Concepts

Introduction

The objectives of this chapter are to:

- Review operational layout or design parameters and operational characteristics
- Identify and evaluate master carton and pallet storage transaction vehicles, storage devices, and rack and transport concepts
- Review methods and technologies to improve employee productivity, and to examine storage capacity, position identification, and inventory control
- Review vendor-delivered UOP and CO piece flow

Awareness and understanding of design parameters and storage operation characteristics are key factors in designing and operating a facility. Improvements in master carton or pallet storage and transport activity increase company profits and improve customer service.

For quick and easy reference to key topics covered in Chapter 3, the following outline is provided.

- Storage area design parameters
  - Peak, average, and most frequent handling volumes
  - Three-month average or three-month moving average storage volumes

- SKU position projection
- Inventory stratification
- SILO inventory scrap concept
- Facility design considerations
- Storage area design factors
- Master carton and pallet
  - Dimensions
  - Pallet types
  - Pallet materials
  - Block and stringer designs
  - Components
- Slip sheet
  - Material
  - Designs
  - Forklift truck attachments
- How to secure cartons on a pallet
- Master carton and pallet storage components and concepts
  - Position identification location
  - Floor stock or block and stacking frames
  - Decked or hand-stacked in standard pallet racks
  - Standard or slotted angle iron shelves
  - Standard pallet rack components, dimensions, and designs
  - Double deep racks
  - Bridge or bridge and aisle racks
  - Drive-in racks
  - Drive-through racks
  - Mobile or sliding shelves or racks
  - Gravity flow racks
  - Air flow racks
  - Push-back racks
  - Cantilever racks
  - AS/RS high-rise racks
  - Car-in or mole racks
  - Sort link
- Master carton or pallet location and layout
  - The 80/20 rule, abc theory and family group
  - Pallet size
  - Manufacturer lot number and product rotation
  - Aisle lengths
- Aisle and position identification concepts
- Master carton or pallet storage vehicles
  - Wide-aisle group and features
  - Narrow-aisle group

- Very narrow-aisle group
  - AS/RS cranes and stacking AGVs
- Deposit transactions
- Manual-controlled forklift truck routing patterns
- Master carton or pallet storage transaction verification concepts
- In-house transport concepts
- Inventory tracking concepts
- Pick position replenishment

Small-item or flatwear order-fulfillment in a store-and-hold operation completes a CO for SKUs. This is a demand pull product flow, and in many industries the order-fulfillment operation offers a large number of SKUs with an inventory level that is projected to satisfy the customer demand. The large number of SKUs requires a large number of pick positions. To accommodate pick position requirements and inventory volume, an order-fulfillment operation requires a remote storage area for SKU master cartons or pallets. In a small-item or flatwear order-fulfillment operation, there is a need for an efficient and cost-effective remote storage operation as well. The remote storage operation has storage shelves, material handling equipment, and storage vehicles to complete storage transactions. In addition there needs to be an in-house transport concept to move master cartons or pallets from the storage area and deliver on time and to the correct pick area, pick position, or fast pack station.

Basic Operational Data

The first step in designing a new storage/pick operation or remodeling an existing facility is to evaluate the proposed volume of master cartons or pallets, vendor UOPs and CO volumes, number of SKUs, inventory characteristics, and other operational parameters for each activity, including:

- The average number of vendor and customer master cartons or pallets per day, the most frequent number of master cartons or pallets per day, and peak vendor and customer pallets per day
- The average number of master cartons or pallets per vendor-delivered UOP and CO, the most frequent number per vendor-delivered UOP and CO, and peak number per vendor-delivered UOP and CO
- The best master carton or pallet volume to use for projecting equipment needs
- Master cartons or pallets by specific inventory classification
- In-house master carton or pallet transport conveyor or vehicle travel speeds
- Vendor UOP and CO shipping method, delivery vehicle size, and loading pattern
- IT-required CO processing and IT download time

- CO priority
- CO service time or CO order and delivery cycle time and vendor unload time
- Profile of the master carton or pallet storage area
- Proposed design of the master carton or pallet storage area, with vehicle aisle and rack layout and simulation of all required activities
- Proposed master carton or pallet storage concept with pallet flow paths, with block, plan view, and detailed view drawings

Completion of this data collection and analysis step assures that the proposed pallet facility and its storage area layout are designed to handle the projected vendor UOP and CO volume and master cartons or pallets per CO. The features ensure a cost-effective and efficient operation that provides the lowest operational cost and accurate and on-time customer service. Vendor-delivered UOP and CO master carton or pallet volumes determine print quantity and per hour scan transactions for master carton or pallet identification, which determine the IT communications network requirements.

Peak, Average, and Most Frequent Volumes

When we review the monthly figures that are used to project a master carton or pallet operation, important volumes to consider include the master carton or pallet volumes and employee or employee-controlled equipment productivity, which determine the number of vehicles and employees required in each functional area. The number of master cartons or pallets represents the potential vendor UOPs and COs that flow through an operation from the receiving area, through the storage area to a pick/pack area and to the shipping area.

The peak, average, and most frequent monthly master carton or pallet volumes (see Figure 3.1) are used to project future storage volume. This is based on the number of master cartons or pallets that were handled by a storage operation over a specific time period, as well as the anticipated growth rate. These are key factors that determine design of the operation, as well as other important operational issues, such as:

- Projecting or scheduling the quantity of labor and equipment that is required to handle the master carton or pallet volume
- Determining annual and other operational expenses for the annual operating budget
- Determining labor quantity, labor costs, labor savings, and other operational expenses that are used to justify the capital expenditure

PEAK, AVERAGE AND MOST FREQUENT HANDLING VOLUMES

WEEK				DAY OF THE WEEK				
NUMBER	MON	TUES	WED	THUR	FRI	SAT	SUN	TOTAL
1	24,963	21,471	17,552	17,515	16,224	19,098	15,492	132,315
2	20,644	19,035	16,882	18,629	18,263	19,874	16,971	130,298
3	24,425	21,140	19,600	17,729	18,498	17,337	14,461	133,190
4	21,020	19,687	16,683	19,308	16,122	17,036	14,603	124,459
5	23,622	19,575	19,670	15,105	16,514	18,443	16,243	129,172
6	23,838	18,721	19,686	21,482	18,979	19,984	17,067	139,757
7	24,858	20,856	20,762	19,697	17,645	19,654	14,284	137,756
8	22,605	18,779	21,308	18,776	17,070	16,575	15,907	131,020
9	21,360	20,455	16,568	17,073	15,552	14,999	13,710	119,717
10	19,656	16,773	19,312	19,588	18,725	19,383	17,310	130,747
11	22,544	22,785	18,381	18,015	22,030	14,961	13,242	131,958
12	23,526	20,572	14,153	18,158	17,054	19,074	15,883	128,420
13	18,989	19,834	20,128	18,954	19,520	15,718	14,335	127,478
14	31,954	24,347	17,756	18,136	15,723	16,024	12,282	136,222
15	21,721	22,042	17,384	21,168	16,624	19,005	15,797	133,741
16	22,152	20,071	22,487	18,455	18,737	17,892	12,336	132,130
17	22,313	21,485	17,816	19,738	18,539	19,786	15,053	134,730
TOTAL	390,190	347,628	316,128	317,526	301,819	304,843	254,976	2,233,110
	17	17	17	17	17	17	17	119
AVG DAY	22952	20449	18596	18678	17754	17932	14999	18,766
PEAK DAY								31,954
MOST FREQUENT RANGE IS 19600 TO 19697								19649

Figure 3.1 Peak, average, and most frequent volumes.

Peak Volume

A peak master carton or pallet volume is the highest volume of master cartons or pallets that was handled by the operation in a specific period, usually during periods of high receiving or shipping activity. There are several implications if the peak master carton or pallet volume is used as the basis for projecting the annual master carton or pallet volume:

■ When used as the basis for an annual operating budget, the actual dollar expense is below the budgeted dollar expense. At the fiscal year end, the annual actual expenses have a significant actual higher variance to the annual budgeted dollar amount.

- When used as the basis to schedule labor and equipment and processing equipment, for most of the fiscal year labor and equipment will be underutilized or not functioning at the design capacity, until the peak master carton or pallet volume is reached. During this peak volume period, the operation does not incur employee overtime and maintains the on-time CO delivery standard.
- When used as the basis to justify an investment, the actual dollar expense compared to the budgeted dollar expense variance means that return on the investment time period is extended longer than the proposed return on investment criteria.

Average Volume

The average master carton or pallet volume is calculated by one of two methods. Either thetotal time period, that is the volume for a day, week, or month divided by the number of days, weeks, or months, or for a specific period, the highest and lowest volume is divided by two. Using the average master carton or pallet volume for planning has the following implications:

- If the average volume is used as the basis for a dollar expense budget, for most of the fiscal year accounting periods the actual dollar expenses will be below the budgeted dollar expenses. At the fiscal year end, there will be a small, if any, variance between total annual actual dollar expenses and budgeted dollar expenses.
- When it is used as the basis to justify capital expenditure, the actual expenses will be below the budgeted labor dollar expenses. This variance means that the return on investment time is considered longer than the proposed return on investment time, but at the fiscal year end, the accumulated total actual labor dollar expenses equal total budgeted labor dollar expenses. This means that the return on investment time period is equal to the proposed return on investment time period.
- When used as a basis for scheduling labor and equipment, for most of the fiscal year the labor and equipment are underutilized or do not function at their design capacity, until the average master carton or pallet volume period is realized at the storage operation. During the average master carton or pallet volume time period, the storage operation does not incur employee overtime and maintains the storage operation customer service standard. During the few peak master carton or pallet volume time periods, the operation incurs some employee overtime and has the potential for some off-schedule customer deliveries.

Most Frequent Volume

The most frequent master carton or pallet volume for a given period is the volume that occurs most frequently at a storage operation. In a bell curve presentation for tracking master carton or pallet volumes, the most frequent master carton or pallet volume typically occurs between the average and peak volumes.

If the most frequent master carton or pallet volume is used for planning, this impacts several financial and operational reports:

- When used as the basis for a dollar expense budget, for most of the accounting periods in a fiscal year the actual dollar expenses are slightly below the budgeted dollar expenses. At the end of the fiscal year, the total annual actual dollar expenses will show a slightly higher actual expense to budgeted dollar expense variance.
- When used as the basis to justify capital expenditure at a storage operation, the actual expenses will be below the budgeted labor expenses. This actual labor dollar expense to budgeted labor dollar expense variance means that during the fiscal year the return on investment time is considered slightly longer than the proposed return on investment time, but at the fiscal year end, the accumulated total actual labor dollar expenses equal the total budgeted labor dollar expenses. This means that the return on investment time period is equal to the proposed return on investment time period.
- When used as the basis to schedule labor and equipment, for most of the fiscal year the labor and equipment are underutilized or do not function at their design capacity, until the most frequent volume period is realized at the storage operation. During the most frequent volume period, the storage operation should not incur employee overtime and maintains the customer service standard. During the few peak volume periods, the storage operation incurs slight employee overtime and there is the potential for minimal off-schedule customer deliveries.

The guidelines for master carton or pallet storage area volumes are:

- The average volume is preferred to project annual operating dollar expense amount
- The most frequent volume is preferred to project a daily storage operation to justify capital expenditure
- The peak pallet volume can also be used to project a daily storage operation to justify capital expenditure when there is no information on the most frequent volume

Projecting Master Carton or Pallet Numbers

When your task is to project the number of forklift trucks or AS/RS cranes needed for a master carton or pallet operation, a hybrid master carton or pallet volume projection is the preferred method. In an operation in which a forklift or AS/RS crane completes a storage transaction, a peak outbound volume may occur with an average inbound volume or an average outbound volume occur with a peak inbound volume. In this situation, a hybrid master carton or pallet volume projection provides a more realistic measure because the storage operation has inbound and outbound peak master carton or pallet volumes that occur at different times of the week or year.

If the average inbound and the average outbound volume are used to project demand, when there is a peak volume work day there is the possibility that the number of forklift trucks or AS/RS cranes are unable to meet the operational requirement, resulting in overtime and poor customer service.

If a peak inbound volume combined with peak outbound volume is used to project demand, for most work days potentially the number of forklift trucks or AS/RS cranes exceeds the requirement.

Storage Volume

For a pallet storage operation, the pallet inventory volume is the volume of master cartons or pallets that can be housed in a storage area. Pallets are stored in a floor stack, conventional forklift truck or AS/RS crane rack or flow rack position. The master carton or pallet storage inventory is separated by master carton or pallet height, pallet size, and special storage conditions. Inventory volumes are calculated either as fiscal year end master carton or pallet volume and an average volume for a number of months or moving average master carton or pallet volume for a predetermined period. The period with the highest average volume is usually used for planning; in most companies this is the last three months of the fiscal year.

Three-Month Average Inventory

To project a master carton or pallet inventory volume for a specific year using the three-month average method, the following data are taken into account:

- Annual master carton or pallet inventory from a base year to the year being considered or the base year inventory plus each year's growth rate and each year's projected inventory.
- Master carton or pallet height, pallet size, and special storage conditions.
- Number of months for the inventory. With this information, to determine a yearly master carton or pallet inventory, the annual master carton or pallet

inventory is divided by 12. The result is multiplied by the number of months for inventory, which provides the total master carton or pallet inventory.

The master carton or pallet inventory is basically the average master carton or pallet inventory. For example, if the annual master carton or pallet volume is 15,000 and the average master carton or pallet is 15,000 divided by 12, which equals 1250 master cartons or pallets, a three-month master carton or pallet inventory = 1250 times 3, which equals 3750 master cartons or pallets.

Three-Month Moving Average Inventory

Projecting master carton or pallet inventory for a given year with the three-month moving average method is based on inventory for a predetermined month. Required information includes:

- Inventory volume for each month of the year
- Inventory volume from a base year to the year under consideration, or the base year inventory plus each year' growth rate and each year's inventory is calculated
- Master carton or pallet height, pallet size, and special storage conditions
- Number of months for inventory

With this information, to determine each year's master carton or pallet inventory, the monthly master carton or pallet volume is grouped into predetermined or moving average months. This shows for a given year the predetermined grouped inventory that provides the highest master carton or pallet inventory number for a year.

The predetermined moving average month master carton or pallet inventory is the highest master carton or pallet inventory quantity. An example for a three-month period or group is 15,000 annual master carton or pallet inventory: Jan. = 1000, Feb. = 1000, Mar. = 1000, Apr. = 900, May = 1100, June = 1100, July = 1400, Aug. = 1400, Sept. = 1400, Oct. = 1700, Nov. = 1700 and Dec. = 2000, with an annual 15,000 pallet total. The moving average inventory groupings are:

- Jan., Feb., and Mar. group = 3000
- Feb., Mar., and Apr. group = 2900
- Mar, Apr., and May group = 3000
- Apr., May, and June group = 3100
- May June, and July group = 3600
- June, July, and Aug. group = 3900
- July, Aug., and Sept. group = 4200
- Aug., Sept., and Oct. group = 4500
- Sept., Oct., and Nov. group = 4800
- Oct., Nov., and Dec. group = 5400

From the example, the moving average or three-month master carton or pallet inventory is 5400 master cartons or pallets. With an 85% utilization factor the master carton or pallet inventory is 6352.

The moving average method reflects a pallet inventory that is similar to an actual master carton or pallet inventory, relationship to growth in sales, and higher master carton or pallet inventory requirement.

Master Carton or Pallet Storage Position Projections

To determine the short and tall master carton or pallet quantity, standard and non-standard master carton or pallet size quantity and number of special-condition storage positions, the required data includes the total master carton or pallet inventory for each group and the percentage needed for short, tall, standard, non-standard, and special storage conditions.

The total number of master cartons or pallets is multiplied by the percentages of short, tall, standard, non-standard, and special-condition cartons or pallets. From the above moving average example the master carton or pallet inventory was 5400 pallets. From the example:

- Standard short master carton or pallet inventory is 1620 = (5400 × 0.3) with an 85% utilization factor equals 1906 master carton or pallet positions.
- Standard tall pallet inventory is 1080 (5400 × 0.2) with an 85% utilization factor equals 1271 master carton or pallet positions.
- Non-standard short master carton or pallet inventory is 1080 (5400 × 0.2) with an 85% utilization factor equals 1271 master carton or pallet positions.
- Non-standard tall master carton or pallet inventory is 1080 (5400 × 0.2) and with an 85% utilization factor equals 1271 master carton or pallet positions.
- Special-condition master carton or pallet inventory is 540 (5400 × 0.1) with an 85% utilization factor equals 635 master carton or pallet positions.

Inventory Stratification Position Projection Method

Components of a SKU inventory stratification storage position projection method (see Figure 3.2) are the number of SKUs for the base year, with SKUs separated by storage characteristics, the yearly rate of movement of SKUs, growth rate, storage concept, and whether the storage position is used for cartons or pallets. After each SKU is allocated to its category, the stratification method allows you to project your inventory requirements for the year. The inventory stratification method is preferred because it takes into consideration the fact that the SKU on-hand inventory

(1)	(2)	(3)	2 × 3 = 4 (4)	(5)	4 × 5 = 6 (6)	4 − 6 = 7 (7)
Range	Items in inv.	Mid point	Total. pal	Type of row	Pal. in pick pos.	Pal. inv. to be put in res.
0–1.0	591	.5	296	H.O.R.S.	295	1
1.01–2.0	336	1.5	504	H.R.O.S.	168	336
0–1.0	197	.5	99	R1 × 5	98	1
1.01–2.0	98	1.5	147	R1 × 4	49	98
1.01–2.0	193	1.5	290	R1 × 5	96	194
2.01–3.0	240	2.5	600	R1 × 5	120	480
2.01–3.0	80	2.5	200	R1 × 4	40	160
3.01–4.0	168	3.5	588	R1 × 5	84	504
3.01–4.0	56	3.5	196	R1 × 4	28	1680
4.01–5.0	111	4.5	500	R1 × 5	55	445
4.01–5.0	37	4.5	167	R1 × 4	18	149)
5.01–6.0	84	5.5	462	DI2 × 5	42	420
5.01–6.0	38	5.5	209	DI2 × 4	19	190
6.01–7.0	60	6.5	390	DI2 × 5	30	360
6.01–7.0	19	6.5	124	DI2 × 4	9	115
7.01–8.0	45	7.5	338	FS3 × 4	22	316
7.01–8.0	15	7.5	113	FS3 × 3	7	106
8.01–9.0	36	8.5	306	2FS3 × 4	18	288
8.01–9.0	13	8.5	111	2FS3 × 3	6	105
9.01–10.0	35	9.5	333	2FS3 × 4	17	316
9.01–10.0	12	9.5	114	2FS3 × 3	6	108
10.1–15.0	87	12.5	1088	2FS3 × 4	43	1045
10.1–15.0	29	12.5	363	2FS3 × 3	14	349
15.01–19.0	45	17.0	765	2FS3 × 4	22	743
15.01–19.0	15	17.0	255	2FS3 × 3	7	748
19.1–Plus	55	19.0	1045	2FS4 × 4	27	1028
19.1–Plus	19	19.0	361	2FS4 × 3	9	352
	2714		9964		1349	9125

(*) The actual pallet positions provided are 544; however, due to the small vertical rack openings, all totals reflect the required pallet inventory.
(†) The actual pallet positions provided from remote reserve are 595; but, since this is drive-in rack, the availability of remote reserve is a restriction to that individual item because of the nature of the rack and in good warehousing practice, should not be used for remote reserve positions for other items.

(8) IF (+) 7/UTIL = 8	(9) IF (−)	(10) 8 − 9 = 10	(11) 8 − 9 = 12	(12)
Pal. pos. for res. at a utilization factor	Rdy. res. pal. pos. provided	Pal. inv. to be stored in remote res.	Pal. pos. to be used for remote res.	Pal. pos. remote res. cumulative avail. (neg) positive
395[1]	544(*)			
1 (85%)	295		294	294
115 (85%)	294		179	473
228 (85%)	289		61	534
565 (85%)	360	205	0	329
188 (85%)	240	52	381	
593 (85%)	252	341	0	40
198 (85%)	168	30	0	10
524 (85%)	166	358	0	(348)
175 (85%)	111	64	0	(412)
560 (75%)	756		196(†)	
253 (75%)	266		13(†)	
480 (75%)	540		60(†)	
153 (75%)	133	20		(432)
421 (75%)	360	61		(493)
141 (75%)	90	51		(544)
384 (75%)	720		336	(208)
140 (75%)	195		55	(153)
421 (75%)	700		279	126
144 (75%)	180		36	162
1393 (75%)	1740		347	509
465 (75%)	435	30	479	
990 (75%)	900	90	389	
997 (75%)	225	772		(383)
1371 (75%)	1375		4	(379)
469 (75%)	399	70		(449)
11,369				

(*) The actual pallet positions provided are 544; however, due to the small vertical rack openings, all totals reflect the required pallet inventory.
(†) The actual pallet positions provided from remote reserve are 595; but, since this is drive-in rack, the availability of remote reserve is a restriction to that individual item because of the nature of the rack and in good warehousing practice, should not be used for remote reserve positions for other items.

Figure 3.2 Inventory stratification.

has different characteristics or is allocated to different storage concepts. In a carton or single-item operation, the inventory stratification method accounts for your SKU quantity in the pick position. With this information, your facility size, storage concept, and storage equipment design are based on more exact calculations.

The inventory stratification calculation method is based on rate of SKU movement and characteristics of the SKUs that are assigned to pick and storage positions. Your design team uses an inventory stratification form or computer program to complete the analysis. The format has 12 columns and as many lines as required to accommodate your SKU movement ranges.

- In column 1 your planner lists the SKU movement ranges in order from the slowest to the highest range. Each movement range has two lines: one for medium- to small-sized SKUs with non-crushable characteristics and the other for high cube/heavy and crushable characteristics.
- Column 2 is for your design year SKU number that is listed in the corresponding movement range.
- Column 3 is the midpoint for the SKU movement range. To obtain the midpoint, the low and high range figures are added, the total is divided by two and the figure is entered on the appropriate line under column 3.
- Column 4 is for the total design year inventory that is entered for each movement range. The figure is determined by multiplying column 2 by column 3. This figure represents the design inventory range for each inventory that is allocated to SKU pick and storage positions.
- Column 5 identifies the pick/storage position type that is assigned for each SKU. The SKU movement rate, cube, and stability characteristics determine the SKU rack configuration.
- Column 6 represents the inventory number that is allocated to the SKU pick position. The inventory in the pick position represents a portion of the on-hand inventory, which is 50% of the maximum capacity.
- Column 7 is the inventory number that remains for placement into a storage position. To obtain the inventory number, subtract the pick position inventory (column 4) from the design year inventory number (column 6). The remainder represents the inventory number that is placed into ready reserve and remote reserve positions. Ready reserve positions are in a pick aisle and remote reserve positions are in another aisle in the facility.
- Column 8 is the inventory number that is increased for a storage utilization factor for the storage position. The inventory quantity is calculated by the storage position type utilization factor that is divided into the inventory quantity and is entered into column 7.
- Column 9 indicates the ready reserve positions that are provided from the storage concept. Each SKU movement range has a quantity that is determined by the storage concept. If your operation handles only pallets, there is no reduction to the pick position. Your design team multiplies the inventory

number per storage concept depth by the number of levels times the SKU number. It should be noted that standard, mobile, and gravity flow racks are not restricted to one SKU. These storage concepts allow multiple SKUs per vertical stack. If a storage concept is floor stack, drive-in, or drive through, racks are restricted to one SKU. With a floor stack your first pallet stack has one pallet high.

- Column 10 is the design year inventory that exceeded the inventory that was assigned to ready reserve positions. The figure is calculated by subtracting column 8 from column 9. When column 9 is greater than column 8, this figure is entered in column 10 as a negative and constitutes the design year inventory that requires remote reserve positions.
- Column 11 is the extra ready reserve positions that are not required for the design year inventory. This figure is calculated by subtracting column 8 from column 9. When column 9 is less than column 8, this figure is entered into column 11 as a positive number and these are vacant positions that are available for SKU inventory units from other movement ranges.
- Column 12 is the accumulative positions that are available for SKU inventry units from other movement ranges. This figure is calculated by subtracting column 10 and adding columns 11 and column 12. When these positions are standard, mobile, and gravity flow racks, they are available for other inventory units. When column 12 is negative or drive-in, drive through, or floor stack storage positions, your design team adds remote reserve positions.

Pick Position Projections

A master carton or pallet storage position is also the SKU in a pick position. In a dense storage concept, the master carton or pallet position that is adjacent to the main aisle is the pick position.

Obsolete or Scrap Inventory Control

In this section, we describe damaged and obsolete physical inventory scrap concepts (see Figures 3.3 and 3.4). An inventory scrap concept allows an order-fulfillment operation to remove damaged and obsolete inventory from the operation storage area, creating vacant storage positions, improving housekeeping, improving employee productivity, and reducing UOP damage.

Scrap inventory is damaged UOPs that do not meet the company quality standards for a CO. Obsolete UOPs are products that have not been ordered by customers. In an order-fulfillment operation, scrap inventory is generally created from products that have been damaged during handling; buy-ins that had low sales and inventory that exceeded a reasonable volume; CO returns, and items not in favor with customers.

SKU	SKU age	Dollar value per SKU	Inventory SKU quantity	Inventory dollar value	Cubic feet per SKU	Total cubic feet of SKU inventory	SKUs to be scrapped	Dollar scrap value	Scrapped cubic feet
A	13 MO	$5	3,000	$15,000	3	9,000	—	—	—
B	16 MO	$10	500	$5,000	4	2,000	—	—	—
C	19 MO	$15	2,000	$30,000	5	10,000	—	—	—
D	22 MO	$20	1,000	$20,000	3	3,000	—	—	—
E	22 MO	$30	1,000	$30,000	3	3,000	1,000	$30,000	3,000
F	16 MO	$40	1,000	$40,000	2	2,000	1,000	$40,000	2,000
G	19 MO	$5	2,000	$1,000	6	12,000	—	—	—
H	13 MO	$10	2,000	$20,000	2	4,000	—	—	—
I	13 MO	$50	1,000	$50,000	2	2,000	1,000	$50,000	2,000
TOTAL			13,500	$220,000		47,000	3,000	$120,000	7,000
							22.2%	54.5%	14.9%
							3 SKUS		

SKUs to be scrapped	Dollar scrap value	Scrapped cubic feet
3,000	$15,000	9,000
500	$5,000	2,000
—	—	—
—	—	—
1,000	$30,000	3,000
2,000	$10,000	12,000
1,000	$10,000	4,000
1,000	$50,000	2,000
8,500	$120,000	32,000
63.0%	54.5%	68.1%
6 SKUS		

SKU	SKU age	Dollar value per SKU	Inventory SKU quantity	Inventory dollar value	Cubic feet per SKU	Total cubic feet of SKU inventory	SKUs to be scrapped	Dollar scrap value	Scrapped cubic feet
A	13 MO	$5	3,000	$15,000	3	9,000	—	—	—
B	16 MO	$10	500	$5,000	4	2,000	500	$5,000	2,000
C	19 MO	$15	2,000	$30,000	5	10,000	2,000	$30,000	10,000
D	22 MO	$20	1,000	$20,000	3	3,000	1,000	$20,000	3,000
E	22 MO	$30	1,000	$30,000	3	3,000	1,000	$30,000	3,000
F	16 MO	$40	1,000	$40,000	2	2,000	625	$25,000	1,250
G	19 MO	$5	2,000	$1,000	6	12,000	2,000	$10,000	12,000
H	13 MO	$10	2,000	$20,000	2	4,000	—	—	—
I	13 MO	$50	1,000	$50,000	2	2,000	—	—	—
TOTAL			13,500	$220,000		47,000	7,125	$120,000	31,250
							52.8%	54.5%	66.5%
							6 SKUs		

Figure 3.3   Scrap inventory disposal.

How to Identify Scrap Inventory

SKUs in inventory that are candidates for scrap are UOPs in the storage area with accumulated dust on the exterior, UOPs with old receiving date tags, and UOPs that show no volume on a year-to-date movement report. There are different ways to determine scrap inventory: the book inventory method, random selection, the age scrap method, and the SILO method.

SKUs to be scrapped	Dollar scrap value	Scrapped cubic feet
3,000	$15,000	9,000
500	$5,000	2,000
2,000	$30,000	10,000
500	$10,000	1,500
1,000	$30,000	3,000
—	—	
2,000	$10,000	12,000
2,000	$20,000	4,000
—	—	
11,000	$120,000	41,500
81.5%	54.5%	88.3%
	7 SKUs	
	(d)	

Analysis factor	Age of SKU	Typical book	Random selection	S.I.L.O.	S.I.L.O. advantages
Number of items to be scrapped	6	3	6	7	Maximum SKUs are scrapped
Inventory dollar value	$120,000	$120,000	$120,000	$120,000	Equal total dollar value scrapped
Percentage of inventory scrapped	54.5%	54.5%	54.5%	54.5%	Equal percentage of
Number of cartons to be scrapped	7,125	3,000	8,500	11,000	Maximum number of cartons
Percentage cartons to be scrapped	52.8%	22.2%	63.0%	81.5%	Maximum percentage
Total number of cubic feet scrapped	31,250	7,000	32,000	41,500	Maximum cubic feet
Percent of total cubic feet scrapped	66.5%	14.9%	68.1%	88.3%	Maximum percentage
Storage openings based on 64 cu. ft. per pallet load	488	110	500	650	Maximum number of pallet openings
Pallet storage cost savings based on $35 per sq. ft. for each 4-pallet high stack	$68,320	$15,400	$70,000	$91,000	Maximum cost saving

Figure 3.4   Obsolete inventory disposal.

The Book Inventory Scrap Method

The book inventory scrap method is based on data generated from the accounting and purchasing departments. From the annual projected sales, the accounting department allocates a specific dollar amount or percent to sales as inventory scrap expense. From the inventory records, the purchasing department identifies the SKUs as scrap candidates.

The accounting and purchasing departments jointly ensure the dollar value for scrap equals the budgeted inventory scrap expense. The SKU list and each SKU quantity are given to the order-fulfillment manager, who scraps the UOPs from the inventory. The accounting and purchasing managers make necessary entries to remove the SKU and SKU inventory from the inventory books and to increase the scrap expense line item.

The disadvantages are that this method does not create the maximum number of vacant positions possible, does not scrap the maximum cube feet, and it is likely that only a few SKUs will be scrapped. The advantages are that it is easy to implement and scraps the inventory dollar amount.

The Random Selection Scrap Method

A random SKU selection method has the scrap SKU list randomly selected from the inventory. The disadvantages and advantages are the same as the book inventory scrap method.

The Age Scrap Method

The oldest or age scrap method identifies the oldest SKUs in inventory and places them on the scrap list. When the SKU has limited life, the age of the UOP is an excellent scrap method. The disadvantages are the same as the book scrap method but the advantage is that the oldest SKUs are scrapped from inventory.

The SILO Scrap Method

The small-in and large-out (SILO) inventory scrap method has the largest cube SKUs scrapped from inventory and the smallest cube SKUs remain in inventory. The method is based on SKU movement, the approved dollar scrap amount, cubic feet for each SKU. The calculations determine the SKUs that will create the maximum number of vacant positions. The disadvantages are accurate SKU dimensions and management discipline and control. The advantages are that it creates the maximum number of positions, scraps the maximum number of cartons, scraps the dollar scrap amount, and uses the age of UOP method.

Facility Design Considerations

In designing and operating a master carton or pallet storage facility the building floor size and shape is determined by:

- Land size, shape, or site location along with access roads
- Square footage required for the projected storage inventory, the number of master carton or pallet storage positions, and other operational space requirements
- Soil condition and ability to support imposed dynamic and static building (usually the greatest loads are the master carton or pallet storage area loads)
- Local building codes, restrictions, seismic, wind, rain, and snow loads
- Available funds
- Number of female and male employees per shift along with employee automobile parking requirements
- Average and peak delivery truck requirements
- Electrical and other utility requirements
- Fire protection and employee emergency exits requirements

Other factors to consider are:

- Building shape and expansion capability and direction of expansion.
- Local code allowable building clear ceiling height and operational acceptance for a single-floor facility and elevated floor sections for office and administra-

tive areas. Included in the building height and floor area are exits, stairways, and an elevator, with appropriate length for run-out and pit.

- Master carton or pallet in-house transport concept requirement for a pit in the floor along with personnel protection along the pit perimeter, fire wall, or elevated wall or elevated floor penetration for a master carton or pallet in-house transport concept that has fire protection, employee emergency exits, and walkways to each exit.
- Lighting fixtures and light level or lumens at 30 in. above the floor surface, light fixtures turned on/off by photo-eye or activity sensor, and the various lighting levels for each workstation or storage aisle or area except an AS/RS concept.
- Ceiling fans or floor-level fans to circulate the air. This feature is a consideration for a geographic area that has high humidity; the moving fan blades improve mobile vehicle safety by minimizing the moisture build-up on the floor surface and employee working environment. This feature reduces potential employee injury.
- Uninterruptible power supply or UPS that provides sufficient electric power during a brown-out or electrical power failure to prevent a computer crash and to permit the facility to operate for a portion or all of the shift. The UPS may be battery-powered, and connected to specific electrical equipment, and designed to operate for a short time; or a diesel generator that is designed with capacity to provide power for all the electrical equipment for a long time. The UPS may be activated either by a manual start-up or on-line start-up. Because most UPS systems are designed to protect computers, they are started by an on-line concept.
- Communications between two micro-computers or between a host computer and micro-computer-controlled equipment, which may involve either radio frequency or hardwire technology. With radio frequency, the desired frequency is approved by the local and material government and is tested in a building that has a high metal content. With a hardwire communication over a long distance, the communicated wire could have short-haul modems or relays to assure clear communication. In many hardwire communication applications between two computers, the design factors are an electric spike protector, low electrical voltage protector, an electrical power filter to minimize undesired noise on the electrical power line, and dedicated communication line between two computers. To minimize the total investment and assure the standards many electrical power supply considerations are justified to have the electrical power supply sent through the UPS. This feature achieves most or all the electrical power supply design factors.
- Pallet in-house travel path through a facility. Objectives are to (a) have the shortest travel distance and time between locations, (b) have the fewest handlings with minimal master carton or pallet, building, or equipment damage or personnel injury, and (c) assure a master carton or pallet is delivered accurately, to the correct location and on-schedule.

Storage Area Design

Design parameters for a master carton or pallet storage concept include:

■ Pallet dimensions (length, width, height, and weight) and WMS master carton or pallet identification location and line of sight
■ If the product overhangs the pallet, the product overhang becomes the pallet dimensions
■ Pallet description
■ Product shape and ti and hi on a pallet
■ Average, most frequent, and peak pallet receiving and shipping volumes
■ Average, most frequent, and peak pallet number for each activity
■ Average, most frequent, and peak master carton or pallet inventory
■ Unique storage requirements

If a master carton or pallet inventory does not have uniform design features, SKUs with like design characteristics are placed in one storage area and unique SKUs in a separate storage area with the storage positions and forklift trucks. Planning to achieve an efficient, accurate, and on-time storage operation involves designing the rack and aisle layout that can handle your projected volume, scheduling equipment and employees, establishing procedures and practices, implementing a WMS program, and organizing vendor deliveries. Activities can be grouped as follows:

■ Those that precede storage:
  ■ Unloading
  ■ Receiving, checking, and QA
  ■ WMS master carton or pallet identification and inventory control
  ■ In-house transport
  ■ Deposit in a storage position and WMS update
■ Storage activities
■ Withdrawal from storage and WMS update
■ Post storage activities: in-house transport, staging, securing, and WMS manifest and load

Master carton or pallet storage area location design focuses on the physical position of a master carton or pallet or SKU in a storage aisle. SKU location in a storage aisle is determined by guidelines that depend on whether the storage activity uses a manual forklift truck or an AS/RS crane concept. They include:
■ The physical height, width, length, and weight of the master carton or pallet. The heavy and short master cartons or pallets are located at the lowest rack or floor level positions. The light weight and tall master cartons or pallets are located at the highest rack positions.

- SKU rate of movement—sales or physical movement—for one fiscal year. In a master carton or pallet storage aisle application, fast-moving SKUs are located at the aisle start, medium-moving SKUs are located in the middle of the aisle, and slow-moving SKUs are located at the end of the aisle.
- SKU value. High-value SKUs are allocated to high positions that have limited or controlled access.
- SKUs that require specific environmental storage conditions, such as temperature-sensitive product, are located at the lowest rack position or in a separate storage area.
- SKU hazard classification. By most local codes, hazardous SKUs are required to be housed in a master carton or pallet position that restricts SKU flight or has a barrier or pit connected to a containment chamber that restricts flammable liquid run-off.
- Master carton or pallet dimensions. When a master carton or pallet operation has pallets or storage devices with one or more dimensions or types, specific storage areas or aisles are allocated.

Approaches to designing the layout of storage rack rows and forklift truck or AS/RS crane movement patterns involve:

- Projecting for the year, the number of SKUs number and associated master carton or pallet inventory, vendor UOP and CO volumes, transport pallet flow pattern, and CO master carton or pallet flow pattern
- Reviewing how each master carton or pallet storage concept interfaces with inventory control or WMS product identification and storage position concept and selecting the preferred concept
- Identifying building column clear spans, column sizes, clear ceiling height; locating passageways through walls, fire walls and floor levels; identifying utilities, and the locations, quantity, and pattern of light fixtures
- Developing and refining block layouts and drawings for rack rows and aisle layout; vendor and CO master carton or pallet flow patterns; and building items
- Developing and finalizing line drawings for the building and storage area
- Completing a simulation that is based on pallet flow patterns, CO flow patterns, and master carton or pallet volumes and CO volumes
- Reviewing building layout and rack row and aisle layout drawings for compliance with local codes and company policies and practices
- Completing a bid process and selecting the preferred storage equipment vendor
- Developing building construction and storage equipment installation plans and schedules
- Installing, testing, and debugging all building equipment and master carton- or pallet-handling equipment and completing a building and master carton or pallet storage equipment punch list

■ Start-up and turning the building and pallet equipment over to the operational department

Master Carton or Pallet Rack Rows and Vehicle Aisle Design Parameters

When designing a manual forklift truck or AS/RS crane master carton or pallet storage area, SKU locations in a rack row or aisle are determined by:

■ The ability of the forklift truck or AS/RS crane to complete a storage transaction
■ Vehicle type and travel path
■ Average and peak master carton or pallet storage inventory and pallet number per SKU
■ Number and frequency of UOP and CO delivery and delivery vehicle pallet capacity
■ Master carton or pallet physical characteristics

Master carton or pallet and SKU physical characteristics are determined by:

■ Physical dimensions: (a) short height master carton or pallet length, width and height and inventory, (b) tall height master carton or pallet length, width and height and inventory, and (c) master carton or pallet or bottom support type
■ Product classification: (a) toxic or nontoxic, (b) tall or short, (c) heavy or light weight, (d) edible or nonedible, (e) hazardous, (f) price or high value, (g) family group component, and (h) temperature

Important Rack and Facility Dimensions

To design a master carton or pallet storage operation, important dimensions are:

■ The space between two building columns.
■ The space between the floor surface and the bottom of the lowest ceiling obstruction.
■ Complete master carton or pallet dimensions and bottom support device.
■ Clearances or open space between two master cartons or pallets and between a master carton or pallet and the rack structural member. In a facility with pallet racks, clearance is between a pallet and the upright post base plate of the rack.

Clear Space between Two Building Columns

The open space between the bases of two building column bases determines the floor space between the four building walls that is available for pallet storage. The two important building column dimensions are center line to center line or the dimension between two building column mid points and the open span between the column stands or base plates.

During the design process, the clear or open dimensions between all building column bases are verified because some facilities have intermediate columns for an elevated floor support. To fit a facility on a site, sometimes building column span varies. If required to adjust a building column span most facility designs have two building column spans for interior building columns and columns along exterior walls. Alternatives are, whenever possible, to ensure that the interior building column span matches the pallet storage rack design and adjust the building column spans along the exterior wall; or to_ adjust the interior building column span to match pallet storage rack design and maintain the building column span along the exterior wall. With both alternatives building column spans that are different are kept to a minimum.

Clear Space between the Floor Surface and Lowest Ceiling Obstruction

The clear space between the floor surface and the lowest ceiling obstruction or the bottom of elevated floor support members determines the number of pallet positions in a floor (block) stack or rack bay, or number of load beam levels and the forklift truck or AS/RS crane lift requirement to complete a storage transaction. This open space includes master carton or pallet dimensions, forklift truck or AS/RS crane operational dimension, fire sprinkler space, and storage rack structural members such as load beams and overhead aisle ties. When an exceptional situation occurs in a clear space, such as a heater, the exception is calculated in the master carton or pallet storage capacity.

Pallet Dimensions

Pallet dimensions are determined by (1) the UOP dimension, that is the product length, width, and height, and the master carton ti and hi on a pallet (the ti is the number of master cartons per layer and the hi is the number of layers per pallet); and (2) the pallet bottom support device. During the receiving activity, a vendor delivery tally sheet shows the suggested ti and hi or master carton quantity per pallet. If a master carton ti is within a pallet (bottom support device) perimeter, there is no master carton overhang and the pallet dimensions are the storage pallet dimensions. If a master carton ti overhangs a pallet (bottom support device), the master carton overhang length and width become the storage pallet dimensions.

The pallet length and width dimensions determine (a) number of floor-stack pallet positions per area, (b) rack frame, bay, or lane depth and (c) pallet number per rack load beam, rack bay, or between two rack upright posts. Two pallet and rack dimensions along with clearances determine the floor stack pallet number, rack row, and rack bay number that fits between two building columns.

The pallet height is a key factor that determines the number of pallets in a floor-stack, rack bay or in a vertical stack. The pallet height is the overall height between the bottom deck board and the top of the highest master carton. During the receiving activity, a vendor delivery tally shows the suggested ti and hi or master cartons per pallet that does not exceed the pallet storage position height. The pallet weight includes the pallet, securing material, and total master carton weight. During the receiving activity, a vendor delivery tally shows the suggested ti and hi or master cartons per pallet that does not exceed the pallet storage/pick position weight capacity. The pallet weight determines:

- Number of pallets per floor stack.
- Number of pallets or load beam level per rack bay design.
- Rack posts base plate thickness, foot print, and design.
- The forklift truck wheel size and weight determines the floor thickness and rebar characteristics.

The Pallet Bottom Support Device

The pallet length, width, and fork entry opening (unit load bottom support device) dimensions determine the rack structural component location that are:

- Rack load beam length.
- Rack upright post and load beam depth.
- Load arm or rail. In a forklift truck facility, a pallet overhangs each load beam or extends into an aisle or flue space by two to three in. Key factors and the forklift truck stacking requirement determine the storage aisle width.

A pallet bottom fork entry opening is the opening between the pallet bottom of the top deck board and the top of the bottom deck board. The fork entry opening determines the fork length or platen, type, or attachment.

Other Important Clearances or Open Spaces

In addition to the clearances already described, to minimize UOP or building damage, and employee injury and increase employee productivity, other important clearances are:

- For fire sprinklers
- Between a master carton or pallet and rack support member
- For a straddle forklift truck
- Allowance for a rack supported elevated floor or facility
- Flue or open space between back-to-back rack rows or between a rack row and wall
- Tall rack facility could require fire baffles and additional fire sprinklers
- For a food facility, a white space along the wall
- For conveyor travel paths adjacent to a building column

Ceiling Clearance for Fire Spinklers

When a master carton or pallet storage concept is designed for a facility, by local code, the company risk management policy or insurance policy underwriter require fire sprinklers. In a conventional facility, a ceiling fire sprinkler design has an 18-in. clearance or open space between the top of the highest pallet and the sprinkler head. In a facility that is 40 ft. high, the ceiling ESFR fire sprinkler design has a 3- to 5-ft. clearance space between the top of the highest master carton or pallet and the sprinkler head. Clearances or open spaces have: sufficient space for heat queue and proper sprinkler head activation to allow the water spray to cover the area, and minimizes sprinkler damage and accidental water discharge that damages product.

Clearance between Two Master Cartons or Pallets and Rack Support Members

For a forklift truck to complete a storage transaction, a rack bay width has horizontal open space between the master carton or pallet and the rack upright post structural support member or base plate. An open space or clearance between two master cartons or pallets and/or between the rack upright post structural support member, load beam or base plate and a master carton or pallet is 3 to 6 in. Clearance or open space varies depending on the forklift truck type and AS/RS crane and permits a forklift truck or AS/RS crane to complete a storage transaction with low UOP and equipment damage and high employee productivity.

Clearances for a Straddle Forklift Truck

In a pallet storage operation that uses a straddle forklift truck with a pallet on the floor as the first pallet position, the clearance between two pallets and rack upright post structural support member base plate is at least 5 to 6 in., or a straddle width plus 1 in. An open space between two pallets or a pallet and rack upright post structural support member base plate permits a forklift truck straddle to enter a rack bay.

This feature permits a straddle forklift truck to complete a storage transaction with low UOP and equipment damage and high employee productivity.

A Rack Supported Building Means a Wider Rack Base Plate

In a rack supported or tall rack building with a very narrow forklift truck, a larger upright frame post base plate is required to spread the vertical rack stacked load weight. If the vertical pallet rack stack uses the floor as the first pallet position, the base plate is in the rack open clearance. If a base plate is not in a rack open clearance, the options are to raise the bottom pallet level onto load beams or anticipate low forklift truck driver productivity and potential rack damage.

Flue or Open Space between Back-to-Back Racks or Walls

In a conventional building pallet storage rack rows are designed with back-to-back rack rows or a single rack row against a wall. There is a 6- to 12-in. open space between pallets and between the pallet and the wall.

In a facility with an ESFR fire sprinkler concept with back-to-back rack rows or a single rack row against a wall, there is a 4- to 6-in. open space between pallets and between the pallet and the wall.

The open space is defined as the flue space that permits heat from a fire to flow upward to activate a fire sprinkler head and permits fire sprinkler pipe installation.

Floor-Stacked Pallet Clearance

A conventional pallet storage operation that uses floor-stacked pallet positions with a conventional forklift truck has a 2- to 3-in. open space between adjacent pallets. In a straddle forklift truck operation, the open space between pallet lanes is at least 5 to 6 in. The open space permits a forklift truck to complete a pallet transaction without product damage and with good forklift truck driver productivity. If slip sheets are used in a pallet floor stack storage concept, a slip sheet tab (nominally 6 in. long) extension is added to the open space between pallet lanes or adjacent pallets.

Baffle Levels and Additional Sprinklers in a Tall Rack Facility

Tall very narrow aisle (VNA) or AS/RS facilities, per local code, company risk management policy, or insurance policy underwriting, require fire baffle(s) or additional fire sprinkler levels in the vertical pallet racks that have additional clearances.

White Space along a Wall in a Food Facility

When a facility is used for a food storage operation in a floor stack or rack position, most local codes require an 18-in. minimum open space between any wall and pallet. This open space is required to ensure proper sanitation and is painted white.

Travel Paths and Conveyors Adjacent to a Building Column Require Open Space

In an AS/RS crane facility that is designed with a conveyor pick up and delivery (P/D) concept or a conventional facility with conveyor travel paths, all conveyor travel paths have a 6- to 12-in. minimum clearance from any building column, building obstacle, or other fixed building or product-handling equipment.

Other Factors That Affect Employee Productivity

Employee productivity, low UOP quantity, equipment and building damage, and logistics operating costs affect a storage operation.

The design factors are:

- Adequate floor space or conveyor length for master carton or pallet queue
- Master carton or pallet storage concept
- Pallet design
- Method to secure master cartons on a pallet
- Pallet WMS or inventory control system that includes pallet deposit and withdrawal
- Pallet transaction instruction method
- WMS pallet identification concept
- WMS or inventory control pallet position identification concept
- Pallet transaction verification system
- Number of hours per work day and vendor UOP and CO transactions

Pallet Storage

Pallet storage components are:

- WMS or inventory control pallet identification, which may be either pallet or slip sheet type
- Pallet handling vehicles for (a) in-house transport and (b) storage transactions
- Pallet storage/pick position and employee instruction

- WMS or inventory control program activity
- Building or facility

In a pallet storage operation, pallet handling method and pallet support device are important design factors. In this section we define pallet terms, look at pallet designs and WMS or inventory control identification location, and identify pallet materials and how they affect securing pallet identification.

Pallet Types

The type of pallet used in a storage operation is a key factor that allows a pallet to flow smoothly through a facility, support master cartons, and provide a WMS or inventory control pallet identification location. In the pallet storage industry, a pallet is a master carton support device. Wooden pallets are the most common. They are handled by all forklift trucks, fit into virtually every storage structure, can be handled in a company's supply chain segments, and accept a self-adhesive, stapled, taped, or sleeve identification.

Pallets are available in many sizes and shapes and are manufactured from a variety of materials. A pallet has a rigid base, with stringers and blocks and top and bottom deck boards. A pallet provides support for master cartons and has fork openings for a forklift, AS/RS crane, or pallet truck forks and a block or stringer for attachment of identification.

Pallet storage operations use standard or specially designed (engineered) pallets. A pallet supports a wide product mix and is easily handled by forklift, AS/RS crane, or pallet trucks in a company supply chain.

Master Carton and Pallet WMS Identification

In a storage operation, each master carton and pallet has a unique identification. In a storage operation at a receiving dock, the master carton identification is attached to the corner of the master carton and the pallet identification is attached to the pallet stringer or block, or the master carton on the bottom layer. The location of the master carton or pallet identification permits line of sight for an employee hand-held scanner or a fixed-position scanner or RF tag reader. Options for location of the identification are:

- Permanent or disposable identification with (a) hand-held scanner, (b) fixed-position fixed-beam scanner, (c) fixed-position and moving-beam scanner, or (d) RF tag reader
- Ladder orientation
- Picket fence orientation
- Hybrid

- Face one direction
- Wrap around a corner, stringer, or block
- RF tag
- Pallet component or bottom layer master carton
- Pallet stringer or block attachment that is taped, stapled, self-adhesive, or placed in a sleeve
- Barcode height or length and quiet zone

Basic Pallet Types

Basic pallets (see Figure 3.5) are:

- Standard pallets that are (a) captive, (b) throw-away, or (c) exchange pallets
- Take-it or leave-it pallets
- Specially engineered or designed pallets

Standard Pallet

A standard pallet is used in a storage operation that has a manual pallet truck, powered forklift truck, or AS/RS crane. The standard pallet types are:

- Captive. A captive pallet is used by a vendor and is captive, that is, it remains in the storage operation. Pallets are based on specifications and standards and are used in an AS/RS operation. They are designed to accept identification that is attached to a pallet stringer or block front with self-adhesive glue, tape, staple or sleeve insert.
- Throw-away pallet. A throw-away pallet is purchased from a vendor and is delivered with master cartons to a storage operation. After the operation has used a throw-away pallet, it is thrown in the trash. A throw-away pallet is less expensive to manufacture than an exchange pallet. A throw-away pallet has thin top deck boards, narrow stringers, and a few, thin bottom and deck boards. The pallet has identification stapled, self-adhesive glued, or taped to the pallet stringer front.
- Exchange pallet. An exchange pallet is used by a vendor and storage operation. Both operations have the same pallet specifications and standard that assures the pallet accepts identification, which is self-adhesive glued, taped, or stapled to the front of the pallet stringer. Upon receipt of a pallet from a vendor on an exchange pallet program, the receiving department verifies vendor pallet quality and trades one empty pallet for each pallet under master cartons.

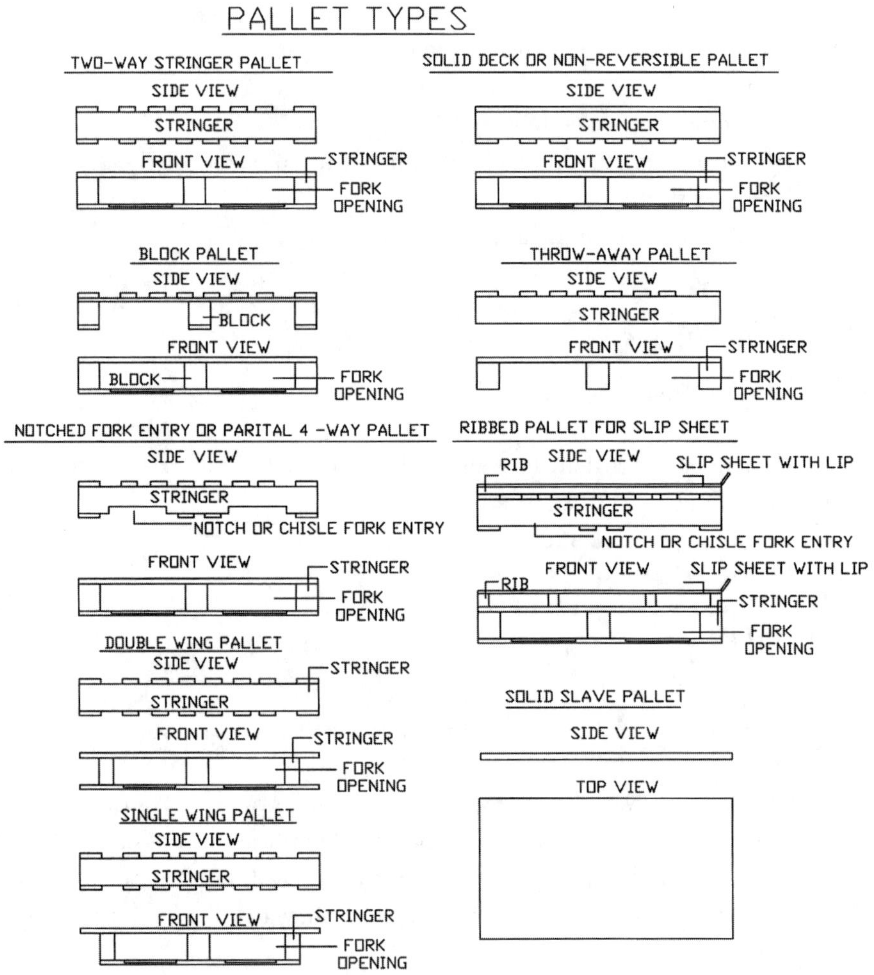

PALLET TYPES

Figure 3.5 Pallet types: two-way stringer, block, notched four-way, double wing, single wing, solid deck, throw away, and ribbed and solid slave.

Take-It or Leave-It Pallet

A take-it or leave-it pallet is basically an exchange pallet that is used by customers and a storage/pick operation. In a storage operation, it is a slip sheet unit load on a pallet that has a specially designed top deck. The unit load combination allows a storage operation to handle a unit load as a slip sheet unit load or as a pallet with a pallet and slip sheet combination. A customer is given the opportunity to purchase a slip sheet and pallet combination or to purchase a slip sheet unit load. If the customer takes the slip sheet option, with a push-pull tine device a slip sheet is pushed and or pulled (removed) from a pallet and a slip sheet unit load is placed onto the customer delivery vehicle floor. If the customer takes a pallet and slip sheet combination option, then with a standard forklift truck they remove the pallet from the storage/pick position and the pallet with slip sheet is placed onto the CO delivery vehicle floor. In a pallet and slip sheet combination option, a pallet storage/pick operation and customers have a pallet exchange program. Pallet identification is self-adhesive glued, taped, or stapled to the front of the pallet stringer.

Specially Designed or Engineered Pallet

A specially designed or engineered pallet is considered a slave or captive pallet that is manufactured to the storage operation specifications. This is required when company storage policy or forklift/AS/RS specifications call for a high-quality pallet. Criteria for a high-quality pallet are no broken or loose top or bottom deck boards or stringers; nails do not protrude above or below deck boards; no crooked, cupped, or bowed members; nails are flathead threaded or screw type; there are knot-free wood members; and the pallet identification is self-adhesive glued, taped, stapled, or sleeve inserted to the pallet stringer or block front.

Pallet Materials

Pallets are manufactured from a wide variety of materials. Factors that determine which pallet material is used are:

- Master carton characteristics: length, width, height, and weight and total weight and ti and hi
- Allowable pallet deflection or bowing
- Combined product and pallet weight that is acceptable to the storage position and forklift truck or AS/RS crane concept
- Storage and safety considerations, such as spark-free, non-corrosive, freezer, or high-humidity environments
- Pallet investment plan
- Pallet exchange program or a captive pallet
- Pallet material as permitted by local codes
- How pallet identification is attached to the stringer or block
- Other operational factors

Pallet materials are:

- Wood: (a) hardwood, (b) medium hardwood, (c) soft hardwood, (d) softwood
- Plastic
- Corrugated

- Pressboard or fiberboard
- Rubber
- Metal and metal clad

Hardwood Pallets

A hardwood pallet has a class C type wood as defined by the National Wooden Pallet and Container Association (NWPCA). The pallet components are nailed together; usually nail holes are predrilled for easy nail entry. A hardwood pallet ensures that the nails are held, provides the greatest support strength, resists shock or damage or splinters, and is the heaviest. Since the hardwood pallet is the heaviest pallet, it is preferred as a captive pallet for an AS/RS or forklift truck storage/pick operation. A hardwood pallet block or stringer provides a good WMS pallet identification location. Prior to selecting a WMS pallet identification attachment method, tests are conducted on a hardwood pallet. Hardwood types include white ash, white beech, red oak, birch, rock elm, hickory, hard maple, hackberry, and oak.

Medium Hardwood Pallets

Medium-density hardwood is a class B wood. To prevent wood from splitting, nail holes are predrilled in wood members. The pallet is medium strength, dries easily, and has medium weight. The characteristics make a medium-density wood pallet preferred for use in a forklift truck or AS/RS crane storage operation and a pallet stringer provides a good pallet identification location. Prior to selecting a pallet identification attachment method, tests are conducted on a medium hardwood pallet. Some medium-density woods are ash but not white ash, soft elm, tupelo, butternut, soft maple, yellow poplar, chestnut, sweet gum, sycamore, walnut, and magnolia.

Soft Hardwood Pallets

A soft hardwood is class A wood. A soft hardwood pallet has medium support weight, is likely to split, and is light weight. A softwood pallet is preferred for a forklift truck storage operation and the pallet stringer provides a good pallet identification location. Prior to selecting a pallet identification attachment method, tests are conducted on a soft wood pallet. Soft hardwoods include aspen, basswood, buckeye, cottonwood, and willow.

Softwood Pallets

A softwood pallet has wood that tends to split easily and is very light weight. A softwood pallet has low support strength and dries easily. Pallet components are nailed, stapled, or glued together and the pallet stringer provides a good pallet identification location. Prior to selecting a pallet identification attachment method, tests are conducted on a softwood pallet. A softwood pallet is low cost and easily handled by employees, preferred in a forklift storage operation and is considered a throw-away pallet. Softwoods are coast or mountain douglas fir, western helmlock, southern pine, and western birch.

The disadvantages are loose nails and wood splinters damage product, the wood easily becomes dirty and infested, and pallets are difficult to use in a spark-free environment. The advantages are repairable and reusable, used in many industries, and interfaces with all forklift truck types and storage/pick concepts.

Plastic Pallets

A plastic pallet is a preformed molded plastic piece. They have similar characteristics to hardwood pallets. Optional metal rods placed in the plastic during the molding process increase the pallet support strength. Plastic pallet use labels manufactured to ensure that the self-adhesive label or sleeve attachment and identification remains on the stringer or block. Plastic pallets can be used with any identification attachment method except staples.

The disadvantages to using plastic pallets are that they are difficult to repair and are likely to bow. Some local codes restrict their use or require additional fire protection. The advantages are that they are washable, have a long life, infestation problems and splinters are minimal, they can be used in a spark-free environment, and interface with all forklift truck types and storage concepts except test arm or rail pallet positions.

Corrugated Pallets

A corrugated pallet is treated heavy-duty corrugated board that is preformed as one piece with its components glued, nailed, or stapled together. Corrugated pallets are used in a forklift truck storage operation and a pallet stringer provides a good pallet identification location. The top deck board of a corrugated pallet provides a good pallet identification location and the self-adhesive, sleeve, staple, or tape identification attachment methods can be used. Prior to selecting a pallet identification attachment method, tests are conducted on a corrugated pallet. Some corrugated pallets have plastic cups for support legs.

Disadvantages are that they are not durable, can be used only in low-humidity areas, and some local codes restrict use. Advantages are they are low cost, can be used in a spark-free environment, are light weight, and available in many shapes and sizes.

Pressboard or Fiberboard Pallets

A pressboard or fiberboard pallet is manufactured from a preformed molded mixture of wood fiber and pressboard. A pressboard pallet block or stringer provides a good pallet identification location using the staple identification attachment method.

Disadvantages are that they are not durable, can only be used in low-humidity areas, and some local codes restrict their use. Advantages are that they are low cost, can be used in a spark-free environment, are light weight, and available in many shapes and sizes.

Rubber Pallets

A rubber pallet is made from one preformed molded polyethylene rubber piece and has the same disadvantages and advantages as a corrugated pallet. The pallet can be used in a spark-free environment. A rubber pallet block or stringer provides a good identification location using a self-adhesive label, sleeve insert, or tape identification attachment method. Optional metal rods placed in the rubber during the molding process increase the rubber pallet strength.

Metal or Metal Clad Pallets

A metal or metal clad pallet has aluminum or steel members. It is the heaviest pallet material. A metal clad pallet has aluminum or steel deck boards with wood blocks or stringers. A metal pallet block or stringer is the pallet identification location using a self-adhesive label, sleeve insert, or tape identification attachment method. Metal pallets are used in a forklift truck or AS/RS crane storage operation.

Disadvantages of metal pallets are that they are difficult to handle, are heavy, and expensive. Advantages are that they are fire resistant and have a long life.

Important Pallet Dimensions

Pallets are available in a wide variety of sizes. Factors that determine the appropriate pallet size for an operation are:

- Storage position and type of fork on the forklift truck or AS/RS crane
- Master carton dimensions, weight, hi and ti, or master carton palletizing pattern
- Supply chain requirements
- WMS identification label dimensions

There are three important pallet dimensions: (1) rack position, depth, bearer, or stringer; (2) width, down aisle, or fork-opening dimension; and (3) height. The length of the pallet stringer determines the length of the forks required on the forklift truck, AS/RS crane, or pallet jack. The forklift truck opening height is determined by a pallet forklift truck, pallet truck set of forks, AS/RS carrier device, and storage concept. Forks on a counterbalanced or straddle chisel forklift truck or pallet truck require a 3- to 5-in. high pallet opening. Most gravity pallet flow storage concepts have a solid slave pallet that is approximately 1 to 2 in. high. AS/RS crane carrier forks have a 3- to 5-in. high pallet opening but an AS/RS carrier with platens handles a pallet underside and a standard four-way pallet is placed onto a slave pallet.

Basic Pallet Designs

Basic pallet designs are block; leg or honeycomb; captive, solid or slave; stringer; and flue.

Block Pallets

A block pallet has equally spaced blocks along its length and width and is a four-way pallet. Deck boards are attached to the blocks. A pallet block provides a good pallet identification location, and permits a front face or wrap-around identification label. In an arm or rail pallet storage position (drive-in, drive-through, or AS/RS carrier with a platen concept) a pallet spans two arms or rails in a pallet position. The block pallet is tested prior to implementation. The pallet material determines the identification attachment method. During label attachment tests, if the block material does not accept one of the identification attachment methods, an option is to have the identification attached to a master carton on the pallet bottom layer.

Leg or Honeycomb Pallets

A leg or honeycomb pallet has a solid deck with equally spaced legs attached to the deck along the length and width. An open space between the top surface and base is the pallet fork opening. The pallet leg provides a pallet identification location, and permits a front face or wrap-around identification label. A self-adhesive label, sleeve insert, or taped pallet identification method is used, with a hand scanner or fixed position scanner.

During label attachment tests, if the block material or shape does not accept one of the above identification attachment methods or provide good scans, an option is to have the identification attached to a master carton on the pallet bottom layer. In an arm or rail pallet storage position (drive-in, drive-through, or AS/RS carrier with a platen concept) a pallet spans the two arms or rails of the pallet position.

Captive, Solid, or Slave Pallets

A solid, captive, or slave pallet is made up of one solid piece of plywood or other material on the top and bottom surfaces. This type of pallet does not have legs, blocks, or stringers. A slave pallet is flat or placed directly on the floor. In a storage concept, a solid pallet is used as a support device for another pallet such as a gravity flow rack or AS/RS platen carrier. Some solid or slave pallets are designed with four holes in the deck corners for easier handling.

When a forklift truck handles a slave pallet, the pallet is set on a support device that permits the forklift truck forks to extend under it. This feature allows the forklift truck to complete a pallet storage transaction. A side of the slave pallet provides a location for identification, using a front face or wrap-around label, but the label height is limited to 1- to 2-in. height of the the slave pallet. The location permits hand scanner or fixed-position scanner line of sight and takes a self-adhesive label, sleeve insert, stapled or taped pallet identification attachment method or the identification is attached to a master carton on the pallet bottom layer. If a standard pallet is placed onto a slave pallet and the standard pallet remains on the slave pallet top until the pallet is sent to a customer, pallet identification is attached to a standard pallet stringer or block that is on the slave pallet.

Stringer Pallets

A stringer pallet has two exterior stringers and one interior stringer for top and bottom deck board attachment. Exterior stringers either are solid or have two notches (openings) in the stringer side. Notches are an additional set of fork entry openings that are 5 to 5½ in. high that permit a forklift truck with chisel forks to handle a pallet from all four sides. A pallet stringer provides a good pallet identification location, and permits a front face or wrap-around pallet identification label. The pallet material determines whether a self-adhesive label, sleeve insert, stapled or taped label is used as the identification attachment method. If a stringer material does not accept one of the above identification attachment methods, an option is to have the identification attached to a master carton on the pallet bottom layer.

Flue Pallets

A flue or rippled pallet is a solid piece pallet or perforated material that has every other ripple sitting directly on the floor. The flue space depth and width between the ripples are the openings for forklift truck fork entry. This pallet is a two-way entry pallet. Identification attachment method is to hang a taped label from the top.. During label attachment tests, if the pallet material or shape does not accept an identification attachment method or provide good scans, an option is to attach the label to a pallet bottom layer master carton.

Pallet Components

Components of a wood stringer and wood block pallet (see Figures 3.5 and 3.6) are top deck boards, bottom deck boards, edge boards, and other deck boards, chamfered edge boards, fork-entry opening, notches, stringer board, block or stringer or bearer, blocks, and pallet component connecting devices.

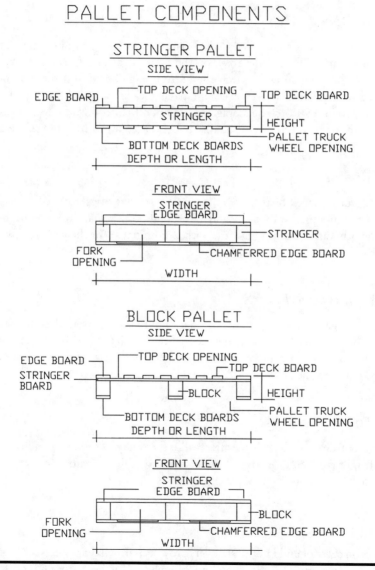

Figure 3.6 Pallet components: stringer pallet and block pallet.

Top Deck Boards

Top deck boards are flat wood, plastic, or material slat or solid piece members that are attached to pallet stringers or to a block pallet stringer board. Top deck boards provide master carton support. When top deck board slats are used on a pallet, the slat width varies from 4 to 8 in. and the open space between slats varies from 1 to 3 in. The open space is called the deck opening and allows air circulation.

Bottom Deck Boards

Bottom deck boards are attached to the bottom of the stringer or block and provide a flat, rigid surface for a pallet to sit on the floor or rack load beams. In a storage operation, the open space between the bottom edge board and the next bottom deck board permits the front wheels of a pallet jack to raise or lower a pallet from the floor. In an AS/RS crane storage operation, all pallet bottom deck boards are reviewed for quality. Only good-quality pallets enter an AS/RS storage operation.

Edge Boards

Top and bottom deck boards are at the fork entry openings. The top and bottom boards are a nominal 6 in. wide and ½ in. high. The remaining top and bottom board width varies from 4 to 8 in. with a nominal ½ in. height and the space between the deck openings varies between 1 and 8 in.

Chamfered Edge Boards

On some pallets, the two bottom edge deck boards and bottom deck boards adjacent to the open space of the pallet jack wheel are chamfered for easy truck fork entry. The chamfered edge on the bottom deck board has the edges cut at a 35 degree angle and the angle cut is 12 in. wide.

Fork-Entry Openings

Top and bottom deck boards are attached to the stringers or blocks and the overall pallet height is 5 to 6 in. with a nominal ½ in. high deck board. This arrangement creates a 4- to 5-in high open space that allows fork entry.

Notches

Notches on exterior and interior stringers provide additional openings for fork entry The notch is a nominal 1½ in. high and 9 in. wide. Notches permit a forklift truck with a set of chisel-type forks to handle a pallet from the stringer side.

Stringer Board

A stringer board is 1 in. thick and is used on a block pallet for top deck board attachment to the blocks.

Stringer or Bearer

A stringer or bearer runs the entire pallet depth. A pallet with a short face or fork opening has two stringers and a pallet with a wide face or fork opening has three stringers. The stringers are 2 to 3 in. wide and serve as top and bottom deck board attachments, create fork entry openings, master carton support and, in a storage operation, the stringer face provides a good location for a front or wrap-around pallet identification with any attachment method.

Blocks

Blocks are square or rectangular parts of a pallet that are placed under the four corners and in the middle of the pallet. Blocks serve for stringer pallet attachment. A small pallet has six blocks and a large pallet has nine blocks, which vary from 3 to 4 in. and are full pallet length and width. In a storage operation, a pallet block face provides a good location for a front or wrap-around pallet identification with any attachment method.

Pallet Component Connecting Devices

Stringer and block pallet components are attached to each other by helically threaded (spiral) nails, staples, nuts and bolts, or glue. The fastener concept is determined by cost, pallet material, and operational requirements.

Pallet Types

Pallets are either flush or wing pallets. A flush pallet has top or bottom deck boards that are flush with the stringers or blocks. A wing pallet has a top deck board width or sometimes both top and bottom deck boards that overhang exterior stringers.

Standard flush and wing pallets have the following features:

- Open decks with 1- to 3-in. wide openings or spaces between two top deck boards or slats that are 4 to 8 in. wide.
- Closed or solid decks with a solid one-piece surface as the top deck.
- Two-way entry, that is, a stringer pallet with two external stringers and two normal high fork openings.

- Four-way entry, that is, a block pallet with four normal high fork entry openings, one on each side.
- Partial four-way or notched entry, that is, a stringer pallet with two normal high fork entry openings at both pallet ends and an additional chisel fork entry opening on the stringers. The two additional chisel fork entry openings are in exterior stringers and are in a middle stringer. Notches are 9 in. wide by 1½ in. high and the open space permits a forklift truck with a set of chisel forks to enter the openings.
- Single wing that has top deck boards that extend beyond the two exterior stringers or blocks. In a straddle forklift truck operation, there is sufficient clearance between the floor and the wing bottom for the straddles to encircle a pallet.
- Double wing that has top and bottom deck boards that extend beyond the top and bottom stringers or blocks. In an operation, a straddle reach forklift truck extends the set of forks outward to handle a pallet.
- Reversible that has two solid deck boards with no pallet truck wheel space on the top or bottom. A pallet deck side that faces up provides master carton support and the other side of the pallet sits directly on the floor. Two fork entry openings permit a forklift truck to handle the pallet, but a pallet truck cannot handle this type of pallet.
- Non-reversible that has the top deck boards to provide master carton support. The bottom deck boards are spaced to permit both pallet jacks and forklift trucks to handle the pallet.
- Take-it or leave-it, that is a two-way pallet that has a normal two pallet fork-entry opening and a top deck surface with ribs that run the entire stringer dimension and elevate a slip sheet unit load (slip sheet and master cartons) above the top deck and provide clear space openings. The openings permit a forklift truck with a tine push-pull device to enter under a slip sheet unit load between the rib spaces. This feature allows a slip sheet unit load to be removed from a take-it or leave-it pallet or a pallet with a slip sheet to be handled as a pallet.
- Nestable pallets have bottom deck boards that span the pallet width. An open space along the length of the pallet bottom allows unused and empty pallets to be nested for storage.
- Specially designed or engineered, that is manufactured to handle a specific product or to interface with a specific device. Examples are (a) round, (b) beer keg, and (c) barrel pallets.

Slip Sheet

A slip sheet is a unit load support device that is used primarily in a transport activity between two facilities. A slip sheet (see Figure 3.7) is designed to minimize unit load height and weight on a delivery vehicle and to permit a forklift truck

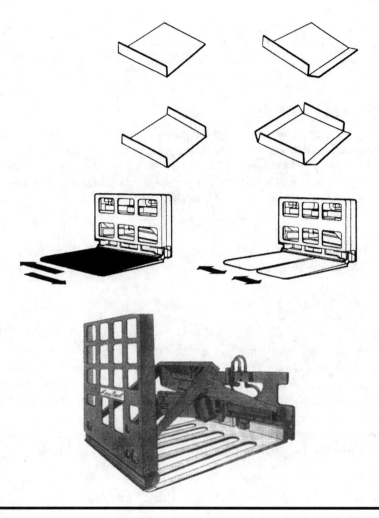

Figure 3.7 Slip sheet types: slip sheet devices, single platens, dual platens, and tines.

with a push device to handle the unit load in the facility. With a pallet or a forklift truck with a push-pull device under a slip sheet, the slip sheet structural or beam strength supports the master cartons. A slip sheet is manufactured from solid fiberboard, corrugated materials, or plastic (polypropylene) shaped to the same length and width as a pallet. A 4- to 6-in. lip extends beyond the slip sheet side. Most slip sheets have a single lip extending forward and facing upward from the slip sheet base. Some slip sheets have two, three, or four lips, which permits a forklift truck push-pull device to clamp onto a lip for lifting a slip sheet unit load from a delivery vehicle floor or pallet. A slip sheet reduces weight and space in a delivery vehicle. When a slip sheet unit load is placed in a storage area, clear space is needed for the tab extension. When a slip sheet is used in a storage concept, it is placed onto a pallet and the pallet carries the identification. When used with no pallet, a master carton on the lower layer carries the identification.

Slip Sheet Designs

Slip sheets are available in sizes that match various pallet dimensions. Slip sheets that are used on delivery trucks or ocean-going containers have one lip. They are loaded onto the delivery truck with all tabs facing the delivery truck door. This arrangement gives a forklift truck push-pull device direct access to the slip sheet lip, which permits the slip sheet to be handled with efficiency and minimal damage. Slip sheets can also come in the following designs:

- Two-lip slip sheet with the lips on opposite sides
- Two-lip slip sheet with the lips on adjacent sides
- Three-lip slip sheet
- Four-lip slip sheet

When a multi-lip slip sheet is used in an operation, to reduce product damage and unit load space requirements, a lip that is not being used is secured to the side of the unit load or removed from the slip sheet unit load. When using slip sheets on a rail car or on a delivery vehicle, a multi-lip slip sheet enables a push-pull device to handle the slip sheet from all four sides and permits various slip sheet placement options onto the delivery vehicle floor. This flexibility permits a slip sheet pattern in a delivery vehicle to maximize the delivery vehicle cube space, weight, and permits an efficient unloading and loading activity with minimal UOP damage.

Forklift Trucks Used to Handle a Slip Sheet

When a slip sheet is used, the operation also requires a forklift truck equipped with a push-pull device and gripper bar.

Types of forklift or pallet trucks that are adapted with push-pull devices include sit-down counterbalanced forklift trucks that can handle the weight lift requirement and can enter and exit delivery vehicles; low-lift, powered rider pallet trucks that have the ability to enter and exit delivery vehicles, travel over the floor to the assigned location, and set a slip sheet onto the floor.

Slip Sheet Attachments

A push-pull attachment has a pantographic arm that extends forward from the forklift truck mast with a backrest and gripper bar. This arm is hydraulically controlled by the operator to extend over a platen, set of platens, or a chisel tine. When an operator activates the gripper, it grips the slip sheet lip and holds it firmly between the backrest bottom and gripper bar. The operator controls a hydraulic system that orders a pantographic device to lift the slip sheet unit load upward. This lifting action raises the front of the slip sheet. In this raised position, the pantographic device and slip sheet are pulled over a platen(s) or tines to the forklift truck mast and it is transferred to another facility location, onto a pallet, or onto the floor of a delivery vehicle. The forklift truck operator positions the slip sheet directly in front of the pallet. A hydraulic system raises the slip sheet to an elevation slightly above the pallet, which is positioned on a slip sheet transfer device, with its far side against the backstop of the transfer device. In this elevated position the slip sheet is moved forward until the slip sheet and platens are over the pallet and the push-pull device and slip sheet are extended forward until the slip sheet touches the backstop. In this action the far side of the slip sheet touches the pallet. When this happens, the forklift truck operator makes a hydraulic system gripper bar release the lip. The backrest remains firm and extended over the near side of the pallet. As the forklift truck is moved backward, the slip sheet rests on the pallet and the transfer is complete.

A platen-type slip sheet attachment can be either single or double. A single platen has one tapered, hardened metal platen 1 to 2 in. narrower than the slip sheet. A double-platen type has two 15- to 18-in. wide platens and resembles a flat shovel. A 4-in. open space between two platens provides at the minimum a 34-in. wide surface to support a 40-in. wide (maximum) slip sheet. The platens are 48 in. long for a 48-in. long slip sheet.

A tine or chisel slip sheet attachment has several full-tapered, hardened metal extensions from the base of the forklift truck mast. The tine length varies from 36 in. minimum up to 48 in. maximum. Each tine is 4 to 5 in. wide, with a 4- to 5-in. open space between tines. The overall width (tines and open spaces) provides adequate support for a 40-in. wide slip sheet.

How to Secure Master Cartons on a Pallet

To unitize master cartons onto a pallet ensures an efficient and cost-effective storage operation with minimal product and equipment damage. Master cartons on a pallet are further stabilized by such factors as the ability to interlock master cartons, master carton physical characteristics and weight, overall height of the pallet, master carton stack alignment, and scanner or RF tag reader clear line of sight to the pallet identification.

A good pallet stabilization concept minimizes master carton damage, improves space utilization, improves security, improves forklift truck productivity, improves master carton appearance, and facilitates scanning the pallet identification.

Pallet Stabilization Concepts

Pallet stabilization techniques used in a pallet storage operation include master carton ti and hi, tape, plastic or steel bands, string, stretch wrap, shrink wrap, netting, glue or adhesive, and industrial rubber bands.

Master Carton Ti and Hi

A basic component of stabilization of a pallet is the master carton-stacking arrangement. It ensures that the number of master cartons on a layer (hi) and each master carton layer (ti) are interlocked on a pallet. Master carton alignment should be straight and there should be no overhang beyond the pallet edges. A master carton unitizing or palletizing pattern for a 48- by 40-in. pallet with an allowable overhand is 52 by 44 in. To achieve a good master carton ti and hi pattern, the master carton varies from 5 to 21 ½ in. width, and 6 ½ to 43 in. length with ½ in. increments from the smallest to the largest dimension.

In a conventional forklift truck storage operation, the pallet storage position allows a 2 in. maximum master carton overhang on a pallet. Allowable master carton overhang is eyeballed by the forklift truck driver. In an AS/RS concept, all pallets that enter an AS/RS concept are reviewed by a laser beam size and weigh station. If master carton overhang on a pallet exceeds the allowed overhang, the pallet is rejected by the AS/RS storage concept. In a storage operation, a receiving tally sheet indicates the number of master cartons per pallet, which helps a receiving clerk determine master carton ti and hi on a pallet. Based on the master carton and pallet dimensions, computer programs are available to determine the optimum master carton ti and hi pattern.

The disadvantage is potential master carton shifting on a pallet. Advantages are low cost, no trash problems, easy master carton removal, master carton visibility, and clear line of sight to the pallet identification.

Tape

An employee applies tape around master carton layers on a pallet and assures that tape is on the top and bottom layers. The tape is plastic or fiberglass-reinforced and is available in various widths, from ¼ to 2 in., and is cut with an industrial knife.

The disadvantages are that this creates a trash problem, the tape difficult to apply on lower levels, and creates a maintenance problem on vehicle wheels. The advantages are low cost, it is easy to apply, can be used at any location, and provides clear line of sight to the pallet identification.

Plastic or Steel Bands

Steel or plastic bands are placed horizontally around one or several master carton layers and over the pallet top and bottom. The bandwidth ranges from ½ to ¾ in. and requires a band stretcher and clamp gripper. To protect the pallet from damage, corrugated board material is used at the corners and top edges. The pallet banding is done manually or automatically with a machine.

Disadvantages are that it is labor intensive, cartons can be damaged, and when cutting the bands, potential employee injury. The advantages are that it permits circulation, allows good visibility, stabilizes the top and bottom location, and provides clear line of sight to the pallet identification.

String

The same application method is used for string as for tape, except that string is tied by an employee around the pallet. String is taken from a spool or cut to predetermined lengths with a loop or metal S-hook.

The disadvantages are that it is difficult to apply at lower carton layers, creates a trash problem, and creates a maintenance problem with vehicle wheels. The advantages are low cost, it can be applied anywhere, and provides a clear line of sight to the pallet identification.

Stretch Wrap

Stretch wrap is a pallet-stabilizing concept that has plastic film wrapped tightly around the four sides of a pallet. The plastic film is applied by a manual or semi-automatic method. Some mechanical stretch wrap methods use prestretched film, which provides a more secure pallet. To wrap a pallet, the stretch wrap film is tucked under a master carton and wrapped around the pallet and tied under another master carton. The stretch wrap should not cover the pallet identification. If stretch wrap material covers a pallet identification, it is cut to assure a clear line of sight to a pallet identification. Stretch wrap application can be manual or automatic.

A manual portable stretch wrap concept has a hand-held device that has a handle grip and a spool. After a pallet is in position, an employee tucks film under a master carton, walks around the pallet a number of times to secure it, and cuts the film and tucks it under another master carton.

A second manual concept employs a portable stretch wrap machine that is made of welded steel and mounted on 3-in. high heavy-duty casters and a wheel bed. A forklift truck sets a pallet adjacent to the machine, which has an attached film holder. An employee tucks the film under a master carton, walks one complete turn around the pallet, activates the brake and continues to walk around the pallet, which stretches the film wrap. A manual portable machine handles 10 to 15 pallets per hour.

The automatic wrap machine is a steel machine with a film roll attached to a single post in the turntable front with a spring-loaded top. After the forklift truck or conveyor places a pallet onto the turntable, an employee tucks the film under a master carton or mechanical device holds the film end and the machine is activated. With the wrap activity complete, the employee or machine cuts the film and tucks it under a master carton or secures the film. The automatic concept handles approximately 25 pallets per hour. A second machine type is a standard model that is similar except that it has a brake and handles about 35 pallets per hour. A third automatic machine concept uses the automatic multi-roll machine, that has the same features as the automatic standard machine except that the multi-roll machine has two to three film rolls and handles 30 to 40 pallets per hour.

Disadvantages are waste or cost due to errors, trash problem, and investment. Advantages are it handles all pallets, creates a moisture barrier, improves security, and provides clear line of sight to pallet identification.

Shrink Wrap

The automatic shrink wrap concept has a shrink wrap tunnel with in-feed and out-feed conveyors. To operate, a pallet is placed onto the in-feed conveyor. A plastic sheet is draped over the pallet and it is moved into the shrink wrap tunnel where heat is applied to the plastic material, which causes the plastic to shrink around the pallet. If the shrink wrap material covers the pallet identification, it is cut to assure a clear line of sight to the pallet identification.

With a hand-held gun plastic shrink wrap concept, a plastic sheet is draped over the pallet and an employee uses a hand-held heat gun to direct heat against the plastic. This heat causes the plastic sheet to conform to the pallet.

Disadvantages are investment cost, the equipment requires floor space, as well as energy for heat. Advantages are protection against moisture, dirt and dust, less wrap waste, and clear line of sight to the pallet identification.

Netting

A net is placed over a pallet and is pulled tightly around the pallet base and top. The netting material is made from plastic and the application to the pallet is similar to the stretch wrap method. If the netting hangs over the pallet identification, the netting is cut to provide a clear line of sight to the pallet identification.

Disadvantages are that the netting permits dust or dirt and creates a trash problem. Advantages are air circulation, good visibility, stabilization of the top and bottom, and pallet identification clear line of sight.

Glue or Adhesive

Glue and adhesives are used to stabilize master cartons on a pallet. With modern glue or adhesive and new application technologies and equipment, glue is applied to the tops of master carton on a pallet. As master cartons with glue are placed on top of each other to form a pallet, they stick to each other, forming an entire pallet that is stabilized with glue. When an employee requires a master carton from the pallet, the bond between the master cartons is broken with a simple bump or pull on a master carton.

Disadvantages are moisture, dirt or dust protection, and possible damage to master carton. Advantages are low cost, no trash problem, good visibility, and pallet ID clear line of sight.

Industrial Bands

An elastic band concept has bands placed around master carton layers on a pallet. Elastic bands are available in various widths, thicknesses, and lengths and are manufactured in a circular form to specific lengths that are precut and joined by a metal clip. This elastic band type secures the pallet top and bottom.

The disadvantages are high cost, difficulty in applying bands to the lower layer, and sometimes difficult to reuse. The advantages are reusable and clear line of sight to the pallet identification.

Inventory Control

The inventory control activity ensures that a master carton or pallet with identification is transferred from the receiving area to a specific storage position from which it is transferred when needed to another location and in the correct quantity. Pallets and storage positions are tracked in the inventory files.

Other inventory program concerns are:

- Proper SKU or master carton or pallet rotation.
- Accurate pallet, master carton, and piece counts.
- Ensuring minimal "stock outs" or "out of stocks" at a pick position. A "stock out" is defined as when, per a customer order pick instruction, an employee or automated pick device arrives at a pick location and the pick position is depleted although the inventory file indicates an on-hand SKU inventory quantity. An "out-of-stock" is defined as when, per a CO pick instruction, an employee or automated pick device arrives at a pick position and the pick position is depleted and the inventory file has no on-hand inventory quantity.

- Tracking pallet flow through the supply chain logistics strategy or through each segment.
- Ensuring that master carton or pallet transactions in each logistics segment is completed by scanning master carton or pallet identification and storage position identification, and sending the scans transaction to the inventory control computer.

Storage Position Concepts

A master carton or pallet storage position concept ensures that identified master cartons or pallets are accessible, that the master carton or pallet identification and pallet position identification have line of sight for a hand-held scanner or RF tag reader device, and that there is maximum storage position utilization. The master carton or pallet storage concept is a key factor that determines the ability of the storage operation to satisfy both company objectives of lowest possible logistics cost and best customer service. The design parameters that influence the preferred master carton or pallet storage concept are:

- Operation type
- Average and peak handling and pallet inventory
- Master carton or pallet dimensions and weight
- Product type, rotation, master carton or pallet number per SKU, SKU type, and shape
- Whether the master carton has a bottom support device, as well as the pallet type or bottom support device
- Type of forklift truck or AS/RS crane
- Storage conditions and building codes
- Master carton or pallet identification and master carton or pallet storage position identification line of sight or RF tag reader

The master carton or pallet storage position concepts are:

- Floor stack or block that are (a) 90 degree angle and (b) 45 degree angle
- Tier racks, stacking frames, or containers
- Shelves or slotted angle racks
- Double-deep racks
- Standard pallet racks that include two pallets on the floor
- Decked standard pallet rack
- Aisle bridge
- Drive-in racks
- Drive-through racks
- Gravity and air flow racks
- Push-back racks
- Cantilever racks

- High-rise or tall racks
- Mole, car-in, and sort link racks

Storage Position Identification

In an operation with standard racks and a forklift truck, each master carton or pallet position has a unique identification. The type of identification is determined by forklift truck type, whether man-up or man-down, load beam face or front flat surface, upright post, and load beam location.

Man-Up Very Narrow Aisle or High Rise Order Trucks

With a man-up VNA forklift truck at the P/D station or high rise (HROS) truck, the operator completes an identified master carton or pallet scan or RF tag read transaction. With a good scan RF tag read, a forklift truck travels in an aisle to a microcomputer-assigned or employee-directed identified storage position. As the forklift truck travels down the aisle, the operator platform and forks rise to the level of the assigned storage position. Each position has identification on a rack load beam or upright post. At the proper elevation, forklift truck driver completes the transfer and scans the pallet identification and the storage position identification.

Man-Down WA, NA, or VNA Forklift Truck

With a man-down WA, NA, or VNA forklift truck at the receiving dock or P/D station the operator completes a pallet identified scan transaction. With a good scan or RF tag read, a forklift truck travels in an aisle to a microcomputer-assigned or driver-directed identified pallet storage position. In the aisle, the forklift truck operator platform remains at ground level, scans or reads the pallet identification, and the forks are elevated to the assigned or employee-selected pallet storage position. At the correct elevation, the forklift truck driver completes the pallet storage transaction. With the forklift truck driver at the ground level, the storage position identification is attached to a low-level upright frame post location or load beam that permits a forklift truck driver line of sight to scan or read the pallet position identification. Storage position identifications are sequenced in the same sequence as the load beam levels. To ensure easy and accurate pallet identification relationship to the elevated load beam level, each load beam level is identified as level 1, 2, 3, 4, 5, or 6 and has a color-coded label or RF tag surface.

Selected label surface colors are tested in a storage area aisle with a hand-held scanner device to assure maximum goods reads from a scanner device depth of field, RF tag reader or forklift truck driver operator platform. After completing a pallet storage transaction, a forklift truck driver hand scans or reads a RF tag corresponding pallet storage position identification.

Load Beam Face Storage Position Identification

In a forklift truck operation that places the storage position identification label on a load beam, the preferred load beam location is as a forklift truck driver faces the storage position to complete a transaction, with two pallets per bay on the right-hand side. In this position, with the master carton or pallet identification on a right-hand stringer or block, the forklift truck driver has clear line of sight to both the master carton or pallet identification and the storage position identification. In a WA, NA, or VNA forklift truck operation, two master carton or pallet storage position identification locations increase productivity due to the fact that driver platforms on most forklift trucks have access on the right side. Most drivers are right handed and hold the scanner in the right hand to complete a scan or RF tag read transaction. This position does not require the forklift truck driver to transfer the hand-held scanner or RF tag reader and therefore minimizes hand-held scanner damage with less chance of it being dropped from the forklift truck.

A master carton or pallet storage position identification on a load beam has a 5 to 6 in. vertical dimension. In a VNA man-up forklift truck operation with the bottom pallet level on load beams, the forklift truck driver has clear line of sight to the identification. Otherwise, the bottom load level identification is placed on, next to, or above the load beam level. In this arrangement, the load beam level has two master carton or pallet storage position identifications for quick and easy forklift truck recognition on one load beam, and the storage position identification is attached to the lowest master carton. Alternatively, the pallet storage position identification is on the lowest load beam face and the top or highest pallet storage position identification is on the top load beam face. If master carton or pallet position identifications are placed onto a load beam, the labels are arranged in a step pattern, with each bar code or RF tag identification on a different elevation above the floor. During a scanning or read transaction, the step arrangement minimizes the possibility that a hand-held scanner light beam reads an incorrect bar code or two bar codes, which would create a "no read." The forklift truck driver uses color-coded labels, as previously described, and places direction arrows before or after the pallet position identification label.

In a forklift truck and WMS program application with one master carton or pallet storage position identification per load beam, the flat front and indented load beam front design has sufficient flat surface for a standard WMS pallet storage position identification label. If a forklift truck and WMS program application has two WMS master carton or pallet storage position identifications per load beam on a flat front load beam, a flat front load beam design offers the entire load beam front (5 to 6 in.) for two standard master carton or pallet storage position identification labels. A 5- to 6-in. flat front is sufficient space for two standard master carton or pallet storage position identification labels that are arranged above each other with no line of sight or RF tag read problems. When there are two master carton or pallet storage position identification labels on one load beam, to assist the forklift truck driver to understand the relationship between a master carton or pallet storage position identification label and master carton or pallet storage position, directional arrows are used before or after the label bar or RF tag code and/or color-coded bar code label face.

With a WA, NA, or VNA forklift truck and WMS program application that uses two master carton or pallet storage position identifications per indented front load beam, the load beam design has a smaller flat front that offers 3 to 4 in. for two standard master carton or pallet storage position identification labels. The 3- to 4-in. flat front does not provide sufficient space for two standard master carton or pallet storage position identification labels and can create potential line of sight problems.

Upright Post Position Identification

In a forklift truck operation that attaches a master carton or pallet storage position identification label to upright posts, the preferred upright post is the post as the forklift truck driver faces a rack bay with two pallets. A master carton or pallet storage position identification is on the the right-hand upright post for a pallet position on the right, and the left-hand upright post for a pallet position on the left.

With two storage position identification labels on one upright post, left- and right-hand pallet positions are consistently labeled on the same upright post location. Identifications labels on a middle upright post are arranged in a step pattern, with each barcode identification on a different elevation above the floor. During a scanning transaction, the step arrangement minimizes incorrect bar code reading.

Directional arrows are used before or after the master carton or pallet storage position identification to identify the master carton or pallet storage position. Options are color-coded barcode labels and master carton or pallet storage position directional arrows to minimize forklift truck problems.

Storage Concepts

Master carton or pallet storage concepts (see Figure 3.8) include a single deep-storage location or dense storage. The preferred storage concept for a small-item order-fulfillment operation is determined by SKU number, storage units or UOPS per SKU, inventory requirement, and required UOP rotation. The storage concept is a major factor in determining the square footage required for the operation.

Single Deep-Storage Concepts

A single deep-storage concept has shelves or racks that are serviced by an employee, employee-controlled forklift truck, or AS/RS crane. With an employee concept, master cartons are stacked one, two, or more high in one storage position. If the building has at least a 40- to 60-ft. high ceiling clearance, a VNA lift truck is used in the operation and with a 60- to 80-ft. high ceiling clearance an AS/RS crane

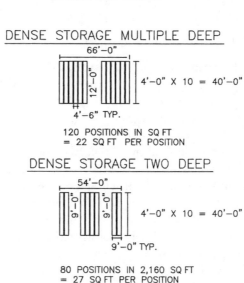

SINGLE DEEP AND DENSE STORAGE COMPARISON

SINGLE DEEP STORAGE

63'–0"

12'–0"   12'–0"   12'–0"   4'–0" X 10 = 40'–0"

9'–0" TYP.

66 POSITIONS IN 2,529 SQ FT
= 42 SQ FT PER POSITION

DENSE STORAGE MULTIPLE DEEP

66'–0"

12'–0"   4'–0" X 10 = 40'–0"

4'–6" TYP.

120 POSITIONS IN SQ FT
= 22 SQ FT PER POSITION

DENSE STORAGE TWO DEEP

54'–0"

9'–0"   9'–0"   4'–0" X 10 = 40'–0"

9'–0" TYP.

80 POSITIONS IN 2,160 SQ FT
= 27 SQ FT PER POSITION

Figure 3.8 Single deep and dense storage concepts: single deep, dense, and two deep.

is used. A single deep-storage concept provides maximum access to all SKUs or UOPs, low storage density, or fewer UOPs per square foot, FIFO (first in, first out) product rotation, and a large-square-foot facility.

Dense Storage Concepts

A dense storage concept means that there are two or more master cartons or pallets deep per storage position and the storage position has one or two aisles. In a low bay building, the dense-storage concepts are floor stack, two-deep, drive-in racks, drive-through racks, stacking frames/containers/stack racks, mobile shelves or racks, gravity flow racks, or sort link.

In an AS/RS storage operation, the dense-storage options are gravity or air flow racks, car-in or mole racks, and two-deep racks.

The dense-storage concepts have few aisles, more UOPs per square foot, wider range of SKUs. Some do not provide FIFO product rotation. Dense storage concepts require a smaller-square-foot facility.

Floor Stack or Block Storage Concepts

A master carton or pallet floor stack or block storage concept (see Figures 3.9 and 3.10) has a pallet placed directly onto the floor or onto another master carton or pallet. A floor-stacked concept is a dense storage concept that provides a maximum of six to ten master cartons or pallets deep per storage lane. There can be either a single lane or back-to-back storage lanes. Due to the master carton or pallets leaning and forklift truck placement variance on the floor and on another master carton or pallet, longer or deeper storage lanes reduce the ability of manually controlled forklift truck to complete transactions with good employee productivity. To utilize the air space, additional master cartons or pallets are stacked on other master cartons or pallets to a maximum height of three to four master cartons or pallets. To assure good employee or forklift truck productivity and minimal damage, in a floor stack concept 3 to 5 in. is needed between storage lanes. If a slip sheet is used in a floor stack concept and the slip sheet lip remains on the unit load, at least 6 in. is needed between storage lanes. Master cartons should be capable of supporting the stacked weight. The master carton (SKU) at the aisle position should be the same for the entire storage lane. The floor stack storage concept, when full, offers the highest storage density with the lowest cost, but poor pallet accessibility.

A 60 percent utilization factor is used to determine the number of storage positions required. A 60 percent utilization factor is used due to honey combing (vacant master carton or pallet positions in a vertical stack) and vacant pallet positions in a storage lane depth. These situations occur from normal employee or forklift truck storage transaction activity. In a floor stack or block concept to make a pallet transaction, from aisle A an employee or forklift truck enters a storage lane, travels to a master carton or pallet position, completes a storage transaction, and backs out from the storage lane to the same aisle A. When a floor stack storage concept is designed, master carton or pallet depth per storage lane is varied to bury building columns in or between storage lanes. The concept has a LIFO (last in, first out) product rotation, handles a high-throughput volume, and interfaces with wide-aisle or narrow-aisle forklift trucks.

A common floor stack concept is a 90-degree stack concept that has master carton or pallet fork openings and the WMS master carton or pallet identification face the main aisle. The concept provides the highest number of SKU openings and master carton or pallet positions per aisle. A 45-degree floor stack concept allows a minimum right-angle forklift truck turning aisle, one-way forklift truck traffic, and fewer master carton or pallet openings per aisle. To complete a master carton or pallet storage transaction, from a main aisle a forklift truck enters a storage lane

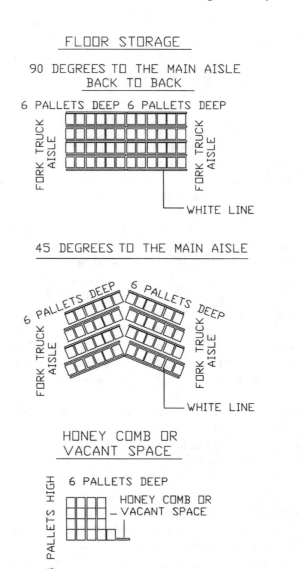

Figure 3.9 Floor stack storage: back to back; angled; and honey comb.

with the forklift truck driver facing the master carton or pallet fork opening and pallet identification.

In a floor storage operation, each master carton or pallet and storage position has a discreet identification. To track a master carton or pallet in a floor-stacked storage location, as the master carton or pallet is placed into a storage position, the forklift truck driver hand scans the identification label on both the master carton or pallet

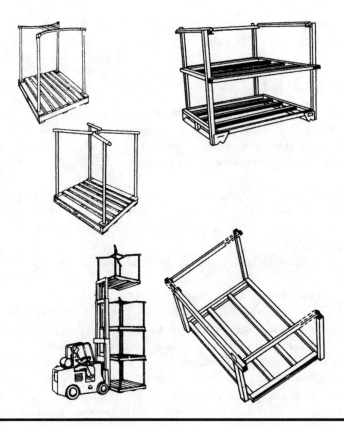

Figure 3.10 Stacking frames and tier racks.

and the storage position. The scans are sent on-line to update the inventory files. If the labels are read by the employee, the inventory files are updated later on.

Portable Containers, Tier Racks, and Stacking Frames

A portable container, tier rack, or stacking frame (portable rack) is a dense storage concept. When master cartons that are placed in a floor storage concept are crushable (not self-supporting or not square), a container, tier rack, or stacking frame makes stackable and uniform unit loads that optimize the cube (air) storage space. Containers and stacking frames have fork openings and a tier rack is placed onto a standard pallet. With the storage devices, the common practice has one master carton (SKU) per unit load lane and stack. Local fire protection codes are reviewed for storage height, sprinkler requirements, and depth restrictions.

In a storage operation, a pallet (unit load) identification is permanently or temporarily attached with a sleeve, tape, or self-adhesive label to a container or stacking frame or tier rack to ensure that the forklift truck driver's hand-held scanner has clear line of sight. In a storage lane, each unit load identification faces a main aisle.

Decked or Hand-Stacked Master Cartons in Standard Pallet Racks

Decked or hand-stacked master cartons in a standard pallet rack bay concept have the same rack components plus a wire-meshed or solid deck in the rack bay. Solid deck material is metal, wood, pressed board, or plywood. To assure minimal deck material bow, front-to-rear or cross-members are placed on or between load beams.

The rack bay may be divided, with one pick position facing one pick aisle and the second pick position facing another pick aisle. The barrier in the center of the rack is made of wood, rope, or pipe. Alternatively, all pick positions in the rack bay face one pick aisle.

Decked or hand-stacked master cartons in a standard pallet rack bay are used for slow-moving, lightweight, and small-cube SKUs. Pick position replenishment is done by an employee hand-stacking cartons onto a deck.

The concept is appropriate for medium SKU hit concentration and hit density, provides medium number of faces per aisle, additional time for pick position replenishment, and limited master carton quantity in a position.

If a decked rack concept has master cartons stacked, during the identified master carton deposit transaction activity, the first identified carton is placed at the stack bottom or the deepest into a rack position. To complete an identified master carton pick transaction activity, the bottom or the deepest identified (oldest or by manufacturer lot number) is the suggested master carton.

Multiple Levels of Shelves or Slotted Angle Iron Shelves

A standard shelf or slotted angle shelf concept has upright posts, shelving or decks and bracing, and connecting components. It is designed with open or closed sections and single or back-to-back rows (see Figures 3.11 and 3.12). The open section has cross bracing along the sides and back and the closed section has solid sides and a solid back section. The slotted angle shelf concept is an open section.

A shelf or slotted angle iron concept is used to handle non-stackable cartons. This concept has limited application in a carton storage operation, handles minimal weight, provides fair SKU hit concentration and hit density, cost is low, and provides a small carton quantity in the pick position.

If the shelf or angle iron shelf concept is one master carton deep and high, a standard program is used in the storage and pick concept. If the shelf or angle iron shelf concept has stacked or multiple deep master cartons, the inventory control program requires an enhancement or a directory concept.

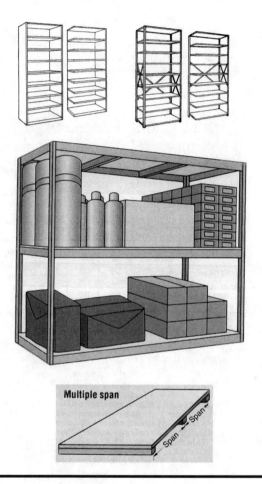

Figure 3.11   Shelves and hand stack concepts: slotted angle shelves, deck material, closed and open shelves, and deck.

Standard Pallet Racks

A standard pallet rack concept has one, two, or three pallets (see Figure 3.13) in a rack opening, level, or bay. The number of pallets per rack bay is determined by the actual pallet width and load beam length. Usually pallets in the first vertical bay are placed on the floor and in the other vertical bays, pallets are placed onto load beams. In a WA forklift truck operation, there is a 3- to 6-in. clearance between a pallet and upright pallet, between pallet and pallet and between pallet top and load beams. If a NA straddle forklift truck is used in a storage rack concept with three pallets or if pallets are placed in the rack opening length-wise (48 in. down aisle), the bottom rack bay is raised above the floor onto load beams. This feature provides clearance for the forklift truck straddles to travel under the load beams. If

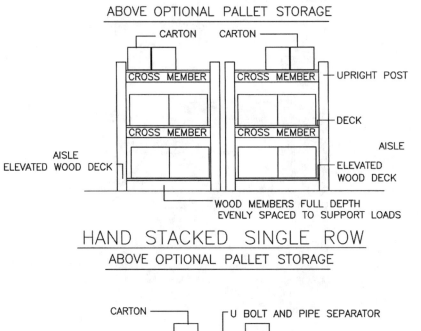

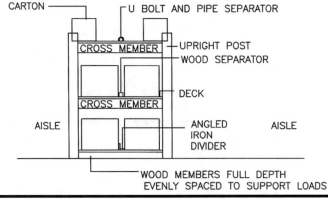

Figure 3.12 Hand stack separators.

a straddle forklift truck is used with the pallet on the floor in the 40 in. dimension, at least 5 in. is allowed between the pallet and rack upright post or pallet to pallet. In a VNA forklift truck operation, both options are considered for a rack and forklift truck design.

A rack bay has two vertical upright frames and for each rack bay above the floor two load beams support the pallets and are attached to upright frames. At the minimum, one load beam pair is at the first level and a second load beam pair is at the top of the upright frame, which provides good stability. Many conventional

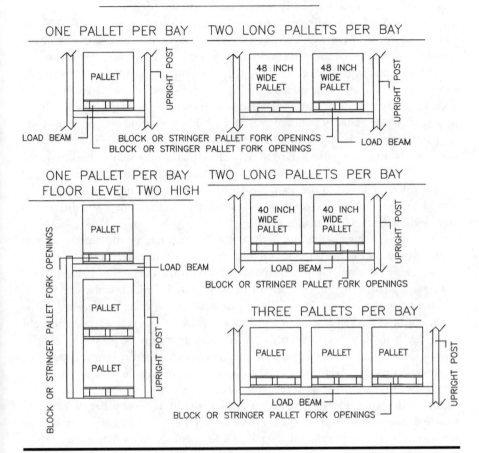

Figure 3.13 Pallet rack concepts: high pallets on the floor, one pallet high, two long pallets wide, and three short pallets wide.

rack installations are three or four pallet bays high. In a tall building, rack bays are at least four or six levels high and an AS/RS rack is ten or more pallet bays high.

Upright frames are designed by a manufacturer to hold all rack bay pallet weight plus rack member weight. A load beam pair is designed to hold a rack bay pallet weight. It is common for an upright frame post and the load beam connection method to permit 2 to 3 in. of adjustability in vertical height. A standard upright frame width is 4 to 5 in. and faces a forklift truck transaction aisle. A standard load beam vertical height is 4 to 6 in. For two pallets wide, the length is a nominal 8 ft. and the solid surface faces a forklift truck transaction aisle.

In a pallet storage facility, a standard pallet rack concept is designed as single rows or back-to-back rack rows. A pallet position utilization factor is 85 percent. With access to all pallet positions, pallet rotation is FIFO and it handles a medium to high volume. Whenever possible, all building columns and fire sprinklers are located in a flue space (8 to 12 in. open space) between back-to-back rack rows or a wall. If required, a building column is designed to occupy one pallet position in a rack bay, but not in an aisle. With an aisle between each rack row, most conventional standard rack layouts have good density but excellent pallet accessibility.

During rack installation, the upright frame bottom diagonal and horizontal support structural member intersection faces the aisle. If a forklift truck impact occurs against a frame post, this arrangement allows the bracing to withstand the impact. To meet stability requirements and seismic conditions, the back-to-back upright frames are tied together and single rows along a wall are tied to the wall or overhead. With heavy pallets, upright frame base plates are widened to disperse the weight. When a standard pallet rack is installed in a facility, the first front upright frame post is anchored to the floor at each row start and the next anchor is on the second upright frame post rear. The anchor pattern ensures rack stability but is approved by the rack manufacturer.

In a pallet storage operation, if there is concern about a pallet becoming dislodged from a rack bay, a solution is to have front to rear members installed in each pallet storage position. Front to rear members span the open space between load beams and provide support for the pallet bottom exterior deck boards.

An alternative to secure the pallet in a rack bay is to use the load beam and the pallet bottom exterior deck boards to lock the pallet in the rack bay. The upright frame depth and load beams are equal to the pallet depth less the pallet two exterior bottom deck boards depth. After a forklift truck places a pallet into the rack bay, the two bottom deck boards are on the two load beams outside. This arrangement has the two or three stringers of the pallet resting directly on two load beams. Since the deck boards are at least ½ in. high, in a rack bay this reduces pallet forward and reverse movement. This arrangement does not interfere with forklift truck storage transaction performance due to the pallet fork openings being maintained at the standard height.

The vertical rack bay is made up of upright frames that have:

■ Two upright posts with a base plate under each post. To hold four pallets in a rack bay, each upright post is 4 to 5 in. wide and has holes for load beam attachment. The upright post has sufficient width to hold a pallet storage position identification.
■ Horizontal brace members between upright posts with connection members.
■ Diagonal brace members between upright posts with connection members.

Upright frames are designed to structurally support a projected total pallet weight on both frame sides that is determined by the rack manufacturer from the design parameters. The upright frames can be straight-legged or canted. With the straight upright post design, the front or aisle upright post is a straight structural member from the base plate to the frame top. The cant leg upright post design has the aisle post from a location interior to the rack position extend on an angle upward and outward toward the aisle and connect with the straight upright frame post. In a WA forklift truck operation, the cant leg design provides additional clearance that allows the forklift truck driver to make a right-angle turn without hitting the upright frame leg with a pallet or a counterbalanced forklift truck rear chassis. Pallet storage identification is placed on the flat surface of an upright post that faces the aisle.

Basic designs for the metal upright frame post are C-channel, round shaped, and H or structural shaped.

Load beam designs are step, structural, or H and rectangular or box that are (a) flat front that has 4 to 5 in. height with a smooth surface and (b) indented front with an overall 4 to 5 in. height with a 2 to 3 in. height with a flat and smooth surface.

A flat and smooth surface has sufficient space for a pallet storage position identification or stacked pallet storage position identifications. Each rack manufacturer has holes that are punched into a upright frame post on 2- to 3-in. centers for load beam connection and possible future rack opening height elevation changes. The hole designs are unique and make it difficult to mix one set of manufacturer racks with another.

Load beams provide a pallet support structure. From pallet weight and pallet number per bay design parameters, a rack manufacture designs a load beam internal and center line to center dimensions, metal guage, and vertical height to support design weights. The load beams are connected to the upright frame using bolts, washers and nuts, clip-ons, and safety locks.

In designing a pallet storage area, the correct rack dimensions ensure that the desired storage design fits into the building and that forklift truck operators meet projected transaction productivity. Important dimensions to consider are rack depth, height, and length.

The rack depth is determined by the pallet stringer dimension that you plan to have supported by load beams and to permit your forklift truck to complete a transaction. In a standard one- or two-deep rack design, the rack depth dimension is 4 to 6 in. less than the pallet stringer dimension.

Rack height is the height of the upright frame or post that is sufficient to support a top pallet and to permit sufficient horizontal and diagonal brace members and as-required overhead ties. In a standard pallet rack design, an upright frame height extends 2 to 6 in. above the top load beam. The additional upright post height serves as a guide for the lift truck driver to complete a pallet transaction. In a drive-in or drive-through rack concept, the front or aisle upright post extends upward and is tied at the top. If a standard rack installation is in a seismic location, per the local code the upright frames extend above the top pallet position to top tie the rack structure.

Rack load beam length dimensions are internal dimension or ID, and center line to center line or C/C dimensions.

The rack bay opening dimension horizontally holds pallets and permits a fork-lift truck to complete a transfer transaction. A load beam length or ID dimension is the rack length that handles the combination for the width or fork opening for the storage pallets and the required clearances or open space. Another internal dimension definition is the open or clear space between two upright frames or posts.

A centerline to centerline dimension is the rack length that is a load beam length and one upright frame post or width combination. To calculate a storage aisle length, use a centerline to centerline dimension plus one upright post or frame width. To determine a facility aisle, count the number of rack bays or openings and multiply by the centerline to centerline dimension plus a nominal 3 to 5 in. post width. With a tall rack, rack-supported facility, or heavy loads, the rack post has a minimum 5 to 8 in. width. To calculate the storage area aisle with the load beam length or internal dimension (ID), the storage area aisle is short by the upright post width in the aisle length.

A standard pallet rack position has two load beams per elevated rack bay connected to each upright frame post. Per the design parameters, load beams support the pallet number and pallet load weight. In a storage operation, a standard flat load beam face has a 4- to 5-in. high flat and smooth surface and an indented load beam has an overall 4- to 5-in. dimension with a 2- to 3-in. high flat and smooth surface. In a WA or NA man-down VNA forklift truck operation, load beams have pallet position identifications that are attached to the face with hand-held scanner line of sight.

If a pallet storage operation is a tall warehouse (above 30 ft.) and a change in pallet vertical height is not expected, in your rack specification state that a load beam spans two or three rack bays. This unique load beam design has a load beam attached to an upright post face or aisle side. For stability, load beams are staggered (front load beam spans two bays and rear load beam spans three bays). Compared to standard upright frame depths, the upright frame depths are narrower. In comparison to standard rack installation costs, this arrangement has the potential to reduce rack material and installation costs. With a staggered load beam feature, if a rack bay vertical height is adjusted, an entire rack row requires adjustment.

Two-High Pallets on the Floor Level

If a storage operation has a large number of pallets for one SKU, pallets with the same SKU are stacked two high on the floor. When a standard, drive- in, or drive-through rack design has two-high pallets on the floor, the rack bays are designed for one pallet height. This rack feature is required in the rack specifications. With this arrangement, the upright frames and upright posts are manufactured with additional strength for the increased bottom-level vertical height. This installation design is approved by the rack manufacturer. Rack post protectors are placed on the upright frame post aisle side to reduce forklift truck damage.

In a two-pallet-high stack in the bottom rack position with a WA or NA forklift truck operation, pallet position identification locations are on the upright frame and additional pallet storage position identification labels are placed onto an upright post. Pallet storage position identifications face the aisle and colored labels and directional arrows are used in the pallet position identification to minimize forklift truck problems. As a forklift truck driver faces the rack bay, the left-hand pallet position to the upright post left location has the pallet storage identification, and the right-hand pallet position to the upright post right location has the pallet storage identification. Since there is one upright post with identified pallets on each side, pallet position identifications do not have to have pallet storage position identifications adjacent to one other. This pallet storage position identification adjacent arrangement could create no hand-held scanner reads or an incorrect scan.

Double-Deep or Two-Deep Racks

A two-deep or double-deep rack concept (see Figure 3.14) is considered dense storage with a two-deep NA forklift truck concept. A two-deep pallet rack components and design characteristics are similar to those for a standard pallet rack concept, with a few exceptions. A rack upright frame is determined by total pallet weight. To design flexibility to reuse the upright frames at another location, options are to use one long frame with four load beams and two upright frames; or a standard upright frame with four standard load beams and four upright frames.

A two-deep rack concept has the same SKU for each interior and exterior pallet position and has an 85 percent utilization factor for the interior pallet position and 70 percent for the exterior pallet position. A double-deep rack row design has one rack bay that is placed directly behind another rack bay. The space between exterior and interior pallet positions varies between 1 and 4 in. determined by a forklift truck stroke or fork extension. A two-deep rack design allows a rack opening to hold four conventional pallets. A two-deep rack concept has a two-deep reach forklift truck that has a stroke or set of forks with a 43 to 51 in. forward extension.

In a two-deep rack concept, the first pallet deposit is made to an interior pallet storage position and the second pallet deposit is made to the aisle or exterior pallet storage position. The first pallet withdrawal transaction is made from an aisle or exterior pallet position and the second pallet withdrawal transaction is made from the interior pallet position. This feature and the time required to extend the set of forks provides a LIFO product rotation and the ability to handle a medium volume. With four pallets per rack bay, a two-deep rack concept provides medium position density and fair pallet accessibility. The two-deep rack designs are known as up and over, or raised above the floor, and bottom pallets set on the floor.

In an up-and-over or raised-above-the-floor concept, the bottom rack opening is raised above the floor. An open space permits forklift truck straddles to pass under the bottom load beams and to turn in a narrow aisle. The aisle allows easy pallet transactions at any rack level due to the forklift truck straddles not being required

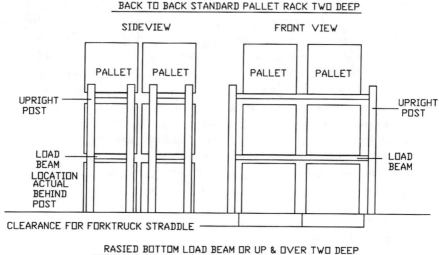

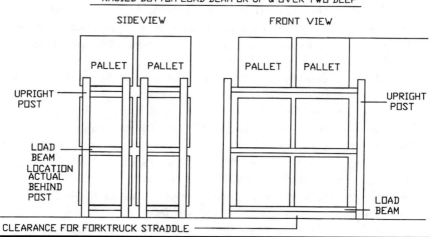

Figure 3.14 Two deep pallet rack: standard and up and over.

to straddle the floor-level pallet. A raised bottom pallet level concept handles pallets with narrow widths or pallet widths that exceed an open distance between the forklift truck straddles and with a narrower rack opening but increases the rack cost, height, and building height.

In the bottom pallets set on the floor concept all pallets in the elevated rack positions are in direct alignment with the floor pallet. This feature reduces forklift truck operator transaction productivity due to the additional time that is required to line up a forklift truck straddles between the pallets. This concept reduces the rack investment, rack height, and building height, but provides a wider rack (load beam length) and fewer pallets per aisle. A two-deep pallet storage concept is a dense storage concept that has the same SKU in the interior and exterior pallet storage positions. To complete pallet storage position identification and pallet identification scan transactions, the forklift truck driver substitutes the indicated identified pallet (same SKU) for another identified pallet (same SKU) or uses the floor stack and pallet identification location directory concept.

To assure proper pallet storage position identification, the pallet position identifications are placed onto the load beams that permit line of sight, or onto upright posts. Color labels or directional arrows are used on the pallet position identification face. In a two-deep storage rack concept, the pallet position identifications are placed on the load beams or upright posts that are adjacent to the aisle and the identification faces the aisle. As the forklift truck driver faces the rack bay, a left-hand pallet position to an upright post left location has a pallet storage identification and a right-hand pallet position to an upright post right location has a pallet storage identification.

Bridge Racks or Bridge across the Aisle

A standard pallet rack concept is designed to bridge the pallet operation (see Figure 3.15) and interface with a WA or NA forklift truck aisles and standard pallet racks. This is considered a storage pallet design that takes advantage of the total available space within a facility and has same rack design and operational characteristics as a standard pallet rack concept. Load beams with front-to-rear (cross) members span an aisle and are connected to the appropriate rack row upright frame that forms a bridge across an aisle. For each rack bay the bridge across the aisle provides one or two additional pallet openings. A very important bridge design consideration is to allow sufficient space for the forklift truck mast (collapsed) or clearance between the floor and bridge load beam bottom. Due to pallet storage position identification line of sight problems color labels and directional arrows minimize forklift truck problems. Identified pallet and pallet storage position scan transactions are similar to standard pallet rack scan transactions. As the forklift truck driver faces the rack bay, a left-hand pallet position to an upright post left location has a pallet storage identification, and a right-hand pallet position to an upright post right location has a pallet storage identification.

Drive-In Racks

A drive-in rack concept (see Figure 3.16) handles non-self-supporting master cartons on pallets. Drive-in racks have pallet rack storage lanes and are a dense storage concept. With drive-in racks storage lanes are used for one SKU. The rack components are upright frames, upright posts, support arms, guide rails, support rails and side, top and back bracing. The drive-in rack bottom lane has the bottom pallets set on the floor and the elevated storage lane pallets are set on rack structural members. Each pallet position level has two support arms and rails. Support arms are attached to each upright post. A drive-in rack lane holds two to ten pallets deep and three to

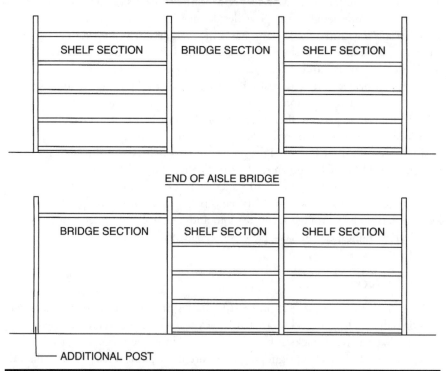

Figure 3.15 Aisle bridge.

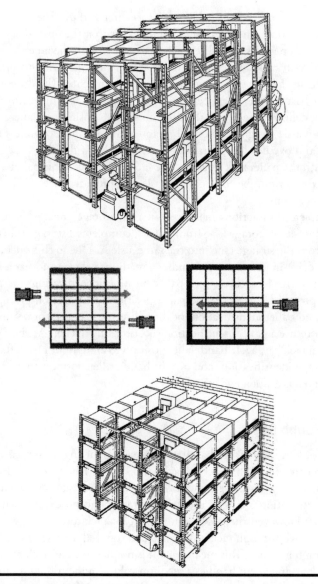

Figure 3.16   Drive-in rack and drive-through rack.

four pallets high per storage lane. This rack layout provides medium to good storage density, but poor pallet accessibility. Product rotation is LIFO. Drive-in racks are designed with a single row or back-to-back rows. A SKU that is in the storage lane floor level or aisle position is in rack storage lane pallet positions. This feature provides a 66 percent pallet utilization factor. In a drive-in rack design, the distance between a rack second-level storage lane and the structural members and floor have sufficient strength and height to permit a WA or NA forklift truck overhead guard or collapsed mast and as-required straddles to enter and exit the floor-level storage lane. The second drive-in rack design parameter is that the open distance between a lane's two rails permits a forklift truck mast to travel in the lane.

For a drive-in rack concept, the number of pallets is varied to place building columns within the flue of the back-to-back drive-in space. The drive-in rack storage lanes are best designed between two building columns. This feature means that the building columns are not in the drive-in storage position or lanes. In all drive-in rack down aisle designs, structural members require fire sprinklers and forklift truck mast and overhead and straddle clearances. Most drive-in rack concepts are designed with the pallet fork opening side facing the aisle. This pallet arrangement has the maximum number of faces per aisle and provides excellent pallet stability in the rack positions.

To complete transactions in a drive-in rack storage concept, a forklift truck enters the floor-level storage lane from aisle A, completes a storage transaction, and backs out from the storage lane into the same aisle A. Due to this operational characteristic, a drive-in rack concept handles a medium volume and has a LIFO product rotation. Drive-in rack concept pallet position identifications are placed onto an upright post and have the same pallet position identification arrangement as a floor stack and pallet identification location directory concept or permits a forklift truck driver to use a pallet identification substitution concept. As the forklift truck driver faces a rack bay, a left-hand pallet position to an upright post left location has a pallet storage identification, and a right-hand pallet position to an upright post right location has a storage identification.

Drive-Through Racks

A drive-through rack concept handles non-self-supporting master cartons on pallets. With a drive-through concept, racks and storage lanes are used for one SKU. Drive-through racks have the same rack components, design characteristics and a 66 percent utilization factor as the drive-in rack concept, except that there is no back bracing. However, there is a requirement for rack top bracing. With no back bracing, the drive-through rack is designed as a stand-alone rack row with a forklift truck aisle on both sides. This means that it is not designed as back-to-back rows. A drive-in rack arrangement handles a medium volume and product rotation is LIFO or FIFO. With a LIFO product rotation, a WA or NA forklift truck enters a rack lane from aisle A, completes a storage transaction and backs out into aisle A. With a FIFO product rotation, a forklift truck enters a rack lane from aisle A, completes a storage transaction to the elevated lane and drives through the rack lane into aisle B. After all elevated lanes are full, the floor-level lane is handled as a drive-in rack lane. An alternative procedure in a FIFO product rotation is to exit a storage lane by driving without a pallet through a storage lane and exiting into aisle B. To retrieve a pallet, a forklift truck enters a rack lane from aisle B, retrieves a pallet and backs into aisle B.

A drive-through rack has medium storage density and poor pallet accessibility. Pallet position dimensions have the same characteristics as these of drive-in racks. A drive-through rack concept has pallet position identifications on the upright post. As a lift truck driver faces the rack bay, a left-hand pallet position to the upright post left location has the pallet storage identification, and the right-hand pallet position to the upright post right location has the pallet storage identification.

Mobile or Sliding Shelves or Racks

A mobile or sliding rack concept (see Figure 3.17) is similar to a standard shelf or rack concept, except that it is a dense storage concept, requires fewer employee or forklift truck aisles, and shelf or rack rows move to create an employee or forklift truck aisle.

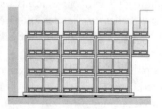

Figure 3.17   Mobile or sliding rack.

Mobile shelves or racks are standard single-depth shelf or pallet rack rows or back-to-back rack rows that are placed onto movable bases. At the end of a module are fixed shelves or rack rows. All shelves or rack rows are placed perpendicular to the main traffic aisle. The mobile shelf or rack is designed with a nominal six back-to-back movable rows, one employee or forklift truck aisle, and at the end of each movable section is a single deep shelf or rack sections or rows. A mobile shelf rack concept has a ratio that is four to five master cartons or pallets high to one master carton or pallet deep. If sprinklers are required in shelves or racks, the mobile shelf rack manufacturer designs sprinklers in the concept.

For access to a shelf or pallet position, the mobile rack moves to the side and creates an employee, WA, or NA forklift truck transaction aisle between the required rack rows. An employee or forklift truck enters an aisle, performs the required storage transaction and exits into the main traffic aisle. After the transaction, as-required mobile shelf or rack sections are moved to create a new employee or forklift truck aisle between two different rack rows. Sensing devices on the movable shelf or rack base bottom section sense an object or employee in the mobile shelf or rack section travel path and if there is an object then the mobile shelf or rack section stops its movement. This feature prevents the movable shelf or rack from causing equipment damage or employee injuries.

With one access aisle, the mobile shelf or rack provides high pallet position density and good accessibility. The mobile shelf or rack concept has an 85 percent utilization factor and handles a low to medium volume due to slow shelf or rack movement to create an employee forklift truck aisle. An inventory control program and batched transactions per row improve employee or forklift truck transaction productivity. With a mobile or sliding shelf or rack concept to assure employee, WA, or NA forklift truck line of sight to pallet position, identifications have the same arrangement as standard pallet racks. A master carton or pallet storage position identification location, and master carton or pallet identification and master carton or pallet storage position identification scan transactions are the same as standard shelf or pallet rack concept.

Gravity Racks

A gravity or flow-through rack concept (see Figure 3.18) is designed as a single or stand-alone rack concept that has aisle A for pallet in-feed and aisle B for pallet out-feed. Each pallet flow lane is allocated to one SKU. This means that for a four-level gravity flow rack, options are one SKU per pallet flow lane or one SKU in all pallet flow lanes in a stack.

The pallet gravity flow rack concept has upright frames, upright posts, braces, brakes, end stops, and skate wheel or roller conveyors that make individual flow or pallet storage lanes. In a WA or NA conventional forklift truck operation, the pallet flow lanes are three or four levels high. In a VNA forklift truck or AS/RS crane operation, pallet lanes are designed five to eight pallets high. Pallet weight and

Transportation pallet:
Specially designed
polyethylene compliant
runners are attached to a
customer specified deck

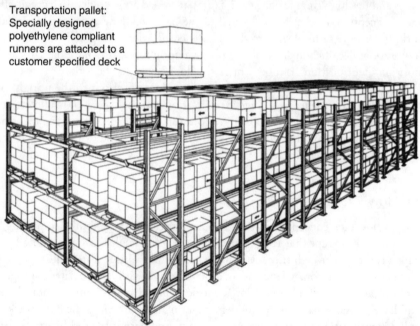

Figure 3.18   Gravity flow rack and air flow rack.

height determine the slope and pitch to the flow rack lanes. A pallet unit height and length ratio is 3 to 1. If pallets exceed this ratio, there is a potential for uneven flow or hang-ups in a storage lane. A gravity flow rack concept is designed with 3 to 20 pallets per lane. In most pallet flow rack systems, to ensure smooth flow through a storage lane, a pallet is placed onto a slave pallet. In some flow rack concepts, conveyor rollers have flanged wheels that act as guides for a pallet as it flows through the rack. To prevent rack damage, lane entry guides, upright post protectors, forklift truck stops, and sufficient WA or NA forklift turning aisle widths are recommended at the entry and exit positions.

A pallet gravity flow rack concept has a WA or NA forklift truck that permits access to a flow rack deposit side. After a forklift truck in aisle A places a pallet onto a conveyor, gravity and the pallet weight on rollers allow the pallet to flow through the storage flow lane to the storage lane exit end. The pallet is removed by a forklift truck in aisle B from the storage lane exit end. This activity permits the next pallet in the lane to flow, move or index forward to the withdrawal position. In a long concept, to reduce line pressure and product damage, pallet brakes and a pallet separator at the exit position are installed in the flow lane.

The pallet gravity flow concept indexing a pallet movement from the deposit (entry) position to the withdrawal position allows each storage flow lane to accommodate one SKU per storage lane. This feature permits gravity flow racks with two aisles and has high storage density and fair pallet load accessibility. In a gravity flow rack concept, as a pallet is transferred to the flow rack lane entry end, the charge forklift driver has line of sight to the pallet identification. The gravity flow rack pallet position identifications are placed onto an upright post and have the same pallet position identification arrangement as the floor stack. As the fork-lift truck driver faces the rack bay, the left-hand pallet position identifications to the upright post left location have the pallet storage identification and the right-hand pallet position identifications to the upright post right location have the pallet storage identification. At the charge or exit end, the forklift driver double handles an identified pallet to have the pallet identification face the forklift driver's hand-held scanner.

Air Flow Racks

The air flow rack concept operational characteristics are very similar to pallet gravity flow racks except that the pallets are indexed through the storage lane by air, which is forced through tiny holes in the pallet rails. This feature requires a higher pitch or slope to a storage lane and an air compressor with its associated piping. The air flow rack concept pallet position identifications are placed onto the upright post and have the same pallet position identification arrangement as the floor-stack and pallet identification location directory concept and permit the forklift truck driver to use a pallet identification substitution concept. As the forklift truck driver faces the rack bay, a left-hand pallet position to the upright post left location has a pallet storage identification, and a right-hand pallet position to the upright post right location has a pallet storage identification.

Push-Back Racks

A push-back rack is a stand-alone rack concept with one aisle. A push-back pallet rack concept (see Figure 3.19) has the same rack structural components and design characteristics as a pallet gravity flow rack with some exceptions. A push-back rack concept is designed as a single rack row that is installed along a building wall, in a building location that permits one aisle for a forklift truck to perform entry and withdrawal transactions or as back-to-back rack rows. The concept is designed to hold three to four pallets deep and three to four loads high with one SKU per lane. To complete a pallet transaction, a forklift truck places a pallet against an existing pallet that is in the storage lane aisle position. When the existing pallet is sufficiently pushed back into the storage lane, it creates the required new pallet space. To withdraw a pallet, a WA or NA forklift truck slowly raises the pallet 2 to 3 in. high above the storage lane conveyor and backs out the pallet from the storage lane. As the pallet is removed from the storage lane, gravity moves the next pallet into the storage lane exit or aisle position. This feature permits 66 percent lane utilization and a LIFO product rotation. The push-back rack provides good storage density and fair accessibility and handles a low to medium volume. Each lane handles one SKU. The push-back rack concepts can be standard or telescoping. In a standard pallet push-back flow rack, the conveyor does not require brakes but requires end stops at both ends of the flow lane. A telescoping carriage push-back rack concept has pallet carriages that ride on a set of tracks. The carriage design permits the carriage that travels into the interior pallet position when not in use (without a pallet) to nest over the empty carriage that is adjacent to the aisle.

In a push-back concept, the pallet position identifications are placed onto the upright post and have the same pallet position identification arrangement as the floor stack and pallet identification location directory concept or permit a forklift truck driver to use the pallet identification substitution concept. As the forklift truck driver faces a rack bay, a left-hand pallet position to the upright post left location has a pallet storage identification and a right-hand pallet position to the upright post right location has a pallet storage identification.

Cantilever Racks

A cantilever rack concept is designed to handle long pallets with SKUs such as pipe. A cantilever rack has upright posts, support arms, legs, and braces. At several levels sufficient space is designed between pallets and metal members for fire sprinklers. With solid or wire-mesh decks on the support arms, the cantilever rack is designed to handle all pallet types. The cantilever rack is designed as a single arm row or double arm rows. The arms extend outward from the upright post and create pallet

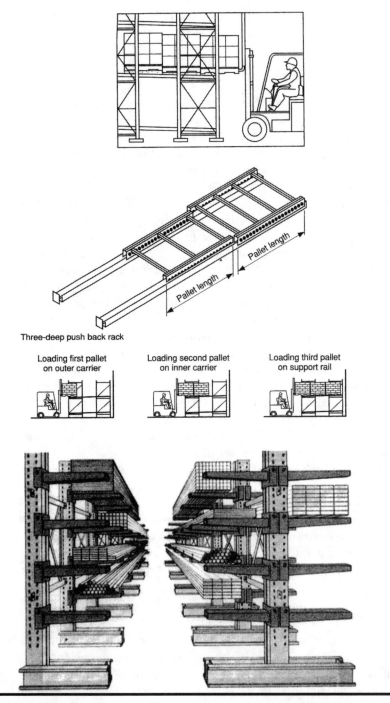

Three-deep push back rack

Loading first pallet on outer carrier

Loading second pallet on inner carrier

Loading third pallet on support rail

Figure 3.19 Standard push-back rack, carriage push-back rack, and cantilever rack.

positions. Pallet positions are serviced by a WA or NA forklift truck. A cantilever rack permits a FIFO product rotation, excellent storage density, excellent product accessibility, and 85 percent utilization. With a cantilever rack concept to assure WA or NA forklift truck line of sight to a pallet position, identification has the same procedure as standard pallet racks.

AS/RS High-Rise Racks

An AS/RS high-rise rack facility has very tall rack rows and an aisle between two rows. The rack and aisle design uses a vertical air space rather than a horizontal footprint. A high-rise rack concept is designed in a conventional building or in a rack-supported structure. A rack-supported structure has rack upright frames and posts as the support and attachment members for in-rack sprinklers, walls, or skin, and the roof. These systems require a dead level or F75/F100 level floor and a vehicle guidance system. It is good practice, prior to rack installation, to assure with a laser light beam that the floor is level. The structure is designed for the snow, seismic, and wind forces, fire protection (walls, barriers, and sprinklers) and AS/RS crane mast clearance at the top of the racks. In an AS/RS high-rise facility, all pallets pass through a size and weigh station and a scanner to verify that pallet dimensions and weight conform to the system design standards and have a readable pallet identification. After passing a pallet identification reading device, the identified pallet transactions are controlled by MHS and the MHS communicates and updates the inventory files. With this feature the AS/RS rack positions do not have rack position identifications. The high-rise rack facilities are designed with single deep racks, two deep racks and flow racks as part of the structure.

A high-rise single deep rack openings can be standard or four-arm openings. A standard AS/RS pallet rack bay has two or three pallets per set or load beam pair. With this rack opening concept, an AS/RS crane has one set of forks. To complete a transaction, the crane forks enter the pallet fork openings as with a normal forklift truck. In a typical high-rise facility, the first pallet level sets 30 to 36 in. above the floor. Advantages are fewer upright posts per aisle, greater face number per aisle, and shorter building. The four-arm pallet openings has four arms and two pallet rails that support one pallet per opening. This four-arm rack opening has a captive or slave pallet and an AS/RS crane has a set of platens that go under the pallet to complete a storage transaction. The first rack level is approximately 17 to 24 in. above the floor, but at each rack level an additional 3 to 6 in. clearance is designed for the platens between a pallet top and the next pallet position level support arms.

A standard pallet position is on pallet support arms or load beams that are support by two upright posts. An additional pallet position has additional structural support members that are attached to the upright above the P/D station. Basically, these additional pallet positions are similar to a cantilever rack position that has AS/RS complete pallet storage transactions.

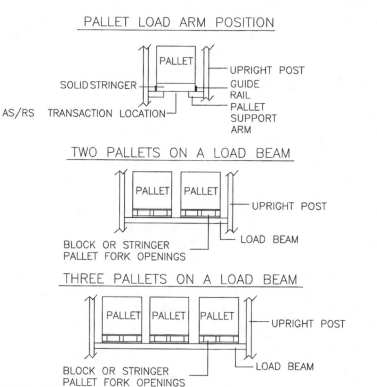

Figure 3.20 Pallet AS/RS rack, arm rack, two-wide pallet rack, and three-wide pallet rack.

As an identified pallet travels to an AS/RS crane storage area, barcode scanners or RF tag readers along a pallet travel path to an AS/RS in-feed station reads each pallet identification, and in the AS/RS storage area, the MHS have assigned a pallet to a pallet storage position. After the AS/RS crane places the identified pallet into a pallet storage position, MHS update the inventory file.

Car-In or Mole

A car-in rack concept has upright frames, posts, car rails, pallet support rails, car-in rack vehicle, slave pallets, and an aisle travel vehicle. Sufficient space is designed between a pallet in rack members or storage lane for fire sprinklers and a car with an elevated pallet travel in a storage lane. A car-in rack concept is designed with 3 to 8 pallet levels high and 10 to 20 pallet positions per storage lane. Each storage lane has one SKU. From an aisle P/D station, an AS/RS crane carries a car-in rack

Figure 3.21 Attached to the front posts.

vehicle between rack rows to the storage lane and performs a pallet storage transaction. As a WMS-identified pallet travels to an AS/RS car-in or mole storage area, barcode scanners along a pallet travel path read each pallet identification, and in an AS/RS storage area, the MHS have assigned a pallet to a pallet storage position. After the AS/RS crane places an identified pallet into a pallet storage position, MHS updates an inventory file.

In a one-aisle arrangement, the car-in rack vehicle leaves the AS/RS crane from aisle A, enters and travels into a storage lane, arrives at the required pallet storage lane position, performs a pallet deposit transaction, and exits the storage lane to the same aisle A onto the AS/RS crane. This arrangement provides a LIFO product rotation, 66 percent pallet position utilization, and handles a low to medium volume. The one-aisle car-in rack concept provides high pallet density and excellent security, but poor pallet accessibility.

A two-aisle car-in rack concept requires two AS/RS cranes and two car-in rack vehicles. In the two-aisle concept, from the inbound aisle A, all pallet deposits are made by the crane and car-in rack vehicle on that aisle., From an outbound aisle B, all pallet picks are completed by the aisle B crane with its car-in rack vehicle.

A two-aisle car-in rack concept provides FIFO product rotation, excellent security, and high pallet density, but poor pallet accessibility. If pallets in a storage lane are indexed forward by a car-in rack vehicle toward the withdrawal aisle, a two-aisle car-in rack concept has an 85 to 90 percent pallet position utilization factor and handles a higher volume. With a two-aisle concept an identified pallet is withdrawn in the same sequence as entered into the inventory file and pallet identification is not oriented for forklift truck pick-up. To assure a pallet identification has the correct orientation, a conveyor travel path has a pallet turntable that turns a pallet to have the pallet identification in the correct direction of travel for a forklift truck to hand scan a pallet identification.

Sort Link

A sort link concept is an automatic or computer-controlled concept, with a captive or slave carrier and travel path, rail-guided horizontal four-wheel pallet cart and travel path, multiple storage level facility, vertical cart carrier, dense cart storage lanes, and computer controls.

The slave carrier is manufactured from hardened, coated steel and is designed to handle a pallet. Under each metal frame corner is a flanged wheel. The flanged wheels move the carrier in two travel directions, between the pallet carrier and the storage lane or discharge lane. To improve the ability to move carriers, a carrier lead end has a special coupler and hook device. When one carrier hook is engaged in another carrier coupler, the motor-driven parent device moves one or four carriers between two locations. The carrier is moved over a dual rail guide path that interfaces with the carrier flanged wheels. The rail guide travel path assures that the carriers are in a straight line.

Each storage facility level is designed to take a slave pallet carrier on board. The four-wheel devices are motor driven to travel over the rail guide path and, at a pick-up location, to take an inbound carrier on board. In the storage area, the hooking device engages or disengages the carrier. A guide rail travel path is a straight travel path that runs the entire length of the storage facility. Depending on the facility design and storage lane configuration, there is one or several guide paths to serve each facility level.

The storage facility has one level for in-bound and out-bound transactions and required pallet staging area. The facility floor provides the guided travel path for the required four-wheel devices to complete a pallet carrier transfer to the storage lane. As required each floor level is serviced by a vertical in-house pallet carrier device.

The number of faces and the depth of the storage lanes is determined by the land and building design parameters. On the proper floor level, a forklift truck performs transfer transactions to feed pallets to the storage facility or to pick a pallet from the storage facility to complete a customer order. The vertical cart carrier is an in-house transport concept that moves the pallet carrier between the computer-assigned floors.

A dense cart storage lane has two flanged wheel rails per storage lane. The storage lanes hold the pallet carriers. As required by the operation, the motor-driven guided carrier moves a pallet carrier from one storage lane to another. This activity is performed to provide access to the assigned pallet or to increase the storage space utilization. Computer controls and inventory control ensure smooth and constant pallet carrier movement through the facility, assign a pallet to the proper storage area, and track the inventory. As an identified pallet travels to a sort link storage area, barcode scanners along the pallet travel path read the WMS identification on each pallet and, in the AS/RS storage area, the MHS has assigned a pallet to a pallet storage position. After the sort link carrier places the identified pallet into a pallet storage position, MHS update the inventory file. The sort link carrier concept can rearrange the identified pallet position in a storage lane to have a suggested pallet identified withdrawn from a storage area.

Layout Considerations

The factors that determine layout in a storage facility are the need to control operational costs, earn a profit, and satisfy and increase the number of customers. In this section, we will look at:

- SKU popularity or Pareto's law (the 80/20 rule)
- The ABC theory
- The unloading and loading ratio
- Family group
- Product rotation
- Rack row and aisle direction
- Aisle length
- Pallet size
- Value
- Temperature-sensitive products

SKU Popularity or Pareto's Law

When a master carton or pallet storage operation has a layout that is based on SKU popularity (see Figure 3.22), it is based on Pareto's law (after Vilfredo Pareto, 1848–1923; Italian economist). This law states that "85 percent of the wealth is held by 15 percent of the people." In the master carton or pallet storage industry, this law indicates that 85 percent of the volume shipped to customers is derived from 15 percent of the SKUs. Many studies have indicated that another 10 percent of the volume that is shipped to customers results from another 30 percent of SKUs

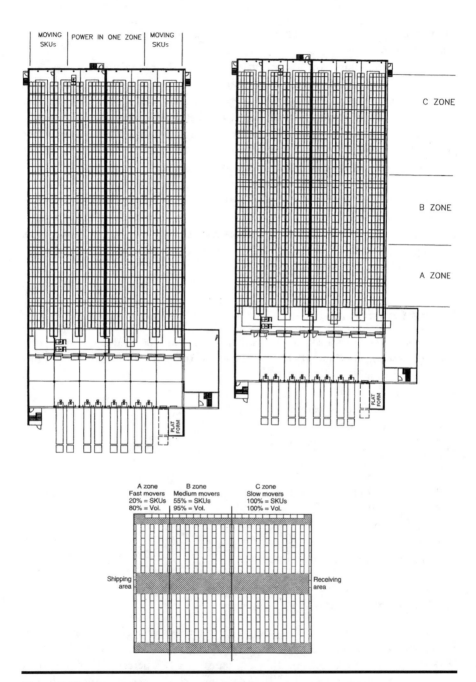

Figure 3.22   Power in one zone and ABC theory.

and that an additional 5 percent of the volume shipped to customers is attributed to 55 percent of the SKUs. In a catalog or direct mail business, 90 to 95 percent of your business is from 5 percent of the SKUs because two to four catalogs are introduced within a year. Each catalog has a wide SKU mix. In recent studies, results show that 95 percent of the volume shipped to the customers is obtained from 55 percent of the SKUs. This is referred to as "Pareto's law revised."

The ABC Theory

The ABC theory is a reference to the three zones of Pareto's law. The A pick zone is allocated to fast-moving SKUs. There are few different types of SKU the and a large inventory quantity per SKU. The B pick zone is allocated to the normally moving SKUs. The number of different SKUs is medium and there is a medium inventory quantity per SKU. The C pick zone is allocated to the slow-moving SKUs. There are a large number of different SKUs and a small inventory quantity per SKU. In a storage operation layout in which the receiving and shipping docks are located at the front of the facility and the SKU location is based on the ABC theory, fast-moving SKUs are located at the front of the facility. If the receiving and shipping docks are located on opposite sides of the facility, the fast-moving SKUs are located by the SKU unloading and loading ratio.

The Unloading and Loading Ratio

The unloading and loading ratio compares the number of trips that unloading employees make with the number of trips loading employees make to handle a vendor delivery. When the number of trips for loading and unloading are equal, SKU master cartons or pallets (positions) are located near the shipping docks or any location in the rack row. When employee unloading trips are more than loading trips, the SKU pallet (position) is located near the receiving docks. This feature reduces the total travel distance for employees.

Family Group

The family group philosophy is dictated by a requirement that SKUs are located in a storage aisle with other SKUs that have the same inventory classification. With this philosophy, by a predetermined criterion, SKUs are assigned to specific locations (areas, aisles, or positions) within a facility. This layout philosophy has facility and racks and aisles concepts that are designed to accommodate SKUs that have similar characteristics: dimensions and weight, components, are located in the same retail store aisle, require the same normal, refrigerated, or freezer conditions, require the same level of security, form either toxic or nontoxic materials, are edible or non-edible substances, flammable or non-flammable materials, stackable or non-stackable products, and crushable or non-crushable packages.

Product Rotation

The master carton or pallet storage rotation philosophy is dictated by the carton life cycle and requirements for specific master cartons to be picked for a customer order. In a FIFO UOP rotation the master carton that is received first is shipped out first. This is appropriate for a UOP that has a predetermined life before it spoils. After a specific date, the SKU is not withdrawn from inventory for COs. An operation that is designed to have SKUs with a FIFO UOP rotation requires access to all master carton or pallet positions in the storage area and assures that the oldest pallet is withdrawn first. Master carton or pallet flow logic and MHS in a standard rack, drive-through two aisle, gravity flow rack, air flow rack, two-aisle car-in rack, or sort link ensures that master carton or pallet with the oldest receiving date is at the withdrawal position. In the inventory files, the program ages the identified master carton or pallets and suggests the oldest identified master carton or pallets are withdrawn first. The two-deep pallet rack and drive-in rack concepts have difficulty in achieving a FIFO UOP rotation.

In LIFO SKU rotation, a pallet that is received last is shipped out first from the facility. This is used with master carton types that do not have a specific shelf life. The storage facility design does not provide access to the oldest master carton or pallet. This feature allows a storage operation layout to use dense storage positions that reduce the required building size. With LIFO UOP rotation and an inventory program, the master carton or pallet position identifications are placed onto the upright post and have the same master carton or pallet position identification arrangement as the floor stack and master carton or pallet identification location directory concept or permits the forklift truck driver to use the master carton or pallet identification substitution concept.

Rack Row and Aisle Direction

With a WA or NA forklift truck operation, the layout is based on the rack rows and aisle direction, that is, the flow of traffic to the pick aisles. With the rack rows and aisle direction parallel to the pick aisles there are at least two turning aisles, at the ends of the storage aisles and rack rows. The aisles lead to the pick aisles. This arrangement increases WA or NA forklift truck travel time between the storage area and pick aisles.

With the rack row and aisle direction straight from the storage area to pick aisles, each storage aisle provides access to a pick aisle and the main traffic aisle serves as a turning aisle for the WA or NA forklift truck. To optimize forklift truck productivity, and perform an identified master carton or pallet transaction in an adjacent aisle, a middle traffic aisle is required in this layout.

Aisle Length

Storage aisles may be either long or short (see Figure 3.23). A short storage aisle layout has rack rows and storage aisles run in the short facility dimension or short width of a rectangular facility. This concept has turning aisles at the end of each rack row and storage aisle. It provides less density and lower employee productivity because of an increased number of non-productive turns at the ends of the storage aisles.

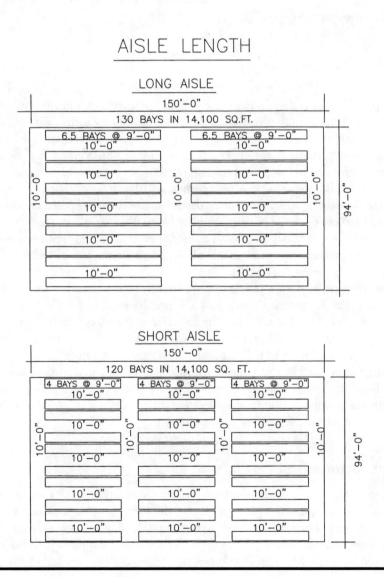

Figure 3.23 Aisle length: long aisle and short aisle.

A long storage aisle layout has rack rows and storage aisles that are arranged to flow in the long direction of the facility or rectangular building. The long storage aisle concept has a cross aisle in the middle of the rack row or storage aisle to provide easy WA or NA forklift truck transfer to another storage aisle. The concept provides greater density and fewer non-productive forklift truck turning aisles.

A WA or NA forklift master carton or pallet facility is designed with 300 to 400 ft. nominal length, to accommodate 75 to 100 pallets. In a VNA forklift truck or an AS/RS facility, the nominal aisle length is 400 to 600 ft., to fit from 100 to 150 pallets, with no cross aisle.

Pallet Size

The pallet storage/pick operation layout determines how high pallets are stacked. The pallet height is dictated by the company specifications and the ability of the vendor to palletize master cartons to a predetermined height. In a very tall storage rack operation, tall pallets (on which master cartons are stacked high) are located at higher pallet positions that compensate for forklift or AS/RS crane clearance. Short pallets are located in low pallet positions to increase the rack structural strength. If an AS/RS pallet storage operation returns partially depleted (short) pallets to the storage area, the hybrid pallet storage design improves storage position utilization.

Tall pallets fit the maximum number of master cartons on a delivery truck, which means fewer dock positions required. Fewer handlings mean less potential damage to product, equipment, and building. There is an increased requirement to secure master cartons on a pallet.

When a pallet storage/pick operation receives pallets that are short pallets, the storage pallet floor stack or rack positions are designed to handle a short pallet. With this concept, the pallet has the fewest master carton number per pallet. A short pallet has the maximum master carton layer number without crushing the cartons on the bottom layer. The short pallet offers a greater number of transactions and trips completed by the storage vehicles, increased potential product, equipment and building damage due to increased handlings, with a short position opening minimum position flexibility and a greater load beam number, which increases the rack investment and lowers the cube utilization.

Manufacturer Lot

In a master carton or pallet storage operation, the ability to track a specific manufacturer lot or production run is similar to the FIFO product rotation. As part of the receiving activity, the manufacturer lot or production run is attached to a pallet identification. The inventory program tracks the master carton or pallet identification similar to the FIFO product rotation.

Pallet Identification

Factors that determine how master cartons or pallets are identified include:

- WA, NA, HROS truck, or VNA forklift truck or AS/RS crane operation
- Identification scan options (a) hand-held or (b) fixed-position bar code scanner/RF tag
- Vendor pallet identification

Master carton or pallet identification methods are classified as no method, manually printed methods, or machine-printed or produce methods.

No Method

This is when a company relies upon the vendor's master carton or pallet identification. The identification method does not require the company to place an identifier onto the exterior of the pallet or master carton. The majority of the companies that use this concept have only one SKU in an aisle and a limited number of SKUs. This storage system is usually two or three master cartons or pallets deep, floor stacked, and there is a large number of master cartons or pallets for one SKU. The "no method" is not considered for a storage operation with an inventory program. Disadvantages are it handles a limited number of SKUs, possible errors, and low employee productivity. Advantages are no expense or cost and less time spent to attach the label.

Manually Printed Methods

An employee writes the pallet identification onto the side of a master carton or pallet or onto a label that is then placed on the side of the master carton or pallet. The pallets are identified with crayon or chalk marks directly on the master carton or pallet, crayon or chalk marks on a label, or crayon or chalk marks on a color-coded label.

Machine-Printed Methods

Machine-printed labels are human-readable or human/machine-readable and are printed directly on the master carton or pallet or onto a label that is placed on the master carton or pallet.

Human-Readable Labels

With this method, a computer-controlled printer prints the master carton or pallet identification onto a label. Each label has alpha characters or digits that are preprinted prior to delivery truck arrival at the dock, or printed on demand at the receiving dock. A receiving clerk places the label onto the master carton or pallet. The disadvantages of this method are the additional costs for the labels, as well as the printer and computer, office space, and print time. Advantages are uniform and clear identification, low labor requirement, no transposition errors, and preprinted labels.

Human/Machine-Readable Labels

With this concept, the computer prints human-readable characters and digits and machine-readable symbology (bar code/RF tag) on each label. This is referred to as a master carton or pallet license plate. The disadvantages and advantages are similar to those described above for human-readable labels but the additional disadvantage is slightly more time needed for printing. An additional advantage is barcode scanning/RF tag reading capability. Human/machine-readable labels can be self-adhesive or non-adhesive, black print on a white background, or color-coded.

Self-Adhesive Labels

A self-adhesive label is widely used in the pallet storage industry as both a disposable master carton or pallet identification and a permanent storage position identification concept. The white face provides a surface to print the black bars that create the bar code/RF tag, as well as alpha characters or numeric digits.

The self-adhesive surface allows the identification label to be secured to a pallet block, stringer, or master carton. This feature ensures that an identification label remains with a master carton or pallet, allows an employee to recognize the label, and permits a hand-held or fixed-position scanner line of sight. After master cartons are depleted from a pallet, the self-adhesive master carton or pallet identification is removed from the pallet and thrown in the trash.

Non-Adhesive Labels

The non-adhesive label is printed on white paper or thin cardboard front. The printed label is slipped into an insert that is secured to a master carton top or pallet stringer or block or stapled to a pallet stringer or block. The sleeve or insert has a self-adhesive backing and a clear (transparent) front and permits an employee to recognize the label, permits a line of sight for a hand-held or fixed-position scanner/RF tag reader, and is used in an operation that uses a disposable pallet identification method. After the master cartons are depleted from the pallet, the pallet identification is removed from the insert or sleeve and thrown in the trash and the insert or sleeve remains on the pallet.

Black Print on a White or Colored Background

The paper used for the labels can be plain white or a light color. In a WA, NA, HROS, or VNA man-down forklift truck operation with an inventory program, the labels are used to identify pallet storage positions on different rack levels (see Figure 3.24). For example, the first-level storage positions have labels with a white background, the second-level positions have labels with a light beige background, and the third-level positions have labels with a light blue background. Prior to implementation in a storage operation with an inventory program, colored labels should be tested to ensure good reads from a hand-held scanner.

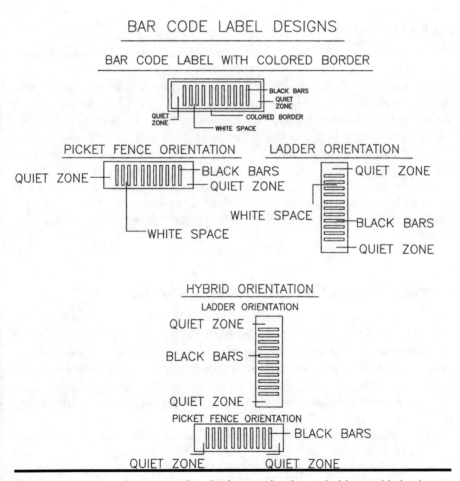

Figure 3.24   Bar-code types: colored edges, picket fence, ladder, and hybrid.

Color-Coded Labels

A system that uses color-coded labels has a unique color for each month of the year or for a specific product group. In a storage operation that has a man-down forklift truck with elevated storage positions that are above a hand-held scanner line of sight or a conventional forklift truck and an AS/RS concept, one colored label is for identified master carton or pallets that are directed to an AS/RS storage section or another colored label for identified master carton or pallets that are directed to a standard rack section. The color code is the label borders or color ink. The color-coded label provides a quick indication of the month that a master carton or pallet was received at the facility. Prior to implementation in an operation, complete hand-held and fixed-position scanner tests on the colored printed labels.

Radio Frequency Tags

The radio frequency (RF) tag identification concept has a tag (symbology) that is attached to each master carton or pallet UOP. Each storage forklift truck has a receiver (antenna) or it is situated as a fixed position device. The tag has a transponder that transmits a radio signal. The radio signal is produced from an antenna grate circuit that is located inside the tag. Each tag has a unique transmission that discreetly identifies the UOP. The storage forklift truck or fixed-position receiver (antenna) picks up a radio signal from the tag and the reader decodes and validates the signal for transmission to the microcomputer that identifies the UOP storage position or SKU. Tags are either (1) passive, permanently coded, able to store up to 20 alpha-numeric characters and relatively inexpensive, or (2) programmable (recodable), able to store up to 2000 characters, contain a battery, and are more expensive.

Storage Transaction Instruction Concepts

A storage transaction instruction concept directs a WA, NA, HROS, or VNA forklift truck driver- or computer-controlled AS/RS crane to complete a master carton or pallet deposit or withdrawal. With an employee-based concept, the employee has to read and clearly understand the instruction. The instruction options are:

- To direct an employee with characters or numeric digits that are printed on a paper document
- Paperless directions that are either an RF device with a visual display screen or voice directed
- Computer-controlled directions to an AS/RS crane

Paper Document

The computer-printed or manually printed concept is used in any operation that has an employee control a forklift truck to complete a pallet storage transaction. The paper document is a printed page (single sheet or multi-carbon sheets) that lists all available master carton or pallet positions and SKU descriptions. Each master carton or pallet handled has a corresponding column or open block. After completing a transaction, the forklift truck driver places a mark in the block to verify the transaction completion. With all required transaction completed, the document is given to an office clerk to update the inventory files. In a master carton or pallet operation, the paper document is not the preferred option. The disadvantages are transposition or entry errors, clerk and an employee to read, inventory update is a delayed transaction, and paper document can be lost. The advantages are no investment, little employee training, able to use in all manually controlled or operated systems, and in a small operation.

Paperless Directions

A paperless transaction directs a WA, NA, HROS, or VNA forklift or a computer-controlled AS/RS crane to complete a master carton or pallet storage transaction.

Radio Frequency and Digital Display

An RF terminal and digital display screen concept has an employee obtain an identified storage position and identified master carton or pallet from the alpha characters/digits that are shown on a display screen. For a deposit transaction, an employee uses a barcode scanner/RF tag reader to read the pallet barcode label/RF tag, the identified master carton or pallet is assigned to an identified storage position that appears on the RF display screen, the employee/forklift truck operator travels to the assigned master carton or pallet position, and completes the deposit transaction and scans the master carton or pallet identification label and the storage position identification label. The RF device sends the scan-identified storage position and identified master carton or pallet transaction information directly to the computer or it is held in memory for later transfer to the computer.

For a withdrawal transaction, the employee is directed to an assigned master carton or pallet position that has the suggested identified master carton or pallet. At the assigned position, the forklift truck driver scans the pallet identification and storage position identification to verify that he or she is completing the directed master carton or pallet storage transaction, and removes and transports the identified master carton or pallet to the assigned location. Scan transactions are sent to the microcomputer to update the inventory and master carton or pallet storage position files.

Disadvantages are the investment required, as well as the need for identification on each master carton or pallet and position and for on-line communication to verify that the RF operates in a facility and employee training. The advantages are accurate transaction completion, on-line transaction verification, and requires few employees.

Voice Directed

A voice-directed pallet transaction concept has a computer system and an employee with a microphone and earphones (headset). With a voice-directed concept (voice recognition and speech synthesis), an employee has on-line communication with the host computer. In this concept, each employee talks to the computer through a microphone and receives verbal transaction instructions from the computer through the headset. Disadvantages and advantages are similar to a RF paperless concept, except the concept requires additional employee training.

Computer Directed

A computer-controlled paperless concept is used in an AS/RS facility. In an automatic computer-controlled concept, the computer communicates to each AS/RS crane computer to complete identified master carton or pallet deposit or withdrawal transactions between an assigned master carton or pallet position and a pick-up/delivery (P/D) station. To verify each identified master carton or pallet transaction completion, the HMS sends the information to the computer and a barcode scanning/RF tag reading device along the master carton or pallet travel path reads each master carton or pallet identification and verifies the transaction completion.

Disadvantages are cost, on-time COs, master cartons or pallets have identification, requires a computer program, and communication between the HMS and computer. The advantages are accurate transaction completion, on-line transaction verification, requires few employees, and requires a building with a small footprint.

Aisle and Position Identification Formats

In a storage operation, each master carton or pallet aisle has identified master carton or pallet storage positions that have alpha characters, numeric digits, and a bar code/RF tag. With a WA, NA, HROS, or VNA forklift truck or AS/RS crane operation with a master carton or pallet rack concept, the identification information sequence is: warehouse, warehouse aisle, bay within an aisle, level within a bay, and master carton or pallet position within a level.

A storage operation with a WMS program that has a floor stack, drive-in, drive-through, pallet flow rack, or car-in rack storage concept, the format sequence is: warehouse, warehouse aisle, bay within an aisle, and level within a bay.

The pallet storage concepts are dense storage concepts and have the same SKU.

Warehouse

The warehouse is used to identify one warehouse from another warehouse in a company supply chain logistics strategy or in a company that has a main warehouse and a remote warehouse. In a concept this component is an alpha character.

Warehouse Aisle

A warehouse aisle is an alpha character or digit that identifies a storage aisle in a warehouse. In a dynamic storage operation 26 alpha characters are sufficient to identify all storage operations.

Bay in Aisle

A rack bay is identified with numeric digits that have an arithmetic progression from the aisle entrance to the end. The bay identification uses digits that match the pallet number or rack bay number in a nominal building length.

Level in a Bay

A load beam or rack level is within a rack bay and has an alpha character. In a WA, NA, HROS, or VNA or AS/RS or tall-rack operation, 26 alpha characters are sufficient to identify all pallet levels in a rack structure. With an alpha character between two numeric digits, a forklift truck driver is less likely to become confused than with three consecutive numeric or digit components.

Master Carton or Pallet in a Bay

Master carton or pallet position in a rack bay is a significant component in the forklift truck driver instruction format. At a master carton or pallet position in a bay, the forklift truck driver or AS/RS crane completes a directed master carton or pallet transaction. The forklift truck driver scans the identified master carton or pallet position in a bay that assigns a master carton or pallet identification to the master carton or pallet position identification. The master carton or pallet position is identified with numeric digits and the print is the easiest to read on a digital display or paper document.

Aisle and Position Identification

To ensure an accurate and on-line transaction in a pallet storage operation, aisle signs clearly identify each aisle and identified pallet storage position.

Characters, Digits, and Bar Code

The aisle and position identification process determines how a storage transaction instruction appears at an aisle entrance, at a rack bay (level), and master carton or pallet position. One option is to use only alpha characters and numeric digits. This basic concept is not used in a dynamic operation because it is difficult to detect errors and there is no system for checking. A better option is to use alpha characters, numeric digits, and a bar code.

With barcode label scanning by a hand-held or fixed-position scanning device, storage scan transaction information is transferred on-line and accurately to the computer files.

Storage Aisle Identification

Options for storage aisle identification (see Figure 3.25) are:

- Placard with one-way vision. The one-way vision or placard is placed flat against an aisle end upright frame or is hung from the ceiling. In this design, the aisle identification faces outward, allowing a forklift truck driver entering the aisle to identify the aisle. The one-way vision placard is used in a forklift truck driver operation and is used to identify each AS/RS aisle on an AS/RS front wall.
- Two-way vision placard against the rack. A two-way vision concept uses either two or three placards. If two placards are used, one is placed on each upright frame and extends outward from the upright post into the main traffic aisle. At the proper elevation the placard is easily recognized and does not get damaged. If three placards are used, two extend outward into the main traffic aisle and a third is placed flat between the other two. With a two-way sign, as a forklift truck driver travels in the main traffic aisle, each aisle number is easily recognized.
- Four-way aisle placard. A four-way aisle identification concept is a ceiling-hung placard that has four sides. Each side has the aisle identification. With a four-way sign, as a forklift truck driver is traveling in the main traffic aisle, each aisle number is easily recognized from the main traffic aisle.

AISLE IDENTIFICATION SIGNS

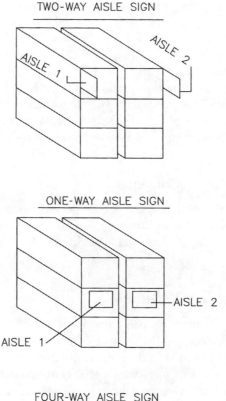

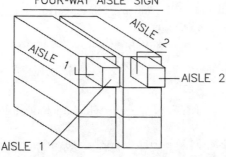

Figure 3.25 Aisle identification concepts.

Identification Concepts

In a master carton or pallet storage operation, the concept used to identify a master carton or pallet position has a direct impact on forklift truck driver transaction productivity and transaction accuracy. Concepts include those that are used with floor stacks: (a) ceiling-hung or (b) floor-embedded, and those that are used with racks.

Floor Stack Identification Concepts

A master carton or pallet floor or block storage position identification has (1) a placard hung from the ceiling with a storage lane position identification. With a placard concept, the clearance between the placard and a forklift truck mast is a factor, (2) position identification is embedded in the floor. With this concept, an identification durability in the floor is a factor or (3) rack post between two floor stack lanes. As an employee or forklift truck driver faces the upright post, the left-hand master carton or pallet lane is on the upright post left side and has the master carton or pallet storage identifications and the right hand master carton or pallet lane to the upright post right side and has the master carton or pallet storage identification. The rack upright post maintains clearance between storage lanes.

Rack Identification Concepts

A storage master carton or pallet position identification is attached to a rack bay metal structural member and assures a hand-held scanner line of sight to the storage master carton or pallet position identification.

Storage Facility Vehicles

The type of vehicle used in a pallet storage facility is a key factor that determines design of aisle widths and the number of master carton or pallet positions (load beam levels) above the floor surface. This influences facility storage utilization, land and building/equipment investment costs, and annual facility operational costs. The storage area vehicle assures a driver hand-held scanner line of sight to the master carton or pallet identification and master carton or pallet position identification to complete the scan transactions.

Master carton or pallet storage vehicle types are (see Figures 3.26 and 3.27):

- Wide aisle (WA) group
- Narrow aisle (NA) group
- Very narrow aisle (VNA) group which includes HROS, car-in rack, and AS/RS vehicles
- Mobile-aisle (MA) or transfer or bridge car (T-car) group
- Captive-aisle (CA) group

Wide-Aisle Vehicles

The group of vehicles classified as wide-aisle (WA) includes forklift trucks (see Figure 3.28) that require a 10- to 13-ft. wide clear aisle between two floor-stacked pallets or two or three pallets in a rack bay. This aisle distance allows a manually

FORKLIFT TRUCK AISLE WIDTHS

WIDE AISLE FORKLIFT TRUCK

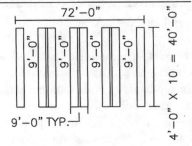

60 POSITIONS IN 2,520 SQ FT
= 42 SQ FT PER POSITION

NARROW AISLE FORKLIFT TRUCK

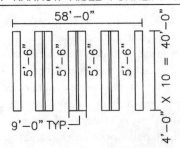

80 POSITIONS IN 2,880 SQ FT
= 36 SQ FT PER POSITION

VERY NARROW AISLE FORKLIFT TRUCK

80 POSITIONS IN 2,329 SQ FT
= 29 SQ FT PER POSITION

Figure 3.26 Forklift truck aisle width: wide aisle, narrow aisle, and very narrow aisle.

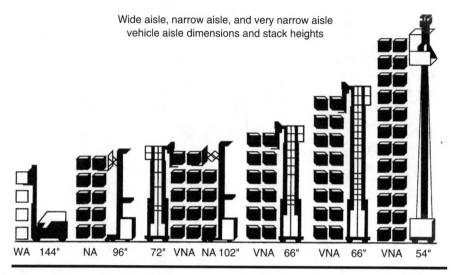

Figure 3.27 Aisle widths and forklift's fork heights.

controlled forklift truck to make a right-angle (stacking) turn and to complete a pallet storage transaction. The aisle distance allows an operator to achieve the desired storage transaction productivity with minimal product, equipment, and building damage. The exact aisle width is based on the stated manufacturer standard for pallet dimensions plus 6 to 12 in. To complete a storage transaction, a forklift truck driver completes a pallet storage transaction at the pallet position and has a clear line of sight with a hand scanner to scan the pallet identification and pallet storage position identification. To assure operator access to a hand-held scanner and display screen vision, a scanner holder and display screen are attached to the forklift truck.

WA vehicles are mobile aisle vehicles that are powered by an electric rechargeable battery, gasoline, liquid propane gas, or a diesel engine. The diesel- and gasoline-powered vehicles are preferred for outdoor activities. Some vehicles have wide, hard rubber tires, and can be used to transport pallets between two facility locations. Because of the long wheel center to center distance and a high undercarriage, the vehicle easily travels up or down grades. A WA vehicle is equipped with a set of forks and as required, a slip sheet or other attachment to an elevating mast. The forks lift a pallet to a 20-ft. high pallet position. Tilt and side shift options improve operator productivity and reduce product and equipment damage. Free lift permits the set of forks to move upward without the mast moving upward. This feature permits the forklift truck with free lift to operate inside a delivery vehicle or under a low ceiling.

Different WA vehicles are walkie stacker forklift truck, straddle or counterbalanced walkie stacker forklift truck, straddle or counterbalanced walkie stacker

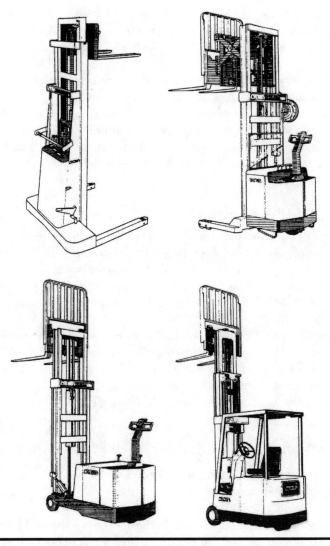

Figure 3.28 Wide aisle forklift trucks: manual walkie stacker, powered walkie stacker, powered counterbalanced stacker, and sit-down counterbalanced forklift truck.

forklift truck with a fork reach feature, sit-down three- or four-wheel counterbalanced forklift truck, and stand-up rider counterbalanced forklift truck.

Walkie Stacker Forklift Truck

A walkie stacker forklift truck is a manual push or electric-powered drive wheel vehicle that handles a 2500 lb. maximum load up to a 10 to 12 ft. pallet position height. In an operation, the hand-held scanner is attached to the operator's waist or the forklift truck chassis. All pallet storage position identifications are placed onto a floor-stack ceiling-hung placard, floor level or rack post, or pallet rack load beam or upright post.

In an inventory program concept, the forklift truck driver picks up a pallet with the pallet identification facing the driver. This permits the forklift truck driver to have pallet identification clear line of sight. A forklift truck driver scans the pallet storage position identification and the pallet identification and the scanner device holds the scan transactions in memory or RF sends the scan transactions to the computer for a SKU inventory and storage position update.

Straddle or Counterbalanced Walkie Stacker Forklift Truck

A straddle or counterbalanced walkie stacker forklift truck has two stabilizing straddles that extend forward from the vehicle base and a space between the straddles to pick up or deposit a pallet. A hand-held scanner is attached to the operator's waist or the forklift truck chassis. All pallet storage position identifications are placed onto a floor-stack ceiling-hung placard, floor level or rack post, or pallet rack load beam or upright post.

A counterbalanced forklift truck wheel base determines the aisle width required. The wheel base is the distance between the front and rear wheels. A forklift truck with a short wheel base requires a narrow right-angle turning aisle and makes the forklift truck very maneuverable. A forklift truck with a long wheel base requires a wide right-angle turning aisle, but has an improved ride, steering traction, and stability.

Counterbalanced Walkie Stacker Forklift Truck with Reach Forks

A counterbalanced walkie stacker forklift truck with a fork reach feature has two stabilizing straddles that extend forward from the vehicle base, a space between the straddles for a pallet transaction, and the set of forks extends outward. In an operation, the hand-held scanner is attached to the operator's waist or the forklift truck chassis. All pallet storage position identifications are placed onto a floor-stack ceiling-hung placard, floor level or rack post, or pallet rack load beam or upright post.

This type of forklift truck has a set of forks on a telescopic mast that elevate and lower a pallet at 10- to 12-ft. high storage positions, two load-carrying wheels, one drive and steer wheel and an operator handle, which carries all vehicle movement controls. The operator, who walks behind the vehicle, has easy access to the fork movement controls. The forks are free lift, which allows them to rise while the mast remains at a low elevation. This feature permits the vehicle to operate in low ceiling areas. The vehicle has a slow travel speed, which means low transport employee productivity. It receives its power from a rechargeable electric battery. The walkie stacker forklift truck is used in a low-volume pallet operation or on an elevated floor.

Stand-Up Rider Counterbalanced Forklift Truck

A stand-up rider counterbalanced forklift truck has similar design features and operational capabilities as a sit-down rider counterbalanced forklift truck. The major difference between the forklift trucks is that a stand-up rider vehicle has the operator stand in an operator platform area. This feature allows a shorter wheel base, which permits right-angle stacking turns in a 9- to 10-ft. wide aisle and places a 2000- to 3000-lb. pallet into an 18-ft. floor or rack position. A 24 or 36 V electric rechargeable battery-powered mobile aisle vehicle is very maneuverable and versatile, performs all activities on a normal facility floor, and enters and exits most delivery vehicles. The forklift truck is used in the receiving, in-house transport, floor stack or rack storage, and shipping areas or activities. One model has a retractable overhead guard option that permits delivery truck entry. In an operation, a hand-held scanner is attached to the operator's waist or a forklift truck overhead support member. All pallet storage position identifications are placed onto a floor-stack ceiling-hung placard, floor level or rack post, or pallet rack load beam or upright post.

The storage position identification location assures WMS storage position identification line of sight for the forklift truck driver. In an inventory program concept, a forklift truck driver completes a pallet transaction with the pallet identification facing the driver.

Sit-Down Three- or Four-Wheel Counterbalanced Forklift Truck

A sit-down counterbalanced forklift truck with a low mast and overhead guard is a very maneuverable and versatile pallet forklift truck. This type of truck is used as a receiving dock, in-house transport, floor-stack or rack storage or shipping dock area vehicle that requires a normal floor and is able to travel up a 15-degree grade or ramp. The forklift truck is designed with a set of load carrying forks that elevate and lower on a telescopic mast. From an operator seat, the operator has access to all horizontal vehicle and vertical fork movement controls. Some models have see-through masts that provide the operator with a complete view of the vehicle travel path. The vehicle counterweight is located in the rear chassis and the vehicle has a long wheel base that provides the forklift truck with the ability to off-set the pallet weight.

An indoor forklift truck is powered by a rechargeable electric battery. When a forklift truck is used as an outdoor vehicle, the power source is diesel, LP gas, or gasoline. The vehicle has three or four wheels, with one wheel as a drive and steer wheel. The wheels are fitted with pneumatic tires for outdoor use and cushion and solid polyurethane or rubber tires for indoor use.

Counterbalanced forklift trucks are available with one-, two-, three-, or four-stage masts that provide the forklift truck with a 16- to 18-ft. stacking height. A counterbalanced forklift truck is used to deposit, pick, or withdraw a 2000- to 4000-lb. pallet from a floor or rack position. Some forklift trucks have a counterweight and wheel base to handle a heavier load at a lower height. Fork side shift and mast tilt devices are options that increase operator productivity and reduce product and equipment damage. An experienced operator makes 20 to 25 storage transactions per hour. All pallet deposits and picks are controlled or eyeballed by the operator. The sit-down rider counterbalanced forklift truck with a low mast and low overhead guard is a very maneuverable and versatile pallet truck. With free lift and a wheel base that has good grade clearance, a forklift truck easily loads and unloads a delivery vehicle, transports a pallet through a facility, and performs standard floor and rack storage transactions. With limitations, a standard counterbalanced forklift truck handles a wide pallet variety over a normal floor.

A four-wheel counterbalanced forklift truck with its long wheel base and chassis weight is fitted with various attachments to perform other tasks. For any forklift truck, you verify with the forklift truck manufacturer that the forklift truck accepts the attachment weight, new load center, and reaches the position elevation. In an operation, the hand-held scanner is attached to the operator's waist or the forklift truck overhead support member.

Lift Truck Capacity and Stability

A forklift truck capacity is the weight that a forklift truck safely lifts. Each forklift truck has a manufacturer-specified capacity. This weight capacity decreases with a longer load. The forklift truck handles shorter loads up to the limit specified by the manufacturer. As the unit load length increases from the standard 24-in. center, the forklift truck handles less weight. If a pallet is heavier than or over the standard dimensions, a bigger, heavier and more expensive forklift truck is needed for the operation or additional chassis weight is added to the existing forklift truck.

The term forklift truck load capacity with stability means when a counterbalanced forklift truck picks up a pallet, the fulcrum (balance point) for the counterbalance action is the front wheel centerline. A pallet weight plus any attachment is counterbalanced by the forklift truck counterbalance weight and wheel base dimension. The forklift truck weight includes the battery weight and counterbalance weight (see Figures 3.29 and 3.30).

Lift Truck Mast

Forklift truck mast parameters are the overall extended or lift height that allows a forklift truck to complete a transaction from the highest floor stack or rack position and lowered, down, or collapsed mast height.

This height is the clearance for a forklift truck to pass through the facility doorways, enter and exit a delivery vehicle, and interface with most conventional storage position concepts. For maximum forklift truck height, a forklift truck height is determined by the backrest height or pallet height to the lowest ceiling (building) obstruction. The clearance height for a forklift truck is affected by an overhead

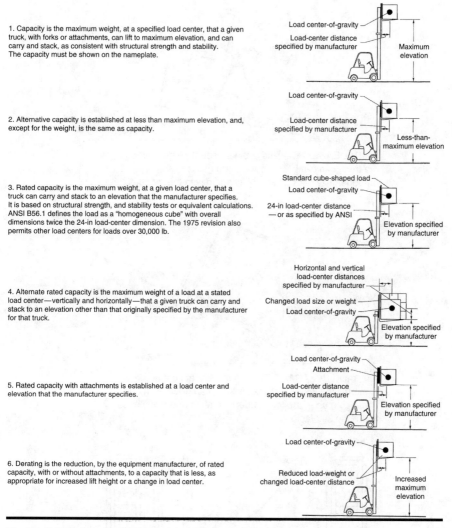

1. Capacity is the maximum weight, at a specified load center, that a given truck, with forks or attachments, can lift to maximum elevation, and can carry and stack, as consistent with structural strength and stability. The capacity must be shown on the nameplate.

2. Alternative capacity is established at less than maximum elevation, and, except for the weight, is the same as capacity.

3. Rated capacity is the maximum weight, at a given load center, that a truck can carry and stack to an elevation that the manufacturer specifies. It is based on structural strength, and stability tests or equivalent calculations. ANSI B56.1 defines the load as a "homogeneous cube" with overall dimensions twice the 24-in load-center dimension. The 1975 revision also permits other load centers for loads over 30,000 lb.

4. Alternate rated capacity is the maximum weight of a load at a stated load center—vertically and horizontally—that a given truck can carry and stack to an elevation other than that originally specified by the manufacturer for that truck.

5. Rated capacity with attachments is established at a load center and elevation that the manufacturer specifies.

6. Derating is the reduction, by the equipment manufacturer, of rated capacity, with or without attachments, to a capacity that is less, as appropriate for increased lift height or a change in load center.

Figure 3.29 Counterbalanced forklift truck features.

guard height that may be higher than the mast. In this situation the forklift truck ability to pass through a low ceiling area, doorway, rack lanes, and delivery vehicle is a restriction.

Forklift truck masts are:

- Single, non-telescopic masts that lift a pallet to a 7- to 8-ft. high floor or pick position. In a storage operation, a onestage mast has very limited application.
- Two-stage non-telescopic masts that have little or no free lift. No free lift means that as the set of forks start to rise, the mast also rises. With this mast, a forklift truck completes a pallet transaction at 12 to 13 ft. high.

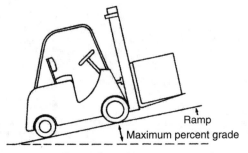

GRADEABILITY—This is the steepest percent grade a truck can climb with load. If you plan to use the ramp frequently, check gradeability again with the manufacturer.

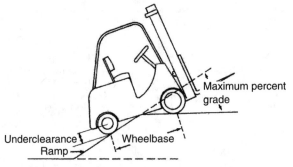

GRADE CLEARANCE—This is the steepest grade, expressed in percent, that a truck can go up without hanging up at the top. Underclearance and wheelbase are the key factors.

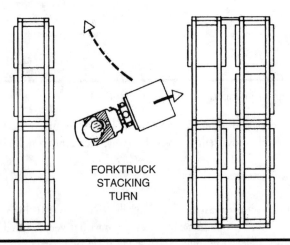

FORKTRUCK
STACKING
TURN

Figure 3.30 Counterbalanced forklift features: forklift truck gradability, clearance, and stacking turn.

- Standard two-stage masts with low free lift enables a forklift truck to complete a transaction at 15 to 20 ft. high. This low free-lift mast feature conserves energy. This type of forklift truck is used as a multipurpose vehicle in a low-ceiling area.
- Two-stage masts with high free lift permits a forklift truck to complete a pallet transaction at an elevation that is equal almost to the half the mast height. This feature permits a forklift truck to operate as a dock forklift truck to enter and exit delivery vehicles and to complete a transaction at a 20-ft. high position and operate in low-ceiling areas.
- Three-stage masts with full free lift allow a forklift truck to complete a transaction at 20 to 25 ft. above the floor. The full free-lift feature permits a forklift truck to operate in a low-ceiling area and to enter and exit most delivery vehicles.
- Four-stage masts with full free lift permit a forklift truck to complete a pallet transaction from an elevated rack position that is 30 to 40 ft. above the floor. This type of mast is used on a vehicle that has aisle guidance and is used in a storage area.
- Rigid masts with or without free lift permit a forklift truck to complete a transaction to an elevated rack position that is 40 to 60 ft. above the floor. A rigid mast is used on a hybrid or high-rise vehicle that is captive to the storage area.

Narrow Aisle Vehicles

A narrow aisle (NA) forklift truck (see Figure 3.31) operates in an aisle that has a 7- to 10-ft. clearance between pallets. Most NA forklift truck manufacturers suggest a minimum aisle width plus 6 to 12 in. To ensure good NA forklift truck driver productivity and assure that the forklift truck straddles enter a pallet rack bay, a rack layout has 5 to 6 in. between pallets or between a pallet and upright post.

With a raised load beam or the lowest pallet rack bay on load beams concept, the NA truck operates in an aisle that has a narrow 7 ft. dimension, but it provides lower employee productivity. With the raised load beam concept, the first pallet storage level is raised 12 in. above the floor. With a 6 in. high load beam, the truck straddles can travel under the load beam and into the rack position. This means that the forklift truck straddles do not travel between pallets or between a pallet and upright post. This concept permits the forklift truck to complete a pallet storage/pick transaction without striking a pallet but increases the top pallet height.

The NA forklift truck group is made up of mobile aisle forklift trucks that travel between two storage area aisles. The forklift truck is powered by an electric rechargeable battery and has a set of forks that elevate on a mast to raise a pallet to a storage position that is 25 to 28 ft. above the floor. The operator stands on a platform under an overhead guard and has access to all forklift truck controls and the steering wheel. An option on the NA forklift truck is a safety bar on the operator platform rear side, which is attached to the compartment metal side and extends

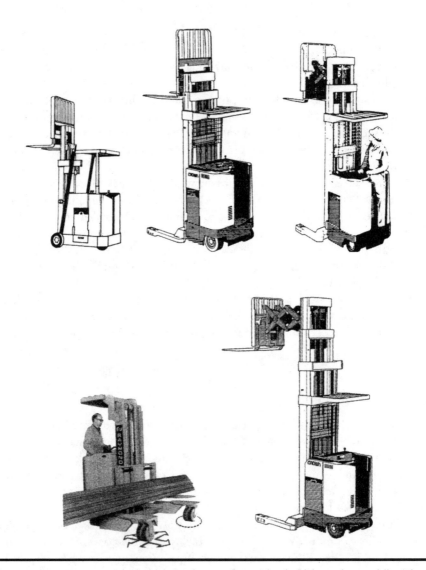

Figure 3.31 Narrow aisle forklift truck: stand-up rider forklift truck, straddle rider forklift truck, and straddle reach forklift truck.

upward to the overhead guard. In an aisle and when backing a forklift truck after completing a pallet storage/pick transaction, the safety bar minimizes the chance of an operator backing into a rack load beam.

Stand-Up Rider Straddle Forklift Truck

A stand-up rider straddle forklift truck is powered by a rechargeable electric battery. If there is no deflection on a dock lever, the forklift truck is able to enter and exit a delivery vehicle. The forklift truck has two load-carrying wheels under straddles (outriggers) and one drive and steering wheel, a set of forks that elevate and lower over a telescopic mast, an overhead guard, and an operator platform area. The operator platform area has both horizontal truck and vehicle fork movement controls and WMS hand-held scanner and display screen. To elevate, lower, and transport a pallet, straddles assume most of the pallet weight. Because most of the pallet rests between the straddles (outriggers), the forklift truck can operate in a 7- to 8-ft. wide aisle. On a normal floor surface, the straddle forklift truck completes a transaction up to a 20-ft. high pallet position and the operator completes 18 to 20 transactions per hour. In a pallet operation, the forklift truck driver hand-held scanner is attached to the driver's waist or forklift truck overhead support member and the display screen is attached to a lift truck overhead guard member. All pallet storage position identifications are placed onto a floor-stack ceiling-hung placard, floor level or rack post, or pallet rack load beam or upright post.

Stand-Up Rider Straddle Forklift Truck with Reach

A stand-up rider straddle forklift truck with reach has similar features to a straddle forklift truck, except that its forks are attached to a pantographic reach device. The pantographic reach device allows the set of forks to extend out beyond or forward of the straddles to pick up and retract a pallet. The pantographic device permits a reach forklift truck to handle a pallet that is wider than the interior space between the straddles. This reach feature increases the right angle turn or stack requirement to 8- to 10-ft. wide aisles and allows an operator in a normal aisle length to complete 15 to 18 transactions per hour.

Stand-Up Rider Double-Deep Forklift Truck

A stand-up rider double-deep forklift truck is also called a two-deep or deep-reach forklift truck. Its basic design features and operational characteristics are the same as the straddle reach forklift truck except that the pantographic device has a greater extension. It extends forward out into the second deep interior pallet position of a two-deep rack concept. A deep-reach forklift truck has an extension (stroke) that is fixed at 43 to 51 in. Operators make 13 to 16 transactions per hour in a 9- to 10-ft. wide aisle.

Four-Directional Forklift Truck

A four-directional forklift truck has similar operational features to a regular forklift truck with reach, but all four truck wheels turn to the direction of travel. One load wheel hydraulically shifts to forward or reverse or lateral for travel sideways. The other load wheel is the free-swiveling wheel that gives the forklift truck its name.

While traveling in a main aisle with a long load in front and for fork entry into a narrow aisle to perform a transaction, the driver stops and shifts the load wheel direction. This permits the forklift truck to travel sideways into a narrow aisle and complete a transaction. Narrow aisle is a nominal 8 ft. 6 in. wide.

Very Narrow Aisle and AS/RS Trucks

The HROS, VNA, and AS/RS forklift truck group (see Figure 3.32) includes:

- ■ VNA forklift trucks, which are battery powered and manually operated vehicles with a 35- to 37-ft. fork height above the floor
- ■ AS/RS cranes that are DC electric powered and manually or microcomputer controlled to a storage position that has an elevation range up to 110 ft. above the floor

The VNA forklift trucks and AS/RS cranes operate in a 5 ft. 6 in. to 7 ft. 6 in. wide aisle product to product and complete pallet storage transactions. The pallet storage transactions (see Figures 3.33 and 3.34) are completed by a set of forks or platens that complete a pallet storage transaction without a vehicle turning in the aisle. This ability to complete a pallet storage transaction from the vehicle side permits the very narrow storage aisle width.

HRSO, VNA forklift truck, or AS/RS crane characteristics are rail or wire-guided aisle travel, master carton or pallet storage transaction is performed by a set of forks or a platen, and with HROS and VNA master cartons are hand stacked; in most applications, the lowest master carton or pallet position is raised 16 to 34 in. above the floor, the operator is located in a protective station with a tether line, some vehicles handle two master cartons or pallets per trip and complete pallet transactions to a two-deep pallet location or pallet flow rack lane, many facilities have the rack structural members support the walls and roof and master carton or pallet position height range is from 30 in. to 110 ft. above the floor

HROS and VNA forklift truck groups are:

- ■ Man-controlled and man-down group that includes (a) rider side-loading forklift truck, (b) stand-up rider double-deep forklift truck, sit-down rider counterbalanced side loader with a swing mast, (c) operator-down counterbalanced side-loading forklift truck, and (d) operator-down side-loading turret forklift truck
- ■ Man controlled and man-up group that has (a) operator-up side-loading forklift truck, (b) stand-up outrigger side-loading rising cab forklift truck or a turn a load or rack loader, (c) operator-up side-loading turret forklift truck, (d) counterbalanced side-loading forklift truck with rising cab and auxiliary mast lift, and (e) high-rise order picker or HROS master carton handling truck

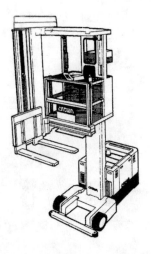

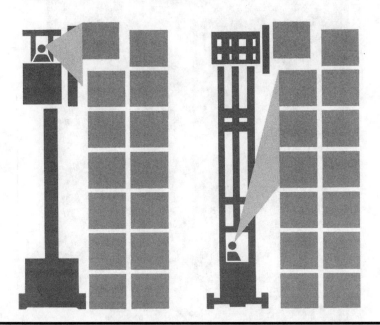

Figure 3.32 Very narrow aisle forklift trucks: man-up, man-down, and fixed mast man-up top and bottom guide.

- Man-controlled storage retrieval or M/SR: (a) captive aisle cranes and (b) mobile aisle cranes
- Computer-controlled storage retrieval or AS/RS cranes or stacking AGV vehicles: (a) sort link, (b) mole, or (c) car-in rack concepts and options for captive aisle machines and mobile aisle machines that use a transfer or bridge car

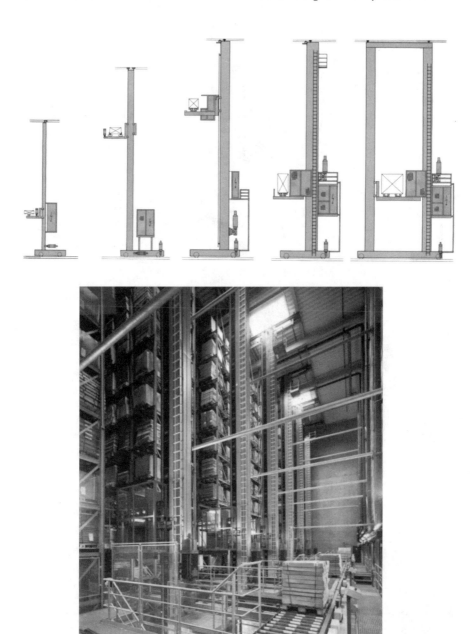

Figure 3.33 Pallet AS/RS: P/D station and single and dual mast.

Man-Controlled and Man-Down Forklift Truck

VNA forklift trucks have the capability to complete a pallet transaction to a tall rack pallet position from a very narrow aisle with a forklift truck operator at the floor level. In a man-down operation, each pallet storage position identification

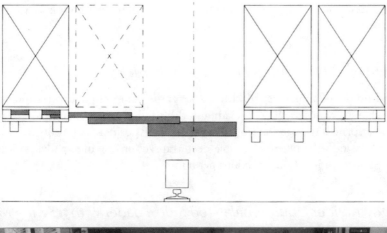

Figure 3.34 Two deep and two pallet carriers.

has a line of sight by the forklift truck operator. With this design parameter, the elevated pallet storage position identifications are arranged in an understandable manner attached to an upright post or load beam.

Rider Side-Loading Forklift Truck

A rider side-loading forklift truck has a set of load-carrying forks that are attached to a pantographic device that elevates and lowers on a telescopic mast. It has an overhead guard and an operator area at one end of the vehicle with full machine controls and a steering device. The vehicle has two steering wheels and two load-carrying wheels under the two decks or straddles. There is sufficient width between the decks for a pallet that is taken onto the set of forks into the well. Prior to travel the pallet is lowered and rests on the front and rear decks. The two decks become the stabilizing platform for a pallet. The forklift truck operates in a 5 ft. 6 in. wide aisle and completes transactions to positions 30 ft. above the floor. A side-loading forklift truck takes a pallet onto the set of forks with the pallet operator facing the rack position. Since the set of forks face one aisle side, all transactions are performed from one aisle side. Without a computer inventory control program to sequence the storage transactions, the transaction number per hour is 17 to 19 in a normal aisle length.

Stand-Up Rider Double-Deep Forklift Truck

A stand-up rider double-deep forklift truck has the same design features and operational characteristics as the side-loading forklift truck except that a pantographic device performs a transaction from a two-deep rack position. A forklift truck has a 6- to 7-ft. wide aisle. Due to a double-deep transaction at a high pallet position, a driver performs 15 to 17 transactions per hour.

Sit-Down Rider Counterbalanced Side-Loader with a Swing Mast

The features and operational characteristics of a sit-down rider counterbalanced side-loader forklift truck design are similar to a regular counterbalanced forklift truck. The truck requires a dead level floor and 10- to 12-ft. intersecting aisles. To make a storage transaction to a pallet position, the telescopic mast swings 90 degrees to the right side of the aisle. The mast movement gives the forklift truck its name. An electric rechargeable battery is the power source and the truck operates in an 8-ft. wide wire-guided aisle with a minimum aisle width that is a nominal 1 ft. wider than a pallet. The swing mast forklift truck makes a pallet transaction from floor to rack positions 25 to 30 ft. above the floor. In an aisle of normal length, the operator throughput capacity is 17 to 20 pallet transactions per hour for a dual cycle. When transactions are not on the same side of the aisle side (single cycle), the productivity is 15 to 17 transactions per hour due to the need for two trips down the aisle. If a swing mast forklift truck functions as a transport and dock vehicle, a facility and delivery vehicle floor has the strength to support a forklift truck and swing mast truck mast to enter a delivery vehicle. For maximum forklift truck driver or aisle trip productivity, the computer inventory control program to sequence storage transactions for each forklift truck trip in an aisle has the forklift truck perform all storage transactions on an aisle side.

Operator-Down Counterbalanced Side-Loading Forklift Truck

An operator-down sit-down rider counterbalanced side-loading forklift truck is an electric rechargeable battery-powered forklift truck, with the set of forks attached to a telescopic mast. The set of forks moves from the right side of the aisle to the left. This feature permits the forklift truck to complete pallet transactions at 30- to 40-ft. high rack positions. The forklift trucks have side shifts to assist a pallet transaction. This type of truck is also called a turret or swing-reach forklift truck. Aisles are a nominal 20 in. wider than a pallet or 6 in. wider than the pallet diagonal dimension to turn a pallet in the aisle as the forklift truck travels in an aisle. The operator transaction activities are 20 to 23 transactions per hour in a guided normal aisle length. With 40-ft. high transaction capability, the mobile aisle forklift truck requires a dead level floor. The intersecting aisle is a 17-ft. wide minimum.

Operator Side-Loading Turret Forklift Truck with Operator Down

With this forklift truck, the set of forks is attached to the carriage top. The carriage is attached to a telescopic mast that permits the set of forks to handle a pallet load transaction from a rack position on either aisle side. For best operational results, the forklift truck has an aisle guidance system, end of aisle pick up and delivery stations, and a dead-level floor. A forklift truck travels in a 5 ft. 6 in. to 6 ft. 6 in. wide aisle that is a nominal 1 ft. wider than a pallet. To change aisles, a forklift truck has a 17 to 18 ft. wide intersecting aisle. A forklift truck receives its power from an electric rechargeable battery.

Man-Controlled and Man-Up Forklift Truck

A man-controlled, man-up forklift truck has the capability to have an operator rise with a pallet to a rack and position. In this elevated position, the operator has vision to complete a pallet transaction. In a man-up forklift truck operation, each pallet position identification and each pallet identification has a forklift truck operator line of sight and permits pallet position identification attached to any load beam level. In the elevated position, the operator has line of sight to the WMS pallet identification and permits the operator to complete a pallet identification scan. In an inventory program concept, a forklift truck driver picks up a pallet with the pallet identification facing the forklift truck driver. With this design parameter, each elevated pallet position identification is placed directly on the load beam or upright post that supports the pallet.

High-Rise Order Picker or HROS Truck

A carton order picker truck is a manually controlled order picker truck that has the ability to travel vertically and horizontally through a guided aisle to the assigned pick position. To assure maximum employee productivity, fast travel speed, and minimal equipment damage, a guided aisle is between two rack or shelf positions. Vehicles have an electric rechargeable battery, operator platform with all vehicle steering, forward and reverse, and fork-elevating controls. An operator tethering line, load-carrying forks, and securing device, three or four wheels with one wheel as the drive and steering wheel, electric-powered drive motor and drivetrain, chassis, elevating mast, aisle guidance system, and options and accessories.

The operator platform has an overhead guard, a rubber mat surface, wire-meshed or plastic barrier between the operator and the elevating masts, safety gates, and a rail that surrounds the operator platform. The safety gates and rail devices are engaged to move the order picker truck or platform. The controls area steering device, operator, and load-carrying device elevating control handle and forward and reverse control handle. To assure employee safety, a tethering line is used to connect the operator to the overhead guard section. Since the HROS handles a pallet or cart, the HROS truck has a set of forks and a claw that grabs the pallet stringer and secures the pallet to the platform. In a cart application, a chain secures a cart to the operator platform. An elevating mast is located in front of the operator platform, and the mast, operator platform, and load-carrying device are raised or lowered.

To have high HROS productivity, fast travel speed, and minimal equipment damage, HROS applications travel in a guided aisle. HROS truck aisle guidance concepts (see Figures 3.35 and 3.36) are:

- Rail or mechanical
- Wire
- Tape
- Paint
- Laser beam or electric

Mechanical guidance systems are used with some modifications on existing vehicles or in an existing or new facility. If an aisle guidance system is considered for an existing or new facility, careful attention is given to the floor (surface, loading, metal object location, and re-bar) prior to a system purchase and implementation. This is especially true for a wire guidance system.

In a HROS concept, the inventory program has an identification on each master carton and on each storage or pick position. During the master carton or pallet put-away physical transaction and master carton/pallet identification and pick position identification scan transactions, the inventory and pick position status is updated in the files. The HROS concept uses master cartons hand stacked in the pick position.

VNA and HROS Rail Guidance

The basic rail HROS aisle guidance system has one or two angle iron rails, while some rack manufacturers have guide rails. The angle iron is 4 in. by 3 in. or 4 in. by 4 in. and a manufacturer rail has a 3-in. base extension. The bases are anchored to the floor on both sides for each vehicle pick aisle full length. In one rail guidance system, the guide rail is on the aisle right side as the vehicle travels through the aisle. The guide rail is matched to the vehicle guide rollers that are attached to the vehicle side or a roller device is attached to the vehicle right side. The vehicle roller riding against the rail or rails provides down aisle vehicle travel guidance. To assist the vehicle operator with aisle entry and to reduce rack and vehicle damage, entry

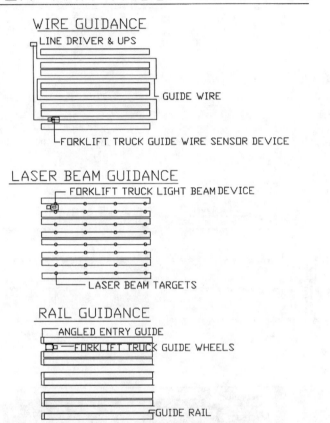

Figure 3.35 VNA and HROS guidance: wire guidance, rail guidance, and laser beam guidance.

guides are installed on both openings to the aisle entrance. The aisle end angled entry guides occupy a portion of the main traffic aisle.

At the main traffic aisle, angled entry guides have a wider opening (nominal at 7 to 10 degrees from the aisle) and become narrower as each entry guide meets the aisle guide rail. The aisle entry guides are rounded, spring loaded, curved, and straight. An entry extension outward from the rack creates a 27-in. open space. When this area is covered with a deck, it provides a good location for a trash container or a document control table that improves facility appearance, safety, and housekeeping.

In most guide rail applications, the angle iron 4-in. side extends upward (leg-up) and comes in contact with the vehicle guidance rollers. To secure the guide rail onto the floor, in the typical installation the rail is anchored on 8- to 9-in. centers for the entry guides and for the first 12 ft. into the aisle. After the first 12 in., the anchors do not receive as much pressure due to proper truck alignment. This per-

Type of guidance concept	Number of trucks									Total	%
	1	2	3	4	5	6	7	8	9+		
Rail	66	26	15	3	1	1	-	1	2	109	33%
Wire	127	55	12	7	7	6	1	4	2	221	67%
Total	187	81	27	10	8	7	1	5	4	330	100%

Type of guidance concept	1984	1985	1986	1987	1988	1989	1990	Total	%
Rail	5	9	8	11	20	48	10	111	35%
Wire	6	31	25	24	45	51	30	212	65%
Total	11	40	33	35	65	99	40	323	100%

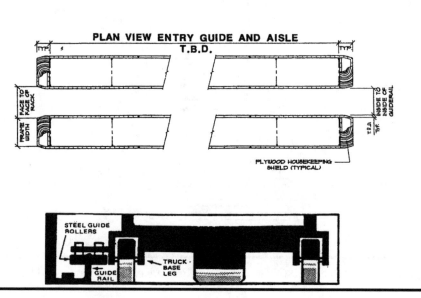

Figure 3.36 Guidance usage, single rail guidance, and rail guidance plan view.

mits the anchors on nominal 24 in. centers. To provide smooth travel in the aisle and reduce guide roller wear, at the locations where two angle iron pieces are joined together, the angle iron is ground smooth and painted with a safety-approved colored coating. The rail guidance systems are either single or double-rail systems that are elevated and floor-mounted or rack-mounted.

A double-rail guidance system has a nominal 4 ft. 6 in. to 4 ft. 8 in. clear space between rails. The clear space between the rails is determined by the vehicle roller width, pallet or cart dimensions, and elevation for the highest pick level. The HROS truck manufacturer for each application calculates specific dimensions.

An elevated rail aisle guidance system has two rack rows on an elevated floor section. The elevated floor section aisle side has a full-length angle iron face. The vehicle two side-mounted guide rollers ride against the two metal sides. This method is used for concepts that handle pallets or hand-stacked cartons. An elevated double-rail system has a replenishment vehicle that operates within the same aisle. To perform the required transaction at these elevated carton pick position above the floor, the vehicle turns and extends the pallet into the rack opening or the order picker truck operator hand-stacks cartons onto pick positions.

The disadvantages are it is difficult to relocate or reuse the facility floor area, capital investment, and potential employee tripping hazard. Advantages are first pick position is elevated above the floor, which reduces housekeeping problems and does not require floor anchors.

In a floor-anchored double-rail concept, two guide rails are anchored full length on both sides of the aisle. In this concept the first storage level is on load beams that are elevated above the guide rails. Pallet replenishments are made from the same aisle or from a separate aisle and the pallets are set on the floor or load beam levels. The latter is used in a carton order pick system and has the same operational characteristics as the elevated rack system. The vehicle guide rollers are side-mounted or there is a guide roller device on the vehicle sides. Disadvantages are with one concept, a separate replenishment aisle and entry guides create a housekeeping and employee-tripping hazard. Advantages are that it is easy to relocate and the floor can be reused for another activity.

In a double rail anchored to the rack concept, the guide rails are attached to the rack structure for the full length to both aisle sides. This concept provides a 6 in. open space between the floor and the bottom of the guide rail. In a carton pick system, the vehicle has a guide roller device on both truck sides or the side guide rollers are set at a height to come in contact with the guide rails. Pallet or hand-stack carton replenishment is made from the pick aisle or from a separate aisle with a narrow-aisle forklift truck to the pick position rear.

Disadvantages are load beam and rail costs and separate replenishment to complete a pallet transaction. Advantages are it reduces housekeeping problems, is easy to relocate, and provides good air circulation on the bottom level.

With a single rail anchored to the floor concept, a single rail is attached to the floor for the full length of an aisle and the rail is used to guide the picker truck. The rail is on the right side and has a vehicle with a guide roller device on the right side. With this rail and guide roller arrangement and replenishment is a hand stack carton activity and is performed by a VNA vehicle that extends the pallet into the pick position or the replenishment is made from a separate aisle.

Disadvantages are housekeeping problem, replenishment vehicle with a guide roller device, and with most applications, a separate vehicle replenishment aisle for a replenishment transaction completion. The advantages are lower rail and lower load beam costs.

A single rail anchored to the rack concept is used specifically to guide an order picker truck. A guide rail is attached for the full length of an aisle on the right side at the first load beam level height. Pallets are replenished with the up-and-over rack concept and performed with a stand-up rider straddle forklift truck. There is a 6 in. clear open space between the floor and guide rail bottom and the aisle is wide enough to complete right-angle stack transactions. This concept is used in a carton pick operation from SKUs that are hand-stacked into a pick position or from pallets. The concept conserves facility floor space because there is one aisle for pallet and carton replenishment and carton pick transactions.

Disadvantages are load beam cost, the vehicle has a guide roller device, rail attached to the rack, and straddle forklift truck that has a set of forks at 14 ft. to 18 ft. elevation. Advantages are less facility floor space, reduces housekeeping problems, and permits good air circulation on all pallet levels.

VNA and HROS Electronic Guidance

A HROS vehicle electronic aisle guidance options are:

- Wire
- Magnetic tape
- Magnetic paint
- Laser beam

A HROS truck requires an aisle guidance wire, tape, or paint to extend into the intersecting aisle. This feature provides the necessary aisle and distance for a HROS truck to pick up the guidance signal. Entry guides are optional for a HROS truck that is guided by an electronic system.

A wire guidance concept is the most popular electronic concept for all HROS truck systems. The wire guidance concept has a wire that is buried (nominally ⅜ to ⅝ in. deep) in a saw-cut path in the order pick aisle center. After the wire is embedded in the floor, the saw-cut path is filled in with an approved substance. The wire runs the full length of the order pick aisle and is a closed loop for an approximate 4000 to 5000 ft. length. The loop starts and ends at a line driver (electric impulse creator). At the floor joints and especially at the expansion joints, the wire is looped to compensate for movement. The electric impulses are sent from the line driver, through the wire, and are picked up by the sensing device on the HROS truck's undercarriage. This sensing device is attached to the undercarriage in the middle of the operator platform. On a HROS truck, to prevent equipment damage and for safety, most wire guidance systems have a short-duration UPS capability and an off-wire stop feature.

After the HROS truck is on the wire guide path and the system is activated, if the HROS truck leaves the wire guide path by 1 to 2 in., truck travel is stopped in the pick aisle. Floor levelness, metal object location, content of metal hardener, and guide wire to re-bar (wire mesh) depth have specific standards to ensure good operation. After the last pick aisle exit and to have the guide wire return or exit to the main aisle, options for completing the loop design are to have it intersect with the existing wire path in the main aisle, which saves on wire installation cost and wire material, or to have its own wire path that is parallel to the existing main wire path.

This concept has a separate floor cut, installation cost, and serves as a vehicle test travel path. If a guide wire path cannot exit in a pick aisle, the guide wires path to the main aisle under an adjacent pick position row.

Disadvantages are floor cut, it creates a slurry that requires disposal, specific floor tolerances, investment, difficult to add length, and capacity length is 4000 to 5000 ft. per line driver and requires a UPS system. Advantages are reduces house keeping and tripping problems, floor positions use, used by both HROS and VNA storage vehicles, and preferred for a multi-vehicle and multi-aisle system.

The magnetic tape strip is applied to the floor surface for the pick aisle full length. With the tape aisle guidance concept, a sensing device is located on the HROS truck undercarriage and directs a light beam downward onto the floor and onto the magnetic tape. The light beam is reflected back from the reflective tape to the vehicle-sensing device that ensures that the HROS truck is on the tape guide path. The disadvantages are floor level is similar to the wire guidance concept, low durability due to the tape tearing easily, and wears with cross vehicle traffic that becomes a problem in the main aisle for HROS truck aisle entrance and to replenish the pick positions, a separate replenishment aisle or a vehicle that extends the pallet into a pick position or the cartons are hand-stacked into a pick position. The advantages are easy to add length and less expensive.

A magnetic paint order picker truck aisle guidance system has similar design and operational features to a tape guidance system.

The laser beam HROS truck aisle guidance concept is new technology. A laser beam guidance concept has a light beam source that is sent from the HROS truck to reflective targets. Reflective targets are strategically located in the pick aisle to reflect the light beam (from the truck) back to a receiver on the HROS truck. If the truck does not receive the light beam, aisle travel is stopped.

Disadvantages are capital investment, HROS truck line of sight, and replenishment vehicle to have the same guidance concept or to use a separate replenishment aisle.

When to Use Rail or Wire Guidance

If a building floor conditions accept either vehicle aisle guidance system, the question is which vehicle aisle guidance system is selected for implementation in a master carton storage or pick operation. For wire guidance, the downtime and investment are factors that influence the decision. If an operation is in a remote area and the factory-qualified wire guidance and vehicle service center is not close to the facility and could create an unacceptable downtime, a rail guidance concept is considered for an operation. If an operation is close to a factory-qualified wire guidance and vehicle service center, a wire guidance concept is considered for the operation. With a mobile HROS system (one truck per two or more aisles), wire guidance represents the lowest investment. If a HROS truck is captive to an aisle or is a mobile vehicle between two pick aisles, rail guidance offers a good return on cost.

VNA and HROS Aisle End Vehicle Slowdown Devices

Most newly constructed master carton storage and order-fulfillment operations have tall racks with very narrow pick aisles. The tall rack structures have VNA or HROS trucks that permit an employee to complete master carton or pallet storage and pick transactions at elevated positions and operate within a very narrow pick aisle. The vehicles have minimum clearances between the two rack rows. Some employees refer to the aisle travel as traveling in a tunnel. To achieve the anticipated savings, return on investment and VNA and HROS employee productivity, the VNA and HROS truck travels at maximum travel speed and gives an operator access to all master carton storage or pick positions. In many operations, to meet these objectives, the VNA and HROS trucks transfer between two aisles and while in the aisle the VNA and HROS truck requires a guidance system.

Per a master carton storage or pick area design, the VNA and HROS truck performs:

- Transfer between two aisles, prior to an aisle transfer to obtain a very slow travel speed. This feature reduces potential accidents and to complete a storage/pick transaction at the aisle last pick position.
- With dead-end aisles (one turning aisle) prior to aisle transfer or to perform an aisle end pick transaction, the VNA and HROS truck travel stops or slows down.

This feature reduces potential accidents and permits storage and pick transaction completion at the last position in an aisle or in a dead-end aisle stop to avoid hitting the building wall. The aisle end VNA and HROS truck travel stop or slow down concept (see Figure 3.37) provides the operator with a signal or stops the VNA and HROS truck to prevent uncontrolled travel into a wall at a dead-end aisle or exits the pick aisle at a high travel speed with potential to hit another truck or employee.

VNA and HROS Slowdown Concepts

The VNA and HROS truck slowdown concepts can be rail and rack, mechanical, electromagnetic or automatic. The VNA and HROS truck slowdown or stop concepts are used with some modifications on an existing vehicle or in a new facility. If an electromagnetic system is considered for an existing facility, prior to the system purchase and implementation, careful attention is given to the floor surface, loading, metal objects, and re-bar or wire mesh location, and depth.

Vehicle slowdown options are:

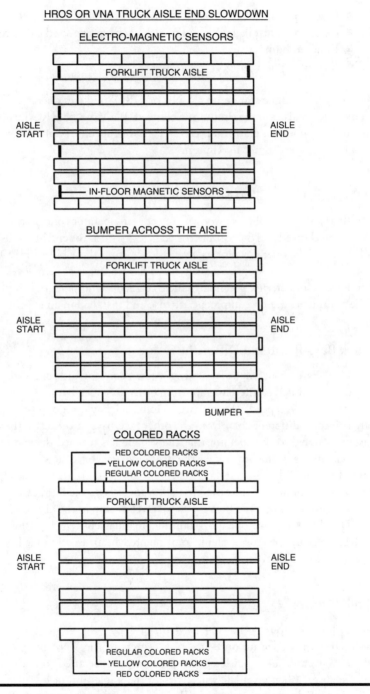

Figure 3.37 End of aisle slow down devices: colored racks, bumper, and electromagnetic.

- Operator controlled
- Operator controlled with end of aisle rack bays painted a different color
- Rail and bumper stop
- Electromagnetic stop

A wire guidance system on a VNA or HROS reduces operator effort to steer a vehicle down aisles at maximum travel speed. If a VNA or HROS truck load-carrying forks or operator platform is elevated above a specific point, the VNA or HROS truck travel speed is not maximum.

Manually Controlled Slowdown

In a manually controlled slowdown concept, an operator controls a VNA or HROS truck travel speed and determines the vehicle arrival at the end of the aisle. At the end of the aisle, the operator proceeds very slowly from the VNA or HROS aisle to the main traffic aisle or stops at a dead-end. Disadvantages are employee training, low employee productivity, potential vehicle accidents, and potential employee injuries. The advantages are no equipment expense and installed in an existing building.

Rack Bays Painted a Different Color

In the operator-controlled aisle end slowdown concept, aisle end rack bays and load beams are painted a different color from other aisle rack bays. The operator controls the vehicle travel speed. The last two to four rack bay upright posts and load beams are painted a different color from the other rack bays in the aisle. The different colors are a signal to the operator that the vehicle has reached the end of an aisle and the operator is trained to stop at the dead-end aisle or to slow the vehicle for entry into the main aisle. The standard rack component colors are (a) green for high travel speed in the middle of the aisle, (b) yellow for slow to medium travel speed prior to the aisle ends, and (c) red for very slow travel or stop at the aisle ends.

Disadvantages are operator training, manual concept, and low productivity. Advantages are low-cost system, implemented in an existing facility, and used in combination with other systems.

Rail Bumper

A rail bumper concept has a 4 by 4 in. or 4 by 5 in. angle iron member that is secured with the leg up on 8- to 9-in. centers to the VNA or HROS truck aisle floor. With a dead-end aisle concept, a bumper is secured to the floor at the end of the aisle. When a vehicle wheel strikes the angle iron or bumper at slow travel speeds, vehicle travel is stopped. The rail and bumper are painted with a safety-approved color. The concept is used in pick aisles that have dead-ends. When used in conjunction with the colored rack concept, it provides the advantages of both concepts and reduces equipment and increases floor maintenance problems.

Electromagnetic Guidance and Stop Concept

The electromagnetic guidance and stop concept has a sensing device that is attached to the VNA truck undercarriage and magnets that are set in the floor on both sides of the aisle at a specific distance to the aisle end. As the VNA or HROS truck travels over the magnets, the sensing device network automatically slows the VNA or HROS truck travel speed and warns the operator that the vehicle is approaching the aisle end. With this slow travel speed, the VNA truck leaves the aisle and enters the main traffic aisle. The concept is best used in an aisle that has two turning aisles and a truck wire guidance system. When an electromagnetic concept is installed with the colored rack concept, both concept advantages are realized by the operation.

Before deciding to install an aisle end slowdown or stop concept in an operation, review the facility layout, floor quality, and operational procedures. Design factors are reviewed with a vehicle manufacturer, managers, and employees.

Master Carton HROS Trucks

The carton order picker truck is a manually controlled rider (high-rise or multi-level) HROS truck that has the ability to have guided travel vertically or horizontally in an aisle to the assigned pick position. Arriving at the assigned pick position, the employee scans the pick position identification and carton identification and transfers the ordered carton quantity from a pick position onto a truck pallet or cart. With all pick transactions completed, the truck travels forward or reverses to exit the pick aisle to travel to an adjacent aisle for additional pick transactions or to deposit picked cartons into an assigned area.

Various HROS trucks (see Figure 3.38):

- Handle cartons
- Handle a pallet or cart, including (a) counterbalanced truck, (b) straddle truck, and (c) platform truck
- Handles a pallet as a storage vehicle and permits an operator to pick cartons

HROS Truck with a Carton Load-Carrying Surface

A HROS truck with a carton load-carrying surface is an electric battery-powered truck that has an operator platform and a carton-carrying surface. Both elevate to a pick position and travel in a guide aisle.

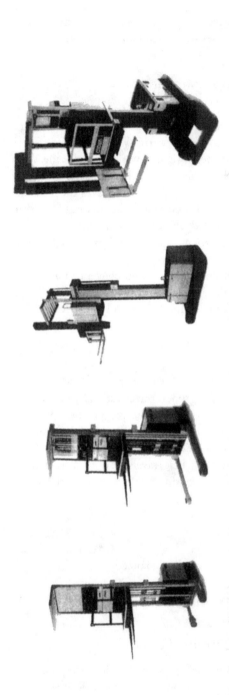

Figure 3.38 HROS vehicles: counterbalanced HROS, straddle HROS, VNA man-up, and VNA man-up.

Counterbalanced Truck

A counterbalanced HROS truck has a set of pallet forks that extend beyond the truck short straddles. Truck motor and battery weight is forward or ahead of an operator platform and offsets a pallet or cart weight. This feature allows the counterbalanced truck to handle a typical 48-in. long pallet or cart with a 3000-lb weight to a 20-ft. height above the floor.

Straddle Truck

A straddle HROS truck is similar to the counterbalanced truck. Its unique feature is that the straddles extend beyond the set of forks. The operator platform is between the drive motor and forks. With an elevated pallet or cart, the drive motor and battery and the two straddles stabilize or support a 48-in. long pallet that weighs 3000 lb. up to 20 to 30 ft. above the floor. The straddle truck operational characteristics are the same as the counterbalanced HROS truck.

Platform Truck

A HROS platform truck is similar to a counterbalanced truck. The platform truck is designed to elevate the operator platform and cart up to 20 ft. above the floor with a maximum total 500 lb. or less weight.

Man-UP VNA Truck

The man-up VNA truck is very similar to the straddle HROS truck. A VNA or HROS truck is designed to travel horizontally or vertically in the aisle and elevate the operator platform up to 30 to 40 ft. above the floor. The man-up VNA truck is designed to handle a pallet or cart and perform pallet storage and carton pick transactions. Prior to operating a straddle or counterbalanced HROS truck in an aisle that is designed for a VNA truck, the truck manufacturer assures that the aisle width, and truck guidance system meet the truck safety and operational features.

HROS Truck Master Carton-Carrying Devices

The master carton-carrying devices (see Figure 3.39) permit a picker to transfer the required carton quantity from the pick positions onto the HROS truck and provide the support for the picked cartons as the HROS truck travels in an aisle. The master carton-carrying devices are:

- Load-carrying surface is a solid sheet metal deck that has the structural strength to support a specified master carton quantity and weight. Most load-

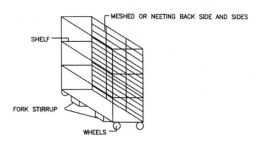

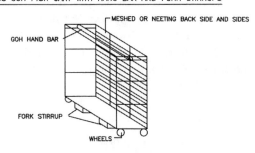

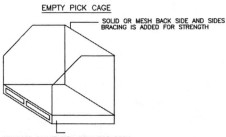

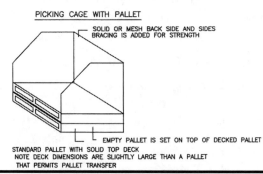

Figure 3.39   HROS picking devices: picking cart and picking pallet.

carrying surfaces are welded or connected with nuts and bolts to the HROS truck frame.

■ Pallet dimensions determine the HROS truck set of forks length and aisle width. A claw device is located at the operator platform base to grab and secure the pallet middle stringer to the HROS truck and permits an employee to step onto the pallet.

■ Multi-shelf or single shelf cart. The aisle width determines the cart width. In most applications, a chain secures the cart to the HROS truck and the cart bottom shelf has two sets of fork sleeves that permit the HROS truck set of forks to be inserted full length.

Pick Cage

A HROS pick cage is used with a HROS truck that has an employee pick cartons onto a pallet. The HROS pick cage helps reduce a picker's fear of high levels and product damage. The pallet cage has two sides, a rear wall, and a bottom. The material is wood, metal, or plastic structure that has a bottom support device with a middle stringer. The picking cage cavity or dimension holds a pallet with the pallet stringer facing the open side. The picking cage open side is designed to permit a picker to transfer cartons from a pick position onto a pallet and to easily remove a pallet from the pick cage.

The height of the pick cage sides is predetermined by the pick computer cube program. With the picking cage middle stringer clamped by the vehicle claws and during elevated vehicle travel in the pick aisle as the elevated vehicle starts and stops in the aisle, the pick cage sides and rear wall prevent cartons from falling to the floor and the clamp securely holds the pallet. If the HROS truck does not have a clamp device, the pick cage is secured by a chain and lock to the HROS vehicle overhead guard or other member.

HROS Truck-Routing Concepts

A HROS truck-routing pattern is a specialized routing pattern that is used in a HROS carton picker concept. A HROS picker routing pattern directs a picker who rides on board a HROS vehicle to an assigned pick position. The vehicle travels in a vertical and horizontal direction down the aisle between two pick position rows.

The HROS routing patterns (see Figure 3.40) are:

■ One-way truck or vehicle traffic through an aisle
■ Two-way or forward and reverse directional travel in an aisle

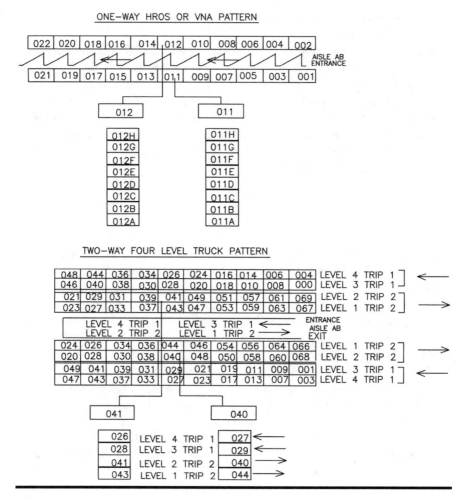

Figure 3.40 HROS picking pattern: one-way and two-way four-high pick levels.

One-Way HROS Truck-Routing Pattern

The one-way HROS truck traffic or routing pattern through a pick aisle directs an order picker to make one trip through an aisle. With this routing pattern, as the picker enters a pick aisle, the first pick position or lowest number is at the first rack bay bottom and the last pick position or highest number is at the last rack bay top.

This routing pattern guides the order picker to enter the aisle at one end and to exit at the opposite end. Each next rack bay bottom rack position has the bay or stack lowest number. At the bay bottom, the pick position number sequence is from low to high in a vertical direction. With this pattern the bottom pick position is numbered 0010 and the above pick position is numbered 0011, that is in the same rack bay on the second level and so on. In a rack bay after the pick positions reach the maximum rack position number, the next arithmetic pick position is the aisle rack bay bottom position on the aisle opposite side. With the first bottom bay pick position number 0010, the next bottom bay pick position number is 0020. With this order picker routing pattern, the order picker picks from both pick positions in the aisle and transfers to the aisle to complete a CO.

Disadvantages are lower productivity due to increased up and down vehicle movement, double travel with a complete order or full load, turning aisles at pick aisle ends, and increased product damage from cartons falling from a full pallet at high vehicle travel elevations. Advantages are easy to implement and trips are in one direction.

Two-Way HROS Truck-Routing Pattern

The two-way HROS truck-routing pattern directs the picker and truck to start at the highest level pick position. At the end of the pick aisle, the operator lowers to the same pick position bottom level and travels (picks) from the aisle end to the aisle front. With trip back to the aisle front (entrance) and the two-way routing pattern has an aisle end for HROS truck entrance and exit. The basic HROS routing patterns (see Figure 3.41) are:

- Four-level pick positions
- Six-level pick positions
- Eight- to ten-level pick positions

Four-Level HROS Truck-Routing Pattern

In a four-high rack bay pick position aisle with a maximum 36 in. height per opening, the picker travels down the aisle to withdraw cartons from rack 3 and 4 (highest) pick levels. At the end of the pick aisle, the operator lowers the HROS truck to the floor and travels (reverse direction) down the aisle to the pick aisle entrance. During this aisle trip, the picker withdraws cartons from rack 1 and 2 pick positions.

Six-Level HROS Truck-Routing Pattern

If the vertical rack structure has six rack bay levels, to complete the down-aisle trip, one HROS truck trip is made in the elevated position, thus permitting a HROS truck driver to withdraw cartons from the 4, 5, and 6 rack levels. At the pick aisle end, the HROS truck lowers to the floor and makes the reverse down-aisle trip to withdraw cartons from the 1, 2 and 3 rack levels.

TWO—WAY SIX LEVEL TRUCK PATTERN

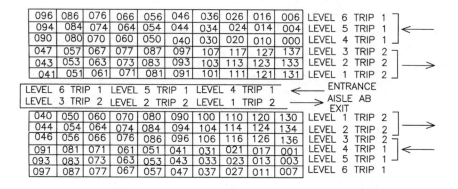

096	086	076	066	056	046	036	026	016	006	LEVEL 6 TRIP 1
094	084	074	064	054	044	034	024	014	004	LEVEL 5 TRIP 1
090	080	070	060	050	040	030	020	010	000	LEVEL 4 TRIP 1
047	057	067	077	087	097	107	117	127	137	LEVEL 3 TRIP 2
043	053	063	073	083	093	103	113	123	133	LEVEL 2 TRIP 2
041	051	061	071	081	091	101	111	121	131	LEVEL 1 TRIP 2

LEVEL 6 TRIP 1	LEVEL 5 TRIP 1	LEVEL 4 TRIP 1
LEVEL 3 TRIP 2	LEVEL 2 TRIP 2	LEVEL 1 TRIP 2

ENTRANCE
AISLE AB
EXIT

040	050	060	070	080	090	100	110	120	130	LEVEL 1 TRIP 2
044	054	064	074	084	094	104	114	124	134	LEVEL 2 TRIP 2
046	056	066	076	086	096	106	116	126	136	LEVEL 3 TRIP 2
091	081	071	061	051	041	031	021	017	001	LEVEL 4 TRIP 1
093	083	073	063	053	043	033	023	013	003	LEVEL 5 TRIP 1
097	087	077	067	057	047	037	027	011	007	LEVEL 6 TRIP 1

TWO—WAY 12 LEVEL TRUCK PATTERN

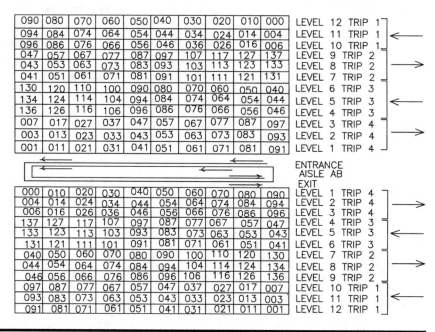

090	080	070	060	050	040	030	020	010	000	LEVEL 12 TRIP 1
094	084	074	064	054	044	034	024	014	004	LEVEL 11 TRIP 1
096	086	076	066	056	046	036	026	016	006	LEVEL 10 TRIP 1
047	057	067	077	087	097	107	117	127	137	LEVEL 9 TRIP 2
043	053	063	073	083	093	103	113	123	133	LEVEL 8 TRIP 2
041	051	061	071	081	091	101	111	121	131	LEVEL 7 TRIP 2
130	120	110	100	090	080	070	060	050	040	LEVEL 6 TRIP 3
134	124	114	104	094	084	074	064	054	044	LEVEL 5 TRIP 3
136	126	116	106	096	086	076	066	056	046	LEVEL 4 TRIP 3
007	017	027	037	047	057	067	077	087	097	LEVEL 3 TRIP 4
003	013	023	033	043	053	063	073	083	093	LEVEL 2 TRIP 4
001	011	021	031	041	051	061	071	081	091	LEVEL 1 TRIP 4

ENTRANCE
AISLE AB
EXIT

000	010	020	030	040	050	060	070	080	090	LEVEL 1 TRIP 4
004	014	024	034	044	054	064	074	084	094	LEVEL 2 TRIP 4
006	016	026	036	046	056	066	076	086	096	LEVEL 3 TRIP 4
137	127	117	107	097	087	077	067	057	047	LEVEL 4 TRIP 3
133	123	113	103	093	083	073	063	053	043	LEVEL 5 TRIP 3
131	121	111	101	091	081	071	061	051	041	LEVEL 6 TRIP 3
040	050	060	070	080	090	100	110	120	130	LEVEL 7 TRIP 2
044	054	064	074	084	094	104	114	124	134	LEVEL 8 TRIP 2
046	056	066	076	086	096	106	116	126	136	LEVEL 9 TRIP 2
097	087	077	067	057	047	037	027	017	007	LEVEL 10 TRIP 1
093	083	073	063	053	043	033	023	013	003	LEVEL 11 TRIP 1
091	081	071	061	051	041	031	021	011	001	LEVEL 12 TRIP 1

Figure 3.41 HROS picking pattern: two-way six-high and two-way twelve-high pick levels.

Eight- or Ten-Level HROS Truck-Routing Pattern

In an eight to ten vertical rack pick position concept with a 30-in. maximum height opening, the HROS truck travels down the aisle and withdraws cartons from the highest four or five levels. In one aisle trip, a picker picks from levels 10, 9, and 8 and in the second trip from levels 7 and 8. The third pick aisle trip permits a picker to withdraw cartons from the levels 5, 4, and 3. In the final pick aisle trip, the picker withdraws cartons from levels 2 and 1.

HROS Aisle Design

Important aisle design considerations are to:

- Have sufficient ceiling height.
- Have sufficient aisle width to permit pallet or hand-stack carton replenishment to a pick position.
- Provide sufficient in-feed and out-feed staging area in the HROS front area.
- Have sufficient aisle length and WMS program that minimizes the vehicle to transfer between two pick aisles. With a VNA vehicle the additional major considerations are the floor levelness and ability to install a guidance concept.

Disadvantages are investment in equipment, installation of a guided aisle, one pallet or cart with cartons per trip, operator to wear a safety harness, carton replenishment is made from the same aisle or from a separate replenishment aisle, at least a 20- to 25-ft. clear ceiling height above the floor and good picker routing pattern. Advantages are improved space utilization, good picker routing pattern means high picker productivity per aisle, permits SKU hit concentration and hit density, with guide path, reduces equipment and building damage, and permits high vehicle travel speed, minimizes building area, and handles pallets or carts.

Operator-Up Side-Loading Forklift Truck

Features of an operator-up side-loading forklift truck are the same as a standard side-loader forklift truck, except that the operator elevates and lowers with a pallet to a position. The driver performs one to two additional transactions per hour more than with the standard side-loader forklift truck.

The truck forks are placed on a set of straddles, which permits a vehicle to pick up and deliver a pallet from the forklift truck aisle. As the set of forks rises, the operator cab rises with the pallet. In one model, at the rack position, the set of forks rotate on a turntable to perform a transaction from either aisle side. In a second model, at the pick-up station the set of forks is flipped to the proper side to make the transaction at the rack position. This feature gives the forklift trucks their nickname of turn-a-load or rack loader.

A forklift truck is powered with an electric rechargeable battery and operates in an aisle that is 10 to 20 in. wider than the pallet. The bottom pallet rack position is required at a 14- to 16-in. elevation above the floor and the top rack position is 20 to 21 ft. high. An operator performs 17 to 18 transactions per hour.

Side-Loading Turret Forklift Truck with Operator Up

This forklift truck forks are attached to the carriage top. The carriage is attached to a telescopic mast that permits the forks to handle a pallet transaction from a rack position on either aisle side. For best operational results, the forklift truck has an aisle guidance system, aisle-end pick up and delivery stations and dead-level floor. The forklift truck travels in a 5 ft. 6 in. to 6 ft. 6 in. wide aisle that is a nominal 1 ft. wider than the pallet. To change aisles, a forklift truck requires a 17 to 18 ft. wide intersecting aisle. The forklift truck receives its power from an electric rechargeable battery.

Counterbalanced Side-Loading Forklift Truck with Rising Cab and Auxiliary Mast Lift

A multi-mast forklift truck carries a pallet in the operator cab front and has all vehicle movement and set of fork travel controls. As the cab moves up or down, a pallet on an auxiliary mast moves with the front cab. This feature provides a forklift truck with a free set of forks that allow a pallet transaction to a 40-ft. high rack position with minimal damage to product or equipment due to the operator having a complete view of the pallet transaction. In a carton storage/pick operation, the operator hand stacks cartons. A forklift truck has 6 to 20 in. clearance from the ceiling bottom. Compared to other forklift trucks, operator productivity is increased by one to two pallets per hour.

Outrigger Stand-Up or Sit-Down Truck with Rising Cab and Fixed Mast

This type of vehicle is a hybrid between an AS/RS crane and a side-loading forklift truck. The forklift truck handles a pallet at a 60-ft. high rack position. With some vehicles, the set of forks is attached to the cab front and the cab elevates and declines with a pallet on one rigid mast to a pallet rack position. Other forklift trucks have dual masts with shuttles or a set of forks between them. The first storage position is a nominal 16 in. above the floor. Some vehicles carry a pallet between the two outriggers and are guided by a rail in the aisle middle. Other vehicles have guide wheels mounted on both vehicle sides. In an aisle, a truck obtains its power from a ceiling DC power system and recharges the rechargeable battery. Normal vehicle aisle transfer time ranges from 1 to 2 minutes and the transfer aisle width ranges from 24 to 27 ft. Due to the electric hook-up, top guidance system, and wide transfer aisle, in an operation a forklift truck enters from one aisle end. With proper electric devices at the dual mast vehicle top, a forklift truck entry is designed at both aisle ends. A storage aisle is 4 to 6 in. wider than a pallet. To perform a transaction at an aisle end, an 8 to 10 ft. run-out is required at another aisle end. Operators perform 25 to 30 transactions per hour in a normal aisle.

Turning a Pallet in the Aisle

In a VNA forklift truck operation, to turn a pallet in the storage aisle the forklift truck driver pick up a pallet from a P/D station (see Figure 3.42). During aisle travel from the P/D station to the pallet position, the forklift truck driver turns the pallet so the pallet identification faces the aisle when the pallet is placed in a pallet position in either rack row. In an inventory program to turn a pallet in an aisle permits maximum flexibility to assign or let an employee determine pallet storage positions. At the P/D station, the pallet has the pallet identification face the aisle to assure forklift truck driver pallet identification line of sight.

In a VNA forklift truck operation, not to turn a pallet in the storage aisle requires the forklift truck driver to pick up a pallet from a P/D station with a 1 to 2 in. narrower aisle width. In a directed put-away concept, at the P/D station the forklift truck driver determines the assigned rack row that is on the left or right aisle side, and turns the pallet at the P/D station to have the pallet identification face the aisle as the pallet is placed into the pallet position. During aisle travel from

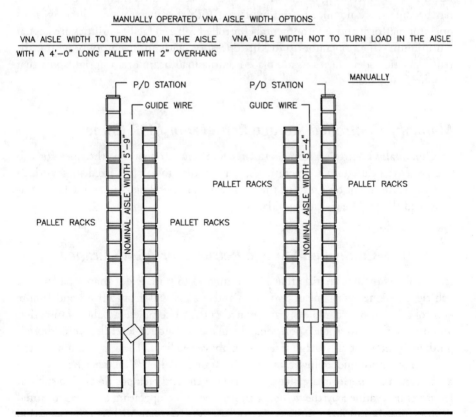

Figure 3.42   VNA forklift truck aisle: turn a load and do not turn a load.

a P/D station to an employee-directed pallet position, the forklift truck driver has the pallet remain with identification facing the aisle as the pallet is placed in a pallet position in one rack row to face the aisle. In a program, not turning a pallet permits an employee to select or assign an identified pallet to a pallet position to complete a storage transaction. At the P/D station, the pallet has a pallet identification face the aisle to assure forklift truck driver pallet identification line of sight. The preferred P/D station concept is the sawtooth.

When to Consider Operator-Down or Operator-Up Forklift Truck

If an operation pallet rack stacking height is three to four pallets high and there is cost difference between operator-down and operator-up vehicles, consider an operator-down vehicle. In an operator-down forklift operation with a program, the pallet position identifications are placed onto the upright frame or load beam to assure pallet storage position identification line of sight. The pallet position identifications are arranged with arrows, numbers, or color coded to correspond to the load beam level. For stacking heights above three to four pallets high, the operator-up has become more popular in the industry. With the operator-up vehicle with its operator platform and environmental controlled cab and facility utilities are kept at a minimum. In an operator-up forklift operation with an inventory program, the pallet position identifications are placed onto a load beam or an upright post that supports the pallet.

Manually Controlled Storage Retrieval or MSR Crane

In a manually controlled storage and retrieval crane operation, the operator cab rises with the cab and the operation is very similar to a VNA fixed mast forklift truck operation. For best results, the vehicle has aisle-end pick-up and delivery stations and the middle aisle rail guidance.

Computer-Controlled Storage Retrieval or AS/RS Crane

An AS/RS crane has an aisle that is a nominal 3 to 6 in. wider than a pallet. The vehicle is designed with one or two masts and as a manual (from a front-end remote control station) or a completely automated crane to complete a pallet transaction at a 40- to 80-ft. high rack position. The crane has a dead-level floor, middle rail guidance, and the first storage level 34 in. above the floor. There is clearance at the ceiling for a mast and ceiling attached DC system. All AS/RS cranes have a 15 to 20 ft. run out to perform a pallet transaction at the end rack positions. The run out length is determined by the vehicle length. The crane operating area has minimal utilities and the tall racks support the roof and walls. In an AS/RS crane operation with an operation, the pallet position identification is not required due to the fact that the actual AS/RS crane operation is microcomputer controlled. To assure that the assigned pallet has arrived at the assigned AS/RS aisle, each pallet on an AS/RS conveyor or shuttle car in-feed and out-feed system has a pallet identification, and has a fixed-position scanner clear line of sight. Good operational practice and due to employee-controlled forklift trucks, in activities that require a pallet identification scan transaction, a pallet identification is attached to each pallet.

Sort Link, Mole or Car-In Rack

Sort link, mole or car-in rack concepts are dense or multiple deep pallet storage positions and a mobile cart or car. A DC electric-powered mobile cart or car transports a pallet over a fixed travel path between the storage area and the pallet pick and delivery station. In a sort link, mole, or car-in rack storage operation, the pallet position identification is required due to the bar scanner tracking as the cart travels on the storage area and is updated in the material handling system (MHS). To assure that an assigned pallet has arrived at an assigned storage aisle, each pallet in a sort link, mole, or cart-in rack storage concept has the MHS update identified pallet and has a clear line of sight to a fixed-position scanner that is located along the travel path. For additional information, see sort link, car-in rack, or mole in the rack concept section of this chapter.

AGV or Automatic Stacking Guide Vehicle

The AGV or automatic stacking guided vehicle is a counterbalanced or straddle vehicle. The power source is several rechargeable electric batteries. The AGV vehicle has three to four wheels (at least one is steering), a fixed mast, a set of load-carrying forks, a manual or remote-controlled on-board control device, a wire guidance system, and a very flat floor.

After the AGV vehicle picks up a pallet from the receiving area, it travels over the main traffic aisle to an assigned aisle, enters the assigned aisle, travels in the aisle to an assigned pallet position, and deposits the pallet to an assigned position. In the pallet position, the WMS pallet identification faces the aisle. To pick a pallet for a customer order, the computer directs the AGV to the assigned pallet position in an aisle. At the assigned pallet position, the AGV takes the pallet on-board, travels in the aisle, and exits to a main traffic aisle and travels to the shipping dock area. In the shipping dock area, the pallet is transferred to a customer-assigned staging area. An AGV completes a pallet transaction to a rack position that is 180 in. above the floor and performs 10 to 12 pallet transactions per hour. Pallet transaction aisle width is 10 to 12 ft. wide for a counterbalanced truck and for the straddle type truck it is less. Due to floor-stacked pallet instability, the majority of the AGV vehicles stack pallets one high on the floor. In a storage operation, AGV or MHS system updates the MHS as to the pallet storage identification location.

Computer-Controlled AS/RS Crane Considerations

To have an efficient and cost-effective AS/RS pallet storage operation, AS/RS crane factors are:

- Pallet handling device. A pallet-handling device is an AS/RS crane component that carries one or two pallets between a pick-up and delivery station, storage position, and completes a storage transaction. The pallet handling device is either a set of forks or platen. The AS/RS crane has a set of forks that are inserted into a pallet fork opening and complete a pallet transaction to a pallet rack position that has minimal vertical clearance in a pallet position, exact tolerances, design approved by a seismic engineer, lower pallet position cost with two to three pallets per bay, and maximum pallet number per aisle. A pallet bottom has a slave pallet or has structural strength to span a rack position pallet support arms/rails and rack opening has two rails, two support arms and sufficient top and bottom pallet position clearance to complete a transaction. This option has a higher cost per pallet due to one pallet per bay, design approved by a seismic engineer, with an increase in the upright number per aisle, fewer pallets and pallet openings per aisle, and tolerances are not as tight.
- Pallet handling capacity. Pallet capacity options are:
  - Single shuttle or carrier. A single carrier has the capacity to handle one pallet per aisle trip. This feature has less run-out space, less weight, and lower overall productivity.
  - Dual carriers or shuttles. A dual carrier concept has the capacity to handle two pallets per trip into the aisle. An AS/RS crane is designed with the carriage area to take two pallets on board. The feature has additional run-out space, additional crane weight, and an additional 20 to 25 percent productivity due to additional starts and stops and slower travel speeds in the aisle to complete the pallet transactions.
- AS/RS crane commands. Crane commands refer to computer commands that direct an AS/RS crane to complete a pallet storage transaction. AS/RS crane command options are (see Figure 3.43):
  - Single command mode. A single command mode has a microcomputer direct the crane to make one down-aisle trip. During an aisle trip, a crane completes one pallet storage transaction. With a single command mode, a crane completes 18 to 24 transactions per hour. A single command mode does not require inbound and outbound pallet transaction balance.
  - Dual command mode. In a dual command mode, an AS/RS crane makes an aisle trip. During an aisle trip, the microcomputer directs a crane to complete a pallet deposit transaction and on the aisle exit to complete a pallet withdrawal transaction. This represents two pallet transactions per trip and the crane completes 22 to 26 dual commands per hour. The dual command mode does not double the single command productivity rate

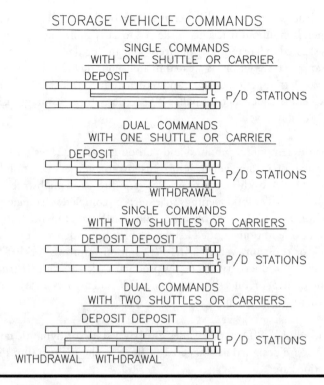

Figure 3.43 AS/RS crane or VNA commands: one carrier single command, one carrier dual commands, two carriers single command, and two carriers dual commands.

due to additional aisle travel time and set of fork (platen) activity time. To achieve dual commands, there is a requirement for a balance between the inbound and outbound transactions and aisle-end P/D stations.

Pallet Pick-Up and Delivery (P/D) Stations

An aisle end P/D station is designed as a static or dynamic pallet station and is a temporary holding pallet position for a pallet that is transferred between the VNA forklift truck or AS/RS crane storage area and an in-house transport vehicle. In the pallet storage industry, recently many companies have increased their storage space utilization and storage forklift truck productivity by a VNA forklift truck or AS/RS crane storage concept implementation. In these storage concepts, a VNA forklift truck and AS/RS crane perform the storage transactions to tall rack positions, operate within a very narrow aisle, and perform the maximum pallet transaction number per hour. The VNA forklift trucks and AS/RS cranes remain in an aisle and perform the maximum dual or single pallet transaction number per hour.

The VNA trucks and some AS/RS cranes are mobile aisle vehicles. A typical pallet storage operation inbound activity is skewed to early morning hours and outbound activity is skewed to late morning or afternoon, an imbalance in the projected pallet storage and pick transactions is created in an 8-hour day. To obtain the desired pallet storage/pick transaction balance and return on investment, a VNA forklift truck and AS/RS crane concepts have pallet P/D stations at each storage aisle end.

A P/D station:

- Provides temporary inbound and outbound pallet queues for pallets that are assigned for a storage deposit or transport to another operation location.
- Gives a VNA forklift truck or AS/RS crane direct and unobstructed access from an aisle to pallet fork openings that starts or completes a storage transaction.
- Ensures that pallet fork openings and WMS pallet identification are in the correct orientation or line of sight for pick-up or delivery by a VNA forklift truck or transport vehicle. The P/D station improves forklift truck or AS/RS crane productivity, decreases product, rack and forklift truck or AS/RS crane damage, reduces transaction errors, and assures a clear line of sight to a pallet identification.

An end-of-aisle P/D station is designed as a static or dynamic concept. Static P/D stations are designed as staggered or flush to the main traffic aisle. The P/D station options are:

- Static type that include (a) floor, (b) structural stand, or (c) standard pallet rack
- Dynamic types are (a) four-wheel carts and (b) on-board transfer car
- Gravity-powered conveyor
- Powered roller conveyor
- Powered shuttle car

Static Staggered P/D Stations

A staggered P/D station (see Figure 3.44) has a pallet fork entry facing the storage aisle. A staggered or saw-tooth rack layout has four rack rows and two storage aisles. In a pallet storage area plan view, the racks have the interior two rack rows shorter by a pallet position number at a P/D station and the exterior two rack rows extend outward toward the main traffic aisle by an equal P/D station number. The extend rack bay design provides 10- to 15-ft. aisle width between the two exterior rack rows. The open space permits any in-house transport vehicle to complete a transport transaction. The staggered P/D station concept is used with a manually operated VNA forklift truck concept, provides pallet queue, and assures that the pallet identification faces the forklift truck driver and the driver has pallet identification line of sight.

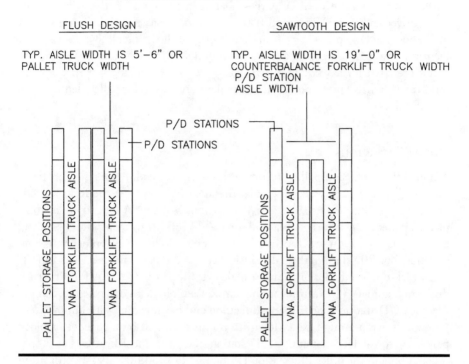

Figure 3.44 VNA forklift truck P/D stations: flush design and sawtooth design.

Static Flush P/D Stations

In a static flush P/D station concept, the P/D stations are at each rack row end and have a pallet transaction location face the main aisle. In the receiving area an in-house transport truck picks up an identified pallet with the pallet identification facing the pallet truck driver. In a flush P/D station, the pallet identification and pallet opening face the main aisle. With this pallet orientation, the pallet identification does not face the storage aisle and the VNA forklift truck driver does not have pallet identification line of sight. This means low VNA forklift truck and in-house transport employee productivity due to pallet double handling to have the pallet identification face the VNA forklift truck or transport vehicle operator. The concept handles a low volume, minimal pallet queue, interfaces with a VNA forklift truck, and is difficult for a dynamic operation, and is not preferred for an inventory program concept.

VNA Aisle Middle P/D Station

In some VNA forklift truck operations, an in-house transport pallet truck places a pallet in a VNA forklift truck aisle middle. After a pallet truck picks up a pallet in the receiving area, the identification faces the pallet truck driver. As the pallet truck driver enters the VNA aisle, the pallet identification faces the pallet forklift truck driver and if the VNA forklift truck is in the aisle, the VNA forklift truck driver does not have pallet identification line of sight. It is difficult to obtain a pallet identification light of sight for a scan transaction. A VNA aisle middle P/D station is not preferred for a VNA forklift truck operation due to the fact that a VNA forklift truck driver performs additional pallet transactions for pallet identification line of sight.

Floor P/D Stations

A floor P/D station is a pallet position at the rack row end. After a forklift truck picks up a pallet in the receiving area, the identification faces the forklift truck driver. In this floor P/D station, a pallet truck or WA or NA forklift truck completes a pallet deposit at the VNA end for a VNA forklift truck pick-up by having the in-house transport truck driver face the racks. With an inventory program, and at a floor P/D station, the pallet identification does not face the VNA forklift truck pick-up. The pallet orientation makes it difficult for a VNA forklift truck driver to obtain a pallet identification scan transaction. With one pallet level high, the floor P/D station is poor space utilization and has no cost. With an inventory program, a wrap-around pallet identification is preferred on a stringer or block pallet due to the fact that at one P/D station, the pallet identification has proper orientation and the other P/D station to obtain VNA forklift pallet identification line of sight, it requires an in-house transport truck to double handle the pallet.

Structural Frame P/D Stations

Structural frame or stand P/D stations are placed at the rack row ends. After a forklift truck picks up a pallet in the receiving area, the identification faces the forklift truck driver. In the storage area, a WA or NA forklift truck deposits a pallet at the rack row end P/D station. With an inventory program and structural P/D station, each pallet identification does not face the VNA forklift truck pick-up. The pallet identification orientation makes it difficult to obtain a pallet identification scan transaction. With one pallet level high, the structural P/D station is poor space utilization and has low cost. With an inventory program, a wrap-around pallet identification is preferred on a stringer or block pallet due to the fact that in one P/D station, the pallet identification has proper orientation and in the other P/D station an in-house transport truck is required to double handle the pallet to obtain VNA forklift pallet identification line of sight.

Standard Pallet Rack P/D Stations

Standard pallet rack P/D stations are two standard pallet rack bays that extend beyond the first pallet storage positions and the VNA forklift truck aisle is between the rack rows. In the receiving area a transport truck picks up a WMS-identified pallet and in a tight aisle width, a manual pallet truck completes a pallet deposit to the floor level P/D station. With an identified pallet P/D deposit, in a rack row a wrap-around stringer or block pallet identification faces the VNA forklift truck aisle. To have an efficient in-house transport concept, it requires a powered pallet truck or forklift truck to transport the pallet and a hand pallet truck to complete the P/D station pick-up or delivery transaction.

The rack bays above the P/D stations provide additional pallet storage positions. With the potential low pallet transfer employee productivity and rack and product damage, the manual pallet truck P/D and a standard pallet rack P/D station are not preferred for a VNA forklift truck operation.

Staggered Standard Pallet Rack P/D Stations

A staggered standard pallet rack P/D station concept is designed with four standard pallet rack bays. The two interior pallet rack bays are short by one rack bay and the two exterior standard pallet rack bays extend beyond the interior pallet rack bays. The standard rack P/D station design forms a U shape and creates a wide aisle that permits a manual or powered pallet truck or a WA or NA forklift truck to complete a P/D station pallet deposit. After a pallet deposit with a one face or wrap-around stringer or block pallet identification, the pallet identification faces the VNA forklift truck aisle. The rack bays above the P/D stations provide additional pallet storage positions. With the high pallet transfer employee productivity, low rack and product damage and proper pallet orientation, the standard pallet rack P/D station is preferred for a VNA forklift truck operation with a pallet identification. P/D station transfer options are:

- Rub bars between upright frames, with two posts. The bar height does not cause pallet hang-up.
- Angle iron pallet back stop with the base facing the rack interior and is anchored to floor. The angle iron height does not cause pallet hang-up.
- Elevated rack bay has front-to-rear members.

Dynamic P/D Stations

A dynamic P/D station (see Figure 3.45) uses pallet conveyor lanes, on-board transfer car, shuttle car/transport vehicle, or four-wheel carts. Most four-wheel cart P/D

Figure 3.45 P/D station concepts: conveyor and AGV.

stations interface with a manually operated VNA forklift truck concept. Pallet conveyor lanes, shuttle car, and on-board transfer car P/D stations interface with an AS/RS concept.

Four-Wheel Cart P/D Stations

A four-wheel cart P/D station has a pallet on its load-carrying deck and a deck push handle that permits a VNA forklift truck to complete a P/D station transaction. The P/D station is a rack row extension and has sufficient height for an employee or guide rails to align a cart at the P/D station. The cart and pallet orientation have a pallet identification face the VNA aisle for VNA forklift truck pallet pick-up or deposit. A wrap-around identification on a pallet stringer or block is preferred to assure pallet identification line of sight for the VNA forklift truck. In a dynamic storage operation with the additional employee activity and time, the four-wheel cart P/D station is not preferred in a VNA forklift operation.

On-Board Crane P/D Stations

On-board AS/RS crane P/D stations (see Figures 3.46 and 3.47) are powered roller or drag chain conveyor sections that are located full length on an AS/RS crane aisle

Figure 3.46 P/D station conveyor.

Figure 3.47 P/D stations: forklift truck and stand, transport vehicle, and conveyor on crane.

front and both sides of a crane. One side is for in-feed pallet flow to the AS/RS crane and the other side is for out-feed pallet flow to the in-house transport forklift truck. A WA or NA forklift truck completes a pallet transfer from an AS/RS P/D station. A wrap-around pallet identification is preferred for a storage operation to provide easier forklift truck scanning of a pallet identification. With MHS communication to a pallet storage transaction, a pallet identification storage deposit by an AS/RS crane is not required to read the pallet identification.

Gravity Conveyor P/D Stations

Gravity conveyor P/D stations are non-powered conveyor travel path sections that are located on a VNA forklift truck or AS/RS crane aisle front both sides. One side is for in-feed pallet flow to an AS/RS crane and the other side is for out-feed pallet flow to a WA or NA forklift truck. A WA or NA forklift truck completes pallet transfers from a VNA forklift truck or AS/RS crane. A wrap-around pallet identification is best for the operation. The MHS communicates the pallet identification storage deposit by an AS/RS crane that is not required to read a pallet identification.

Powered Roller Conveyor P/D Stations

Powered conveyor P/D stations are powered roller or drag chain conveyor travel path sections that are located on a VNA forklift truck or AS/RS crane aisle both front sides. One side is for in-feed pallet flow to the AS/RS crane and the other side is for out-feed pallet flow to the WA or NA forklift truck. Along the powered roller conveyor travel path, a turntable or right angle turn assures that a pallet identification is facing an in-house transport forklift truck driver's line of sight. A WA or NA forklift truck, AGV, or powered conveyor completes pallet transfers from a forklift truck or AS/RS crane P/D station. A wrap-around pallet identification is preferred for the storage operation. The MHS communicates the pallet identification storage deposit; an AS/RS crane is not required to read the pallet identification.

Powered Shuttle Car P/D Stations

The powered shuttle car P/D station is a powered carrier that travels over a rail-guided fixed travel or wire guide flexible travel path between an AGV or powered conveyor P/D to an AS/RS crane aisle both sides. One side is for in-feed pallet flow to the AS/RS crane and the other side is for out-feed pallet flow to the AGV or powered conveyor travel path. The shuttled car assures that a pallet identification is facing an in-house transport forklift truck driver line of sight. A WA or NA forklift truck, AGV, or powered conveyor completes pallet transfers from the forklift truck or AS/RS crane P/D station. A front or wrap-around pallet identification is used in the storage operation. With the MHS communication the pallet identification storage deposit, an AS/RS crane is not required to read the pallet identification.

Transport Vehicle

A station transport vehicle is a microcomputer-controlled vehicle with an elevating pallet-carrying surface and is a four-wheeled buss bar motor-powered vehicle that travels over a closed and fixed travel path. The transport vehicle travel path has multiple vehicles that travel past each P/D station and in-house transport concept pallet transfer station. Per the microcomputer command, a transport vehicle is assigned to pick up a pallet from an in-house transport transfer station, travel over the travel path, and to discharge the pallet at the assigned P/D station. To complete a transfer transaction, the transport vehicle load-carrying surface raises up or declines to interface with the transfer or P/D station.

The transport vehicle load-carrying surface is set at a 30- to 36-in. high elevation above the floor. The transport vehicle design options are:

- Travel path speed to range from 30 to 48 ft. per min.
- Pallet transfer speed range from three to 10 ft. per min.
- Pallet weight range from 500 to 2500 lb.
- Carrier surface options are (a) powered roller conveyor, (b) chain, and (c) elevated ribs.
- Pallet width dimensions.
- Travel path has 90 or 180 curves.
- Travel on mezzanines.

The features are faster pallet transport speeds, easy to expand, minimal queue, on a per unit basis a lower cost, and easy to add vehicles to handle a high volume.

Pallet Delivery and Pick Up or In-House Transport Vehicles

The VNA forklift truck or AS/RS crane concept pallet delivery or in-house transport vehicle completes a pallet transport transaction between the storage area and receiving/shipping dock area. The in-house transport vehicle objectives are to assure:

- A constant pallet flow
- For a VNA forklift truck pick-up, a pallet with its fork openings and pallet identification properly oriented with a forklift truck driver line of sight as a pallet is transferred to a P/D station

The in-house pallet transport concepts are:

■ Pallet truck or forklift truck. A pallet truck or forklift truck concept inter-faces with a manual VNA lift truck operation that has staggered or flush P/D stations and to an AS/RS concept with flush P/D stations.

■ AGV. An AGV in-house transport concept interfaces with a manual VNA lift truck operation that has staggered P/D stations or to an AS/RS concept with a conveyor P/D lanes.

■ Monorail. DC electric-powered overhead monorail pallet in-house transport concept that has a powered conveyor carrying surface that interfaces with an AS/RS operation that has conveyor lane P/D stations.

■ Powered conveyor. DC electric-powered pallet roller conveyor in-house trans-port concept interfaces with an AS/RS operation that has conveyor run P/D lanes.

Captive Aisle Vehicle

With a captive aisle VNA forklift truck or AS/RS crane operation, there are mul-tiple VNA forklift trucks or AS/RS cranes, and one forklift truck or AS/RS crane remains in one storage aisle and performs all storage transactions. The projected pallet storage transaction activity has each VNA forklift truck or AS/RS crane perform at its maximum transaction activity rate. This means that the storage area computer spreads daily pallet storage transactions as evenly as possible over the VNA forklift trucks or AS/RS cranes aisles.

Mobile Aisle Vehicle

Mobile pallet storage vehicle major groups are:

■ Wide-aisle or WA forklift trucks
■ Narrow-aisle or VN forklift trucks
■ Very narrow-aisle or VNA forklift trucks
■ AS/RS crane with a transfer (T-car) car (see Figure 3.48)

With a mobile aisle forklift truck design, from the main traffic aisle a WA, NA, or VNA forklift truck has access to any storage aisle. To transfer between the stor-age aisles, a WA or NA forklift truck concept has turning aisles at the aisle ends and for a long rack row a middle cross aisle is in the rack row. With a VNA forklift truck concept, options are to have:

■ Turning aisles at the aisle ends.
■ Only one turning aisle at the aisle front and a rear aisle that permits last pal-let position access and employee emergency exit. In a storage aisle, a forklift truck completes an assigned or employee-directed pallet storage transaction.

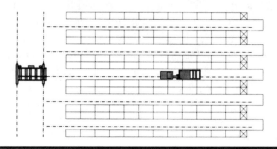

Figure 3.48 AS/RS crane T-car.

With a mobile aisle concept, the storage area computer spreads the pallet storage transactions that are based on an arrangement to maintain the forklift truck productivity standards and service the receiving and shipping area requirements. The mobile aisle forklift truck is powered by a rechargeable battery, has three or four wheels, and is manually controlled. VNA forklift trucks have wire or rail guidance in the storage aisle.

Most AS/RS cranes are restricted to one storage area aisle. With a proper storage area design, T-car, microcomputer controls, and sufficient time, a captive aisle AS/RS crane is transferred by a transfer or bridge car between storage area aisles. With a DC electric-powered transfer (T-car) in a transfer aisle, a T-car has the mobility and capacity to transfer an AS/RS crane from one aisle to another aisle. The T-car is microcomputer controlled to physically receive an AS/RS crane in its cradle or bridge (take on-board) and travel from one aisle, travel in the T-car aisle on its wheels over a rail-guided travel path, and to align itself for AS/RS crane discharge into another aisle. The T-car aisle width is 25 to 40 ft. wide according to the AS/RS crane length and the T-car aisle is located at the rear or non-P/D station AS/RS facility side.

Deposit Activity

A deposit activity has an identified pallet placed into a microcomputer-assigned or operator-directed identified storage position and both pallet and pallet storage position identifications that were involved in the storage transaction are scanned. A scan transaction is sent on-line or delayed to the microcomputer. An accurate and on-schedule identified pallet deposit transaction completion ensures that the right SKU is in the proper place, in the proper quantity, in the correct condition, and at the correct time. An accurate and on-time deposit activity and communication to the computer allow inventory update and on-time pick activity.

Computer-Assigned Pallet Storage Position

A storage operation with an inventory program that assigns an identified pallet to an identified pallet storage position is a directed or assigned pallet storage transaction. In this concept, to complete a deposit storage transaction a forklift truck driver scans a pallet identification, a RF hand-held scanner device visual display indicates an assigned pallet storage position. After a forklift truck driver arrives at the assigned pallet position, the forklift truck driver places the pallet into a pallet position and scans a pallet storage position identification. Pallet identification and pallet position identification scan transactions are sent on-line or delayed to the computer for inventory update. To complete a directed pallet withdrawal transaction, the forklift truck driver reads the RF hand-held scanner device visual display screen. The screen indicates the assigned pallet position for the withdrawal transaction. The forklift truck driver travels to the assigned position. At the position, the forklift truck driver scans the pallet identification and the pallet position identification. The scan transactions are sent to the computer that updates the pallet identification and pallet position status inventory files.

Operator-Directed Pallet Storage Position

This is a storage operation with an inventory program that has a forklift truck driver select an identified pallet storage position for an identified pallet. In this concept, to complete a deposit storage transaction the forklift truck driver scans a pallet identification, the RF hand-held scanner device visual display indicates the pallet is ready for deposit. After the forklift truck driver arrives at a vacant or assigned pallet position, the forklift truck driver places the pallet into the identified pallet position and scans the pallet and the storage position identifications. The pallet identification and pallet position identification scan transactions are sent on-line or delayed to the computer for inventory file update. To complete a directed pallet withdrawal transaction, the forklift truck driver reads the RF hand-held scanner device visual display screen. The screen indicates the assigned pallet position for the withdrawal transaction. The forklift truck driver travels to the assigned position. At the position, the forklift truck driver scans the pallet identification and pallet position identification. Scan transactions update the pallet identification and pallet position status inventory files.

Forklift Truck Patterns

A fundamental rule for a successful pallet storage operation with an inventory program is that a pallet storage transaction instruction follows a pattern through each storage aisle. The instruction form directs a WA, NA, or VNA forklift truck driver or AS/RS crane to an assigned and identified pallet storage position to complete a storage transaction. In a driver-directed put-away operation, the employee selects the put-away identified pallet location and scans the pallet identification and the pallet storage position identification. The scan transactions are sent to the computer inventory file. In a directed put-away activity, an inventory program suggests an identified pallet position.

With a sequential aisle number or routing pattern, a WA, NA or VNA forklift truck driver or AS/RS crane enters an aisle, completes the storage transaction, and exits from the same aisle. As a forklift truck or AS/RS crane travels through a storage aisle, the lowest pallet storage position number starts at the entrance from the main aisle and the pallet position numbers are progressive to the highest number at the aisle end. To assure maximum forklift truck driver or AS/RS crane productivity, sequential aisle number patterns have an arithmetic progression through an aisle. For aisle progression options (see Figure 3.49):

- Aisle pallet position numbers that end with an even digit are located on the right side of an aisle and position numbers that end with an odd digit are located on the left side.
- The aisle rack row or side has an aisle number and the pallet position numbers are progressive. For example, the right-side rack row has 100 and the left-side rack row has 200.
- Aisles are kept clear and well illuminated, and good housekeeping is maintained.

In most pallet storage operations, a forklift truck or AS/RS crane storage transaction handles one pallet per trip. The aisle routing pattern has the forklift truck or AS/RS crane enter the aisle from the main traffic aisle, complete a transaction, and travel (back-out) to the main traffic aisle. Some AS/RS cranes have the ability to

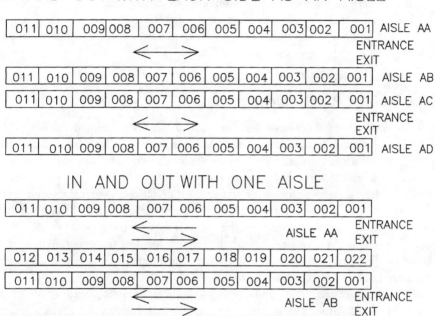

Figure 3.49 Forklift truck routing pattern: each row as an aisle and both rows are an aisle.

carry two pallets and to complete dual pallet storage transactions in one aisle. With this option, an AS/RS crane follows the same procedure.

Pallet Transaction Verification

A pallet storage operation with an inventory program requires that a forklift truck driver verify a storage transaction completion to ensure that an identified pallet has been deposited into an assigned pallet storage position; when needed the pallet is withdrawn from the storage position; and the transaction scan is sent to the computer and updated in the inventory file.

The storage transaction verification completion concepts are:

- Manual or memory
- Manual written document
- Automatic or barcode scanning method

Human Memory Concept

A forklift truck operator mentally remembers the position of a pallet. This is a basic transaction verification method that should not be considered.

The disadvantages are low productivity, possible errors, low volume and few SKUs, and difficulty controling over two shifts or a large area. The advantages are low or no cost.

Handwritten Paper Document

The handwritten method has a forklift truck driver use a printed document to record each storage transaction. The printed document is a four-column activity sheet that has a space for a forklift truck operator's name and the date. The four columns are separated into two groups each with two columns. The two columns under the deposit heading for listing all deposit transactions. Under the withdrawal heading, the two columns for listing all withdrawal transactions. After a forklift truck driver completes a transaction, the operator lists the pallet position and SKU identification number. At the shift end, the activity documents are sent to the office. An office clerk performs an inventory update. The handwritten document concept should not be considered for an operation.

Disadvantages are it can handle only low volume, requires clerical effort, adds to a forklift truck driver's work, and creates transposition errors. Advantages are no investment, and it can be used in two-shift operations and in a large facility.

Barcode Scanning or RF Reading

To verify an identified pallet storage transaction, a forklift truck driver uses a hand-held barcode scanning/RF reading device to scan an identified pallet and identified pallet storage position. Pallet or pallet storage position barcode/Rf tag labels are human/machine-readable and are discreet. After a label is attached to a pallet and at each identified pallet storage position, a hand-held barcode scanner with line of sight reads an identified pallet label and identified pallet position label and to hold the transaction in memory or with a RF device to send the transaction on-line to the computer.

In a storage operation with an inventory program that directs all transactions, prior to picking up an identified pallet a lift truck operator scans the identified pallet label and per the storage transaction instruction travels to the assigned or employee-selected identified pallet position. Arriving at the assigned pallet position, an identified pallet is placed into an identified and assigned position and the identified pallet position identification and identified pallet are scanned by the operator. This scanned information is relayed or sent by on-line transmission to a computer. In an employee-directed operation with barcode scanning/RF tag reading, a forklift truck driver scans the identified pallet, picks up the pallet and selects the storage aisle. In a storage aisle, a forklift truck driver selects and scans an identified pallet storage position, scans an identified pallet and places the pallet into an identified pallet storage position. In an AS/RS crane operation, as the identified pallet travels over a travel path, fixed-position fixed-beam barcode scanners or RF tag readers read each identified pallet. This identified pallet information and MHS tracking devices update an identified placement into a computer file.

Disadvantages are investment, employee training, position and pallet identifications, and management control and discipline. The advantages are accurate information, on-line or delayed information transfer, and high employee productivity.

In-House Transport Concepts

A pallet storage operation has several different locations where activity takes place. To complete the receiving or shipping activity, identified pallets are transported between the various locations, using pallet-handling vehicles that travel over a variable or fixed travel path. The power sources for in-house transport vehicles include manually powered, DC electric motor, electric battery, and fuel engine.

The in-house transport concepts (see Figure 3.50) are manual pallet trucks, powered pallet trucks, forklift trucks, powered vehicles with cart trains, AGVs, towlines, monorails, and overhead towveyors.

Manual Pallet Trucks

A manually powered pallet truck is used to transport a pallet over a variable and short travel path. Due to employee physical effort required, a manually powered pallet truck is used only in the dock area and cannot complete an elevated pallet storage transaction. In this type of operation, each driver has a hand-held scanner device that is attached to the employee's waist and communicates a scanned pallet identification and pallet position identification on-line or delayed to the microcomputer.

Powered Pallet Trucks

An electric battery-powered pallet truck is available in walkie, walkie/rider, and rider models and have the ability to transport one or two pallets over a long distance at a fast travel speed over a variable travel path, but cannot complete an elevated pallet storage transaction. With a powered pallet truck operation, each employee has a hand-held scanner device that is attached to the employee's waist. The scanner communicates a scanned pallet identification and pallet position identification on-line or delayed to the microcomputer. These were discussed in more detail earlier in the chapter.

Figure 3.50 Pallet in-house transport concept: overhead towveyor, forklift truck, transport vehicle and conveyor.

Wide- or Narrow-Aisle Forklift Trucks

The electric battery- or fuel engine-powered WA or NA forklift truck has the travel speed and capacity to transport one or two pallets between locations in a facility. The travel path can be variable and this type of forklift truck completes elevated pallet storage transactions to an identified pallet position or onto a VNA or AS/RS P/D station. With a WA- or NA-powered forklift truck operation, each employee has a hand-held scanner device that communicates a scanned pallet identification and pallet position identification on-line or relayed to a microcomputer. A hand-held scanner and display screen are attached to an overhead guard upright member.

Powered Tugger or Wide-Aisle Forklift Truck with a Cart Train

Powered tuggers or WA forklift trucks have the capability to tow a cart train. The powered tugger, most powered WA forklift trucks, and AGVs use a draw bar to tow a cart train over a fixed or variable travel path. Each cart has the capacity to handle one or two pallets. With the cart train concept, consideration is given to the intersecting aisle widths and towing vehicle capacity and a forklift truck or machine to unload and load pallets at the assigned locations. With a powered tugger or WA forklift truck with a cart train operation, each employee uses a hand-held scanner device that communicates on-line or relayed to the microcomputer. The hand-held scanner is attached to the employee's waist, the tugger chassis or to the forklift truck overhead guard upright member.

AGV or Driver-Less Vehicle

The AGV is an electric battery-powered driver-less vehicle that has a carrying surface for one or two pallets or the ability to tow a cart train. The AGV follows a fixed, closed-loop travel path between two locations. Stop or address locations and spur (branch) lines are segments from the main traffic line and run parallel to the main traffic travel path. The return line has a separate travel path or runs parallel to the main travel path. The travel path spur and return line design permits a vehicle to unload or load a pallet at a stop while other AGVs are traveling over the travel path. All AGV types are equipped with safety bumpers and have multiple vehicles on one system.

With an identified pallet on a fixed-position carrying surface, a fixed-position, fixed-beam scanner along the AGV travel path reads the pallet identification and communicates to the MHS and microcomputer. The AGV as a pallet in-house transport concept requires:

■ Transport activity performed by a forklift truck, pallet truck, or tugger with a cart train
■ Delivery between two locations

- Fixed delivery path.
- Frequent, on–schedule deliveries.
- Automatic or forklift truck–assisted loading and unloading of pallets.
- Standard pallets.
- Smooth or level floor and gradual grade.
- Delivery travel path is a long distance and is closed loop.

AGV Control Methods

The AGV is a driverless vehicle and AGV movement is controlled with one of the following:

- Basic controls
- Advanced controls with microprocessor
- Microprocessor controls that interface with another system

The basic control method is the simplest method and is implemented in a system that has one or two vehicles, has less than four address locations and has a total guide path of less than 5000 ft. The advanced control method with microcomputer controls is used for more complex systems that have a guidance path of more than 5000 ft., a vehicle travel path that is more complex or requires blocking, if the activity requires more than four vehicles and more than four automatic unload and load stations. The microcomputer control method can interface with another automatic system. The AGV performs a pallet pick-up and delivery to and from another automated system. The pick-ups and deliveries require on-time performance and have at least the same operational design parameters as the advanced control method.

AGV Dispatch Methods

The AGV dispatch options are:

- Fundamental human dispatch method. The manually AGV dispatch concepts are:
  - Toggle switch
  - Thumb wheel switch
  - Pushbutton numeric pad
  - Via remote communications to a host computer or automated system

The AGV vehicle efficiency and safety features are:

- A safety bumper on the AGV lead end. When the bumper comes in contact with an object or employee, this stops the AGV.
- Running lights are illuminated when an AGV is traveling on the guide path. This is a colored light that is readily noticed by an employee near the travel path.
- A control pad and mimic display panel that show each AGV operational status on the guide path and permit you to start or stop the system. The panels indicate where each AGV is located on the guide path and the safety stops and they contain other devices that control the AGV system.
- An off-line stop device is an emergency device that reduces damage to the pallet, equipment, and the building. When an AGV deviates or strays from the guide path by more than 2 in., the device is activated and the vehicle is immediately stopped.
- A UPS (uninterruptible power source) that provides temporary electric power to the AGV wire guidance system during a brown-out or electric power failure. It provides power long enough to ensure that all AGVs on the travel path are returned to the main dispatch station. This reduces damage to the pallet, equipment, and building.

AGV Anticollision Methods

The most sophisticated AGV systems have anticollision controls that are a blocking system. The blocking system permits several AGVs to travel on one guide path. The anticollision system is on-board the AGV or built along the guide path. The two methods are the optical anticollision method and the zone blocking methods (including point-to-point blocking, continuous blocking, and computer zone blocking).

Optical Anticollision Methods

The optical controls on each AGV are a light beam (source), receiver, and reflective target. On the AGV front end a light device produces a light beam at a fixed level and directed to the AGV front. A receiver on the AGV front is at a fixed level to receive the reflected light. A reflective target at the same elevation as the light beam is located on the rear of each AGV. On a multiple AGV transport system, when the second or trailing AGV light beam is returned (reflected) from the first or lead AGV target, the controls on-board the second AGV stop its travel. This anticollision system is used with a pallet or towing AGVs that have a structural member on the rear, or a cart for the light source, light receiver, and reflective target attachment. The system is not preferred on a driverless pallet truck because there is not always a reflective target on the AGV rear. If a reflective target is temporarily attached to the pallet rear and backstop, the anticollision system functions as a normal system.

Zone-Blocking Method

In the zone-blocking method, the guide path is divided into various zones or segments with sufficient length to contain an AGV or a cart train plus a safety margin. The anticollision control program permits only one AGV or cart train per zone, at a time. If an AGV occupies a zone and another AGV enters the zone, the entering AGV must wait until the first AGV exits the zone. The zone-blocking methods are (a) point to point, (b) continuous, and (c) computer.

The point-to-point zone blocking method has zone-sensing devices that are mounted on a physical object or column adjacent to the guide path. As an AGV passes from zone A into zone B, the AGV activates the sensing devices in the zone B control package to prevent another AGV from entering zone B and permits another AGV to enter zone A.

The continuous blocking method has an on-board AGV-blocking signal transmitter and receiver, and an auxiliary multi-loop wire that is buried in the finished floor under the travel path. Each loop wire section becomes a zone. As each AGV travels on the guide path, it sends a signal into this looped wire that is detected by the blocking devices on the next AGV. When an AGV is in zone A and a second AGV sends a signal on the loop wire, the sensing device on the second AGV picks up the signal and stops (queues) until the first AGV leaves or clears the zone or travels from the loop wire.

In the computer zone-blocking system, each AGV is equipped with an on-board microcomputer and on-board transmitting and receiving antenna and has the ability to communicate with a host computer or another smart AGV. The finished floor area under the travel wire guide path contains magnets or plates that identify each zone. As an AGV passes over the magnet or plate, it sends a signal through the guide wire or by RF radio link. This method links the AGV zone status to a second AGV and to the host computer. When one AGV is in a zone, a second AGV that wants to enter the occupied zone is restricted by its on-board microcomputer or host computer.

AGV Design Parameters

When considering an AGV transport system in a facility, you are required to define the following design parameters:

- Good traction for the wheels.
- Floor hardness.
- Metal is not within 2 in. of the guide wire.
- Guide wire path avoids expansion joints.
- If expansion joints are passed, the wire is looped.
- All grades are 10 percent or less in slope.
- UPS for the system.

- ■ Determine and identify the number of turns in the travel path that are uncompensated or tangential turns, which have a shorter width to turn, and the number of turns that are compensated turns, which have a wider width to turn and that mitered or 90 degree turn.
- ■ Calculate a vehicle travel requirements. Basic is used for one- or two-directional AGV system with a simple manual dispatch method. Advanced is used for two-directional smart AGV with a host computer or automated dispatch system. Simulation is used for the sophisticated system with a host computer or automated dispatch method that interfaces with another automated system.
- ■ Calculate the required AGV number on the loop, travel time (travel from a start location and return to the start location), an 80 percent allowance for AGV utilization, and dispatch and on-line placement time (includes load and unload time).
- ■ Determine the guidance system: which can be (a) inductive—the most common method, it has a floor-embedded wire that carries a low (40 V) electric current and sensors on the AGV undercarriage; (b) chemically treated tape or paint and an ultraviolet light beam from the AGV underside that stimulates fluorescent particles in the path and transfers the light beam back to the AGV underside sensor; or (c) laser beam that has red light sent from a light on the AGV mast and the light is reflected back to the AGV from strategically placed reflective targets.

AGV Types

The pallet AGV types are (a) towing AGV, (b) driverless pallet truck, and (c) pallet load AGV.

AGV Towing Vehicle

When a pallet operation has frequent deliveries (high volume) over a long travel path or distance (more than 3000 ft.) and several carts or pallets are assigned to one facility location, a towing AGV is an option. An AGV with a cart train drops or picks up pallets at multiple locations. This feature has an employee couple and uncouple the carts and program the towing AGV for travel to the next location. If a towing AGV with a cart train is to make a turn, consideration is given to the cart train-turning requirement. A loading and unloading spur or siding has sufficient length to handle an AGV and cart train. This design feature permits other towing AGVs on the main travel path to pass the AGV that is on the siding. The AGV has an employee hook-up/unhook cart and load/unhook pallet activity. When a forklift truck or mechanical device places an identified pallet onto the AGV tow cart, a wrap-around pallet stringer and block has pallet identification face the tow cart side. In this tow cart position as an AGV tow cart travels on a travel path, barcode scanners have a line of sight to an identified pallet that is used as a divert bar code on the transport concept and at each delivery location there is pallet identification clear line of sight.

Driverless Pallet Stop and Drop Vehicle

The driverless pallet truck carries one or two pallets to a drop location and is a driverless vehicle that travels in one (forward) direction. It is manually controlled for the pallet truck to travel in the reverse direction for pallet pick-up. After the pallets are on the pallet truck set of forks, an employee enters the dispatch code or codes in the vehicle push button control panel and dispatches the vehicle. Arriving at the assigned location, a driverless pallet truck travels onto the siding, stops, and drops the pallets at the location. The driverless pallet truck travels on the main travel path to perform, if one pallet was dropped, another drop function or continues to the dispatch location. The vehicle is battery-powered and has an operator control area, safety light, and safety bumper. With a wrap-around pallet stringer or block has pallet identification facing a pallet fork opening and a side and has on a driverless pallet truck each pallet face the direction of travel and the travel path side. When used on a driverless transport concept, the pallet identification orientation is in a consistent location at the load and unload delivery station, the forklift truck has a clear line of sight to each pallet identification. With the pallet identification facing the travel path side, barcode readers use the pallet identification in the transport concept.

Pallet AGV

A pallet AGV is used to transport one or two pallets between two locations. The pallet AGV has a battery compartment, three to four wheels (one at least for steering), safety bumper, control device, path sensor device, and pallet-carrying surface. The important non-vehicle components are vehicle guidance system, pick-up and delivery stations, and control panels.

On-board rechargeable batteries power the AGV on a guide path. The most common guide path uses a wire embedded in the floor. This is the inductive guidance method in which an electric charge travels through the wire. The sensor devices under the AGV maintain travel control via electric impulses that direct the AGV travel on the guide path. Other guidance systems are the optical method, which uses chemically treated tape or paint on the floor, and the laser light beam method, which uses a light beam sent from the vehicle to reflect from reflective targets. With three or four wheels, an AGV travels in the forward and reverse direction over the travel path. In some newer models, the pallet-carrying surface is designed to handle a wide pallet variety. When a lift truck or mechanical device places a wrap-around pallet stringer or a block identification is placed onto the AGV-carrying surface, the identification faces the AGV tow cart side. In this AGV-carrying position as the AGV travels on a travel path, barcode scanners have a line of sight to an identification and is used as a divert bar code on the transport concept and at each delivery location, the pallet identification has a clear line of sight.

Various AGV Pallet-Carrying Surfaces

An important pallet AGV feature is the pallet-carrying surface or how the AGV handles the pallet. The first concept is to tow the pallet-carrying cart, as reviewed earlier in this section. An AGV pallet-carrying surface is equipped with any device type that is placed on a pallet or forklift truck. Pallet-carrying surfaces are:

- A fixed-position single pallet carrier that has a forklift truck or fork device to deposit or lift a pallet from an AGV surface
- A single pallet-carrying surface that can be lifted and lowered for unassisted and assisted loading or unloading transactions
- A gravity roller surface for side assisted loading and unloading transactions
- A powered roller surface for automatic loading and unloading
- A set of forks to elevate and lower a pallet

AGV Pick-Up/Delivery Stations

An AGV pick-up/delivery (P/D) station is an important part of the design and is matched to the AGV pallet-carrying surface and travel-direction capability. AGV P/D station types are:

- Manual type that has a WA or NA forklift truck deposit or pick up a pallet
- AGV automatically loads or unloads onto a structural stand that requires the AGV to turn and back up to the stand
- Conveyor stand that has a gravity or powered roller conveyor section to interface with the AGV
- Slave-driven stand that has the AGV provide the power to transfer the pallet
- Floor or rack position that has an AGV with a set of forks

Towline

The in-floor towline concept has each towline cart with a pallet travel at 60 to 90 ft./min. over a fixed travel path. An in-floor towline has a fixed path system that is designed with multiple automatic divert and charge locations (spurs or sidings) and it is installed in a facility with inclines and declines having a slope of 10 percent or less. A towline cart is transferred from one chain travel path to another. On a towline system, a tow cart is manually removed from a spur or placed onto a chain. A major component of the in-floor towline is the towline chain, which is enclosed in a hardened metal travel path and is driven by an electric-powered motor and towline cart.

The towline concepts are:

- Conventional heavy duty. A conventional towline chain is in a pit 7 in. deep, 6 in. wide at the top, and 3 in. wide at the bottom, with clean-out pits that are 4 in. deep. The clean-out pits are periodically checked and cleaned out. Along a chain travel path there are automatic oilers at various locations to ensure proper chain lubrication.
- Low-profile or light-load towline. The low-profile towline chain concept is a pit 3 in. deep and 2½ in. wide. A low-profile chain has a lower installation cost and is installed in a thin or average deep floor.

A towline has a good preventive maintenance program and employees are prevented from sweeping trash into the chain slot trench. Inside, the chain travel paths have bottom and two side wear bars that are hardened metal stripes. The wear bars are the tracks for the towline chain. As a chain travels across a wear bar, the wear bar prevents excessive chain wear. The towline chain is pulled by an electric motor-driven sprocket that is located on a straight run and is in a pit. In this pit is a take-up device that allows the towline slack chain condition to be taken-up. Slack chain results from wear and weather changes. A tight chain ensures good performance from the towline.

When a forklift truck or mechanical device places a wrap-around pallet stringer or block-identified pallet is placed onto the AGV tow cart, the identification faces the tow cart side. In this tow cart position as the AGV tow cart travels on a travel path, barcode scanner has line of sight to the identified pallet label and is used as a divert bar code on the transport concept and at each delivery location.

Chain Links and Link with a Cavity or Dog

A towline has motor-driven chain links that travel in a C-channel track and it is designed to travel over 90- or 180-degree turns or to incline or decline over approximately 10- to 15-degree grades. It has numerous chain links and every 20 ft. there is a link with a cavity or dog. For long-lasting wear, the chain components are hardened steel. At every 20 ft., or as determined by the towline cart length, a dog link is located and is the towline chain component that pulls the tow line cart. The dog is a specially designed link with a cavity to hold the cart tow pin. The motor, sprocket, and take-up device is located in a pit. The motor and sprocket pull the chain forward through the travel path and the towline is designed to pull the maximum number of carts on a chain. A take-up device provides the means to increase or decrease the chain length that adjusts the chain tension.

The spur is a left or right non-powered track that intersects the chain travel path at 45-degree angles. A branch line permits non-powered cart travel to a location. The chain sensor device or pad is in the floor prior to the spur. After the spur is activated by the cart-coded sensor, the spur diverter (floor-mounted lever) slides across the chain travel path. The lever being in this position and the chain pulling the cart forward together force a cart pin to become disengaged from the chain link dog and to flow into the spur open travel path. The spur allows a diverted cart to clear the main travel path that permits other carts to travel on a main line. All spurs or branch lines are designed for a specific length and have full line sensors. When activated by a towline cart, a sensor does not allow a diverter mechanism to divert additional carts onto a spur. To purchase used carts, test the carts to assure a match of the towline with the present system specifications.

Towline Cart

A towline cart is a four-wheel vehicle that has a pallet-carrying surface and a bar code on its front end. The standard carts are 6 ft. and 10 ft. long and 3 ft. to 4 ft. wide. The pallet-carrying surface is a flat wooden surface with hardened metal borders. This feature requires the pallet transfer transaction to be completed by a forklift truck. A towline cart features are (a) tow pin, (b) selector pin rack, selector pin, or barcode label and barcode scanner device, and (c) location and wheel types under a cart.

These features make the towline cart different from the other carts.

Tow Pin

To prevent excessive wear, the tow pin is a hardened steel rod that is set inside a sleeve. The pin sleeve is attached to the lead end and the tow cart center, and the tow pin is removed from a tow cart pin sleeve.

The tow pin positions are non-engaged or engaged. The non-engaged position is maintained by a cap or retainer on the end of the tow pin. This prevents the pin from mistakenly being placed in the tow chain. When a cart is not being towed on the system, this is the preferred position to prevent pin damage. In the engaged position, the tow pin is in the secured position that keeps the pin in a raised position above the floor. An engaged position is in a lowered predetermined position.

The exact pin length is a specification that is based on the track and dog depth. The rod retainer or cap permits a tow pin to reach this predetermined depth. When a tow pin is in an engaged position and in a track, as an empty dog arrives at the cart, the pin automatically slides into the dog (open link), and the tow chain pulls the cart forward to an assigned location.

Selector Rack, Pins, or Barcode and Scanner

The selector rack and selector pins are at the cart lead end and the options allow an employee or WMS pallet identification to assign a divert address to the cart. A selector rack(s) is (are) on each tow pin side and at a cart center. Holes (12) permit the selector pin to hang in a vertical position which is approximately level with the floor surface. Each position has an alpha character in the front. The pin hanging arrangement from the specific selector rack hole corresponds to a discrete series of sensors in the spur (finished floor mounted) sensor device. One selector pin rack handles up to 12 address locations, two selector pin racks handle up to 150 locations, and three pin racks handle several hundred locations. The selector pin or probe is a metal rod with a flexible or spring-loaded magnetized tip and metal cap or retainer. To prevent pin damage, the pins are attached to the selector pin rack with a string or chain. When not in use, pins are placed in a tray that is behind the selector pin rack. The metal cap and selector pin rack permit a pin to hang vertically at specific heights to have the magnetized tip at approximately floor level. This is the proper height to activate the floor spur-sensing device which activates the diverter to transfer the cart from the towline travel path.

An alternative to a selector pin is a bar-coded label or RF tag on a cart lead end or front side. A bar code or RF tag is read by a scanning or RF tag reader device communicating with a microcomputer that triggers a divert device to direct the cart from a main travel path to a spur or to another travel path.

When a forklift truck or mechanical device places a wrap-around stringer or block pallet-identified pallet is placed onto the tow-line cart, the identification faces the towline cart side. In this towline cart position as the towline cart travels on the travel path, barcode scanners have a line of sight to the identified pallet label and can be used as a divert bar code on the transport concept and at each delivery location, the pallet identification has a clear line of sight.

Tow Cart Wheels

The towline cart has four wheels. The two wheels on the rear are rigid casters or wheels and are located on the two corners under the rectangular load-carrying surface. The two front casters or wheels are the swivel type and are located under the two front corners. With a cart underside plan view, the four-wheel locations are interior to the rear wheels. This feature reduces drag as a tow cart is pulled through a towline travel path curve.

During towline operations, to prevent injuries to employees and damage to material-handling equipment or to the building, a cart is not pushed or towed by another cart that is being pulled by a towline dog. To operate a towline, an employee-operated forklift truck places the pallet onto a cart load-carrying surface and an employee pushes the cart onto the towline track. At the towline track, the selector pin is inserted into the assigned selector rack hole. The employee waits for an open link or dog on the towline every 20 ft. and pushes the cart over the track.

In this position, the tow pin is released and rides on the chain top and the dog automatically engages the tow pin. When this occurs, the tow line cart is moved by a tow chain to the new warehouse location that corresponds to a selector pin arrangement or barcode label.

As the tow cart approaches the assigned spur, the magnetized selector pin passes over the floor-mounted sensing device (the bar code passes a reader) and activates (trips) a floor-mounted divert device (lever) which extends across the cart path. When the cart tow pin strikes the lever, the two pins slide from the dog mouth or cavity and the cart travel onto the spur away from the main travel path. At this location, the employee removes the cart from the towline travel path, removes the pins from the engaged rack position, and a forklift truck removes the pallet from the cart.

Overhead Towveyor

An overhead towveyor concept is used to transport pallet loads over long travel distances. The components are:

■ An overhead, powered or electric motor-powered, rivetless chain travel path and a free trolley travel path. The free trolley has a pendant or arm that extends downward toward the floor. At the end of the pendant is a loop or strap.
■ A four-wheel tow cart with a rigid tow mast that extends upward and has a T-shaped top or end, or a non-powered pallet truck with a T-shaped handle.
■ A tow cart travel path across the floor.

After a forklift truck or mechanical pallet-lifting device transfers a pallet onto the four-wheel tow cart-carrying deck, or a non-powered pallet truck with a T-shaped handle picks up a pallet on the dispatch station siding or spur travel path, an employee attaches the free trolley pendant loop or strap onto the T-shaped end of the mast on the tow cart, or raises and inserts the T-shaped handle of a non-powered pallet truck into the loop or strap. With the tow cart or pallet truck attached to the free trolley the employee enters the pallet dispatch destination or delivery address (code) onto the trolley code device, then transfers the tow cart or pallet truck by a rail switch device onto the free trolley main travel path, which is directly under the rivetless powered chain travel path. At this location, the free trolley with the tow cart or pallet truck is queued or waits for a rivetless powered chain pusher dock to engage the free trolley pusher dog. The four-wheel tow cart or pallet truck is pulled over the towveyor tow-cart floor travel path to the assigned location.

The overhead towveyor design parameters are:

■ A four-wheel pallet truck or cart travel path. Because in the overhead powered towveyor concept the four wheels of the pallet-carrying cart come in contact with the floor and the pallet-carrying cart is pulled forward over the travel path, the finished travel path features include (a) pallet-carrying cart wheels and floor surface with a low coefficient of friction, (b) free and clear of debris, cracks, holes, and joints, (c) considered a high traffic aisle and is coated to minimize dust and floor re-bar is designed for the tow cart combined pallet and cart weight, or the cart wheel load factor, and (d) closed-loop travel path feature and requirement for the cart to be returned to home base, the return travel path has the same design travel path window criteria as the outgoing or fully loaded cart travel path window.

■ Tall floor or ceiling-hung structural support members. With the overhead powered towveyor transport concept, the combined pallet and cart weight is borne by the four wheels and the floor surface. With this feature the structural support members have less load weight to support than the typical overhead powered towveyor transport concept.

■ A P/D station area. A free trolley with tow cart or pallet truck is diverted onto a P/D station spur with the following design features: (a) a divert spur travel path and loop that is sloped to the P/D spur, or at the P/D station end an employee ensures that the diverted tow cart or pallet truck has cleared the main travel path and is at the P/D spur end; (b) full line controls on the spur travel path to control the P/D station divert mechanism; and (c) a rail switch that allows an employee to transfer the free trolley onto the powered rivetless chain and free trolley main travel path.

After a tow cart or pallet truck is on the divert spur, to ensure an efficient in-house transport concept, a P/D station employee removes the tow cart or pallet truck from the divert spur. At this location, the employee removes the free trolley pendant strap or loop from the T-shaped component and pushes the tow cart or pallet truck to an inbound staging area. An employee transfers an empty or full tow cart or pallet truck to the main travel path in-feed line. On this in-feed line, an employee places a free trolley pendant loop or strap onto the tow vehicle T-shaped component, enters a dispatch code onto the free trolley and moves the free trolley back onto the main travel path, where it waits for pick-up by a rivetless powered chain pusher dog. This concept requires sufficient floor space to stage inbound tow carts or pallet trucks, sufficient floor space to stage empty or outbound tow carts or pallet trucks, and adequate floor space for fork lift trucks to complete pallet transfer transactions and travel from the staging area to the activity area.

When a forklift truck or mechanical device places a wrap-around stringer or a block WMS-identified pallet is placed onto the tow-line cart or a pallet truck, the WMS identification faces the tow-line cart or pallet truck side. In this cart position as the tow-line cart or pallet truck travels on the travel path, barcode scanners have a line of sight to the WMS-identified pallet label which is used as a divert bar code on the transport concept and at each delivery location.

Monorail

A monorail pallet transport concept has an electric motor-powered pallet-carrying surface that moves over a fixed travel path. The overhead monorail travel path or track has straight and curved sections and is suspended from the facility beam, rafters, joists, or other structural members. The overhead monorail concept is used to move pallets between the receiving dock and the storage area and from the storage area to the shipping dock. The monorail pallet transport concept is a hybrid mix concept that has AGV and towveyor features and the best services of an AS/RS concept. When a forklift truck or mechanical device places a wrap-around stringer or block pallet-identified pallet is placed onto the monorail carrying surface, the identification faces the side of the monorail carrier. In this position, as the carrier travels on the travel path, barcode scanners have line of sight to the identified pallet label and at each delivery location.

Carton or Pallet Replenishment

A SKU replenishment to a depleted carton pick position is a requirement for a carton order-fulfillment operation that uses fixed carton pick positions. A SKU replenishment activity ensures that the correct SKU is removed from an assigned storage position on schedule, in the proper quantity, and is placed into the correct SKU pick position.

SKU replenishment activities include the pick position listing that indicates a replenishment on a paper document or into an RF device, appropriate SKU withdrawal quantity from the storage position, SKU transfer into the pick position, and verification of completion. In an inventory program, each pallet replenishment transaction has an employee with an RF device scan each pallet identification and pick position identification and enter the master carton quantity in the files. In the various replenishment concepts, a warehouse employee transfers product from a random storage position to a fixed pick position. In an order-fulfillment operation that has a floating slot or pick position concept, a product is assigned to several discreet pick positions entered into the computer. When pick activity depletes the product from one pick position, the computer transfers the next pick activity to another pick position that contains the product.

Transaction Verification and Inventory Tracking

A very important order-fulfillment operation activity is the product deposit and withdrawal transaction verification. The replenishment activity involves SKU transfer from the receiving dock to the storage location and the replenishment transaction (transfer) from the storage location, at the correct time and to the correct pick position. The replenishment transaction verification options are (a) manual or visual memory, (b) manual written, and (c) automatic.

Human Memory

The employee keeps the storage and pick positions for the SKU replenishment transaction in memory. When there is a demand for a SKU replenishment to the pick position, the employee remembers the storage location and completes the replenishment transfer transaction. The human memory concept is not used for an inventory program. Disadvantages are low employee productivity, possible errors, low volume and few SKUs, difficulty controling a large area or two shifts, difficult to handle a FIFO product rotation, and storage location is not always the optimum position. Advantage is low cost.

Handwritten Paper Document

The replenishment employee uses a printed form to record the replenishment transaction. A printed form is a four-column activity sheet that has a space for an employee's name and date. The four columns are separated into two subcolumns: one is the storage deposit heading, with two columns to list each deposit transaction; the other is the replenishment heading, with two columns to list each replenishment transaction.

After the employee completes a transaction, the operator lists the storage position and SKU identification number or the pick position and SKU identification number. At the end of the shift, the activity forms are sent to the warehouse office where a clerk performs an inventory control update. The handwritten concept is not appropriate for an inventory program.

The disadvantages are that it handles a low volume, requires a clerk to make an inventory entry, adds another task to the employee's work, and creates possible transposition errors. The advantages are no capital investment, and can be used over two shifts and in a large facility.

Manual File

The manual bin file concept uses reserve position cards. Each card corresponds to a reserve position in a facility aisle and the cards are placed in sequential order in a cardholder. In a carton order-fulfillment operation, the cardholders have a slot for each reserve position or pick position in a rack bay. The cardholder is attached to the rack upright post.

The card has the reserve position number printed on the top left side and has three columns. The employee who performs the replenishment transaction completes each column. One column lists the SKU identification number that was involved in the transaction. The other two columns are to indicate a deposit or withdrawal transaction. After the replenishment transaction, an employee obtains the appropriate storage position card from the holder, and lists the SKU identification number and places a mark in the appropriate in or out column that reflects the transaction. The completed card is returned to the cardholder for future reference.

The manual bin file memory concept is not considered for an inventory program. Disadvantages are unclear handwritten statements, possible transposition errors, card can become lost, and requires another employee activity. The advantages are low cost, low capital investment, information in the warehouse area, and easy to implement.

Barcode Scanning/RF Tag Reading

A barcode scanning/RF tag reading concept has the following features:

- Each carton handled in an operation has a human/machine-readable carton identification.
- Barcode label/RF tag (human/machine-readable) identification for each SKU position.
- Human-held barcode scanner/RF tag reading device that holds or transmits the replenishment transaction and quantity to the computer.

In the barcode scanning/RF tag reading replenishment operation, each product movement or replenishment transaction or activity is barcode scanned/RF tagged and the information is transmitted to the host computer for inventory update. Considerations for designing and implementing this concept are:

- Scanning/reading devices and transaction number
- Data transmission distance and clean lines
- Host computer capability to handle on-line transactions and other activity transactions
- Scanner device that is finger, wrist, or cord held
- Whether RF communications can be performed within the building
- Whether there are government regulations regarding wave use in the facility area

Disadvantages are investment, employee training, management control, and possible requirement for a temporary data-holding capacity to handle all the transactions. Advantages are handles a high volume, accurate information transfer, provides accurate transaction record, permits on-line transaction transfer, and high employee productivity.

Replenishment Concepts

A replenishment transaction moves master cartons from a storage position to a fixed pick position. This replenishment activity occurs in a carton operation with a fixed-position concept and ensures that the correct SKU quantity is in the correct pick position for the order picker to complete the customer order. With any replenishment concept, the replenishment employee with a RF device scans each master carton identification, the pick position identification and the date are transferred to the inventory files. SKU replenishment methods are (a) random, (b) slug, and (c) sweep.

Random Replenishment

The random pick replenishment concept has an employee make replenishment transactions that are based upon the employee's thoughts. To reduce travel distance and travel time, the order-fulfillment facility aisles are separated into zones. The disadvantages with this concept are replenishment transactions do not complement the order picker transactions, management control is low, employee productivity is low due to unscheduled work activity, and if the pick position is not completely depleted, the extra cartons are not placed in the best location. The advantage is fast-moving SKUs are handled first.

Slug Replenishment

The slug replenishment concept has an employee complete all replenishment transactions within one order pick aisle before moving to another aisle. The disadvantages are increased stock-out potential and lower employee productivity with carton hand-stacking into a pick position. The advantage is decreased employee travel time and distance between replenishment positions.

Sweep Replenishment

In the sweep replenishment concept, the pick area aisles are separated into zones. A replenishment employee starts in one zone and performs all replenishment transactions in sequential order in one order-fulfillment aisle prior to moving to the next aisle. An additional advantage is that the concept closely complements the order pick activity.

SKU Allocation to a Pick Area

The basic SKU pick position location principle is to keep the order picker and replenishment employee's travel distance as short as possible between two transaction locations. SKU allocation to the pick position involves (a) no concept, (b) the ABC concept, or (c) the family group concept.

No Method

With the no method concept, SKUs are randomly assigned to the pick positions within the pick area or pick aisle. With this method, a pick position that has fast-moving SKUs is separated by several slow-moving SKU pick positions; therefore, an employee increases walk or ride time and distance to complete a replenishment or order pick transaction. Disadvantages are replenishment and picker productivity is low, mixes SKUs from different product groups in one area, and does not maximize space. The advantages are easy implementation and does not require management control and discipline.

ABC Method

In the ABC method, the SKU pick positions in the pick aisle are separated into three major zones, A, B, and C, which are then subdivided into micro-zones, a, b, and c.

Each zone is restricted for particular SKUs that have a specific annual movement. The A zone positions are reserved for fast-moving SKUs. The positions within the B zone are for medium-moving SKUs. The C zone positions are for slow-moving SKUs. Eighty percent of the SKU movement (based on Pareto's law) is from zone A. The ABC concept improves the pick aisle SKU hit density and concentration.

Disadvantages are management control and discipline, SKUs from different family groups are mixed in one aisle, and accurate projections for SKU movement are needed. Advantages are improves replenishment and order picker productivity and handles a high volume.

Family Group

In the family group, SKUs with similar characteristics are assigned to pick positions within a pick aisle, zone, or area. The order-fulfillment operation management team determines the characteristics. Disadvantages are increases management control and discipline, additional positions are required for new SKUs, fast- and slow-moving SKUs are mixed in one pick aisle or zone, and lower replenishment and order picker productivity.

Replenishment Quantities

The carton replenishment activity is planned to handle one SKU quantity per pallet, one layer of a pallet, or less than a pallet layer.

Pallet

With a pallet replenishment concept, the entire pallet is transferred from the storage area to the carton pick position. To maximize space and labor, this method has a pick position depleted. With this situation, the entire pallet is placed into the pick position. The pallet replenishment concept replenishes approximately 50 to 75 cartons per unit load and is used for high-cube or fast- to medium-moving SKUs. With the pallet replenishment concept, an employee with an RF device scans the pallet and pick position identification.

Disadvantages are if cartons remain in a pick position, it lowers replenishment productivity, if a replenishment activity is not scheduled this creates a stock-out. Advantages are minimizes the SKU number, high replenishment employee productivity, and used for medium- to fast-moving SKUs.

Replenishment of One Layer of a Pallet

A pallet layer carton or SKU replenishment concept has one or two carton layers from a pallet removed and transferred from the reserve position to a pick position. This means that the pick position is a hand-stack rack position, case flow rack position, or shelf position. The method is used for slow-moving to medium-moving SKUs and activity to remove the layer(s) is performed in the storage area or pick area. If the activity is performed in the storage area, the cartons are removed from a pallet and transferred onto a conveyor system or vehicle load-carrying surface for transport to a pick aisle. In a pick aisle or in the replenishment aisle, cartons are transferred into a pick position. With this concept, an employee with an RF device scans the pallet and pick position identification, and enters a master carton quantity that was transferred to a pick position and the data is sent to the files.

The disadvantages are transfer equipment cost and required floor area, product is handled twice, and increases vehicle traffic in the facility aisles. The advantages are: minimizes the number of pallets moved and reduces the potential product and equipment damage, has FIFO product rotation, and handles slow- to medium-moving SKUs.

If the layer removal activity is performed in a pick area, an employee removes the pallet from the storage area and transports it to the pick area. From the pallet, an employee transfers cartons (layers) to a pick position. After the replenishment transaction completion, the employee removes the partially depleted pallet from the pick area and returns it to the assigned storage position.

Disadvantages with this concept are pallet double handling, increased product and equipment damage potential, and increased number of forklift truck or pallet truck trips in the aisles. Advantages are large inventory quantity is transferred to a pick position and employee physical effort to bend and reach for cartons is minimized.

Replenishment of Less than a Layer of a Pallet

This method is used for very slow-moving SKUs. The employee basically order picks the slow-moving cartons from the storage area, and transports the cartons (SKUs) from the storage area into the appropriate pick position. With this concept, an employee with an RF device scans the pallet and pick position identification, and enters the master carton quantity that was transferred to the pick position, and the data is sent to the files.

The disadvantages and advantages are similar to the previous replenishment method except that the product movement is one way from the storage area to the pick area.

Timing Methods for Replenishment

The SKU or carton fixed pick position replenishment activity is performed when the pick position becomes depleted and with a SKU quantity that maximizes use of the pick position space and employee transport activity. To achieve this objective, for best results all replenishment transactions and quantities are made by a microcomputer. The next carton or SKU replenishment activity is to determine the appropriate time for a replenishment transaction to occur. An employee is directed to move a specific SKU from a storage position to a fixed pick position. SKU replenishment activity can be controlled either manually or by computer.

Manual Concept

The manually controlled replenishment transaction concept relies on an employee to determine the time at which the cartons are moved from a storage position to a pick position. This random concept is based on employee experience and is not sequenced with the other order-fulfillment operation activities. On some occasions, the replenishment transaction is made to a partially depleted pick position. This condition requires the replenishment employee to waste time transferring the cartons.

The disadvantages are minimal replenishment employee activity control, low employee productivity, pick position poor spatial utilization, and noncoordination with the order pick activity. The advantages are low cost, no capital investment, and does not require employee training.

Computer Concept

With the computer-controlled replenishment transaction concept, the computer directs an employee to perform a product replenishment transaction from a storage position to a pick position. The replenishment is determined by:

- Customer-ordered SKUs
- Inventory quantity in a pick position
- Storage concept type

The methods to direct the employee to complete a replenishment transaction are:

- Paper document that lists all the replenishment transactions to occur on the employee shift. On this form the SKUs are listed by SKU number and the description in a sequential order that is based on the anticipated time that a pick position becomes depleted. The first column states the SKU reserve position and the second column is for the replenishment employee mark that indicates and verifies that the transaction withdrawal portion was completed on schedule. The third column states the SKU pick position and the fourth column is for the employee mark to verify replenishment transaction completion.
- Paperless or RF device replenishment transaction has an employee with an RF device scan the pallet and pick position identification, and enter the master carton quantity that was transferred to the pick position, and the data is sent to the files. The RF barcode scanning device transmits the replenishment information to the host computer for inventory position update.

The disadvantages with this concept are that it requires management control and discipline, capital cost, and employee training. The advantages are accurate transaction records, excellent employee productivity, reduction of lost inventory, improved spatial utilization, and enhanced management control.

Chapter 4

Small-Item and Flatware Apparel Pick and Pack Activities

Introduction

A small-item and flatware order-fulfillment operation is a very complex operation, with customer order pick and pack activities at the heart of the operation. The factors that contribute to the complexity of facility design, equipment layout, and management control for vendor-delivered units (UOPs) and customer order (CO) flows are:

- SKU characteristics
- Number of SKUs with (a) different UOP classifications, (b) large inventory along with a store and hold requirement, and (c) pick position number
- Large number of COs, wide CO lines, CO pieces, and mix per CO; COs have (a) a wide product mix, (b) short CO order/delivery cycle time, and (c) wide fluctuations in CO volumes, pieces, and lines per order and total daily pieces
- Potential for value-added activities
- Maximum number of work stations that can fit in the facility and the potential to have automated pick concepts
- Number of employees
- Potential for a multi-level facility

- Requirements for vendor delivery trucks or ocean-going containers, including dock receiving areas and delivery truck areas
- Local building codes
- β Operational costs

The objectives of this chapter are to:

- Identify and evaluate the characteristics of a small-item and flatwear order-fulfillment operation, and design parameters for a facility. These guidelines apply whether you are remodeling an existing operation or developing a new one.
- Review manual, mechanized, and automatic pick concepts and design parameters.
- Introduce manual picker routing patterns.
- Describe pick area equipment and layout.
- Identify activities that can ensure efficient, accurate, and on-time UOP and CO flows through a small-item order-fulfillment facility.

Facility or Building Pick Area, or Pick Aisle and Pack Area

The first factor that has an influence on design of a small-item or flatwear apparel manual order-fulfillment operation is the building area for the pick and pack activities. This area houses the most important order-fulfillment activities. The receiving, storage, shipping, IT, and other administration areas are support functions to the pick and pack activities. The size and shape of the site dictate facility design. The best design for a small-item or flatwear manual order-fulfillment operation is in a square or rectangular facility. The buildings have 10 to 13 ft. between the floor surface and ceiling support structure. The pick and pack area layout should be designed so that pick aisles and pick position rows are parallel or perpendicular to the pack stations, and there is a queue area for CO picked pieces, as well as a transport path from the pack area to the carton manifest and shipping area.

Facility Site Shape

The shape of a small-item order-fulfillment facility has significant impact on the arrangement of the pick position, pick line and pick aisles, pack area, space utilization, and the flow of vendor-delivered UOPs and CO pieces through the facility. When the shape of a facility site is not square or rectangle, there is low utilization of land space and internal building space. A square or rectangular facility is set on a site with the land angle sections providing the best land utilization, green area, or set back from the property line.

Square-Shaped Order-Fulfillment Facility

A square-shaped order-fulfillment facility provides the best balance between wall area and floor area. It creates an efficient facility structure for internal transport from receiving docks to all storage areas and for CO shipping carton transport from the pick area to the pack area and then to the shipping area. A square-shaped building is best for a new order-fulfillment operation with a store/hold piece inventory flow that handles square or rectangular SKUs. It is normal as a facility expands that a square-shaped building becomes a rectangular building.

With a manual pick concept in a square building, the recommended layout of the pick position and order-fulfillment equipment is:

- Receiving and shipping docks along one wall
- A storage area that is located directly behind the receiving docks
- Value-added activity, pick, transport, sort, and pack areas are located adjacent to the storage area with the shipping dock in the front, so that a vendor-delivered UOP and CO flow pattern has a U shape through the facility

Pick aisles and rows are short and either parallel or perpendicular to the pack and shipping areas.

Aisle length and flow direction to the pack and ship areas are determined by the building dimensions, number of pick positions, projected store/hold inventory, and number of pack stations.

Rectangular Facility

A rectangular order-fulfillment facility provides an increase in the ratio of wall area to floor space. This structure provides additional facility wall space that can accommodate an increase in the number of docks available for delivery vehicles and permits efficient transport from the receiving dock area to the storage area. A rectangular facility is an excellent option for an operation that provides service to CO pack stations or handles a store and hold inventory piece flow concept.

With a manual pick concept in a rectangular building, the locations of pick rows and aisles places include:

- Receiving docks and shipping docks along a long wall
- The storage area directly behind the receiving docks
- Value-added activities, storage, transport, pick, sort and pack activities adjacent to the storage area and in front of the shipping docks, with pick aisles and position rows:
  - Perpendicular or direct flow from the storage and pick area to the pack area and shipping docks. This means that pick aisles and pick position rows are short.

- Parallel flow along the docks, which means that pick aisles and pick position rows are long and located along a short wall, with pick areas that are located directly behind the receiving docks.
- Value-added activity, pick, transport, sort and pack areas that are located directly behind a storage/pick area and before the shipping docks
- Shipping docks along the other facility short wall that is directly in front of the pack area

With this option, the vendor-delivered UOP and CO flow pattern is a straight line through the facility. The length of the pick aisles and pick position rows is determined by the number of positions, building dimensions, and store and hold inventory.

Single or Multiple Floors in an Order-Fulfillment Facility

In a metropolitan area, available sites are high-priced. Buildings have ceilings that are 30 to 40 ft. above the ground-level floor. With this height the facility can be designed as a low bay or ground floor building, or with floors above ground-floor level.

Low-Bay or Single-Floor Facility

A low-bay or single-floor facility has all order-fulfillment activities, value-added activities, and employee support activities on the ground-level floor. This facility design has:

- Large square footage
- Highest building construction cost per square foot
- Low space utilization
- Longest travel distance and time between two locations

High-Bay or Multi-Floor Facility

Design options for a high-bay or multi-floor facility are as follows.

- Slow-moving, light-weight/small-cube SKUs, value-added activities, pack areas, and administrative support offices can be located on an elevated floor.
- Storage areas and high-volume, heavyweight/high-cube SKUs are located on an elevated floor.
- Receiving and shipping dock areas, fast-moving and heavyweight/ large-cube pick and pack areas are on the ground floor.
- Heavyweight/high-cube SKUs are located on the ground floor and pack stations are located on an elevated floor with direct flow to the manifest/shipping area.

In this design, the storage area, high-volume and heavyweight/high-cube SKUs are located on the ground floor. The office, administrative, employee support, low-volume, lightweight/small-cube order-fulfillment activities and value-added activities are located on elevated floors. To assure maximum facility flexibility, elevated floors are designed to support the dynamic and static forklift truck loads, one to three high pallet storage racks, and automatic or other pick line.

Depending on the design parameters and other planned activity in a high-bay building, the pack area can be located either on the ground floor or an elevated level.

Mezzanine or Second Floor

To reduce total land and building costs consider designing an elevated level in a new facility or when designing expansion in an existing facility that has sufficient ceiling height and ground-level floor structural support for the weight of an additional floor. This second level increases your operation's space for storage, pick, or other process areas.

The additional level is planned within a 15- to 16-ft. minimum clear space between the ground-level floor and the ceiling or in an 8-ft. space above an existing structural area. To achieve this design, lights and fire sprinklers fit within the ceiling joists or mezzanine support members.

A typical low-bay floor area in a building 25 ft. high has a clear ceiling height for a mezzanine (see Figure 4.1) above many locations, including receiving, pack, picking except HROS, shipping, ticketing, office, processing, and two-pallet-high storage.

The cost justifications for adding a mezzanine floor area is based on additional construction costs compared to additional land and building construction costs for a ground-level floor area. Approximate construction costs for a ground-level floor area are $50,000 an acre for land and $35 per square foot for building construction costs. Estimated costs (that include lighting and fire sprinklers) for an additional elevated floor level has a range from $10 to $25 per square foot. From an option cost comparison, we conclude that an additional elevated floor level is more economically attractive to provide additional floor space.

Other advantages are the opportunity to expand small-item and flatwear storage and pick area, increased security, increased inventory capacity, and reduced horizontal travel distances between two process locations.

With an elevated level, there is some additional expense for employee stairways and unit of product vertical incline and decline travel paths.

Options for designing a mezzanine or elevated floor are:

- Constructing an additional floor during building, which is not really classified as a mezzanine
- A mezzanine that can be
  - A fixed mezzanine with lightweight aggregate concrete filled on waved deck
  - A freestanding or equipment support mezzanine
  - A custom-engineered post-supported mezzanine with a solid deck

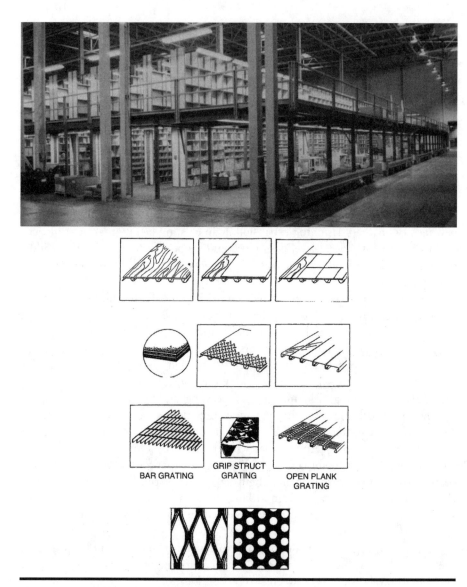

Figure 4.1 Mezzanine and floor material.

In a new facility design that includes an additional floor you allow sufficient ground floor to ceiling height to have an additional floor constructed in the initial construction or have the ground floor designed and constructed for future additional floor expansion. Before elevated floor construction, your elevated floor specifications are:

- Static and dynamic loads that are based on your processing or material-handling equipment static and dynamic loads
- Floor thickness to assure equipment anchor bolt installation
- Vertical material-handling elevated floor penetrations and personnel and fire protection
- Building column spans and floor structural support members and the column fire protection

Basic mezzanine designs are fixed and free-standing. Either design can be installed during an initial building construction phase or in an existing building shell. For a fast-track mezzanine installation and less impact on your operations, a mezzanine is installed in the initial building construction phase. This feature is especially true for a concrete floor mezzanine type.

Additional design factors are:

- Sufficient floor to ceiling space to permit a mezzanine.
- The installation area is free and clear of all operational activities.
- At the additional elevated level there are emergency exits in compliance with existing codes.
- The ground floor has the structural strength to support the new loads.
- If attached to existing racks or shelves, the existing upright posts and base plates have sufficient strength to support the new load.
- Upright posts have sufficient height and strength to hang light fixtures.

When considering a mezzanine, the following factors should be considered:

- Equipment layout
- Vertical UOP and CO movement between the two levels
- Horizontal UOP and CO movement on a mezzanine level
- Local building codes and standards
- Fire protection

Concrete-Filled Mezzanine

A concrete-filled mezzanine has a metal waved roof deck supporting the concrete filler. The mezzanine design has upright structural members that are attached to building columns. Attached to upright structural members are the cross beams and joists. On top of these components is placed the roof deck. The concrete is poured onto the roof deck and the depth is determined by your architect or structural engineer and your seismic location. The depth factors are:

- Anchor bolt depth
- Imposed equipment loads
- Allowed by structural support members

A concrete-filled mezzanine floor covers the entire elevated floor area and is very flexible, to allow equipment re-arrangement. It permits processing and material-handling equipment changes. Expansion to have another mezzanine floor elevation that exactly matches the existing mezzanine is very difficult.

Rack-, Shelf-, or Equipment-Supported Mezzanine

A rack- or equipment-supported mezzanine has the rack or equipment upright posts structurally support the additional positions and aisle between two pick position rows or pick position row and wall. The most frequently used rack types are a standard pallet rack with hand stack or carton storage/pick positions on the elevated level, carton or pallet flow racks, or standard shelves.

An equipment-supported mezzanine has:

- Upright frames or posts
- Standard rack load beams or shelf cross bracing
- Specially designed elevated floor or aisle support load beam or aisle support members
- Connection nuts, bolts, and washers

On the floor level, the upright rack frames/posts are spaced on relatively close centers with load beams or decks to hold pallets of hand-stacked cartons or totes. The space between upright frames and load beam horizontal members are on 8 ft. to 9 ft. 6 in. centers. Additional standard or specially designed mezzanine support beams are attached to the tops of upright frames or posts to provide structural stability and support (cross aisle ties) for the metal roof and elevated floor surface.

A solid deck on the elevated floor permits use for processing activities or a pick position layout that is different from that on the ground level. This feature permits maximum flexibility to accommodate changes in pick, storage, and processing equipment and product mix. Mezzanine or elevated floor expansion is easily achieved by the addition of upright frames, load beams, and deck material to the starter rack bay.

The shelf or carton flow rack-supported mezzanine with a solid deck has the same features and components as the standard pallet rack-supported design. The shelf-supported design has shelf upright posts that are on very close centers (nominally 2 or 3 ft. deep and 3 to 4 ft. horizontally). The carton flow rack design has posts or frames that are on close centers (nominally 5 to 10 ft. deep and 5 ft. horizontal centers).

With the rack- and shelf-supported mezzanine group, with sufficient ceiling clearance and floor structural strength, a double-decked type is designed with an additional elevated floor. In the equipment-supported mezzanine designs, upright frame or shelf posts extend upward to have structural stability and rigidity from the ground floor through the first elevated floor to support a second elevated floor. A double-deck concept is used to provide the same pick positions on the ground floor, first elevated level, and second elevated level. In most buildings with a 28- to 30-ft. high ceiling, a double-deck mezzanine concept is considered.

Attached between the vertical structural members are cross-aisle ties that support the elevated floor decking and load beams or shelves that provide storage or pick positions. An employee walkway and aisles are solid or grated deck material. The deck material is secured to the load beam or cross-aisle ties.

An equipment-supported mezzanine has similar characteristics and handles limited change in UOP mix and weights. The designs are restricted to the specific use for pick walkways or aisles due to storage and pick position designs being the same on the ground level and elevated levels. With an add-on to an existing bay, upright rack frames, shelf posts, load beams or shelves, cross-aisle ties, decking are added to a starter bay. This feature with sufficient ground floor space permits easy expansion.

Free-Standing Mezzanine

A free-standing (custom-engineered) beam and post support mezzanine design has support columns and horizontal structural members that are bolted to columns. The horizontal members are connected together to provide support for long spans or heavier loads on a metal or wood deck. This structural design permits, on the elevated floor levels, flexible equipment design and layout. Additions to a free-standing mezzanine are planned in any direction with sufficient ground floor space.

A free-standing mezzanine design has flexible ground and elevated floor areas, handles product mix changes, equipment layout redesigns, and minimal problems for expansion, large ground level base plates, with heavy elevated loads, large columns, and with a solid deck additional lighting and fire sprinklers.

The mezzanine deck material provides employee walkways/aisles and support for material-handling equipment that is not supporting the elevated floor. The most common deck material is formed steel roof deck or waved deck that is 16 gauge steel with 1½-in. deep ribs that are spaced on 6 in. centers. This roof deck is secured to the structural members (cross-aisle ties, load beams, or beams) with self-drilling or self-tapping screws. If heavy equipment is anticipated on the elevated level, the area is identified to the vendor who provides additional elevated floor-support members directly under the equipment locations and a solid metal or wood floor deck under each equipment leg.

The elevated deck or floor surface major types are solid and non-solid. The deck, floor, or aisle materials are:

- Solid group that includes (a) plywood, (b) plywood with a coating, (c) plywood with asonite, (d) plywood with tile, (e) polytexture, (f) metal plate, and (g) solid grates.
- Open group that includes (a) grates that are bar preformed deck grates, open plank or grip struct type, and (b) perforated metal or expanded metal.

A plywood deck surface is solid material that covers the entire elevated aisle. Disadvantages are when there is high employee or wheeled vehicle traffic on the surface, it tends to wear quickly and bare plywood can splinter and become difficult to clean. The advantage is that it is inexpensive to install.

A plywood deck with a coated surface has the same operational characteristics as a bare plywood surface. The additional advantages are that it is easier to clean and reflects some light.

A plywood deck with a Masonite surface in high traffic aisles provides a surface that resists wear from high employee or wheeled vehicle traffic. If the Masonite does not cover the entire elevated floor area, at the intersection between the plain plywood surface and Masonite floor area surface there is a slight elevation change that creates an employee tripping hazard.

A plywood deck with tile surface provides an even surface for the entire area and is long-lasting. The tile cover permits easy housekeeping and minimal employee tripping hazards. There is a mastic or glue seal between the roof deck and the plywood deck. The mezzanine requires additional installation time.

With the plywood deck types, a deck is fastened to the steel roof with flathead self-drilling or self-tapping screws that are spaced on nominal 12–in. centers in lateral and longitudinal directions. Plywood deck material is a minimum ¾ in. thick, interior, APA class 1 or 2 plywood, and tongue and grooved on all edges with the C side up. Fire-treated tongue and grooved plywood with its high potential for uneven edges creates installation problems and results in an uneven mezzanine floor surface.

The polytexture deck material has a textured high-density polyethylene overlay pre-attached to a plywood sheet. This feature creates a smooth and even mezzanine surface with less installation time. The surface provides a good surface for flexible equipment arrangement. As part of the installation work, any gaps between two polytexture deck sections are filled with an epoxy filler and are smoothed even with the floor edges. Each screw top is coated with a coating to match the polytexture coating color.

The metal plate deck has a solid metal plate that has diamond shapes on its surface to improve employee safety.

Plank grate mezzanine decks are used for aisles or walkways. For rigidity, additional cross-aisle members used in the aisle and planks are crimped or welded together.

The grated open mezzanine deck concept is bar grates, grip struct grates, or open plank deck grates. Each open-deck grate type provides sufficient support for an employee aisle in a rack-to-rack or shelf-to-shelf design. When using these deck types, the concept has additional cross-aisle ties and the grates are crimped or welded together.

The perforated metal or expanded metal deck is a metal sheet that has pre-punched holes in it. This design provides a walkway between two pick rows.

Protection Concepts for Open Spaces under an Elevated Pick Position

Decking material is used under elevated pick positions to improve employee safety under the open pick position. The decking material under the elevated pick position protects employees from items falling through a vacant or empty pick position. Options are:

- To have a deck that covers the elevated floor area
- To secure the lowest pick position deck to the upright structural members
- To use wire cloth or wire mesh, nylon netting, wire screen, or expanded metal

Choosing an Open or Closed Elevated Floor

When you must decide between using a closed or open deck or mezzanine, the following situations may provide some insight. A solid deck is preferred when:

- Carts or other wheeled vehicles are used on the mezzanine level.
- Small-items are handled on the elevated floor.
- Elevated floor equipment layout is different from that on the lower floor.
- There is high activity on the mezzanine floor level.
- There is a large mezzanine floor area.
- Women in dresses work on the elevated level.
- The elevated area is allowed to exceed the standard mezzanine area with additional smoke detectors and fire sprinklers.

An open-decked mezzanine is considered when:

- There is a short span for the walkway or aisle.
- The mezzanine is rack- or shelf-supported.
- Pick positions are the same on the ground and elevated levels.
- Few employees work on the elevated level.
- Heat and ventilation conditions are the same on both levels.

Product Movement between Elevated and Lower Levels

A major component of planning for an elevated floor is the vertical small-item transport concept that is used to transfer small items between the two levels. The vertical transport concept that is used is determined by the physical characteristics of the small items and available building space. Vertical transport concepts include:

- Master cartons or totes on a pallet with a forklift truck. The pallet is transferred through a safety gate onto or from the elevated level. To protect the elevated floor edge, there is a metal angle plate along the front edge.
- A powered belt conveyor travel path.
- A powered overhead trolley with baskets that pass through a kickplate and handrail corral.
- A vertical lift.
- A whiz lift.
- Decline transport only, which can be gravity chutes or slides or a non-powered roller or skatewheel conveyor.

For additional information on the vertical transport, the reader is referred to the vertical transport section in this chapter.

Other Considerations

A major component is compliance with building codes and standards. Some are:

- Kickplates and handrails on the perimeter and floor penetrations/openings. If a conveyor or overhead trolley breaks the handrail or kickplate perimeter, the handrail and kickplate extends inward to from a corral of at least 6 to 7 ft. length. At the corral interior, there is a trip bar or kickplate in the product travel path with no handrail.
- Sufficient employee exits or stairways.
- Floor penetrations are protected with employee guarding and have sufficient fire protection, that is, a deluge or dog house.
- Base plates, structural upright and horizontal members are designed for the seismic location and for the imposed static and dynamic loads.
- With the additional floor, upright posts or columns may have to be fire protected with additional fire sprinklers or fire proof sprayed mineral fiber or dry walls.
- On both levels, sufficient lighting and head clearances.
- High traffic areas have additional floor supports and on the lower level with powered mobile equipment all columns or posts have guards.
- In warm and humid climates, sufficient air movement on both levels.

Equipment arrangement, which is determined by the weight of the equipment, elevated floor design, and UOP flows.

Building Columns

During building construction, the number and size of building columns are determined by the dynamic and static loads supported by an elevated floor and facility geographic (seismic) location. The building column loads are:

- For the ground-level building columns to support the elevated floors order-fulfillment equipment dynamic and static loads
- Highest elevated floor building columns to support the roof load

The building columns that support the roof loads are few in number and size but the roof support beams could have a larger size. Building column characteristics are that the highest elevated floor building columns support only roof rain and snow loads. Fewer building columns mean an increase in useable elevated floor area, increased equipment layout flexibility, and lower building construction costs.

The important building column characteristics are:

- Base size that includes length, width, and height
- Internal span between two column bases
- Center to center span between two column centers

To locate the pallet storage area and fast-moving, heavyweight/high-cube SKUs on a building elevated floor has high construction costs with a thick floor, due to the heavy dynamic and static loads and geographic location. Lower construction costs result with a building elevated floor that is designed for the lightweight and small-cube SKU storage/pick area, and pack and return activities.

In a densely populated area and especially in high seismic zone areas, an order-fulfillment building design has a large number of building columns, as well as larger columns to support elevated floor dynamic and static loads from the pallet inventory, forklift truck, and storage rack static and dynamic loads. The building columns have a 20- by 30-ft. pattern and have a size range from 1 to 3 sq. ft. Multi-floor building columns add 10 to 15 percent space requirement, and thus decreases your useable space. Building columns are unused space that is included in a facility construction costs. Building columns are included in the facility square footage and with many, large columns, facility space utilization is lower than a building with fewer, smaller columns.

With any multi-floor facility, elevated floors have vertical personnel, vendor-delivered UOP or CO shipping carton transport equipment travel paths, and protected access to elevated floors. A multi-floor facility design with the office, administrative, employee support, low-volume, lightweight/small-cube order-fulfillment activities, pack area, and value-added activities on an elevated floor characteristics are:

- Small site area (in square feet).
- Low building construction cost per square foot.
- High cube or space utilization.
- Personnel stairways and vendor-delivered piece vertical transport equipment travel paths.
- Each elevated floor has fire sprinklers, light fixtures, and employee support areas.

■ From an elevated floor level pack station to a shipping sortation concept induction station a CO shipping carton decline transport conveyor travel path

UOP or CO Flow through a Facility and Value-Added Activities

The vendor-delivered UOP or CO flow through a facility and valued-added activities has an impact on the order-fulfillment process. It is understood that a building's physical shape and size, the number of floors, and the value-added activities determine the location of the pick area and pattern of flow of vendor-delivered UOPs or COs through a facility. The design team determines the most cost-effective and efficient flow pattern and sequence of value-added activities.

Options for vendor-delivered UOP or CO flow patterns are:

■ For a single-floor facility, a horizontal flow pattern that is:
  ■ One-way or straight flow
  ■ Two-way patterns
■ A vertical or up-and-down flow pattern that is used in a multi-floor facility

One-Way or Straight Flow Pattern

A one-way or straight vendor-delivered UOP or CO flow pattern through a small-item or flatwear order-fulfillment facility is also referred to as an "in one side and out the other" pattern. Vendor-delivered UOPs enter the facility from one side and COs exit the facility from the opposite side. In this one-way flow pattern, vendor-delivered UOPs travel through the entire facility. In a store/hold inventory flow pattern, there are increased transport costs because storage area forklift trucks do not perform dual-cycle storage transactions. When the order-fulfillment operation has a low volume, few SKUs and low pallet inventory, and this pattern is used, facility construction and landscape costs are higher due to the need for more vendor and customer delivery truck roadways and yard area.

Two-Way Flow Pattern

In a two-way pattern, vendor-delivered UOPs enter a facility from one side and COs exit on the same side of the facility. This pattern is excellent for a store and hold inventory requirement. It improves transport and storage area employee productivity due to the fact that the employees have a high potential to make dual-cycle transactions. Advantages are lower land and construction costs for delivery truck yard and roadway, and high storage area productivity with storage area dual cycles. The two-way flow can be either a U or a W pattern.

The U Pattern

In the U flow pattern, vendor-delivered UOPs are received on one side of the facility, transported to the storage or pick area in the middle of the facility, and withdrawn from the storage/pick/pack area and transported to a shipping dock. With the receiving docks on the right side of the facility and the pack stations and shipping docks on the left side, vendor-delivered UOP and CO movement through the facility makes a U pattern.

The W Pattern

With the W flow pattern, vendor-delivered UOPs are received in the middle of the facility, transported to the middle for storage and pick/pack activities, and transported to the shipping dock. With the receiving docks on the left and right sides and the pack stations and shipping docks in the middle, the vendor-delivered UOP and CO movement through the facility makes a W pattern.

The Vertical or Up-and-Down Pattern

A vertical or up-and-down flow pattern through an operation is specific for a multi-floor facility. In a multi-floor operation, receiving and shipping functions, pallet storage positions, and high-volume, pick positions for heavyweight/high-cube SKUs are located on the ground floor. The pick positions for low-volume, small-cube/lightweight SKUs, pallet storage positions, pick area and pack stations, offices, and value-added activities are located on elevated floors.

With the order-fulfillment activities located in this arrangement, vendor-delivered UOP flow pattern and SKU allocation in the facility optimizes small-item flatwear vertical and horizontal transport activities. If the 80/20 SKU movement rule is true, 20 percent of the SKUs and 80 percent of the volume remain on the ground floor and follow a U or W vendor-delivered UOP flow pattern. Other UOPs or 80 percent of the SKUs and 20 percent of the volume flows on a vertical travel path to an elevated floor. Vendor-delivered UOPs and COs follow a vertical or up-and-down flow pattern. This design has the following features:

- Reduced site investment
- Increased investment in construction for an elevated floor and vertical transport equipment
- Slightly increased expense for additional vertical and horizontal transport

To assure that the suggested pattern fits into the proposed building area, your design team establishes the required square footage for each order-fulfillment activity and each value-added activity. In a small-item order-fulfillment operation, the value-added activities are:

- Pre-pick and pack value-added activities
  - Price ticket
  - Repackage
- Post pick and pack value-added activities
  - Monogram
  - Gift wrap
  - Product repair
  - Return goods
  - CO pick-up
  - Out-of-season

Value-Added Activities

The first step is to list and define the small-item or flatware apparel order-fulfillment activities that are performed to complete a CO. An activity list assures that the design team and operations manager have included all the required order-fulfillment and value-added activities to provide accurate and on-time customer service at the lowest possible operational cost. Small-item order-fulfillment activities include:

- Delivery truck or ocean-going container yard control and rail car spotting at receiving docks for UOPs, and CO delivery truck spotting at the shipping docks.
- Receiving, unloading, SKU quantity and quality control, individual small-item packing, piece identification, price ticketing, and pre-pick value-added activities.
- Internal piece transport between small-item order-fulfillment activities.
- Storage deposit and withdrawal transaction and inventory control.
- IT CO processing, cubing, and download to the pick area microcomputer.
- Per CO requirements, replenishment quantity from a SKU storage position to a pick position.
- Per a CO fulfillment operation, CO carton make-up with bottom flaps sealed and carton introduction to a pick line and transfer to a pick vehicle or automatic pick machine; other pick area or line start activities include CO discreet identification and pack list insert into a CO container.
- Picker or automatic pick machine SKU quantity transfer from a pick position onto a load-carrying surface or into a CO captive or shipping carton, including SKU identification with a CO discreet code.
- For batched small-item COs, SKUs are sent to a sortation area where they are separated and packed into a shipping carton.
- After pick or pick and pack activity, a CO shipping carton is sent to a pick quantity and quality check activity.
- On a small-item order-fulfillment pick line, most CO shipping cartons require filler material to fill void spaces and to seal carton.

- Manual or automatic manifest a CO discreet identification.
- CO cartons are placed into a temporary storage area until required at the shipping dock, unitized onto a pallet, four-wheel cart, or slip sheet, or directly loaded onto a CO delivery truck.
- Trash removal.
- Handling CO product returns, out-of-season, damaged or obsolete SKUs, which includes the entire processing, transport, and storage.
- CO return SKU activity to ensure the returned piece was received.
- Security, risk management, sanitation, and maintenance assure that a building equipment and order-fulfillment concept performs and protect assets and inventory from damage.

Block Drawing

After your design team has established the required square footage and has sequenced the order-fulfillment activities and value-added activities, the next design step is to develop a block layout drawing. A block drawing is a basic drawing that shows the total square footage for the ground floor, elevated floor, and the location of each activity inside the facility. The block drawing also shows building exterior walls, interior load-bearing and fire walls, building columns, receiving and shipping docks, and the location of stairways, toilets, offices, and emergency exits.

The advantages of doing this are that it ensures that all pick and value-added activities fit inside the facility and allows you to trace vendor-delivered UOP or CO flow to ensure that it optimizes the in-house transport activity.

Plan View Drawing

Another important step in understanding UOP and CO flow through the facility is to develop a detailed plan view drawing. This plan view is drawn to scale and shows each order-fulfillment activity area, including material-handling equipment and all building items. The detailed drawing shows the relationship between the order-fulfillment activities and the floor space.

Pick Line, Pick Aisle, or Automatic Pick Machine Design Parameters

In designing a manual, mechanized, or automatic pick line or pick area for small items, the following factors apply:

- Ability for replenishment and pick employees to complete transactions
- Load-carrying surface or conveyor travel path to move vendor-delivered UOPs or COs, available electric or air power to move a transport load-carrying surface and travel path window

- CO piece and pick line volume, CO volume, and CO shipping carton volume
- Physical characteristics of the SKUs, as well as their rate of movement and value
- How a SKU is picked from a pick position onto a conveyor travel path, into a carton, or into an employee's hands

The design of the pick area will be determined by the characteristics of the SKUs, including cube (volume) and the actual length, width, height, and weight of each SKU. Other physical characteristics include UOP classification, SKU exterior packaging and the ability of the packaging to protect the SKUs, support the weight of another SKU stacked vertically, and withstand horizontal line pressure on a conveyor travel path.

UOP classifications are:

- Fragile
- Crushable
- Toxic or nontoxic
- High cube or small cube
- Heavyweight or lightweight
- Edible or nonedible
- Hazardous
- Price or value
- Family group or kit component

Exterior packaging factors are:

- Sealed or nonsealed
- Low or high coefficient of friction
- Ability to accept a self-adhesive pick label
- Whether exterior SKU ends are susceptible to damage
- SKU retail surface
- Nature of the pack material: plastic bag, paper bag, loose, chipboard box, cardboard box, or sealed with a rubber band, plastic band, or plastic wire
- SKU shape

The physical characteristics of the SKU determine its ability to move over a belt conveyor travel path and interface to an automatic pick machine.

Another SKU characteristic is how it is picked from a pick position onto a take-away transport concept. SKU transport options are:

- Individual, each, piece, single, or loose item from a master carton or sleeve, with no additional exterior packaging.

- As a sleeve, SKUs are connected together by a manufacturer package. As a group, SKUs are removed from a vendor master carton, with no additional exterior packaging.
- Individual, each, piece, single, or loose item from a master carton, sleeve, or inner pack, placed into a small bag or box.

Pick Line or Pick Aisle Activity Sequence

Pick activities to complete a CO need to be efficient and cost-effective. Each pick activity should take place in sequential order to ensure accurate and on-time CO completion. These activities include:

- Carton make-up and carton introduction to a pick line, pick aisle, placement onto a pick vehicle load-carrying surface or introduction to an automatic pick machine
- CO discreet identification code attached onto a CO pick tote exterior surface or release from a microcomputer instruction to an automatic pick machine
- CO pack list or manifest insert into a CO pick carton
- SKU pick position on a pick line, in a pick aisle or in an automatic pick machine
- Loose picked SKU or completed CO carton take-away transport network from a pick line or automatic pick machine to the next order-fulfillment operation activity
- SKU pick position replenishment activity
- CO-picked SKU quantity or quality check activity
- Single CO pack activity
- Batched CO-picked SKU sortation and pack activity
- CO shipping carton void fill and CO delivery label activity
- CO carton seal activity
- CO carton manifest, load, and ship activity
- Trash removal from pick line, pick aisle, or automatic pick machine

This is a complete list of the activities in an order-fulfillment operation. Your specific operation determines which of these are carried out in your operation. The four common types of pick operation are:

- Single CO pick and pack or direct transfer from a pick position into a CO shipping carton
- Batched CO loose SKU pick or transfer from a pick machine onto a belt conveyor travel path, into a cart, or into a captive tote for transport to a sort concept and pack into a CO shipping carton
- Batched CO loose SKU pick and sort for transport to a pack station for sort and transfer into a CO shipping carton

- Single CO pick into a captive CO container or between two cleats on a belt conveyor travel path surface and packed into a CO shipping carton at a separate station

SKU Subassembly Activity

In a small-item order-fulfillment operation, there may be SKU subassembly activity prior to the order-fulfillment activity. These activities are designed to:

- Minimize time to complete or pick COs
- Improve SKU handling
- During low-volume periods, perform an activity that is required at a later time
- Protect SKUs from scratches

SKU subassembly activity is performed in the receiving dock area, special work area, or in a pick area. An example of this type of subassembly activity is placing a SKU into:

- A chipboard box, plastic bag, or paper/plastic bag, depending on the transport concept:
  - If the item is picked as a single SKU, it is packed in a chipboard box, plastic box, paper bag, or plastic bag.
  - If picked in bulk, 25 or 50 SKUs, or the most frequent SKU count, are placed into a plastic or paper bag.
  - For small items such as jewelry, rings are placed into a foam rubber pad with slits, or chains are placed on a plastic loop.
- A shipping carton for COs with a high SKU volume.
- A presentation case, for individual SKUs.

SKU Pick Position

The SKU pick position is a physical location on a pick line, pick area, or in a pick machine that temporarily holds SKUs to satisfy a CO. In response to a CO pick instruction, SKUs are removed by a picker or automatic pick machine from a pick position and transferred into an employee's hands, directly into a CO carton/tote, or onto a belt conveyor travel path for travel to a sortation area or pack area.

Small-item CO SKU pick activities may be manual, mechanized, or automatic.

For maximum picker and replenishment employee productivity, high accuracy, and on-time CO completion, small-item pick positions are arranged in an arithmetic progressive sequence that routes a picker or automatic machine pick device through a pick aisle, along a pick line, or along a pick machine front. SKU allocation to a pick position impacts the efficiency and cost effectiveness of an order-fulfillment operation and are determined by the physical characteristics of the SKU and CO volume.

Pick position locations can be described as (see Figure 4.2) fixed, variable, golden zone, fixed zone, variable zone, ABC theory, pairs, kit or family group, high value, UOP rotation, and pick line profile. These are described in more detail later in the chapter.

Fixed or Variable Pick Positions

Pick positions may be either fixed or variable. In a fixed pick position, SKUs are permanently allocated. During the receiving process, a specific quantity of each SKU is allocated to a pick position; additional quantities are allocated to storage

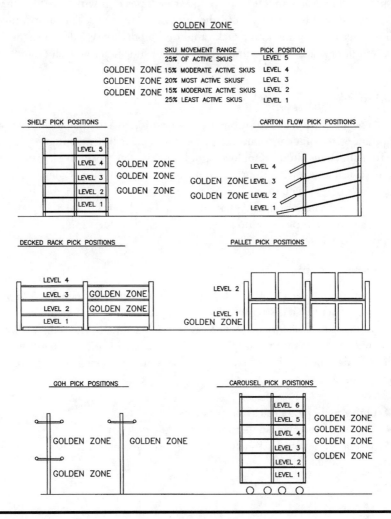

Figure 4.2 Golden zones: shelf, flow rack, pallet, and GOH.

positions. After CO quantities deplete SKUs in a pick position, a MWS program creates SKU replenishment transactions from a storage position to a pick position. The fixed pick position maintains pick line profile productivity, best picker productivity, family grouping, less floor space, replenishment errors, stock-outs, and on-time and accurate pick position replenishment.

A variable, random, or floating pick position means a SKU is temporarily allocated to a pick position. After the receiving process, a WMS allocates the SKU quantities to several pick positions in a pick area; no SKUs are allocated to storage positions. SKUs remain in each pick position until the SKUs in the first pick position are depleted by COs. Then a pick area microcomputer assigns the next CO picks to another pick position with SKUs. In a variable pick position application, the pick area microcomputer floats the next pick position within a "SKU present" pick zone. With this concept, it is difficult to maintain a pick line profile; increased walking distance means lower pick employee productivity, additional pick positions and floor space, and minimizes the replenishment activity, labor and errors.

Golden Zone

The physical elevation and location of a pick position in a manual pick concept is an important factor that contributes to picker and replenishment employee productivity. To achieve maximum employee productivity, pick position elevation reduces employee reaching or bending to complete a pick or replenishment transaction. This means that a shelf, decked pallet rack, carton flow rack top, and bottom pick positions are the least desirable pick positions. Pick levels between 2 ft. and 5 ft. 6 in. are the preferred pick positions, referred to as the golden zone.

Zone Area Picking

Zone area picking separates a pick line or pick area into defined fixed or variable pick zones. Each picker is responsible to complete CO picks for one pick zone. A fixed zone means that for each work day a picker works in the same pick zone or has a predetermined number of pick positions. A variable pick zone means that for each work day the number of pick zones that a picker is assigned to changes. Pick zone options are:

- One gravity pallet or carton rack bay
- Two or three gravity pallets or carton flow rack bays
- Several static standard shelves and hand-stacked onto decked pallet rack bays

The ABC Theory

When order-fulfillment professionals refer to the three zones of a pick area, they are referring to the ABC theory, which states that A-zone pick positions are allocated to fast-moving SKUs that are few in number with a large inventory quantity per SKU. B-zone pick positions are allocated to medium-moving SKUs, medium in number, with a medium inventory quantity per SKU. C-zone pick positions are allocated to slow-moving SKUs, large in number with a small inventory quantity per SKU.

If an order-fulfillment operation has receiving and shipping docks in the front of a facility and pick locations are based on the ABC theory, fast-moving SKUs are located at the front of the facility. If the receiving and shipping docks are located on opposite sides of the building, the fast-moving SKUs are located by their associated travel distance theory.

The Golden Highway Theory

Fast-moving or power SKUs are located in one area, referred to as the golden highway. All promotional, seasonal, special sales, and fast-moving SKUs are located in one area or pick aisle, near the packing stations or shipping docks.

The arrangement of SKU pick positions increases your picker hit concentration (hit number per aisle) and density (hit number per SKU). High hit concentration and density means high pick and replenishment employee productivity due to short walking distance between two active pick positions.

The Pairs or Family Group Theory

The family group theory for pick area layout is dictated by your company or local code requirements. Specific SKUs are allocated to pick positions in specific locations, zones, or aisles within a facility. This layout is designed to accommodate SKUs that complement each other, have similar dimensions, are located in the same customer retail or plant aisle, require the same environmental conditions, are high security risks, toxic materials, edible or nonedible, hazardous or flammable, and slow moving.

The Cell Pick Line Theory

The cell pick line concept is used in an order-fulfillment operation that has a unique SKU mix and a large number of SKUs. The cell pick line theory is similar to the family group theory. Within the SKU mix there are separate groups, such as CO groups with a common language on a SKU. This situation means that a pick area has pick positions that are allocated for common SKUs that are available for all COs, as well as separate cell pick positions that are allocated for unique SKUs that are for a particular CO group.

A cell pick line layout uses a four-wheel cart or zero- or low-pressure powered conveyor travel path that is strategically located along each cell pick front. The cell pick line layout has several pick modules that have SKUs that are common for all COs. This layout has the standard equipment with gravity carton flow racks facing a pick conveyor or cart travel path, standard static shelf rows, and aisles that are perpendicular to a pick conveyor or cart travel path at the static shelf aisle end with a standard shelf bay on casters. This feature means that a shelf at the end of the pick aisle is mobile. For each language or SKU group, there is one pick cell or as many cells as required. The pick cell has standard gravity carton flow racks, standard static shelves, and a mobile standard shelf at the end of the aisle.

Jewelry or Spare Parts by Family Group or Cell Group

To optimize order-fulfillment activities, SKU location in a pick area is a very important factor. SKU location has an impact on picker productivity and fast CO SKU return to a pickable location.

The family group or cell SKU location philosophy separates a pick area into distinct pick aisles or cells that have a unique SKU group. Each unique SKU group is assigned to separate pick positions. SKU family groups or cells are allocated by their unique description or last digit of the inventory number.

A family group approach has:

- One aisle for gold watches
- One aisle for other watches
- One aisle for silver watches
- One aisle for gold bracelets
- One aisle for silver bracelets
- One aisle for silver chains
- One aisle for gold chains
- One aisle for gold rings
- One aisle for silver rings
- One aisle for gold earrings
- One aisle for silver earrings
- One aisle for other items

Similarly, for auto parts, separate aisles would be allocated for Ford, GM, BMW and Toyota spare parts.

Typical family group allocations for jewelry or spare parts are as follows:

- The high-cube SKU group is located nearest to a pack area. With fewer totes per trip, this location minimizes receiving put-away and picker travel trips.

- The smallest cube SKU group is located the greatest distance from the pack area. With the number of totes per trip, this location minimizes receiving put-away and picker travel trips.
- SKU groups are assigned to an aisle according to physical description of the SKU. This feature means that employees easily recognize a SKU-assigned aisle.
- Because most sales programs are per SKU group, each aisle has A SKU movers that improve picker productivity.

SKU family group allocation to a pick aisle improves picker productivity by increasing CO SKU hit concentration and hit density and decreasing the number of picker trips between the pick area and the pack area.

Improvements in processing CO returns results from a continuous CO return piece flow from a process station to a pickable location. To achieve this objective, each CO SKU return process station has a tote for each product group. With this arrangement, a returns process station employee physically identifies a good-quality SKU and quickly and accurately places it into the appropriate container. Full totes are transferred from the returns area to a pick area. In a pick area, with one SKU group in a tote, employees returning SKUs to a pickable position remain in one aisle and are not traveling between aisles.

SKU Rotation

SKU or UOP rotation is either first in, first out (FIFO) or last in, first out (LIFO). In a FIFO UOP rotation concept, an SKU that was received first in an operation is the SKU that is shipped first from the operation. A FIFO pick area philosophy typically indicates that a SKU has a limited shelf life. After a specific date, if a SKU is not withdrawn from a pick position, it is not acceptable for a CO. SKUs that have a FIFO UOP rotation have a gravity carton flow rack pick position.

In a LIFO UOP rotation concept, SKUs have no specific time for rotation. SKUs do not have a predetermined shelf life and are allocated to a shelf, hand-stack decked pallet rack, or pallet rack pick position.

Pick Line, Pick Aisle, or Automatic Pick Machine Profile

An important factor in designing pick activity is to have a SKU profile, which allows your design team to assign each SKU to a pick position. If you do not complete a SKU pick line profile there is the potential for low picker productivity. The problems are:

- With a manual or mechanized pick line, lower than anticipated picker rates or low picker productivity due to insufficient daily pick quantity per pick zone.

- In one pick aisle, increased nonproductive picker walking time and distance between two picks.
- With a manual single-item pick line, low picker productivity to transfer a SKU from a pick position into a CO carton.
- With any pick concept, potential SKU damage from random placement of heavy, crushable, fragile, or lightweight SKUs into a CO carton/tote.
- Fast-moving SKU slugs are discharged onto an automatic pick machine conveyor belt or traveling over a powered conveyor surface travel path.

Factors in allocating SKUs to a pick line, pick aisle, or automatic pick machine are:

- Physical characteristics, cube, and weight of each SKU
- Peak and average SKU movement
- CO container cube at a pack station, in a CO carton or vehicle load-carrying capacity
- Cube and load capacity per pick position
- Number of pick positions per rack, shelf, or gravity carton flow rack bay, bin, or automatic pick machine side
- Pick area layout and detailed view drawings
- Anticipated pieces per hour and pick rate per hour
- Manual or personal computer with profile spreadsheets

After you have collected and reviewed the above data, you allocate, on paper or using a PC, each SKU to a pick position on a pick line, pick aisle, pick area, or in a pick machine. SKU allocation to a pick position assures that you have a physical location for each SKU and that each pick zone has sufficient daily SKU pick quantities to achieve your budgeted picker productivity rate. Completing a pick line profile gives your design team the opportunity to review the profile with order-fulfillment managers.

Pick Area Layout Considerations

In a manual picker concept, the direction of flow to SKU pick locations may be perpendicular or parallel to the pack stations (see Figure 4.3). Pick position rows and aisles perpendicular to a pack area have turning aisles at the ends of the aisles, which permit a picker to travel from one pick aisle to an adjacent pick aisle. A middle aisle decreases picker travel distance and time to move from a pick area to a pack area or to another aisle. Turning aisles and the middle aisle lead to a pack area. A pick area objective is to maximize picker productivity and if pick position rows and pick aisles have a long length, there is good possibility to obtain budgeted picker productivity. The feature means fewer picker turns and turning aisles. The advantages are maximize picker travel, high picker productivity with a high hit concentration and hit density, and good pick area space utilization.

PICK POSITION & ROW FLOW DIRECTIONS

PICK POSITIONS PERPENDICULAR TO PACK STATIONS

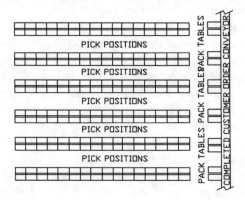

PICK POSITIONS PARALLEL TO PACK STATIONS

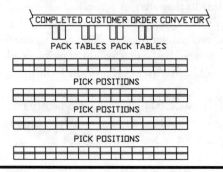

Figure 4.3 Pick position layout: perpendicular to pack stations and parallel to pack stations.

Pick position aisles that are parallel to the pack areas have a direction of flow that is direct from a pick area to a pack area. Each pick aisle provides access to a pack area. With short aisles there are additional picker turns and turning aisles. The features create low picker productivity with a greater pick aisle end turn number and fair pick area space utilization.

Guidelines for deciding on SKU pick positions include:

■ SKU cube and weight. Heavyweight and high-cube SKUs are located at the start of a pick line, medium-weight and medium-cube SKUs are located in the middle of a pick line, and lightweight and small-cube SKUs are located at the end of a pick line.

- SKU velocity, sales, or movement for one fiscal year. Fast-moving SKUs are located at the start of a pick line, medium-moving SKUs are located in the middle, and slow-moving SKUss are located at the end.
- Whether the SKU is crushable or fragile. SKUs with crushable or fragile characteristics are located at the end of a pick line.
- SKU value. High-value SKUs are allocated to a microcomputer-controlled pick machine, to a pick position that has limited or controlled access, or to a pick position with a security camera that is directed at the pick area.
- SKU hazardous classification. Hazardous SKUs are placed in a pick position that restricts SKU flight and liquid run-off.
- Maximum customer service or a pick line with SKUs allocated by family group or kit. SKUs with similar characteristics are located in sequential pick positions on a pick line. Reasons for using family groups are (a) SKUs are located in the same retail aisle, for example shoes or clothing SKUs that have one color with various sizes, or (b) SKUs are picked for a specific customer group such as SKUs in different languages.
- Review of historical SKU movement to identify SKUs with high movement per COs as candidates that are picked/packed off-line.
- Pick position elevation on a pick line. For example, the golden zone described above is the preferred manual pick position for fast-moving, heavyweight, or high-cube SKUs. Given the average height of pickers and the physical characteristics of most SKUs, in a five-level pick shelf, the golden zone is two, three, or four pick levels.
- Final pick aisle factors are (a) pick position type, (b) pick position identification that impacts your picker and replenishment employee productivity and accuracy, and (c) high- and low-volume SKUs.

Pick position types and pick position identification concepts are reviewed later in this chapter. The high- and low-volume SKUs were reviewed previously in this section.

Pick Line Profile

When a pick line is separated into pick zones and pickers are assigned to a pick zone, a proper pick line profile is a major step to ensure that you realize your budgeted picker productivity. A pick line profile is created using a form or PC spreadsheet (see Figure 4.4). Information required to create a pick line profile includes:

- Budgeted picker productivity rate or pieces per hour
- Historical SKU movement and characteristics
- Pick position type and physical characteristics
- Pick position identification

PICK LINE PROFILE SHEET AS PICKER FACES PICK FACES

PICKLINE PROFILE (SEE PICK LINE LAYOUT)
PICK LINE 1 - PICK SIDE
BAY 01

		PICK LINE LAYOUT		
		PICK AND TAKE-AWAY		
		CONVEYOR TRAVEL PATH		
		PICKER DIRECTION >>>>>>>>>		
		PICK SIDE		
		BAY 01		
TOP LEVEL 4	D05	D04	D03	D02	D01
SKU #				
PICK QUANTITY				
GOLDEN ZONE	C05	C04	C03	C02	C01
SKU #				
PICK QUANTITY				
GOLDEN ZONE	B05	B04	B03	B02	B01
SKU #				
PICK QUANTITY				
BOTTOM LEVEL	A05	A04	A03	A02	A01
SKU #				
PICK QUANTITY				

PICKLINE PROFILE (SEE PICK LINE LAYOUT)
PICK LINE 1 - PICK SIDE
BAY 02

		PICK LINE LAYOUT		
		PICK AND TAKE-AWAY		
		CONVEYOR TRAVEL PATH		
		PICKER DIRECTION <<<<<<<<<<<<<<<		
		PICK POSITIOCK SIDE		
		BAY 02		
TOP LEVEL 4	D05	D04	D03	D02	D01
SKU #				
PICK QUANTITY				
GOLDEN ZONE	C05	C04	C03	C02	C01
SKU #				
PICK QUANTITY				
GOLDEN ZONE	B05	B04	B03	B02	B01
SKU #				
PICK QUANTITY				
BOTTOM LEVEL	A05	A04	A03	A02	A01
SKU #				
PICK QUANTITY				

PICKLINE PROFILE (SEE PICK LINE LAYOUT)
PICK LINE 2 - PICK SIDE
BAY 02

		PICK LINE LAYOUT		
		PICK AND TAKE-AWAY		
		CONVEYOR TRAVEL PATH		
		PICKER DIRECTION <<<<<<<<<<<<<<<		
		PICK POSITION SIDE		
		BAY 02		
BOTTOM LEVEL	D05	D04	D03	D02	D01
SKU #				
PICK QUANTITY				
GOLDEN ZONE	C05	C04	C03	C02	C01
SKU #				
PICK QUANTITY				
GOLDEN ZONE	B05	B04	B03	B02	B01
SKU #				
PICK QUANTITY				
TOP LEVEL 4	A05	A04	A03	A02	A01
SKU #				
PICK QUANTITY				

PICKLINE PROFILE (SEE PICK LINE LAYOUT)
PICK LINE 2 - PICK SIDE
BAY 01

		PICK LINE LAYOUT		
		PICK AND TAKE-AWAY		
		CONVEYOR TRAVEL PATH		
		PICKER DIRECTION <<<<<<<<<<<<<<<		
		PICK POSITION SIDE		
		BAY 01		
BOTTOM LEVEL	D05	D04	D03	D02	D01
SKU #				
PICK QUANTITY				
GOLDEN ZONE	C05	C04	C03	C02	C01
SKU #				
PICK QUANTITY				
GOLDEN ZONE	B05	B04	B03	B02	B01
SKU #				
PICK QUANTITY				
TOP LEVEL 4	A05	A04	A03	A02	A01
SKU #				
PICK QUANTITY				

Figure 4.4   Pick line profile sheet.

A pick line profile is completed by an employee. The spreadsheet or paper profile document shows:

- Each pallet or carton gravity flow rack bay or static shelf bay pick position
- Estimated pick volume for each SKU or pick position and pick zone
- Picker routing or pick position number sequence or picker direction across a pallet or carton gravity flow rack bay or shelf bay face

Benefits of using a pick line profile spreadsheet are:

- SKUs can be allocated to pick positions within a pick zone for maximum picker productivity.
- Paper picker instruction form to attach pick position numbers to a pallet or carton gravity flow rack bay or shelf bay pick position face. With pallet or carton gravity flow racks, the labels are attached to a pallet or carton gravity flow rack bay replenishment side.
- Diskette or paper form allows pick position and SKU identification entry into a microcomputer.
- For initial set-up, as a replenishment employee instruction to transfer SKUs to a pick position.

Working with a pick line profile form, the considerations are:

- A profile document that shows each pallet or carton gravity flow rack or shelf bay pick position number by width and height.
- With a dual pick line, a picker routing pattern is different for pick aisle side A and side B.
- As an employee faces a pallet or carton gravity flow rack bay replenishment side, the pick position number sequence is different from the pallet or carton gravity flow rack bay pick side.

Pick Area Design

The design team fits the development of pick aisles and UOP and CO flow patterns into the existing facility, taking into account the building characteristics, such as walls, columns, and the roof.

Factors in designing a pick line or pick aisle are:

- Projected number of SKUs to be handled during the year, associated SKU or piece inventory as master cartons or pallets, CO piece volume, CO number, CO lines, vendor-delivered UOP pattern, and CO order flow pattern
- Selection of the preferred pick concept
- Identifying the building column clear span, column size, clear ceiling height, and locating the passageways through walls and elevated floors, location of utilities and light fixtures
- Block layouts and plan view drawings that are used to develop and refine pick line or pick aisle layout, vendor-delivered UOP flow pattern, CO flow pattern, and other material-handling and building items
- Developing and finalizing building drawings and pick and pack area line drawings
- Completing a simulation that is based on vendor-delivered UOP flow patterns with travel paths from a storage area to a pick area, CO flow patterns

from the pick area to the pack and shipping areas, and CO piece or pick volumes, and CO volumes

- Reviewing building and pick line layout drawings for compliance with local codes, company policies, and operational acceptance
- Complete a request for quote and select the preferred small-item pick concept vendor
- Develop building construction, remodel or new pick line equipment, and WMS program installation plans and schedules
- Install, test, and debug all building and pick-line equipment and complete a building and pick line punch list
- Start up and turn over of the building and pick line to the operational department

Pick Position Set Up and Replenishment

SKU set up occurs as new SKUs enter a pick line or as a pick line is organized for the next work day. SKU pick position replenishment activity occurs in both a fixed pick position concept and a floating pick position concept. In a fixed pick position concept, prior to SKU depletion in a pick position, a replenishment activity ensures that the correct SKU in the proper quantity is transferred on time from a SKU storage position to the correct pick position.

In a floating pick position concept, during the receiving process SKUs are transferred to several vacant pick positions. When the first SKU pick position is depleted, the pick area computer automatically has the next pick transaction occur at another SKU pick position.

SKU Storage Positions

In small-item order-fulfillment operations, SKU storage positions are:

- Off-site, in a separate facility. This feature means (a) all pick area and building storage positions are at capacity, (b) it is difficult to control inventory, (c) the operation incurs delivery truck shuttle costs between the two facilities, and (d) SKUs are obsolete.
- Remote storage, which is located in the same facility but is not adjacent to a SKU pick line or pick aisle. A remote storage position means (a) in-house forklift truck or pallet truck transport costs from a remote storage area to a pick area, (b) minimal delivery truck shuttle costs, and (c) improved inventory control.
- Ready storage, which is a SKU storage position that is adjacent to a pick position. Ready storage adjacent locations are (a) in a hand-stacked decked pallet rack bay, gravity flow rack, or standard shelf bay pick concept, positions that

are beyond a picker's normal reach or directly behind a pick line; (b) in a carousel, car trac, Robo pick, A-frame or Itematic pick concept, a ready storage position is in hand-stacked decked rack, carton gravity flow rack or standard shelf bays that are arranged parallel or perpendicular to the pick concept replenishment side.

■ On-line storage positions are additional SKU pieces beyond a workday's pick volume that are held in a pick position. In a gravity pallet or carton flow rack, slide, or chute pick concept, an on-line storage SKU piece quantity is located in master cartons that are queued behind a master carton in the pick front.

In a small-item or flatwear order-fulfillment operation, pick line replenishment activity involves:

■ Either fixed or floating pick positions
■ SKU replenishment quantities that are (a) maximum quantity, (b) optimum quantity, (c) next required replenishment position, or (d) pick clean
■ SKU replenishment transaction creation and verification that are (a) manual or (b) computer-controlled
■ Pick position identifications that are (a) alpha characters and numeric digits, (b) bar code, (c) human- and machine-readable codes, and (d) with or without SKU numbers

These replenishment concepts are reviewed later in this chapter.

Multiple Small-Item or Flatwear SKUs for Customer Orders

A small-item order-fulfillment operation has to pick multiple SKUs for COs, which may be either single COs or COs that are batched or grouped.

Single Customer Orders

If your operation picks multi-line or multiple SKUs as single COs, there is one picker who picks all SKUs for a CO. Pick options are:

■ To pick one CO per trip
■ To pick multiple COs on each trip, and place them into a separated container
■ To pick *en masse* each SKU for a CO group and sort each piece into a separate container, carton, or tote

The first method has low picker and packer productivity, additional pick cart and tote cost, less than optimum completed CO flow from a pick area to a pack area, increased pickers per aisle with wider pick aisles, no sortation labor, does not require a computer program or advanced pick instruction formats.

The second pick method has one picker with a four-wheel shelf cart pick multiple CO SKUs onto one shelf or into a multiple compartment tote. Depending on volume, a shelf has several totes or a tote is subdivided into smaller compartments. Each shelf, shelf section, tote, or tote compartment has a CO identification. With this order-fulfillment concept, a picker uses a pick and sort instruction format that is a paper pick document, paper self-adhesive labels, pick (sort) to light, or RF device. When a shelf, shelf section, tote, or tote compartment is full or all assigned COs are completed, the tote or cart is sent from the pick area to a pack station.

The features are good picker and packer productivity, few pick totes or carts, computer program to assign a CO to a shelf or tote section, and on each shelf tote section a CO discreet human/machine-readable identification, queue at a pack station, few pick cart trips in a pick aisle, no sortation labor, and any pick instruction format.

The last manual single CO pick philosophy has an order picker with a four-wheel shelf cart to pick each SKU *en masse* for a CO group and place the SKUs into a tote. A CO human/machine-readable discreet identification is attached to the front of each tote and a picker instruction matches this discreet identification.

Batched or Grouped Customer Orders

A multiple batched or grouped CO pick/sort/pack philosophy has a computer group several CO mixed SKUs into a batch or group. Each picker completes CO picks and places SKUs as mixed CO picked SKUs into a tote, picks or labels each CO SKU into a tote, cart load-carrying surface, or directly onto a belt conveyor travel path for transport from a pick area to a CO sortation area.

The features are few picker aisle trips, constant CO picked flow to pack stations, low pick cart and tote cost, computer program, each SKU has a human/machine-readable CO discreet identification, sufficient table top surface at a pack table, sortation area with additional sort and pack employee time, and handles high CO volume.

Order Picker Concept

Work flow for order pickers may be:

- A single CO is completed by a single picker.
- A single CO is completed by several pickers.
- Batched or multiple COs are completed by a single or multiple pickers.

Single Customer Order and One Order Picker or Pick Machine

A single CO that is completed by a single picker is used in a manual pick and pack or automatic pick machine operation. The picked SKUs are transferred from a pick position to a CO tote. One picker travels all pick aisles or the entire pick area to complete all SKU and piece pick transactions for a CO.

With a manual picker concept, the disadvantages are low employee productivity due to greater travel distances. The manual concept handles a low CO volume and small CO size or cube, employs the greatest number of pickers, longer CO order and delivery cycle time, and with an automatic pick concept, a high capital cost. The advantages are maintains CO integrity selection process, easy to identify problem pickers, easy to control, and with an automatic pick concept, accurate picks.

Single Customer Order with Multiple Order Pickers or Pick Machines

A single CO that is completed with multiple order pickers, mechanized pick concept, or automatic pick machines is used in an order-fulfillment operation that has a pick and pack operation. Picked SKUs are transferred directly from a pick position into a CO discreetly identified tote and transported through the pick area.

Instructions for a CO can take one of the following forms:

- Paper pick instructions and a picker who manually picks and transfers SKUs directly into a CO discreetly identified carton
- Pick-to-light or RF device instructions with a picker who manually picks and transfers SKUs directly into a CO discreetly identified carton
- Paper or pick-to-light instructions with a cell pick line, or pick area layout for specific CO SKUs, and a picker who manually picks and transfers SKUs directly into a CO discreetly identified carton

To handle one CO with multiple pickers, a manual pick line, mechanized pick concept, automatic pick machine, or pick area layout features CO discreet identification on each carton, and a carton transport travel path that links all the pick positions together. The transport concept moves a CO discreetly identified carton to the travel path past all manual or mechanized pick positions or automatic pick machine discharge locations. The pick instruction includes the CO discreet identification, CO SKU quantity, and SKU pick position identification.

The results are an increase in picker productivity due to shorter travel distances, a decrease in CO order and delivery cycle time, an ability to handle a greater CO number, COs that have few pieces per CO, and COs with several lines or SKUs, medium picker number, management controls, and a well-designed pick area with a CO carton transport concept, and difficult-to-identify problem pickers.

Multiple Customer Orders with Multiple Order Pickers or Automatic Pick Machines

Multiple COs with multiple pickers are also known as batched COs with multiple pickers. CO SKUs are grouped by pick position and for all COs, each CO SKU is printed onto a pick instruction. SKUs are picked and transferred from a pick position into a picker's hand or apron, tote, or onto a belt conveyor as mixed CO SKUs. The mixed CO picked SKUs are sorted or separated in the pick area, with a paper document sort instruction that directs a picker to transfer a CO picked and identified SKU into a CO assigned sortation location or carton. Alternatively, a mixed CO group with SKUs that are discreetly identified is sent from a pick area in a tote or as loose SKUs over a transport conveyor from the pick area to a sort area, where an employee or machine sorts each CO picked and identified SKU to each CO identified sortation location.

A high-volume batched CO pick operation has separate pick and sortation areas. In a batched CO pick concept, a picker instruction format is a machine-printed self-adhesive label. The front of the label has human- and machine-readable symbologies. For each SKU, a label is printed and a as a picker completes a pick transaction, a label is placed onto the SKU exterior surface. With a batched CO pick concept and self-adhesive label pick instruction, a picker travels through pick aisles and at the assigned pick position, transfers a SKU from the pick position, labels it, and places it into a tote or loose onto a belt conveyor travel path surface. The tote or belt conveyor transports the SKUs from a pick area to a sortation area.

After all pick transactions are completed, the mixed CO picked and labeled SKUs in totes or as loose SKUs on a conveyor belt are transported from a pick area to a sortation area. In the sortation area, a human/machine-readable discreet label on each SKU serves as the sortation employee instruction or is read by barcode scanner as a mechanical sort instruction. A sortation instruction permits a sortation employee or mechanical device to sort the mixed or batched SKUs by each CO discreet identification or SKU identification into a CO sortation location. The sortation activity completes the CO pick, transport, and sort activity.

Multiple Customer Orders with a Single Order Picker

For slow-moving SKUs in a high-volume operation, a picker with a paper document or RF device as a pick and sort instruction format, picks and sorts a CO picked SKU from a pick position into a CO tote.

Picked SKU Transport Concept

The SKU transport concept that is used depends on how SKUs are sorted. With SKUs that are sorted in a pick area, the transport concept moves picked and sorted

SKUs as individually sorted or complete COs from a pick area to a pack area where it is packed into a shipping carton.

In an operation in which mixed COs are picked and labeled, the transport concept moves mixed CO SKUs as a mixed group in totes or as loose SKUs on a belt conveyor travel path from the pick area to a sortation and pack area. In the sort and pack area, SKUs are sorted into a discreet location and packed into a shipping carton.

Sortation Concepts

There are different concepts for sorting batched COs. CO sortation location design parameters are:

- Determine the peak, most frequent, and average CO SKU or piece number and cube.
- Determine the sortation location percent utilization and associated SKU number and cube. Fifty percent sortation location utilization is a reasonable factor with 50 pieces per sortation location.
- Design each sortation/pack location with an end stop, a pack station surface for easy SKU flow onto a pack table, and sufficient space to perform final CO sort activity and pack activity.
- Determine whether the sortation activity is manual or mechanized and ensure that the pick and sort instruction satisfies the pick area and sortation area requirements.

Manual Sortation Concepts

Manual sortation concepts (see Figure 4.5) are:

- In a pick area with a paper document, RF device, or light pick/sort instruction, a picker picks and sorts customer-ordered SKUs to a cart shelf or into a separated tote.
- In a pick area with a paper document, RF device, or light pick/sort instruction, a picker picks and sorts CO SKUs to a mixed tote or bin location.
- In a sort area with a discreet label pick/sort instruction from a tote, basket, or a cart, an employee sorts CO picked and labeled SKUs to a pigeonhole location.
- In a sort area with a discreet label pick/sort instruction from a tote, basket, or a cart, an employee sorts CO picked and labeled SKUs to a slide or chute location.

Manual sort concepts are reviewed in detail later in this chapter.

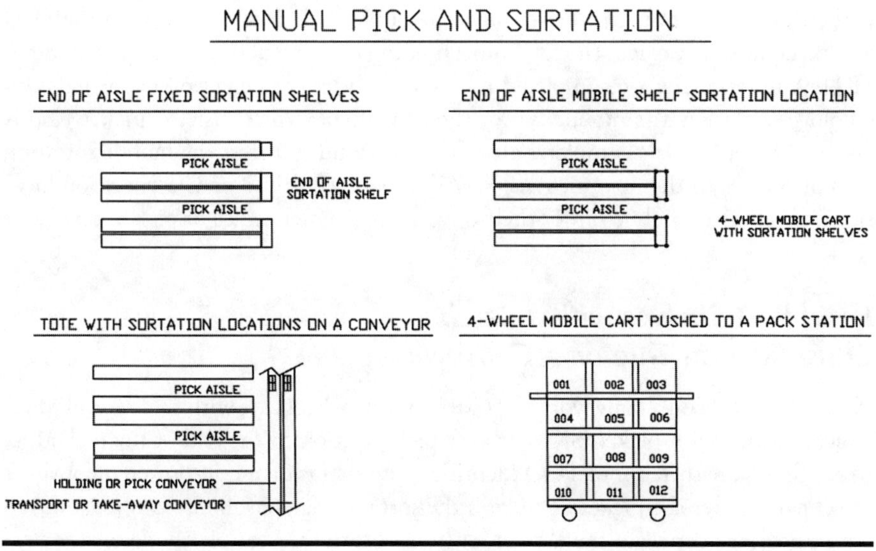

MANUAL PICK AND SORTATION

END OF AISLE FIXED SORTATION SHELVES

```
PICK AISLE
                          END OF AISLE
                          SORTATION SHELF
      PICK AISLE
```

END OF AISLE MOBILE SHELF SORTATION LOCATION

```
         PICK AISLE

         PICK AISLE
                          4-WHEEL MOBILE CART
                          WITH SORTATION SHELVES
```

TOTE WITH SORTATION LOCATIONS ON A CONVEYOR

```
      PICK AISLE

      PICK AISLE

HOLDING OR PICK CONVEYOR
TRANSPORT OR TAKE-AWAY CONVEYOR
```

4-WHEEL MOBILE CART PUSHED TO A PACK STATION

001	002	003
004	005	006
007	008	009
010	011	012

Figure 4.5 Manual sort concepts: end of aisle fixed shelves, end of aisle mobile shelves, tote with separators on a conveyor, and multi-shelf cart.

Batch Consolidation Tote with Fixed Shelves

This concept has fixed sort shelves at the end of a pick aisle end where a picker sorts SKUs into a CO assigned bin or shelf. After all CO SKU picks and sorts are completed, a consolidation person transfers SKUs from the fixed bin or shelf section at the end of each pick aisle into a mobile batch consolidation or master tote. This master tote is separated into multiple CO sections and each CO section is identified with a CO identification. A batch consolidation tote travels past all pick aisle fixed sortation bins or shelves. From each pick aisle sortation location, CO picked and sorted SKUs (pieces) are transferred into a CO identified section in the batch consolidation tote. Each master tote has the capacity to hold 4 to 6 totes and a 4-wheel cart holds 8 to 12 CO totes.

A picker walks with batched CO picked SKUs from a pick position to the end of the aisle to complete individual CO sortation activity to a CO identified sortation location. Depending on the SKU pick quantity or cube, the picker uses a box, apron, or hand to transport one or several SKUs. The concept is used with a paper pick document or self-adhesive label, or an RF device.

Batch Consolidation Tote with Mobile Shelves

With mobile bins or shelves at the end of a pick aisle that are discreetly identified for a CO, a picker verifies that the SKU pick/sortation instruction matches the master consolidation tote identification. SKUs are sorted into the appropriate bin or shelf.

If there is no match, the picker ensures that a pick tote is identified with the SKU and batch number, places the tote into a holding location and starts to pick another SKU. After all picks are completed in one pick aisle, the picker or a consolidation employee transfers the mobile bin or shelf to the next pick aisle. A picker who is assigned to a pick aisle completes all picks, sorts, and transfers the mobile sortation bins or shelves to the next pick aisle. With a batched group or CO sortation locations, this master mobile bin or shelf travels past all pick zones or pick aisles.

Fixed Sort Shelves with a Batch Consolidation Tote on a Conveyor

With a large tote on a conveyor travel path or four-wheel cart with shelves and totes, a picker walks with picked SKUs from a pick position to the end of the pick aisle, where SKUs are sorted into a CO identified tote that is on a non-powered conveyor travel path. After the picker completes the sortation, a consolidation tote is moved onto a powered conveyor travel path for transport to the next pick zone. At the next pick zone, the tote is diverted onto a pick zone conveyor travel path. This concept has a paper pick document, self-adhesive label, with an RF device.

A Mobile Pick Cart

CO discreetly identified bins or shelves on a mobile sortation vehicle (pick cart) is moved into a pick aisle by a picker, who picks and sorts SKUs from pick positions directly to a CO identified tote or section on the cart. After all required picks in one pick aisle, the picker transfers the mobile sortation vehicle to the next pick aisle, where the picker assigned to that aisle completes picks and transfers the cart to the next pick aisle. This four-wheel sortation cart with CO shelf sections or totes travels through all pick aisles.

An option with a four-wheel cart that has multiple CO sections is to have one picker with pick and sortation instructions for entire CO SKUs push a mobile sortation cart through all aisles and complete all CO SKU pick and sortation activities. This concept is used with a paper pick document, self-adhesive label, light to sort, or with an RF device sort instruction.

Large Tote or Self-Dumping Cart and SKU Transfer onto a Belt Conveyor

With a large tote or a self-dumping cart, a picker transfers each picked and labeled SKU (all pieces for a batched CO group) into a tote or cart cavity. A self-adhesive pick label is applied to each SKU and a full tote or cart is pushed to a transport concept. At this location, the tote or cart contents are dumped onto a belt conveyor travel path surface or the tote is transferred onto the conveyor travel path. The conveyor moves the mixed CO SKUs from the pick area to a sortation/pack area.

An option has a picker transfer a CO picked and labeled SKU directly onto a belt conveyor travel path. All individual pick belt conveyor travel paths merge or fall onto a master belt conveyor travel path. CO picked and labeled SKUs are transported from the pick area to a sortation/pack area, where an employee or mechanical sortation concept sorts SKUs into a CO assigned sortation location.

How to Determine Piece Number per Batch

With a batched or grouped CO concept, activity in the pick and sortation areas are designed according to the quantity of SKUs that are handled. Methods that are used to determine the number of small items or flatwear per batch are either picker driven or packer driven.

The picker driven method determines customer ordered SKU pieces per batch by the projected picker productivity rate. The method relies on a computer program to separate COs by projected picker productivity rate. The picker productivity rate determines the number of pickers that is required to handle the CO SKUs. If the picker productivity rate cannot handle the quantity of batched CO SKU pieces, management adjusts the number of small items or flatwear per batch and recalculates the projected picker number. Factors that affect the picker driven batch quantity are the number of COs per pack station, number of SKUs or pieces per CO, anticipated pack employee productivity, total daily pieces and COs, work hours per day, projected picker employee productivity, and budgeted sortation rate.

The packer-driven batch quantity projection method uses the number of SKUs that are handled by each sortation and pack station to determine small-item or flatwear pieces per batch. A computer program ensures that there is sufficient SKU quantity and CO quantity to keep each pack station active. CO picked, labeled, and sorted SKUs are evenly distributed to each pack station.

In a catalog or direct mail business, the picker driven method is an acceptable method to determine the batch small-item or flatwear piece quantity due to the computer allocating approximately 50 pieces per pack station with the average CO piece quantity ranging from two to five pieces. However, this SKU quantity does not exceed piece number per pack station and the active pack station number. Factors in determining packer-driven batch quantity are sortation/pack employee productivity rate, number of COs per pack station, number of pieces per CO, total daily piece number and COs, work hours per day, projected picker productivity, and budgeted sortation rate.

Batch Release Control

Batch release (see Figure 4.6) controls the transfer of picked and labeled SKUs to a transport concept. With a paper document pick instruction format, a picker picks

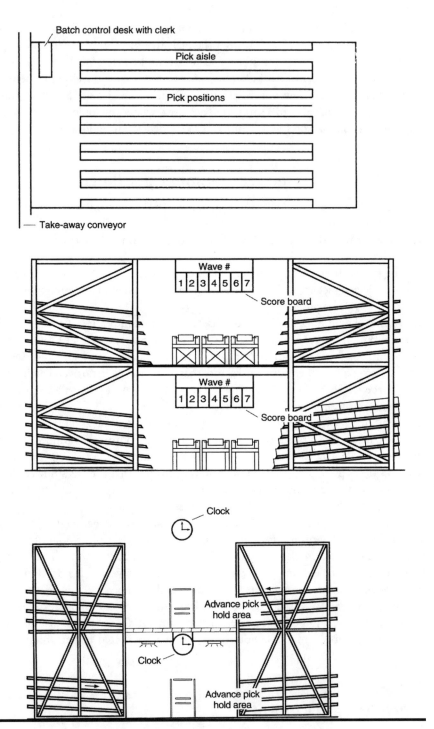

Figure 4.6 Batch release concepts: clerk, score board, and clock.

and sorts the SKU directly into a section of a tote or four-wheel cart. Batch release control components are:

- A paper pick/sort document that accompanies each tote or cart. Each tote or cart is identified for each batch. At a pack station, the paper document is used to verify the accuracy of the CO.
- In a pick area, there are sufficient empty totes or carts to handle the flow between pick/sort area and pack area activities.
- Each tote or cart has sufficient and clearly visible space for a batch identification and location of each CO.

With multiple COs and multiple pickers, each picker transfers mixed CO SKUs that have been labeled or coded into a tote that is moved over a conveyor travel path, or SKUs are transferred directly onto a belt conveyor. A batch release or control concept ensures that each picker transfers correctly batched SKUs on time onto a transport concept. If the batched SKUs are not properly sequenced with the sortation area activity, an unscheduled batch release from the pick area to the sortation activity has a high potential to create sortation errors or decrease sortation productivity and accuracy.

There are several guidelines for minimizing an imbalance between pick and sortation activity.

- Determine batched CO number and CO quantity or cube
- CO cube or pieces
- Clear pick and sort instructions
- In a pick area, a device to indicate the proper batch that is scheduled for pick and transfer activity
- In a pick area, sufficient space to queue an out-of-balance occurrence and a predetermined tote or cart queue number
- In a sortation area for each sortation station, multiple sortation locations
- If totes are used to transport SKUs, discreetly identify each tote for each batch or use color-coded totes for a batch and provide tote conveyor queue area

A batch release factor is that each pick label, pick position, and sortation location has a clear and understandable CO discreet identifier that is a human-readable or human/machine-readable code. This identifier method assures a quick and easy employee or machine pick/sort label match to a pick position and to a sortation location.

Batch Release

Each pick area has a device that controls the transfer of batched SKUs from a pick area onto a transport concept. A computer separates SKUs into several groups or waves. A batch release concept is a technique that is used in a pick area to instruct order pickers regarding the time for a specific batch or CO group to be transferred onto a transport concept. This is very critical, because in a workday there are several CO batches or waves that are sent to a limited number of sortation/pack stations. A batch release concept ensures that picker activity is coordinated with the sortation/pack station activity. This includes good communication between the sortation/pack area and pick area. The sortation area notifies the pick area regarding receipt of each batch. After proper notice from a sortation/pack area that a batch has been received and sorted into a sortation location, the pick area starts to pick, label, and transfer the next batch of SKUs onto a transport concept. Unless there is a transport or sortation/pack station problem, a computer has estimated the time for each batch release. This batch release is a component of the picker instruction.

To reduce batch overlaps in the sortation/pack area, the guidelines are:

- Ensure that the total picker productivity is equal to your sortation/pack station productivity
- In the pick area provide queue space and totes to hold a CO batch that is out of balance
- In the pick area use a batch release concept to instruct pickers as to each CO batch transfer time
- In the sortation/pack area provide each sortation station with at least three sortation locations

In the pick area, batch release concepts are:

- Control clerk. With a CO batch release desk and clerk in a pick area, at a control desk a clerk issues to each picker their pick paper document, self-adhesive label, or RF device instructions. A control clerk issues CO batch 1 pick instructions to pickers. After all CO batch 1 pick instructions are released to pickers, a control clerk with verbal pick and hold instructions to pickers and a paper hold tag with a batch number hand-printed on the surface, issues CO batch 2 pick instructions. To assure proper control, a clerk lists each picker name who received the pick instructions along with the issue time and picker anticipated return time.
- Clock and printed release time on the pick label. With a pick label CO batch release method, the computer-controlled printer prints each pick zone batch release time onto each CO batch pick label. This CO batch release time on a pick label indicates to a picker the expected time that CO batch 1 SKUs are scheduled for transfer onto a transport concept. In the transfer location, each pick zone has a clock that is visible to pickers. Each pick zone has a slightly different CO batch release time that is determined by the computer due to the travel distance from the pick area to the sortation area.

- Scoreboard. With a scoreboard CO batch release method, each pick zone has a scoreboard that has a symbol for each CO batch. In a pick area at a transfer location and based on sortation/pack activity, a scoreboard shows an assigned time that is scheduled for a CO group to be transferred to a transport concept.

Floor Queue Space

There should be sufficient floor space in the pick area adjacent to a transport concept to serve as a queue area to hold totes or carts with batched CO SKUs that are picked in advance or out of balance with the sortation area. Each tote or cart has a human-readable code that identifies each CO batch. When sortation activity is out of balance with pick activity, picked and labeled SKUs in totes or carts are held in the pick area queue section. When the sortation area signals release, SKUs are transferred from the totes or carts in the queue area onto a transport concept for transport from the pick area to the sortation area.

Mechanical Sortation Lane with Three Sections

With a mechanical sortation concept, each sortation/pack station is designed with at least three sortation sections:

- Number 1 sortation section for the batch 1 SKUs.
- While a sort/pack employee is handling batch 1 SKUs, a number 2 sortation section is designed for the second batch.
- If there is a problem with one sortation/pack activity, the number 3 sortation section is designed to provide a third sortation location.

With a manual sortation concept that has totes travel on a transport conveyor, prior to a sortation station is a queue tote conveyor lane. When loose SKUs are transported on a belt conveyor travel path from the pick area to the sortation area, each sortation location has a chute or belt conveyor travel path. To assure a constant sortation activity, there are one or two extra sortation lanes, which allow the sortation activity to continue if there is a problem in a sortation lane.

The mechanical sortation concepts are flap sorter, gull wing, Bombay drop, SBIR or moving belt, Nova sort, and tilt tray. After CO picked and labeled SKU symbology is read by a scanner device, a mechanical sortation device transfers each labeled SKU from a sortation conveyor travel path into a CO sortation location. Mechanical sort concepts are reviewed later in this chapter.

Other Factors That Influence Efficiency and Cost Effectiveness

Other factors that influence efficiency and cost effectiveness of a small-item pick operation are:

- SKU profile or SKU location on a pick line, pick aisle, or in a pick machine
- Pick tote or carton fill rate
- Pick zone or aisle design
- Pick/pack equipment layout
- Pick position type
- Common or unique SKUs

SKU Profile or Pick Position Location

The first factor that influences pick performance in a small-item or flatwear order-fulfillment operation is the SKU pick position or location on a pick line, pick aisle, or in an automatic pick machine. The SKU profile concept was reviewed earlier in this chapter.

Pick Methodology

One factor that ensures an efficient, cost-effective, accurate, and on-time order-fulfillment operation is to ensure that SKU pick quantity matches CO pick tote/carton or sort pack station capacity. The pick quantities can be projected manually or by computer. In a dynamic order-fulfillment operation, the computer-projected quantity is the preferred option. For a pick and pack operation, the computer determines the SKU cube pick quantity that fits into a CO pick tote or carton. This is determined by the internal dimensions of the carton, actual cube and weight of SKUs, desired carton utilization rate, and CO delivery truck company standard carton weight.

Tote or Shipping Carton Space Requirements

An important factor in the efficiency of the pick line is to ensure that the spatial data on your SKUs is accurate. This includes the actual length, width, height, and weight of each SKU as it is transferred from a pick position or pack table into a CO shipping carton. SKU cube data is not the best criterion to determine a CO pick tote or pick/shipping carton.

In an order-fulfillment operation, accurate SKU spatial information for a CO carton means less nonproductive picker and packer time, reduced UOP damage, computer selection for the best size shipping carton, and less filler material.

The features are realized due to the fact that for each CO the computer allocates a sufficient SKU quantity (length, width, and height) to maximize shipping carton utilization.

Methods for obtaining spatial information on SKUs are (see Figure 4.7):

■ Vendor data. The merchandising department notifies each vendor that the order-fulfillment operation requires SKU data. This data is entered into the computer files and becomes the basis to determine each CO cube. The disadvantages of relying on vendor data are that for loose SKUs, it does not accu-

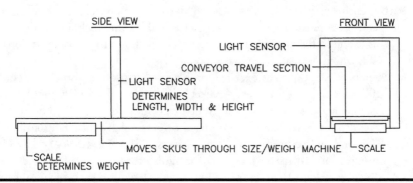

Figure 4.7 SKU cube concepts: cube platform and size/weight machine.

rately reflect the shape of a picked SKU, the data does not change as a SKU flows through an order-fulfillment operation, and there is no management control. The advantages are no data collection expense and no equipment investment.

■ Manual data collection method. An employee is assigned to collect the data, using a tape measure and portable scale. The employee measures and weighs each SKU and enters the identification number, dimensions, and weight on a paper document. The information is then entered into the computer files. For each new SKU received at a facility, a receiving or QA department employee collects the information.

■ Cube platform. A manual cube platform has a bottom surface and two sides that all intersect at one point. The bottom surface and each side have a set of colored lines. Each set of lines represents a shipping carton. An employee places a new SKU onto the platform, using the colored lines that to measure the appropriate shipping carton size for that SKU. The disadvantages of this method are time consuming, labor expense, and prone to errors. The advantages are accurate SKU measurement and the method establishes procedure for new SKUs.

■ Mechanized data collection. An employee uses a scale and a cube device. Each SKU is passed through a cube device which cubes the SKU to 0.04 in. and determines the weight to 0.01 lb. accuracy. The information is sent to a computer file or is manually entered in the computer files. The disadvantages are investment and it requires an employee. The advantages are it cubes all SKUs, is a quick and easy activity, mobile, provides accurate and on-line data, and can be used for new SKUs.

Pick Zone Length

Options for establishing the length of a pick zone in a pick line or pick aisle are to travel:

■ The entire pick line or pick aisle. A dynamic small-item order-fulfillment operation does not require a picker to travel the entire pick line or pick aisle. This is due to the fact that it is a great travel distance to all pick positions for a CO, which decreases employee productivity.

■ Fixed pick zone. With an excellent profiled pick/pack line, a small-item order-fulfillment operation has each picker travel over a fixed pick zone. A fixed pick zone has the same start and end pick position. It is preferred due to high picker productivity that results from a shorter travel distance, familiarity with SKUs, and, with a profiled pick line, increased SKU hit density and SKU hit concentration. To assist a picker to distinguish the limits of a fixed pick zone, an identification placard is placed at each aisle entrance and exit and fixed pick zone limit signs identify each pick zone start and stop pick position.

- Variable pick zone. A variable pick zone means that the pick zone limits are expanded or reduced depending on CO SKU picks or hits. Due to increased potential picker confusion, the variable pick zone is not the preferred option for a dynamic small-item order-fulfillment operation.

Determining the Best Small-Item/ Flatwear Apparel Pick Position

Information in this section helps you to determine the best small-item/flatwear apparel pick position for your order-fulfillment operation. A pick position is a discreet location for one SKU that contains sufficient SKU quantity to satisfy at least a large portion for one day or several waves of your CO SKUs. In most small-item/ flatwear apparel order-fulfillment facilities, SKU inventory is allocated to a pick (forward or active) position, a ready reserve position, or remote storage position. A ready reserve position is adjacent to, above, below, or behind a pick position, which means there is SKU inventory readily available to replenish a pick position. A SKU remote reserve position is in the general storage area and is transferred as required from the remote position to a ready reserve position or to a SKU pick position.

Factors to determine the best pick position are:

- Your picker and replenishment employee reach height
- Pick position type and capacity
- CO handling method such as single CO or batched COs
- Product rotation and inventory control method
- CO SKU demand, CO volume, CO number, lines per CO, and CO SKU mix and size
- Building design
- Physical characteristics and classification of the SKU
- IT processing capability to cube and process COs

Other factors that influence a pick position are order-fulfillment operation type, fire protection, risk management factors and seismic conditions, and pick and replenishment procedures.

Pick Position Objectives

The objectives of properly locating pick positions are:

- To satisfy a predetermined throughput or CO piece pick volume requirement
- To ensure proper product rotation
- To provide the best pick position density per pick aisle
- To provide maximum SKU openings or faces per pick aisle

- To permit easy and quick SKU transfer into or from a pick position to a CO carton/tote or in-house transport concept
- To ensure the lowest logistics operating costs and on-time CO fulfillment activity

To satisfy these pick position objectives, there are numerous pick position concepts that are available for use in an order-fulfillment operation. These pick position concepts are reviewed later in this section.

Pick Position Concept Design Parameters

Prior to selecting a small-item/flatwear apparel pick position concept for purchase and implementation, design parameters are most important and should be clearly defined by your pick area concept design team. The design factors are:

- Average and peak sales volume for each SKU, average and peak number of COs, pieces and lines per CO, and CO order and delivery cycle time
- SKU dimensions and physical characteristics
- Physical features of the building, such as
  - Pick area or size, including clear height between the floor surface and ceiling joist bottom
  - Building column spacing
  - Building column size
  - Passageways, stairways, walls, or floor penetration locations
  - Floor surface and condition
  - Utilities availability and quantity
- Average and peak on-hand inventory, replenishment method, inventory rotation requirements, and pick characteristics such as single or multi-packs
- Existing or proposed order-fulfillment and material-handling equipment that includes fire protection, safety, and seismic requirements
- IT or computer capability to process and download COs
- Labor quantity and availability height or employee reach height
- Other order-fulfillment activities such as carton make-up, packing, seal, check manifest, and load a CO onto a delivery vehicle, and other value-added activities

The pick position concept selected should be the one that best satisfies the majority of your pick concept objectives. Whenever possible, the design parameters are uniform. If these design parameters are not uniform, SKUs with similar characteristics are grouped in one area within a facility. For best results, a pick concept design is made in conjunction with the fundamentals of good SKU allocation profile that dictates SKU pick position in a pick aisle or along a pick line.

Several factors should be considered in determining SKU allocation to a pick position:

- An each, piece, single item, or individual item package. This is an important factor in an automatic pick machine, manual, or mechanized pick concept.
- Sleeve, inner pack, or multi-pack, important in a manual or mechanized pick concept.
- Loose small-item/flatwear apparel SKU in a container, important in a manual or mechanized pick concept.

When we allocate SKUs to a manual pick position, a SKU is allocated as a sleeve, inner pack, or multi-pack, as loose pieces, or as a group in a container. Each SKU has different characteristics that influence selection of a pick position. A container allocation to a pick position has 12 or more medium- to large-size small items.

Small-Item/Flatwear Pick Positions for a Manual Pick Concept

Manual small-item pick position concepts are:

- Containers that are stacked on a floor
- Wood rack structures with cartons or totes
- Standard pallet racks
- Hand-stack containers on a decked pallet rack
- Versa shelving
- Standard shelves, which include:
  - Open type
  - Closed type
  - Wire meshed
  - Mobile
  - Round-shaped
  - Wide-span shelving
- Carton flow racks, which include:
  - Mobile type
  - Stationary type, with straight, tilt, or slant back models
- Push back flow racks
- Slides or chutes
- Peg boards
- Drawers
- Kits

Most small-item/flatwear apparel manual pick concepts handle two SKU types: stackable or nonstackable.

Stackable SKUs

Stackable SKUs have four side walls and two ends, are square or rectangular, with sufficient length, bottom width, and rigidity, side walls, and top surfaces. In a pick position, the structural or rigid characteristics permit containers to be stacked. To ensure that small-items are retained in a container pick position, a container is designed with solid rear walls, solid side panels, and an open front. To maintain discipline and an orderly pick position, stackable SKUs are placed into a carton, bin, or container or remain in the vendor carton. Each master carton or container has a cut or half-open front. In some situations, where pick position structural members do not have sufficient space for pick position discreet identification, the front of the pick position bin provides a location for SKU discreet identification.

Nonstackable SKUs

Nonstackable SKUs have physical shape or package characteristics that prevent stacking in a pick position. Nonstackable SKUs are placed into a carton, bin, or container. The maximum number of SKUs is allocated and retained in a pick position and permits a picker to easily complete a pick transaction. If required by a pick position design, a small-item pick position bin front has a half-cut or open face and has sufficient space for a pick position identification attachment.

Small-Item/Flatwear Pick Position Bin, Carton, or Container Design

A pick position bin or container is designed for placement in any pick position that is used in a manual or mechanized pick concept. In some order-fulfillment operations, a vendor master carton serves as a pick position bin. In many operations, depending on the type of SKU, a pick position bin is a preformed plastic, metal, wood, corrugated, or chipboard bin. Bin length, width, and height are determined by the physical characteristics of the SKU, pick position dimensions, and desired inventory quantity in a pick position.

When designing a pick position bin, two additional important design factors are:

■ The front of the bin should have sufficient clearance or open space between side walls to permit an employee to complete a SKU replenishment or pick transaction. An employee's hand must be easily inserted into the bin or container.
■ The front of the bin should have a surface with width and height that allows for SKU identification or pick position identification. An alternative location is to attach the identification to a structural support member or shelf or deck front lip. If your pick position bin or structural support member does not satisfy these design features, the result is lower employee productivity due to

difficulty in transferring SKUs between a pick position and time to identify a SKU pick position.

An alternative SKU identification concept is place it on a bin side wall. This location involves additional steps (for the employee to pull and replace the bin).

Other important considerations in selecting the type of bin to be used in a pick position are:

- Outside dimensions that determine a the number of bins to fit in a decked rack bay or shelf pick position
- Internal dimension, which determines the number of SKUs that fit into the bin
- Stackable or nonstackable bins
- Dividers in interior
- Permanent or removable bin front or barrier
- How is the bin to be transported in-house between two facility locations?

When allocating pick position bins to a carton gravity flow rack, shelf, or decked rack bay and to have the optimum design, most pick area design professionals use a bin or carton dimension plus an additional 1/8 in. This 1/8 in. safety factor ensures that the projected number of bins fits within a carton gravity flow rack, decked rack, or shelf pick position and assures easy employee transfer of a bin into or from a pick position.

A pick position bin design provides LIFO UOP rotation, SKU control in a pick position, pick position reduces inventory mix-up and inventory loss, and assures maximum face or pick position number per shelf, gravity flow rack bay, or decked rack bay.

With small bins that are 6.5 in. wide by 6 in. high, it is possible to place five bins on a 36-in. wide shelf position. In a vertical stack with 12 pick position levels high, a pick bay has 156 pick facings within a 7 ft. 3 in. shelf location.

The disadvantages are additional one-time cost, bin with a narrow opening, SKU transfer is difficult, and LIFO UOP rotation. The advantages are discipline in a pick position, organization of a pick position, maximum face number per pick aisle, good SKU density, ensures good SKU accessibility, improves in-house transport, and minimizes inventory shrinkage.

Types of Bins

When your consider using a pick position bin box or container in your small-item/flatwear apparel order-fulfillment operation, options are vendor master carton, preformed plastic bin, metal, or wood bin, corrugated box, or chipboard box.

Vendor Master Carton as a Pick Position Bin

When a vendor master carton is used as a pick position bin, the carton is transferred to a pick position. Pick position replenishment options are:

- Transfer a carton with the closed top seal and flaps
- Open the top seal and carton top flaps that are cut from a carton, restrained by a rubber band or clips
- Open the top seal, top flaps, and square, smile, or V cut in a carton front

For additional information on vendor carton replenishment to a pick position, we refer the reader to the carton presentation in a pick position section in this chapter.

The disadvantages are vendor master carton or box dimensions do not optimize your pick position space, in a carton gravity flow rack pick concept, a poor-quality carton causes hang-ups, additional writing on a carton, possible replenishment employee confusion, and trash disposal. The advantages are no material cost, no labor cost to transfer SKUs into a bin box, and easy to inventory.

Preformed Plastic Bin

Preformed plastic bins are polypropylene or corrugated plastic. In a pick position, joints are glued and stapled together and the bottom exterior surface has an additional plastic corrugated layer and top perimeter with a solid hardened plastic rim. Plastic bins are very common in many small-item order-fulfillment operations, for several reasons:

- Durability or ability to reuse a bin.
- In a shelf or decked rack pick position, structural stability and rigidity to permit bins to be stacked.
- Bottom surface permits in-house transport by most concepts.
- Separators permit interior separation into smaller components to handle multiple SKUs per bin.
- Front design has a removable quarter-high or half-high front barrier and open top that permits easy and quick SKU replenishment or pick transfer.
- Relatively lightweight and water-tight.
- Sufficient front surface for pick or SKU identification.
- Available in several colors.

A preformed plastic bin is available in a wide variety of standard sizes and colors and they are manufactured to match your written functional specifications.

Preformed Metal or Wood Bin

Preformed metal or wood bins are manufactured in standard sizes from metal and wood. Because of the weight of this type of bin, it is not preferred in a manual small-item/flatwear apparel order-fulfillment operation.

Chipboard Bin

A chipboard bin is most frequently used for lightweight, slow-moving and small-cube items. A preformed chipboard pick position bin is a very common bin in many small-item/flatwear apparel order-fulfillment operations, for the following reasons:

- Ability to reuse a bin, structural stability, and rigidity to retain SKUs, and fits into most pick position types
- Bottom surface permits in-house transport by most concepts
- Front design and open top permit easy and quick SKU replenishment or pick transaction
- Relatively lightweight
- Sufficient flat and plain-colored front surface for pick or SKU identification
- Lowest cost but requires labor to assemble

A preformed chipboard bin is available in a wide standard size and color variety that are manufactured to match your written functional specifications.

Corrugated Bin

A preformed corrugated pick position bin is very common in many small-item/flatwear apparel order-fulfillment operations that have medium-moving SKUs, medium weight, and medium-cube items. A corrugated pick position bin is used for several reasons:

- Durability or ability to reuse a bin.
- Structural stability and rigidity to retain SKUs and fits in most pick position types.
- Hard and flat bottom surface that permits in-house transport by most concepts
- Front design and open top permits easy and quick SKU replenishment or pick transaction.
- Relatively lightweight.
- Sufficient flat kraft or white-colored front surface permit pick or SKU identification.
- Interior can be divided into smaller compartments.
- Low cost but requires labor to assemble.

An option to increase corrugated bin rigidity and durability is to have it treated or coated with a resin substance. A preformed corrugated bin is available in a wide variety of standard sizes and most bin manufacturers manufacture bins to match your written functional specifications.

In conclusion:

- A vendor master carton is the lowest cost option, but requires labor to open the top and front. A vendor carton is very common in small-item/flatwear apparel order-fulfillment operations that use carton gravity flow rack or hand-stack in a decked pallet rack or shelf pick concept.
- A preformed plastic pick position bin is more expensive and is a preferred pick position bin for stacking on a shelf or for a hand-stack decked rack pick position.
- A preformed metal or wood bin is not favored for use in a manual pick concept.
- A chipboard bin is a low-cost bin that can be used for very small parts and in most shelf pick positions.
- A corrugated bin is a medium-cost bin that is used for medium-sized parts and in shelf or decked rack pick positions.

Carton Presentation in a Pick Position

To have high picker productivity, carton presentation in a shelf, decked rack, or carton flow rack pick position allows a picker to complete a pick transaction with minimal nonproductive tasks and time. To satisfy this objective, a carton has all SKUs ready for pick transaction completion with a full or half open front, and all filler material removed from the carton.

To prepare SKUs for removal from a carton, a replenishment employee has:

- Cut and removed carton top flaps. This is preferred for cartons that are allocated to pick level 1, the lowest pick level, and for all cartons that are allocated to the golden zone pick levels. When an empty carton with removed top flaps is thrown in the trash, on a trash conveyor, a carton with no flaps minimizes jams. Also, if the carton is manually folded into a cart, it means less work and a higher trash carton handling volume.
- Cut the front of the carton, in a square cut, smiley face, or V cut (see Figure 4.8). This practice is used for all cartons on the highest pick level or the highest pick level of the golden zone and when a carton is not recycled in an operation but thrown in the trash.
- If an empty or depleted carton is used in a pack activity, the carton flaps are opened and secured to the carton sides by a rubber band or several clips. Empty or depleted cartons are transferred from a pick position to a transport

MASTER CARTON PICK POSITION PRESENTATION

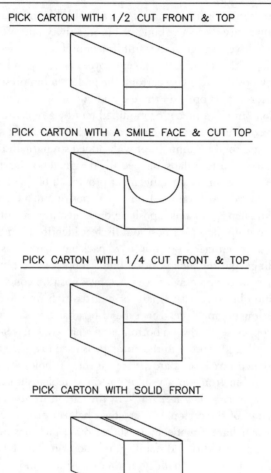

PICK CARTON WITH 1/2 CUT FRONT & TOP

PICK CARTON WITH A SMILE FACE & CUT TOP

PICK CARTON WITH 1/4 CUT FRONT & TOP

PICK CARTON WITH SOLID FRONT

Figure 4.8   Pick position bin presentation: no cut top and front, half cut front, quarter cut front, and smiley face.

concept for travel to the start station or pack station, where rubber bands and clips are removed and the carton is introduced to a pick line or used at a pack station. Rubber bands or clips are returned to the carton gravity flow rack replenishment side or to a shelf or decked rack front.

Filler material is removed from the carton and thrown in the trash.

Pick Position Bin Replenishment

A key to an accurate and on-time small-item order-fulfillment operation is to assure that picker productivity meets the budgeted productivity rate. With this criterion, most small-item/flatwear apparel order-fulfillment professionals feel that a carton or bin box in a pick position ensures that an employee can easily and quickly transfer a SKU from a pick position to a CO carton. The preferred bin presentation in a pick position is a carton that is open on the top and cut or open in the front. An open face pick position bin has a precut or modified (cut by a replenishment employee) standard bin box or vendor box that has two side walls and a rear wall with an open top, open front, and solid bottom. To provide added strength and rigidity to a bin, tape is used to secure bottom flaps, or specially designed box inserts are secured in pre-cut holes. An open bin box arrangement provides a bin with an open face for easy and unobstructed SKU transfer from a pick position to a CO container. A bin or carton with an open front, a pick position deck member has sufficient structural support and a front lip for a pick position discreet identification. Most SKUs that are placed into an open-faced bin box have package characteristics that prevent SKUs from falling from the pick position bin. An open-faced bin box is used in a shelf, decked rack, or carton gravity flow rack pick position concept.

A pick position bin box with a half- or quarter-high front barrier is precut or cut by a replenishment employee. This type of box has two side walls, a rear wall, open or closed top, closed and solid bottom, and a half- or quarter-high open front. To provide strength and rigidity to the bin, tape is used to secure the bottom flaps, or specially designed box inserts are secured in precut holes. To create a half- or quarter-high open bin front on a vendor carton, a replenishment employee uses a knife to cut a square, smiley face, or V in the carton front. With a cut section removed from front of the carton, this creates a half- or quarter-high front barrier to retain SKUs that have a tendency to slide from a pick position. This method provides increased bin rigidity and stability and a location for SKU or pick position identification, increased SKU quantity in a pick position, and open space for easy and quick SKU transfer from a pick position to a CO container. Other half- or quarter-high front barrier bin box considerations are when a low SKU inventory is kept in a bin box with a front barrier, the barrier creates some difficulty to remove a SKU, and there is some potential product damage or employee injury during cutting. Safety gloves minimize employee injuries. A pick position bin box with a half- or quarter-high front barrier is used in a shelf, decked rack, or carton gravity flow rack pick position concept.

Floor Stack Concept

A floor stack concept is considered a basic small-item/flatwear apparel pick concept and requires a large floor area. The floor stack pick concept has a low economic investment because it requires replenishment employees to stack SKU cartons or containers onto a facility floor, onto a pallet, or onto another material-handling device. The floor stack concept requires that a SKU container has sufficient rigidity and strength to stack several containers high. It should be non-crushable and able to support the stack weight. The container stack height should not exceed 5 ft. 6 in., or a picker's reach.

High-cube or high-volume SKUs are stacked one container high and one container deep. With this concept, additional containers are considered ready reserve inventory locations. Alternatively, each stack has one SKU. This container concept provides the maximum pick face or number of positions per aisle but creates a bottom container that is a difficult location to complete a SKU replenishment or pick transaction.

A floor stack concept provides LIFO UOP rotation. Depending on the design, floor stacking provides a low number of pick positions per aisle for slow-moving and small-cube SKUs that are one deep; or a large number of pick positions per aisle for SKUs with a large inventory; and good storage density.

The disadvantages are poor container stack stability, low picker productivity due to low hit concentration and density, LIFO UOP rotation, and additional cost. The advantages are does not require a shelf, decked rack, or carton flow rack investment, used on an in-house transport concept, and handles a wide SKU mix or shapes.

Plastic Bins Clipped or Attached to a Support Member

This design uses plastic bins with an outward extending hook or lip on the exterior rear wall and a stationary or fixed support member. The support members have hooks or lips that extend outward and interface with a bin hook or lip or slots or holes that interface with a bin hook or lip. A bin and support member components have sufficient structural strength to support a fully loaded bin or a bin group.

Most small-item order-fulfillment operations do not use the floor-stack pick position concept. For additional information, we refer the reader to the plastic stackable container pick position concept that is in this section.

Wood Rack Structure with Cartons or Totes

The wood rack structure is 2 × 2 in., wood top and bottom with upright posts on each corner. The length is determined by the desired pick position rigidity, stability, and employee reach. The members are nailed or screwed together. For structural rigidity and stability, several 1 × 11 in. deep members are placed on the top, bottom, and on both ends. An option is to place a 1 × 11 in. deep member in the middle that serves as a shelf. These connected wood members form a rectangular frame.

Corrugated cartons or totes are stapled or taped to the wood racks. SKU pick positions are created from cardboard boxes that are placed inside the wood frame. To ensure rigidity and uniform openings, whenever possible each cardboard carton flap is stapled to a wood member and all cartons are stapled or taped to each other to form one stacked or column unit. Each pick position holds a small SKU quantity. To identify a pick position, a flat cardboard section is attached to a carton or box front.

The disadvantages are difficult to relocate, difficult to use in humid locations, handles lightweight SKUs, fixed pick position opening size, and holds a small inventory quantity. The advantages are low cost, provides a divided pick position, and maximum SKU openings or facing per shelf design.

The wood rack structure with cartons or totes is not preferred in a dynamic small-item order-fulfillment operation because of its limitations and disadvantages.

Standard Pallet Rack Pick Concept

A standard pallet rack has two upright frames and two load beams for each elevated pallet pick level. To prevent rack sway and ensure rigidity, there are at least two load beam levels per rack opening or bay. Load beams provide the support for an elevated pallet pick position and a flat surface that faces a pick aisle for a pick position identification. In a typical pallet rack design, load beam dimension is based on pallet face dimension and horizontal clearances in a rack bay. In most applications, a rack opening bay has two pallet pick levels. A rack bay additional pallet position above these two pick levels is used for ready reserve pallet positions. The pick aisle width between two pallet rack rows has sufficient width for a forklift truck to make a stacking or right-angle turn. In most applications, this aisle width has a range from 8 ft. wide for a straddle truck to 10 or 13 ft. wide for a three-wheel sit-down or stand-up counterbalanced forklift truck.

In most pallet rack applications, the first pallet position is on the floor. In some narrow-aisle forklift truck applications, the first pallet pick position level is raised onto two load beams. This feature permits narrow-aisle forklift truck straddles to travel under the load beams but based on the load beam connected plate raises the first pick position by 6 to 12 in. If fire sprinklers are in a rack structure, they are placed in a flue space between the back-to-back rack rows or located in a rack bay above a pallet position.

In a normal pallet rack pick aisle, the first pick position is the floor surface. With a 54 in. overall pallet height, 5 in. load beam vertical height and 4 in. clearance, the second load beam level is 63 in. high and the highest carton top on a pallet is at 117 in. above the floor.

When you design a standard pallet rack pick row, your first rack bay is a starter bay and has two upright frames and two load beams for each elevated pick level. After a starter bay, each additional rack bay or add on bay has one upright frame and two load beams for each elevated pick level.

A standard pallet rack is designed to handle high-cube, heavyweight and very fast-moving SKUs. Rack design options are:

- Two 48 × 40 in. pallets in a rack bay with a pallet 48 in. dimension down aisle, or facing a pick aisle. This rack bay design has a load beam with a 113-in. center line to center line dimension. This pallet arrangement has
  - Four pick positions per rack bay
  - Eight inches shorter picker reach into a rack bay
  - Fewer number of pick faces per pick aisle
- Two 48 × 40 in. pallets in a rack bay with a pallet 40 in. dimension down aisle or facing a pick aisle. This rack bay design has a load beam with a 96-in. center line to center line dimension. The pallet arrangement provides
  - Four pick positions per rack bay
  - Eight inches longer picker reach into a rack bay
  - Increase in the number of pick positions per pick aisle
- Three 40 × 32 in. pallets in a rack bay with a pallet 32 in. dimension down aisle or facing a pick aisle. The rack bay design has a load beam with a 117-in. center line to center line. The pallet arrangement provides
  - Six pick positions per rack bay
  - Eight inches shorter picker reach into a rack bay
  - Medium number of pick positions per pick aisle
- Four 40 × 32 in. pallets in a rack bay with a pallet 32 in. dimension down aisle or facing a pick aisle. This rack bay design has a load beam with a 156-in. center line to center line. The pallet arrangement provides
  - Eight pick positions per rack bay
  - Eight inches shorter picker reach into a rack bay
  - Greatest number of pick positions per pick aisle due to fewer upright frames

Components of a standard pallet rack are:

- An upright frame that includes the diagonal and horizontal braces. At the base of each upright post is a foot or base plate that spreads a pallet rack load and permits the rack upright to be anchored to the floor.
- Two load beams for each elevated pallet level.

The upright frame has hardened coated metal members that provide support for an elevated pallet load beams and pallets on each load beam level. The pallet number and weight are stated as your rack written functional specification to your rack vendor. Per the facility seismic location, the above information and additional information such as additional pallet storage levels, the rack manufacturer determines the final rack upright frame dimensions, base plate size, rack metal gauge, and load beam dimensions and size.

In most manual small-item pick applications, the rack elevation has one pallet position on the floor surface and three elevated rack levels. This configuration provides a small-item order-fulfillment operation with two pallet levels for pick positions and two pallet levels for ready reserve positions. To ensure rack row rigidity, most rack applications criss-cross the upright post base plate anchor to the floor. The criss-cross upright post base plate anchor pattern has the first upright frame post front base plate anchored to the floor and the second upright frame post front base plate anchored to the floor.

As an option to improve safety, many operations add front to rear or cross members to the elevated pallet position load beams that are above pick positions. The cross members are metal members that span the two load beams and restrict a pallet from falling through a rack bay. Most pallet rack pick applications use two front to rear or cross members per pallet position.

Most rack manufacturers allow a pallet to overhang each load beam by 2 to 3 in. With this overhang feature for a 40–in. stringer pallet in a rack bay, the upright frame depth is 36 to 34 in. deep and for a 48–in. stringer pallet, an upright frame depth is 44 to 42 in. deep.

The last upright frame considerations are the upright post width or down aisle dimension and the upright frame height. When we consider a pallet rack has two pick position levels and two ready reserve position levels, in most rack applications an upright frame width is 3 to 4 in. wide. The final upright frame width is determined by your rack manufacturer and is based upon your rack written functional specifications. Included in your rack written functional specification is the facility geographic location for the rack vendor to determine the seismic zone.

The upright frame post height depends of the number of pallets high in a rack structure, pallet overall height, your company desired vertical clearances that are:

- Four to 6 in. between pallet and load beam.
- Height above the top pallet level. Many logistics professionals consider that the upright post extension above the top pallet level as a guide to a forklift truck operator as a pallet transaction is performed at a pallet position.
- Load beam height and connection to an upright frame requirement.

In most standard rack applications, a load beam height has a range from 4 to 6 in. high. A load beam connection to an upright post is based on the upright post center spacings that are standard for each rack manufacturer. Most load beam vertical settings are adjustable on 2- to 4-in. centers. Some rack manufacturers have a flush load beam end plate and other rack manufacturers have a load beam end plate that extends 2 to 4 in. beyond the load beam edge. This extended load beam end plate has an impact on an upright post height in design options. These are:

- When a floor level pallet pick position is set on two load beams
- At the highest load beam level to assure that the load beam end plate connection devices are connected to an upright frame

Two load beams span the distance between two upright frame posts and provide structural support for two pallets. Load beam designs are the box or step design.

Your seismic location and rack manufacturer determine a load beam metal gauge, length, design and vertical dimension. When we consider a load beam length, the down aisle or length dimension options are:

- Internal load beam dimension that is the distance between two upright posts interior sides. This load beam dimension is considered the rack opening dimension that is determined by your pallet width and required horizontal clearances in a rack bay. If a straddle or straddle reach forklift truck is used an operation, a forklift truck straddles require 5 to 6 in. open space between floor level two pallets that adds 15 to 18 in. to a load beam internal dimension. An alternative for a straddle forklift truck operator and rack design is the up-and-over concept. The up-and-over concept has the first pallet level set on two load beams that increases the pick level height above the floor but decreases the horizontal clearance.
- Center to center dimension that is the distance between two upright frame post middle location. This load beam dimension is your rack layout dimension. To calculate a rack row length, you multiple a rack bay number by the load beam center to center dimension and add an upright frame width. The total is the overall rack row length.

The vertical load beam dimension is determined by the pallet number, pallet weight, span, and allowable load beam deflection per two load beams. The load beam deflection means the degree of bow or cup for a load beam to a parallel plane or floor. A large amount of load beam deflection increases a forklift truck difficulty to complete a pallet transaction to a pallet position. Most rack manufacturers have standard colors for their rack components and have different colors for the upright frames and load beams.

With a pallet rack, a pick position identification is placed onto a load beam surface that faces a pick aisle. In most applications, the first pallet pick position sets on the floor. The first load beam lower section is the location for the floor pick position and the first load beam upper portion is the location for the pallet rack level pick position. To provide additional pick position identification understanding, the number sequence has 1 as the lowest or floor level pick position and 2 as the elevated pick position. Some pick operations have directional arrows at the pick position identification to show a pick position.

The disadvantages are low SKU hit density and hit concentration, large floor area, few SKU faces per aisle, and two pick levels high. The advantages are handles heavyweight, high-cube and fast-moving SKUs, minimizes replenishment activity, large inventory in a pick position, provides good SKU access, and ready reserve positions.

Hand-Stacked Containers on a Decked Pallet

Cartons, containers, or packaged small-items are hand stacked into a standard decked pallet rack opening or bay. This concept has similar upright frame and load beam components as a standard pallet rack. The exceptions are:

- Design parameters have a lighter load that is supported by the upright frames and load beams. The load weight is eight containers in a hand-stacked decked rack opening nominally 8 ft. 3 in. long by 44 in. deep. An example is if the 30 lb. cartons or containers with a load beam opening that is 10 ft. wide by 1 ft. long by 1 ft. high. This means two load beams and rack opening product load weight is 240 lb. With a pallet rack opening decking material and a safety factor, the estimated total load weight supported by two load beams is approximately 240 lb. To have load beam flexibility, many companies use a hand-stack load beam support weight to hold two standard pallets weight.
- Each rack bay assigned as a pick position has decking material that covers the entire opening and provides a smooth and flat surface (space) for the small-item containers. To provide rigidity and to assure the deck supports a carton or container load, some applications have three to four front to rear or cross members span the two load beams.

To select a load beam for your hand-stack pick application, you determine the preferred load beam span between two upright posts. Most rack manufacturers have tables that show center to center upright post dimension, associated weight capacity, stated load beam deflection, or bow to the floor, and vertical load beam height. The load beam vertical height provides the surface for a pick position identification. Later in this section, we review the box and step load beam designs.

Because a hand-stacked load is lightweight, your selected load beams have a longer span and shorter vertical dimensions. One very important rack load beam design consideration is that these lightweight load beams are designed for a hand-stack application, not a pallet pick or storage position application. If a heavy pallet is placed onto these lightweight-designed load beams, the load beam more than likely would collapse and load beam and product become damaged with potential employee injury.

An alternative load beam design option is to have two load beams structural strength designed to have the capacity for two pallets maximum. This feature assures maximum rack use flexibility but has a slightly higher load beam dimension.

The second hand-stack container rack component is the upright frame that has two upright posts. Each upright post has a base plate and diagonal and horizontal bracing members. An upright frame is a hardened coated metal structural rack member that has two load beams attached at the proper elevation above the floor surface. The depth distance between two upright posts determines the hand-stack carton or container number deep per rack bay.

When you design a hand-stack pallet rack pick position row, your first rack bay is a starter bay that has two upright frames and two load beams for each pick position level. After a starter rack bay, each additional rack bay or add on rack bay has one upright frame and two load beams for each pick position level.

With hand-stacked cartons in a decked pallet rack bay, the upright frame load capacity and height options are:

- Decked rack components and hand-stack carton load weight along with the height requirement
- Decked rack components and hand-stack carton load weight along with the height requirement and pallet storage above the highest pick position

When we consider the maximum elevation for a 5 ft. 8 in. tall picker to complete a pick transaction is 5 ft. 3 in., the elevation above the floor surface to the bottom of the top pick position is 5 ft. 3 in. Several factors determine the number of pick position levels:

- Bottom pick position level elevation above the floor surface. Hand stack bottom level options are:
    - 6-in. high pallet that sets on the floor
    - 8-in. high bottom level that has a load beam and deck material
    - 2- to 3-in. high bottom that has deck material on wood slats.
- Per your operation desire to have a bottom pick position level supported by two load beams or by pallets.
- Vertical clearance or open space between a carton top and the next load beam bottom that permits a picker to complete an order pick transaction or SKU transfer from a pick position carton to a CO container. If a carton front in a pick position has an open front, the clearance is a nominal 6 in. high. If a carton front in a pick position has a closed front, the clearance is 9 to 12 in.
- Load beam vertical height to support the load beam design load or weight.
- Deck material vertical height and the deck design.
- Your replenishment and pick employees' reach height.

Open Front Carton on a 3-In. High Deck

From our example, a 12-in.-high open-face carton, 6-in. clearance, and a 5-in. high load beam with deck material and three hand-stacked pick position levels:

- First pick level sets on a deck on the floor, has 3 in. deck material with one 12 in. carton and 6 in. clearance = 21 in.
- Second pick level sets on a load beam with deck material = 5 in., one carton = 12 in. and 6 in. clearance = 23 in.

- Third or top pick level sets on a load beam with deck material = 5 in., one carton = 12 in. and 6 in. clearance = 23 in.

A carton top on the top deck has 61 (67 – 6 = 61) in. elevation above the floor and with an open front carton to access a 12-in. high SKU, an employee pick reach height is 55 (61 – 6 = 55) in. elevation above the floor.

Open Front Carton with Bottom Level on Load Beams and Deck Material

From our example, a 12-in. high open-face carton, 6-in. clearance, and a 5-in. high load beam with deck material and three hand-stacked pick position levels:

- First pick level sets on load beams with 5–in. deck material with one 12-in. carton, and 6 in. clearance = 23 in.
- Second pick level sets on a load beam with deck material = 5 in., one carton = 12 in., and 6 in. clearance = 23 in.
- Third or top pick level sets on a load beam with deck material = 5 in., one carton = 12 in., and 6 in. clearance = 23 in.

A carton top on the top deck has 63 (69 – 6 = 63) in. elevation above the floor and with an open-front carton, an employee pick reach height is 57 (63 – 6 = 57) in. elevation above the floor.

Closed Front Carton with Floor Level on a Deck

From our example, a 12-in. high closed face carton, 6-in. clearance, and a 5-in. high load beam with deck material and three hand-stacked pick position levels:

- First pick level sets on a deck on the floor, with 3 in. deck material, one 12-in. carton, and 6 in. clearance = 21 in.
- Second pick level sets on a 5-in. load beam with deck material = 5 in., one carton = 12 in., and 6 in. clearance = 23 in.
- Third or top pick level sets on a load beam with deck material = 5 in., one carton = 12 in., and 6 in. clearance = 23 in.

A carton top on the top deck has 61 in. (67 - 6 = 61) elevation above the floor and with a closed front carton, an employee pick reach height is 61 in (61 - 0 = 61) elevation above the floor.

To determine the upright frame height and load weight, the options are:

- Upright frames support only the pick position height and weight, you state this fact in your written functional specifications. With a maximum of three pick position levels high, the overall upright frame height is minimum above the base plate or floor surface or the rack manufacturer standard height.
- Upright frames support the hand-stack pick position levels and ready reserve pallet positions, you state the additional load weight, capacity, and height requirements in your written functional specifications.

In your written functional specifications you state your order-fulfillment operation forklift truck type that performs the ready reserve transaction. This allows the rack manufacturer to determine the load beam span that allows a forklift truck to complete a storage transaction to an elevated pallet position.

Hand-Stack Options

The upright frame depth or rack opening or bay depth permits your order-fulfillment operation to stack one carton or two cartons deep into a rack opening. Standard rack depths are 24, 32, 36, 40, 44, and 48 in. Per your carton or container length and your allowable carton or container upright frame depths overhang, you select an upright frame depth that matches your design parameters.

Carton hand-stack pick position design options are:

- Stack containers into a single rack row with pick aisles on both sides of the rack opening. A barrier in the middle of the rack bay divides the deck into two sections. One section faces pick aisle A and the second section faces pick aisle B. The barriers can be:
  - Angle iron with leg up that is secured to the deck material.
  - Rope that is attached to two upright posts.
  - Metal or plastic pipe that is secured to the deck material.
  - Wood member that is secured to the deck material With an 18-in. long carton on each side of the barrier in a typical 44-in. deep rack, the barrier can be up to 2–3 in. wide. The barrier ensures that SKU containers on each side of the barrier are properly aligned in a pick position and as a replenishment or pick employee from aisle A pushes a carton into a rack bay that a carton on the other side (facing aisle B) does not fall into aisle B. This option provides maximum number of pick positions in a pick area but minimal inventory in a pick position.
- Back-to-back rack rows with SKU pick positions or pick faces on each pick aisle or use one rack or shelf opening side. This option permits additional SKU inventory in the pick position or handles long SKUs but provides fewer pick positions in the pick area.

Deck Material

The pallet rack deck design and material depend on rack load beam design. The load beam design options are:

■ Step load beam with a step on the interior side that is half inch deep from the load beam top and ¾ in. inward from the load beam rear side
■ Box or H structural load beam

When you have a step load beam, your selected decking material edges sit within load beam steps. The nominal step depth is half inch deep, which allows a portion of the deck material to be retained or recessed in the step. The feature retains the deck material in a rack bay and has a shorter decking material depth than the span between two load beams. Most step load beams have a ¾ in. step. An example is with a 44 in. frame depth and step load beam, the deck material depth is 42.5 in. deep.

When you have a box or H structural load beam, your deck material sits on two load beam tops. This load beam and deck feature means deck material has length or depth to span two load beams, slightly higher elevation to a pick position, and deck material is secured to both load beam tops. Most load beams are secured with countersunk screws and several S-shaped plates along a load beam.

The hand-stacked box or container in a pallet rack bay has an influence on the selected deck material. The deck material has sufficient rigidity and structural strength to support hand-stack carton or container load weight and provides a smooth flat surface to easily move a carton or container across a deck surface. The deck materials are:

■ Wire mesh and the options are (a) flat or (b) inverted 'V'
■ Wood sheets
■ Metal slats or sheets
■ Gravity conveyors

When you specify deck material for your hand stack application, it is suggested to state maximum load weight and carton/container length, width and height; load beam type and span between two load beams; rack upright frame depth; and maximum allowable deflection to the deck material.

Wire-Meshed Deck

The wire-mesh deck material has two front to rear members with coated wire strands. One wire strand set is on the bottom and the other wire strand set is on the top. These wire strands are welded together in a pattern to form small square or rectangular openings. A wire-mesh deck has the structural support for the hand-stacked carton weight. To assure the wire-mesh deck is retained in the rack opening, the front and rear ends are angled and extended beyond each load beam edge. The wire-mesh end design serves as a lock to retain the deck on the load beams.

The other wire-mesh deck considerations are:

- Wire-mesh top wire strands are in the direction of the rack bay depth. With the top strands in this direction and when moving cartons or containers forward across a deck, this top wire strand design minimizes carton or container bottom hang-up on wire strands.
- Dividers or separators the full depth between two pick positions. These dividers ensure that the SKUs do not fall from a carton or container and become mixed in a pick position.
- Wire-mesh deck shaped in an inverted V form. The inverted V wire-mesh deck is used in a pick aisle layout that has pick positions on the both sides of the rack opening. The inverted V-shaped wire-mesh deck allows gravity to move SKUs forward in a carton or container and gives a picker a better view of the carton or container interior.
- Wire-mesh deck shaped in an angle or with a pitch. The pitch has the high side on the rack bay interior and low side on the rack bay aisle side. The angled or pitched shaped wire-mesh deck is used in a pick aisle layout that has back-to-back rack rows with pick positions on one rack bay side. The angled or pitched wire-mesh deck allows gravity to move the SKU forward in a carton or container and gives an order picker a better view of the carton or container interior.

Openings allow dust and dirt to fall onto the floor. There is minimal increase in elevation, The wire-mesh deck supports a heavier load with minimal deflection, but has a higher cost, and it is difficult to handle loose small-items.

Wood Deck

A wood deck is a sheet that is made of plywood or particle board. The wood is trimmed or cut to fit within your rack opening. After the wood is placed in a rack opening with a step load beam, the wood edges rest in the load beam steps. With a box load beam, wood edges are flush with the load beam edges, rests on load beams, and countersunk screws or S-shaped plates secure a deck to the load beams. If wood has a greater deflection or bow than anticipated or than is acceptable, front to rear cross members are connected between two load beams. With three to four cross members evenly spaced along a rack bay or opening, the wood deck is flat and smooth. Wood deck option features are solid deck surface to handle loose items, less expensive, per code, fire sprinklers are in the rack opening below, solid wood deck protects the sprinkler head, and is the preferred deck material, and with a box load beam design, wood deck adds 1 in. to the load beam height.

Sheet Metal

Metal sheets or corrugated metal slats have the same design parameters as the wood deck option. The metal deck option is slightly more expensive.

Non-Powered Roller or Skatewheel Conveyor

A gravity roller or skatewheel conveyor deck has each carton or container on a gravity roller or skatewheel conveyor lane. With an H or box load beam, each gravity roller or skatewheel lane end is designed to fit onto a securing clamp device to hold a conveyor section onto a load beam. With a step load beam, the gravity conveyor section front is set in a load beam step and the rear conveyor frame is secured with welds or clamps to the load beam. Most gravity conveyor applications have a slope or pitch between the conveyor lane charge (replenishment) end and the discharge (pick) side. This slope or pitch plus the carton or container weight and a gentle push from a replenishment employee permits a carton or container to travel over the conveyor lane. At the discharge end is an end stop device to retain the lead carton or container on the conveyor lane.

In most applications, gravity rollers or skatewheels in a conveyor lane are spaced on 2- to 3-in. centers. The roller width is between the two conveyor lane frames and skatewheels are evenly spaced between the two conveyor lane frames to support the narrowest carton bottom surface. These features ensure that there are three rollers or skatewheels under the full width and length of a carton or container bottom surface. To provide rigidity to a gravity roller conveyor section, conveyor lane frames rests on two or four load beams. With a short gravity roller conveyor section, two load beams support the conveyor frame. With a longer gravity roller conveyor section, there are four load beams that support the conveyor frame. These four load beams are attached to the back-to-back upright frames.

If carton or container tracking or travel over a roller or skatewheel conveyor travel path creates jams or hang-ups, guide rails are used on both sides, the full length of the conveyor lane to minimize hand-ups. A push pole with a hook is used to unjam carton or container hang-ups. The various guide rail options are preformed sheet metal, plastic or metal pipe, and rope or metal wire.

Most non-powered roller or skatewheel conveyor deck applications have a replenishment and a pick aisle. The conveyor lane deck features assure no carton hang-ups on a conveyor lane, each lane has guide rails, pick area has a replenishment aisle and pick aisle, which means fewer pick positions in the pick area, and cost.

Aisle Width

A pick aisle width is the distance or open space between two opposing pick positions or two pick position rows. The pick aisle width options are:

- Pallet reserve positions above the pick positions, UOP to UOP pick aisle width permits a forklift truck to make a stacking or right-angle turn, and complete a storage transaction
- Only pick positions in the two pick position rows, UOP to UOP pick aisle width is the normal aisle width that is required by your pick operation, and is the open space between two rack rows

The aisle has a range from 30 in. wide that permits a picker to walk in one direction through the pick aisle or 6 ft. wide that permits two pickers to walk in different directions through a pick aisle.

The disadvantages are: LIFO UOP rotation, requires double handling, additional cost, and medium pick position number per pick aisle. The advantages are: handles medium to heavy, long and high cube SKUs, ready reserve pallet positions above the pick positions, and are easy to install.

Shelf or Hand-Stacked Rack Bridge at Middle Aisle

With shelf or hand-stack in pallet rack bay as pick positions and the shelf or pallet rack structural members are designed for ready reserve pallet positions, at the middle aisle between two shelf or pallet rack sections, the shelf or pallet rack is designed as a bridge over the middle or cross aisle. A shelf or pallet rack bridge provides additional ready reserve pallet positions. The shelf pallet bridge concept design parameters are:

- Sufficient open space between the floor surface to the lowest load beam that permits a picker or forklift truck mast to travel under the bottom load beam without striking the load beam.
- Load beam length to provide the additional shelf or pallet storage levels and a wide passageway. Most applications use a 10 ft. center or center long load beam.
- Under each pallet reserve position on the bottom load beam level there are two front to rear members.

Versa Shelving

Versa shelving pick position concept has two 2 ft. or 4 ft. deep upright frames, specially designed upper and lower load beams with holes in upright frame posts, shelf support devices, wood shelves with metal side runners, upright posts, and dividing components. These components divide a normal 9 ¼ ft. wide rack opening or bay into smaller compartments. Versa shelves are adjustable on a 1 ft. high by 1 ft. down aisle dimension and have a 1 in. shelf bottom support member. Per your carton or container height or to increase pick position number, versa shelving is vertically adjusted on 2-in. centers. A standard versa shelving bay provides eight pick positions wide and five pick positions high for a 40 pick position total. Versa shelving provides good SKU access, medium pick faces per aisle, and LIFO UOP rotation. It handles medium- to lightweight, slow-moving SKUs and the pick position has a small on-hand inventory. Versa shelf pick position concept is designed as a single rack row with pick positions facing both aisles or as back-to-back rack rows with a pick position facing each aisle. Disadvantages are higher cost and LIFO UOP rotation. Advantages are handles loose or containerized SKUs, good SKU accessibility, handles a wide UOP mix, and vertical adjustable shelf or pick position openings.

Standard Shelves

In many small-item order-fulfillment operations, the standard shelf pick position concept is used as a pick position for medium- to slow-moving SKUs, medium- to lightweight SKUs, and medium- to small-cube SKUs. With the flexibility to easily add or adjust shelf levels, add shelf bays, handle a wide SKU variety, and have ready reserve positions above pick positions, the shelf pick position concept is very common in most small-item apparel order-fulfillment operations.

The standard shelf pick position concept components are upright support post or end panels that provide support to a shelf or deck back panel or sway brace that stabilize a shelf pick position, shelves, or decks that support a SKU, connecting items that attached a shelf deck to the upright post or panel and options and accessories.

Standard shelf pick position concept options are:

- Closed shelf bay
- Open shelf bay
- Wire-mesh or hardened plastic shelf bay
- Fixed track mobile shelf bay
- Four-wheel mobile shelf bay
- Round shelves
- Wide-span shelves

Closed Shelves

A closed shelf bay concept has a solid rear wall and two solid side panels with an open front to a pick aisle. The open front allows an employee to complete replenishment and pick transactions. A closed shelf bay has a solid coated sheet metal back panel, two solid coated sheet metal side panels with welded upright posts, and a solid shelf or deck. After the components are connected together, a closed shelf bay creates a barrier on three sides and an open front. A closed shelf bay concept is preferred for loose small-items, packaged small-items that are stackable, and to provide a clean and organized appearance to a pick area. A closed shelf pick position layout has an aisle along all pick positions or faces. A closed shelf bay pick area is designed with single shelf rows or back-to-back shelf rows.

The standard closed shelf pick position components are welded closed upright assemblies or end panels, back panel, box shelf, or shelf, deck, connecting items, and options and accessories.

Closed shelf welded upright assemblies or end panels form the sides of the closed shelf bay. The welded closed upright assembly or end panel has a solid coated sheet metal piece that is welded to two upright posts. Most upright posts are made from 14 gauge steel, provides support for all load weights, permit shelves or decks attachment, and allows shelf opening vertical adjustability. Most shelf manufacturers have uprights that have shelf or deck adjustability on 1 to 1 ½ in. centers. On closed shelf bay applications, most shelf manufacturers use a T-shaped post, which is sometimes referred to as an angle or off-set post. The off-set post has the T front face a pick aisle and each upright post has one half of the base or T-top extend into a shelf bay. This design reduces the useable shelf space. With a 3 ft. wide shelf and a T-post, the useable shelf opening is reduced by as much as 1 in., which is the sum of the two ½ in. offset posts. This means that a 3 ft. wide shelf opening is reduced to a 35 in. wide opening. The upright post is the full height of the closed shelf pick position and the solid sheet metal is full depth of the shelf bay.

Alternative upright post designs are:

- Delta upright post, which has narrow front or face
- Box post, which has the shelf edge against the post
- Beaded upright post, which has a slightly rounded front or face

These alternative upright post designs do not extend into a shelf bay.

A very important upright post consideration is the future use in your order-fulfillment operation. If your order-fulfillment operation has future plans to increase shelf pick levels and height or to support a catwalk, design parameters are given to the shelf manufacturer. With this information, the manufacturer assures that each upright post and base plate supports the proposed dynamic and static loads, decide to anchor and shim the upright posts to the floor, and allows for a splice plate connection.

Most shelf manufacturers have welded closed upright assemblies or end panels that are available in a variety of sizes: 12, 15, 18, and 24 in. deep and 40, 61, 73, 85, 88, 97, 109 and 121 in. long. Shelf manufacturers use two end panels for upright assemblies or end panels that have a height over 97 in.

A closed shelf back panel is a coated sheet metal piece that is secured to each end upright post, each welded upright assembly, or end panel rear upright post. A back panel is a shelf bay full height and width. The back panel sizes are 30, 36, 42, and 48 in. wide, and 40, 61, 73, 85, 88, 97, 109, and 121 in. high. Most shelf manufacturers use two panel sections for panels that are higher than 97 in.

A closed shelf box shelf or shelf deck is made from 18 to 20 gauge coated steel that is roll formed with welded front and rear edges. A box shelf or shelf deck provides a solid and flat surface for carton or container support in a pick position and provides some stability and rigidity to a shelf bay. A shelf deck front lip extends downward by approximately 1 in., which is sufficient space for pick position identification. An alternative pick position option is a label holder or preformed pick-to-light holder. The label holder is riveted or tek screwed to the shelf deck lip. Standard shelf or deck sizes are 12, 15, 18, 24, 30, 32, and 36 in. deep, and 30, 36, 42, and 48 in. wide.

If you have a requirement to support a heavier concentrated or unevenly distributed load on a shelf or to permit a picker to stand on the first or second shelf level (to use a shelf as a step ladder), the box shelf or shelf deck is designed with reinforced members. Reinforced members add to the cost but minimize possible shelf damage. Most shelf-reinforced members are C-channel members that are the full width or depth of the shelf deck and are factory attached to the underside of the deck. Shelf reinforcement options are:

- Side reinforced option that is used to support a heavier load
- Center reinforced option that is used to support a heavier concentrated load
- Single front or both front and rear reinforced option that is used to support a heavier load on a shelf front or to permit an employee to step onto a shelf deck without damage to the shelf deck

A standard box shelf or shelf deck supports an evenly distributed load weight. A 36-in. wide shelf deck load weight range is:

- Medium or normal capacity that has a 625 to 800 lb. capacity
- Heavy duty reinforcement that has a 1000 to 1050 lb. capacity
- Extra heavy duty or maximum reinforcement capacity that has a 1100 lb. capacity

A 48-in. wide shelf deck load weight range is:

- Medium or normal capacity that has a 550 to 600 lb. capacity
- Heavy duty reinforcement that has a 725 to 750 lb. capacity
- Extra heavy duty or maximum reinforcement capacity that has a 1025 to 1050 lb. capacity

To design the number of shelf pick position levels number in a closed shelf concept, there is one shelf deck or box shelf for each pick position level plus one shelf deck at the top of the shelf bay.

The number of pick position shelf levels that is designed in a shelf bay concept has several important design factors:

- Upright post shelf deck connection locations to an upright post and a shelf deck lip. The first shelf pick position level is 1 to 3 in. above the floor surface.
- Carton or container height.
- Shelf deck lip downward extension.
- Vertical shelf adjustability, which ranges from 1 to 1 ½ in.
- Your pick and replenishment employees' reach or height.

To ensure effective cube utilization and maximum number of shelf pick position levels in the golden zone, most shelf pick position design professionals vary the height for the top or bottom shelf opening. The closed shelf connection components are nuts and bolts, clips or flanges, or slots in the shelf decks and upright posts. The particular connection component is per the shelf manufacturer standard.

In a closed shelf pick position row design, a row starter bay has two upright assemblies or end panels, one back panel, and the appropriate shelf deck and connecting item number. After a starter bay, each additional shelf pick position bay or add-on bay has one welded closed upright assembly or end panel, one back panel, and the appropriate shelf deck and connecting item number.

Most shelf manufacturers' standard closed shelving bay colors are gray, blue, beige, and green. Shelving color selection is determined by your company preference. If a dark color is used for the shelving, consideration is given to light in a pick aisle due to the fact that dark blue makes an aisle look dark.

Open Shelves

An open shelf pick position bay has no rear wall but an X sway brace member and each side has an X or short sway brace or a panel and an open front to a pick aisle that permits a picker to complete a replenishment or pick transaction. An open shelf bay has a factory-assembled back sway brace, two factory assembled X side sway braces or upright post sets that have two welded side panels, four upright posts, and solid shelves or decks. After these components are connected together, an open shelf bay creates a shelf bay that has an open front and is basically open on all three sides. An open shelf bay concept is preferred for loose small-item/flatwear apparel in a carton or container, or medium to large packaged items that are stackable. An open shelf pick position layout has an aisle along all pick positions or faces. An open shelf bay pick area is designed with single shelf rows or back-to-back shelf rows.

The standard open shelf pick position components are four upright posts with two factory assembled X side sway braces or two upright post sets that have two welded side panels, one factory-assembled back sway brace, box shelf or shelf deck, connecting items, and options and accessories.

Two factory-assembled X side sway braces or two upright post sets have two small welded side panels. If your open shelf manufacturer uses four upright posts in a shelf design, each of the two upright posts is field connected together with the factory-assembled X side sway brace. A sway brace has two 1-in. wide hardened metal coated stripes that are factory connected in the middle. Each metal stripe end has a punched hole that permits a field installation employee to connect the sway brace to an upright post. After the side sway brace is connected with nut and bolt to the upright posts, the attached sway brace provides rigidity and stability to a shelf pick position. Sway brace length is determined by the shelf manufacturer standards. If your open shelf uses two upright posts sets that have two welded side panels, each upright post set has one end panel is attached near the top and one panel is attached near the upright post bottom. These end panels provide rigidity and stability to a shelf pick position.

Most upright posts are made from 14 gauge coated steel, provide support for the load weights, permit shelf or deck attachment, and allow vertical shelf opening adjustability. Most shelf manufacturers have uprights that have shelf or deck adjustability on 1 to 1 ½ in. centers. On an open shelf bay, some shelf manufacturers use a T-shaped post, which is sometimes referred to as an angle or off-set post. The off-set post has the T or front face a pick aisle and each upright post has one half of the base or top of the T extend into a shelf bay. The design reduces the useable shelf space. The upright post is the full height of the pick position and the solid sheet metal is the full shelf bay depth. Alternatives include:

- Delta upright posts, which have a narrow front or face
- Beaded upright posts, which have a slightly rounded front or face
- Box posts, which have the shelf edge against the post

Available heights are 73, 85, 88, 97, 109, and 121 in. The upright post metal gauge and base plate size are determined by your design parameters, which include support for additional pick levels or catwalk. With this feature, the manufacturer allows for splice plate attachment to the base or existing upright post.

A back sway brace has the same design characteristics and use as a side sway brace, with a few exceptions: it is longer and is used in an open shelf bay design.

An open shelf box shelf or shelf deck has the same design characteristics, use, and sizes as a closed shelf box shelf or shelf deck. Components are nuts and bolts, clips or flanges or slots in the shelf decks, and upright posts that are per your shelf manufacturer standard.

In an open shelf pick position row design, the starter bay has four upright posts with two sway braces or two upright post sets with welded sway panels, one back sway brace and appropriate shelf deck, and connecting item number. After a starter bay, each additional shelf pick position bay or add-on bay has two upright posts with one sway brace or one upright post set with welded sway panels, one back sway brace, and appropriate shelf deck and connecting item number.

If your shelf pick area has additional ready reserve locations and you do not use mobile material-handling equipment that elevates above the top shelf level, you have an opportunity to increase your ready reserve positions. For additional information, the reader is referred to the closed shelf section.

Most shelf manufacturers' standard open shelving bay available colors are gray, blue, beige, and green. Shelving color selection for your pick aisle is determined by your company preference. If a dark color is used for the shelving, consideration is given to light in the pick aisle. Open shelves allow more light in a pick aisle.

Features of the open shelf pick position concept include lower expense per shelf bay, labor expense to install, SKUs are in a carton or container, low static load on the floor, permits light and air to flow through the pick area.

Standard Shelving Accessories

Many shelf manufacturers offer shelf accessories for a basic closed or open shelf bay. Accessories are used to:

- Handle a unique SKU
- Improve employee productivity
- Improve the safety and appearance of a pick area

Some accessories are:

- Fixed with a nylon clip or sliding shelf dividers. A divider is a coated sheet metal member that is full depth and separates a shelf bay width into smaller sections and helps to maintain SKUs in a pick position.
- A bin front or base that is made of coated sheet metal and is the full width of the bottom shelf bay edge. Attached to the bases of the shelf bay, a bin front minimizes dust and dirt accumulation from the pick aisle under the shelf bay.
- Bin box to hold loose small items.
- Swinging doors that are attached to the front of the shelf bay. With a lockable door, it provides a secured pick position for high value or hazardous SKUs.
- Per your seismic location, shelf manufacturer standards and your floor, shim plates and upright post anchor plates that secure the shelf bay to the floor and assure level shelf pick positions.
- Pick position label holder or pick-to-light holder.
- Post adapters and splice plates to increase the height of an existing upright post.
- GOH rail and side plates to provide GOH pick positions.

◾ Step rail that is attached to the front of an upright post, which permits a picker to step on the rail and reach the top position.

Disadvantages are difficult to handle large and heavy SKUs, medium inventory in a pick position, SKU double handling, provides LIFO UOP rotation, and collects dust and dirt on the decks or shelves. Advantages are low cost, handles medium- to slow-moving SKUs, provides good hit concentration and hit density, handles a wide SKU mix, pick aisle is a replenishment aisle, and supports additional levels.

Wire-Meshed Shelves

A wire-meshed shelf bay has coated upright posts, coated wire-mesh or strand decks, and connection items. The maximum load weight for a wire-mesh shelf is 800 lb. and five shelves per bay means that four upright posts support a 4000 lb. load weight. Per your shelf manufacturer and application, wire-mesh coating options are chrome-plated, galvanized, zinc, and stainless steel. Each coating plus the open deck design make the wire-meshed shelf concept very common in the food or medical order-fulfillment industry. If the SKUs that are handled could cause rust on a metal shelf, a coated wire-meshed shelf is used.

A meshed pick position shelf has four upright posts with one upright post in each rectangular deck corner. The upright post serves the same purpose as a standard shelf upright post. Most wire-meshed shelf upright posts have indentations, round shape, and a height range. For most order-fulfillment operations, the approximate height range is 55, 63, 75, and 87 in. At the base of each upright post is a leveling bolt to compensate for an uneven floor surface. Most wire-meshed applications allow one ready reserve level which is on top of the 87–in. shelf level.

Coated wire-mesh or strand decks provide a flat and smooth surface for a pick position level. A deck has coated wire strands with the top strands perpendicular to the pick aisle, which allows a carton or container to be easily pushed forward on the deck. On the underside of the top strand is a single wire strand or at least two to three double diagonal wire strands, depending on the deck span or length and weight capacity. A long deck span has the greatest number of single and double wire strands. Between the strands is a preformed, wave-shaped wire strand that is welded to the double wire strand hole. Each rectangle-shaped corner has a round shape. The end diameter permits a deck to slip on a round-shaped upright post. At the desired deck elevation above the floor, a split plastic cone, sleeve-shaped shelf or deck retainer two sections are placed onto each upright to form a solid cone or sleeve. The cone or sleeve base diameter is slightly larger than the deck round-shape diameter and has a projection. As the wire-meshed deck is lowered onto the upright post and the round holes come in contact with the formed cone or sleeve base, the cone sleeve diameter stops the deck downward movement, and the projection becomes engaged into the upright post indentation.

Wire-mesh shelf elevations on the upright post are adjusted on 1-in. centers. A 36-in. or less wire-meshed deck supports a 800 lb. total weight and a 48-in. or more shelf deck supports 600 lb. total weight. Available sizes are:

- Depth range is 14, 18, 21, 24, 30, and 36 in.
- Length or face range is 24, 30, 36, 42, 48, 60, and 72 in.

In a wire-meshed shelf pick position row design, a wire-meshed shelf pick position row starter bay has four upright posts with the appropriate shelf deck and connecting cone or sleeve number. After a starter bay with an S-hook, to have each shelf level at the same elevation, each additional shelf pick position bay has two upright posts and appropriate shelf deck, and connecting cone or sleeve number. If your application does not used an S-hook, to have each shelf level at the same elevation, each add on bay or additional shelf pick position bay has four upright posts and the appropriate number of shelf decks and connecting cones or sleeves.

The wire-meshed shelf options are dividers, solid mat overlay, and ledges or barriers along each or any shelf side.

A wire-meshed shelf concept features are most wire-mesh shelf applications do not provide additional ready reserve positions, wire-mesh shelf is slightly more expensive, used in a humid environment, and used to handle food or medical products industry.

Hardened Plastic Shelves

A hardened plastic shelf bay has reinforced polymer upright posts, slotted decks, and connection items. The maximum load weight for the polymer shelf has a range from 250 to 800 lb. and five shelves per bay means that four upright posts support a 1250 to 4000 lb. load weight.

A hardened plastic shelf concept is very similar to a wire-meshed shelf concept, except for the following differences:

- A slightly lower load weight capacity
- A deck that has three horizontal strands with slotted or vented grids that fit on a rectangular deck or one solid hardened plastic deck
- It uses a wedge connection item to hold a shelf at the desired elevation above the floor surface
- Slightly more expensive

Round-Shaped Shelves

A round or revolving shelf concept is used for small items but not flatwear apparel SKUs. A revolving shelf pick position has coated sheet metal or hardened plastic shelves that are circular and attached to a center post. To access a pick position, the appropriate shelf level is turned until the assigned pick position arrives at the pick station.

The center post is 66 in. high and serves as a leg or base to provide stability to the shelf unit. The center post also permits an employee to revolve the shelf around the post.

A revolving shelf with a center hole has a convex slope to the deck that is higher at the center and lower at the end. This feature assures that gravity moves small items to the shelf edge for easy SKU transaction completion. A center hole permits attachment to a center post and allows an employee to turn or revolve a shelf level independent of the other shelf level movement. Circular-shaped shelf available designs are:

- Diameter range is 17, 28, 34, 44, and 58 in.
- Per shelf level with five compartments that has 625 lb. capacity or with ten compartments that have 500 lb. capacity.
- Shelf level number range is four to ten levels. In a center post height, shelf level number permits sufficient clearance between levels to assure easy and quick SKU replenishment or pick transaction.

The adjustable shelf divider bin and pan are designed to divide the existing shelf large compartment into smaller compartments. The pan is designed to handle very small-items.

In a small-item pick area layout, the circular-shaped pick position shelf bay is used as:

- Single circular-shaped shelf row with pick aisles on both sides
- Back-to-back circular-shaped shelf rows with a pick aisle on one side
- At the end of back-to-back standard closed, open, or other fixed shelf row
- At the corner or the intersection of two walls

The circular shelf has limited application in an order-fulfillment operation. Most circular shelves are used in the maintenance shop or on a retail sales floor.

Mobile Shelves

A fixed track mobile shelf and a four-wheel mobile shelf bay (see Figures 4.9 and 4.10) are specialized small-item or flatwear pick positions that are moved by an employee onto a pick line or into a pick aisle. The mobile shelf options are:

- Fixed overhead track
- Two deep overhead or in-floor track
- In-floor track
- Four-wheel platform

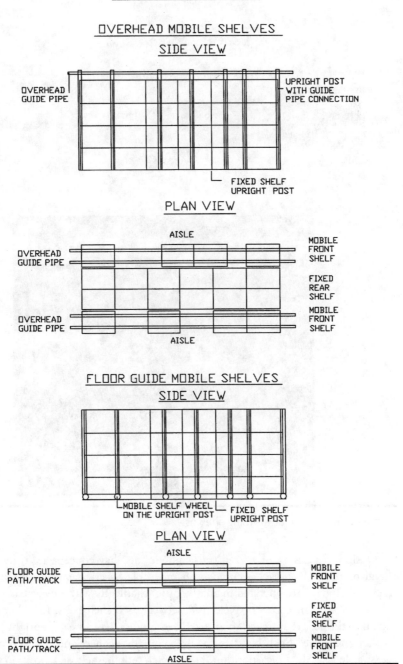

Figure 4.9 Overhead guide plan view and floor guide plan view.

Figure 4.10 Mobile and sliding shelves.

Options for design of a fixed track mobile shelf bay concept are three to four closed or open shelf bays secured onto moveable bases that travel over in-floor tracks. Shelf bays are operated manually or mechanically with an electric-powered motor to create an access or pick aisle between two shelf bays. (b) Alternatively, each upright post is attached to an overhead guide rail, and each upright base has a caster and wheel. Shelf bays are moved manually or mechanically to provide an access aisle to a SKU pick position. Fixed-track mobile shelf pick position concept design options are:

- Overhead track
- Two-deep overhead or in-floor track mobile shelf
- In-floor track mobile shelf
- Four-wheel mobile shelf bay

Fixed Overhead Track

A fixed overhead track mobile shelf concept has one-deep open or closed shelf bays. Shelf bays have rectangular shelves with the long shelf side or pick face perpendicular to the main aisle. The travel direction is perpendicular to or across the main aisle front. When shelf bays are in the closed position, bumpers on the corners of the shelf base create an open space between the upright posts. This permits a picker to easily reach a shelf upright and move the shelf. It also minimizes UOP damage if the UOP overhangs the shelf edge, and minimizes potential picker injury from having a hand become jammed between two shelves.

The top of each upright post is attached to an overhead track that serves as a shelf guide. At the base is a caster and a 5-in. diameter wheel. The wheel cover is a polyurethane or resilient hardened plastic cover to assure easy movement across the floor surface. To create a pick aisle for access to a pick position, shelf bays are manually or mechanically pushed or pulled to the side. In most mobile shelf applications, the two end shelf bays are fixed shelves and there are three to four mobile shelf bays.

Two-Deep Overhead or In-Floor Track

The two-deep shelf bay concept has two shelf rows. The rear shelf row has two shelf bays that are fixed and face the pick aisle. The front shelf row is one mobile shelf bay that is manually moved along the two rear shelf fronts or pick faces. The mobile shelf bay faces a pick aisle. To access a rear shelf pick position, a picker moves the front mobile shelf to the appropriate location that provides access to the rear shelf pick positions. This mobile shelf concept has an in-floor or overhead fixed track with end stops for the front shelf travel path.

In-Floor Track

The in-floor fixed-track mobile shelf bay concept has one shelf bay or back-to-back shelf bays that is secured to a moveable platform. In most applications, 1 to 2 in. is allowed between each mobile shelf bay. To manually move a mobile shelf bay, a picker pushes or pulls it over the in-floor track. To mechanically move a mobile shelf bay, a picker turns a wheel or handle that mechanically moves the shelf over an in-floor track. To move a shelf bay with an electric-powered motor, a picker controls shelf movement from the control panel.

With an in-floor track mobile shelf concept, the moveable platform has four casters/wheels that are under each corner. Casters and wheels move over tracks that are embedded in the floor. To create a pick aisle, the shelf bays are moved to the side, parallel, or across the main aisle front. In the parallel mobile shelf design, there are three to four moveable bases and each end is a fixed shelf bay. An alternative embedded track and mobile shelf travel direction is to have the shelf bays moved forward, perpendicular, or into the main aisle. With the mobile shelf in the main aisle, a picker has access to the appropriate pick position. In a perpendicular mobile shelf design, there are two to three moveable shelf platforms. In mobile shelf applications, shelf bay designs are:

- Closed shelf type
- Open shelf type with edge barriers
- SKUs are in cartons or containers with a half-high front

Other design considerations are:

- Each rectangular shelf is 18 or 24 in. wide and 36 to 48 in. long
- All manually moved shelves have bumpers

For additional caster and wheel information, the reader is referred to the manual push pick cart section in this chapter.

The design parameters and operational characteristics for a mobile shelf bay concept are similar to those for a regular shelf bay concept, except that a mobile shelf concept is slow moving and is best suited for very slow moving SKUs. To ensure stability as a shelf bay is moved across a floor surface, most have upright posts with a 7 ft. maximum height above a moveable platform surface and the floor surface is clean and clear of debris and obstruction.

In a small-item/flatwear apparel order-fulfillment operation, a mobile shelf concept has some application for very slow-moving, almost obsolete SKU pick positions or for record storage. If these SKUs do exist in your operation, a mobile shelf concept provides high hit concentration and hit density in a small area. To have relatively quick and easy access to a pick position, the two-deep mobile shelf concept is the preferred mobile shelf design. The two-deep mobile shelf bay concept is designed as single rows, back-to-back rows, or as a stand alone unit in a rack bay.

The disadvantages are cost, time to create a pick aisle, and height restrictions. Advantages are increases pick positions per square foot, fewer aisles, and used for slow or obsolete SKUs.

Four-Wheel Mobile Shelf Bay

A four-wheel mobile shelf bay concept has one open, closed, or wire-meshed shelf or carton gravity flow rack bay. A rigid or swivel caster arrangement with a polyurethane-covered wheel is under each upright post or corner. A four-wheel carton gravity flow rack design has a 3- to 4-ft. rack depth. The caster and wheel arrangement permits a picker to push or pull a mobile shelf to the desired location on a pick line. For additional caster and wheel information, the reader is referred to the manual push pick cart section in this chapter.

In a pick area design with the pick aisle perpendicular to the pick conveyor, a four-wheel mobile shelf bay concept is used at the end of the pick aisle or shelf row. Prior to the pick activity, a mobile shelf bay is moved at the end of the pick aisle end, with a mobile shelf pick position facing the pick aisle. In this position, the proper picker routing or number pattern is from a pick conveyor:

- Along a fixed shelf row pick position
- Across the mobile shelf bay pick position
- Along a fixed shelf row pick position back to a pick conveyor

To assure easy movement of a fully loaded mobile shelf bay, lightweight SKUs are allocated to the mobile shelf bay pick positions. Swivel casters are preferred on a manually moveable shelf bay.

In a pick aisle layout, the perpendicular pick aisle width determines the shelf bay width. In most small-item order-fulfillment operations, this perpendicular pick aisle width is 3 to 4 ft. In a manual pick operation, the four-wheel mobile shelf is used to increase the number of pick positions for slow- to medium-moving SKUs.

Wide-Span or Slotted-Angle Shelves

A wide-span, slotted-angle, or long-span shelf concept is a pick position that is a hybrid between a standard shelf concept and a decked pallet rack concept. With a wide-span concept, a pick position or deck span is between two upright posts. The feature is very similar to decked pallet rack load beam span and a wide span shelving load weight capacity is similar to standard shelf weight capacity.

A wide-span or slotted-angle shelving pick position concept has upright posts with slots, load beams with one to three rivets on each end, several front to back members with at least one or two rivets on each end, shelf, or deck, and accessories.

The upright post is a 1 ½ in. by 1 $^{11}/_{16}$ in. by 2 $^5/_{16}$ in., coated 14-gauge angle iron member. Each upright post is prepunched with slots for load beam elevation adjustment at 1 ½ in. centers. The slots are on both sides of each upright post and are used for attachment of load beams and front to back members. The slot configuration permits the rivets for load beams and front to rear members to be easily inserted or removed from an upright. After load beam and front to back members are connected to an upright post, the slot design and rivet design securely lock the components together.

With wide-span shelving, to have each shelf level at the same elevation above the floor surface, each starter bay and add-on-bay or additional bay has four upright posts. If shelf elevation above the floor surface can vary, a starter bay has four upright posts and each additional or add-on-bay has two upright posts. A wide span shelf upright post is available in 1-ft. increments from 315 ft.

A wide-span load beam serves the same purpose as a decked pallet rack load beam. Each shelf level has two load beams. A load beam is a 1 ½ in. by 1 $^{11}/_{16}$ in. by 2 $^5/_{16}$ in. angle iron, J formed or C-channel solid piece member. At the end of each load beam is from one, to three rivets, depending on the load weight and span design and manufacturer standards.

The wide span shelf load beam span has a range in 1 ft. to 1 ft. 6 in. increments from 2 ft. to 8 ft. long. The preferred method of connecting the load beam and upright frame is with the flat leg on the top, which provides support for a deck or shelf. To ensure shelving stability and rigidity, two load beams span the top of the upright posts.

Load beam adjustability on an upright post permits flexibility in the number of shelf levels. A shelf or load beam front to back member has the same characteristics and function as the standard shelf end panel member or decked pallet rack front to back member. Front to back members:

- Span the shelving bay depth or width
- Are connected with rivets
- Provide rigidity and stability at each level
- Do not support any load weight

Standard sizes are 1 to 5 ft. long. The wide span shelf bay length is determined by your acceptable deflection, shelf bay weight capacity, and requirement to have a single row with one pick aisle face or to have a pick aisle face a single wide span shelf row both sides.

The wide-span shelving supports SKU pick positions. Applications include:

- Plywood
- Particle board
- Solid steel shelf
- Grated steel shelf

Per the pick position management, a deck is divided with the same material that was described in the hand stack in standard decked pallet rack section of this chapter.

Wide-span shelving accessories to improve stability are overhead aisle ties, wall ties, and foot plates that anchor the upright post to the floor surface.

The wide-span shelving concept has a lower cost than a standard pallet rack with a deck, and costs about the same as a standard shelf concept. The load beam is slightly shorter than that of a standard pallet rack and slightly longer than a standard shelf. It has the same number of pick faces per aisle, and is considered an open pick position concept. It is deeper than a standard shelf. Wide-span shelving does not require X sway braces or a back panel like standard shelves.

The wide-span shelf is a very flexible pick position concept, and can be designed as a single row with pick faces on both sides of a shelf or as back-to-back shelf rows or a single row with pick faces on one side of the shelf.

For add-on bays to have pick levels at the same elevations above the floor surface, the bay has four upright posts and two load beams with front to back members and the deck for each pick level. If the add-on bays are to have pick levels on different elevations above the floor surface, the add-on bay has two upright posts and for each pick level two load beams with front to back members and the deck. If a wide-span shelf is to support a future catwalk or higher pick levels, the manufacturer compensates the upright post gauge and base plate to handle the additional load weight.

The disadvantages are that it is not as popular as standard shelves, cannot support a pallet, and is slightly more expensive than a standard shelf concept. The advantages are long spans down a pick aisle and improved hit concentration and hit density.

Carton Gravity Flow Rack Pick Concept

The carton gravity flow rack pick position concept (see Figure 4.11) is used in a small-item or flatwear pick operation to provide pick positions for fast- to medium-moving, high-cube, and heavyweight SKUs. These are best allocated to a carton gravity flow rack because of the carton presentation at a pick position, the number of pick faces per bay, and the number of cartons in a lane or in ready reserve positions.

Carton gravity flow rack components are structural support frames, shelves, or pick level frames and connecting clips, bracing member, roller or conveyor lanes, guide rails or dividers, end stop or tray, and replenishment or loading shelf.

The frame is welded and coated 11-gauge steel with a 1 ½-in. wide C-channel that has vertical, angled, and horizontal members that form one unit. The support frame has prepunched slots, which are the locations for attachment for a pick position level, clip, and bracing members. Slots are on 1- to 2-in. centers to ensure that a pick level or pick frame has the correct slope or pitch, which allows cartons to travel over a conveyor travel path from the replenishment side to a pick face or pick side. On the two horizontal members, prepunched holes are used to secure or anchor a structural frame to the floor.

Carton gravity flow racks are available in a wide variety of dimensions and sizes. The width of the rack face determines the number of pick positions per pick line. Carton gravity flow rack face or front presents SKU pick positions and cartons to a pick line. To optimize picker productivity, the maximum number of pick faces is allocated to the golden zone. Standard dimensions are 48, 52, 56, and 60 in. wide.

The height of the pick front or pick face height determines the maximum number of pick positions within easy reach of an employee. Most carton gravity flow rack manufacturers have a 96-in. overall structural frame height. With the first

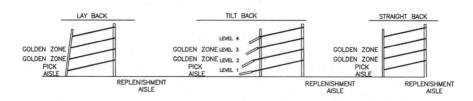

CARTON FLOW RACK DESIGNS

Figure 4.11 Carton flow rack designs: straight front, tilt front, roller conveyor, and separator.

pick level or pick face 3 in. above the floor surface, there is sufficient space to allocate three to four pick position levels high.

The front to rear dimension determines the pick line location and pick aisle in the facility and determines pick aisle and replenishment aisle locations. Most carton gravity flow rack manufacturers have basic design options that include:

- A straight back design, in which the front and rear vertical post members are perpendicular to the floor surface. This perpendicular design has each pick position level extend toward a pick aisle or replenishment aisle at the same length. This means that the next highest pick position level is directly above the lower pick position level. For a picker to complete a SKU pick transaction, this pick position design has a higher clearance between a pick position SKU carton or container and the structural member of the pick position above it. To minimize this clearance between two pick positions, each carton has a partial open front. Standard straight back carton gravity flow rack overall dimensions are 5 ft., 90 in., and 10 ft.
- A layback carton gravity flow rack, which has the front and rear vertical members angled at a 10 degree angle to the floor surface. This layback frame feature means that the lowest pick position level extends outward the furthest distance toward a pick aisle. The next highest pick level extension toward the pick aisle is 1 to 2 in. less. From a side view, the layback feature resembles a set of steps or a stairway. The layback feature is reversed on the replenishment side of the rack. On the replenishment side the highest pick level has the greatest extension into the aisle. An option is for the replenishment side to have a straight design. Standard overall layback carton gravity flow rack dimensions are 6 ft., 8 ft. 6 in., and 11 ft.
- A tilt front carton gravity flow rack, which is a layback frame design with a tilt front. The tilt front feature presents a carton interior or SKUs at an angle, which enhances picker productivity because it is easier to withdraw or remove a SKU (piece) from a carton in a pick position.

Most professionals prefer the layback frame design, because it provides easier access for a picker to complete a pick transaction.

A carton gravity flow rack pick position level frame or shelf with connection clips are welded coated members that from a rectangular frame. The metal frame has two long full-length members or side runners, which

- Provide rigidity to the frame
- Permit attachment of welded front, rear, and middle cross members
- With an exterior flange and a clip inserted into a vertical frame structural member for connection to a structural frame

An add-on bay has one upright frame. Per the frame length or depth, there are:

- Two to three middle cross members that support carton gravity flow rack conveyor or rollers
- Carton gravity flow rack conveyor or roller lane front member attachment and pick position member
- Rear or replenishment member for carton gravity flow rack or roller attachment

To reduce roller or conveyor damage, the rack frame replenishment side has a sheet metal step or section that extends inward from the frame end 1 ft. into the frame depth. During carton replenishment to a carton gravity flow rack conveyor lane, this metal step serves as a solid metal surface to absorb carton impact. If rollers or conveyor lanes are at this replenishment side, over time these rollers or conveyors become damaged from carton impact. The pick frame or shelf is a bed that structurally supports the carton gravity flow rack rollers and conveyors and presents the carton to a pick position.

Designs for a pick position or pick level frame are:

- Straight shelf front. A straight front presents a carton in a pick position on a perpendicular angle to flow lane rollers or conveyors.
- Carton tilt front. A tilted front is manufactured with the end section of the pick frame or pick shelf angled downward toward the floor and forward to a pick aisle or pick line. The tilt front exposes more of the carton top to a picker. The most common tilt front angles are 13, 18, and 21 degrees.

On a small-item pick aisle or pick line, most professionals consider that a carton gravity flow rack with a layback and tilt front frame provides the best picker productivity.

Connection clips are inserted into the front, rear, and middle member side slots of a structural upright frame. After the clips are inserted into slots at the proper elevation above the floor, employees set a pick position frame or shelf exterior flanges onto the clips. The clip design holds a carton gravity flow rack pick frame to a carton gravity flow rack bay.

Additional structural supports and front and rear sway bracing members provide added stability and rigidity to a rack bay.

Most carton gravity flow rack manufacturers install front to rear sway braces on each rack bay. A sway brace is preformed sheet metal with pre-punched holes on each end. After the rack bay is standing on the floor, nuts and bolts are used to secure the sway brace to the rack bay frame. A front sway brace is attached to the upper part of the upright frame structural front member and a flat surface is used to identify a pick bay. Rear sway braces are attached with nuts and bolts to the upper and lower upright structural rear member sections and a flat surface is used to identify a pick bay.

Rollers or conveyor lanes are thin metal C-channels with pre-punched holes in the same location on both sides of the channel. Solid plastic skatewheel axles are factory-inserted into the holes, which run the full length of the C-channel. On the top surface are pre-punched holes or slots. When a C-channel is set on its legs, the length and depth of these slots and the side holes permit the skatewheel conveyor surface to extend a quarter inch upward above the flat or solid base. The plastic skate wheels run the full length of the C-channel section.

In a carton gravity flow rack application, a pick level frame is easy to install and the elevation above the floor is adjustable. To change the pick level elevation above the floor, employees relocate the connection clips and insert the connection clips in new slots on the frames. After the clips are secured in the slots, the pick level frame exterior side flange is placed onto the clip.

For an 11-ft. deep flow rack lane, the elevation change between a carton gravity flow rack pick level frame replenishment side and pick side is 10 to 13 in. This elevation change provides proper pick level frame slope or pitch to assure good carton travel over a conveyor travel path. For optimum carton travel over a conveyor travel path or lane, on-site actual carton gravity flow rack travel tests on the existing pick level frame determines the best pick level frame slope or pitch.

Most carton gravity flow rack manufacturer pick level frame or shelf replenishment and pick side end members have sufficient space for paper pick position identification placement on both the pick and replenishment sides or pick-to-light module attachment on a pick level frame pick side. The front and rear pick level frame member front dimension is 2 ½ in. high and carton gravity flow rack bay front full width.

It is considered good order-fulfillment practice to have paper pick position identification along with the pick-to-light module. This feature assures during a pick-to-light problem or down time that your order-fulfillment operation with a paper pick document provides service to your customers.

With a tilt front carton gravity flow rack pick level frame configuration, design for a carton support in a tilt front is:

■ Separate rollers or conveyor lanes that match a pick level frame conveyors or roller lanes.
■ Coated and solid sheet metal tray that covers the entire front and is supported by the tilt front interior lip or edge. When we compare other front options, a solid sheet metal tray has less cost and maintenance.

After a C-channel is secured with a clip or inserted into the pick frame front member metal clip, the pans under the front and the rear members provide support and retain the carton gravity flow rack rollers or conveyor lanes in the pick frame.

The conveyor roller wheels are the bottom support for the carton and with gravity force from a pick level or frame slope (pitch) assures that a carton travels from the replenishment side, over a conveyor travel path, and to the pick side. To assure good carton travel over a carton gravity flow rack rollers or conveyor lanes, there are a minimum of three C-channel flow lane members under the short carton exterior bottom surface or width. The carton width is based on a replenishment employee carton transfer to a pick position. To assure good carton flow over a carton gravity flow rack conveyor or lane, replenishment employees are instructed to place a carton with the flaps in the carton direction of travel. In many carton gravity flow rack applications, across a carton gravity flow rack pick level frame width or front, roller or conveyor lanes or C channels are located on 2- or 3-in. centers.

In a carton gravity flow rack application, a carton has potential to hang up in a carton gravity flow rack lane. Carton hang up results from a poor carton travel surface, poor carton tracking over the conveyor travel path, or jam against an adjacent carton. To resolve a carton hang-up situation the options are:

■ Carton flow dividers or guides the full length of the conveyor travel path.
■ Replenishment and pick carton gravity flow rack has long poles along the side with hooks on the ends. The pole is used by an employee to move a carton forward on the conveyor lane.

A guide rail or pick level divider runs the length of the pick level frame or conveyor lane full length and is located at specific points across the pick level frame width. A guide rail or lane divider is a preformed coated thin sheet metal member that resembles an inverted T. The inverted T has the top as the guide rail base and the stem extending upward from the base above the top of the conveyor. The extension acts as a guide for a carton to travel over a conveyor travel path and serves to minimize carton hang ups on a carton in an adjacent carton gravity flow rack conveyor lane.

In a carton gravity flow rack application, to permit a wide carton mix in a carton flow rack concept, guide rails or dividers are not used in carton gravity flow rack conveyor lanes. When some companies re-profile a pick line, per a new pick line profile to relocate a guide rail requires additional time and labor expense. Second, if a replenishment employee properly prepares the carton top flaps, there is sufficient open space or a gap between two adjacent cartons and the carton has a good-quality bottom exterior surface, there is minimal potential for carton hang up in a carton gravity flow rack conveyor lane.

The carton gravity flow rack conveyor lane end stop prevents a carton from falling from the conveyor lane onto the floor. Most carton gravity flow rack manufacturers use each pick level frame or shelf front end as the end stop.

The carton flow rack option is a flat coated hardened metal shelf on a replenishment lane or shelf end. During a carton replenishment transaction, a metal shelf absorbs carton impact instead of the conveyor rollers or wheels. This flat metal shelf feature minimizes roller or wheel damage, improves carton gravity flow through a flow lane, and improves replenishment employee productivity.

In most carton gravity flow rack small-item/flatwear apparel order-fulfillment applications with a normal employee reach, a good pick line profile and standard carton heights for a carton gravity flow rack pick position, the maximum pick level number is four or five levels high. The pick level number high is used in the:

- Straight front flow rack
- Lay back flow rack
- Lay back with a tilt front flow rack

Vendor master carton width, carton gravity flow rack bay width, and pick line profile are key factors that determine number of pick positions per rack bay and for each pick level. Most professionals allow 1 to 2 in. of open space between two carton conveyor lanes or pick positions. In most pick line profiles, for a five level high rack bay the golden zone or middle pick levels are pick levels 2, 3, and 4. In a 4-level high rack bay, the golden zone or middle pick levels are pick levels 2 and 3. SKU allocation to a carton gravity flow rack pick levels are:

- Golden zone pick levels have fast-moving, medium-weight, and medium-cube SKUs
- Lowest pick levels have heavy-weight, high-cube, and slow-moving SKUs
- Highest pick levels have the lightweight, medium- to small-cube, and slowest-moving SKUs

SKU pick line profile allocation of a carton gravity flow rack bay zone has SKU movement to match your projected employee productivity rate.

One important carton gravity flow rack pick level feature with a horizontal picker routing pattern across the entire front for one pick bay is that the very fast-moving SKUs are effectively and efficiently allocated to one pick level in the golden zone. This feature means that the rack level can have one fast-moving SKU the full width of the pick frame. With this SKU profile or allocation, there is a large inventory in the pick position to minimize stock outs and assures budgeted picker productivity. With a standard size carton, most pick line profile results have five cartons or pick positions per pick level. With a small size carton, most pick line profiles have seven cartons or pick positions per pick level.

From the above pick line profiles, a carton flow rack pick bay has a pick position range. Maximum pick positions are:

- Five high pick level and five wide pick level flow rack bay has 25 positions.
- Four high pick level and five wide pick level flow rack bay has 20 positions.
- Five high pick level and seven wide pick level flow rack bay has 35 positions.
- Four high pick level and seven wide pick level flow rack bay has 28 positions.

During a pick position replenishment activity an employee gently pushes a carton forward in a carton gravity flow rack lane. To assure accurate picks in a carton gravity flow rack pick concept, each SKU replenishment lane is discreetly identified with a pick position number, SKU number, and SKU description. This identification process minimizes replenishment errors that become CO pick errors.

A carton gravity flow rack pick position concept is designed as a stand-alone pick concept that has pick positions face a pick line or pick aisle. This pick area layout feature means that a carton gravity flow rack pick line has a large bay number. In most small-item pick applications, carton gravity flow rack pick bay features are:

■ Separated into pick zones. Each pick zone is profiled to assure that the SKUs allocated to a pick position zone match your budgeted order picker productivity rates.
■ 30 to 36 in. wide pick aisle between a carton gravity flow rack pick face and a pick conveyor or vehicle travel path.
■ Replenishment aisle that has direct access to a carton gravity flow lane replenishment side and to SKU ready reserve positions.
■ After a predetermined number of pick bays, per local code and company practice, to have a cross aisle. The cross aisle provides picker entrance and exit from a pick line to the replenishment area or to an emergency exit.

Design options for a carton gravity flow pick area are:

■ Single pick line
■ Dual or mirror pick line
■ Two dual or mirror pick line sets with two completed CO take-away conveyor transport concepts
■ With short carton gravity flow rack lanes or standard shelves as a pick cell for unique SKUs
■ Carton gravity flow rack structural support frame, support carton, or pallet ready reserve storage positions, an elevated carton make-up area or other pick line activity, an elevated in-house carton transport concept, or an elevated pick line
■ Pick position holds a larger ready reserve inventory quantity
■ Good hit concentration and hit density
■ Minimizes product double handlings
■ Handles fast- to medium-moving, high-cube, and heavy-weight SKU mix
■ Minimizes picker non-productive walk time and distance
■ Replenishment and pick activities in separate aisles

A carton gravity flow rack pick line has a starter bay and add-on-bays. A starter bay has two upright structural frames, pick position frame, or shelf complete with a carton roller or skatewheel conveyor lane for each pick level, four to six clips for each pick position frame, and front and rear sway braces. Each add-on bay in the series has one upright structural frame, pick position frame, or shelf complete with a carton roller or conveyor for each pick level, four to six clips for each pick position frame and front and rear sway braces.

The disadvantages are higher cost, larger area, and management discipline and control for SKU allocation to a pick line. Advantages are interfaces with a paper pick document, pick label, pick-to-light pick, or RF device instruction, provides good picker and replenishment employee productivity, large ready reserve inventory in a pick position, and minimizes employee physical effort to pick small-items or flatwear.

Picking Fast-Moving SKUs from Pallet or Carton Flow Racks

Most small-item order-fulfillment operations have fast-moving SKUs. The SKUs are enclosed in master cartons and account for a large percentage of the customer-ordered piece volume. In a pick line operation, options for handling very fast-moving SKUs are to use a carton gravity flow rack pick position or a powered or non-powered pallet pick position.

Pick from Carton Gravity Flow Racks

With a carton gravity flow rack pick line arrangement, there are separate replenishment and pick aisles. From the replenishment aisle a replenishment employee opens each master carton, transfers it onto a carton gravity flow rack lane and throws the filler material and empty carton trash into the trash concept. With this arrangement, an order picker does not have non-productive time on replenishment activities that include cutting or opening cartons, throwing the filler material and carton top into the trash, moving master cartons to an accessible location on a pallet, bending and reaching for master cartons on the bottom of the pallet, removing an empty pallet to a holding lane, and moving a full pallet forward to the pick position.

In a carton flow rack pick position, a picker has easy access to SKUs in a master carton in the pick position. After the master carton is emptied, the picker throws the empty carton into a trash concept.

The concept features higher picker productivity due to direct access to the SKUs in a master carton and fewer non-productive activities, and decreased walk distance between two pick locations.

Pick from Non-Powered Pallet Lanes

On a pick line, pick positions for very fast-moving SKUs means that a SKU or pallet is placed onto a pick line. Per a CO, an employee picks a CO SKU quantity from a master carton on a pallet. In a pick position, SKUs in master cartons are presented to a picker on a pallet.

The features are lower picker productivity due to additional non-productive activities such as opening cartons, throwing items in the trash and moving cartons to have access to a specific master carton. and removing empty pallets from the pick line. Alternatively, a replenishment employee performs these activities. This concept has increased distance for the picker to walk between two pallet pick positions, and large SKU piece quantity in a pick position.

With master cartons on a pallet, carton presentation options are:

- Closed master carton
- Precut master carton tops
- Wide pallet front is a position face
- Narrow pallet front is position face

Enclosed or Non-Cut Master Cartons

An enclosed or non-cut master carton option has a picker cut open a master carton top, throw filler material and empty carton into a trash concept, and transfer a CO SKU quantity from a master carton into a CO container. To complete a pick from a pallet a picker has to move an open master carton to the front of the pallet and reach across the pallet to remove SKUs from a cut master carton.

Features of this option are:

- Picker non-productive time to cut and throw filler material and empty carton into a trash concept that lowers pick line productivity
- Additional picker activity
- With no picks from a fast-moving SKU potential uncontrolled CO carton queue on a pick line and completed CO flow interruption
- In a WMS program, faster SKU allocation to a pick position, and CO release

For additional information on master carton presentation in a pick position, we refer the reader to the previous section in this chapter.

Precut Master Cartons

The precut master carton option has a replenishment employee precut each master carton on a pallet. In a master carton cutting activity, a cut master carton is transferred onto an empty pallet. When the pre-cut master carton pallet is full, it is transferred onto a pick line position. The picker continuously picks SKUs, so there is no or minimal queuing on a pick line, and constant CO flow. Master cartons are cut by a replenishment employee, which is more productive.

For additional information, we refer the reader to the previous section in this chapter.

Wide Pallet Front in the Pick Position

The widest dimension of the pallet faces the pick aisle. To pick a SKU from a master carton, a picker has a long reach to the deepest master carton, which causes fatigue and additional time and lower pick productivity.

Narrow Pallet Front in the Pick Position

A narrow pallet in a pick position means that a picker has a shorter reach to the deepest master carton. When completing a pick activity, the shorter reach means less fatigue and time and improved pick productivity.

Handling an Empty Pallet

When a pallet becomes depleted of SKUs, it is transferred from the pick position.

- Empty pallet holding position. With an empty pallet holding position option, a picker has to manually or mechanically lift, remove, push or pull, and transfer the empty pallet into a holding position.
- Pallet flow lane. With a pallet flow lane option, the designs are (a) a flow lane that has the pallet flat as it travels over the travel path or (b) a flow lane that has the pallet vertical or standing up as it travels over the travel path.

Pick Line Design

There are a number of options for designing SKU location or pick position on a pick line.

- SKUs are located on a pick line using the 80/20 rule.
- A single pick line concept in which the entire SKU inventory is located in a pick position or in ready storage positions in the pick area. In a single pick line design, there are (a) a pick conveyor and take-away conveyor located on the opposite side of the pick aisle from pick positions; (b) a pick conveyor travel path along the front of the pick positions, with pick positions over and under the pick and completed CO carton conveyor travel paths; and (c) a pick platform that travels on a rail along the front of the pick positions
- A dual or mirrored pick line concept, in which one pick line faces the pick aisle conveyor and CO take-away conveyor travel paths, and a second pick

line face with its own pick conveyor and completed CO carton travel paths but has an independent or common trash conveyor travel path. A mirrored pick line characteristic is that pick positions for the same SKU are on opposite sides of the pick line. This means that the first pick position on pick line 1and the first pick position on pick line 2 have the same SKU.

- Four pick lines or two mirrored sets. With this pick line arrangement, a front mirrored pick line set has a CO take-away conveyor travel path that bypasses the rear mirrored pick line set. The CO take-away conveyor travel path has lift gates. The conveyor bypass is elevated above the pick positions or located in a floor pit below the pick line sections.

- With a SKU mix for all customers and unique SKUs that are assigned to specific customers, SKU pick positions are separated into sections or cells. The front pick position section or cell is designed for the common SKUs that are available to all COs. The rear pick position section or cell is designed for the unique SKU allocation. A cell is either parallel or perpendicular to a main pick aisle.

- Fixed or floating pick positions. With a SKU allocated to a fixed pick position, unless management reassigns a SKU to a new pick position, each work day, a pick position has the same SKU. A floating or random SKU pick position concept is one in which SKUs are assigned by a computer to several pick positions and replenishment transactions are completed in the concept. After the first pick position is depleted of SKU pieces, the computer assigns the next CO pick transaction to another pick position that contains a SKU inventory.

SKU Allocation to a Pick Line

To ensure maximum picker productivity, accurate picks and on-time CO completion, a carton flow rack pick line has high-demand SKUs and medium- or low-demand SKUs in appropriate pick positions. Some professionals refer to high-demand SKUs as power SKUs, which are promotional, seasonal, special sale, or daily promoted SKUs. These SKUs have a one- to-three day high-demand life cycle. The medium- or low-demand SKUs are regular-inventory SKUs.

SKUs can be allocated to pick positions based on high or low volume, and groups. SKUs are grouped together in one pick zone based on the historical or estimated sales volume or on similar volume characteristics. With the high-volume SKU allocation concept, the number of SKUs allocated to a pick zone matches the budgeted picker productivity rate. This feature controls order picker queue in a pick aisle or zone and increases the completed CO number within a short time, short pick distance, and one pick zone. This allocation group has all low-volume SKUs in one pick zone.

With a low-volume SKU allocation concept, SKUs with a low sales volume are allocated to one pick zone, which maximizes the number of SKUs within a pick zone. This feature minimizes picker travel distance between two slow-moving SKU picks.

For additional information, we refer the reader to the section on SKU hit concentration and hit density later in this chapter.

High to Low Pick Line Profile

The high to low pick replenishment concept is used for an order-fulfillment SKU group with two unique features: a high-volume pick area and a low-volume pick area that follow the 80/20 rule and customer service standard to complete customer orders within two hours.

A high to low replenishment concept has a WMS program sequence a pick line set-up or replenishment that is based on your CO SKU volume. The pick line set-up or replenishment transactions are sequenced with the highest CO SKU volume as the first SKU for the replenishment activity. This means that a storage area completes all UOP withdrawal and transport transactions to move the SKU pieces from the storage area to a pick line replenishment area. The next storage area replenishment transaction is for the second highest SKU. This sequence is followed for each CO SKU.

The high to low replenishment concept requires a WMS program to sequence the CO SKUs and ready reserve positions in the replenishment or pick area.

The disadvantages are the cost of the WMS program, the space requirement for ready reserve positions, and some travel path queue between the storage and pick areas. Advantages are the greatest number of SKU pieces are replenished to the pick positions, fewer trips between the storage and pick areas, and faster CO pick activity.

Pick Position Allocation or Profile for Demand Pull Inventory Flow

One objective of a small-item order-fulfillment operation is to complete a CO order/delivery cycle in the minimum time and at the lowest possible logistics operational costs. For many operations, the CO order/delivery cycle time is within 24 to 48 hours from receipt of a CO. With a high CO number, large SKU mix, and CO mix that includes single-line COs, multi-line COs, and combi COs, and required host, WMS and MHS computer processing time, the demands to have an early or warmed pick line and pack activity start are growing at an increasing rate.

Most catalog, e-mail, direct marketing, and television marketing order-fulfillment operations have a demand pull inventory flow and dual pick lines with the same number of SKUs allocated fixed or different pick positions. For SKUs in a fixed pick position, replenishment transactions move SKUs from a storage area to the pick position. Alternatively, SKUs are allocated to a pick line position daily or periodically.

Fixed-Position Allocation

A fixed-position allocation concept has a SKU allocated to a WMS identified pick position, based on historical sales volume or other CO SKU allocation philosophy. This was discussed earlier in this chapter. As COs deplete SKU piece quantity in a pick position, a WMS program has a MHS complete a SKU UOP replenishment transaction from a storage area to a pick position. With this demand or CO pull inventory flow and large SKU number, a pick line is long and requires a large area, and it is difficult to have a high SKU hit density and concentration. Advantages are that it is easy to implement, SKUs are in a pickable position for early pick and pack activity or start-up, per CO demand requires replenishment, and minimal pick line set-up time.

Daily or Periodic Pick Line Set-Up

If your business has a daily CO demand pull, limited SKU number, high CO number, and a wide SKU mix, each day your pick line is set up with new SKUs that are allocated to pick positions. After a host computer transfers CO information to a WMS computer, a MHS computer directs a WMS-identified UOP transferred from a WMS-identified storage position to a WMS identified pick position. As a SKU is transferred to a WMS-identified pick position, an employee scans the WMS pick identification and enters the WMS-identified piece quantity into a hand-held scanner. Scan transactions are sent to the WMS computer that permits CO release to the pick line. To have the shortest time and travel distance to physical storage, transport, scan and SKU transfer to a pick position, design options are to set up the pick line each day, or to have dual pick lines and set up one pick line ahead of time for the next day's COs.

Set Up a Pick Line Each Day

To set up a pick line each day, from a WMS program an order-fulfillment MHS is provided with the number of UOP SKU pieces that are required to complete all COs. For each CO type, the MHS receives a WMS-identified storage quantity. From a pick line set-up list, UOPs are transported to the pick line replenishment aisle or side. An employee scans each SKU WMS identification and WMS pick position and piece quantity into a RF device. The RF device sends the scans to the WMS computer to update the WMS inventory files as to SKU quantity in a WMS pick position. If the WMS identified SKU quantity overflows a pick position capacity, each WMS-identified UOP quantity is scanned and physically transferred to a WMS-identified ready reserve position. The scans are sent to a WMS computer to update the WMS inventory files. With an updated WMS-identified SKU quantity in a pick position, the WMS releases COs to a pick line or MHS.

With COs that have 500 to 700 SKUs per day, 100 full pallets and 1000 to 2000 master cartons complete replenishment transaction has a two-hour pick line set-up time. To have an active pick and pack area activity at 0900, a pick line set-up starts at 0600/0700 and requires exact storage forklift truck drivers, in-house transport, scan and replenishment employees, and activities completed per the schedule.

Set Up a Pick Line for the Next Day

A pick line set up for the next day's COs means that an order-fulfillment operation has two pick lines and means that pick line 2 is set up on day 1 for day 2 COs. To set up pick line 2 on day 1, the order-fulfillment operation anticipates day 2 COs and SKUs, based on projected sales from the merchandising or sales department. An option is to base the projection on actual sales recorded by the order entry or IT department at predetermined times and the historical percentage for these sales. Based on the estimated SKU sales, the order-fulfillment department estimates each SKU UOP piece quantity that is required to meet the two to four hours of pick and pack activity. These calculations are rounded-up to full pallets or master cartons. With these SKU UOP full master cartons or pallets, the MHS is directed to move WMS-identified UOP quantities from a WMS-identified UOP storage position to a WMS-identified pick position on pick line 2.

Prior to the end of the work day, the MHS concept withdraws WMS-identified master carton and pallet quantities from WMS-identified storage positions, and moves the WMS-identified UOPs over an in-house transport concept to a pick line. In the pick line replenishment area, a replenishment employee completes a scan and SKU physical replenishment transaction to a WMS-identified pick position. The WMS-identified UOP, WMS-identified pick position, and SKU piece quantity are sent to the WMS computer and updated in the WMS inventory files.

With these SKU quantities in WMS-identified pick positions and in the WMS inventory files, on day 1 pick line 2 has SKUs in a WMS pickable location. As a WMS computer allocates SKUs to COs, the WMS allocates SKUs in pick line 2 pick locations. The concept features are warm or on-time pick line and pack activity start at 0600/0700, at 0600/0700 WMS/MHS creates additional WMS-identified UOP move transactions from WMS-identified storage positions to complete day 2 COs, during day 2, all COs are completed and during day 2, pick line 1 is prepared for day 3 COs.

Pick Line Balance

A pick line balance activity is a WMS/MHS computer process that is used in a small-item order-fulfillment operation that has at least two pick lines or areas. Each pick line has the same WMS- identified UOPs. A host computer downloads the CO pieces to a WMS/MHS computer. To ensure each pick line has sufficient WMS-identified piece quantity and CO number to achieve the budgeted picker productivity rate, maintain the customer service standard, or another CO characteristic such as priority order, CO zip code, and CO age are in the WMS wave or CO group planning process. A pick line balance activity takes place either automatic (on-line) or after the host computer CO process close time.

Automatic or On-Line Pick Line Balance

An on-line CO pick line balance concept occurs as soon as possible after COs are received by a WMS computer. The host computer downloads COs to a WMS/MHS computer either as COs are received or at frequent intervals. An on-line CO pick line balance concept has a computer allocate each CO to a pick line as it is received by a WMS computer. The on-time concept requires a high-speed computer and COs allocated by piece quantity or CO number that is randomly spread by piece or CO volume over the two pick lines.

After Host Computer Process Pick Line Balance

Pick line balance after CO entry close time at a host computer concept has a host computer transfer all COs at one time to a WMS/MHS computer. After a WMS/MHS computer has received a day's COs, the CO piece quantity and CO number are allocated between two pick lines. CO piece quantity and CO number are allocated by a methodology to each pick line. The pick line allocation or balance options are:

- Piece quantity
- Customer order number
- Single SKU and multi-line COs by piece quantity or CO number and same multi-line COs or COs with the same SKUs
- Shipping carton size

Piece Quantity

The piece quantity pick line balance concept has a WMS/MHS computer allocate a daily CO piece quantity on a predetermined basis between two pick lines. If an order-fulfillment operation for a workday has 36,000 CO pieces, the WMS/MHS computer pick line balance concept assigns 18,000 CO pieces to each pick line.

The concept features are evenly allocated piece quantity to each pick line, simple computer process, randomly spreads a particular SKU quantity between two pick lines, SKU replenishment is made to both pick lines, and does not optimize replenishment, shipping carton make-up, and pick activity.

Customer Order Number

A CO number pick line balance concept has a WMS/MHS computer allocate a daily CO number on a predetermined basis between two pick lines. If an order-fulfillment operation for a workday has 20,000 COs, a WMS/MHS computer allocates 10,000 COs to each pick line. The concept features are evenly allocates COs between two pick lines, simple computer process, randomly spreads SKU and SKU quantity between two pick lines, SKU replenishments are made to both pick lines, does not optimize replenishment, shipping carton make-up and picker productivity, and with a different piece quantity per CO mix, possible piece and pick imbalance between two pick lines.

Single and Multi-Line SKUs on Two Pick Lines by Piece Quantity or Customer Number

If a small-item order-fulfillment operation has dual pick lines and COs with single and multi-line SKUs, single SKUs and multi-line SKUs for COs are separated by piece quantity or CO number. After a WMS/MHS computer receives COs from a host computer, for each pick line the computer separates COs. For a predetermined piece quantity, the pick line balance concept allocates multi-line COs with the same two SKUs and single-line CO with the same SKUs are allocated to pick line 1. After a predetermined quantity for common SKUs for multi-line and single-line COs is allocated to pick line 1, a WMS/MHS computer allocates other multi-line COs and single SKU COs with different SKUs to pick line 2. The concept features are for a workday budgeted pick productivity, allocates COs to pick line 1 and then allocates COs to pick line 2, a more complex computer process, concentrates CO SKUs to pick line 1 and other CO SKUs to pick line 2, more frequent SKU replenishment is to one pick, which optimizes replenishment and picker employee productivity, similar COs on pick line 1 has fewer shipping make-up carton activity change over number and time and to some degree a piece or CO quantity imbalance between two pick lines.

Customer Order Shipping Carton Size Balance

If a small-item order-fulfillment operation has dual pick lines and COs with single and multi-line SKUs, the CO pieces are allocated by shipping carton size to each pick line. After a WMS/MHS computer receives COs from a host computer, for each pick line a WMS/MHS computer separates COs by multi-line CO SKUs and single-line SKUs that have the same suggested shipping carton size. The concept allocates multi-line and single-line COs with the same suggested shipping carton size to pick line 1. For the next shipping carton size or sizes, multi-line and single-line COs are allocated to pick line 2. When the pick line 2 CO number or piece quantity matches pick line 1, the WMS/MHS computer repeats the allocation process for each shipping carton size between the two pick lines. The concept features are allocates COs to pick line 1 and then to pick line 2, more complex computer process, randomly allocates SKUs between pick line 1 and pick line 2, SKU replenishment is random between pick lines which does not optimize replenishment and picker employee productivity, similar CO shipping carton sizes on pick line 1 has fewer shipping make-up carton activity change over number and time and to some degree a piece or CO imbalance between two pick lines.

Carton Flow Rack Profile Concepts

Carton flow rack pick line profile or set-up concepts are:

- Continuous flow rack pick line with random SKU allocation or profile along a pick line
- Continuous carton flow rack pick line with specific flow rack lanes that are allocated or profiled for high-demand SKUs
- Pick area with three modules that are separated into three sections

A carton flow rack pick line is designed with the number of rack bays, channels, or lane numbers required to provide pick positions for the projected number of SKUs for the year. High demand SKUs have a three- to seven-day life cycle, separated into:

- A or high-demand SKUs
- B or medium- to high-demand SKUs
- C or low- to medium-demand SKUs

Carton Flow Rack Pick Line with Random SKU Profile

A carton flow rack pick line with random SKU profile concept has SKUs allocated to any pick position on a pick line. Pallet flow lanes and carton gravity flow lanes are located along the pick line. SKU allocation to a pick position is completed on a random basis and sku allocation to a pick position is not based on SKU life cycle curve or number of days that the SKU is available for sale. This means a wide SKU mix along a pick line. This feature has high-demand or A SKUs in one pick position with B and C SKUs in pick positions on both sides of the pick line.

The concept increases picker travel distance and time between two pick locations, which lowers picker productivity, increases travel distance and time to complete a pick line set-up, and periodic pick position replenishment transactions that means low replenishment productivity, per a SKU life cycle, increases travel time and distance to rotate or remove SKUs to other pick positions, and difficult to manage or establish picker zones to assure good picker productivity.

Carton Flow Rack Pick Line with Specific SKU Profile

A carton flow rack pick line with a specific SKU profile concept has SKUs allocated or profiled to specific pick position zones or cells on a pick line. A pick line has pallet flow lanes and carton gravity flow lanes that are located along a pick line. SKU allocation to a pick position is completed on a specific basis, such as the SKU life cycle curve or number of days that a SKU is available for sale. Most applications have:

- A SKUs allocated to pallet flow lanes
- B SKUs allocated to a single carton gravity flow rack lane or multiple carton flow rack lanes
- C SKUs allocated to a single carton gravity flow rack lane or shelf position

The concept features slightly decreased picker travel distance and time between two pick locations, which means higher picker productivity, slightly decreased travel distance and time to complete a pick line set-up, and periodic pick position replenishment transactions, per the SKU life cycle, requires travel time and distance to rotate or remove SKUs to other pick positions and improves management to establish picker zones to assure good picker productivity.

Vertical or Horizontal Picker Routing Pattern

In a pick-to-light concept that interfaces with a carton flow rack, static shelves, or a mobile cart pick concept, a picker routing pattern directs a picker to sequentially complete pick transactions from illuminated pick positions. Pick-to-light picker routing patterns (see Figure 4.12) are:

- Vertical: a vertical picker routing pattern on the right side of a pick aisle has the first pick position in the lower right-hand corner of a pick bay, adjacent to the upright support member. For each pick level bay on a pick line, a picker starts in the lower right-hand level position. With a vertical pick position stack, a picker proceeds to complete all pick transactions from the lowest pick level to the highest pick level. After all picks from this vertical stack within

HORIZONTAL PICKER PATTERN

BAY 02				BAY 01			
A04	A03	A02	A01	A04	A03	A02	A01
B04	B03	B02	B01	B04	B03	B02	B01
C04	C03	C02	C01	C04	C03	C02	C01
D04	D03	D02	D01	D04	D03	D02	D01

PICKER DIRECTION OR TRAVEL <<<<<<<<<<<<<<<

VERTICAL PICKER PATTERN

BAY 02				BAY 01			
D04	C04	B04	A04	D04	C04	B04	A04
D03	C03	B03	A03	D03	C03	B03	A03
D02	C02	B02	A02	D02	C02	B02	A02
D01	C01	B01	A01	D01	C01	B01	A01

PICKER DIRECTION OR TRAVEL <<<<<<<<<<<<<<<

Figure 4.12 Carton flow rack pick patterns: horizontal and vertical.

the same bay, the picker moves to an adjacent pick position vertical stack and performs all pick transactions. In each pick bay, the last pick position is in the upper right-hand corner. This picker routing pattern is repeated in each pick line. With a wide master carton mix it is difficult to maintain a vertical carton stack and to handle fast- and very fast-moving SKUs that require an entire level, and it is also difficult to maintain a golden zone.

■ Horizontal: a horizontal picker routing pattern is one in which a picker enters a pick aisle or pick zone and faces to the right at each bay, starts at the upper right-hand first pick position, across the bay level from right to left, drops to the next lower level right-hand position and continues across the bay from right to left and ends at the lowest level at the last pick position on the left of the pick bay. The horizontal picker routing pattern features are easy expansion, more pick faces in the golden zone, and understood by pickers.

Manual Paperless Pick Activity

A manually controlled paperless, pick-to-light or RF device pick method is a digital display pick instruction method. Digital display method options are:

- At a pick line start, a CO discreet identification is placed onto each CO carton exterior surface. A CO carton discreet identification has the same sequence as CO discreet identifications that were downloaded to a pick area microcomputer. On a pick line, a picker matches a CO carton and pick-to-light discreet identification that is shown on a zone controller. At a pick position a CO SKU pick position light is activated to match a CO carton discreet identification. As a CO identified carton travels on a pick line, for a picker each activated pick light represents a pick for a CO.
- At the entrance to each pick zone, a picker with a fixed position, hand-held or finger barcode scanner/RF tag reader reads a label on a CO carton or reads a CO discreet identification with a magnetized card and reader. A read transaction enters a CO identification into a microcomputer. The microcomputer activates each required pick zone pick lights.

As a picker arrives at a lighted pick position, he or she removes a SKU quantity from the pick position and presses a pick button or activates a sensing device that registers a pick. The pick quantity is reduced by one. If another pick is required at the same pick position, a picker repeats the pick activity until the digital display screen indicates zero. After all pick transactions, a picker transfers a CO carton to the next pick zone or onto a take-away conveyor. For the next CO, the picker repeats the pick position activation (barcode scanning, reading a magnetic card, or pressing a button).

The pick-to-light picker instruction format the preferred format for a single CO with a single or multiple pickers in a pick/pack operation. This pick/pack or pick/pass concept has a pick line with flow racks or shelves, a mobile cart, or carousel concept.

Disadvantages are cost, on-time COs, and replenishments, multiple lanes for fast-moving or high-cube SKUs and the need for a microcomputer, a smooth let down, or UPS electronic back-up system. Advantages are increases picker productivity and accuracy, easy training, pickers read numbers not a description, frees up a picker's hands, on-line pick line productivity review, and used on a pick/pack pick line that has one or multiple pickers for one CO.

RF Pick-to-Light Device with a Barcode Scanner/RF Tag Reader

A hand-held RF pick-to-light device with a RF barcode scanner/RF tag reader is a paperless pick concept. Components are barcode scanner/RF tag reader with a hand-held, wrist or finger light device or a spring-loaded cord, CO download to the RF device, bar code/RF tag label on each pick position or piece, and picked SKU-carrying container or transport concept.

After the RF device arm control component receives a CO download from a host computer, a picker places a barcode scanner onto a finger/RF tag reader. As a picker walks with a barcode scanner/RF tag reader in a pick aisle, the arm component shows the pick position. At the assigned pick position, a picker directs a finger barcode scanner/RF tag reader to read the pick position barcode/RF tag label and sends the label data to the memory component. The memory component on the arm registers a correct pick and on a display screen indicates the SKU quantity on a CO.

A picker transfers a SKU quantity from a pick position into a CO carton. After a pick, a display screen directs a picker to the next pick position that is on a CO. After a picker completes a CO or fills a CO carton, the carton is transferred from a pick conveyor onto a CO take-away conveyor travel path or onto a four-wheel shelf cart. With a completed CO, a picker starts another CO.

The features are paperless pick concept, minimizes reading, investment into barcode scanners/RF tag readers and program, single pick concept, and barcode/RF tag label on each pick position or SKU.

Sort-to-Light Customer Order

Sort-to-light is a paperless pick concept that is used to pick COs for A or B SKUs that are less than a master carton and C and D SKUs or SKUs with a small inventory quantity. In a sort-to-light concept, each sort-to-light position has a display screen that shows a pick quantity, next or completed CO. To have an efficient and cost-effective sort-to-light concept, the factors are:

- CO SKU arrival sequence at a pick station
- Sort-to-light position numbers, bay lamps, and sort-to-light cell or position layout
- CO pick or shipping carton make-up and label print and apply area
- CO take-away concept

Customer Order Arrival Sequence

In a small-item order-fulfillment operation, a sort-to-light concept has CO SKUs downloaded from a WMS computer to a MHS computer. A WMS CO download sequence to a MHS printer is to print a bar for each CO. The print sequence is based upon a design team selected SKU transfer sequence. Each printed barcode/RF tag is an instruction for an employee to transfer CO cartons/totes to a pick-to-light cell or position. After a MHS has printed a barcode an employee scans and transfer a CO barcode/RF tag labeled pick carton/tote into a sort-to-light cell or position.

A MHS transfers CO SKUs from a storage area to a sort-to-light pick cell or position and has controlled pick for CO SKUs. In a small-item order-fulfillment operation, the CO types are:

- Single line with one SKU and one piece
- Single line with one SKU and multiple pieces
- Multiple lines with one or more SKUs and pieces
- Multiple lines with one or more SKUs from two separate pick areas or zones

The first step to activate a sort-to-light concept is to have a WMS/MHS concept transfer CO SKUs in a controlled sequence from a storage area to a sort-to-light area in a controlled sequence. The feature ensures maximum CO orders are completed to create vacant sort-to-light positions in a cell for new COs and high picker productivity.

SKU sequence options are:

- SKUs for single-line COs with one SKU and one piece, single-line COs with one SKU and multiple pieces, SKUs for multiple-line COs with one or more SKUs that only have one pick in the sort-to-light pick area
- SKUs for multiple-line COs that are completed with two or more picks in a sort-to-light area
- Combine all CO order types

Single COs with One SKU and One Piece, Single Line COs with One SKU and Multiple Pieces and Multi-Line COs with Only One Pick in a Sort-to-Light Area

In a sort-to-light concept that has a SKU transfer sequence that brings SKUs for single-line COs with one SKU and one piece or single-line COs with one SKU and multiple pieces, as a SKU travels on the in-feed conveyor travel path, a barcode scanner/RF tag reader updates the SKU location and diverts a SKU to a pick cell and is stopped at a pick position. A barcode scanner scans/RF tag reader reads the SKU barcode/RF tag and sends a message to a microcomputer, which activates all sort-to-light positions that require that SKU piece. A picker removes a SKU quantity and transfers a SKU piece into each CO that is in an activated sort-to-light position in a cell. After SKU transfer to a CO carton, a picker presses the sort-to-light position pick button that completes a pick transaction. If a sort-to-light position display screen shows a completed CO, a picker presses the complete button and transfers the CO carton from the sort-to-light position to a CO take-away conveyor travel path. If a sort-to-light position display screen shows another SKU pick quantity, a picker repeats the pick activity until the display screen shows a completed

CO. A carton make-up employee labels, scans, and transfer another CO carton to a vacant sort-to-light position.

At a sort-to-light pick cell, after all picks for a SKU, a new SKU is transferred to the pick cell and the completed SKU is transferred to another pick cell. In another pick cell, a picker completes all pick transactions for COs in the pick cell.

After all CO picks at all sort-to-light pick cells are completed for a CO wave, SKU transport options are:

■ Return to the storage area. If there is available space on a pick area queue conveyor travel path, the SKU is sent to a storage area, which has the front storage positions reserved for these SKUs. For additional information, we refer the reader to the discussion of storage position strategy in this chapter.

■ Route to a queue conveyor for re-entry to the sort-to-light pick area. The design teams Options are (a) if a SKU is not required on a multi-line CO, it is sent from the sort-to-light area to the storage area and (b) if a SKU is required on a multi-line CO, and if there is available space on a queue conveyor travel path, the MHS transfers the SKU onto a queue conveyor travel path. If the conveyor travel path has a recirculation feature with barcode scanners/RF tag readers, the MHS concept ensures that the SKU is in proper sequence to match the WMS/MHS CO demand or download sequence. If there is space on the queue conveyor travel path and there is no recirculation conveyor feature, the SKU sequence cannot be made to match the CO demand or download sequence. This feature could have a slight negative impact on the CO completion.

Multiple Line COs That Are Completed with Two or More Picks in a Sort-to-Light Area

If a WMS/MHS has one CO group for multi-line COs after a single-line CO group is completed, the sort-to-light concept is set up to pick multi-line COs. The storage/MHS area releases SKUs to the sort-to-light area to match the WMS/MHS CO download to the sort-to-light concept. For multi-line COs, SKU release can be:

■ Random: a random sequence does not optimize the pick activity, and creates potential SKU recirculation and slow CO completion

■ The highest-volume SKU in a CO is the first SKU downloaded; the second SKU is the one that is needed to complete the greatest number of COs

An employee attaches WMS/MHS print labels to CO cartons, scans them, and transfers the cartons to a vacant sort-to-light position. As the first SKU arrives in the pick cell, at the entrance a scanner reads the SKU barcode/RF tag and activates each sort-to-light position display screen at each pick position that requires the SKU. One screen shows the SKU number and the other screen shows the pick quantity. A picker removes a SKU quantity, places it into a CO carton, and the SKU pick quantity is decreased by one. If another pick is required for the SKU, the pick quantity screen indicates the quantity. If the SKU pick quantity display screen is zero, a display screen shows next and the CO carton is transferred to a take-away conveyor travel path. After all SKU picks, the picker releases the SKU for travel from the pick cell to the next sort-to-light pick cell. After a new SKU arrives at the pick cell, the SKU barcode/RF tag is read and the sort to lights that have COs for that SKU are activated. The picker repeats the pick activity. After a second SKU is added to a CO carton, if the CO requires another SKU the display screen shows that next. If the CO is completed, the display screen shows completed and the picker transfers the CO onto a take-away conveyor travel path. The set-up employee applies a bar code to the carton and transfers the labeled carton to a vacant sort-to-light position.

Combined or All Customer Order Types

The WMS/MHS transfers SKUs for any CO from a storage area to a sort-to-light area. The concept has labels printed in a SKU sequence in which the highest-volume SKU is transferred first, the second SKU is the one that together with the first SKU completes the greatest multi-line CO number, and the third SKU is the one that completes the greatest number of multi-line COs. This SKU release scheme is repeated for each SKU until all COs are completed. After a WMS/MHS has sequenced the SKU release pattern, the WMS/MHS sequences the CO label print sequence to match the SKU releases.

An employee attaches, scans, and transfers CO cartons to a vacant sort-to-light position. As the first SKU arrives in the pick cell, a scanner reads the SKU barcode, and activates each sort-to-light position display screen. One screen shows the SKU number and the other screen shows the pick quantity. A picker removes a SKU and places it into a CO carton. The pick quantity is decreased by one. If another pick is required for the SKU, the pick quantity screen indicates the quantity. If the SKU pick quantity display screen is zero, a display screen shows next. After all SKU picks, the picker releases the SKU for travel from the pick cell to the next sort-to-light pick cell. After a new SKU arrives at the pick cell, the bar code is read and the sort to lights that have COs for that SKU are activated. The picker repeats the pick activity. When the CO is completed, a display screen shows complete, a picker presses the completed button, and transfers the CO onto a take-away conveyor travel path. The set-up employee applies a bar code to the carton and transfers a labeled carton to a vacant sort-to-light position.

Sort-to-Light Position Numbers, Bay Lamps, and Sort-to-Light Position Layout

Sort-to-light components are SKU and empty tray in-feed conveyor travel path from a storage area to the pick area, carton make-up, label print, label application, and labeled carton transfer to a sort-to-light position, SKU scanner, sort-to-light positions, and bay lamps that are (a) parallel or (b) perpendicular to the conveyor travel paths, which are a SKU return conveyor and a completed CO take-away conveyor.

A SKU in-feed conveyor travel path transfers SKU trays or master cartons from a storage area to a pick area. The number of SKU in-feed conveyors is designed to ensure continuous SKU flow from the storage area to the pick area. If an empty tray is required at a sort-to-light location, the MHS concept releases an empty tray onto the in-feed conveyor travel path for transport to a SKU transfer station. With a multiple SKU and CO pick concept, if SKUs have been picked for a single-line CO group and are also required for a multi-line CO group, they are recirculated if there is sufficient queue conveyor travel path, or they are returned to the storage area for deposit and later withdrawal. With a tray or tote SKU transport concept, a separate conveyor travel path for empty trays or totes is designed in the layout. Prior to a SKU transfer station, a conveyor travel path assures a continuous and queued empty tray or tote area.

Carton make-up, printing and applying labels, and transferring labeled cartons to a vacant sort-to-light position require space for empty carton pallets, a manual carton make-up area, a CO label printing machine, a work area for employees to apply labels, and a vacant position area for labeled cartons. Per a WMS CO download to a label printing machine (same as the SKU release from the storage area), a label is printed for each CO. Each label has a carton discreet human/machine-readable identification or CO identification plus a human-readable CO carton size. An employee makes up a carton of the size indicated by the computer and applies the CO label to the carton, which is scanned and transferred to a vacant sort-to-light position, with the label facing the pick aisle. The scan activity sends a message to the microcomputer that the WMS/MHS can start the SKU release or transfer activity.

Per the available floor area and if multiple carton sizes are used in the operation, the carton make-up options are:

- One carton size per sort-to-light cell. This option requires the largest floor area but provides maximum flexibility and permits the WMS/MHS to print any label at any sort-to-light cell.
- Selected carton sizes in each sort-to-light cell. This option uses the smallest floor area, has limited flexibility, requires the WMS/MHS to download specific carton sizes to one sort-to-light cell, and printer to have memory capacity to hold the WMS/MHS CO download. If there is a high carton volume at

one cell, two employees are assigned to the carton make-up, label, scan, and transfer activity.

If a carton make-up area with empty carton pallets is remote from a sort-to-light cell, design features are a label printing machine for each sort-to-light cell, and a transport device with proper sort-to-light identification for empty carton transfer from the make-up area to the assigned sort-to-light cell. Behind each cell there should be sufficient area for transport device staging and sufficient space for an employee to complete scan and carton transfer activities. With this option, a carton form machine makes up high-volume carton sizes.

The SKU in-feed conveyor for each sort-to-light cell that is a branch of the main conveyor travel path. From the main conveyor, SKU trays, totes, or master cartons are diverted onto a SKU in-feed or pick station conveyor. At the middle of each sort-to-light cell section is a barcode scanner/RF tag reader device that reads SKU bar codes/RF tags. The barcode/RF tag is sent to the sort-to-light microcomputer, which activates each sort-to-light cell bay lamp and sort-to-light display light. An activated bay lamp permits a sort-to-light picker to easily identify each sort-to-light shelf bay that requires a particular SKU. An activated sort-to-light display light indicates to a picker each CO or sort-to-light position that requires a SKU. After all picks are completed for the SKU, the picker releases it for travel from a scan station to travel on a main conveyor travel path to the next sort-to-light cell. From the queue conveyor travel path, a SKU is moved forward to the sort-to-light cell scan station.

The layout of the sort-to-light cell enables a picker to carry out pick transactions easily and quickly, with the shortest possible walking distance between the SKU scan station and sort-to-light cell positions. Options for layout of the shelfs and bay lamps are:

■ A parallel design, in which the shelf positions face the conveyor travel paths and SKU scan station. In this design, there is an aisle between the SKU scan station and sort-to-light shelf positions. The aisle permits an employee to complete a sort-to-light CO pick transaction. The scan station is in the middle of the shelf section, where an employee can easily recognize an activated sort-to-light pick position and has a short walking distance along the position faces between the SKU scan location and the sort-to-light position. The sort-to-light cell carton make-up and label area is behind the shelf locations. With a completed CO take-away conveyor travel path above the in-feed conveyor travel path, a short distance is required to transfer a completed CO from a sort-to-light position onto a CO take-away conveyor travel path.

■ A perpendicular design, in which sort-to-light shelf positions face each other across an aisle. The aisle and shelf position ends intersect with an aisle that is between the SKU scan station and sort-to-light shelf positions. The aisle permits an employee to complete a sort-to-light CO pick transaction. The SKU scan station is across from the midway position, which permits a sort-to-light employee to easily recognize an activated sort-to-light pick position and walk a short distance along the sort-to-light position faces between the SKU scan location and a sort-to-light position. The sort-to-light cell carton make-up and label area is behind the shelf locations. With a completed CO take-away conveyor travel path above the in-feed conveyor travel path, the employee does not have to walk a great distance to transfer a completed CO from a sort-to-light position onto a CO take-away conveyor travel path.

Completed Customer Order Travel Path

A completed CO conveyor travel path is above the in-feed conveyor travel path at an elevation that permits an employee to easily transfer a completed CO from a sort-to-light position onto a conveyor travel path. To optimize completed CO transfer onto a conveyor travel path and SKU transfer from a scan station, the completed CO is set back above the in-feed or SKU scan station.

Pick-to-Light Pick Position or Location Options

In a dynamic order-fulfillment operation, a pick-to-light or RF device CO pick concept is used on pallet flow lanes, carton gravity flow rack, and standard shelf pick positions. During a SKU profile to a pick-to-light or RF device concept, the basic profile criteria have:

- Large-cube and fast- to medium-moving SKUs from groups A and B allocated to pallet flow or carton gravity flow rack pick locations. A SKU is allocated to one carton gravity flow rack level or to one carton gravity flow rack lane, depending on how fast-moving it is.
- Small-cube and medium- to slow-moving SKUs from groups B and C are allocated to pick-to-light shelf pick locations.
- Remaining SKUs from the C group are allocated to shelf non-pick-to-light pick positions or to a pick-to-light concept that has one light per shelf. In the C group, SKUs are paper batched, picked by SKU or CO number sequence and are placed onto a mobile or stationary location.

The pick-to-light concept is used in a pallet or carton gravity flow rack or standard shelf pick position concept that has a light for each SKU. To assure maximum picker productivity and minimum pick concept investment the pick-to-light concept has a pick-to-light and SKU allocation to pick position options based on SKU movement and cube profile for a pick line. For optimum results, a pick line has a pallet flow rack, carton gravity flow rack, and standard shelf pick position combination.

Pallet or Carton Flow Rack or Lane Options

As previously mentioned, a pick-to-light pallet flow rack or lane profile or SKU allocation has one SKU per pallet flow lane, with one pick pallet and two ready reserve pallets. A standard carton gravity flow rack bay has three to four pick levels and is four to five cartons or pick lanes wide. A carton gravity flow lane has a minimum of five long master cartons, seven medium length master cartons, and nine short master cartons in ready reserve. SKUs are allocated to pallet or carton flow rack pick positions based on the following factors (see Figures 4.13, 4.14, 4.15, 4.16, and 4.17).

■ One SKU, one flow rack level. This means that one fast-moving SKU from the A group occupies an entire rack bay horizontal level and has one pick position. Design options are (a) one pick light in the middle of the rack level,

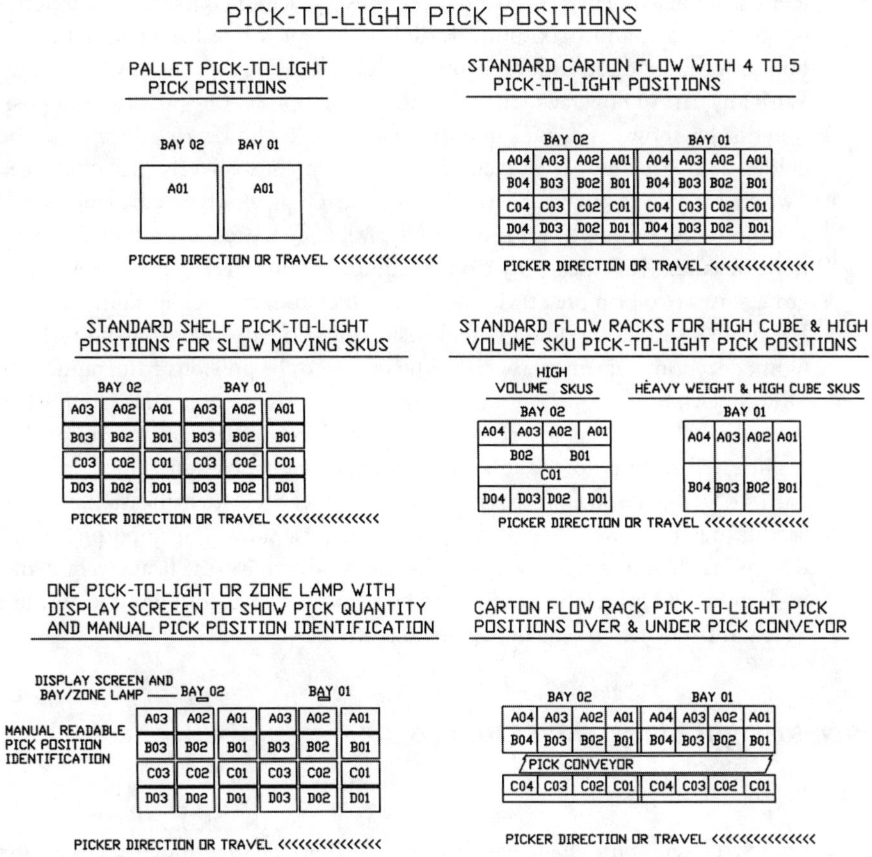

Figure 4.13   Pick-to-light pick positions.

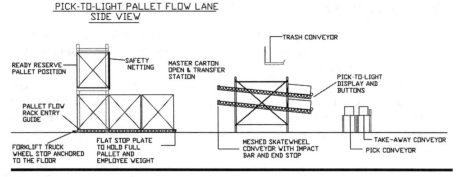

Figure 4.14 Pick-to-light pallet pick position: side view.

which minimizes cost or (b) four to five pick lights for a carton or pallet flow rack level, with a middle light. Inactive pick lights and replenishment position identifications are covered. The concept has a discipline replenishment activity, higher cost, and maximum flexibility. If a pick line has pallet lanes, an option for large-cube and fast-moving SKUs is to use a pallet flow rack lane. With any carton flow rack lane SKU allocation option, one pick-to-light position display shows SKU pick quantity. Fast-moving SKUs are allocated to the golden zone. In a four-level rack bay, the golden zone is levels two and three.

■ Two SKUs on one flow rack level. In most carton gravity flow rack lanes, there are four to five standard cartons; on each level one SKU occupies two pick positions or lanes. This means that one medium-moving SKU from the A and B groups uses two or more gravity flow rack lanes and one pick-to-light position. For more information on SKU replenishment and picker control and pick-to-light concept management, we refer the reader to the previous paragraph.

■ One SKU, one carton flow rack lane or pick position on a horizontal bay level. This concept has one slow-moving SKU from the B and C groups that is allocated to one flow lane and one pick-to-light position. This is a standard SKU allocation concept that has one (same) SKU identification on both carton gravity flow rack replenishment and pick sides. For good pick position management and good replenishment productivity, the heavyweight and high-cube SKUs are allocated to the bottom pick level, and lightweight, tall SKUs are allocated to the highest pick level.

Pick-to-Light Static Shelf Options

Options for SKU allocation to shelf pick positions are:

■ One SKU, one static shelf level. This means that one slow-moving/large cube SKU from the C SKU group occupies one entire shelf level and has one pick position. A pick-to-light position display shows the SKU pick quantity. The

PICKLINE LAYOUT OPTIONS

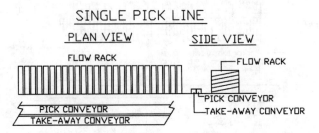

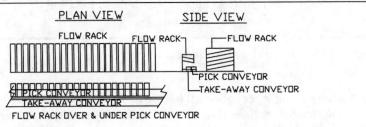

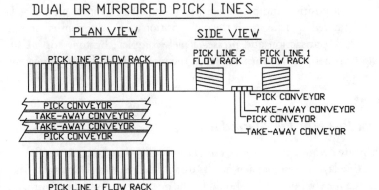

Figure 4.15 Pick line layout options: single pick line; pick line over and under pick conveyors; and dual pick lines plan view.

SKUs are allocated to the golden zone. In a five-level shelf, these are levels 2, 3, and 4.

■ Three SKUs for each shelf level. This means that each SKU from the C group has three pick lights in a 3-ft. shelf width. There is one SKU per pick-to-light position. The pick-to-light position display shows the SKU pick quantity.

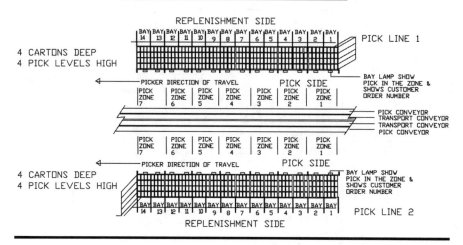

Figure 4.16 Dual pick-to-light lines.

■ One pick light per horizontal level, with three or more human-readable pick positions on each shelf level. This is used with small cube and very slow moving SKUs. This means that each shelf level has one pick-to-light that is for all pick positions on a shelf and when illuminated for a SKU the display shows pick position and pick quantity. When a bay lamp or pick zone lamp is illuminated, a picker reads the pick position and pick quantity, completes a SKU pick transaction and presses a pick complete button. For a CO, if there is another pick for the shelf level, the pick-to-light display shows the next SKU pick position and pick quantity.

Pick-to-Light Pick Line Designs

There are design options for layout of each pallet flow lane, carton flow lane, or shelf pick line layout design; each has its disadvantages and advantages. The one that is selected for your operation should be the one that best satisfies your pick line objectives. The pick-to-light pick line options are:

■ A single pick line opposite the aisle from pick and take-away conveyors
■ A single pick line opposite the aisle from the pick and take-away conveyors, with two pick levels above and one pick level below the conveyors
■ Dual pick lines, each pick line opposite the aisle from a pick conveyor, with one take-away conveyor
■ Dual pick lines, each pick line opposite the aisle from one pick conveyor and two take-away conveyors

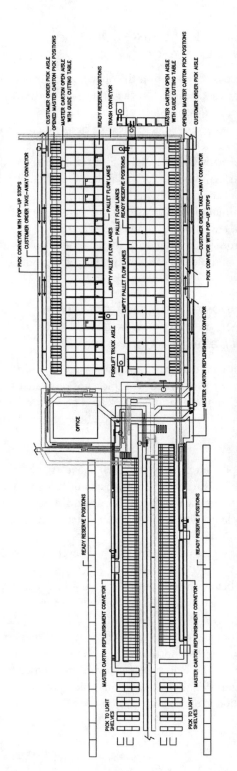

Figure 4.17 Pick line plan view.

Single Pick Line Opposite the Aisle from a Pick and Take-Away Conveyor

A single pick line with a pick conveyor on the opposite side of the aisle is designed with pick positions that face an aisle. On the other side of the aisle (30 to 36 in. wide) is a pick conveyor and adjacent to it a take-away conveyor for completed COs. During the pick activity, a picker walks in the aisle perpendicular to the pick faces and pick conveyor. To complete a pick, a picker identifies an activated pick light, removes a SKU piece, presses the pick light, and transfers the SKU piece into a CO tote or carton. A trash conveyor is located either above the pick and take-away conveyors or on the top of the pick line frame.

The features are minimal physical picker effort, easy to identify a pick, easy to identify pick zone limits, fewest pick positions per square foot, trash conveyor location is flexible, easy to profile a pick line with all pick line set-up, and replenishment transactions to one side, and pickers can achieve budgeted productivity.

Single Pick Line Opposite the Aisle from Pick and Take-Away Conveyors with Two Pick Levels above and One Pick Level below the Conveyors

A single pick line with a pick conveyor on the opposite side of the aisle and pick position above the conveyor is designed with pick positions that face an aisle and two pick levels above the conveyors and one pick level below the conveyor. The pick line layout is very similar to the layout described above, except for three features. These are:

- Two short-length pick levels above the pick conveyor. The elevation of these pick levels permits COs to travel on the pick and take-away conveyor travel paths and the bottom pick level extends outward to the middle of the pick conveyor. To complete a pick transaction, a picker leans against the conveyor frame and reaches outward.
- A trash conveyor is located above the two pick levels or on the standard flow rack frames.
- The pick position below the conveyors extends to the conveyor frame edge. The flow lanes are longer than the flow lanes above conveyors.

The features are additional employee reach effort, more difficult pick line profile due to multiple flow lane lengths, short flow lanes require more frequent replenishments, pick positions above and below the conveyors are profiled with lightweight and small-cube SKUs, additional cost, additional pick positions per square foot, and trash conveyor location is optional.

Dual Pick Lines, Each Opposite the Aisle from a Pick Conveyor and One Take-Away Conveyor

A dual or mirrored pick line with one completed CO take-away conveyor concept has two pick-to-light lines, two pick conveyors, and one completed CO take-away conveyor travel path. Each pick line is basically a single pick line (as previously described). Each pick line has access to a pick conveyor. Both pick line conveyors have access to one completed CO take-away conveyor. A trash conveyor is located above the conveyors or on the top of the flow rack frame.

The features are the same as the single pick line, with two pick lines handles a higher CO number, additional costs for conveyor and pick positions, two pick lines in a small area and one completed CO take-away conveyor could create queues.

Dual Pick Lines, Each Pick Line Opposite the Aisle from a Pick Conveyor, and Two Take-Away Conveyors

A dual or mirrored pick line with two completed CO take-away conveyor concept has two pick-to-light lines, two pick conveyors, and two completed CO take-away conveyor travel paths. The pick line is basically a dual or mirrored pick line (as previously described). Each pick line has access to a pick conveyor and each pick line conveyor has access to a completed CO take-away conveyor. A trash conveyor is located above the conveyors or on the flow rack frame top.

The features are the same as the dual pick line, with two pick lines handling a higher CO number, additional costs for conveyor and pick positions, two pick lines in a small area, and one completed CO take-away conveyor minimizes queues.

SKU Allocation to Pick-to-Light Standard Shelves

A standard shelf bay is four to five pick levels high, three to four carton or pick lanes wide and a master carton deep. Options for SKU allocation to pick positions on a pick-to-light standard shelf are:

- One SKU, one shelf level. This means that one slow-moving/large-cube SKU from the C group occupies one standard shelf level and has one pick-to-light pick position. For pick position management and replenishment and order picker control information, we refer the reader to the discussion of one SKU on a carton flow rack level in the previous section in this chapter. The SKUs are allocated to a golden zone. In a five-level shelf, shelf levels 2, 3, and 4 are the golden zone pick levels.
- Three or four SKUs or pick lights for each shelf level. This means that each SKU from the C group has three to four pick lights on one shelf level width.

There is one SKU per pick-to-light position and a pick-to-light position display shows the SKU pick quantity.

■ One pick light per shelf horizontal level and there are three to four human-readable pick positions on each shelf level. The SKUs are small cube and very slow moving. This means that each shelf level has one pick-to-light that is for all pick positions on the shelf and when illuminated for a SKU the pick-to-light display shows a pick position and pick quantity. When a bay lamp or pick zone lamp is illuminated, a picker reads the pick position and pick quantity, completes the SKU pick transaction and presses the pick complete button. For a CO, if there is another pick for the shelf level, a pick-to-light display shows the CO SKU pick position and pick quantity.

The features are lower investment, higher picker productivity, paperless concept, employee to read a human-readable pick position label and complete a transaction training, mis-picks, and high hit density and hit concentration.

Customer Order Container In-Feed onto a Pick Line

After a CO pick carton has been identified or attached to a CO, an important feature is in-feed to the pick line or start station. Options for transporting CO identified cartons to a pick line are:

■ In-house transport conveyor travel path direct to a pick conveyor
■ Divert from an in-house transport or CO take-away conveyor travel path onto a pick conveyor

When a CO carton travels direct;y from an in-house transport conveyor travel path to a pick conveyor, the design has one CO carton start pick location. Before the first pick station or zone there is sufficient CO carton queue conveyor, the carton transport travel path is direct to a pick line conveyor travel path, carton flow is controlled by a flat stop device or zero pressure conveyor, and the completed CO carton take-away conveyor travel path starts at the first pick position.

To divert a CO identified carton from an in-house transport or CO take-away conveyor travel path onto a pick conveyor layout, the pick area has multiple barcode scanners/RF tag readers and CO start pick locations. With this design, before the first pick station or zone there is sufficient CO carton queue conveyor with a barcode scanner/RF tag reader and divert device to transfer an assigned CO identified carton to a pick line zone. The pick line is separated into pick zones, with each pick zone having a barcode scanner/RF tag reader, and a divert device. The CO carton queue conveyor and completed CO carton take-away conveyor travel path start at the first pick position.

Pick Carton Travel on a Pick Line

A high-volume pick/pack or pick/pass CO identified carton pick concept has a CO carton pick conveyor travel path along carton flow rack bays and shelves, a picker aisle between the pick conveyor and pick positions, and a completed/partially completed CO pick carton take-away conveyor adjacent to the pick conveyor. Pick line alternative designs (see Figure 4.18) are:

- One pick section, at the start of a pick line, which has a CO pick carton entrance with a barcode scanner/RF tag reader and divert device and CO conveyor. All pick cartons are diverted onto one pick line conveyor for travel through the entire pick line (the pick line has "no cell or section skip").
- Two pick cells on one pick line:
  - One pick cell with a barcode scanner/RF tag reader and divert device and CO conveyor at the pick line entrance. If a CO has a pick instruction for a SKU from pick cell 1, the CO pick carton is diverted onto a pick cell 1 pick line conveyor travel path. If a CO has a pick instruction for a SKU from pick cell 1 and pick cell 2, a CO pick carton is diverted onto a pick cell 1 pick line conveyor travel path. After completion of pick cell 1 transactions, the carton is transferred onto the CO take-away conveyor travel path, where a barcode scanner/RF tag reader reads the CO carton identification and activates a divert device to transfer the carton from the take-away conveyor onto the pick cell 2 conveyor.
  - The second pick cell is located in the middle of the pick line, with a CO pick carton entrance with a barcode scanner/RF tag reader and divert device, referred to as "one cell or section skip." If a CO carton does not have a pick transaction in pick cell 1, it travels past pick cell 1 and is diverted from the CO take-away conveyor travel path to pick cell 2.
- Multiple pick cells or sections. A multiple pick cell concept has two or more pick cells on one pick line and each pick cell zone has a barcode scanner/RF tag reader and divert device. The CO conveyor travel path is at the pick line entrance. If a CO has a pick instruction for a SKU from pick cell 1, the CO pick carton is diverted onto the pick cell 1 conveyor. If a CO has a pick instruction for a SKU from pick cells 1, 2, and 3, the CO pick carton is first diverted onto the pick cell 1 conveyor. After completion of all pick cell 1 transactions, the CO carton is transferred onto a take-away conveyor travel path, where a barcode scanner/RF tag reader reads the CO carton identification and activates a divert device to transfer the carton from the take-away conveyor onto pick cell 2 conveyor. The activity is repeated for pick cell 3.

A pick line or pick cell is separated into pick zones, depending on the pick line layout and pick cell or pick section length.

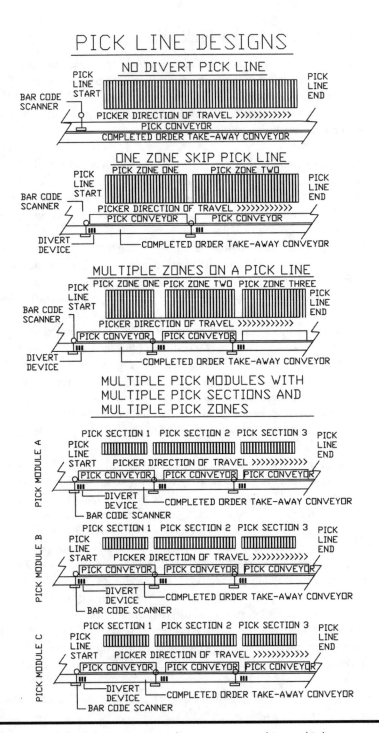

Figure 4.18 Pick line alternatives: no divert, one zone skip, multiple zones, and multiple modules.

No-Divert or No-Zone-Skip Concept

A no-divert or no-zone-skip concept has a pick conveyor travel path that moves past all the pick positions in a zone. In most pick applications, a CO carton train is in a sequence that was entered or read by a pick zone barcode scanner/RF tag reader. The CO numbers are sequenced in the pick zone microcomputer queue and appear on the appropriate pick zone display light and trigger the associated pick position bay lamps.

After a CO container (CO container train) is on a pick conveyor travel path a CO carton is passed by each picker from one pick zone to the next pick zone. The pick conveyor travel path is a non-powered or zero-pressure roller conveyor with a pop-up stop. For best picker control of CO carton train movement, manual stops or pop-up stops are located at the end of each pick cell. These ensure that a picker has ample time to complete all pick transactions and control the transfer of the carton train to the next pick cell. In a section with slow-moving SKUs, it permits one picker to pick across multiple pick cells. If a zero-pressure conveyor is an option, the pop-up stop controls are at multiple locations within easy picker reach in the pick cell.

When a CO carton is completed in one pick zone, a pick-to-light digital display signals to a picker a completed message. With this CO pick completed signal, the picker transfers or pushes the completed CO carton from the pick conveyor onto the take-away conveyor travel path. If a CO carton is not completed in one pick zone, a pick-to-light digital display signals to a picker the next pick message. With this next pick message signal, a picker transfers or pushes a partially completed CO container along the pick conveyor from one pick zone to the next.

Each pick zone has carton gravity flow rack and standard shelf bays. Each rack bay has an individual zone controller and bay lamp. This pick-to-light feature permits a pick line or pick section to have variable pick zones that are expanded or shortened to match CO SKU volumes.

A CO carton train moves or is pushed along the pick conveyor travel path from the high-volume SKU pick zones to the low-volume SKU pick zones. In the high-volume SKU pick section, pick zones are short one to two bays. In a high-volume section or pick line start, a CO carton train has a large number of cartons. In a low-volume SKU pick section, the pick zone is two to four carton flow rack bays or perpendicularly arranged standard shelf bays. With most COs, the majority of SKU pick transactions occur in the high-volume pick zones. As CO cartons arrive in the low-volume area, the CO train is shorter, as completed COs are pushed forward (transferred) from the pick conveyor travel path to the CO take-away conveyor travel path.

One-Zone-Skip or Two Divert Concept

A one-zone-skip or two divert concept has a major pick line divided into two minor pick sections. The first pick section is at the start of the pick line and the second pick section is in the middle of the pick line. On a CO container travel path, a barcode scanner/RF tag reader reads each container symbology and CO containers are diverted from the in-house transport conveyor travel path onto the appropriate pick line conveyor travel path. The pick conveyor travel path is from the first pick zone. If a container is not required at the first pick line, it travels on the transport conveyor travel path and is read by the barcode scanner and diverted onto the second pick line.

After a CO container (CO container train) is on a pick conveyor travel path the container is passed by each picker from one pick zone to the next pick zone. A pick conveyor travel path is a non-powered or zero-pressure roller conveyor. CO handling containers are the same as those used in a no-divert or no-zone-skip concept.

Multiple Zones

The multiple zone concept has a pick line separated into zones, with each zone being a number of flow rack bays. On a CO container travel path, a barcode scanner reads each container symbology and as required the appropriate CO container enters (diverts) from the in-house transport conveyor travel path onto the appropriate pick line zone conveyor travel path. The pick conveyor travel path is only in front of the first pick zone. Upon completion of all picks in a pick zone, the CO container is pushed forward from the pick conveyor onto the co take-away conveyor travel path. If a container is required at the next pick section, a container bar code/RF tag is read by the second pick zone barcode scanner/RF tag reader and is diverted onto the second pick zone. If a container on the in-house transport travel path is not required at a next pick section, the container travels on the in-house transport conveyor and its bar code/RF tag is read by the appropriate pick zone barcode scanner/RF tag reader and is diverted onto the appropriate pick zone or transported to the next pick line activity station.

In each pick zone, a CO container travels on a pick zone pick conveyor travel path and after all picks, a CO container is pushed forward from a pick conveyor path onto an in-house transport conveyor travel path.

Of the three methods described above for CO container flow through a pick line, the one divert concept requires the lowest investment in divert devices and barcode scanners, maximum flexibility to have a variable pick zone length, with a push-off of completed CO containers;the CO container train is easy to push over a low-volume pick line section and improves pick line profile and good picker productivity.

The one or two divert or zone skip concept requires investment in two divert devices and barcode scanners, less flexibility to have a variable pick zone length, with shorter pick line, a CO container train is easy to push over the low SKU section, additional time to complete a pick line profile and good picker productivity and picker training.

In addition to the above features, a multiple pick zone concept requires a higher investment in pick line conveyor divert devices and barcode scanners/RF tag readers; the pick line has the greatest pick cell number and pick line has a requirement for pick cell pop-up end stops.

Multiple Pick Modules Separated into Three Sections

An order-fulfillment operation that has a large number of COs and changes the SKUs on a pick line each day requires a pick area with multiple pick modules. Each module is a pick line that is separated into three sections, and each section has pallet flow and carton gravity flow racks or standard shelves that make up pick zones.

A three-pick module with three sections designs maximum flexibility to match the CO and SKU number. On a day with a large CO and SKU number, this concept permits an order-fulfillment operation to activate all nine pick sections. On a day with a medium CO number and fewer SKUs, the operation is able to activate two pick modules with six pick sections. When there are a few CO numbers and SKUs, one pick module and three pick sections can be activated.

With this concept SKUs are allocated or profiled to specific pick positions in each pick line section. Each pick line section has pallet flow lanes, carton gravity flow lanes, and standard shelves that are located along a pick line. SKU allocation to the pick positions is completed on a specific basis, depending on the SKU life cycle curve or number of days that the SKU is available for sale. Note that the pallet pick area can also be separated into pick zones. . With specific SKU allocation to a pick line, most applications have:

- First day, high-demand or A SKUs allocated to a pallet flow lane
- Second day, medium- to high-demand or B SKUs allocated to a single carton flow rack lane or multiple carton flow rack lanes
- Third day, low- to medium-demand or C SKUs allocated to a single carton flow rack lane

Use of this concept results in decreased picker travel distance and time between two pick locations, decreased travel distance and time to complete a pick line set-up and pick position replenishment transactions. Depending on the life cycle of a SKU, it requires travel time and distance to rotate or remove SKUs to the proper pick position, allowing maximum management flexibility of the pick area and improving management ability to manage or establish picker zones to ensure good picker productivity. It requires a high investment in multiple divert devices and barcode scanners/RF tag readers, and a strict pick zone length with little variation. If there is no pick, each picker pushes off a completed or partial CO container, time to complete a pick line profile, and low picker productivity and picker training.

Push Back Pick Position Concept

A push back pick concept is a unique or modified carton flow rack pick concept (see Figure 4.19). If your order-fulfillment facility floor space is inadequate for a standard carton flow rack replenishment aisle, a 5- or 8-ft. deep carton flow rack is designed as a push back rack, with end stops at the back end and at the front of the carton flow rack conveyor lane.

A push back carton flow rack concept has no replenishment aisle, a steeper slope to the carton conveyor travel path or lane, all SKU replenishment and pick activities are performed from the same aisle, and LIFO UOP rotation.

A push back pick concept is designed as a single carton flow rack row against a wall or as a back-to-back carton flow rack.

Disadvantages and advantages are the same as the standard carton flow rack concept except that the additional disadvantages are LIFO UOP rotation and smaller inventory quantity in a pick position.

Chute or Slide Pick Position Concept

A chute or slide pick position concept has the same structural members as standard pallet rack or shelf pick concepts except one. With a chute or slide rack or shelf pick position, each pick position has a solid deck and side guards to support loose or packaged merchandise. If SKUs are in a container or carton, the solid or wire-mesh deck is a chute. With wire mesh, the top wire strands are one piece and in the pick position depth. The wire-mesh surface is pitched toward a pick aisle. All chute or slide flow lanes have an end stop at a pick face. A slide or chute top surface is coated sheet metal, plastic, or coated wood that is sloped or pitched toward a pick aisle. To assure container, carton, and loose SKU movement over a slide or chute surface, the bottom exterior surface of the carton, container, or SKU and the chute or slide surface have a low coefficient of friction. With most loose small items, the coefficient of friction is very minimal but a front barrier and high side guards retain SKUs in a pick position.

Features of the slide or chute concept are steeper slope, three to four pick levels high for a total of 15 to 20 pick faces per bay and proper pick position identification; either the shelf front lip has a pick position identification that faces the order picker, or a wood or metal surface is placed at the front of the deck. Design options for a shelf slide pick concept are:

- Fixed single deep shelf unit between a replenishment side and pick side a 6 to 7 in. elevation difference above the floor surface
- Mobile single deep shelf unit with casters and wheels under each upright post and between a replenishment side and pick side with the same elevation difference above the floor surface

Figure 4.19   Push back rack, container and drawer.

- Two deep fixed shelf units with a front shelf pick position as a slanted shelf; between the replenishment side and pick side is the same elevation difference above the floor surface
- Double-deep shelf unit that has both shelf sections slanted with the same elevation difference above the floor surface

The slide pick concept handles loose SKUs, containers, and cartons, has a small to medium inventory on-hand, handles medium- to slow-moving SKUs and medium to lightweight SKUs. With separate replenishment and pick aisles, a slide pick position provides FIFO UOP rotation. With one aisle that is a replenishment and another aisle pick aisle, UOP rotation is LIFO. A slide pick position concept is very flexible. Per your design parameters, a slide concept has the same design layout alternatives as standard carton flow racks with a replenishment aisle and pick aisle and as push back carton flow racks with one aisle for both replenishment and pick activities.

The disadvantages and advantages are the same as a standard carton flow rack except for additional disadvantages including smaller inventory in a pick position, fewer pick positions per opening or bay, and steeper slope or pitch.

Cantilever Rack and Pegboard Pick Position Concepts

Cantilever rack and pegboard pick concepts are used for long SKUs that are required to lay flat in a pick position and cannot be handled by one of the previously mentioned pick position concepts. A cantilever rack has two upright posts, top and back bracing members, bases that are 11 in. thick and chisel-shaped arms with a deck or a basket. The arm is 6 to 7 in. thick at the base and 4 to 5 in. at the front. With a solid deck or basket secured to the arms, a pick position handles long SKUs. If your SKU is not stackable on a deck, a two- or three-sided container or open ended basket holds the SKUs in a pick position. Since a cantilever rack arms support a solid deck or shelf, the front of the deck or shelf has sufficient space for pick position identification.

A pegboard concept has pegs that are secured against a flat vertical surface. To form a pick position, the pegs extend outward. A pegboard handles loose or packaged SKUs with a loop, which is slipped over the peg and the SKU is retained in a pick position. The peg permits lightweight SKUs to be flat in the pick position. The best pick position identification location is directly above a peg.

Cantilever racks have a high cost; both concepts have fewer pick positions per square foot, both have a low inventory quantity in a pick position, and both have limited application in a dynamic small-item order-fulfillment operation.

Drawer Pick Position Concept

A drawer pick concept is a large fixed wall container with many smaller interior containers or drawers. Each drawer has the ability to move outward to permit an employee to complete a pick or replenishment transaction. Each drawer can be separated into smaller compartments, each of which has a pick position identifier. The drawer pick concept provides 15 to 20 pick positions per drawer with a 400 lb. maximum load capacity and the large fixed wall container has the height to house eight to ten drawer levels.

The features provide 150 to 200 total pick positions per large container, which is a very large number of pick positions in a small area. The drawer pick concept is best designed for very small loose or package SKUs with a small inventory on hand. The other UOP characteristics are LIFO UOP rotation, very lightweight and slow-moving SKUs, and restricts dust and dirt accumulation on the SKUs.

The drawer pick concept has a variety of sizes:

- Large fixed wall containers that are
  - 24, 30, 48, and 60 in. wide
  - 32, 40, 46, and 60 in. high
  - 24 or 28 in. deep
- Individual drawer sizes are
  - 18, 24, and 28 in. deep
  - 24, 30, 48, and 60 in. wide
  - 3, 4, 5, 6, 7, 8, 9, 10, 11, 12, 13, and 14 in. high

A drawer cabinet is made from 14-gauge sheet metal that is coated with a standard color that is either gray, charcoal, dark blue, red, light blue, or beige.

Other drawer pick concept features are a drawer handle that allows your picker to open or close a drawer, a locking mechanism for additional security, enclosed to minimize dust and dirt collection on SKUs, and has a mobile drawer that has a caster and wheel under each corner.

The drawer pick concept is designed as a single drawer row or as a back-to-back drawer row. The aisle width has sufficient width to open or extend each drawer into the aisle and for a second picker to pass in the aisle as a drawer is open. Alternative layout designs are drawer cabinet in a rack bay or drawers installed in a standard shelf bay.

A drawer concept's features are high investment with a low cost per pick position, but a very small SKU inventory, small area, improved hit concentration and hit density for slow-moving SKUs, and additional security.

Disadvantages are small inventory in a pick position, cost and additional employee activity to open and close a drawer. Advantages are excellent for very small and slow-moving SKUs, larger pick position number per square foot, good SKU accessibility, keeps dirt and dust off the SKUs, and lockable.

Stacked Containers Pick Position Concept

This concept has self-stacking hardened plastic, metal, or wood bins that have a full-height rear wall, two full-height side walls, and half-height front wall. The full-height walls are designed to interlock with the container above and the rear wall and two side walls have the structural strength to support the weight of the stacked containers and the SKUs they contain. The open front permits a SKU pick or replenishment transaction, acts as a front barrier to retain SKUs in a container, and provides sufficient space for pick position identification and sufficient interior cube space to hold an inventory quantity.

A hardened plastic container is a flexible SKU pick position that is used in a floor stack pick layout or a standard decked pallet rack or shelf concept with structural strength to support stacked containers. Each decked rack or shelf level clearance permits an employee to complete a pick or replenishment transaction. Options are:

- Half container front that has a container remain in a position to complete a pick transaction
- Full container front that has an employee remove a container to complete a pick transaction

When you consider a hardened plastic container for your pick operation, in addition to the standard bin considerations that were previously mentioned in this chapter, the other important considerations are number of containers in a stack, maximum SKU load weight for each container, and container actual weight, available space, and how the bin is handled or supported in a pick position. The alternatives are:

- Flat or bottom exterior surface on a decked rack or shelf
- Stacked on another container top with a stacking lip
- From a hook at the rear side and under a rear wall lip
- Pick position identification size and type
- Interior cube and actual internal and external dimensions
- Container front options are full front, half front, no front pull handle, or front pull handle

A self-stacking container pick position concept as a stand-alone floor stack pick position concept is used to handle very small SKUs, slow-moving SKUs, and SKUs that have few COs. As a stand-alone floor stack pick position concept, a container is a single row or back-to-back row. An alternative container single row concept is to hang a container from a wall or from another structural member that is set on four casters and wheels. When a container is used in a hand-stack decked pallet rack, wide span shelf, standard shelf, or carton flow rack pick concept, the container is more commonly used in a small-item order-fulfillment operation with low-volume or small SKUs.

Disadvantages are cost, possible collapse from being struck, for efficiency all container openings are at the same elevation above the floor surface, and difficult to perform replenishment or pick transactions at the lowest and highest pick levels. Advantages are increased pick position number per pick aisle square foot, improves pick area appearance, and long life.

Unique Small-Item Pick Position Considerations

When unique small-item SKUs are allocated to a pick position and need to comply with local codes and company policy, the important considerations are:

- Hazardous or combustible SKUs are allocated to a pick line or pick aisle pick position that is protected with a wire-meshed barrier, sheet metal enclosure, or a fire wall enclosed area with a containment chamber, pit, or pan.
- High-value SKUs are allocated to a pick line or pick aisle pick position that has a lockable wire-meshed cage, enclosed cabinet, or area with controlled employee access and security cameras.
- Toxic SKUs are allocated to a pick line or pick aisle pick position that is separate from other SKUs.

Pick Methodologies

Objectives of pick methods are to:

- Transport a picker to a pick position
- Release a picked SKU from a pick position into a tote or onto a belt conveyor travel path
- Transport a picker and picked SKUs from a pick area

The pick methodologies are:

- Manual pick concepts
  - Picker walks or rides to a pick position
  - Pick position moves to a picker pick station
- Mechanized pick concepts

- Picker moves to a pick position transfers a picked SKU onto a belt conveyor travel path, into a tote that is on a pick conveyor travel path or into a four-wheel cart
- Pick position moves to a picker, who transfers a picked SKU onto a belt conveyor travel path, into a tote that is on a pick conveyor travel path or into a four-wheel cart
- Automatic pick machine concepts
  - A-frame pick machine
  - Itematic pick machine
  - Robo pick machine

Automatic pick machine concepts have a picked SKU released from a pick position into a tote or between two cleats onto a belt conveyor travel path.

Common characteristics of the various pick concepts are design parameters, economic justification factors, and operational factors, including daily average and peak CO number, pieces per CO, lines per CO, SKU volumes, and CO volume and customer growth, SKU growth, SKU physical characteristics and classification, SKU value, labor economics and investment, and customer service standard.

Manual Pick or Order-Fulfillment Concepts

A manual pick concept is a simple and most flexible pick concept. Components of a manual pick concept picker instruction, pick concept, pick position, and transport concept.

Picker instruction formats (see Figures 4.20 and 4.21) are:

- Manually printed paper document instruction
- Machine-printed paper document and self-adhesive label instruction
- Voice-directed instruction
- RF device or pick-to-light digital display instruction

Picker instruction format differences and similarities are:

- Paper document and self-adhesive label instruction forms are paper and voice-directed and RF device to pick-to-light digital display instruction formats are paperless.
- Presentation to a picker. The pick instruction has several large and bold alpha characters or numeric digits that are manual- or machine-printed onto a paper document or self-adhesive label or appear on a digital display. All pick instruction formats have a picker read a pick instruction. A RF device or barcode scanner format has a picker point a hand-held, wrist, or finger scanner device at a bar code and read a pick instruction. A voice communication pick instruction has an employee hear a pick instruction.

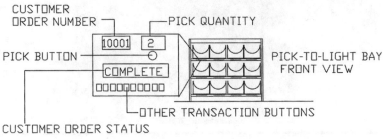

CUSTOMER
ORDER NUMBER ┐   ┌─PICK QUANTITY

PICK BUTTON ─────── │10001│ │2│   PICK-TO-LIGHT BAY
                        ○   FRONT VIEW
              │COMPLETE│
              □□□□□□□□□□
                    └─OTHER TRANSACTION BUTTONS
CUSTOMER ORDER STATUS

Figure 4.20 Manual pick instructions: paper, label, and pick to light.

- Content. With a paper document or self-adhesive label, the information content includes, in addition to the previous information, a CO human- or human/machine-readable discreet identification. With a voice-directed instruction, a pick instruction is a verbal CO identification. With a RF device or pick-to-light digital display instruction, a CO discreet identification appears on a CO carton and on a display device and is read by a picker, a finger/wrist, or hand-held scanner.
- Pick verification. To verify a completed pick transaction (a) with a paper document, a picker places a mark on a paper document, (b) with a self-adhesive label instruction, a picker removes a label from a self-adhesive backing

Figure 4.21 Manual pick instructions: RF and voice directed.

and places it on a SKU, (c) with a RF device or pick-to-light digital display instruction, a picker presses a button or breaks a light beam, and (d) with a voice-directed instruction, there is no pick transaction verification.

Similarities in manual pick instruction formats include:

■ Presentation arrangements that are (a) with a paper document or self-adhesive label pick and voice-directed instruction formats, a SKU pick position first and a CO quantity second, and (b) with a digital display pick instruction for pick-to-light or RF device that shows a CO identification, a pick position and to display a SKU quantity.

■ SKU presentation format. For each CO the SKU presentation format is a pick position numerical sequence and SKU quantity. A random pick instruction arrangement has SKUs in inventory allocated to any pick position on a pick line or in a pick aisle. Pick instructions are presented to a picker in no sequential order. The disadvantages are low picker and replenishment employee productivity due to double travel in a pick aisle or on a pick line between two pick positions and potential employee confusion. A random pick instruction sequence is not preferred for implementation in a small-item order-fulfillment operation. A better option is SKUs presented in a numbered sequence. SKUs in inventory are allocated to the pick positions by a SKU discreet number, and the pick instruction format does have a sequential presentation. The first pick position has a SKU with the lowest discreet number. The SKU with the next highest discreet number is allocated to the next pick position, and the SKU with the highest number is allocated to the last pick position. This format directs a picker along a pick line or through a pick aisle. The disadvantages are arithmetic progression along a pick line or through a pick aisle but with no pick density or pick concentration, no increase in pick and replenishment employee productivity, high potential that the SKUs are allocated along a pick line or in a pick aisle by the family group method, difficulty to expand SKUs on a pick line or pick aisle, does not consider a SKU cube, weight, or other physical characteristics, as picked SKUs are transferred from a pick position to a CO tote or shipping carton, there is potential UOP damage and low employee productivity and with a paper pick instruction format, there is potential lower picker productivity due to SKU numbers varying in width or the alpha characters or numeric digits.

Pick Position Number Pick Instruction Presentation

With the sequential CO SKU presentation, the first possible pick position is the first pick position on a pick line or pick aisle. The features are:

■ Arithmetic progression along a pick line, through a pick aisle, or in an automatic pick machine. The preferred arithmetic progression has
  ■ Pick position numbers that end in odd digits on the left side of a pick aisle and pick position numbers that end in even digits on the right side
  ■ Along a pick line flow rack or shelf bay horizontal pick progression
  ■ With a horizontal pick progression the SKU number per carton flow rack or shelf bay level varies, which decreases the non-productive picker travel time and distance between two pick positions
■ Minimizes picker confusion
■ High potential for SKUs allocated to pick positions to match the projected picker productivity
■ Easy SKU expansion

- SKUs have predetermined allocation to pick positions based on the rate of movement and physical characteristics; transfer of a picked SKU from a pick position to a CO carton has less potential for damage and maintains picker productivity
- Can be implemented in any order-fulfillment operation

The pick position number sequential pick instruction presentation arrangement is preferred for an operation.

Manually Printed Paper Document Instruction

A manually printed paper pick instruction document has manually printed alpha characters or numeric digits a sheet of paper. The printed CO information appears in columns. The first column is for a SKU pick position, the second column is for SKU piece quantity for a CO, the third column is for SKU description, and the other column is for other company information. To verify a pick transaction, a picker places a mark adjacent to a pick position or pick quantity.

The manual pick instruction form is used in a low-volume small-item order-fulfillment operation with a medium SKU inventory and in a single CO pick/pack, single picker for one CO handling concept. It has limited application in a batched CO pick concept but is used for slow-moving SKUs, with a batch of 16 to 20 customers. A picker picks and sorts the SKUs to a CO sortation location. When a picker travels through a pick aisle or along a pick line, a paper pick instruction document is held by a clip board or clipped to a pick tote or carton or attached to a cart holder.

Machine-Printed Pick Position

Computer-controlled printed pick instruction options are:

- Paper document
- Self-adhesive labels that are printed (a) horizontally across a self-adhesive label sheet or (b) on a self-adhesive label roll

Machine-Printed Paper Document Instruction

A mechanically printed pick instruction paper document and a manually printed instruction document are similar. Exceptions are:

- Pick operation has a computer-controlled printer
- Pick instructions are clear and understandable
- Low transposition errors

The similarities are:

- Low picker productivity due to the fact that a picker has to read to find the SKU order line on a pick document
- The picker places a mark on the paper pick document

An order-fulfillment operation uses a machine-printed paper pick instruction document for:

- A single pick/pack low-volume operation with a medium SKU number in inventory
- A single CO pick/pack operation that cubes CO skus into a carton; this feature minimizes the line and page number for a CO, which reduces potential picker confusion
- Slow-moving SKUs, with batched COs
- Back-up pick instruction format for other pick instruction formats

Machine-Printed Self-Adhesive Label Pick Instruction

Machine-printed self-adhesive labels have black or colored alpha characters or numeric digits printed on the face of the label. For a manual pick and mechanical sortation concept, the label has human-readable language plus a bar code/RF tag that has black bars on a white label.

- The pick position is printed in large, bold alpha characters or numeric digits. Other information on the label has a small print size in a different color. The human- or human/machine-readable codes and contrast between the black ink and white surface are read by an employee or a barcode scanning/RF tag reading device.
- The pick position discreet identification is located in the right-hand corner.
- The Label is placed on any SKU exterior surface and the label size is adequate to meet pick and sortation human- or human/machine-readable requirements.
- The label face is easily removed from the self-adhesive backing.
- The self-adhesive coating sticks to the exterior surface of any SKU.
- The labels are printed horizontally on a paper sheet or a roll.

To minimize non-productive picker activities and to control and maintain machine-printed self-adhesive labels in a sequential order is important in a pick activity. The picker carries the labels in a stack or as a roll in a dispenser attached to his or her belt. The advantages of the dispenser are that the picker's hands are free and the dispenser minimizes the risk of dropping the stack of labels or mixing up the sequence. Most dispensers start to remove the self-adhesive label face from the backing material.

During a pick activity, a picker walks between two pick positions and performs the following actions:

- Reads a label for the assigned pick position
- Removes a SKU from a pick position
- Removes a label from the stack or roll and peels the label from its self-adhesive backing
- Places the label onto the exterior surface of the SKU
- Transfers the labeled SKU into a CO carton or onto a conveyor belt travel path
- Places the self-adhesive backing into the trash, apron, or pocket

A picker verifies a pick by placing a label onto the SKU. As a labeled SKU is moved onto a sortation concept the bar code/RF tag on the label is read by a barcode scanner/RF tag reader. All "did not pick" or pick instruction labels are returned to an order control desk.

The features are used in a batched CO pick concept with a manual or mechanized sort concept, used in a medium- to high-volume operation, the retail price can be printed on the label, and it is difficult to identify a problem picker.

Voice-Directed Pick Instruction

The voice-directed or voice recognition pick instruction format is a paperless pick instruction format that has a computer system communicate to a picker through a microphone and headset. The voice-directed format is a speech synthesis method with picker on-line communications to a computer. Each picker talks to a computer through a microphone and receives verbal pick instructions from a computer through a headset. A pick instruction is achieved by a radio transmission that leaves a picker's hands free for SKU transfer from a pick position to the conveyor travel path. A voice-directed pick instruction is used in an order-fulfillment operation that handles single COs that are picked/packed into a tote or carton. With batched COs, SKUs with a human-readable code are picked and sorted into a CO sortation location; SKUs with a machine-readable code are sorted by a barcode scanner/RF tag reader/mechanized sort concept.

The concept should be tested in the facility to ensure that transmission does not interfere with any governmental transmissions. Voice-directed pick instructions minimize paper and print expenses. The disadvantages are investment, picker training, and difficult to identify picker errors. The advantages are reduces transposition errors, no print time and cost, does not require a picker to read, and frees a picker's hands.

Digital Display for Pick-to-Light or RF Device Pick Instructions

The digital display for a pick-to-light or RF device concept is a paperless pick instruction format that is used in pick concepts that have pallet or carton flow racks, static shelves, a mobile cart with shelves or flow racks, deck racks, or at a carousel or multiple carousel pick station. There are many pick-to-light or RF device instruction models that are available for implementation in an order- fulfillment operation. They vary in the method used to introduce a CO pick instruction to a pick zone and verify a pick transaction.

Pick-to-light methods for pick instructions are:

- A computer-controlled label machine prints a label bearing a unique number for each CO, which is attached to a CO carton in the pick area. At the start of the pick line each CO pick is transmitted electronically to all pick zones and pick lights, in an arithmetic progression.
- At the entrance to each pick zone, a barcode scanner/RF tag read scans or reads a CO discreet label and sends the data to the microcomputer to activate a pick zone and all pick lights within a pick zone. An activated pick light represents a pick instruction for a CO SKU or SKU quantity.
- Same procedure as in item 2, except a magnetic strip on a card is read by a reader.
- Other pick buttons include next, cancel, full carton/tote, or other company required activity.

Pick-to-light instruction methods to verify a pick transaction are:

- A CO enters a pick zone and a picker presses the "next" button on the pick-to-light system, activating a digital light display to show a CO number. At the pick position, as the picker transfers a SKU from a pick position to a CO carton he or she presses the pick position light for each picked SKU quantity on a CO. The picker repeats this activity at each illuminated pick position in a pick zone for each CO. When the picker has completed all picks in a pick zone for a CO, he or she presses the "next" button to activate a new CO number on the digital display and activate pick position lights.
- A laser light beam or other sensing device at each pick position. As a picker completes a SKU transfer from a pick position, the picker's hand or the SKU breaks a light beam to register a SKU pick. This activity reduces the pick light quantity by one. The pick transfer activity is repeated until the pick light quantity is zero or blank.

Components of a pick-to-light pick instruction concept are:

- In arithmetic progression, a pick area computer receives COs downloaded from a host computer and transmits pick instructions over a pick line wire or fiber optic run. Options for installing a communication link between a host computer and the pick line computer are

- A diskette that is used when there is a restriction that prevents on-line communications.
- On-line, via a wire run between the host and pick line computers. When using the wire option, a dedicated wire run is preferred in a protective track and with a long transmission distance to use short-haul modems.
- Wire runs or fiber optics for the microcomputer to send a pick message over a pick line to each pick zone and to all pick lights.
- Repeaters or relays, digital display boxes, and other electrical power supply boxes or electrical equipment to assure that a pick message is transmitted to all pick zones and all pick position lights.
- Pick-to-light modules or pick position lights.
- For each pick zone, a microcomputer that establishes the number of rack bays or shelves.
- Bay lamps that indicate a required pick in a picker zone.
- For some concepts, a barcode scanner/RF tag or magnetic card reader that sends a message to a microcomputer to activate a pick zone and pick lights.
- On each CO tote, a human- or human/machine-readable code.
- Back-up electrical supply.

The particular pick-to-light technology used in an order-fulfillment operation is determined by:

- Pick concept
- CO handling methodology
- CO volume, piece pick volume, pick lines, and SKU physical characteristics
- The need to satisfy customer service and investment objectives
- IT/WMS ability to interface
- Manufacturer repair service ability
- Cost

Other Major Considerations

In addition to the concepts described above, other factors that impact picker productivity in an order-fulfillment operation are:

- Pick aisle and pick position identification method
- Pick zone identification
- Picker routing pattern through a pick aisle and hand movement across a pick position face

Pick Aisle and Pick Position Identification

In an order-fulfillment operation, to ensure accurate and on-time pick completion, signs clearly identify each pick aisle and pick position. Signs permit a picker to locate a pick aisle and match a pick position identification to a pick instruction. In a pick-to-light application, human-readable pick position identification is placed at each pick position and on the sides of replenishment aisles. During an electrical power failure, this feature permits paper picking of COs.

Pick Aisle Identification

A pick aisle identification is a placard with alpha characters, numeric digits, or a combination of both. The placard is placed at the entrance and exit of a pick aisle or on a pick line at the start and end of a pick zone. Aisle identification methods are:

- One-way vision placard
- Two-way vision placard against a pick position structural member
- Two-way aisle identification that is hung from the ceiling

One-Way Vision Placard

The one-way vision placard can be made of flat metal, wood, plastic, or cardboard. The surface of the placard is white or yellow, with large, bold-colored alpha characters or numeric digits. The placard is attached securely to a pick position structural member.

The placard is positioned so the pick aisle identification faces outward into the main aisle. As a picker travels in the main aisle, it is possible for the employee to identify the pick aisle.

Two-Way Vision Placard on a Pick Position Structural Member

The two-way vision method can be:

- Three placards that are shaped in the form of a U. Two placards for each pick aisle are placed side by side, facing opposite directions, with the placard end attached to a pick position structural member and extending into the main aisle. Each placard identifies the pick aisle. As a picker enters a pick aisle, the employee easily reads the placard facing him or her. A third placard faces the main traffic aisle. As a picker travels in the main traffic aisle, this placard allows the picker to identify the pick aisle.
- One placard that has a pick aisle identification printed on both sides. The metal or plastic placard is attached to the rack or shelf position post that is at the entrance or exit. The placard extends outward into the main traffic aisle.

As a picker or replenishment employee travels in the main traffic aisle they can easily read the pick aisle identification on the placard.

Two-Way Vision Aisle Identification Hung from the Ceiling

A placard with four sides is hung from the ceiling at the entrance to the pick aisle. This arrangement permits an order picker to identify a pick aisle from a main traffic aisle or as he or she enters the pick aisle.

Pick Position Identification Methods

In a manual order-fulfillment operation, the method that is used to identify a pick position directly impacts employee productivity and accuracy. Pick position identification should be clearly visible, easily understood, permanent, match the pick instruction format, and serve as the back-up pick position identification for paperless or automatic pick machine concepts.

The pick position identification method identifies and differentiates a SKU pick position from the other pick positions. Pick position identification methods include:

■ No method
■ Hand-written paint, crayon, or chalk marks on the pick position
■ Human-marked self-adhesive label or tape
■ Machine-printed self-adhesive labels
■ Human- and machine-readable label in a plastic holder
■ Placard hung from the ceiling
■ Digital display or pick to light

No Identification at a Pick Position

If there is no identification at the pick position the operation relies upon a picker to correctly match a pick instruction document description with the vendor markings on a SKU carton exterior. The disadvantages are ensuring that vendor identification on the carton matches the pick instruction SKU identification, low employee productivity, increase in errors, and pickers to read. The advantage is low cost.

Handwritten Identification on a Pick Position Structural Member

Paint, crayon, or chalk markings are used to identify a SKU pick position. An employee writes a pick position identification alpha characters or numeric digits directly on a the flat surface of a pick position structural member. Disadvantages are identification is unclear or confusing to a picker, not easy to transfer, difficult to have uniform characters or digits, and no uniform position. The advantage is low cost.

Human-Marked Self-Adhesive Label

With the human-marked self-adhesive label method, an employee marks a pick position identification onto a self-adhesive label. After the self-adhesive label backing is removed from a label face, an employee places the label onto the pick position structural member. The disadvantages are unclear or confusing identifications and additional labor expense. The advantages are easy to re-identify a pick position, uniform labels, and labels are in a uniform location, easy to notice, and low label expense.

Machine Preprinted Self-Adhesive Labels or Placard

The machine preprinted self-adhesive labels or placards have a computer-controlled machine print the pick position identification onto a self-adhesive label or placard. After removing the backing, an employee places a self-adhesive label or placard at the pick position. Options are:

- Individual preprinted alpha characters and numeric digits. The labels vary from ½ to 3 in. in height and width. An employee places the labels onto a pick position structural member. Disadvantages are label expense, uneven placement, and extra alpha characters or numeric digit labels. Advantages are clear and easy to read, easy to replace, and uniform identification.
- Labels or placards preprinted with the entire pick position identification. With a barcode scanning/RF tag reading and inventory replenishment or pick concept, a machine-readable code is printed onto each label face. The label or placard is placed directly onto a pick position structural member. The disadvantages are increased cost and costly to replace. Advantages are low label application labor, no label waste, consistent label placement, clear and easy to read, and read by a barcode scanner.

Human- and Machine-Readable Identification Placed into a Plastic Holder

The human- and machine-readable identification is placed into a plastic holder. After an employee places and secures a plastic holder onto a pick position structural member, an employee inserts a paper or cardboard pick position identification into the holder. Disadvantages are potential damage to the plastic holder, additional expense, and potential blurring of the identification. Additional advantages are identification presentation is uniform and identification is easily transferred to another pick position.

Placard Hung from the Ceiling or Embedded in the Floor

The placard that is hung from a ceiling member or embedded in the floor is used for a floor pick position identification. Floor-stack pick positions are rare in a small-item order-fulfillment operation.

Digital Display for Pick-to-Light or RF Device

A digital display for pick-to-light or RF device method is a computer-controlled indicator light that is attached to a pick position structural member. When a CO has a SKU from a pick position, a computer illuminates a pick-to-light or RF device display screen. A pick light or RF display screen directs a picker to a pick position and a display screen indicates a SKU pick quantity. Disadvantages are identification method is part of a total system that has a capital cost, costly to relocate, and human- or human/machine-readable identification, to permit pick activity to operate in an electrical or computer failure. Advantages are minimal reading required and clear and understandable.

Locations for placing pick position identification include:

- Attached to pick position structural member directly below or above a pick position and on the deck facing upward. This is appropriate for golden zone pick positions.
- For a pick position that has SKUs in a permanent tote or has a barrier across the front, the identification is attached to the front of the tote or to the barrier.
- If SKUs are stackable packages and the pick position is the lowest/floor-level pick position, the identification is placed on the pick position deck or shelf edge and faces upward.

Picker Routing Pattern

A picker routing pattern is the sequence for pick positions on a pick line or in a pick aisle and for pick instructions. Pick position numbers guide a picker along a pick line or through a pick aisle to a pick position or directs a picker's hand to an appropriate pick position.

The objective in designing a picker routing pattern is to minimize picker non-productive walking time or hand movements between two pick positions. To achieve maximum picker productivity, an appropriate picker routing pattern is implemented in conjunction with good warehouse practices, including:

- The 80/20, family group principle, and SKUs profiled to pick zones for the budgeted picker productivity
- Use of the golden zone
- Clear and properly lighted aisles
- CO SKU cube
- Starting a picker in the best location in relation to SKUs on a pick line or in a pick aisle
- Clear and understandable pick instructions
- Clear and understandable pick aisle identification and pick position identification

A picker routing pattern is designed to satisfy a company objectives, including:

- Meeting budgeted picker productivity rate
- Accurate picks
- On-schedule activity
- On-budget labor expense activity

Some alternative routing patterns are:

- Non-sequential or non-routing patterns
- Sequential picker routing patterns
  - One pick aisle side with one or multiple pickers
  - One side along a pick line with one or multiple pickers
  - Loop
  - Horseshoe or U-shaped
  - Z-shaped
  - Block
  - Stitch
  - Vertical up or down movement
  - Lateral or horizontal
  - Front to rear movement
  - Fixed or variable pick zones

Non-Sequential or Non-Routing Patterns

A non-sequential or non-routing picker routing pattern (see Figure 4.22) has each picker determine his or her pick path through a pick aisle. This pattern should not be implemented in an order-fulfillment operation. The disadvantages are low picker productivity, because a picker walks the same path at least twice; picker fatigue and confusion, and non-productive time to locate a SKU. There are no advantages.

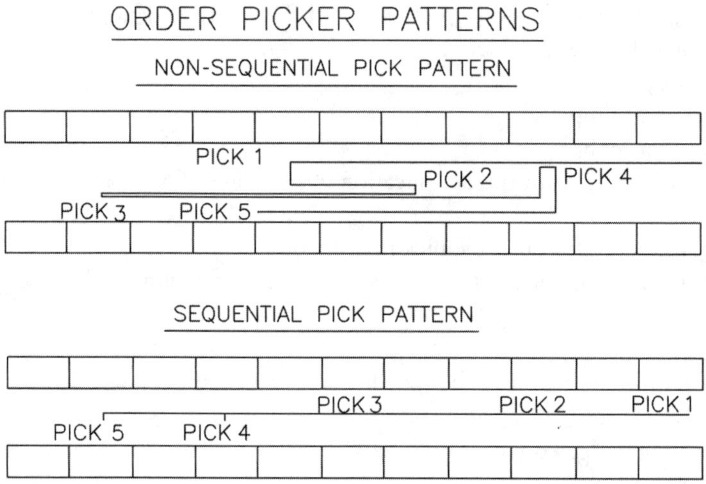

ORDER PICKER PATTERNS

NON-SEQUENTIAL PICK PATTERN

PICK 1

PICK 2   PICK 4

PICK 3   PICK 5

SEQUENTIAL PICK PATTERN

PICK 3   PICK 2   PICK 1

PICK 5   PICK 4

Figure 4.22   Pick position sequence: non-sequential and sequential.

Sequential Picker Routing Patterns

A sequential picker routing pattern is one in which there is an arithmetic progression to pick positions through a pick aisle or along a pick line. This feature means that the lowest SKU pick position is at the entrance to a pick aisle or along a pick line and the highest pick position number is at the pick aisle exit. In a sequential picker routing pattern, a picker starts at the first pick position in a pick aisle or along a pick line. As the picker progresses down a pick aisle or along a pick line the next SKU pick position is as close as possible to the previous pick position. In an order-fulfillment operation, any sequential picker routing pattern provides an efficient and productive picker group. The advantages are that a sequential pattern reduces picker non-productive time and fatigue, minimizes picker confusion, and increases picker productivity.

- Some characteristics of a good picker routing pattern are: As a picker walks through a pick aisle, pick position numbers that end with even digits are on the right side of the pick aisle and those that end with odd digits are on the left side.
- Arithmetic progression in a pick aisle or along a pick line.
- A picker stays in an aisle as long as necessary and "serpentines" through the pick aisle layout.
- Improved SKU hit density and hit concentration.
- Pickers in the high-cube, heavyweight, and fast-moving SKU section.
- Single-item COs have a pick and pass activity.
- Pick activity is cubed.
- Maintain good housekeeping.
- Illuminate pick aisles and pick lines.

Evens on the Right and Odds on the Left

With pick positions on both sides of a pick aisle, to maximize picker productivity, whenever possible all picker routing patterns have pick position numbers that end with even digits on the right side of a pick aisle and those that end with odd digits on the left side of a pick aisle (see Figure 4.23). As a picker travels through a pick aisle, this arithmetic progression reduces picker confusion and is a built-in check system for the pick activity.

Arithmetic Progression through the Pick Aisle

A sequential picker routing pattern routes a picker to each SKU pick position that appears on a CO pick instruction. The picker remove a single piece from a pick position. The pick number sequence is in an arithmetic progression through a pick aisle, from the first pick position to the last pick position. This reduces picker confusion and minimizes the possibility that a picker doubles back as he or she travels down the aisle.

A sequential picker routing pattern keeps a picker in the same pick aisle as long as possible. This feature reduces a picker non-productive travel time between a dispatch station, pick aisle, and pack station.

Figures 4.24 and 4.25 show alternative designs for picker progression through an aisle.

- Straight into and out of each required pick aisle. A picker receives a CO pick instruction and travels in and out of one pick aisle completing all pick transactions. For each pick aisle, the lowest pick position number starts at the main aisle end (001) and progresses arithmetically to the highest pick position number at the other end. After all pick transactions for that CO, the order picker turns the pick cart and walks back through aisle 001 to the main aisle entrance. Leaving aisle 001 entrance, the picker walks in the main aisle to the adjacent aisle 002.The picker then travels through aisle 002, completing all pick transactions. The procedure is repeated through the pick aisles in turn until the CO is complete. The disadvantages are the picker travels twice through each pick aisle; low productivity due to walking past pick positions that do not have picks or that a picker has already completed picks in; requires wider aisles to permit a picker cart to be turned in an aisle; and in one aisle, potential for two pickers traveling in different travel directions.
- Serpentine travel through pick aisles. A picker travels through pick aisle 001, completing picks for a CO. The lowest pick positions are at the entrance to pick aisle 001. After all pick transactions, the picker exits pick aisle 001 at the end or at the highest pick position number location and enters the second pick aisle 002. The picker walks along pick aisle 002 and completes all pick transactions before exiting aisle 002 into the main aisle. From the main aisle, the picker enters the next aisle 003. With a serpentine concept, the lowest

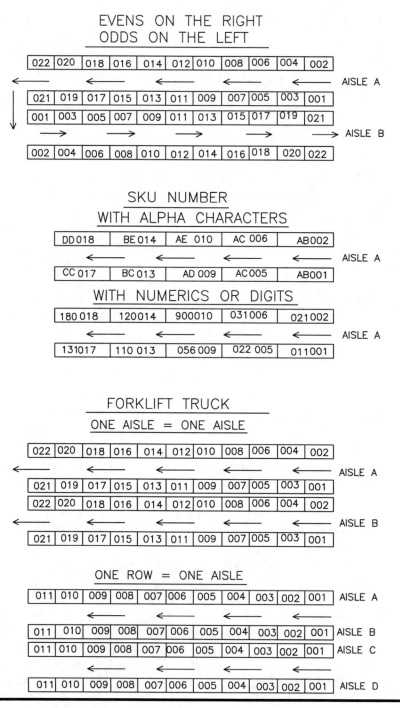

Figure 4.23 Evens on the right, odds on the left: for order picker; for forklift truck; by SKU number.

ORDER PICKER AISLE

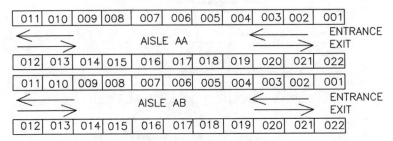

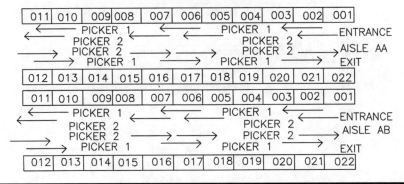

Figure 4.24 In and out one aisle, with one picker and with two pickers.

pick position number starts at the entrance to each pick aisle and arithmetically progresses to the highest pick position number at the exit. In a rectangular pick area with a main aisle and rear aisle, aisle entrances are sequenced so that even numbered aisles have their entrances at the main aisle (front) and odd numbered aisles have their entrances on the rear aisle. The advantages are one trip through each pick aisle, improved productivity, minimal aisle width, and low probability that two pickers occupy a pick aisle at the same time.

Single Side Pick Aisle Pattern

In a single side picker routing pattern pick positions are on both sides of a pick aisle. The options are:

- One picker in a pick aisle. A picker walks down the right side of a pick aisle completing picks for a CO, then turns at the end of the aisle and walk down the left side, completing all pick transactions on that side.

ORDER PICKER AISLE

SERPENTINE

| 022 | 020 | 018 | 016 | 014 | 012 | 010 | 008 | 006 | 004 | 002 |

EXIT ←——— AISLE AB ENTRANCE

| 021 | 019 | 017 | 015 | 013 | 011 | 009 | 007 | 005 | 003 | 001 |

| 001 | 003 | 005 | 007 | 009 | 011 | 013 | 015 | 017 | 019 | 021 |

ENTRANCE ——→ AISLE AC EXIT

| 002 | 004 | 006 | 008 | 010 | 012 | 014 | 016 | 018 | 020 | 022 |

IN AND OUT WITH EACH SIDE AS AN AISLE

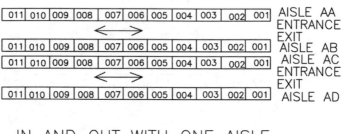

IN AND OUT WITH ONE AISLE

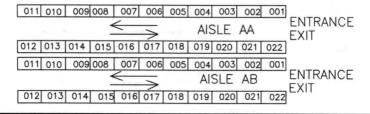

Figure 4.25 Serpentine pattern: in and out each side as an aisle and in and out with one aisle.

- Multiple pickers in a pick aisle. Two pickers complete all pick transactions for a pick aisle. The first picker walks down the right side of pick aisle A and completes all pick transactions, then transfers to pick aisle B. In pick aisle B, the first picker again completes all pick transactions on the right side. The second picker walks in pick aisle A and completes all pick transactions on the left side, then transfers to pick aisle B, and completes all pick transactions from pick positions on the left side.

A single side picker routing pattern is used to complete pick transactions from rack or shelf pick positions and is used with a paper pick document or RF device instruction format. When fast- and slow-moving SKUs are randomly located in a pick aisle, the disadvantage of the single side picker routing pattern is low picker productivity. This is due to a high probability for non-productive travel between two pick positions and to complete a CO, one picker travels in the same pick aisle twice or has two pickers travel in the same pick aisle. This feature has a picker to walk past previously selected pick positions. The advantages are easy to implement in a new or existing facility, minimal training, and pick aisle dead ends into a wall, picker productivity is equal to productivity from the other picker routing patterns.

Single Side Routing Along a Pick Line

A single side routing along a pick line with one or multiple pickers is used in a manual pick/pack or pick/pass concept. To complete pick transactions, a picker walks in a pick aisle and completes all pick transactions from flow rack or shelf pick positions. All CO SKU picks are transferred from the pick position into a CO tote or carton that moves over a conveyor travel path. To ensure optimum picker productivity, the pick line is profiled for your budgeted picker productivity rate and separated into pick zones.

This routing pattern is used with a paper pick instruction document or a pick-to-light or RF device pick instruction. It is used on a pick/pack or pick/pass conveyor and pick to a hanging basket from an overhead rail concept. Disadvantages are good pick line profile, tote transport concept, difficult to identify problem pickers, and pick line is separated into zones. Advantages are improves picker productivity, less picker fatigue with rubber mats on a pick aisle that improve picker safety, handles a high pick volume, high CO volume, and wide range for pieces per CO, handles a wide UOP mix and easy to expand and permits paper pick document, pick-to-light or RF device pick instructions.

Loop Routing Pattern

With a loop picker routing pattern (see Figure 4.26), a picker walks in a pick aisle to complete pick transactions from pick positions on the right side of pick aisle A. At the end of pick aisle A, the picker transfers to pick aisle B and completes all pick transactions from pick positions on the left side. The picker then returns to pick aisle A and completes all pick transactions from the left side of pick aisle A. After completing all pick transactions in pick aisle A, the picker enters pick aisle C and completes all pick transactions from the left side of that pick aisle. This picker aisle entry and exit pattern is repeated in the other pick aisles until a picker completes a CO.

A loop picker routing pattern is used in an order-fulfillment operation that has a pick aisle between two rack or shelf pick position rows and a large number of SKUs. When fast- and slow-moving SKUs are randomly located in an aisle, disadvantages of a single side picker routing pattern are low picker productivity because of non-productive travel distance between two pick positions; to complete a CO,

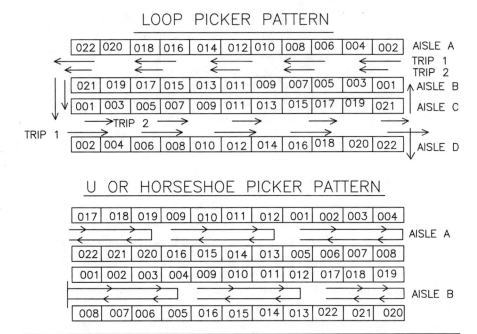

Figure 4.26 Loop and U or horseshoe picker routing patterns.

one picker travels in the same pick aisle twice or two pickers travel in the same pick aisle. This feature has a picker walk past selected pick positions and requires additional training.

Horseshoe or U Routing Pattern

In a horseshoe or U picker routing pattern, a picker walks in a pick aisle and stops at predetermined pick aisle locations in the middle of the pick aisle. At each stop, the picker has the opportunity to complete pick transactions from the adjacent pick positions on both the right and left sides of the pick aisle. On each side of the pick aisle, there are from one to two rack bays or two to four shelf bays. The picker follows a loop through the various pick aisles until a CO is completed.

The advantages and disadvantages are the same as those described above for a loop picker routing pattern. An additional disadvantage is low productivity due to time spent by a picker stopping at predetermined pick aisle halt locations and passing positions that have SKUs not on a CO.

Z Routing Pattern

In a Z picker routing pattern (see Figure 4.27), a picker walks once in a pick aisle that is between two pick position rows that are rack or shelf bays. A Z picker pattern

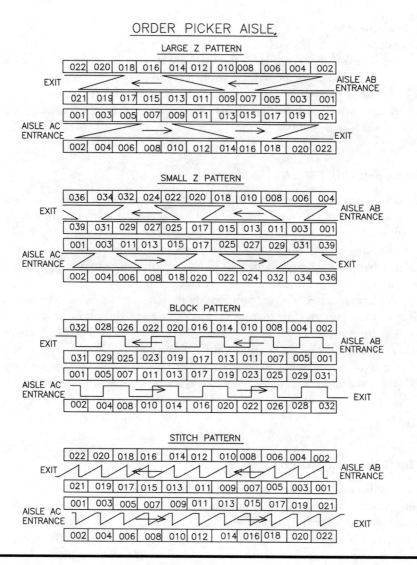

Figure 4.27 Large Z, small Z, block, and stitch picker routing patterns.

directs a picker to complete all pick transactions from the first four pick positions on the left of pick aisle A, then complete all pick transactions from the first eight pick positions on the right side of pick aisle A, then all pick transactions from the next eight pick positions on the left side of the pick aisle. The picker follows this Z pattern until the CO is completed.

The Z picker pattern can be modified to shorten a pick zone length. In this shortened Z picker routing pattern, the picker completes all pick transactions from the first two pick positions on the left side of a pick aisle, then completes all pick transactions from the first four pick positions on the right side, then the next four pick positions on the left side.

The disadvantage is additional training. Advantages are with a good pick aisle profile, excellent picker productivity; with an arithmetic progression through a pick aisle, a reduction in picker time to locate a pick position; and, if slow-moving SKUs are on one side of a pick aisle and fast-moving SKUs are on the other side, there is good picker productivity.

Block Routing Pattern

A block picker routing pattern is similar to the Z pattern. The similarities are that both patterns direct a picker through a pick aisle once, completing pick transactions from both sides of the pick aisle. The difference between the Z and the block picker routing pattern is the block pattern pick zone has two pick bays.

In a block picker routing pattern, a picker walks once in a pick aisle that is between two pick position rows that are in rack or shelf bays. The picker completes all picks from the first two pick positions on the left side of pick aisle A, then complete all picks from the first four pick positions on the right side before completing all picks from the next four pick positions on the left side. The picker follows this pattern through the pick aisles until the CO is complete.

The disadvantage is the additional training required. Advantages are, with a good pick aisle profile, excellent picker productivity; with arithmetic progression through a pick aisle, a reduction in picker time to locate a pick position; and, if slow-moving SKUs are on one side and fast-moving SKUs on the other, good picker productivity.

Stitch Routing Pattern

The stitch routing pattern is similar to the block pattern. The similarities are that both patterns direct a picker through a pick aisle once and has a picker complete pick transactions from pick positions on both sides of the pick aisle. The difference is the pick zone length due to the fact that the stitch pattern zone has one bay.

In a stitch picker routing pattern, a picker walks once in a pick aisle that is between two pick position rows that are racks or shelves. A stitch picker pattern directs a picker to complete all picks from the first pick positions on the left side of pick aisle A, then complete all picks from the first pick positions on the right side of pick aisle A, then complete all picks from the next pick positions on the left side. The picker follows this pattern through the aisles until the CO is complete.

The disadvantage is the additional training required. Advantages are, with a good pick aisle profile, excellent picker productivity; with arithmetic progression through a pick aisle, a reduction in picker time to locate a pick position; and, if slow-moving SKUs are on one side of the pick aisle and fast-moving SKUs on the other side, good picker productivity.

Vertical Hand Movement Routing Pattern

A vertical, or up and down, picker routing pattern is used on a carton flow rack, shelf, drawer, or carousel pick position concept. A picker is directed in a pick aisle or along a pick line to a SKU pick position, or a carousel transports the assigned pick position to a pick station, and the picker moves his or her hand across pick positions on a carton flow rack, shelf, drawer, or carousel.

With a vertical pattern, at a pick bay the first possible pick position is located at the bottom or top and a picker is directed by the pick instruction to proceed up or down the first pick position stack. After all pick transactions are completed in the first pick position stack, the picker proceeds to the second pick position stack. At the next pick bay, the up or down pattern is repeated, until the CO is completed.

This is not a preferred routing pattern because:

- The golden zone does maintain an orderly SKU arrangement in a pick bay.
- It does not ensure good replenishment.
- It is difficult to add SKUs to a pick bay.
- It is difficult to add faces for fast-moving or high-cube SKUs.
- It makes for low picker productivity.

Horizontal Routing Pattern

A horizontal picker routing pattern is used on a carton flow rack, decked rack, or shelf pick concept. The pattern has:

- A single side routing pattern that directs a picker in a pick aisle or along a pick line to the SKU pick position
- A horizontal picker routing pattern at an assigned flow rack bay that directs picker activity across pick positions in a carton flow rack

The horizontal routing pattern starts as a picker enters a pick aisle that is along a carton flow rack bay front. At the first carton flow rack bay, the start pick position is the upper first flow rack bay and the last pick position ends at the lower rack bay.

The Picker Faces Pick Positions on the Right Side

A picker enters a pick aisle on the right side (single or mirrored pick module) and faces the first pick zone, with pick positions are on the right of the pick aisle. As the picker proceeds along a pick aisle, at each carton flow rack bay, the first pick position is in the upper-level right-hand corner. Beginning at the upper right-hand corner, pick transactions proceed horizontally across each rack bay level. The last pick position is in the lower-level right-hand corner.

The Picker Faces Pick Positions on the Left Side

The picker enters the pick aisle on the left side (single or mirrored pick module) and faces the first pick zone. At each flow rack bay, the first pick position is in the upper-level left-hand corner. The picker starts in the upper left-hand corner, moving horizontally across each carton flow rack bay level, and the last pick position is in the lower-level left-hand corner.

With a horizontal picker routing pattern, at each pick bay the first possible pick position is at the top left corner. The pick positions progress along the top level. After all pick transactions on that level, the picker proceeds to the next lower pick level, in the left-hand corner. The horizontal routing pattern is repeated at each carton flow rack bay until the CO is completed.

If a picker entered an aisle on the right side, the upper right-hand corner is the start pick position. The horizontal picker routing pattern is preferred for a pick/pass order-fulfillment operation that has carton flow rack bay pick positions, because it

- Maintains an orderly arrangement in the golden zone in each pick bay.
- Ensures good replenishment activity.
- Is easy to add SKUs to a pick bay.
- With fast-moving or high-cube SKUs requiring additional faces, it allows high picker productivity because the SKU is allocated to one carton flow rack bay pick level.

Front-to-Rear Picker Routing Pattern

The front-to-rear picker routing pattern is used with drawer pick positions. The routing pattern (see Figure 4.28) sequence is:

- A picker is directed into a pick aisle or along a pick line to the drawer pick position.
- Front-to-rear picker routing pattern at the drawer pick position directs a picker hand movement to a pick position in a drawer.

With the front-to-rear picker routing pattern, at a drawer the first pick position is at the front or rear of the drawer. After all pick transactions in the first drawer, the picker proceeds to the next drawer. At each drawer the front-to-rear picker routing pattern is repeated until the CO is completed. The front-to-rear picker routing pattern is preferred for implementation in an order-fulfillment operation that has drawer pick positions and very small-cube SKUs with a very small inventory.

Fixed Zone Picker Routing Pattern

A fixed zone picker routing is used along a pick line to separate it into a fixed number of carton flow rack or shelf pick bays. One order picker is allocated to each

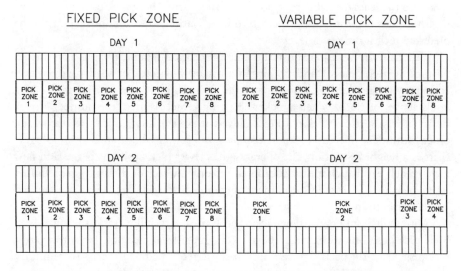

Figure 4.28   Front-to-rear, fixed pick zone, and variable pick zone routing patterns.

pick bay zone. With a well-profiled pick line, an order- fulfillment operation has each picker travel over a fixed pick zone. A fixed pick zone provides high picker productivity that results from a shorter travel distance between two pick positions, familiarity with the SKUs and increased SKU hit density and hit concentration. Along a pick line aisle, to assist a picker to distinguish the fixed pick zone limits, a pick aisle has fixed pick zone limit signs to identify start and stop locations. Limit signs are solid colored banners, alpha characters, or numeric digits on a placard, which extends downward and slightly into the pick line aisle.

Variable Zone Picker Routing Pattern

A variable zone picker routing pattern is used along a pick line to separate a pick line into a number of carton flow rack or shelf pick bays in each pick zone. A variable pick zone has for each work day or CO download order wave. Pick zone limits are expanded or reduced depending on CO SKU picks or hits. Due to potential confusion and difficulty to maintain a good pick line profile, a variable pick zone or pick area is not a preferred option for an operation.

Pick Position Elevation and Location

Pick position elevation and location on a pick line or in a pick aisle influences manual picker productivity. To achieve maximum picker and replenishment employee productivity, pick position elevation above the floor in a carton flow rack, shelf, or drawer pick bay reduces a picker reach or bend to complete a pick or replenishment transaction. This means that the top pick level and bottom pick (pigeonhole) levels are the least desirable pick positions. In a carton flow rack or shelf pick bay, pick positions between 20 in. and 5 ft. 6 in. elevation above a floor surface are the preferred pick positions or levels.

Pick position elevation above the floor or location in a pick bay has an impact on pick area equipment and layout. In a carton flow rack or shelf pick position concept with standard pick openings 18 to 20 in. high, there are four to five pick levels above the floor. When a picker-elevating vehicle is used, the number of pick positions in a vertical stack is 10 to 12 levels above the floor. With this feature, an order-fulfillment operation is more compact and is housed in a smaller building. If there is a height restriction, the operation has fewer pick positions per carton flow rack or shelf bay, which means additional carton flow rack or shelf bays are required in the pick area, your order-fulfillment operation covers a larger area, and is housed in a larger building.

With a tilt-front gravity flow rack or push back carton flow rack, the lower-level SKU pick positions are extended beyond the upper-level pick positions or toward a pick aisle. This tilt back feature minimizes the required open space. To minimize the open space requirement in shelf pick positions, the options are:

- To have the elevated pick position sloped downward toward the aisle and the bottom pick position lane flat
- Whenever possible to have the long dimension of a SKU lie flat on a pick position deck
- When the SKUs are in totes or cartons, to have the fronts cut open to allow a picker to pull a SKU forward and out

How to Increase Picker Reach

Picker access to an elevated pick position can be made easier by:

- Attaching a step to an upright post along a shelf or rack pick position
- Reinforcing the pick position shelf or deck
- Using a step ladder

A hardened metal step is located on the structural support posts of the bottom two pick positions and supports a picker's weight. The step runs the full length of a pick aisle.

Reinforced pick position shelves or decks are the lowest or next to the lowest shelf or deck level from the floor. A shelf or deck has heavy-duty hardened metal that is reinforced with hardened metal slats and runners to support a picker's weight.

Another alternative is to use a safety step ladder in a pick aisle. Options are:

- A safety step ladder that is attached to the back of an order pick cart as a picker push handle. This requires no non-productive picker time to locate a ladder in a pick aisle.
- A three- or four-wheel mobile safety stool. A mobile safety stool is designed with a spring-loaded deck. As an employee steps onto the deck, the bottom surface comes in contact with the floor, which acts as a lock mechanism to secure the stool. A mobile stool has a picker use non-productive time to locate the stool in a pick aisle. The stool has no hand rails, which requires a picker to maintain his or her balance. A stool increases a picker reach by 8 to 10 in.
- A four-wheel mobile independent safety ladder. Designs are:
  - A stationary top handrail structure
  - A collapsible top handrail structure
  - Front structure members that are designed to carry cartons or totes. With the safety step ladder concepts, there is some non-productive picker time to relocate a pick ladder in a pick aisle. The safety step feature and handrails improve picker safety.
- A ladder attached to the top of a structural member at a pick position. To ensure mobility, the ladder has two top wheels that are interlocked onto a travel rail or path and two bottom wheels that run on the floor. The ladder is angled into the pick aisle so a picker can easily climb it. Each ladder has a safety lock mechanism that becomes activated as a picker steps onto it. A spring-loaded safety lock mechanism has component with a rubber bottom that comes in contact with the floor surface and secures the ladder to the floor surface.

Cubing the Pick Activity

A CO pick activity includes the use of a computer program to cube SKUs into the number of pieces that fit into a CO pick carton/tote. The computer cube program ensures that a picker is able to handle the quantity of SKUs in one CO pick carton/tote without additional trips to obtain another pick carton/tote. If the number of pieces in a CO requires two pick cartons/totes, the computer cube program adds another CO pick carton/tote on the pick line.

If an order-fulfillment operation pick instruction method "cubes out" a CO into separate pick cartons/totes, the pick area computer separates the total number of SKUs for a pick line or pick aisle sequence into one or several pick cartons/totes. This feature reduces picker non-productive aisle travel time and UOP damage.

If an order-fulfillment operation does not "cube out" picker activity, an order control clerk estimates and issues instructions to each picker that fill one pick carton/tote or directs the picker to add an empty carton to a pick line or vehicle or make additional trips. When estimating a CO cube requirement, a clerk uses an average cube for each SKU. The average cubic feet change each season. When there is no match between the number of SKUs in a CO and carton/tote capacity, there is a high probability that a picker will have pick instructions that exceed one pick carton/tote, making it necessary for the picker to make unnecessary and non-productive travel trips to complete the CO portion or to add a new carton to a pick vehicle.

Where to Start Pickers

A small-item order-fulfillment operation has a pick aisle, or pick line layout with fast-moving, heavyweight, or high-cube SKUs in one area and slow-moving, light-weight, or small-cube SKUs in another area. To obtain high picker productivity, the question for a pick area manager is where to start a small-item picker. In a small-item order-fulfillment operation, the fast-moving SKUs or 20 percent of SKUs account for 80 percent of the pick volume, and 80 percent of the SKUs or medium- to slow-moving SKUs for 20 percent of the pick volume.

On a small-item flatwear pick line or in a pick aisle, a pickers may start either in the fast-moving, heavyweight, or high-cube SKU area or in the slow-moving, lightweight, or small-cube SKU area.

Starting in the High-Volume, High-Cube, and Heavyweight SKU Pick Area

If your small-item picker starts in the fast-moving, heavyweight, or high-cube SKU area, the following advantages result:

- Improved pick position replenishment.
- Higher pick volume handled by a pick operation and constant completed CO flow from a pick area to a check and pack station.
- With slow-moving, lightweight, or small-cube SKUs at the end of a pick line or aisle, it is easier for a picker to top-off or place other SKUs into a CO pick carton/tote.
- Fewer CO carton/tote are handled.

- High picker productivity and SKU transfer to a carton because the SKUs with high hit density and hit concentration, and the heaviest or largest are placed first into a carton. With a paper pick instruction document, these SKUs are the first SKUs on the pick instruction, which make it easier to read.

If your pick operation handles each carton for a multiple carton CO as a cubed-out carton and your pick line has a completed CO take-away transport concept, the fast-moving, heavyweight, or high-cube SKU area is the preferred location to start your picker or CO.

Starting in the Low-Volume, Small-Cube, and Lightweight SKU Pick Area

If you start your small-item pickers in the slow-moving, lightweight, or small-cube SKU area, the results are:

- With a paper pick instruction document, it is more difficult to read because most picks are fast-moving SKUs that will be located at the bottom of the document.
- It is difficult to move partially full CO carton/tote or lightweight cartons/totes over a conveyor travel path.
- Increased non-productive time to transfer large-cube, heavyweight, or fast-moving SKUs into a CO carton that has the slow-moving, light weight, or small-cube SKU area.

Where to Start Customer Orders with Captive Totes

An order-fulfillment operation that uses a captive tote concept also must determine where to start a CO. With a captive tote pick concept, SKUs are transferred from a pick position into a captive tote and at a pack station the SKUs are transferred from the tote into a CO shipping carton. Other design factors are that this low-volume SKU pick concept is computer directed and uses a carousel, carton AS/RS or mini-load, or shelf, with pick-to-light and sort-to-light pick concepts. The slow SKU pick section has the customer orders batched into predetermined groups. The 80/20 SKU movement philosophy applies.

COs are started either in the high-volume SKU pick zone or in the low-volume SKU pick zone.

Starting in the High-Volume SKU Pick Area

When COs are started in the high-volume SKU pick zone, each partially completed CO tote or 20 percent of the CO totes are transported to the low-volume SKU pick zone. The features are:

- Most COs are completed in the high-volume area and the SKUs are placed in a pick tote and transferred to a CO take-away conveyor.
- Twenty percent of the CO totes are sent on to a low-volume SKU pick zone.
- There is the potential to have a picker waiting for a tote to arrive in the low-volume SKU pick area. With a carousel or carton AS/RS concept and totes traveling on a powered conveyor travel path, the tote travel sequence has a first-in first-out concept.
- In the high-volume area, each CO pack slip is transferred into a CO tote.
- Simple CO transport concept and controls.
- Empty pick totes are introduced in the high-volume pick zone.

Starting in the Low-Volume SKU Pick Area

When CO totes are started in the low-volume SKU pick area, the features are:

- Both the high-volume and low-volume pick locations require a CO pack slip transfer station.
- At each location empty totes are introduced to a pick line.
- With a CO download and batched CO concept, a carousel or AS/RS carton is more effective to have a pick position arrive just in time at a picker location. This means minimal non-productive picker waiting time.
- From a low-volume area, an estimated 20 percent of the CO totes require introduction to a high-volume SKU pick zone. The design feature requires additional controls and conveyor travel paths.
- Maximum CO number are started and completed due to two pick or CO start locations.
- If a CO requires only SKUs from a low-volume SKU pick area, there is less transport.

Picking into a Shipping Carton or Captive Tote

In a pick/pass order-fulfillment operation a picker transfers a CO picked SKU from a pick position into a CO captive tote or shipping carton. For both concepts, SKU cube and weight for a CO are determined by a computer and do not exceed the tote or shipping carton internal cube capacity. If SKUs for a CO exceed shipping carton internal cube capacity, the microcomputer separates CO SKUs into the appropriate number of shipping cartons or totes.

Shipping Carton

In an order-fulfillment operation that uses a CO shipping carton on a pick line or in the pick activity, a picker transfers a SKU directly from a pick position into a CO shipping carton. This application involves:

- Accurate, updated information on each dimensions and weight of each SKU and interior dimensions and weight of the shipping cartons.
- Cartons dispensed from a machine or employees to make-up shipping carton mix and apply tape to the bottom flaps, or a pop-out carton that does not require tape on the bottom.
- With a wide carton mix, the lighter-weight cartons may jam on a powered conveyor travel path or cause problems with the conveyor controls.
- A wide carton mix increases the difficulty a picker experiences to push a carton train queue over the pick conveyor travel path.
- The need for CO discreet identification on a carton side. It may be difficult to use a mechanical label applicator; a short carton may demand additional non-productive picker time to locate and scan or read the CO discreet label; and some small carton flaps do not have sufficient space for a label.
- For the tallest cartons the picker may have difficulty transferring picked SKUs over the carton flaps.
- If there are elevation changes over a pick line conveyor travel path, a wide carton mix creates carton transport problems.
- If there is a 90 or 180 degree curve in a pick line conveyor layout, with a wide carton mix there is potential for carton jams.
- If sales literature or a CO pack slip is required in a CO shipping carton, with a wide carton mix it is difficult to use mechanical insert equipment. With some carton sizes, an employee or machine folds a CO pack slip prior to insertion into a carton.
- If a COD, bank transfer, or hazardous label is required on a carton, a location is required on the conveyor travel path for label attachment to a carton flap.
- If an order-fulfillment operation pick line has an activity that creates a gap or open space between two cartons, a gap is pulled between two cartons.
- CO identification label is placed on the side of a carton or the identification is printed directly onto the carton.
- Does not require empty totes to be returned to a pick area or pick line start station.
- At a pack station, improves packer productivity.
- Improves shipping carton and seal material inventory control and use.

Captive Tote

When an order-fulfillment operation uses a captive tote on a pick line, a picker transfers a SKU directly from a pick position into a captive tote. Features of this concept are:

- Accurate, updated SKU dimensions and weight and interior dimensions and weight of the captive totes.
- A standard size tote minimizes controls, conveyor equipment, and investment.

- A picker transfers a SKU from a pick position into a tote.
- At a pack station, SKUs are transferred from the tote into a shipping carton, the tote identification is scanned, and empty totes are removed to a queue area or take-away conveyor travel path.
- At each pack station, there are tables for carton make-up activity or cartons that have been made up are transported to the pack station.
- CO label is at a constant and visible location.
- All CO address labels, COD, or bank transfers are handled at a pack station.
- With a standard tote length, a standard gap between two totes.
- Sleeve or tote indentation to hold a CO identification or unique identification.
- Transport concept to return empty totes to a pick area or to a pick line start station.

Shipping Carton in a Captive Tote

When an order-fulfillment operation uses a captive tote with an assigned CO shipping carton inside, there is one additional employee pick activity station, where an employee or machine transfers an open empty carton into a tote or along a pick line to transfer a picked SKU directly from a pick position into a shipping carton.

Totes used in this concept may have one of the following designs (see Figure 4.29):

- Full height and four-sided or standard tote, which has a side wall with space for CO identification that permits line of sight for a barcode scanning/RF tag reading device. At a pack station, a packer lifts a carton over the side wall. All totes are returned to a pick line start station.
- Three-sided tote with CO identification located on the cartons. With a wide carton mix, CO identification is placed on the side of the carton, allowing line of sight for a barcode scanning/RF tag-reading device. With a three-sided tote, the open side faces the pack station, which minimizes packer effort to transfer a carton from the tote onto a pack table. All totes are returned to a pick line start station.
- Standard shipping carton used as a tote. CO identification is a label that can be peeled off at a pack station and transferred to a smaller shipping carton. This feature permits line of site for a barcode scanning/RF tag-reading device. At a pack station, a packer lifts the small carton over the larger carton's side wall and transfers the label from the standard carton to the small carton. The standard size shipping carton is returned to a pick line start station and reused. An option is to place a non-formed small shipping carton into a large shipping carton and at a pack station to form a small size shipping carton.

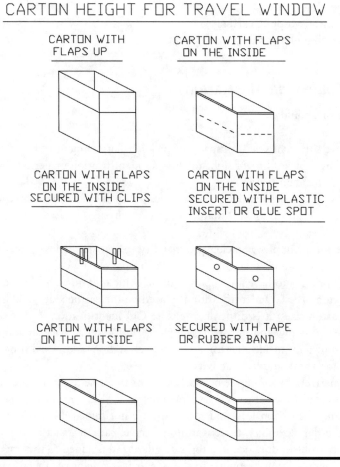

CARTON HEIGHT FOR TRAVEL WINDOW

CARTON WITH FLAPS UP

CARTON WITH FLAPS ON THE INSIDE

CARTON WITH FLAPS ON THE INSIDE SECURED WITH CLIPS

CARTON WITH FLAPS ON THE INSIDE SECURED WITH PLASTIC INSERT OR GLUE SPOT

CARTON WITH FLAPS ON THE OUTSIDE

SECURED WITH TAPE OR RUBBER BAND

Figure 4.29 Carton travel options: flaps-up, flaps on the outside, flaps on the inside, flaps clipped, flaps with rubber bands, and flaps with glue or inserts.

Carton Travel Path Window

Tall cartons or cartons with flaps up create a tall window on a vehicle load-carrying surface or conveyor travel path. This results in low picker productivity because the picker must lift a SKU up and over a carton flap and an employee or machine must then fold down the carton top flaps. The tall window also reduces the possibility to have pick positions above the conveyor travel path. Options for minimizing carton height and lowering the travel path window include:

- Folding the flaps inside the carton
- Taping the flaps by hand or banding the flaps to the carton
- Applying reusable rubber bands or clips to hold the flaps to the side of the carton

- Applying glue to the carton flaps
- Using a two-piece carton
- Inserting plastic stays

Flaps Folded into the Carton

Each carton should:

- Have sufficient side wall space for pick label attachment
- Allow a picker line of sight to the CO identification or permit a scanner to read the identification
- Have a CO delivery label on the side, which should be reuseable and placed onto the carton for CO manifest/scan activity

The concept improves picker productivity to transfer picked pieces into a tall carton but takes additional time to complete or pack-out a CO.

One option is to have an employee or machine fold down all four carton flaps into a carton. The CO identification is placed on the carton side wall. With a wide mix of carton sizes, it is difficult to ensure CO identification line of sight, short travel path window, and narrow travel path, additional activity station or time to fold the four flaps inside a carton and at a pack station, additional time to prepare a carton for a seal, and label activity.

Alternatively, an employee or machine folds two side flaps into the carton interior. This creates a lower carton side for transfer of SKUs from a pick position into the carton. The carton lead and trail flaps are not folded down. With front and rear carton flap flexibility, there is minimal travel path obstruction or hang-ups. If a hang-up problem does occur, a top side guard rail is added to the travel path top and CO identification is applied to the side, if it is a delivery label, then a re-label activity or peel-off label is used.

Another option is to have an employee or machine fold down two or all four carton flaps only for those cartons with flaps that extend above a predetermined height. CO identification is placed on the exterior side wall of the carton. In an operation with little variation in carton heights, it is easier to ensure consistency in the location of CO identification on the carton sides, and easier to ensure CO identification line of sight, short travel path window, and narrow travel path.

Plastic Band or Tape

An employee or machine folds the carton flaps against the exterior side and places a plastic band or tape around the flaps. After the pick/pass activity, as a carton arrives at a fill station, an employee cuts the plastic band or tape and ensures that the carton flaps are ready for sealing. This concept requires additional investment for banding or tape machine, or an employee activity station, as well as additional cost for the bands or tape. It creates a short travel path window with no hang-up problems, CO identification is applied to the side, if it is a delivery label, a re-label activity or peel-off label is required in the concept. It results in good packer productivity due to less time to prepare a carton for a pack activity, and wide carton size mix increases difficulty and additional investment.

Reusable Rubber Band or Clips

An employee or machine folds the carton flaps down and places a reusable rubber band or clips to secure the carton flaps to the carton exterior. After the pick/pass activity, as a carton arrives at a fill station, a fill employee removes the rubber band or clips and assures that the flaps are ready for the seal activity. The rubber band or clips are placed in a container for transfer to the apply station for reuse on a pick line. The features are additional or employee activity station, some additional plastic band or tape cost, but is less due to reusable bands or clips, creates a short travel path window with no hang-up problems, CO identification is applied to the side, if it is a delivery label, a re-label activity or peel-off label is required in the concept, and good packer productivity due to less time to prepare a carton for a pack activity.

Glue

The employee or machine applies glue to the carton side and folds the flaps onto the exterior walls. At a pack station or fill station, an employee releases the glue and ensures that the flaps are ready for the seal activity. The glue concept has the same features as the rubber band or tape concept.

Plastic Inserts or Stays

The employee or machine inserts one or two inserts or stays through two or four sides of a carton to hold the flaps onto the exterior walls. At a pack station or fill station, an employee releases the plastic inserts or stays and ensures that the flaps are ready for the seal activity. The plastic inserts or stays concept has the same features as the rubber band or tape concept.

Two-Piece Carton

The two-piece carton concept uses a machine that forms the carton bottom section. The bottom section is transferred to a pick line and receives a CO identification.

The two-piece carton concept features are the same as the previous concepts but the additional features are carton-form machine investment and carton seal method.

Single Customer Order Selection for Multiple SKUs from Different Pick Zones

In conclusion, to have an effective, cost-efficient operation and to assure good customer service, your SKU mix or size, CO number, and CO pieces (volume) or cube dictate that the order-fulfillment area has different pick concepts to match the design parameters. This means that most small-item order-fulfillment operations have one pick concept for fast-moving and high-cube SKUs and another for slow-moving and small-cube SKUs.

Other considerations are:

- Capability of the warehouse management system or WMS to split a CO between two different pick zones or sections
- A pick process that involves (a) picking SKUs into a CO shipping container or captive tote
- Pick concept or picker instruction method for each pick zone or section

Improving Picker SKU Hit Concentration and Hit Density

To improve CO SKU hit concentration and hit density, SKUs are allocated according to Pareto's law or the 80/20 rule, which states that 80 percent of the pick volume is derived from 20 percent of the SKUs.

The SKU hit concentration means the number of SKUs ordered by your CO lines (stops or hits) within a particular pick aisle, pick zone, or the number of pick positions within a particular pick aisle or pick zone for one CO. The SKU hit density is the number of times or SKU quantity that a CO has for one SKU or line to complete a line on a CO.

On a pick line or in a pick aisle, high SKU hit concentration and hit density dramatically improves picker productivity because it minimizes travel distance between two picks. For best results in a single-item pick and pack operation or pick line, a picker is assigned to a pick zone, pick area, or pick aisle. Each pick zone, pick area, or pick aisle has projected SKU hit concentration and hit density to match your budgeted picker productivity and other pick activity design factors that were previously mentioned in this section.

How to Light or Illuminate Your Pick Aisle

Light fixtures are hung directly above the center of the pick aisle or perpendicular to the pick aisle and above the pick positions. With both alternatives, the light fixture program has to specify a desired lighting or lumen level that is 30 in. above the floor.

With the light fixtures directly above the center of the pick aisle, the light fixtures hang from the ceiling joists, the second floor support members, or cross aisle members that are supported by the pick position structural members. This light fixture arrangement has light fixtures that start at a pick aisle entrance and end at a pick aisle exit, illuminating the entire pick aisle length. When required to replace a light fixture or tubes or bulbs, the maintenance employees have easy access to the light fixture. This is the arrangement that is more familiar to pickers.

The light fixtures are hung perpendicular to a pick aisle and above all pick positions, which are racks or shelves. The light fixtures hang from the ceiling joists or the second floor support members or supported from members that are attached to pick position structural members. With the light fixture arrangement a light fixture row begins at the first pick aisle, strung above other pick aisles, pick position structures, and ends at the last pick aisle. The center spacing between the light fixture rows ensures that the light or lumen level is per your specification. To replace light fixtures, tubes or bulbs, your maintenance employees have a more difficult task due to the fact that the light fixture is less accessible. Your pickers are less familiar with this light fixture arrangement.

A machine-printed label or paper document instruction concept has a picker read print, pick and replenishment aisles have a high lighting level. To read a pick-to-light or RF device pick instruction, a pick-to-light or RF device pick instruction has a picker read from a light or background that is bright, an aisle lighting has a low lighting or lumen level.

Maintain Clear Pick Aisles and Have Good Housekeeping

Pick aisles should be obstacle free to allow a picker to move along a pick line to pick positions without unintentional stops. Many professionals have stated that good housekeeping in a pick area enhances picker productivity.

Handling Single-Item Customer Orders

An order-fulfillment operation that has a high pick volume for a single item or a few items for a large number of COs uses a pick and pack activity that adds one or two additional steps to the normal pick activity, which results in a slight decrease in picker productivity. The picker picks, packs, and seals each CO carton. For high-volume SKUs, a pick and pack activity dramatically improves the total order-fulfillment productivity, because COs bypass a regular pick activity and pack station; SKUs are picked *en masse* and transferred to a specific pick and pack station. Preformed cartons can be delivered to the pack station.

The Pick/Pack Concept

A small-item/flatwear operation with a pick/pack activity can improve customer service, lower costs, and increase CO handling volume. The opportunities are determined by:

- SKU volume
- SKU physical and pack characteristics
- CO number and pieces per CO
- Pack stations that have access to a transport conveyor

A pick/pack activity is used when there is a single SKU or a pair with a high sales volume. The exterior packaging that is used allows a SKU to be placed inside a CO shipping carton or delivered from the vendor as a ready-to-ship carton.

If the SKU is fragile and requires fill material in the shipping carton, a pick/pack line has separate top and bottom fill stations. If a SKU is in a vendor ready-to-ship carton, the pick/pack line has a work station where the CO label is applied and CO packing slip inserted.

The third pick/pack line design factor is the CO number and pieces per CO. If a pick line handles a large single SKU or SKU pair per CO, the layout has a completed CO take-away conveyor. The pick/pack line options are reviewed later in this section.

Pick/pass stations perform the following activities:

- Carton make-up and introduction to the pick line
- Bottom fill
- SKU pick
- CO pack slip insert
- Top fill
- Carton seal
- Carton label and scan manifest

Pick/pass layouts are:

- For regular SKUs, a complete pick/pack line
- For vendor ready-to-ship cartons, label and invoice pick/pack line with
  - Individual CO label
  - Slapper label

Pick/pass activity stations are determined by:

- Fill requirements
- Whether the CO pack slip is inserted into or attached to a CO shipping carton
- How a CO shipping label is placed onto a shipping carton exterior surface
- SKU size and associated CO shipping carton

A "slapper envelope" is a clear plastic envelope or a paper envelope with a plastic window or clear plastic front that has a self-adhesive or glue-coated back that is applied to a carton exterior surface. The CO carton is either a made-up carton that contains a SKU or is a vendor ready-to-ship carton.

- The label window permits a barcode scanner to read the label.
- The size of the envelope is selected to match the smallest carton size.
- The envelope is able to contain the CO pack slip and sales literature.
- The envelope glue or self-adhesive backing adheres to the carton exterior surface.
- The envelope is not damaged or removed from the carton.

At a CO pack slip insert station, an employee or machine transfers a CO pack slip into a CO carton. As an empty carton arrives at the insert station an employee attaches a CO identification to the carton exterior surface and places the associated CO pack slip into the carton interior.

Options for shipping COs are to place a SKU into a:

- Preformed bag
- Formed bag
- Chipboard box
- Corrugated box
- A vendor ready-to-ship carton

Each type of CO shipping bag or carton affects the pick/pass line design and operation.

Pick/Pass into a Bag

With a single, low-profile, non-crushable, or fragile SKU, an employee or machine transfers a SKU, CO pack slip, and sales literature into a preformed bag made of corrugated paper or plastic. Options are:

- To place a CO SKU, sales literature, and customer order pack slip into a bag and attach a CO shipping label to the exterior. This option requires additional employee effort, which means lower employee productivity.
- To insert the CO SKU and sales literature into a bag, which is then closed manually or mechanically and a CO pack slip and shipping label in an envelope is placed onto the exterior surface.

The label application options are:

- A self-adhesive label that is applied by an employee or machine to a sealed bag or envelope.
- Machine-sprayed glue is applied to a bag or envelope and a plain label attached.

In a pick/pass concept that uses a preformed bag the workstation layout is as follows:

- SKUs are picked in bulk and transferred from a storage area to a pick/pack workstation. At the first workstation an employee packages each SKU and transfers it onto a powered conveyor travel path.
- At a second station, the SKU, sales literature, and individual CO pack slip are transferred into a bag and the bag is labeled and sealed.
- At the last workstation, a CO delivery label is scanned for manifest and the bag is transferred into a BMC, side wall cart, or direct load conveyor travel path.

If the operation uses a machine-formed bag, after bulk-picked SKUs are transferred to a pick/pack line start station, an employee places a CO pack slip, sales literature, a SKU, or the appropriate number of SKUs before a finger or between two cleats on a belt conveyor travel path. If required to have solid bottom surface, a customer pack slip, sales literature, and SKU are placed onto a chipboard insert. A CO pick slip has a CO discreet identification code facing in the proper direction for a barcode scanner.

After a CO discreet identification is read by a barcode scanner, a SKU for a CO, sales literature, and CO pack slip moves forward from the first powered belt transport section (finger or cleated) onto a form bag machine-powered conveyor travel path and into a corrugated or plastic bag-forming machine, where the material is placed between two corrugated or plastic bag sections that are square cut and sealed to formed a sealed CO delivery bag.

The sealed bag continues to travel on a powered conveyor travel path to arrive at a CO delivery label print and apply station, where a corresponding delivery label (to a CO discreet identification) that was read by a barcode scanner/RF tag reader is machine printed and applied to the bag. With a CO delivery label on the bag, the sealed and labeled bag travels on a powered conveyor travel path to a manifest barcode scanner/RF tag reader station.

Another option is to pick/pack the SKU into a preformed bag with a slapper envelope label.

- Bulk-picked SKUs are transferred from the storage area to a pick/pack workstation, where an employee transfers the SKU and sales literature into a preformed plastic or corrugated bag and seals the bag. The sealed bag is transferred onto a powered belt conveyor travel path for transport to a slapper label station.
- At this station, an employee removes the sealed bag from the powered belt conveyor travel path and applies a slapper envelope label to exterior surface of the bag. A slapper envelope has a CO pack slip inside and has the CO delivery address facing upward through a transparent window or clear plastic front.

- The completed envelope is placed onto a conveyor travel path for travel to the manifest and load station.

This requires additional stations, but results in high employee productivity, high CO volume, increased management control and discipline, and low cost.

Pick/Pack into a Box

Small or low-profile SKUs are picked and packed in a corrugated or chipboard box. There are several options:

- CO pack slip, sales literature, and SKU are placed into a box and a CO shipping label is manually applied or machine placed on the carton exterior surface.
- CO pack slip, sales literature and label are enclosed in an envelope that is placed on the carton exterior surface.
- SKU and filler inside the shipping carton. A CO pack slip and sales literature are placed into an envelope that is attached to the carton exterior.

A chipboard or pressed board box concept has a two-piece CO shipping carton and is used for SKUs that are low weight, non-fragile, non-crushable, with flat profile characteristics.

The first pick/pack station is a chipboard box bottom section entry station. At a chipboard box bottom station, an employee or machine places one chipboard box bottom section onto a finger, smooth top belt and cleated powered belt conveyor travel path. This conveyor travel path assures a constant travel speed and controlled carton forward movement.

The second station receives the bulk-picked SKUs. At this station, an employee or machine transfers one SKU from a pallet, carton, or tote into the chipboard box bottom section. With the SKU properly arranged in the chipboard box, the box moves on a conveyor travel path to a customer pack slip insert, fill material transfer station, and chipboard box top section apply station.

At a CO pack slip insert station, an employee or machine transfers (drops, slides, or inserts) a CO pack slip into the chipboard box and if required, filler material is added to the carton.

At the chipboard box top section apply station, an employee or machine places a CO delivery label on the carton top. After a cover is properly placed onto a bottom section, at the seal station one or two plastic bands or tape strands secure the CO delivery carton. The plastic bands or tape strands are machine applied as two bands/strands in the same direction or as two bands/ strands in different directions (criss-cross). Alternatively, an employee applies the tape strands. If the operation has a slapper envelope label applied to the carton after the band/tape station at a label station, an employee or machine applies a CO delivery label onto a chipboard box.

A corrugated box is used for crushable, fragile, heavier weight, and larger-profile SKUs. The pick/pack line design is determined by the carton type.

If a two-piece carton is used the operation has a carton make-up station. The pick/pack line sequence is as follows:

- Place a CO pack slip into an envelope
- Transfer a machine-made or preformed carton onto a conveyor travel path
- Bottom fill the shipping carton
- Transfer the carton onto the next workstation and place a SKU into the shipping carton
- If not using a slapper label, place a CO pack slip and sales literature into the carton and add top fill over the SKU
- Place the top or cover on the carton and seal with tape strands or plastic bands
- Place a slapper envelope or CO delivery label onto the carton exterior surface and scan the label

The pick/pack line design for a one-piece carton is designed to handle a CO shipping carton with top flaps that extend upward and are sealed at the end of the pick line.

The first station is a carton make-up station, where an employee or machine forms a carton and applies tape to secure the bottom two flaps. With a pop-out carton no tape is required on the bottom flaps. The empty carton travels over a powered roller or belt conveyor travel to the bottom fill station. At the bottom fill station, an employee or machine momentarily stops the carton traveling and an employee or machine adds bottom fill to the carton interior. With a pre-determined filler material to cover the carton bottom surface, the carton is transported to the SKU pick station, where a SKU is transferred by an employee or machine from a pallet, flow rack, or conveyor lane into the carton. At the next station an employee or machine transfers sales literature and top filler to fill any void space inside the shipping carton. An employee or machine folds the carton top flaps. The carton is transported to the carton secure station, where an employee ensures a CO pack slip is inside the carton, t applies a CO delivery address label, and the employee or machine applies one or two tape strips or plastic bands to seal the carton. The carton length and company practice determine the tape strands or plastic bands in a criss-cross pattern or two bands in the same direction on the carton. After the carton top flaps are secured, the carton moves forward to the CO delivery label station. If a customer pack slip is not in the carton, an employee or machine applies a CO delivery slapper envelope label onto the carton.

Per your delivery company requirements, the label is applied to the carton top or side. Per your equipment layout and label material, the label is self- adhesive, an envelope label with a self-adhesive back or envelope with a plain back that requires a glue sprayed onto the carton surface or glue pot.

Pick Line SKU Pick Station

At a SKU pick station, an employee or machine transfers a SKU from bulk-picked SKUs into a CO shipping carton. Cartons travel on a powered conveyor travel path that runs along the SKU pick station front. The conveyor is a zero-pressure queue conveyor or the conveyor travel path momentarily stops the carton at the SKU pick station. The pick station options are non-powered or powered conveyor travel paths, pallet or four-wheel cart, and carton flow rack.

Non-Powered or Powered Conveyor Travel Path Pick Station

Bulk-picked SKUs are transferred to a non-powered or powered conveyor charge end, where an employee opens a master carton and transfers it onto a conveyor in-feed travel path, along which the SKU master carton flows from the open station to a pick station. At the pick station, a non-powered or powered conveyor travel path intersects with the pick line conveyor travel path. At the intersection, the carton is stopped while a picker or machine transfers a SKU from the master carton into a CO carton.

The disadvantages are investment in non-powered or powered conveyor travel path, facility floor space, employee or machine effort to transfer SKUs from an opened master carton, master carton queue is determined by a conveyor travel path length, powered conveyor travel path requires electric power, and time to change SKU. The advantages are permits a high pick volume, minimizes product damage with fewest handlings, and handles a wide SKU mix.

Pallet or Four-Wheel Cart Pick Station

An employee opens a master carton and transfers it onto a pallet or four-wheel cart. When the cart is full, it is placed adjacent to a SKU pick station. At the pick station, the open end of the pallet or cart faces the pick line conveyor travel path. The open section of the master carton faces the pick line conveyor travel path, and a picker transfers a SKU from the master carton into a CO shipping carton.

The disadvantages are employee effort to transfer SKUs from an opened master carton, physical effort to transfer cartons between the pallets and carts, no master carton queue and time to build a pallet or cart with opened master cartons and to change SKU. The advantages are high pick volume, minimizes UOP damage with fewest handlings, minimal floor space, no electrical power, minimum cost, and handles a wide SKU mix.

Carton Flow Rack Pick Station

An employee opens a master carton and transfers it onto a gravity carton flow rack charge end. The SKU master carton flows from the master carton open station to the SKU pick station, where the gravity carton flow rack discharge end faces a pick line conveyor travel path. At the location where the gravity carton flow rack and pick line conveyor travel path intersect, as a CO carton momentarily stops on a pick line conveyor travel path, a pick employee transfers a SKU from a master carton into a CO carton.

The disadvantages are investment and flow rack space. The advantages are high pick volume, less product damage with few handlings, minimal floor space, no electrical power, minimal time to change SKU, master carton queue, and handles a wide SKU mix.

Pick/Pack Line Conveyor Travel Paths

A customer order carton conveyor travel path ensures a constant and controlled customer carton flow over a pick/pack line with minimal product damage, and appropriate carton queue prior to a workstation.

Selection of a pick/pack conveyor travel path is influenced by:

- Carton physical characteristics. This factor influences a conveyor travel path surface and dimensions.
- CO carton volume. This factor has an influence on a conveyor travel path type and conveyor travel speed.
- Number of pick/pack stations and the time to complete the activity.

These characteristics determine the length of the conveyor travel path, whether the pick/pack line is manual, mechanized, or automatic, and the preferred queue distance between work stations.

Conveyor travel path options are:

- Smooth, rough, or cleated top belt conveyor
- Finger conveyor
- Non-powered roller or skatewheel conveyor
- Powered roller or skatewheel conveyor

Smooth or rough top belt pick/pack line conveyor travel path features are controlled and constant travel path speed, wide SKU mix with different physical features, pick/pack line design options, no carton queue on a pick/pack line travel path, and service to a manual, mechanized, or automatic pick/pack line.

For the other smooth or rough top belt conveyor travel path features, we refer the reader to the belt conveyor section in this chapter.

A cleated belt pick/pack conveyor travel path has a smooth top belt conveyor travel path surface with upward extended projections. Projections are evenly spaced on a belt surface full-length and a nominal ⅛ to ¼ in. high.

The concept provides maximum carton control, spacing on a conveyor travel path, slight cost increase, and interfaces to a manual, mechanized, or automatic pick/pack line.

A powered finger pick/pack conveyor travel path concept has:

- Very narrow powered belt conveyor travel path
- At pre-determined intervals hardened plastic fingers that are attached to a belt conveyor surface. A plastic finger extends upward by 1.5 to 2 in. above and in the middle of the travel path surface.
- Along both sides of the plastic finger are a CO pack slip, sales literature or chipboard sheet travel path surface. The surface options are:
  - Series of skatewheel rollers
  - Hardened plastic surface
  - Coated metal surface

The finger belt conveyor travel path concept has similar operational features and design parameters as the cleated belt conveyor travel path concept, but has a slightly higher cost and some SKUs require a chipboard sheet.

The non-powered skatewheel or roller pick/pack conveyor travel path concept has each roller and several skatewheels that are attached to axles that are connected between two frames. As an employee pushes a CO pack slip or chipboard sheet forward, rollers or skatewheels on the axles permit the rollers or skatewheels to turn and move a CO pack slip or chipboard sheet forward to the next pick/pack station. A pack slip or chipboard sheet width determines a non-powered skatewheel or roller conveyor travel path width.

At a workstation, an employee completes a pick/pack activity and pushes the CO pack slip or chipboard forward over a roller or skatewheel conveyor travel path to the next pick/pack line workstation.

The concept features minimal queue between two workstations, it is difficult to assure constant and controlled flow, handles a wide SKU mix but for maximum efficiency has three rollers or skatewheels under a CO pack slip or chipboard sheet bottom surface, CO pack slip or chipboard sheet has a flat bottom surface, low pick/pack CO volume, and low investment.

The powered roller or skatewheel pick/pack conveyor travel path concept has similar design parameters and operational features as non-powered roller or skate-wheel concept except a powered conveyor travel path:

- Propels a CO pack slip or chipboard sheet forward over a travel path
- Constant and controlled CO pack slip or chipboard sheet flow
- CO pack slip or chipboard sheet queue between workstations
- Slightly higher cost
- An electric power and air supply

Use of Vendor Master Cartons as Ready-to-Ship Cartons

A ready-to-ship vendor carton pick/pack line operation has a CO pack slip placed (a) inside the carton or (b) inside an envelope that is placed onto the carton exterior surface.

Customer Pack Slip inside the Carton

With a CO pack slip placed inside a carton, the pick/pack activity stations are:

- Transfer the carton to a pick/pack line station and open a vendor carton
- Place a CO pack slip and sales literature inside the vendor carton
- Seal the vendor carton
- Place a CO shipping label on the carton exterior surface
- Scan the barcode/RF tag shipping label

The concept features additional labor expense to open and seal the carton, tape expense, label labor expense.

Customer Pack Slip inside a Slapper Envelope

A CO pack slip placed inside a slapper envelope (paper envelope with a plastic window or a clear plastic front) is placed onto a vendor ready-to-ship carton exterior surface. The pick/pack activity stations are:

- Transfer a carton to a pick/pack line
- Place a CO pack slip and sales literature inside a slapper envelope
- Employee or machine places a slapper envelope on the carton exterior surface
- Scan the barcode/RF tag shipping label

The concept features are low labor and label expense.

Company Sales Literature Insert Station

A company sales literature pick/pack station has the company sales literature transferred into a CO shipping carton. The sales literature transfer options are:

- At a pick/pack line workstation. With this option, the sales literature is transferred at the same workstation that transfers a CO pack slip into a carton. This application handles a wide literature mix but has an additional employee activity.
- When a CO pack slip and sales literature is placed inside an envelope, the activity is either a manual or a mechanical activity.

The CO volume, carton size, and cost determine the option. An important factor is a CO delivery envelope window for a CO shipping bar code, paper or plastic envelope thickness, ability to adhere to a carton, and envelope back glue surface.

Picked SKU Transport Concepts

The next important manual order-fulfillment component is the method that a picker uses to transport picked SKUs and completed COs from a pick area to a transport concept or pack station (see Figure 4.30). Options include:

- The picker carries SKUs or COs in their hands, tote, or apron.
- A four-wheel push cart with a cavity, multiple shelves, self-dumping hopper, or step ladder.
- A rolling ladder with four wheels on a floor surface or two wheels that are attached to an overhead guide rail.
- A manual pallet truck or platform truck with shelves.
- A overhead trolley with a basket or bag on ceiling-hung or floor- supported travel path.

A Picker Carries Single Items

In a picker carry concept, a picker walks with a pick instruction to a SKU pick position, withdraws a SKU from a pick position, verifies pick transaction completion and transfers the picked SKU into his or her hands. To increase picker productivity by reducing walking trips, time, and distance, the picker places SKUs into a tote, carton, sack, or apron. The picker continues to pick SKUs into the apron or tote until he or she is unable to carry additional weight; then the picker walks with the SKUs and deposits them into an assigned CO tote or onto a take-away transport concept or pack station surface.

This method is preferred for a pick/pack single CO or batched CO pick/sort or batched CO pick with a self-adhesive label order handling order-fulfillment concept. The order-fulfillment operation has a pick line or short pick aisles. The pick aisles and pick position rows are three to four flow rack bays or shelf bays long.

The disadvantages are picker walk distance and time, low picker productivity, greatest picker number, handles a low pick volume, CO volume, and has a limited load weight or cube, pick errors, and UOP damage, and picker injury. The advantages are low cost, minimal training, low maintenance, handles all SKUs, can be implemented in any type of facility, and interfaces with a single CO pick/pack concept batched CO pick/sort concept.

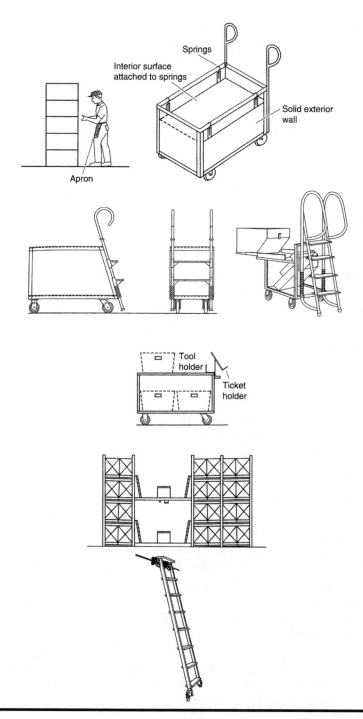

Figure 4.30 Picker options: apron, spring-loaded bottom tote, self-dumping cart, cart tool holder, and ladders.

Manual Four-Wheel Cart Concept

In a manual push or pull four-wheel cart concept, a picker pushes or pulls a four-wheel cart with one load-carrying surface, a cavity, or several load-carrying surfaces through a pick aisle. The four-wheel cart has the same order-fulfillment design parameters, operational features, disadvantages, and advantages as the manual carry concept. Exceptions are increases in carrying load weight and cube, cart expense, two carts travel an aisle, an aisle width matches the requirement, per cart type used in all pick concepts, or a cart is pushed to a pack station, carts have a storage area, and some maintenance.

With a tote or several totes on a four-wheel cart, an order picker moves through a pick aisle. The CO pick instruction has an arithmetic sequence to pick positions. As a picker arrives at the first pick position, the pick instruction indicates SKUs for a CO tote. The picker transfers a quantity of SKU pieces from a pick position to a CO tote. After all pick transactions, a picker transfers a completed tote or totes to a pack station.

At a pack station, a packer, per a paper documentation or visual display screen, identifies CO picked SKUs. With the pack instructions, a packer identifies the appropriate identified SKUs in a tote or spreads the SKUs on a table top and a packer completes a CO pack activity. With the SKUs spread over a table top, SKUs for a CO are more easily identified by a pack/sort employee. Many order-fulfillment professionals refer to this small-item order-fulfillment process as a "low cost sortation" concept.

A four-wheel non-tilt manual push or pull cart concept is a variable path in-house transport concept. A non-tilt cart means that all four wheels are in contact with the floor surface. A picker with pick instructions pushes or pulls a four-wheel cart through a pick aisle. At a pick position a picker completes a pick by transferring a picked SKU into a load-carrying shelf section or cavity on the cart. The four-wheel cart components are:

- Load-carrying surface and structural support members
- Four casters and wheels that are under each load-carrying corner
- Push handle or two handles
- Optional items such as
  - Attached step ladder
  - Document, ticket, or tool holder
  - Locking wheels
  - Brake mechanism

The advantages are that this type of cart handles a wide product mix, larger UOP volume and SKU quantity, larger cube or SKU, and greater weight or capacity.

After receiving pick instructions, the picker pushes or pulls a cart from a dispatch station, through pick aisles to the various pick positions and with a completed CO or full cart to an assigned pack station.

Cart Structural Support Members

Structural support for a four-wheel cart is provided by two or four upright (vertical) posts and two cross or horizontal members that are connected to the upright posts. In a small-item pick operation, this type of cart is used for loose, small SKUs that are placed onto a shelf, into a cavity, or into a container that sits on a shelf.

Four-wheel carts with four upright members or posts support two or more shelf levels or one cavity. Each shelf level is connected to the upright posts. With a two shelf minimum, one shelf level is connected at the upright post top and the second shelf level is connected at the post bottom. These shelf locations are required to ensure sufficient load-carrying capacity and rigidity.

The cross-structural members or shelves are metal or hardened plastic members that are connected to the upright post at the top and bottom. They are welded, hinged, or secured with a nut and bolt to ensure cart rigidity and stability as the cart is moved across the floor surface.

All cart upright posts and cross members are metal and a coating is applied to the exterior surface to reduce UOP damage that occurs when a picker handles a non-coated metal member surface, which dirties the picker's hands. When the picker handles a SKU with dirty hands, the dirt or rust is transferred from the picker's hands to the SKU.

Rigid or Collapsible Shelf Designs

The basic shelf designs are:

- Rigid shelf — all structural support members, comprising upright and horizontal posts and shelves, are welded or connected with nuts and bolts as one unit. The structural support members are permanently attached or fixed to upright posts. With a fixed shelf, a four-wheel cart occupies a large fixed floor area whether a cart is being used in a pick activity or is empty.
- Collapsible shelf — upright posts or vertical support members and load-carrying shelves have a hinge or other non-permanent method for shelf attachment. Shelves are locked into position onto upright posts to support loose small-items or containers of items on the cart shelf.

Compared to a rigid shelf cart, a collapsible shelf cart has an equal load-carrying capacity and occupies the same floor space as a rigid shelf cart, except that a collapsible shelf cart is used for loose, small SKUs in a tote. When a collapsible shelf cart is not needed to carry UOP, flexible structural members and load-carrying shelves are disengaged and collapsed into a small, one unit cart. In the collapsed mode, when compared to the rigid shelf cart, the collapsed shelf cart occupies a smaller area. Depending on your order-fulfillment requirements and handling volume, keep in mind that this collapsible cart feature does require labor to collapse and assemble the cart.

Load-carrying shelves may be made of solid metal, wood, plastic (solid or slats), or wire-mesh. The load-carrying shelves or surfaces are placed at an elevation that permits a picker to perform UOP load and unload functions. For maximum efficiency, the elevation between the shelf levels is set for your tallest SKU height or container height plus a few inches for clearance. The clearance or open space allows a picker to easily and quickly perform UOP loading and unloading transactions without injury or UOP damage. For flexibility, with most rigid cart designs, elevation settings of middle shelves are adjustable to accommodate future SKUs that are taller in height. In all cart designs the top and bottom shelf levels are fixed, which provides cart rigidity and stability.

If your pick concept has SKUs picked into a container, an option is to have one side of the shelf with no lip. The one-lip feature permits easy and quick container transfer from the cart shelf. This no-lip feature minimizes your pick and pack employee physical effort to complete a container transfer transaction between the cart and a workstation.

A four-wheel cart shelves have four sides. Each shelf side has a lip along the perimeter and the lip extends downward toward the next shelf level opening. With this shelf lip arrangement, when you determine each shelf elevation setting, the shelf lip bottom is the top limit for the higher shelf level opening.

The maximum total cart weight that an employee can effectively and efficiently handle is approximately 1200 to 1400 lb. per cart. This cart load weight includes approximately 300 to 400 lb. for the cart weight which means that the UOP weight per cart can range from 900 to 1000 lb.

General written functional specification for a shelf cart is determined by cart use in a pick operation. A typical four-wheel cart load-carrying surface or shelf supports a total of 200 to 300 lb. This UOP load weight is the combined UOP and container weight that is evenly distributed over the load-carrying surface or the shelf. The upright frames or posts equally support the total weight for the UOP and containers plus the weight of the shelf. From your cart design factors, the shelf weight capacity is determined by your cart manufacturer. For best performance, low picker injuries, and minimal product damage, the combined UOP, containers, and cart weight handled by a picker is 1200 to 1300 lb. per cart. The combined cart weight is equally supported by each of the cart's four casters and wheels.

If loose, small items are picked and transferred onto the cart shelf, a solid load-carrying shelf with solid sides, front lip, and rear walls is the preferred shelf design. During in-house transport across the floor between two facility locations, these solid sides, front lip, and rear walls, and solid shelf features reduce the risk of loose, small SKUs falling from the shelf. If the small items will not fall through the wire-mesh openings, wire-meshed sides, shelves, and rear walls are used on cart.

If containers are on the cart shelf, a solid or wire-mesh load-carrying surface with open sides and rear wall is considered for the small-item order pick area. If your cart small-item pick activity or in-house transport activity has loose, small items or containers fall from a cart shelf as the cart is moved across a floor surface, your cart shelf design has a load-carrying surface or shelf lip that extends upward from the shelf on all four sides. With this shelf lip arrangement, the shelf lip acts as a barrier to the SKUs or container bottom. This lip reduces the risk of UOP falling from the cart shelf. When this shelf lip arrangement is used on a cart shelf, a shelf opening is the open space between the shelf lip and the next higher shelf bottom. Another lip feature to consider is that the shelf lip extends upward and during the unloading and loading activities, your picker is required to lift UOPs or the container up and over the shelf lip to perform a SKU or container transfer transaction.

If dust and dirt accumulation on a cart shelf presents a problem to your small-item order-fulfillment operation, your cart design has a wire-mesh shelf or plastic shelf with slats as a cart load-carrying surface. Wire-mesh or slat openings in the shelf permit the dust and dirt to fall through onto the floor. The wire-mesh and slat opening design has a narrow width and length so that the smallest loose SKUs remain on the shelf. When small, loose SKUs are handled by a wire-mesh or slat shelf cart and the SKUs become hung up on the openings or fall through the openings or slats, to reduce this occurrence you specify your shelf wire-mesh opening or slat size to ensure that the smallest SKU cannot fall through and that SKUs do not become hung up.

Method to Prevent SKUs Falling from a Cart Shelf

If during the pick activity or the in-house transport activity, loose, small SKUs fall from the shelf, modifications to the shelf design can reduce the risk. Options are:

- As previously mentioned, shelf lips are turned upward.
- A removable barrier that extends between two upright posts across the entire shelf front opening or UOP transfer opening. The end of each barrier is attached to one upright post at the proper height above the shelf surface. The height of the SKU or container determines the height of the removable barrier or the use of multiple barriers on the shelf opening front. This feature creates an adjustable and removable front barrier across the shelf opening that reduces the risk of the loose, small items or containers falling off the shelf and damaging UOPs.
- Load-carrying surfaces or shelves angled at a 5 degree slope or pitch to the shelf rear wall, side, or middle. With a wide cart that has product transfer transactions performed from both sides, the shelf is shaped in a wide V which creates an incline toward the middle so loose SKUs or the containers are retained on the shelf. This feature minimizes the risk of items falling from the shelf.

Methods to Handle Multiple Discreet SKUs per Shelf or Opening

If your order-fulfillment operation handles batched COs, consider using either dividers on each shelf level or containers or cartons on the shelves.

The first option is to have fixed or movable dividers either permanently or temporarily attached to a shelf, to separate the load-carrying surface or shelf into smaller compartments to handle your CO UOP mix. Each shelf has a lip with sufficient space for each CO discreet identification.

An alternative method to separate the cart shelf for discreet SKUs is to use a captive container or CO shipping carton as a customer order location. Each container or carton is positioned on the shelf so that the picker easily transfers UOP from a pick position into a container.

With either concept, for best performance the computer determines an opening size which is based on a CO cube. With a container or carton application, some cart shelf designs have a shelf angled upward on the back side (inverted V-shape) and a high lip on the front to retain a container or carton.

Push Handle or Push Bar

A four-wheel cart that is pushed or pulled by a picker over a floor surface through pick aisles has push or pull bars, grips, or handles on each of the four upright posts or shelf rear/front sides. The locations are designed to minimize picker injury, permit a picker to push or pull the cart, and allow a picker to steer the cart.

Rubber or plastic bumpers are attached to the ends of the push handles. When several carts are queued in a line, a bumper permits an employee easy access to a cart push handle.

Depending on a cart written functional specifications and drawings, operational requirements, and caster/wheel location and selection, push handles, grips, or bars are attached to one or both cart ends.

Wheel/Caster Location and Other Design Parameters

Other design components of the four-wheel cart are the axle, axle bearing, axle bracket, caster mounting to the undercarriage, wheel/caster location, wheel type, wheel cover, wheel height, and caster type.

The caster design directly affects an employee's ability to move, steer, and perform loading and unloading SKU transactions between a cart and pack station. Important caster/wheel factors are:

■ How casters are attached or mounted to the underside or bottom frame of the cart
■ Wheel diameter and wheel covering or tread
■ Lockable or non-lockable casters
■ Wheel location and wheel number
■ Axles, axle bearings, and axle brackets
■ Cart load rating or weight-carrying capacity

Casters can be rigid or swivel-type.

- Rigid caster. As a cart is pushed or pulled across a floor surface through a pick aisle, a rigid caster serves to:
  - Maintain a wheel in a fixed travel direction.
  - Permit easy cart turning at a pick aisle end or at corners. With most manual inverted T-shaped or regular-framed push carts, a rigid caster is located in a cart load-carrying surface rear and trails in the direction of the cart travel.
- Swivel caster. When a cart is pushed or pulled across a floor surface through a pick aisle, swivel casters permit the wheels to rotate in all directions. This feature permits a picker to easily maneuver and align the cart at a pack station. With most picker push carts, swivel casters are located in the front or lead section of the load-carrying surface or direction of cart travel.

The caster type you select is determined by your cart operational movement requirements, combined cart and UOP weight, cart load-carrying surface or frame length, floor surface and debris conditions in a pick aisle, and economics.

Caster Type and Caster Mounting Locations

Basic caster attachment locations are:

- Two swivel casters in front and two rigid casters in the rear. This concept permits a picker to easily turn a non-tilt cart at the ends of a pick aisle. The advantages are handles lightweight loads, easier to steer through turns and wider aisle space to turn or align the cart at a pack station.
- Two rigid casters in the front and two swivel casters in the rear. This arrangement increases effort and time by your picker to turn the cart because the front wheels are aligned in the direction of cart travel prior to moving a cart into a pick aisle or at a pack station. The advantages are handles heavy loads and are good for long and straight travel paths.
- Four swivel casters in all four corners. A swivel caster under each cart corner permits a picker to easily move a non-tilt cart in any direction with minimal difficulty to turn a cart. Advantages are handles lightweight loads and easy to align a cart at a pack station. The disadvantage is that during cart travel across a floor surface or in a pick aisle, the cart is difficult to steer in a straight travel path.
- Two swivel casters at the two end corners and two rigid casters in the middle of the cart. The rigid casters have a higher diameter than the swivel casters. During cart movement across a floor or in a pick aisle, this caster arrangement permits a picker to steer a cart from either cart end. This arrangement is used with a tilt cart.

Cart Wheel Diameter and Wheel Cover (Tread)

A cart wheel diameter and wheel cover or tread is basically the wheel size or height. The wheel diameter is determined by floor surface and conditions and type, and amount of debris on a floor surface or pick aisle, load bar or shelf height for the employee to complete a SKU loading and unloading transaction, combined product, and cart load weight that is handled by the cart, and economics.

A typical cart wheel diameter for a manual push or pull four-wheel cart has a diameter range from 3 to 8 in. high. A selected wheel diameter affects a shelf level elevation above a floor and directly affects a picker's ability to perform SKU pick and transfer transactions. Cart wheel diameter options are:

- Small diameter. A small-diameter wheel has a small height or low profile. It handles a lighter-weight load, the bottom shelf level or load-carrying surface is closer to the floor surface, and improved cart stability with a lower center of gravity. These factors result in a more stable and safer load as a cart is pushed or pulled across a floor surface or in a pick aisle.
- Large diameter. A large-diameter wheel has a high height which means cart travels more easily over a floor surface with cracks and joints, handles a heavier combined UOP and cart load weight, and bottom shelf level or load-carrying surface is at a higher elevation above the floor surface.

A wheel cover or tread is the wheel portion or surface that comes in contact with a floor surface or pick aisle surface. A wheel cover is a key factor that permits your picker to easily push or pull a four-wheel cart across a floor surface or through a pick aisle. The wheel cover can be a cushion rubber tread, hard rubber cover, or metal cover with plastic or hardened metal. Most cart applications do not use hardened metal wheels due to the noise level that is created as a cart moves across the floor.

Wheel Location and Wheel Number

The wheel location and wheel number that are used on a four-wheel cart undercarriage refers to a caster/wheel attachment or mounting location to a cart leg, upright post, or bottom support frame. Cart wheel design options are:

- Front wheels are set inside the rear two wheels.
- Two front and rear wheels are in a straight line from a cart front to the rear for a cart direction of travel.
- One small-diameter wheel set under the middle of each cart direction of travel ends and one large-diameter wheel set under a cart middle.

With the front two casters/wheels locations set inside the two rear casters, your employee easily moves a cart through turns. The caster is easily moved through a turn because the rear wheels have less drag and a low wheel resistance.

When two front and two rear casters are set in a straight line between two ends of a cart for the direction of travel, the cart is more stable as your picker moves a cart across a floor surface or through a pick aisle.

A non-title cart is the most common in a small-item order fulfillment operation. When casters are located under each cart direction of travel end and have a small-diameter wheel and are swivel-type casters and the two other casters/wheels under each side and in the middle have a high-diameter wheel, this wheel height and location difference gives the cart tilt characteristic. This cart feature provides your picker with the ability to transport a heavier load and to improve cart steering. Wheel number under a cart options are:

- Four wheels. A cart with four wheels under the load-carrying surface has a wheel under each rectangle-shaped cart frame corner. This wheel design supports a typical manual push or pull cart for a normal pick area load.
- Six wheels. A cart with six-wheels under the load-carrying surface or frame has a wheel under each cart corner rectangle-shaped load-carrying surface or frame and one wheel under the cart side surface or frame middle. This six-wheel arrangement gives the cart a higher load-carrying capacity but increases cost.

Axles, Bearings, and Axle Brackets

The axle, bearings, and axle brackets permit the cart wheels to turn as the cart is moved along the floor surface. A cart has four axles, each of which has a hardened metal shaft through each wheel center. Each axle supports one quarter of the combined cart and UOP weight. As a cart is pushed across the floor each wheel revolves on an axle bearing. The axle thickness and length are determined by your cart manufacturer. Axle design factors are wheel selection, caster selection, combined cart and product load weight, and your floor surface and condition.

An axle bearing in each wheel center permits the wheel to revolve on the axle as the cart is pushed. The axle bearings can be oil-less, roller and ball, ball bearing, and plastic sleeve. An axle bracket is a preformed hardened metal bracket that is attached to the cart underside. It has two sides that extend downward on both sides of a wheel. At the bottom of each bracket extension is a hole that secures the axle in a constant position and maintains the proper wheel position. Locking caps or other devices secure an axle in the proper position. Each axle bracket holds approximately one quarter of the combined UOP and cart weight.

The load rating for each caster and other wheel components is determined by your cart manufacturer. The load rating is a manufacturer-stated total cart load-carrying weight capacity. The load-rated weight is the combined product and cart weight plus your cart manufacturer safety factor.

Cart Bumper

A cart bumper is an option for a manual push or pull cart. Cart bumpers are designed for attachment to the four corners of a cart. The bumpers reduce cart, material-handling equipment, and building damage that may occur as the cart is pushed across a floor or through a pick aisle. The bumpers are attached to the cart's load-carrying deck surface. Upon impact, the cart bumper cushions the impact. Cart bumpers are manufactured from material that does not leave marks on objects, such as a wall or workstation that it may strike.

Cart Options

Other optional equipment increase order picker productivity and can added to the cart as a single option or a combination.

Step Ladder

A step ladder increases the ability of a picker to complete picks from elevated pick positions that are above normal reach. Safety ladders have rubber-cushioned legs. The cart and step ladder should be tested to ensure that they remain stable when a picker climbs the ladder.

Self-Dumping Hopper

A self-dumping hopper has four side walls and a bottom surface. The four side walls slide forward across the load-carrying surface. This causes picked loose SKUs in a cart cavity to slide forward and fall onto a belt take-away conveyor. A self-dumping hopper that discharges SKUs onto a belt conveyor travel path improves picker productivity.

Document, Ticker, and Tool Holder

An order picker document, ticker, or tool holder provides a location on a cart to hold all the necessary pick tools, pick instructions, and other items within easy reach, allowing a picker to perform pick transactions with minimal effort. The holder design and location permits unobstructed deposit of picked SKUs into the cart.

Wheel Locks or Floor Surface Lock

A caster/wheel locks or floor surface lock prevents accidental carton movement over a floor surface at a pack station or other workstation. A picker engages or disengages the lock device.

In most small-item/flatwear apparel order-fulfillment operations, wheel brake options are:

- Brakes that are attached to the side of the push handle on a cart. This feature minimizes a picker employee effort to activate and deactivate the wheel brake, which is a levered device that in an active position has a rubber pad that presses against the wheel cover. Pressure from this rubber pad prevents the wheel from revolving on its axle and stops the cart. For best results, a caster/wheel lock device is used on a wheel with a rubber cover.
- A floor surface lock device is attached to the bottom structural frame of the cart and extends outward slightly beyond the edge of the frame. It is activated by a lever located on the side of the push handle. When a picker activates the lever, a floor surface lock friction face or pad extends downward and comes in contact with the floor surface. An engaged floor lock device has a friction face pressed against the floor surface, locks the cart in place. To ensure that a floor surface lock device matches your cart frame elevation, the cart manufacturer installs the device. You specify the mounting height above the floor surface, retracted height above the floor surface, bottom frame elevation above the floor surface, and caster/wheel diameter

Disadvantages of the manual four-wheel cart concept are the physical effort by a picker to push to cart and carry out loading and unloading activities, the limited volume and weight that it handles, and the need for a smooth floor surface that is crack- and debris-free. Advantages are that it handles a wide UOP mix, carton, and containers, variable path concept, normal floor surface, low investment, low maintenance, minimal picker training, operates in a narrow aisle, used in any pick philosophy or methodology type, and no fuel, storage or battery-charging area.

Retail Shopping Cart

A retail shopping cart has four wheels, with two swivel casters on the front and two rigid casters on the rear, a push handle and formed wire strands that create the SKU-carrying cavity. A picker transfers SKUs from the pick position into the retail shopping cart. With a completed CO or a full cart, the picker pushes the cart to a pack station.

This type of cart is not preferred for a dynamic small-item order-fulfillment operation because of low picker productivity resulting from additional picker pick area and small load-carrying capacity. If picked SKUs are not in a container, they may become hung up on the wire strands. A retail shopping cart is used in a single CO pick concept, in a batched CO pick concept with a label pick instruction and in a pick/pack order-fulfillment concept that has a CO container in a cart cavity. A retail shopping cart moves between two rows in a 48– to 72–in. wide pick aisle.

Disadvantages are to complete a load and unload transaction a picker lifts SKUs over the side walls, small load-carrying capacity, and some SKUs hang up on the strands. The advantages are easy to push and steer through pick aisles, is nestable and thus when not being used in an operation requires less storage space, and most pickers are familiar with the cart.

Data Mobile or Pick-to-Light Cart

The data mobile or pick-to-light cart (see Figure 4.31) is an in-house transport concept for small-item picked SKUs or completed COs. The data mobile or pick-to-light cart has similar design parameters for structural support members, shelves, and wheels, and the same operational characteristics as a four-wheel cart.

A data mobile or pick-to-light four-wheel cart is a manual, single-item pick cart that has an on-board microcomputer and two to three shelf levels for CO containers. At each pick position is a pick-to-light instruction device. Some pick-to-light cart models have an on-board printer, top rail aisle guidance concept, either an on-board rechargeable electric-powered battery or an overhead DC (direct current) rail, and receives a CO pick requirements via wireless infrared interface from a microcomputer. The pick-to-light cart has the shelf locations and pick-to-light instruction modules to handle a single CO or 8 to 16 batched COs.

To pick SKUs, options are to have each pick-to-light cart:

■ Make one pass through a pick aisle between two shelf pick position rows. A picker enters a pick aisle from a start location and travels through the entire pick aisle to the end pick location. Then the picker pushes the cart through the rear traversing aisle to the next or adjacent pick aisle and travels through the aisle to the pick area front. This travel path is repeated until all required

Figure 4.31 Pick-to-light cart.

picks or a CO container is full. The picker routing pattern is used for long pick aisles that are between long pick position rows. This feature means that the small-item order-fulfillment operation has a large SKU quantity.

■ Remains in the main traffic aisle and a picker walks from the cart to a pick position, completes a pick transaction, and returns with SKUs to a pick cart, where the picker places SKUs into the appropriate CO container. This picker routing pattern is used for short pick aisles that are between pick position rows. The feature means that a small-item order-fulfillment operation has a small to medium SKU quantity.

In a pick-to-light cart concept, a pick instruction is a microcomputer-controlled pick light digital display screen that indicates the appropriate SKU pick requirement for each CO. With some models a pick instruction includes the pick location, SKU discreet identification, and CO pieces, sleeves, or cartons. With a pick instruction indicated at each CO location, a picker completes all required picks for SKUs on a CO batch.

With a pick-to-light cart, a CO is cube-out for a container, which means that the computer separates a CO per the pick aisle sequence into the appropriate tote number.

Pick-to-light carts or data mobile vehicles handle 8 to 16 COs. Each CO container front faces the cart rear or side, to ensure efficient SKU transfer from a pick position to a CO container. The design considerations are:

■ With 16 CO cart, to have picker access to both sides of the cart, a pick aisle is 5 ft. to 6 ft. wider than the normal cart pick aisle
■ With a rear access cart, the CO number is 8 with a 3-ft. wide aisle
■ Cube COs
■ To minimize equipment damage and ensure good travel speed, guided cart travel through a pick aisle

The disadvantages are since the vehicle travels over a fixed travel path and if the vehicles are not properly dispatched to a pick aisle, potential vehicle queue in a pick aisle, capital investment, management discipline and control, and for best performance, the vehicles are guided through a pick aisle and some applications have a wide pick aisle. The advantages are accurate pick activity, higher picker productivity, reduced UOP and equipment damage, batched or single CO handling option and handles all SKU types.

For additional pick light information, we refer the reader to the section on pick-to-light instructions in this chapter.

RF Pick Cart

RF pick cart technology is very similar to the pick-to-light cart, except for a few features. The similarities are:

- COs are batch picked.
- A picker pushes or pulls the pick cart through a guided pick aisle.
- It is a paperless pick concept.
- CO picked SKUs are transferred from a pick position to a CO container or shipping container.
- Picker accuracy and productivity are improved.
- A scanner or RF device holder leaves a picker's hands free.

The RF pick cart features a digital display screen that indicates the total CO SKU quantity and the required SKU quantity per each CO.

Disadvantages are that because the vehicle travels over a fixed travel path, if vehicles are not properly dispatched to a pick aisle, potential vehicle queue in a pick aisle, capital cost, management discipline and control, and for best performance, vehicles are guided through a pick aisle or have a wide pick aisle. Advantages are accurate pick activity, high picker productivity, reduced UOP and equipment damage, batched or single CO pick options, and it can handle all SKU types.

For added RF pick information, we refer the reader to the section on RF pick instructions in this chapter.

Manual Pallet Truck or Platform Truck with a Shelf Pick Concept

A manual pallet truck or platform truck with a shelf or totes is similar to a manual pallet truck or platform truck carton pick concept. The hand pallet truck concept has a picker operate a non-powered pallet with a multiple tote or carton shelf carrying device through a pick aisle. At a pick position, a picker transfers a SKU into a tote or carton on the shelf. The completed CO or full carton/tote is pulled to the pack area. A hand pallet truck is designed with:

- Operator handles that activate a hydraulic elevating mechanism.
- Hand- or foot-operated hydraulic pressure-release lever.
- One or two steering wheels and two load-carrying wheels to carry a 3000-lb cart or pallet; each load-carrying wheel is under a hardened and coated metal fork.
- Options are double length pallets and load back rest.

For best results, a hand pallet truck concept has:

- Smooth and even floor surface with no debris in a pick aisle.
- Pallet with chamfered (angled cut edges) bottom deck boards to minimize fork entry problems.
- Cart with a chain and lock to a back rest and a cart undercarriage that has the structural strength to support the full weight of the cart and has two sleeves full length. The width and depth permit HROS truck forks to enter

the C-channel opening and prevent a cart from moving side to side. Cart wheel diameter or fork raised elevation assures that a raised cart wheels do not contact the floor surface.

As a picker with a hand pallet truck picks up a pallet with a shelf device or a shelf cart, a picker sets the hydraulic lever in an elevate position to elevate the pallet or cart. A picker pulls or pushes a hand pallet truck through pick aisles and assembles SKUs into a carton on a shelf. With a pick activity complete, the pallet or cart is pushed or pulled to a pack area and to deposit the pallet or cart a picker sets the hydraulic lever in the lower position.

Electric Battery-Powered Walkie Pallet Truck with a Shelf or Walkie Tugger with a Cart Train

An electric battery-powered and -controlled walkie pallet truck with a shelf pallet or cart is a manually controlled and electric-powered tugger with a cart train that has the capability to travel horizontally through a pick aisle. The electric-powered pallet walkie truck or tugger reduces a picker's physical effort to move through a pick aisle.

For additional information on the electric battery pallet truck with shelf, we refer the reader to the manual pallet truck with shelf pallet or cart or electric battery-powered pallet rider truck or electric battery-powered rider tugger with a cart train section in this chapter.

Overhead Manual Trolley with a Basket or Bag

The overhead manual trolley with a basket or bag for picked SKUs or CO transport is very similar to an overhead rail and trolley hanging garment or carton pick concept. For additional information, we refer the reader to the GOH overhead rail and trolley section and to the overhead rail and trolley with a bag or basket section in this chapter.

Mechanized Pick Concepts

Mechanized pick concepts are those in which a picker rides on a vehicle through a pick aisle or walks in a pick aisle between two pick position rows to an assigned pick position, where he or she completes a pick transaction. The mechanized pick group options include:

■ A powered vehicle that a picker rides through a pick aisle. Loose SKUs are placed onto a carrying surface or into a tote or carton for transport to a pack or sortation station.

- A powered conveyor from a pick area to a pack or sortation area. A picker walks or rides through a pick aisle, placing picked SKUs in totes or cartons which are then transported on the conveyor.

There are similarities between a manual pick concept and a mechanized pick concept:

- Both concepts involve (a) a picker who picks SKUs for a single CO directly into a CO carton or tote; (b) a single CO with a single or multiple pickers who pick and transport SKUs to a pack station, (c) batched COs, with one picker who picks, sorts, and transports SKUs to a pack station, and (d) batched COs that has one picker or multiple pickers pick, label, and transport SKUs to a sortation/pack station.
- Both have a pick area in which pick aisles and pick position rows fit into a rectangular or square-shaped facility. A rectangle is the preferred shape because it allows longer pick aisles.
- In both, Pick aisles and pick position rows have a parallel or perpendicular layout with pickers traveling to pack stations. The perpendicular arrangement is preferred. Between a pick area and pack/sortation area is a main aisle for a pick vehicle to traverse between pick aisles.
- There is arithmetic progression to the pick position numbers or a sequential picker routing pattern through a pick aisle.
- Each CO is cubed to optimize vehicle trips in a pick aisle.
- A picker uses a paper pick document, self-adhesive label, pick to light, or RF device instruction.

The differences are:

- Picker ride concepts are difficult to implement on a pick line but are preferred in a pick aisle concept.
- With a mechanized concept, there is a need for a fuel storage or battery-charging area.
- A mechanized vehicle requires a wider turning aisle.
- The load-carrying surface on a powered vehicle handles a larger SKU capacity, more totes or cartons, or more COs.
- A powered concept travels farther in a shorter time.
- Some vehicles utilize a facility entire cube or open space between a floor surface and ceiling joists. This expands the golden zone.
- For best picker productivity and minimal equipment damage some vehicles have a wire or rail aisle guidance concept.
- With a pick label, paper document, RF device, or pick and sort concept, a mechanized concept handles batched COs.

Figures 4.32 and 4.33 show types of powered vehicles. These include:

■ Burden carriers
■ Picker carts
■ Work assist vehicles
■ Pallet trucks with shelves
■ Tuggers with cart trains
■ High-rise picker trucks
■ Decombes
■ Pick cars

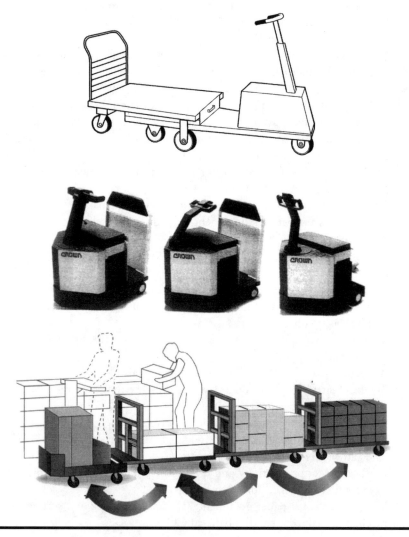

Figure 4.32   Mechanized pick concepts: tow cart and tugger and cart train.

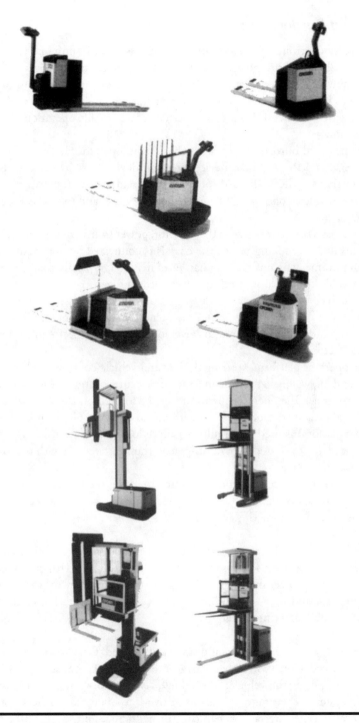

Figure 4.33   Pallet trucks and HROS trucks.

Powered Burden Carrier

A manually controlled powered burden carrier is a liquid propane (LP) gas, gasoline, diesel, or electric-powered battery rider vehicle. An electrical rechargeable battery-powered burden carrier is preferred for indoor order picking operations because of low noise level and no fume emission. An internal combustion engine vehicle, which does emit fumes, is preferred for an outside operation and for long travel distances.

The powered burden carrier has a chassis and operator platform with forward and reverse controls, a brake device, steering device, a solid deck load-carrying platform that handles a divided tote or shelves and a tow draw bar, three or four wheels with at least one wheel being the steering wheel and two wheels being drive wheels.

The chassis or shell encloses a drive motor, power source, or fuel, drivetrain, and other mechanical moving parts. The chassis is connected by nuts and bolts to the structural frame, which is made of hardened material, typically coated sheet metal, fiberglass, plastic, or aluminum, that is connected together to form one piece.

The choice of chassis material depends on the vehicle use, whether it will be used mainly as a pick vehicle or as a maintenance vehicle, whether it will be used indoors or outdoors, be required to travel long distances, trip frequency, and load-carrying factor.

The operator platform, steering device, and vehicle controls are located on the chassis and allow a picker to ride in a safe and secure location with access to all the vehicle controls. The operator platform may be located on the rear, which is used on a single-person transport vehicle with a load-carrying platform in the front. A mid-seat platform is used on a multi-person or load-carrying surface vehicle; and a front-seat platform is used on a multi-person transport vehicle with the load-carrying surface in the rear.

The steering device allows the operator to guide the burden carrier from the dispatch station, through the pick aisles, and to deliver a completed customer order to the assigned location. The steering devices may be:

- A tiller steering device, which is the most common on the rear-stand, three-wheel vehicle types or on some mid-stand vehicles that have a T-shaped handle
- A steering wheel method, which is the most common on the front-seat or mid-seat and four-wheel vehicle and is a round handle that is similar to an automobile steering wheel

The vehicle controls for stop and start, travel speed, forward and reverse vehicle movement, and a key switch that activates the internal combustion or draws from a battery to power the vehicle. The vehicle also has forward or reverse brakes or braking system. The brake options are drive line brakes and wheel brakes. The wheel brakes types are (a) mechanical, (b) hydraulic, and (c) drum or four-disk brake type.

The two power sources are:

- LP gas or gasoline internal combustion engine. The internal combustion engine vehicle is preferred for outdoor operations to carry heavy loads, for travel over grades, and for long travel distances. The fumes and fuel storage are disadvantages.
- Electric-powered battery. With no fumes, less noise, and a battery-charging area, the electric battery-powered burden carrier is preferred for indoor order-fulfillment activities.

The load-carrying deck surface, load deck capacity, and tow bar are factors that determine the number of SKUs, totes, customer orders, and the maximum load weight a carrier has the capacity to handle. Standard deck sizes are 30 in. wide by 48 in. long; 32 in. wide by 44 in. long; 24 in. wide by 44 in. long; and 41 in. wide by 77 in. long. Most powered burden carrier manufacturers recommend carrying totes that have a height that is equal to or less than ½ in. of the deck width or length. For the above deck sizes, the load weight capacities are 500 to 1000 lb., 1000 to 3000 lb., 500 to 1000 lb., and 500 to 6000 lb., respectively.

A powered burden carrier payload or tote-carrying feature is the ability to tow a four-wheel cart. To tow a four-wheel cart, a powered burden carrier needs a tow bar (hitch) and a coupler. Standard power train draw bar pull capacities are 1000, 1100, 2200, and 2600 lb.

The wheels on a powered burden carrier come in contact with the floor surface. The wheel types are:

- Pneumatic. A pneumatic tire has a tube, tubeless or pneumatic tube option, that is foam-filled urethane liner or vinyl sealant.
- Extra cushion. An extra cushion tire wears longer but is more costly and gives a passenger a harder ride.
- Solid cushion. A solid cushion tire is a solid rubber tire that is pressed onto an iron wheel and is used for hauling heavy loads and gives the passenger a rough ride.

Tire selection factors are the load weight and floor surface.

Optional accessories include headlights and turn indicators, cab or cargo box, hitch to tow four-wheel carts, and roll-out or lift-out battery compartment.

Disadvantages of the powered burden carrier are that it is limited to handling SKUs in totes and has a limited weight capacity, high cost per small-item, fuel storage or battery-charging area, and increased maintenance. Advantages are reduces picker physical effort, travels over longer distances at faster travel speeds, carries one or more totes, can also be used as a tour or maintenance vehicle, and operates over grades and outdoors.

Powered Work-Assist Vehicle

A powered work-assist vehicle is an electric rechargeable battery-powered vehicle. Totes or cartons are carried on the vehicle load-carrying surface. A powered work-assist vehicle has:

- Four wheels, each with a rubber cover with the front two wheels located under the load-carrying surface, being the two steering wheels. The two rear rigid-type wheels are under the chassis and allow the vehicle to turn in a very small area.
- An operator platform permits an employee access to all vehicle controls and steering device. Vehicle controls include vehicle forward and reverse movement, and elevate and lower an operator platform. Attached to the operator platform is a tote-carrying surface. In an elevated position, this feature makes it very easy for an order picker to transfer a picked SKU into a tote or carton and to activate the E-stop device. As a safety feature, the operator platform has a waist-high handrail. The load-carrying surface has the ability to elevate and lower on a collapsible mast to an elevation height that is 13 ft. above the floor surface.
- A fixed load-carrying surface that is above the wheels and the previously mentioned vertical moveable load-carrying surface that handles one tote/carton.
- An electric battery-powered motor with a drivetrain propels the vehicle through a pick aisle.

The features are capability to elevate and lower the operator and tote-carrying surface to a 13 ft. elevation above the floor surface, short wheelbase, which means a shorter turning aisle requirement, and carries one tote or carton.

Electric-Powered Cart with a Towed Cart

An electric battery-powered cart is a manually controlled vehicle that is a variable travel path order pick vehicle. The electric battery-powered cart design:

- Handles one cart
- Limited drag pull
- Operates in a narrow-order pick aisle
- Aisle turns have a shorter radius
- Small electric rechargeable battery
- Low combined vehicle, cart, and product weight
- Low cost
- Does not elevate a picker
- Greater SKU- or tote-carrying capacity
- Wider turning aisle
- Faster travel speed

Components of an electric battery-powered cart are an operator platform with vehicle forward and reverse travel controls and a steering mechanism connected to a steering device, electric rechargeable battery, electric-powered motor connected to a drive wheel, and four-wheel cart with a tow hitch or hook, and a roller frame.

The operator platform is located inside the chassis so that an operator is located in a position to have access to all forward and reverse movement controls and the cart steering mechanism. For operator comfort and safety, the platform has a rubber mat surface inside the chassis that has a welded coated steel or hardened plastic frame.

Controls are mounted on a handle bar and are a T-handle steering device, brake device, and a release pedal at the rear of the platform, which disengages the hitch and releases a towed cart with shelves from the roller frame. This feature allows a cart to be pushed from the vehicle and allows the cart to travel forward without a towed cart with a coupler. When a pick cart is carried on a roller guided frame, the load-carrying wheels have sufficient diameter to have a cart wheels above a floor surface. The two front wheels are swivel-type and the rear two wheels are rigid. Other controls include:

- Forward travel speed controls.
- Reverse speed control.
- Horn.
- Ignition switch.
- Flashing operating light switch.
- Three wheels with rubber covers or tread; one is the steering wheel and load-carrying wheels under each rear roller frame corner are rigid.

An electric-powered four-wheel cart with a tow cart is used to pick loose small-items as a single CO into one tote, pick and sort batched CO SKUs into an assigned tote on a shelf, or batched CO SKUs loose into a container.

Disadvantages are one cart per trip, battery-charging area, fixed trips per battery, cart matches a powered picker cart requirements, and limited order number. Advantages are very similar to a powered burden carrier concept except that a powered cart with a multi-shelf cart has a large capacity.

High-Rise Order Picker and Man-Up Very Narrow Aisle Vehicle

A man-up very narrow aisle (VNA) vehicle or high-rise order picker (HROS) is a manually controlled vehicle that has the capability to travel horizontally and vertically through a pick aisle. Both a picker and load-carrying surface elevates and lowers to the proper height above a floor surface to complete a deposit or pick transaction. For best picker productivity and minimal equipment damage, a HROS vehicle has a rail or wire guidance travel concept, a locking device to hold a pick pallet or cart, and a picker routing pattern. In a small-item flatwear HROS order-fulfillment operation, a picker picks to a pallet or cart that has multiple shelves and dividers.

The vehicles have:

■ An electric rechargeable battery
■ A picker platform with all the vehicle steering, forward and reverse and fork-elevating controls and an operator tethering line
■ Shelves on a pallet or cart with a securing device
■ Three or four wheels with one wheel as the drive and steering wheel
■ An electric-powered drive motor and drivetrain
■ A chassis
■ An elevating mast
■ An aisle guidance system
■ Accessories

The first unique feature is a picker platform with the entire vehicle steering, forward and reverse, and fork-elevating controls and an operator tethering line. The picker platform has a rubber mat surface, wire-meshed or plastic barrier between the picker and the elevating masts, safety gates and rail that surround the picker platform, and an overhead guard. The safety gates and rails devices are engaged to move a picker truck or platform. Controls are a steering device, picker and load-carrying device, elevating control handle, and forward and reverse control handle. To ensure picker safety, a tethering line is used to connect a picker to an overhead guard section. The picker truck has a set of forks and a claw that grabs a shelf pallet stringer and secures the pallet to a platform, a shelf cart has a middle stringer for claw security or a chain to secure a shelf cart to a picker platform. Elevating masts are located in front of the picker platform, and a picker is able to raise or lower the mast, picker platform, and SKU-carrying device above the floor.

To have high picker productivity, fast travel speed, and minimal equipment damage, most picker truck applications travel in a guided aisle. The aisle guidance concepts are:

■ Rail or mechanical
■ Wire
■ Tape
■ Paint
■ Laser beam or electric

The mechanical guidance systems are used with some modifications on existing vehicles or in an existing or new facility. If a guidance system is considered for an existing or new facility, prior to the purchase and implementation of the system, careful attention is given to the floor (surface, levelness, loading, metal objects, and rebar location). This is true for a wire guidance system.

For additional HROS guidance, slow-down device, truck types, routing patterns, and order pick devices information, we refer the reader to the Storage HROS section in Chapter 3.

Electric-Powered Pallet Rider Truck with a Shelf Pallet or Cart Concept

A manual pallet rider truck with a pallet shelf or shelf cart concept is referred to as a low-lift powered pallet truck. The truck is used in other operation activities, requires minimal picker training, and is easy to operate.

The following types of pallet trucks are included in the group:

- Walkie/end rider pallet truck
- Mid-control rider pallet truck
- Remote-controlled pallet truck
- Elevating single pallet truck
- Operator step single pallet truck

To improve picker safety and productivity, most pallet trucks are available as double pallet length trucks or side-rider pallet trucks.

Walkie/End Rider Single Pallet Truck

The distinguishing feature of a walkie or rider pallet truck is the picker riding platform. A walkie or rider pallet truck has a handle bar with controls to raise and lower the forks, and a horn device. Most manufacturers have stabilizing casters under the front rider platform.

With the controls and steer handle in the center of the truck and rider platforms on both sides or the front, the vehicle can be operated by a right- or left-handed picker.

The control bar provides a handgrip for the pickerso that he or she can stand and maintain his or her balance on the platform. This feature improves picker safety.

During a pick-up or deposit activity, the picker activates forks by pushing on the control button located on the bottom of the handle bar.The picker is able to raise or lower the pallet truck forks without releasing his or her grip on the steering handle. This feature improves picker safety and productivity.

The end rider pallet truck stabilizing casters are under each corner or at both rider platform ends. Each stabilizing caster is a swivel-type caster with a 4-in. diameter hardened plastic wheel on an axle. These stabilizing casters ensure that as a pallet truck makes turns the end ride platform remains balanced in relation to the floor surface. This feature reduces accidents and improves picker safety and minimizes damage to the pallet truck or floor surface.

To operate an electric battery-powered end rider pallet truck as a pick vehicle, picker activities are similar to picker activities with a non-powered walkie pallet truck except that during empty pallet pick-up or full pallet deposit, the picker remains on the vehicle, and during pick aisle travel between two pick positions, the picker walks in the aisle or rides on the vehicle platform.

The disadvantages are increased cost, increased weight, picker training, and wider right-angle turning aisle. The advantages are improves picker productivity and safety, faster travel speeds, greater travel distance, used as a walkie or rider truck, and operates in a narrow aisle.

Mid-Control Rider Single Pallet Truck

The picker-controlled and electric battery-powered mid-controlled rider pallet truck is available as a single pallet truck. A mid-control rider pallet truck has a platform in the middle of the chassis, between the drive motor and forks. It is equipped with a standard full-height pallet backrest and fixed T-shaped control and steering handle.

The picker platform is located between the pallet backrest with forks and battery compartment. This open space is approximately 15 to 16 in. wide and is the full width of the chassis. From the platform, a picker has direct access to the pallet truck steering and control handle. The truck has protective bumpers on the pallet backrest side. The mid-controlled pallet truck requires a wider right angle or pick-up/deposit transaction aisle and wider intersecting aisles. The intersecting aisle permits the pallet truck to travel from one pick aisle to an adjacent pick aisle.

A full-height pallet backrest is located between the forks and the picker platform, and has welded and coated hardened metal members. During pallet truck transport from the pick area to a CO staging area, the pallet truck has quick stop travel in a pick aisle, a pallet backrest restricts tote, cartons, or cart moving forward into a picker platform. This feature improves picker safety and reduces carton damage.

The T-shaped steering and control handle extends upward from the drive motor chassis and is angled toward the picker platform. At the pallet truck steering handle are fork raise and lower controls, complete pallet truck forward and reverse movement controls, and a horn.

An electric battery-powered mid-controlled rider platform pallet truck has the longest overall length and longest wheelbase dimension. An electric battery-powered mid-controlled rider platform pallet truck has the longest right-angle empty pallet pick-up or full pallet deposit turning radius and longest intersecting aisle width.

An electric battery-powered mid-controlled and rider pallet truck is operated as a pick vehicle similar to the electric-powered walkie pallet truck except:

■ During empty pallet pick-up or full pallet deposit, the picker remains on the vehicle.

- During pick aisle travel between two pick positions, the picker rides on the vehicle platform.

The disadvantages are cost, increased weight, additional picker training, to pick up an empty pallet the truck is driven in reverse, widest right angle turning and intersecting aisle, and low productivity due to walking from the picker platform and pick position. The advantages are picker productivity and safety, faster travel speeds, greater travel distance, and if travel is quickly stopped in an aisle, minimizes carton damage from cartons falling from the pallet or cart to the floor.

Remote-Controlled Single Pallet Truck

A remote-controlled electric-powered single pallet truck is a sophisticated vehicle that has similar controls to a mid-controlled/rider single or double pallet truck, except that the vehicle forward and directional movements are controlled by a picker from a pick aisle. A transmitter sends a signal to a receiver on the vehicle. By controlling the manually remote-controlled single pallet truck forward movement and steering, this communication network and device allows a picker to walk in a pick aisle between the pallet truck and pick positions.

Elevating Single Pallet Truck

The vehicle design and operating characteristics are the same as the mid-controlled/ rider standard pallet truck, with two exceptions:

- One model has a picker platform on which a picker stands to reach an elevated pick position.
- Another model has both a picker platform and pallet-carrying set of forks that elevate. The ability to elevate approximately 30 to 48 in. above the floor reduces the need for a picker to reach for a case from the second-level pick position.

During travel in a pick aisle on a mid-controlled pallet truck with elevating picker platform and forks, a picker stops the pallet truck at the required pick position. To follow a pick instruction for a carton from an elevated pick position, the picker elevates the platform and pallet.

Operator Step Single Pallet Truck

The vehicle design and operating characteristics are the same as a mid-controlled/ rider standard double pallet truck with one exception: a step platform that allows a picker to step onto an elevated step to reach a SKU from a second-level pick position.

During travel in a pick aisle on a mid-controlled pallet truck with a step platform, a picker stops the pallet truck at the required pick position. If the pick instruction is to pick a carton from the elevated pick position the picker steps onto the platform.

Side Rider Single Pallet Truck

A side rider picker-controlled and electric battery-powered pallet truck is available with fork lengths that handle one pallet. Features are:

- A protected picker platform to the side of the pallet truck behind the battery compartment
- A swivel steering handle that extends upward and is angled toward the picker platform
- A separate control device that extends upward and is angled toward the picker platform

The picker platform has sufficient space to allow the operator to sit or stand. The platform has a nonskid surface and padding on the protective wrap around side. These features improve picker safety.

With the truck control handle angled toward the platform area, a picker has easy access to all pallet truck movement controls, fork-elevating and -lowering controls, and a horn.

The side rider pallet truck has an overall length of 88 in., with a 36-in. wide chassis and a 51-in. long wheelbase.

Disadvantages of the side rider pallet truck are picker acceptance, separate pallet truck controls and steering device, and vehicle steering stick or device. Advantages are protective picker area and picker sits or stands.

Pallet Truck Options

To improve a picker-controlled and electric battery-powered rider pallet truck pick concept (see Figure 4.34) options are:

- Chamfered or bottom deck boards angle cut pallet that improves empty pallet pick-up and full pallet deposit.
- Entry wheel that is a hardened plastic wheel set that has an axle and bearing and is located directly under each load-carrying fork tip. As a picker drives a pallet truck forks with fork entry wheels into a pallet fork opening, fork entry wheels come in contact with a pallet bottom exterior deck board and easily roll over bottom exterior deck boards. This results in improved picker productivity and lower potential building and equipment damage.

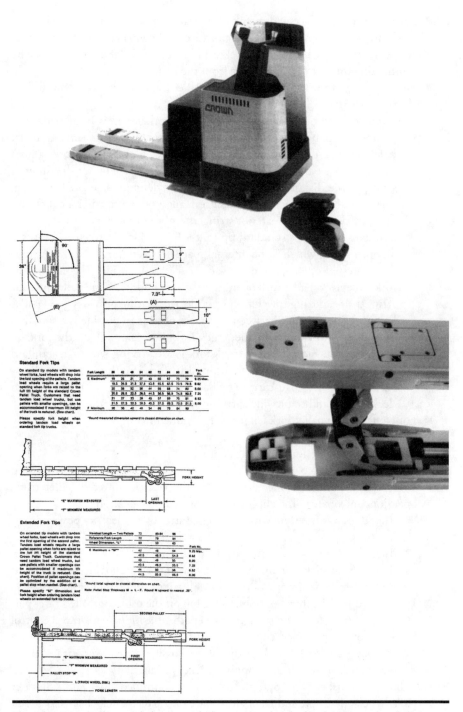

Figure 4.34 Pallet truck features: forks, entry wheels, back rest, and stabilizing casters.

- Backrest is a welded-coated metal structural member that is attached to the load-carrying fork base. As a picker pushes or pulls a pallet truck with a pallet across the travel path, the high backrest reduces potential carton damage by minimizing totes or cartons falling from the pallet.
- Fork tips. The fork length handles the pallet depth or stringer length. The electric-powered pallet truck fork designs are:
  - Extended fork tip design. The extended tip fork design is 10 in. wide with an open space between the two forks. The open space between two pallet truck forks is 4 to 8 in. wide. This open space between two forks allows them to enter pallet fork openings and provides space for a pallet interior stringer. The distance to the two forks outside to outside is 23 to 28 in. The overall pallet truck fork dimension allows them to enter a pallet fork opening between two exterior stringers and to support a pallet.
  - Standard fork tip. A standard tip fork is 9 in. wide. Between two forks is an open space, which is the internal space between the two forks. A common internal open space dimension is 4 ft. 9 in. wide. This allows the forks to enter a pallet fork opening. With each fork on one pallet interior stringer side, during travel in a pick aisle, this feature provides good pallet balance. The distance between two forks outside to outside is a standard 22 to 27 in. This overall pallet truck fork dimension allows the forks to enter a pallet. From a pallet truck fork load-carrying wheels, a standard tip forks end extension is 9 to 10 in. This fork length dimension allows each pallet truck load-carrying wheel to set between two pallet bottom deck boards and to provide carrying support for a pallet.
- Fork length. The pallet fork length is determined by:
  - Pallets carried per trip
  - Pallet stringer length or pallet depth
  - Corresponding spacing for a pallet bottom and top deck boards opening
- Clipboard document holder attached to the electric-powered pallet truck operator control area. The clipboard is attached in a location
  - That does not interfere with a picker control of an electric- powered pallet truck
  - That allows a picker to read pick instructions
  - That allows a picker an unobstructed view of the travel path. The clipboard document holder improves picker productivity and enhances safety
- High-speed travel control option is a button on the mid-controlled rider pallet truck. During long and straight travel paths, the high-speed travel control button allows a picker to engage the control button and to ensure that an electric-powered truck travels at the fastest possible speed over a travel path. A high-speed control button improves picker productivity.
- A fire extinguisher is an option for the mid-controlled rider and side rider pallet trucks that have sufficient clear space on a pallet truck chassis to attach a fire extinguisher. A fire extinguisher is placed in a location that permits a

picker easy access to it and does not interfere with a picker on a pallet truck. A fire extinguisher improves the safety of your order-fulfillment operation.

■ An audible travel horn is available on all electric-powered pallet truck types. When the electric-powered pallet truck control handle is lowered or when the electric-powered pallet truck is moving across the floor surface, the audible travel horn becomes activated and makes an audible sound that alerts a picker and other employees in the pick area or aisle that the electric-powered pallet truck is ready to move or is moving across a travel path. When considering an audible travel horn option, you evaluate the work area noise level and the horn noise level. The audible travel horn option reduces picker injury and improves safety in the pick aisle and work area.

■ Traveling flashing light. When the electric-powered pallet truck control handle is lowered or when the electric-powered pallet truck is moving in a pick aisle, it activates a flashing light that acts as a signal to a picker and other employees in a pick aisle that an electric-powered pallet truck is ready to move or is moving in a pick aisle.

When considering a flashing travel light option, you should evaluate light conditions in a pick aisle and mounting location on an electric-powered pallet truck. The flashing light location on the electric-powered pallet truck should not interfere with a picker or become an item that is frequently damaged. The travel flashing light option reduces employee injury and improves the safety in the pick aisle and work area.

Other key components include:

■ An electric rechargeable battery as the power source and battery charger. The important battery and battery charger features are:
  ■ Volts. The battery voltage is a battery electric power. A pallet truck with a higher voltage or volts travels at higher travel speeds, up grades and lifts and lowers pallets at a faster speed. A pallet truck with a lower battery volts or voltage has less electric power, slower pallet truck travel speeds, and slower pallet lift and lower speeds. If your order-fulfillment operation has long travel distances between a pick area and CO pack area, the 24-V battery is the preferred battery for your pallet truck. If an order-fulfillment operation has short travel distances between pick area and CO pack area, a 12-V battery is the preferred battery for a pallet truck.
  ■ Ampere hours. A battery with a high ampere hour rating provides electric power to operate a pallet truck for a longer period of time. A battery with a low ampere hour rating provides electric power to operate a pallet truck for a shorter period of time. Pallet trucks that operate for four to five hours of an eight-hour shift and travel short distances are candidates for a low ampere hour battery. Pallet trucks that operate for seven to seven and a half hours of an eight-hour shift and travel short distances are candidates for a high ampere hour battery.

■ Connector types. When your electric-powered pallet truck and forklift truck fleet has batteries with several different voltages, each battery voltage group (12, 24, and 36 V) should have a different type, shape, and colored battery connector. This reduces battery-charging errors (a low voltage battery hook-up to a high voltage battery) and damage (cooking) to batteries from the incorrect battery hook-up to a battery charger with a different voltage.

■ Battery charger cable length. A standard battery charger cable length is 10 ft., which is sufficient length to connect the battery charger to the battery in the pallet truck compartment.

In the previously mentioned pallet truck electric battery-charging options, when your battery charger cable is not connected to the battery in the pallet truck, your battery-charging station should have an open, flat shelf space or hook to hold the unused cable. To minimize risk of battery damage, a common practice is to have battery chargers elevated and secured on a shelf, stand, or decked pallet rack, and have one employee assigned the responsibility to connect the battery charger cable to the pallet truck battery cable.

Important Operational Features

Features include:

■ Overall vehicle length
■ Wheelbase
■ Maximum fork height
■ Right angle or transaction turning radius
■ Grade clearance
■ Maximum gradeability
■ Load weight and carrying capacity
■ Travel direction

Overall Vehicle Length

The overall length of the powered pallet truck is the distance between the fork tip and the pallet truck front chassis. The length varies depending on the type of truck.

■ A walkie pallet truck ranges from 65 to 89 in.
■ A walkie/end-rider pallet truck ranges from 87 to 138 in.
■ A side-rider pallet truck that has a standard fork length ranges from 76 to 136 in. and a truck with an extended-tip fork ranges from 123 to 136 in.
■ A mid-controlled rider pallet truck that has a standard fork ranges from 89 to 149 in. and with an extended-tip fork from 137 to 149 in.

The overall length is an important factor that determines the electric battery-powered pallet truck turning transaction or right-angle turn radius. A shorter overall length means a shorter turning radius, which means a narrower right-angle turning aisle and intersecting aisle. A longer pallet truck has a wider turning radius, which means a wider right angle turn area and intersecting aisle.

Wheelbase

The wheelbase length is the dimension between the steering/drive wheel center and the fork load-carrying wheel center. The wheelbase is important because it influences the pallet truck's turning radius and grade clearance. Typical wheelbase lengths are:

- For a walkie pallet truck the wheelbase ranges from 45 to 69 in.
- A walkie/end-rider pallet truck wheelbase ranges from 50 to 110 in. for standard-tip forks and 82 in. for extended-tip forks.
- A side-rider pallet truck wheelbase ranges from 50 to 110 in. for standard-tip forks and 82 to 84 in. for extended-tip forks.
- A mid-controlled rider pallet truck wheelbase ranges from 68 to 128 in. for standard-tip forks and 100 in. for extended-tip forks.

A pallet truck with a shorter wheelbase has a higher grade clearance. The pallet truck grade clearance is the steepest grade percentage that the pallet truck climbs without underclearance problems. Pallet truck underclearance problems occur at two locations:

- At the top of a ramp or dock leveler where a pallet bottom deck becomes hung up
- At the bottom of a ramp or dock leveler bottom as the load-carrying forks or pallet bottom deck boards strike the floor surface

Pallet truck underclearance is the distance between the floor surface and the lowest pallet truck undercarriage part or pallet bottom deck boards. A single pallet truck with a short wheelbase travels without hang-ups over standard dock plates and ramps. Most dock levelers or ramps have an angle peak at the top, thus a double pallet truck with a long wheelbase has a greater potential for hang-ups on the ramp or dock leveler travel path by one of the pallet truck underside or pallet bottom deck boards.

Fork Height

Important fork heights are:

- Elevation of the lowered fork above the floor surface

■ Maximum elevation of the raised set of forks above the floor surface. For most electric battery-powered pallet trucks, the elevation of lowered forks above the floor surface is 3.25 in. and the maximum elevation of the raised forks above the floor surface is 9025 in. The minimum and maximum set of fork height elevations along the electric battery-powered pallet truck wheelbase influence the electric battery-powered pallet truck grade clearance.

Turning Radius

A pallet truck's turning radius is the minimum distance between two racks or work stations that is required in an aisle for the pallet truck to make a turn and complete a pallet pick-up or delivery transaction. The turning radius is the distance from the end of the load-carrying wheels to the front end. Turning radius depends on truck design and fork length:

■ A walkie pallet truck ranges from 58 to 82 in.
■ A walkie/end-rider pallet truck with standard-tip forks ranges from from 71 to 130 in. and for extended-tip forks 103 in.
■ A side-rider pallet truck ranges for standard-tip forks from 70 to 130 in. and for extended-tip forks 100 to 104 in.
■ A mid-controlled rider pallet truck ranges for standard-tip forks from 82 to 142 in. and for extended-tip forks 100 in.

The walkie pallet truck has the shortest turning radius and the mid-controlled pallet truck has the widest turning radius. This means that the walkie pallet truck operates in the narrowest aisle and that the mid-controlled pallet truck requires the widest aisle.

Maximum Gradeability

Maximum gradeability is the steepest grade percentage ramp or dock plate slope that an electric battery-powered pallet truck climbs without the chassis underside becoming hung up. Maximum gradeability depends upon the pallet truck length.

■ For a walkie pallet truck with a 5000 lb. pallet, the grade is 5 percent
■ For a walkie-end rider pallet truck with a 4000 lb. pallet, the grade is 5 percent
■ For a side-rider pallet truck with a 4000 lb. pallet, the grade is 10 percent
■ For a mid-controlled rider pallet truck with a 6000 lb. pallet, the grade is 10 percent and with no pallet or empty the grade is 15 percent

Load Weight

The pallet truck load weight includes pallet weight, pallet truck weight, battery weight, and operator weight for a rider truck. The load weights to handle a stringer pallet or pallets are:

- For a standard fork walkie pallet truck, from 1210 to 1286 lb.
- For a walkie-rider pallet truck:
  - From 1326 to 1540 lb. with standard forks and 1345 to 1605 lb. to handle two pallets
  - From 1323 to 1619 lb. for one pallet truck with extended-tip forks and 1583 and 1679 lb. to handle two pallets
- For a side-rider pallet truck:
  - From 1685 to 1899 lb. with standard forks and from 1704 to 1964 lb. to handle two pallets
  - From 1765 to 1823 lb. for one pallet with extended-tip forks and from 1825 to 1823 lb. to handle two pallets
- For a mid-controlled rider pallet truck:
  - From 1685 to 1899 lb. with standard forks and from 1704 to 1964 lb. to handle two pallets
  - From 1882 to 1978 lb. for one pallet with extended-tip forks and from 1942 to 2038 lb. to handle two pallets

Load- or Weight-Carrying Capacity

The weight-carrying capacity is the maximum pallet weight that the electric-powered pallet truck hydraulic pump and system is able to raise and support. The pallet weight-carrying capacities are:

- For a walkie pallet truck, 5000 lb.
- For a walkie-end rider single pallet truck, 6000 lb. and for a double pallet truck, 8000 lb.
- For a side-rider single pallet truck, 6000 lb. and for a double pallet truck, 8000 lb.
- For a mid-controlled single pallet truck, 6000 lb. and for a double pallet truck, 8000 lb.

Most electric-powered pallet truck manufacturers have an overload valve in the hydraulic pump system, which prevents a picker from lifting a pallet that exceeds the design load weight. This feature minimizes the risk of equipment damage.

Direction of Travel

Options for the direction of travel (see Figure 4.35) are:

■ The set of forks leads, or the picker is in the trail direction of travel. When a picker trails the powered pallet truck, the battery compartment trails the picker. During travel in a pick aisle, a picker has minimal control of the vehicle, which results from the fact that as the picker turns the steering handle to the right, the pallet truck turns to the left. This pallet truck steering method is not natural for a picker and has the potential to result in slower travel speeds and increased product, equipment, and building damage and picker injury.

■ The picker or steering/drive wheel leads, or the picker is in the lead direction of travel. The picker is located in the front of the pallet truck and the battery compartment trails the picker. During travel in a pick aisle, the picker has maximum control, which results from the fact that the steering is similar to the automotive steering method. As a picker turns the steering to the right, the pallet truck turns right. This results in faster travel speeds and minimal UOP, equipment, and building damage, and picker injury.

With operator on platform forks are in the lead
◄—— Direction of travel

With operator on platform forks are in the trail
Direction of travel ——►

Figure 4.35 Pallet truck travel direction: forks in the lead and forks in the trail.

Smooth and Even Floor Surface

For best results, the electric-powered pallet truck requires a smooth and even floor surface with no debris in the pick aisle and no tape or string on the floor. Tape on the travel path can become wrapped around a truck wheel and is a maintenance problem.

Options or Accessories

Optional accessories on a pallet truck serve different functions, included to:

- Improve picker productivity
- Reduce building, equipment, or UOP damage
- Improve picker safety
- Make the pallet truck more flexible and equipped to perform other functions

Optional accessories include clipboard document holder, pallet stop, high-speed control, quick pick handle, fire extinguisher, audible horn, and flashing travel light.

Pick Activity

When your pick operation uses the electric-powered pallet truck as a pick vehicle, the pick instruction options are voice-directed, paper-printed, and self-adhesive labels, and RF device. After your picker is given CO pick instructions, the picker uses the pallet truck to pick up an empty pallet on the truck forks, then drive to the first assigned pick position. The employee travels through the pick aisle and stops at the appropriate pick positions transferring the required number of SKUs from each pick position onto a shelf on the pallet. With the totes or cartons on the shelf filled, the picker travels from a pick aisle to a CO pack area.

Electric Battery-Powered Rider Tugger or Tractor with a Cart Train

The electric battery-powered tugger with a cart train is a manually controlled vehicle that has the capability to travel horizontally through a pick aisle and to pull several carts. An electric tugger with a cart train is a variable-path small-item or flatwear apparel pick concept (see Figure 4.36). Tugger components are a draw, tow bar or hitch, and an on-tilt four-wheel cart train. Each cart has a coupler and hitch.

Powered tugger drive options are:

- Electrical rechargeable battery. A power motor, drivetrain, steering, and other features are similar to those on an electric battery-powered pallet truck. An electric-powered tugger is preferred for indoor operations.

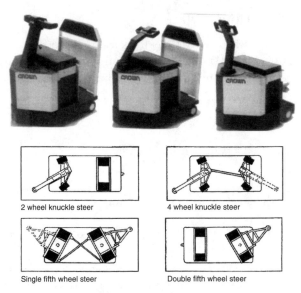

2 wheel knuckle steer | 4 wheel knuckle steer

Single fifth wheel steer | Double fifth wheel steer

TRAILING INFORMATION

Chart I gives the aisle widths (A) required to make a 90° turn with a number of trailers in specific sizes; Chart II, the space (B and C) required to make a 180° turn for a number of trailers in specific sizes. Conversely, both charts can be used to determine the number of trailers that can be used in existing situations. Please contact us for any additional information required . . . . particularly regarding unequal aisles and/or unusual situations.

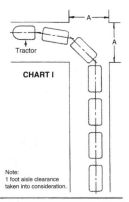

Tractor

CHART I

"A" = aisle width required for 90° turn

	Number of trailers					
	1	2	3	4	5	6
Deck size	Dimension "A" (caster steer with automatic couplers)					
30 × 60″	6′-6″	7′-6″	8′-3″	9′-0″	9′-6″	10′-0″
36 × 60″	7′-2″	8′-0″	8′-10″	9′-5″	10′-0″	10′-6″
36 × 72″	7′-8″	8′-6″	9′-3″	9′-10″	10′-5″	10′-10″
36 × 92″	8′-6″	9′-6″	10′-5″	11′-2″	11′-10″	12′-5″
Deck size	Dimension "A" (4 wheel knuckle steer)					
36 × 72″	6′-9″	7′-9″	8′-6″	9′-0″	9′-6″	10′-0″
36 × 96″	7′-5″	8′-6″	9′-5″	10′-2″	10′-10″	11′-4″
48 × 96″	8′-5″	9′-6″	10′-5″	11′-2″	11′-10″	12′-4″

Note:
1 foot aisle clearance taken into consideration.

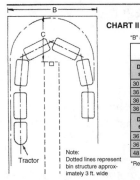

Tractor

Note:
Dotted lines represent bin structure approximately 3 ft. wide

CHART II

"B" and "C" = space required for 180° turn

		Number of trailers					
		1	2	3	4	5	6
Deck size	*Dim. "B"	Dimension "C" (caster steer with automatic couplers)					
30 × 60″	17′-6″	6′-10″	7′-8″	8′-5″	9′-1″	9′-8″	10′-2″
36 × 60″	18′-0″	7′-4″	8′-2″	8′-11″	9′-8″	10′-3″	10′-10″
36 × 72″	19′-9″	8′-0″	8′-10″	9′-9″	10′-7″	11′-1″	11′-6″
36 × 92″	24′-6″	8′-10″	10′-2″	11′-6″	12′-7″	13′-7″	14′-6″
Deck size	*Dim. "B"	Dimension "C" (4 wheel knuckle steer)					
36 × 72″	17′-6″	6′-6″	7′-3″	7′-11″	8′-6″	9′-1″	9′-8″
36 × 96″	21′-0″	7′-0″	7′-9″	8′-7″	9′-4″	10′-1″	10′-10″
48 × 96″	22′-3″	7′-9″	8′-6″	9′-2″	9′-10″	10′-6″	11′-1″

*Regardless of number of trailers

Figure 4.36 Tugger and cart train: tugger types, cart wheels, and aisle.

- Internal combustion engine power source: (a) LP gas, (b) gasoline, and (c) diesel. A fuel-burning engine produces fumes and is best for an outdoor operation.

The tugger has a coated metal or hardened plastic chassis that encloses the power source and cart-coupling device at the rear with three or four wheels to support the chassis. Most powered tuggers have an operator platform on which the vehicle forward and reverse controls, stop, and speed controls are located.

The tow cart or cart train steering is designed so it follows a tow tractor or lead tow cart. The front wheels can be either:

- Two-wheel caster steering with the two casters/wheels in the rear rigid and two casters/wheels in the front swivel-type. This wheel design has lead and trailing characteristics, avoids whipping action, and has easy manual movement.
- Knuckle steering, which is (a) two wheels in the lead are swivel-type and two rear wheels rigid, which provides trailing characteristics and (b) four wheels, with the two wheel sets connected by a draw bar. This feature has front wheels turn in one (right) direction and the rear wheels turn in the opposite (left) direction that has good trailing characteristics.
- Fifth wheel steering which is (a) single wheel that has two wheel sets and each has one axle that is on a turntable with two draw bars and all wheels are rigid and (b) double wheel that has rigid rear wheel set on one axle and front rigid wheel set on an axle that is on a turntable.

Powered Tuggers or Tow Tractors

Model types include:

- A walkie model, where a picker walks through a pick aisle with the tugger following a picker. The walkie tugger is a low-cost vehicle and has a slow travel speed that is set by the picker.
- A walkie-rider, which features are (a) moveable control and steer handle, (b) operator platform that permits a picker to stand in a protective area, (c) faster travel speed, (d) greater draw bar pull or cart pull capacity, and (e) less physical effort.
- A mid-controlled rider has a picker control and steer a tugger from an operator platform
- A remote-controlled has forward and directional movement controls that are directed by a picker transmitter. The transmitter sends a signal that is received by a receiver on the tow tractor. While controlling the tugger forward movement and steering, this communication network and device allows a picker to walk in an aisle between a cart and pick positions. The picker stops the remote-controlled tugger at a pick position, transfers SKUs from the pick

position into a tote or carton on a cart shelf and walks to the next pick position while the tugger moves to the next pick position.

The vehicle features are large vehicle number has a higher remote control cost, additional maintenance, approval of the RF wave and test the system prior to the purchase, and higher picker productivity.

SKUs Transported by a Belt Conveyor

A picker walks or rides in a pick aisle and picked SKUs are transported by a powered conveyor travel path to a pack or to a sortation/pack station. At a pick position, an order picker completes a pick transaction by transferring a SKU from a pick position into a tote or placing loose SKUs onto the conveyor. If picked small-item flatwear SKUs are transferred into a tote, when a CO is completed or the tote is full, the picker transfers the tote onto a powered roller conveyor. The conveyor transports the SKUs from the pick area to the pack station or to the sortation/pack station.

The mechanized pick concepts are:

- Pick to a conveyor travel path
- Decombe
- Pick car

Alternative Employee Pick to Conveyor Layouts

Alternative designs for a pick to conveyor concept are:

- For loose SKUs, a belt or, for totes, a roller take-away conveyor with a pick conveyor on one or both sides of a main aisle and perpendicular pick position rows and aisles
- Belt or roller conveyor take-away travel paths on both sides of a main aisle with pick positions that are parallel to the pick conveyor

Employee Pick into a Tote on a Conveyor

Alternative designs are:

- A powered roller take-away conveyor travel path in a main aisle with a perpendicular pick module (pick aisle between two pick position rows) on one or both sides of a pick conveyor. In a pick aisle, a picker transfers a SKU from a pick position into a hand-carried tote or onto a four-wheel cart shelf. The four-wheel cart tote shelf has a slightly higher elevation than the roller conveyor and has a removable side or three side walls. This feature permits easy tote discharge from a cart shelf onto a powered roller conveyor travel path.

■ A pick to tote conveyor travel path that is adjacent to or parallels pick positions and full totes are transferred to the conveyor travel path. A pick area is separated into separate pick zones and a picker passes a partially completed CO or full tote to the next pick zone. Perpendicular pick modules are on one side of a pick conveyor. A picker walks from a pick conveyor into a perpendicular pick aisle between two pick position rows to a pick position, removes and labels a SKU, returns to the pick conveyor, and transfers the SKU into a tote. For easy tote transfer from a pick conveyor onto a take-away conveyor, a gap plate is the full length between the pick conveyor and the take-away conveyor.

■ A pick to tote conveyor travel path that is adjacent to and the full length of a powered roller take-away conveyor travel path. A pick line or aisle is separated into separate pick zones and a picker passes a partially completed CO or partially full tote from one pick zone to the next. Pick positions are parallel along a pick conveyor and take-away conveyor travel path. A picker walks in an aisle or along a pick line to a pick position, removes and labels a SKU, and transfers it into a tote. For easy tote transfer from a pick conveyor onto a take-away conveyor, a gap plate is the full length between a pick conveyor and take-away conveyor. The gap plate serves to bridge the gap between two conveyor travel paths and as a guard rail.

Employee Pick to Belt Conveyor

With a pick to belt conveyor concept, a picker transfers picked, labeled, or coded SKUs loose from a pick position directly onto a powered belt conveyor. To increase picker productivity, an employee uses an apron, bag, tote, or four-wheel self-dumping cart to transfer picked and labeled (coded) loose SKUs from a pick position to a belt conveyor travel path.

A pick to belt conveyor design has pick area layouts similar to other conveyor concepts. Loose SKUs are picked to a belt conveyor that transports them through the pick area to a sortation/pack area. To have a long belt conveyor travel path, between two belt conveyor sections have gap plates/rollers or waterfall sections.

The design includes a batch release or control concept, so that if an imbalance occurs between a pick area and a pack area, there is space for empty totes to be queued. Top and side guards on the belt conveyor permit easy SKU transfer from a picker or cart onto the belt. There are E-stop devices to ensure that loose SKUs do not fall through gaps between two belt conveyor travel paths, and there are solid side guards on both sides of the travel path.

Decombe

A decombe high-rise picker vehicle is a manually controlled vehicle that travels horizontally and vertically through a pick aisle with shelf or decked rack pick positions on both sides (see Figure 4.37). A picker is raised or lowered on a platform

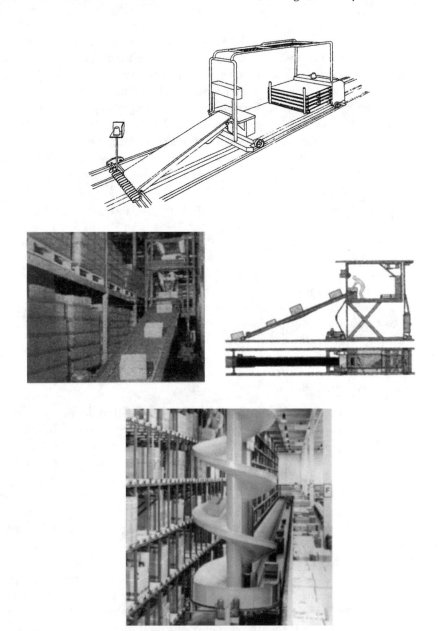

Figure 4.37 Pick and decombe vehicles.

to the proper height above the floor surface to complete a pick transaction. After a picker picks and labels each CO SKU (coded SKUs are not labeled), he or she transfers a SKU into a non-powered spiral chute on the front of the platform. The spiral chute extends from the ground-level powered belt conveyor travel path surface to the highest pick position. SKUs slide down the chute to a discharge end, and fall onto the ground-floor level belt conveyor travel path. The belt conveyor is located on the right side of the aisle and runs the full aisle length, and discharges the picked and labeled SKUs onto the main conveyor travel path, which transports them to the sortation/pack area. This allows an order picker to pick SKUs from pick positions on the left side of an aisle and makes it difficult for an operator to pick SKUs from the right side.

While driving a decombe in a pick aisle, a picker stops at a pick position on the left-hand side of the aisle. In a batched CO pick methodology, as each SKU is transferred from a pick position, it is labeled and placed in the spiral chute To improve picker productivity and minimize damage, a decombe vehicle has a rail guidance travel concept and floor-level take-away belt conveyor travel path.

Pick Car

A pick car HROS vehicle is a manually controlled vehicle that has the capability to travel horizontally and vertically through a pick aisle with shelf or decked rack pick position on both sides. A picker is raised or lowered on a platform to the proper height above the floor surface to complete a pick transaction. A decline powered belt conveyor travel path transports picked and labeled loose SKUs or SKUs in totes from the operator platform to a floor-level powered belt conveyor travel path, which transports SKUs from the pick area to a pack or sort/pack station.

For best picker productivity and minimal equipment damage, a pick car HROS vehicle has a rail or wire guidance travel concept and a floor-level take-away belt conveyor travel path.

As a pick car is driven in a pick aisle, an operator stops at a pick position between the floor and the top position in a rack or shelf structure. In a batched CO pick methodology, a pick car design options are:

- To pick, label (coded SKU), and transfer SKUs directly onto a powered belt conveyor travel path. In this design, as each SKU is transferred from a pick position, it is labeled and placed onto a decline powered belt conveyor travel path for transfer to the floor take-away powered belt conveyor travel path.
- A picker transfers picked and labeled SKUs into a tote or a tote section that is located on the operator platform. When the tote is a full or a CO completed, the picker transfers the tote onto a decline powered belt conveyor travel path for transport to a pack or sortation/pack station. An alternative batched CO to a single tote process has several totes on a picker platform, with one tote for each CO. This concept has a picker pick and sort the batched CO SKUs into discreetly identified CO totes.

Mechanized Stock Transfer Concepts

In a mechanized stock or pick position concept the stock or the pick position is transferred to a manual pick station, where an employee transfers a SKU from a pick position in one of the following ways:

- Loose, labeled or coded SKUs picked in bulk into a tote
- SKUs picked into a tote CO section or shelf
- Loose, labeled or coded SKUs picked directly onto a powered belt conveyor travel path

Depending on the pick methodology, a picker removes a SKU from a pick position, labels it(or coded SKU), presses a pick-to-light button, and marks on a paper document or RF device to verify the pick transaction. SKU pick position transfer is repeated until all SKUs for a CO are picked and transferred to a transport concept.

The mechanized stock or pick position transfer concepts include:

- Horizontal or vertical carousels
- Carton AS/RS or stacker cranes
- Cartrac, which is a mini-load or pallet type with shelves

Carousel Concepts

A carousel pick concept (see Figures 4.38, 4.39, and 4.40) has bins, baskets, or totes that are attached to a powered chain. The powered chain is an endless loop with necks, pendants, or open chain links that interfaces with a motor-driven sprocket wheel. A sprocket wheel has teeth that interface with the endless chain loop necks, pendants, or open chain links. As the tooth sprocket wheel turns, the chain cavity or pendant interfaces with the sprocket teeth. As the sprocket turns, it moves the chain and basket forward or in reverse over a fixed travel path. Forward and reverse movement is controlled by a control panel that receives instructions from an operator or by a microcomputer. The movement instruction causes the carousel to rotate until a pick position is halted at a pick station, where a picker completes a pick transaction by removing a SKU from a bin, placing it into a CO tote or placing a labeled or coded SKU onto a CO take-away conveyor travel path, and then programs the carousel to advance or move the next bin to the pick station.

SKU replenishment activities to a pick position are performed at a pick station or at a replenishment station. If pick and replenishment stations are at different locations, for picker safety, a pick and replenishment interlock network is designed not to have pick and replenishment transactions that are performed at the same time. If there is no interlock, replenishment transactions are completed on a different shift or during pick breaks. Carousels may be horizontal or vertical.

The carousel type used as a pick concept determines the chain and bins travel direction.

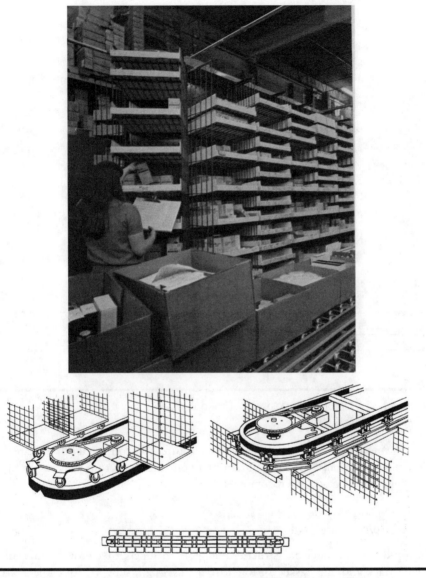

Figure 4.38 Carousels: pick location, bottom drive, top drive, and plan view.

Horizontal Carousel Concept

A horizontal carousel concept moves bins, baskets, or totes across a horizontal fixed travel path that is above the floor surface. The bins travel past a replenishment station and past a pick station. A horizontal carousel is top- or bottom-driven to move pick positions past the pick location. Options are to have one SKU per large tote or to have a large bin with separators or dividers to make smaller compartments. With three to five SKUs per bin, one carousel handles a large number of SKUs.

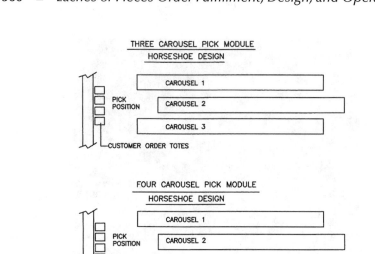

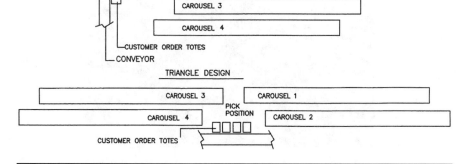

Figure 4.39   Carousel three and four modules.

The mechanical design and operational features of a horizontal carousel are that it is top- or bottom-driven, has intermittent or continuous run, is manually operated or controlled by a microcomputer.

Top-Driven Carousel

A top-driven carousel has structural support members that provide an endless chain track and bin, basket, or tote carrier travel path support. An endless chain is pulled by an electric-powered motor-driven tooth sprocket. The opening or distance between each sprocket tooth interfaces with a carrier neck or open link. As a tooth sprocket revolves, the sprocket teeth and carrier neck or open chain link interfaces and movement propels the carrier over a travel path. This travel path has straight sections and two 180 degree turns. One turn has a motor-driven sprocket and the other turn has a non-driven tooth sprocket. The non-driven tooth sprocket ensures accurate chain travel through a turn. The top-driven carousel is used for lightweight and small-items in baskets and GOH.

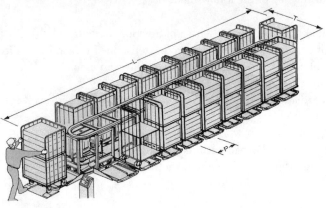

Figure 4.40 Cart carousel: transfer station and plan view.

Bottom-Driven Carousel

A bottom-driven carousel concept has a motor-driven tooth sprocket, endless chain track, and wheeled carrier travel path at the floor level. A bottom-driven carousel has structural support members and guide rollers on the upper-level travel path. The bottom-driven carousel concept is used to handle heavier items or heavier loads in a bin stack.

Intermittent-Run Carousel

With an intermittent-run carousel, a SKU pick position carrier travels the shortest travel distance to a pick station in either a forward or reverse direction from the active bin pick position to the next pick position. Control options are manual or microcomputer.

Continuous-Run Carousel

With a continuous-run carousel concept, the carousel bins are constantly moving over a travel path. A continuous run carousel is manually controlled and has an electric motor-driven tooth sprocket that has a variable speed drive. A continuous-run carousel moves bins at slow travel speed past pick stations, where pickers perform pick transactions.

Manual Carousel

A manually controlled carousel has a picker at a pick/control station pick and transfer a SKU from a carousel bin to a transport concept. At a pick position, a picker completes replenishment transactions to a pick position and controls the carousel movement using one of the following devices:

- A hand-controlled device with a bidirectional toggle switch
- A foot-controlled switch with two buttons, one for each travel direction
- A number dial and push button that allows an employee to set the dial and activate a carousel
- A keypad and digital display unit with several push buttons
  - 0 to 9
  - Revolve
  - Cancel and keypad permits a picker to enter the bin number onto a keypad that has the carousel move the bin to a pick station

Computer-Controlled Carousel

A computer-controlled carousel concept has the capability to handle approximately 500 commands from six carousel units. A multi-carousel concept revolves several carousel pick positions past a pick station. This results in higher picker productivity due to minimal picker waiting time for an assigned pick position to arrive at a pick station. A computer-controlled carousel has the capability to revolve carousel bins in a forward or reverse travel direction to bring an assigned bin to a pick station in the shortest time. During the pick activity, a computer commands other carousel units to revolve in forward and reverse directions to bring an assigned SKU to a pick station. While one carousel unit, A, is stopped at a pick station and a picker completes a pick or replenishment transaction, carousel B or carousels B and C revolve bins to another pick station.

Components of a horizontal electric-powered carousel are an electric motor-driven unit with a tooth sprocket and non-powered tooth sprocket, track support structure and frames, bin, and controls.

Top- and bottom-driven carousel types have the same electric motor and tooth sprocket design. The motor-driven tooth sprocket device is located on a 180-degree turn inside. Carousel design parameters and manufacturer standards determine the number of sprockets. Tooth sprocket spacings create an open space between two teeth. When this tooth and chain neck or open space center lines spacing match, as a motor-driven tooth sprocket revolves, it permits a carrier neck or open chain link to fall into an open space and a tooth sprocket to move a chain and bin over a travel path. The carrier movement is the same as the motor-driven tooth sprocket turning direction. The teeth per sprocket and sprockets per carousel are determined by your carousel manufacturer standards and your design parameters. The design factors are motor horse power, spacing between two carrier necks, combined weight product and bin, travel path length that is determined by the number of SKU pick faces, and pieces per pick position and carousel application.

The carousel track or travel path allows bins, totes, or baskets to be carried past the pick stations. The track device is moved by tooth sprockets and the components are:

- A straight track with at least two 180 degree turns, with metal tubular pipe sections that form an endless loop. Standard carousel length ranges from 14 to 110 ft. with a track on the top or bottom.
- A carrier with two wheels or a grooved wheel that travels over a metal track. The hardened carrier metal or plastic wheels have a low coefficient of friction that provides the wheel with a smooth, continuous movement over the metal track.

With a top-driven carousel, metal brackets extend outward from the wheels and the brackets are connected together at the bottom by hardened metal links or necks. The link thickness or metal gauge matches the opening space between two sprocket teeth and the other carousel design parameters. A bottom-driven carousel has the brackets extend upward and are connected together at the top of the carrier wheel by a hardened metal link.

A carousel carrier link, arm, or neck is made of hardened metal and serves to interface with the opening space between two sprocket wheel teeth and provide a connection between bins, totes, or baskets.

The links and bins are connected together to form an endless chain. In a top-driven carousel unit, a carrier bin is attached to the link bottom and the bottom-driven carousel unit, a carrier device is attached to the link top. The shape of the link is determined by the carrier shape, top or bottom connection location, and combined weight.

Carousel support structure design and configuration is determined by the carousel type and application. A top-driven carousel support structure has a coated metal frame that is attached along the full length of the tubular track. A-frame is supported by at least two upright posts that have wide base plates to spread the load and have predrilled holes for base plate anchor bolt locations. The number of support frames, size of the base plate, and metal gauge are determined by your manufacturer standards, carousel application, and geographic (seismic) location. A bottom-driven carousel support structure has the same design parameters as a top-driven carousel.

Carousel carrier type is determined by SKU characteristics and pick volume, pick position number, inventory quantity for each SKU, and manufacturer design. Types of carousel carriers include:

- Wire-meshed basket. The wire-meshed basket is made from preformed zinc or galvanized or coated steel wire strands, and handles loose small-items that are in bins, totes, or cartons. Depending on the SKU size and volume, the bins have separators and dividers to make small compartments to handle a SKU with a small inventory and increase the number of SKUs per basket. The basket has lighter weight and four to five per pick position carousel stack.
- Bin or tote. The bin is designed with two solid sides, a solid rear wall, and solid bottom surface and has separators and dividers to make smaller compartments to handle more SKUs in each carrier. Bins in a carousel stack 8 ft. 10 in. maximum above the floor surface. The number of bins per stack is determined by SKU characteristics, inventory volume, and employee reach. Standard bin dimensions range from 21 to 30 in. wide and from 14 to 21 in. deep.

Other bin and basket carrier device features are that each carrier bottom surface has a 5 percent slope or pitch to the rear, a permanent or removable front barrier, and optional solid or wire-meshed sides and rear.

During revolution of a carousel over a travel path, the last two features minimize the risk of loosing small-items that fall from the carrier.

Carousel Designs

A standard carousel concept design increases space utilization, improves picker productivity, has a 100 ft. or less overall track length, has a 24 in. wide carrier device, and has approximately 50 carrier devices or stacks.

Horizontal carousel concept designs are:

- Multiple carousels that service one pick station with a stationary ladder to permit a picker easy access to elevated pick positions.

- Multiple stacked carousels that revolve to service one pick station. A picker has the ability to move up or down on a picker-controlled vertical-moving platform or stationary ladder to access elevated pick positions.
- Carousels that are installed on elevated mezzanines.
- Carousels that interface with a mechanical device with bin injector and extractor.

Other design requirements are:

- Separate replenishment and pick stations with corresponding interlocked carousel movement controls. When a pick transaction is being performed, interlocked controls prevent a replenishment station from directing the carousel to move.
- A transport concept for empty CO totes and completed CO totes.
- A CO tote location at each pick station.
- A replenishment product travel path, queue area, and trash removal concept.
- A back-up CO pick method to be used during mechanical break-down or electric failure, or a 3 ft. wide aisle between carousels.
- Pick instruction at a pick/control station.
- Safety fencing to enclose a carousel.

Carousel Pick Concept

A small-item flatwear order-fulfillment carousel concept revolves stock or a pick position to a pick station. At a pick station, a picker ensures that the identification on a bin matches a paper document, self-adhesive label, pick to light, or RF device pick instruction. With a correct match, a picker completes a pick transaction, removes SKU from a carousel bin and transfers it (labeled, coded, or non-labeled) into a CO tote or transport concept. After all picks have been completed from this carousel bin stack, the carousel is directed by a picker or microcomputer to revolve and bring the next carousel bin to the pick station. After all picks for a CO have been completed or the tote is full, it is released from the pick area to a transport concept for delivery to the sortation/pack station.

Per your order-fulfillment operational practice, SKU replenishment activities are performed on a separate shift at a replenishment station or at the pick station.

Alternative procedures are for a picker to handle:

- Single COs at a pick station with 3 or 4 ft. aisle between carousels, and a CO tote on a cart shelf, fixed shelf, flow rack, or conveyor travel path.
- Batched COs that are picked into a tote on a conveyor travel path, static or cart shelves, or carton flow rack. A carton flow rack discharges totes onto a powered conveyor travel path.

- Completed batched COs that are picked, labeled, or coded and transferred into a tote or directly onto a powered belt conveyor travel path that transports SKUs to a sort/pack station area. In this carousel pick station design from a 3- or 4-ft. aisle between the conveyor travel path and pick station, picked and labeled or coded SKUs are placed into a tote or transferred directly onto a belt conveyor path. The carousel picker instruction is a paper document, self-adhesive label, RF device, or pick to light.

A carousel batched CO philosophy has SKUs that are picked and sorted by a picker direct to a CO tote. With this concept, CO totes are introduced by a transport concept to a carousel pick station. If totes are transported by a cart or picker, a picker at a pick station scans a CO tote identification and obtains pick instructions. With CO totes on a conveyor travel path, carousel pick instruction sequence options are:

- Picker scans each CO tote that has the computer sequence for pick instructions to match a CO tote discreet identification.
- Based on the first-in-first-out CO tote travel over a conveyor travel path that has each tote travel past a barcode scanner/RF tag reader. The barcode scanner RF tag reader sends the tote discreet identification to a microcomputer-controlled printer, RF device, or pick-to-light microcomputer that creates a pick instruction in the scanning sequence.

Options for rotating a horizontal carousel past a pick station are:

- One-way direction. A horizontal carousel with a one-way pick position travel direction has all carousel bins or baskets rotate in a forward direction past a pick station. The forward direction rotation means that if carousel C bin number 10 is at a pick station and the next CO pick position is from bin number 2, the carousel rotates the bins in a forward direction until the bin number 2 arrives at a pick station. This forward direction has all the carousel bins rotate from bin number 10 through bin number 50 and bin number 1 until bin number 2 arrives at a pick station. This results in non-productive picker time as bins that do not have a pick transaction travel past a pick station. With a multiple carousel concept that has several carousels interface with a microcomputer and travel past one pick station, this non-productive picker time is minimized due to carousels A and B rotating bins while a picker is completing a pick transaction from carousel C.
- Two-way reverse direction. The two-way or forward or reverse carousel bin rotation is microcomputer controlled and has all CO SKUs entered into the computer file. With a two-way carousel pick position rotation concept, a computer controls the horizontal carousel rotation for the bins in a forward or reverse direction past a pick station. This dynamic bin rotation minimizes

non-productive picker waiting time for an assigned carousel bin to arrive at a pick station. This pick position two-way rotation feature means that if carousel C bin number 10 is at a pick station and the next CO pick is from carousel C bin number 2, a horizontal carousel rotates the bins in a forward or reverse direction that brings the next required carousel bin by the shortest travel time to a pick station. With the next pick position in bin number 2, the preferred carousel movement is a reverse direction. This carousel pick position reverse direction rotation is from bin number 9 through bin number 3 and the carousel rotation stops at bin number 2.

The advantages of a single carousel concept are a reduction in picker non-productive time by a 30 basket factor. With the multiple carousel concept interfacing with one pick station, the non-productive picker time is probably reduced to zero. This is due to the fact that while a picker is completing a pick transaction from carousel C, carousels A and B have completed a pick position rotation, and at a pick station these bins have SKUs that are on COs.

At a carousel pick station, a picker routing pattern directs the picker's hands across a carousel bin or basket stack. With a carousel pick position stack, there is a golden zone. A carousel pick position stack has four or five baskets high within an 8- to 10-ft. elevation. With this feature and an average picker height, the golden zone has two, three, and four baskets. To assist a picker access the top pick level, a pick station has a stationary ladder or a hydraulic lift.

Picker routing patterns are:

- With one SKU per basket, there is one motion to complete a pick transaction
- With multiple compartments on a basket, picker routing patterns are (a) vertical up or down for each basket stack and (b) horizontal across the top-level basket row and to next lower basket row

A small-item flatwear order-fulfillment operation that has a powered carousel pick concept for B- and C-moving, small-cube SKUs and a second flow rack pick concept for high-cube, A-moving SKUs, has two separate CO start locations. For additional information, we refer the reader to the previous section.

The disadvantages are investment, small inventory quantity, specific design and layout requirements, separate UOP replenishment shift or location with interlock for carousel pick and replenishment stations, microcomputer controlled with pick-to-light or RF device pick instructions, each CO tote has a human/machine-readable CO identification, and for best picker productivity, batched COs. The advantages are: handles a wide product mix, reduces UOP and building and equipment damage, minimizes employee injury, requires fewer employees, good picker productivity, improves space utilization, and reduces aisles.

Vertical Carousel Concept

A vertical carousel concept (see Figure 4.41) moves trays or pans by dual motor-driven, endless loop chains over a fixed travel path. The chains follow a closed loop, vertical, elliptical fixed travel path and are propelled by an electric-powered motor and tooth sprockets. Options for a vertical carousel are:

- One pick position on one floor that completes pick and replenishment transactions.
- Pick positions that complete both replenishment and pick transactions. With a dual-position concept, the tray travel path is past lower and upper-level pick stations.

With interlock carousel movement controls sequenced so that only one work-station has access to a vertical carousel tray, at each pick station a picker completes a pick or replenishment transaction. Another vertical carousel travel path feature is that the entire travel path and trays are enclosed in a shroud with a carousel access door at each pick station and a maintenance access door to the drive motor and chain area. Per local code a shroud is wire-meshed material, solid sheet metal, or fire-resistant material with fire doors, that surround the entire vertical carousel and access door entrance and exit.

Key Components

Components of an electric motor-powered vertical carousel pick concept are divided tray, two endless closed loop chains and tray attachment devices, electric-powered motor, drives and tooth sprockets, start and stop controls, carousel chain travel command entry device, key pad and tray access doors, structural support upright post, horizontal support members and protective shell or shroud, and on each floor a pick station.

Carrier options are:

- A tray that acts as a small-item container and is attached to the vertical carousel chain. If a mix of small SKUs is placed into a tray with solid sides and bottom, dividers or separators are used to separate the tray into smaller compartments to provide maximum pick faces. To permit a pick and replenishment transaction, a tray has one side or top open. Standard tray features are determined by your UOP characteristics and vertical carousel manufacturer standards.
- A two-component carrier that has each carrier end attached to a powered chain. This concept has (a) a carrying surface that is designed with a perimeter barrier on four solid sides and bottom surface to retain and support totes and (b) solid wall land bottom tote that holds a SKU or has dividers to hold several SKUs and fits securely onto a carrier-carrying surface.

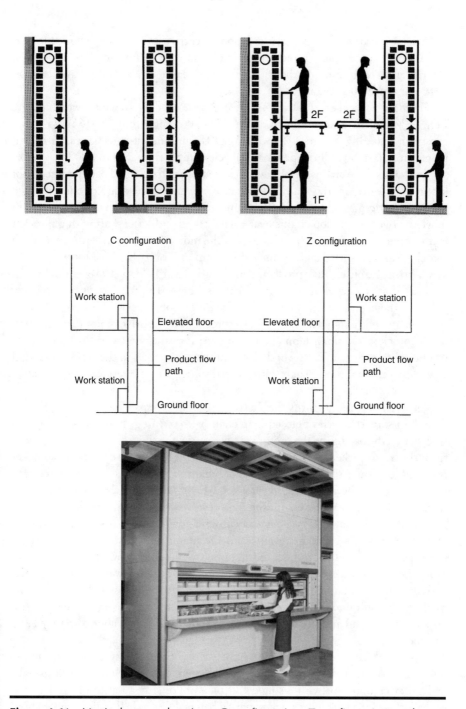

Figure 4.41 Vertical carousel options: C configuration, Z configuration, and one pick front.

A vertical carousel tray or two-component carrier/tote front is designed to permit a picker to complete a pick transaction, provide surface for a pick position identification on a tote side that faces a vertical carousel access door and contain maximum inventory quantity.

Standard vertical carousel and tray features are overall pan width, which ranges from 33 7/8 to 130 in., overall pan depth that has a range from 121/16 to 25 1/2 in., pan pitch that ranges from 3 7/8 to 22 1/2 in. The number of pans per carousel unit ranges from 6 to 50 and the pan load capacity ranges from 200 to 800 lb.

Two closed loop chains move trays or carriers over a fixed travel path between floor locations. The exterior tray side load surface is attached to the chain interior side.

An electric-powered motor, drive, and tooth sprockets provide the power to move the two closed-loop chains over a fixed travel path. The shaft with a sprocket is connected to an electric drive motor. As the motor turns the drive and the tooth sprocket turns, a small chain is moved forward and over the drive tooth sprocket and vertical carousel tray sprocket. This small chain is looped over a dual tooth sprocket that is located at the vertical carousel base. To propel trays or pans over the fixed travel path, a short closed-loop chain is looped between the drive motor sprocket and one pan or tray chain tooth sprocket. As the drive motor turns the first tooth sprocket, the small loop chain is moved forward, which in turn moves the second tooth sprocket. The second tooth sprocket interfaces to the vertical carousel tray or pan chain and to a shaft with a tooth sprocket that interfaces with a second carousel tray chain.

As an electric motor turns the shaft, the gear at the shaft has a tooth sprocket. Due to the fact that a short closed-loop chain with openings in the links matches a sprocket teeth spacing, as the motor sprocket turns the short chain, the short chain turns the tray sprocket and tray chain threads over its teeth. As a drive tooth sprocket is turned forward, a sprocket pulls the short chain forward. The short chain forward movement turns the second tooth sprocket, which pulls a tray chain forward. Since trays are attached to a long chain, the trays revolve over the travel path.

The motor is matched to a vertical carousel application, based on the number of trays, tray load-carrying capacity, and the manufacturer standards. A vertical carousel has guide rollers and sleeves that improve movement over a travel path and minimize vibration. A vertical carousel controls the tray movement over a travel path, E-stops, manual movement tray chain up or down, and switches and indicators, such as off/on and tray number. A vertical carousel tray's movement control options are key pad and microcomputer controlled and accept approximately 1250 tray moves.

A vertical carousel pick station features are lockable vertical carousel access door, light fixture, pick instruction paper document or label printer or pick-to-light or RF device indicator with a single CO pick philosophy a flat top or roller conveyor surface for a pick location, and with a batched CO a shelf or flow rack to hold multiple customer order totes.

Design options for a vertical carousel dual pick station are:

- A C configuration that has pick stations on each floor, located on a travel path on the same side. In this C dual-access door configuration, the upper-level door is above the lower-level door.
- A Z configuration that has a pick station on each floor, located on different sides of a vertical travel path.

Components of a vertical carousel are:

- Structural support upright posts and horizontal support members. Hardened and coated structural upright posts and horizontal support members provide structural support for an the travel path, trays or pans, guide track and workstation or counter.
- A solid sheet metal or dry wall shell or housing that encloses the entire vertical carousel moving parts between the floor and ceiling. The enclosure has an access door to each workstation and a maintenance door. The enclosure minimizes employee injury, reduces the noise from the carousel moving parts, and with fire-resistant material minimizes fire risk. An alternative with local code approval is a wire-mesh shell which minimizes employee injury but does not reduce noise or fire hazard.

The picker routing pattern for a vertical carousel directs a picker's hand to a SKU pick position. The routing pattern is determined by several factors, which include:

- CO tote location options.
  - Directly below the access door. This arrangement is to use a single CO pick philosophy that has a picker pick SKUs into one CO tote/carton.
  - To have several CO totes or cartons on a multilevel shelf or carton flow rack that is directly behind a pick station. In this arrangement, there is a 30 in., 3 ft. to 4 ft. aisle between the access door and CO tote or carton location. This arrangement handles a pick philosophy that has a picker pick/pack single COs, a batched CO pick philosophy that has a picker pick and sort into a CO tote/carton, or a batch CO pick philosophy that has a picker pick, label, and transfer the SKUs into a tote or carton or directly onto a belt conveyor travel path.
  - Control station location.

The two factors determine whether the routing pattern is horizontal across a tray from right-to-left or left-to-right.

A vertical carousel tray rotation over the vertical carousel travel path options are:

- One-way rotation that is sequential from the lowest bin number to the highest bin number

- Two-way or forward or reverse rotation that has carousel trays rotate in a direction that has the shortest travel time between a bin number at the pick station and the next required bin number. For additional information, we refer the reader to the discussion above of horizontal carousel one-way and two-way rotation.

A picker enters the vertical carousel command to revolve a tray to a pick station. At the pick station, the picker transfers a SKU from the tray to a CO tote or carton. With a completed pick transaction, the picker activates the carousel to revolve the next tray. This activity is repeated until the CO is complete or the tote or carton is full. During a pick activity on the lower-floor level, a carousel interlock shuts the upper-floor access door.

The disadvantages are limited SKU quantity in a pick position, investment, one picker, limited UOP mix and at one time, only pick position accesses UOP. The advantages are small area, storage area, and vertical transport concept, uses vertical space, and excellent UOP security, and minimizes employee injury.

Carton AS/RS, Mini-Load or Mini-Stacker Crane

A carton AS/RS, mini-load, or mini-stacker crane bin or carton concept (see Figure 4.42) has a SKU in a tote or carton or multiple SKUs in smaller compartments within a large tote. Operationally, a carton AS/RS crane is a captive-aisle crane that travels to a pick-up or delivery (P/D) station at the end of an aisle, takes a tote or carton on board, travels down the aisle between two storage position rows to an assigned location and deposits the tote or carton into the assigned storage position. Per your small-item design parameters, transactions per hour and facility design, a T-car is used to transfer a carton AS/RS crane between two aisles. To withdraw a SKU, a carton AS/RS crane travels to an assigned position, withdraws a tote or carton from a storage position, takes it on board, travels down the aisle to the output (P/D) station and places a tote or carton onto a take-away conveyor. Totes/cartons are transported via a powered conveyor or are placed onto carts for transport from the carton AS/RS pick area to a CO pack station.

Crane control options are:

- Employee-controlled, with an operator at a control station at the end of an aisle. An operator enters each P/D transaction into a microcomputer and the computer directs a carton AS/RS crane to perform a transaction.
- Computer-controlled. A carton AS/RS crane control concept is based on economic savings from the labor, on-time, and accurate transactions.

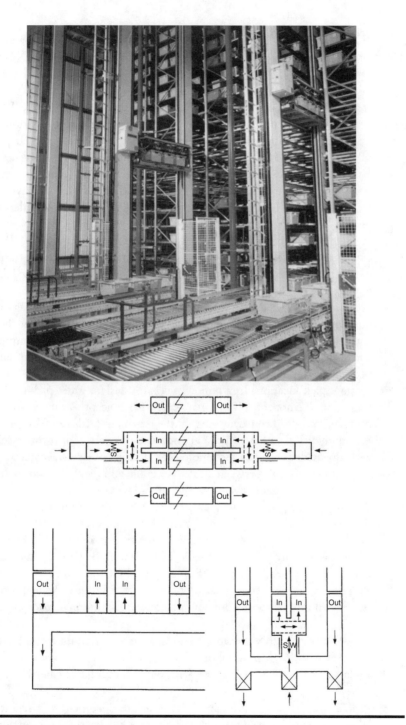

Figure 4.42 Carton AS/RS: three carrier crane and layout options.

Carton AS/RS Pick-Up and Delivery and Pick Station

Design options for a carton AS/RS pick station (see Figures 4.43 and 4.44) are:

- Front-end pick location. The carton AS/RS front-end pick location has powered roller conveyors that are placed in the front of a carton AS/RS crane aisle. One powered roller conveyor direction of travel is to the carton AS/RS crane aisle and provides totes to an aisle input station. A second powered roller conveyor travel direction is from a carton AS/RS crane aisle and is an aisle output station. The third powered roller conveyor section is a pick station that has a direction of travel from an output powered conveyor, over a pick station, and to the input powered conveyor. The powered roller conveyors are low- or zero-pressure queue conveyor with photo eyes to send messages to a microcomputer to stop and start conveyor travel path. The front end carton AS/RS pick location is preferred for a small-item order-fulfillment operation that has a low to medium pick volume, few to medium SKUs, and few pieces per CO.
- Front to rear pick location. A front to rear carton AS/RS crane design has the same operational characteristics, disadvantages, and advantages as the front-end carton AS/RS crane design except that both aisle ends have output, pick, and in-feed powered conveyor network. This dual activity design handles a higher volume and requires additional investment.
- Remote pick location. In a remote pick area design, order pick stations are located in a remote area from a carton AS/RS concept. A carton AS/RS concept has an in-feed conveyor travel path and an out-feed conveyor travel path. A remote pick area has additional and queue powered conveyor travel paths with divert devices to service more pick positions. A conveyor network is a tote travel path link between an AS/RS crane and pick area and to any pick stations. This output conveyor travel path allows a tote diverted at a pick station and queue prior to any pick station. At a pick station, a picker removes the assigned SKU from a tote and transfers it into a CO tote or carton. After a pick transaction is completed, a picker transfers a carton AS/RS tote to a conveyor travel path for travel to the next pick station or return to a carton AS/RS concept in-feed station. With a completed CO or a full pick tote/carton, a picker transfers the tote or carton onto a powered roller take-away conveyor travel path for transport from a pick area to the sortation or pack area.

A remote carton AS/RS concept pick location is preferred for a small-item pick operation that has a high pick volume, large number of SKUs, several pieces per CO, and large CO number. A carton AS/RS pick concept has the flexibility to use:

- Pick instructions that are paper pick that are document and self-adhesive labels and paperless that are pick-to-light or RF device pick instruction

Figure 4.43 Carton AS/RS concept.

Figure 4.44   Crane carriers: side belt, fork, and three totes.

- A pick philosophy that is (a) single CO with a single picker, (b) single CO with multiple pickers, (c) batched COs with a single picker who picks and sorts CO picked SKUs, or (d) batched COs with multiple pickers who pick and label CO picked SKUs
- A pick methodology with mixed SKUs into a tote, SKUs into a separated tote, and mixed SKUs onto a powered belt conveyor travel path

Carton AS/RS Concept Components

Components of a carton AS/RS concept are a stacker crane with a tote/carton-carrying surface and handling device, tote/carton for loose small parts, tote/carton storage positions along the sides of a stacker crane travel aisle, stacker crane operational or control station, tote/carton in-feed and out-feed stations, and order pick stations.

Carton AS/RS Stacker Crane

A stacker crane is a two- or four-wheeled captive aisle crane with a tote/carton-carrying surface. A stacker crane travels horizontally down the aisle and carries a tote/carton-carrying surface on either a single mast or dual masts. A stacker crane aisle has either:

- An in-feed station at an aisle end and an out-feed station at the other end
- Both in-feed and out-feed (P/D) stations located at one aisle end

As determined by a stacker crane written functional specifications, tote/carton storage positions begin at a specific elevation above the floor and end a specific level below the ceiling steel. Tote/carton positions that are elevated above the floor are determined by the clearance for a stacker crane wheel or wheels and the tote-carrying/handling surface or device. The first storage position elevation is 1 to 2 ft. above the floor surface, depending on the manufacturer's standards.

The maximum height for the tote/carton storage area is the clearance from the ceiling steel for a stacker crane overhead section and a stacker crane tote-handling device top. The highest tote/carton storage position is located at least 1 to 2 ft. below the building overhead structural steel.

Stacker crane wheels travel on a smooth concrete floor surface or on a hardened steel rail. A carton AS/RS crane installation has F-100 or dead level rated concrete floor as a minimum. This floor is a flat or very flat floor and ensures smooth crane operation and a level rack and crane travel rail installation. If a stacker crane concept is a floor rail-guided vehicle, the rail is properly anchored to the floor. Rail anchor bolts are on 2 ft. centers for an entire aisle length. During guide-rail installation, shim plates are used to ensure a flat and even crane wheel travel path.

A tote/carton-carrying and -handling surface or device is referred to as the injector and extractor. A tote/carton-carrying and -handling surface is designed to pick-up, carry, and deposit a tote/carton to complete a mini-load transaction. The tote/carton transactions are performed at a P/D station at the end of an aisle and at assigned storage positions in the aisle.

With these operational requirements, a stacker crane tote/carton-carrying and -handling surface or device is designed to handle a specific tote/carton size, tote/carton front design, carton top and bottom, and total tote/carton weight. A total tote weight includes the empty tote and divider sections. The sections combined become a tote tare weight and the maximum quantity or weight for your product mix in the tote.

A tote/carton-carrying and -handling surface or device specifications are for a maximum 600 lb. combined product and tote/carton weight. Tote/carton dimensions vary depending on product mix characteristics and crane manufacturing standards. Tote/carton exterior dimensions match a carton AS/RS crane tote/carton-carrying and -handling surface or device characteristics. The standard tote sizes are 48 to 72 in. long by 18 in. wide and 36 to 72 in. long by 24 in. wide. Tote/carton height, length, and width dimensions are determined by your UOP height and manufacturer standards.

The tote/carton-carrying/handling surface or device handles and travels with the tote/carton long dimension perpendicular to an aisle storage position. This tote/carton on-board carrying feature gives a stacker crane concept the maximum number of totes or cartons or storage openings for a given aisle length.

A mechanical tote/carton-carrying and -handling and tote grab method uses a device to grab and push a tote or carton or lifts it by the top or grabs the sides. The method allows a carton AS/RS crane concept to handle nonmetal totes.

A traversing magnetic carton AS/RS crane tote-carrying and -handling device completes a tote/carton transaction by the magnetic device engaging the steel tote end or a steel plate on a tote end. The mini-stacker with a magnetic tote-carrying and -handling device has each tote to have a metal end or plate and side wall structural strength to withstand the magnet pull pressure.

A carton AS/RS crane tote/carton carrying and -handling device transactions take place at in-feed and out-feed stations and at each storage position. A tote or carton is:

■ Deposited at an aisle P/D station or inserted into a storage position
■ Withdrawn from an aisle P/D station or storage rack position

To complete a tote/carton transaction, a carton AS/RS crane tote/carton-carrying and -handling device has a side traversing feature that allows it to transfer a tote or carton into a storage position or P/D station or to withdraw a tote or carton from a storage position or P/D station.

The motor and drive mechanism is a DC electrical component that receives electrical power from an overhead bar that serves to guide AS/RS crane travel in an aisle. The motor and drive mechanism moves a carton AS/RS crane in both horizontal and vertical directions:

■ Horizontally through an aisle
■ Tote/carton-carrying and -handling device vertically along the crane mast to a P/D station or storage rack position. At this elevation, electric power moves the tote/carton carrying or -handling device to perform a tote/carton transaction.

In a dual-entry carton AS/RS crane concept design, the motor, drive, control platform, and tote/carton-carrying and -handling device length dimension determines a carton AS/RS crane run-out length at both ends of an aisle. A carton AS/RS crane run-out length is the distance a carton AS/RS crane tote/carton-carrying and -handling device needs to access P/D stations and all the storage rack positions.

A carton AS/RS crane mast and top guide-rail functions as a vertical travel path for the tote/carton-carrying and -handling device. Mast height is determined by a height of the storage positions above the floor or P/A station. The width and size are determined by a total tote/carton-carrying and -handling weight, seismic location and conditions and travel speed, mast height, and manufacturer's standards.

A top guide device is an overhead rail that is attached to the top of the storage structure and performs to:

- Guide a mast top section as a carton AS/RS crane travels horizontally through an aisle
- Supply electrical power to a carton AS/RS crane motor
- Act as a structural support member that provides rigidity and stability to the storage structure

With a carton AS/RS concept, the UOP is a vendor or company carton that has the structural strength to satisfy your AS/RS concept design parameters. These design parameters include:

- Ability to be queued and moved by the in-house transport concept
- Ability to ensure UOP line of sight to barcode scanners/RF tag readers and employees
- Capability to be handled by the crane carrier
- Able to be transferred at a storage position
- With possible humid storage no corrugated carton side wall or bottom deflection that would obstruct a storage transaction (if there is a humidity problem, a tray could be used as required)
- Ability to handle a carton with open or closed top

The features are no tray or tote cost, load and upright post storage concept, slightly narrower travel path window, some cartons could require trays and each carton has an identifier.

Carton AS/RS Tote/Carton Carriers

A carton AS/RS crane carrier performs:

- At a P/D station to take on-board and deposit totes or cartons onto a conveyor or station

- On a carton AS/RS crane to carry totes or cartons and deliver each to an assigned storage position
- At a storage position to complete a deposit or withdrawal transaction

The number of totes or cartons that can be handled one ach AS/RS crane trip in the aisle depends on the type of crane. A standard carton AS/RS crane has a mast for a transporting a carrier vertically. The first storage position is at a limited height above the floor, and at the P/D station has less run-out to complete a transaction. This type of crane requires a lower investment.

A dual mast crane vertical travel of the carrier guided by a front and/or rear mast. A dual mast crane features storage positions that are at a maximum elevation above the floor and at a P/D station has a greater run-out distance to complete a transaction. The dual mast crane requires a higher investment than a standard crane.

Standard carrier options are:

- A single tote carrier. For each crane trip in an aisle, a single carrier handles one tote. A single carrier crane handles approximately 50 to 70 totes per hour.
- A multiple tote carrier. On each crane trip into an aisle, a multiple carrier stacker crane handles two or more totes. It is capable of handling approximately 100 totes per hour.

The dual mast crane has a carrier with the capability to carry multiple totes or cartons. Carrier options are:

- One level with two side-by-side totes/cartons or trays that handles an estimated 100 totes/cartons or trays per hour
- One level with three side-by-side totes, cartons, or trays that handles an estimated 130 totes, cartons, or trays per hour
- Two levels with two side-by-side totes, cartons, or trays that handles an estimated 150 totes, cartons, or trays per hour
- Two levels with three side- by-side totes, cartons, or trays that handles an estimated 170 totes, cartons, or trays per hour
- Three levels with two side- by-side totes, cartons, or trays that handles an estimated 190 totes, cartons, or trays per hour
- Three levels with three side-by-side totes, cartons, or trays that handles an estimated 180 totes, cartons or trays per hour

A single-deep carton AS/RS crane carrier is designed to complete a carton, tray, or tote storage transaction to one UOP deep storage position. To complete a storage transaction, the feature has a single-deep crane carrier device extend outward into a storage or P/D station position one UOP deep. The single-deep crane carrier features are slightly short storage transaction time, fast transactions, slightly short storage position window, FIFO product rotation, access to all UOPs, and fewer UOPs per storage aisle.

A two-deep or double-deep carton AS/RS crane carrier is designed with a pantographic device that has the capability to extend outward two UOPs deep into a storage position. To complete a storage transaction, the device extends at a P/D station to one UOP outward into a P/D station position and at a storage position to complete a two-deep storage transaction to extend outward two UOPs deep and to complete a one deep storage transaction to extend outward one UOP deep. The two-deep crane carrier features are two-deep storage transaction time, slightly slower transactions, slightly taller storage position window, product rotation which is to minimize with crane UOP shuttle, with crane UOP shuttle access to all UOPs, and greater UOP number per storage aisle.

Factors that influence the design of container-handling devices are:

- How a carton AS/RS crane completes a P/D station or storage position transaction
- Clearance space in a storage position between a tote, carton, or tray and structural support members
- Tote, carton, or tray characteristics
- Time to complete a storage transaction
- How a carton AS/RS crane carrier device interfaces with a tote, carton, or tray

Carrier handling devices include:

- A telescopic grab device with an extendible and retractable arm that handles a tote from the top with two fingers. The two fingers are moveable and are set at an angle. As the fingers are extended over a tote, they move upward or downward and interface with the tote top lip. To complete a storage transaction, the telescopic grab device slightly elevates and transfers a tote between a stacker crane solid platform carrier and storage position. The features are an open top tote that is made of hardened plastic, hardened and coated metal, or wood; the top lip of the tote is designed to interface with the two moving arms of the carrier; solid or stranded smooth and continuous bottom surface in a storage position and crane carrier; permits one- or two-deep storage positions. A crane can have multiple carriers, crane has two masts, to telescope into a storage position, a slight increase in the storage transaction time, in a storage position, for a telescopic arm to complete a transaction, additional open space above a tote, standard or mini-stacker crane wide platform, tote physical characteristics are good for conveyor travel and queue and standard crane carrier handles up to 3–25 lb. tote to a 48 ft. elevation with a 720 ft. horizontal travel speed and a 100 ft. vertical travel speed.
- Telescoping platen or table, platform, or plank. To complete a storage transaction, the telescoping table concept has

- ◼ A tote, carton, or tray storage position with storage rails. The side of each storage position creates an open space between the rails. Storage rails support a tote, carton, or tray.
- ◼ A telescopic platen or plank that extends between the two rails. The telescopic platen extends fully to perform a storage transaction.
- ◼ The tote, carton, or tray bottom surface has structural strength to support the load weight. A tote, carton, or tray bottom surface is supported by two rails. This feature means that a tote, carton, or tray bottom surface has the structural strength to support the load without deflection. Excessive tote, carton, or tray bottom surface deflection or concave bowing restricts a storage transaction completion. If cartons are stored on the two rails you assure that humidity does not create bottom carton structural strength problems or deflection.
- ◼ For travel and queue on a conveyor travel path, a tote, carton, or tray front and rear side wall and bottom surface physical characteristics are conveyable.
- ◼ Interface with a tote, carton, or tray.
- ◼ Gripper device has two arms that extend into or retract between a stacker crane carrier and storage position to complete a storage transaction. The gripper device handles a tote from two sides. The tote side walls are designed with indentations that permit stacker crane arms to interface with the tote. After a stacker crane gripper device arms are extended into a storage position both arms are moved to grip the sides of a tote. To complete a withdrawal transaction, the arms grip the tote and raise it before moving it on-board the stacker crane. To complete a deposit transaction, the two arms grip a tote side walls, raise the tote, and move it from the stacker crane, extend into a storage position and deposit the tote onto a storage position two rails. The features are tote is designed to interface with the stacker crane grippers, tote has an open or closed top, tote material is hardened and coated metal, plastic, or wood, storage position has a solid or strand bottom surface, permits one- or two-deep storage locations, stacker crane has multiple carriers, to extend and retract into a storage position, gripper arms have a slight increase in transaction time, storage position has additional open clearances along both sides, tote indents reduce a tote interior storage space, standard crane carrier handles up to four carriers. Each tote holds an estimated 25 lb. A crane performs storage transactions at 5 ft. high above the floor, a crane has a 80 ft. horizontal travel aisle speed and a 57 ft. vertical travel speed.
- ◼ A telescoping blade gripper has a similar design as a gripper device except
  - ◼ Each tote side wall has a location for a blade gripper
  - ◼ Storage position is single deep
  - ◼ Carrier handles a 5 lb. load at a maximum 10 ft. height, single mast crane, and up to four carriers

- A rough top belt conveyor container handling device has very similar crane and storage position design characteristics and operational features as a telescoping table handling device. The exceptions are
  - Storage position is one deep
  - Carrier handles a 5 lb. load at a maximum 10 ft. height, single mast crane with a 80 ft. horizontal travel speed, and a 40 ft. vertical travel speed
  - Crane has four carriers
  - Maximum 10 ft. storage height above the floor
  - Interfaces with any tote or tray type
- An extractor tote- or tray-handling device has one or two lips or a fish hook-shaped lip. The lip is located at the front of a tote or tray and has sufficient open space between a tote or tray side wall and lip. This open space permits a carrier extractor fish hook to engage or interface with a tote or tray. A crane extractor fish hook-shaped device secures or interfaces with a tote or tray lip. With the hook engaged on a tote or tray lip, the interface permits a stacker crane to complete a storage transaction. This tote or tray and transaction feature has a hardened and coated plastic or metal tote or tray with the structural strength to perform a transaction.

To perform a storage transaction, the extractor fish hook-shaped component engages or interfaces to a tote or tray lip and extends outward into the proper location, slightly below a tote or tray lip and directly below an open space between tote or tray side wall and lip. With an extractor in the proper location, a stacker crane carrier moves the extractor upward to properly secure or engage a tote or tray. With a tote or tray properly engaged, a crane carrier moves the tote or tray slightly upward to reduce any friction between a tote or tray bottom surface and a storage position or crane carrier component. This permits a stacker crane to complete a storage transaction. With a storage transaction completion, a stacker crane carrier fish hook moves slightly downward to disengage or release the tote or tray lip. In this position, a crane carrier fish hook is ready to perform another storage transaction for another tote or tray.

The features are similar to a telescoping device with the following exceptions:

- With a lip on a tote or tray front end in the direction of travel on a conveyor travel path, a conveyor travel path has angled photo eyes
- Tote or tray material is hardened and coated plastic or metal
- Storage position is single deep
- Stacker crane features are single mast, two carriers, 10 to 30 ft. transaction height and 75 ft. horizontal travel speed and 30 ft. vertical travel speed and interfaces with a tote

A vacuum grab device has an extendible and retractable moving arm with one or two vacuum suction cups that are located at the end of the arm. From a vacuum device and through a tubing network, cups receive a constant and sufficient vacuum pressure to lift and move a tote to complete a storage transaction. The vacuum feature has a tote or carton top surface that is solid, smooth, secured to the side walls and dirt-free to ensure proper interface between vacuum suction cups and a tote. To complete a storage transaction, a vacuum device holds a tote or carton by the top surface.

A vacuum device has similar design characteristics and operational features as a telescoping device. Additional features are:

■ Tote or carton has a solid, smooth upper surface secured to the side walls, dirt-free, and with side walls and bottom structural strength
■ One deep storage position
■ Stacker crane features are single mast, four carriers per crane, 10 ft. to 30 ft. transaction height above the floor, handles 5 lb. tote weight, and 80 ft. horizontal travel speed and 40 ft. vertical travel speed

A powered roller conveyor concept has a storage position to have a powered roller conveyor bottom surface with an end stop and a stacker crane carrier to have a powered roller surface with an end stop. A powered roller conveyor concept has similar design features and operational aspects as rough top belt conveyor concept, with the exception that a tote, carton, or tray bottom surface assures proper traction for interface with a powered roller conveyor travel path surface.

Another option is a stacker crane with forks that are extended outward and inserted into a tote or tray fork openings. The tote or tray is deposited into a storage position or taken on-board the stacker crane carrier. This concept has similar design characteristics and operational features as a belt conveyor device with the following exceptions:

■ The tote or tray front has bottom fork openings
■ Storage position is one or two deep
■ Stacker crane features are tote or tray material is hardened plastic, wood, or metal, single mast, four carriers per crane, 10 to 30 ft. transaction height above the floor and 75 ft. horizontal travel speed and 30 ft. vertical travel speed

A tote, carton, or tray (see Figure 4.45) is the material-handling device that ensures the SKUs are retained in a storage position and permits a stacker crane to transport and transfer SKUs between locations.

Each stacker crane tote, carton, or tray handles one SKU or has interior separators and dividers to make smaller compartments. With multiple smaller compartments, a tote or tray can carry a number of small SKUs.

When you design a stacker crane concept, a tote, carton or tray width or down aisle dimension determines a stacker crane aisle length, crane tote-carrying and -handling device width dimensions, stacker crane transaction stroke, storage posi-

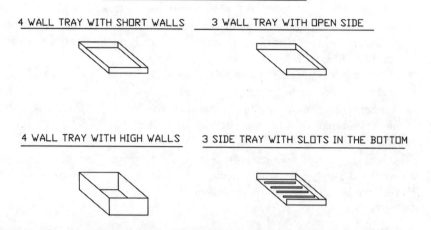

Figure 4.45 Tray types: four walls, three walls, long tray, and three walls with bottom slots.

tion depth, and P/D station depth. A crane transaction stroke is the distance a tote, carton, or tray-carrying and -handling device needs to move into a storage position or P/D station to complete a tote, carton, or tray deposit or withdrawal transaction.

The tote or tray can be made of metal, plastic, wood, or treated corrugated material. The preferred tote or tray material meets your stacker crane concept design parameters and product characteristics. Tote or tray design parameters are:

- Ability to interface with and be handled by a stacker crane tote, carton, or tray-carrying and -handling device, P/D station, and storage position
- A bottom surface that holds maximum number of SKUs and weight and ensures that SKUs are secure in a tote, carton, or tray with minimal bottom surface deflection
- Maximum storage area utilization
- Compatible with a tote, carton, or tray transport concept
- Identification code attachment
- Picker access to SKUs or allow an automatic device to transfer a tote or carton onto or from a tray

In addition, operational guidelines are:

- Long tote, carton, or tray dimension is the down aisle travel width
- From a rack storage or P/D delivery station, a narrow tote, carton, or tray dimension faces the aisle and is the side that is handled by a stacker crane-carrying and -handling device

- Totes, cartons, or trays with a short height are at the rack stack base and all tall totes, cartons, or trays with a tall height are at a rack stack top.
- Tote or tray interior has the ability to have dividers or separators. Dividers or separators make smaller compartments and permit the tote or tray to handle multiple SKUs with a picker routing pattern.
- Handles maximum SKU quantity and weight.
- Each small compartment within a tote or tray has a discreet identification that is either (a) a machine-readable code that is a bar code or RF tag, (b) an alpha/numeric human-readable code, or (c) a human/machine code combination.

Tote, carton, or tray features provide maximum storage position number per aisle length and storage area, sturdy storage structure, move on a powered conveyor travel path, and at a pick station, easy and quick access to a SKU.

Carton AS/RS Tray

In a carton AS/RS crane operation, a tray is a material-handling device that is used to handle a wide carton mix, and is designed to minimize carton bottom surface deflection in a storage position, ensure quality conveying bottom surface, and in humid areas and in storage position to maintain the carton bottom strength.

One design option is a tray with a solid bottom and four side walls. To transfer cartons a picker physically transfers a carton, a mechanical device transfers a carton by grabbing the two sides, or a vacuum lifts a carton from the top or cover. With the latter two carton transfer options it is essential that the carton side wall, top/cover, and bottom section has sufficient structural strength to withstand the pressure. With the vacuum lift option, the carton cover or top is secured to the side walls. The concept handles a carton with sufficient side wall, bottom, and cover structural strength; accepts all cartons that have a conveyable bottom surface; interfaces with many AS/RS crane carrier or handling device options; and is employee based.

An alternative tray design is to have a tray with three side walls and slots in the bottom. To transfer a carton from a conveyor travel path as a carton arrives at a pop-up stop, the carton is square and the tray bottom surface slots are aligned above pop-up belts. When a carton is stopped on the conveyor travel path, the pop-up belts rise through the slots and come in contact with the carton bottom surface. The forward movement of the pop-up conveyor belts moves the carton forward through the open tray side wall onto a take-away conveyor travel path. With the carton square and totally on the take-away travel path, it travels from the carton AS/RS area to the pick area. The concept (a) handles a wide carton mix with or without covers, (b) accepts all cartons that have a conveyable bottom surface, and (c) interfaces with many AS/RS crane carrier or handling device options.

The number of totes or trays designed to fit in a storage position (see Figure 4.46) are referred to as one-deep or two-deep storage positions.

LOAD ARM CARTON AS/RS STORAGE POSITION

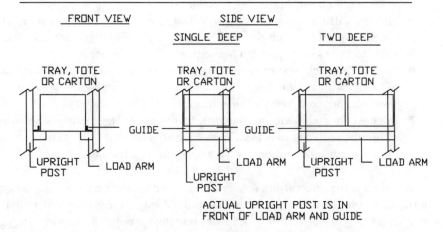

LOAD BEAM CARTON AS/RS STORAGE POSITION

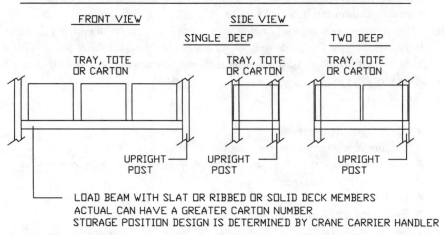

Figure 4.46 Carton AS/RS storage positions: load arm front view, load arm side view, load arm two deep, load beam front view, load beam one deep, load beam two deep.

- A one-deep tote, carton, or tray storage position concept offers maximum flexibility, good storage position utilization, simple operation, and permits the selection of almost any carrier device type
- Two-deep tote, carton, or tray storage positions provide good storage density per aisle, has a special carrier, and low utilization per storage position

Carton, tote, or tray storage positions options are:

- One carton, tote, or tray per two uprights, which has a high cost per storage position but minimal possible deflection.
- Multiple cartons, totes, or trays on a load beam. Each UOP has wide rails and is designed to match the AS/RS crane carrier. With two uprights it has a lower cost per storage position but some possible deflection.
- Installed in a rack support facility that has the upright posts support the walls and roof. With a facility over 20 ft. high, a rack support facility option has a lower cost per position and a shorter project schedule.
- Installed in a conventional building with the storage positions installed within the walls and roof shell. With a facility under 20 ft. high, the concept has a lower cost per position and a longer project schedule.

A stacker crane tote, carton, or tray storage position has a narrow tote, carton, or tray dimension face a stacker crane aisle and the long dimension is into a storage position (see Figure 4.47). A tote, carton, or tray storage position provides a storage location for small parts and structural support for an overhead guide-rail.

Cartons in a storage position have no additional bottom support tray or tote. The storage position can be a decked rack, or a ribbed or slated storage structure. To complete a storage transaction, cartons on a decked rack are handled from the sides with (a) side clamps, (b) a side belt, or (c) a side clamp with rubber pads. With a ribbed or slated storage position cartons are handled from the bottom with a comb or ribbed fork device.

The features are:

- No tray or tote expense
- Quality check for carton entering the AS/RS
- In-house transport path is powered roller, skatewheel, or belt conveyor
- Less weight on the storage members
- Poor-quality cartons are placed onto a tray or in a tote
- Each carton has an identifier

In-Feed and Out-Feed or Pick-Up and Delivery Stations

The in-feed (pick-up) station is the location where totes, cartons, or trays from a receiving department or a return tote, carton, or tray from a pick area are entered into an AS/RS crane concept. With this operational feature, a tote, carton, or tray conveyor travel path from the receiving department or a pick area is direct to the stacker crane in-feed (pick-up) station.

An out-feed (delivery) station is the location where totes, cartons, or trays exit the AS/RS crane concept storage area for delivery to the pick area. With this operational feature, a tote, carton, or tray conveyor travel path is a direct travel path from the stacker crane concept to a pick location. This pick location is on a conveyor

Figure 4.47   Carton AS/RS SKU containers: tote, carton, and tray.

section that is in the front of a stacker crane or is remote from the stacker crane concept. For best operational results the P/D stations have queue conveyor sections.

Stacker Cranes as a Small-Item Pick Concept

A stacker crane used as an order-fulfillment concept (see Figures 4.48 and 4.49) transfers the stock or pick position from a storage area to a pick station. At a pick station, a picker reads a tote, carton or tray discreet identification and matches the discreet identification to a pick instruction format (paper document, self-adhesive label, pick to light, or RF device). With a correct match, a picker completes a SKU pick transaction and removes a SKU from a stacker crane tote, carton, or tray and transfers it into a CO tote or carton. After all picks from a tote, carton, or tray, it is released from the pick station onto a transport conveyor travel path for transport to another pick station or to a stacker crane P/D station. After a CO has been completed or when a container is full, it is released from the pick area for travel to the sortation or pack station.

Stacker Crane with a Dynamic Conveyor Pick Front End

A stacker crane concept with a dynamic conveyor pick front end has (see Figure 4.50):

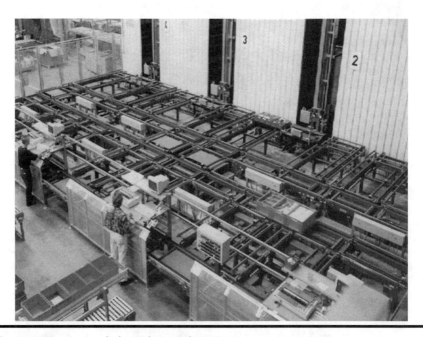

Figure 4.48 Sort-to-light with a stacker crane.

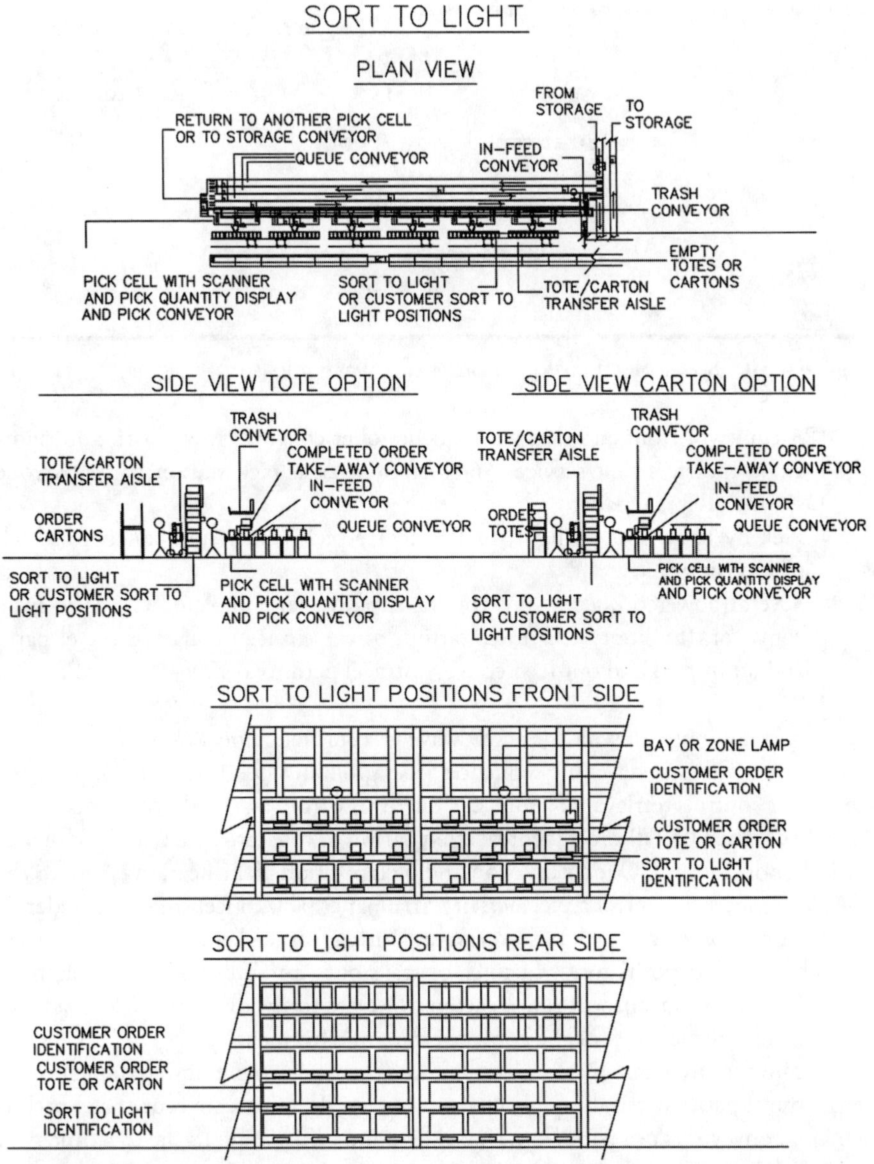

SORT TO LIGHT

PLAN VIEW

RETURN TO ANOTHER PICK CELL
OR TO STORAGE CONVEYOR

QUEUE CONVEYOR

FROM STORAGE

TO STORAGE

IN—FEED CONVEYOR

TRASH CONVEYOR

EMPTY TOTES OR CARTONS

PICK CELL WITH SCANNER
AND PICK QUANTITY DISPLAY
AND PICK CONVEYOR

SORT TO LIGHT
OR CUSTOMER SORT TO
LIGHT POSITIONS

TOTE/CARTON
TRANSFER AISLE

SIDE VIEW TOTE OPTION

TRASH CONVEYOR

COMPLETED ORDER
TAKE—AWAY CONVEYOR

IN—FEED CONVEYOR

QUEUE CONVEYOR

TOTE/CARTON
TRANSFER AISLE

ORDER CARTONS

SORT TO LIGHT
OR CUSTOMER SORT TO
LIGHT POSITIONS

PICK CELL WITH SCANNER
AND PICK QUANTITY DISPLAY
AND PICK CONVEYOR

SIDE VIEW CARTON OPTION

TRASH CONVEYOR

TOTE/CARTON
TRANSFER AISLE

COMPLETED ORDER
TAKE—AWAY CONVEYOR

IN—FEED CONVEYOR

QUEUE CONVEYOR

ORDER TOTES

SORT TO LIGHT
OR CUSTOMER SORT TO
LIGHT POSITIONS

PICK CELL WITH SCANNER
AND PICK QUANTITY DISPLAY
AND PICK CONVEYOR

SORT TO LIGHT POSITIONS FRONT SIDE

BAY OR ZONE LAMP

CUSTOMER ORDER
IDENTIFICATION

CUSTOMER ORDER
TOTE OR CARTON

SORT TO LIGHT
IDENTIFICATION

SORT TO LIGHT POSITIONS REAR SIDE

CUSTOMER ORDER
IDENTIFICATION

CUSTOMER ORDER
TOTE OR CARTON

SORT TO LIGHT
IDENTIFICATION

Figure 4.49 Sort-to-light views.

- Several microcomputer-controlled stacker cranes with a single or multiple tote, carton, or tray handling carrier
- A tote, carton, or tray that holds one or multiple SKUs; each tote, carton, or tray has a discreet code

SORT TO LIGHT PIECE AND ONE WAVE CUSTOMER ORDER COMPLETION

ORDER PROFILE FROM HIGH TO LOW VOLUME ORDERED SKUS	PIECE NUMBER	SINGLE LINE ORDERS		MULTI-LINE ORDERS		WITH SKU A MULTI-LINE ORDER COMBINATIONS					
		ORDER NUMBER	PIECES	ORDER NUMBER	PIECES	220A	220A 200B	220A 200B 300C	220A 200B 300C 400D	220A 200B 300C 400D 500E 600F	220A 200B 300C 400D 600F
100A	50	25	25	25	25	25					
200B	34	12	12	22	22		8	5	5	3	1	22
300C	24	10	10	14	14			4	4	4	2	14
400D	21	10	10	11	11				5	3	3	11
500E	15	5	5	10	10					5	5	10
600F	2	1	1	1	1						1	1
		63	63	83	83	8	9	14	15	12	58

Figure 4.50 Sort-to-light customer order one wave release.

■ A conveyor front end that has a queue roller conveyor travel path and right-angle transfers that service a pick station; each pick station has a barcode scanner/RF tag reader

■ Pick by light, sort by light, or RF device pick instruction concept at each pick position

■ Several powered queue conveyor travel paths: (a) completed CO take-away conveyor, (b) empty CO tote, carton, or tray in-feed conveyor travel path, and (c) in-feed and out-feed conveyor travel path to a crane

A host computer downloads a CO wave to a stacker crane pick area microcomputer. Per a CO stacker crane sequence SKU grouping (see Figure 4.51), a microcomputer controls retrieval of a tote, carton, or tray from assigned storage positions. A picker scans a CO discreet identification carton or tote into a pick-to-light or sort by light position. A CO identification and pick position identification is sent to the microcomputer to control a pick activity. In a paper pick concept, each CO identifier is written onto each pick position front. With a full stacker crane carrier device, a stacker crane deposits totes, cartons, or trays onto an outbound conveyor travel path that serves to transport totes, cartons, or trays to the dynamic pick area front end and as a queue zone for the conveyor dynamic front end.

A conveyor dynamic pick area front end has three or more powered queue conveyor travel paths and right-angle chain transfers. The first conveyor travel path is a major conveyor travel path from a stacker crane P/D station to the dynamic pick area. Other conveyor travel paths are queue conveyor travel paths with right-angle transfers that sequence the totes, cartons, or trays on a conveyor travel path to a pick station. The next conveyor travel path is a pick conveyor travel path that goes past each pick zone. Prior to a pick station, each tote, carton, or tray code is read by a barcode scanner/RF tag reader device. At a pick station, a tote, carton, or tray stops on a pick conveyor travel path and the pick instruction (sort by light, self-adhesive label, or RF device) shows a SKU pick quantity for each CO that is in an active or scanned pick position. After a pick transaction, if a SKU is required for another CO at another pick station, the tote, carton, or tray is sent from the pick position to

SKU TRANSFER TO SORT TO LIGHT PICK AREA

FIRST SKU		SINGLE ORDERS		SINGLE COMPLETED ORDERS	NEW SINGLE ORDERS	NEW MULTI-LINE ORDERS	POTENTIAL MULTI-LINE ORDERS	MULTI-LINE COMPLETE ORDERS	MULTI-LINE NON-COMPLETE ORDERS
100A	50	25	25	25	0	25	25	0	25
SECOND SKU									
200B	34	12	12	12	12	22	47	8	39
THRID SKU									
300C	24	10	10	10	10	14	53	9	42
FOURTH SKU									
400D	21	10	10	10	10	11	53	14	39
FIFTH SKU									
400E	15	10	10	10	5	10	49	15	34
SIXTH SKU									
400F	2	10	10	10	1	1	35	12	23

Figure 4.51 Sort-to-light customer order completion.

another pick station queue conveyor travel path. If a SKU is not required for a CO, the tote, carton, or tray is returned to the in-feed conveyor.

With a one pick station concept, as a CO pick activity is completed at a pick station, conveyor queue travel path A releases a tote, carton, or tray onto a pick station conveyor travel path. This feature ensures a constant and continuous flow of totes from queue conveyor A to a pick station. After all A conveyor totes are transferred, queue conveyor B releases totes, cartons, or trays to a pick station conveyor travel path. After all B conveyor totes, cartons, or trays are released to a pick station, queue conveyor C releases totes, cartons, or trays to a pick position.

After each queue conveyor travel path releases the totes, cartons, or trays and the conveyor is empty, the concept directs other totes that contain SKUs required for the wave or the next CO wave to the vacant positions on the conveyor travel path.

With multiple pick stations on a pick conveyor travel path, the conveyor network is the same as the single pick station concept, except that:

■ The pick zone conveyor travel path moves totes, cartons, or trays from a pick station to pick station 2.
■ Each pick station has a transfer device to transfer totes, cartons, or trays from the conveyor travel path to pick station 2.
■ With no required picks, totes, cartons, or trays by-pass the pick station and travel on a conveyor travel path for return to a stacker crane P/D station.
■ Pick area in-feed queue conveyor travel path sections that serve the same functions as the queue conveyor travel path that feeds pick position 1.

The features are pick to light/sort-to-light or RF device concept, permits high picker accuracy and productivity, investment, few pickers, two or more pick stations per pick area, permits one or more pickers for maximum picker productivity and CO number, de-box area for SKU transfer to a stacker crane tote, carton, or tray, and small area.

A carton/tote/tray AS/RS crane concept can have one of a number of designs (see Figure 4.52) to deliver and retrieve totes/cartons/trays between a pick position and AS/RS crane storage position. These pick concepts handle a smaller volume and have a lower investment than the sort-to-light with a crane. These pick concepts are:

- C configurations
- Left and right configuration
- Picking drawers
- Picking loop
- U-shaped picking station

A carton AS/RS crane pick line set-up and replenishment concept is an automatic carton pick line set-up or replenishment by carton AS/RS crane carrier to a flow rack pick position. In this concept each carton receives a discreet identifier and each carton travels past a preparation station where an employee prepares each carton front and top for the pick activity and ensures that the carton dimension and physical characteristics match the concept design parameters.

If the carton physical characteristics are not acceptable, the employee transfers the carton onto a tray/tote and completes the necessary scan transactions to update the inventory files and WMS/MHS computers. Past the preparation station, each carton travels past a barcode scanner/RF tag reader that sends the carton identifier to the WMS/MHS computer. Per the WMS/MHS computer, the computer directs each carton to an AS/RS crane aisle. As the carton arrives at the P/D station, the AS/RS takes the carton on board and travels down the aisle to the suggested pick position. At the pick position, the AS/RS crane transfers the carton to the pick position and notifies the MHS that a pick line set-up or replenishment transaction is completed. With a completed transaction, the pick line is ready for the pick activity and the AS/RS crane returns to the P/D station and picks up another carton. If the AS/RS crane cannot complete a pick line set-up or replenishment, the crane signals the MHS computer that directs the crane to another pick position or to place the carton into a ready reserve position. The MHS identifies the pick position problem and suggests management corrective action.

The features are AS/RS crane cost, MHS and WMS system and computer cost, protective area, additional roller lanes in the pick position, carton preparation area, good-quality cartons, and in-house carton conveyor travel path.

With an AS/RS crane automatic carton replenishment to carton flow rack concept, the carton, tray, or tote has to travel from the carton flow rack charge (AS/RS crane) end to the carton flow rack pick side. After a carton AS/RS crane transfers a carton, tray, or tote onto a flow lane pick position charge end, the carton, tray, or tote travels by gravity force over or through the flow rack conveyor lanes. The flow rack conveyor lanes are strands of plastic skatewheels on 1- to 2-in. centers and there are two to five strands per carton flow rack pick position. The gravity force is created by the elevation change between the charge end and discharge end and

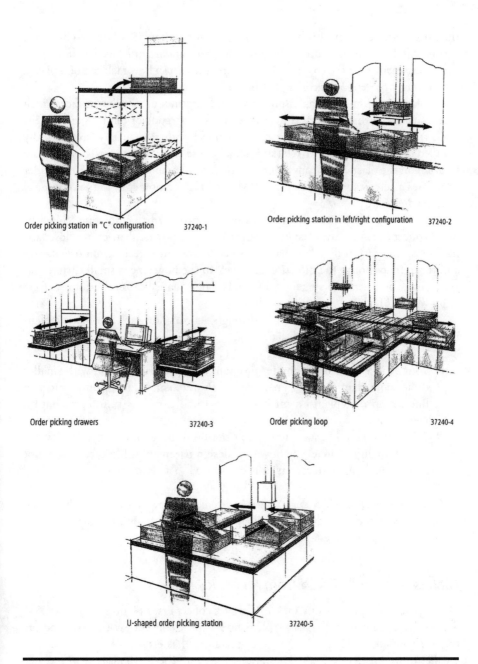

Order picking station in "C" configuration   37240-1

Order picking station in left/right configuration   37240-2

Order picking drawers   37240-3

Order picking loop   37240-4

U-shaped order picking station   37240-5

Figure 4.52 AS/RS crane pick locations: C configuration, left/right configuration, picking drawer, picking loop, and U-shaped picking station.

the carton weight. With this standard elevation for all flow rack lanes, lightweight cartons can become hung-up on the flow lane and standard or heavyweight cartons travel very easily over the conveyor travel path. The carton flow rack can be a plastic tote or tray, or a vendor or company carton.

A plastic tote or tray has a solid, flat, and continuous bottom surface which ensures that the tote can travel over three or four strands of plastic skatewheels. Each tray has a discreet identifier on each travel path side that is automatically scanned by a barcode scanner/RF tag reader. Per the barcode scanner/RF tag reader location on the plastic tray conveyor travel path, the barcode scanner/RF tag reader completes a zero scan or updates the WMS or MHS computer as to the tray travel location and status.

With a vendor carton or company cardboard carton in an automatic replenishment concept, some carton characteristics have the potential to create flow lane hang-ups. The carton flow lane hang-ups occur from incorrect carton orientation and travel (flaps not in the direction of travel) in the flow lane; a small carton that gets caught between two skatewheel strands; poor-quality carton bottom surface; light–weight SKU.

For maximum carton flow control through a carton flow rack automatic replenishment concept, it is suggested that:

- Your WMS or MHS computer program identifies each carton with smaller dimensions than the design parameters. Prior to entry into an automatic pick line set-up or replenishment concept, a small carton is transferred onto a plastic tray.
- All cartons travel past a preparation station to have an employee verify the carton quality matches the flow rack design parameters. If a carton does not match the design parameters, it is transferred to a plastic tray.

The features are AS/RS cost, control system, or computer cost, AS/RS crane area requires protection, additional roller lanes in each pick position, carton preparation area, and employees and conveyor travel path.

Eaches or Pieces Picked from an AS/RS Pallet

Picking eaches or pieces from an AS/RS pallet concept (see Figures 4.53 and 4.54) is considered a no forklift truck replenishment pick concept. With this concept, the AS/RS crane or pallet pick line set-up concept has each AS/RS crane deposit a pallet onto a pallet delivery (D) station conveyor. The method used to transfer pallets from the AS/RS to the pick area is determined by the AS/RS front structural support bracing members. If there is minimal structural support and there is sufficient space between the members for a D station conveyor travel path, each D station conveyor travel path ends at a pick station. If the AS/RS front end structural members allow one or two D station conveyor travel paths to interface with a pal-

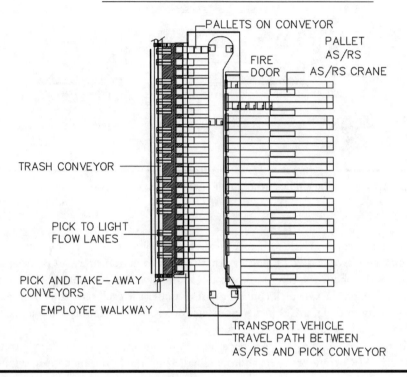

Figure 4.53   Picking eaches from pallet AS/RS lanes: plan view.

let transport device, the pallet transport device transfers a pallet from a D station conveyor travel path end to a pick position.

After the host computer has downloaded the CO SKU requirements in pallet quantities to the MHE concept computer, the MHE computer directs AS/RS cranes to move pallets from the storage area to the special D station conveyor travel path. Each D station conveyor travel path has a forward travel path direction of travel from the AS/RS area to the pick area. The pallets travel over a pallet conveyor travel path through a fire wall opening to the pick area. In the pick area, the first pallet is moved to the pick position and other pallets are queued behind an employee walk-way on the conveyor travel path. With a pallet in a pick position, a pick station employee by walking over a walk-way removes the plastic wrap from a pallet. Next, a pick station employee hand scans or an automatic scanner reads the pallet identification. The hand scanner sends the information to the pick line computer that a pallet has arrived at a pick position and the pick line computer enters the SKU into the pick inventory. Next, the pick station employee removes a master carton from the pallet, prepares the master carton for a pick position and transfers the prepared master carton onto the CO eaches pick conveyor. The pick conveyor

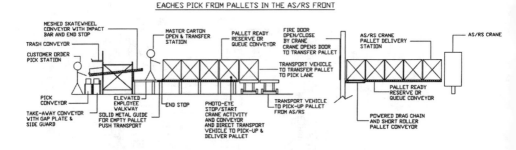

EACHES PICK FROM PALLETS IN THE AS/RS FRONT

Figure 4.54 Picking eaches from pallet AS/RS lanes: side view.

has sufficient lanes and length to provide a reasonable master carton queue. At the pick conveyor end is a photo-eye that registers a carton in a CO eaches pick position. The photo-eye communicates this to the pick area computer that has CO orders travel on the CO pick conveyor past or under the each pick conveyor. Per a CO pick instruction, a CO picker transfers the CO SKU quantity from a master carton into a CO carton or tote.

In this high-volume pick area, per a vacant position on the pallet conveyor travel path, an AS/RS crane transfers additional pallets on the D station conveyor travel paths until the total CO demand is complete. The master carton picker continues to open and transfer master cartons from a pallet onto the master carton conveyor lanes. Per a CO pick instruction, a CO picker continuously transfers SKU pieces from master cartons in a pick position into a CO carton or tote.

To ensure constant trash removal, a trash conveyor is located above the master carton pick conveyor. In this location, the CO picker has easy access to transfer empty master cartons and from a walk-way the master carton picker has trash conveyor access to dispose of carton tops and fronts as trash.

Components of the eaches or pieces concept are the pallet AS/RS crane, specially designed D station conveyor travel paths with one direction of travel, fire wall penetration and protection, pallet queue conveyor with a walkway around a pallet in a pick position, master carton conveyor lanes with an elevated trash conveyor, pick-to-light instruction concept with a pick conveyor, and completed CO take-away conveyor and CO pick carton or tote conveyor feed concept.

The features are no forklift truck investment or employee activity, constant replenishment with no fixed ready or shadow reserve positions, few employees, handles high-volume SKUs and a wide product mix, can be located in a small area and no forklift truck aisles, good WMS or inventory control concept, minimal UOP or equipment damage or employee injury, and pick-to-light as the CO pick instruction.

Stacker Crane Pick Tunnel

A stacker crane pick tunnel concept (see Figure 4.55) has:

- Microcomputer-controlled stacker cranes with single or multiple tote, carton, or tray-handling carrier
- Bar-coded/RF tagged tote, carton or tray that holds one or multiple SKUs
- One deep or two deep tote, carton, or tray storage/pick positions
- Code reader, a pick to light, sort to light, or RF device instruction concept at pick stations

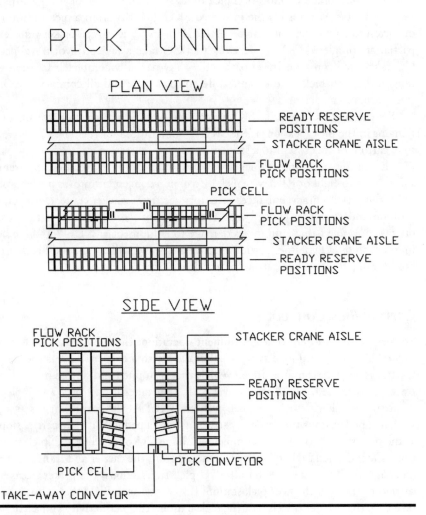

Figure 4.55   Pick tunnel.

■ In-feed empty tote, carton, or tray travel path to each pick station, partial CO tote/carton transport conveyor travel path, and a completed CO tote/carton take-away conveyor
■ SKU de-box or transfer to a tote location and a trash disposal concept

COs are downloaded from a host computer to the stacker crane microcomputer. Per a CO sequence, the stacker crane retrieves a tote, carton, or tray from a storage position and deposits it at an assigned pick position. Pick positions are between a stacker crane aisle and pick aisle. A stacker crane is captive to a storage aisle and a pick aisle has a specific number of pick positions.

When a SKU in a pick position is required for a CO, the microcomputer activates the appropriate CO location pick light, sort light, or RF device pick instructions and a picker transfers the indicated SKU quantity from a pick position tote or carton to a CO pick tote/carton. After all picks for a wave, a picker transfers all partial or completed CO totes/cartons to a CO take-away conveyor travel path. A CO take-away conveyor travel path moves any partially completed CO totes/cartons to the next pick position in a different pick aisle and all completed COs are transported to a pack area.

For the next CO wave, if a SKU is required on the wave, the tote, carton, or tray remains in the pick location. If the SKU is not required for the next CO wave, the tote, carton, or tray is sent to a stacker crane pick-up station.

The features are minimal picker walk distance, low picker reading requirement, sort or pick-to-light or RF device pick instruction concept, improves picker productivity and pick accuracy, enhances inventory accuracy, stacker crane, conveyor, pick position and computer control cost, wide SKU mix, minimal picker tote, carton or tray movement that minimizes damage to equipment and UOP and employee injuries, high CO volume, optimizes space and cube utilization, and SKUs are de-boxed into a tote, tray, or carton for a pick activity.

Cartrac Pick Concept

For use in a small-item order-fulfillment operation, the S.I. Cartrac concept (see Figure 4.56) is a mini-load type or pallet type with a carrier that has shelves. The Cartrac is an electric motor-driven vehicle that is stopped and started manually and travels over a fixed, closed-loop travel path, that takes it past pick and replenishment stations. Each carrier or shelf has one SKU. As the carrier arrives at a pick station, a picker stops the carrier by stepping onto a pedal. With the carrier stopped at the pick station, the picker completes a single CO or batched CO pick transaction. The indicated SKU quantity is removed from the carrier and transferred onto an assigned CO location. After all SKU pick transactions, the picker releases the carrier for travel to the next pick station. If a carrier has a SKU that is not required on a CO, that carrier travels past the pick station. The pick activity and carrier stop and start activity are repeated for all SKUs on a CO order or batched COs.

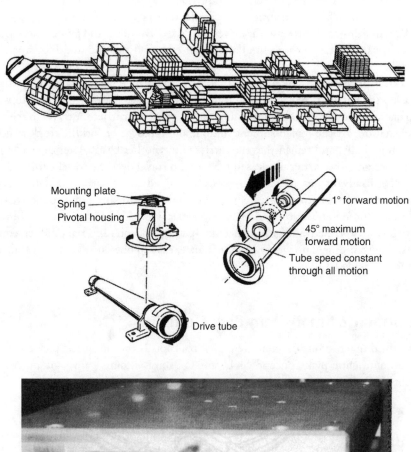

Mounting plate
Spring
Pivotal housing

1° forward motion

45° maximum
forward motion

Tube speed constant
through all motion

Drive tube

Figure 4.56 S.I. Cartrac.

Components of the S.I. Cartrac are track supports, a drive tube and drive motor, turntable curves, transfer and parallel travel paths, carriers for a pallet with shelves or mini-load carrier with shelves, chassis, drive wheel and guide wheel, travel wheels, cam switch with accumulation lead end, and tail end and foot stop/start pedal and controls.

After an employee or forklift truck has set-up or replenished a SKU onto a carrier, the carrier is released for travel to the first pick station. A picker stops the carrier and completes pick transactions and releases the carrier for travel to the next pick station. At the replenishment station, an employee or forklift truck removes an empty pallet and replenishes the carrier with another UOP. After the replenishment activity, the carrier is released for forward travel over the travel path.

The disadvantages are capital investment, fixed travel path, singulated carrier over a fixed travel path can create queues on a travel path and non-productive time at a pick station and mini-load or pallet carrier with shelves carries a small inventory quantity. The advantages are machine-paced concept, minimal product, equipment and building damage, low maintenance, easy to operate, easy to expand, and handles a wide UOP mix.

Automatic Small-Item Pick Concepts

The third major group of small-item pick concepts is the automatic pick concepts. In an automatic pick concept, SKUs for COs are automatically (computer controlled) released from a machine pick position. Picked SKUs are:

- Released onto a rough top belt conveyor travel path for transport to a CO pack station
- Released onto an open space between two cleats on a rough top belt conveyor travel path for transport and transfer into a CO tote/carton
- Released into a discreet CO carton or tote for transport to the next pick area station to a pack station

Key automatic pick concept design parameters are:

- SKU characteristics, including length, width, height, weight, and exterior packaging
- CO cube and pick line profile

Design parameters for an automatic pick concept are similar to those for a manual pick line design. The exceptions are restrictions for round or odd-shaped, crushable or fragile, and very slow- and fast-moving SKUs.

Where to Locate the Automatic Pick Machine to Have the Customer Order Carton or Tote Interface with a Manual Pick Area

In a small-item order-fulfillment operation or pick area layout, an automatic pick area has several CO pick concepts. An important concern is then, where do you locate an automatic pick machine in relation to a manual pick concept?

An automatic pick machine handles SKUs that are small- to medium-cube, light- to medium-weight, medium- to fast-moving, square or rectangular, non-crushable or fragile, and expensive SKUs. A manual pick concept handles heavy-weight, large-cube, less-valuable, very slow and very fast-moving, odd-shaped, fragile or crushable SKUs. An automatic pick machine that is 30 to 50 ft. long has a fast pick rate that completes a three piece CO within three to five seconds.

During the manual pick activity, picked SKUs with the above-mentioned characteristics are manually transferred from a pick position into a CO carton or tote. To minimize UOP damage and to maintain good picker productivity and on-time CO completion with an automatic pick machine, a pick area design requirement question is where to start a CO pick process and to complete a CO pick process. In most applications that pick into a barcode identified carton, the manual section sends partially completed COs to the automatic pick machine area.

For additional information, we refer the reader to the previous review of where to start your picker in a low- or high-volume area section in this chapter.

SKU Allocation to an Automatic Pick Machine

SKU allocation to an automatic pick machine is based on the physical characteristics, velocity, and family group of the SKU. Pick activity in an automatic pick machine is in an arithmetic progression beginning at home base from the first pick position to the last pick position. Per your design criteria and automatic pick machine manufacturer standards, home base is a pick position that is nearest the picked SKU transfer location or at the take-away conveyor belt travel path start. The microcomputer release or pick impulse is along one side of an automatic pick machine and down the other side. An automatic pick machine has large-cube, heavy, and fast-moving SKUs located at the first pick position zone and the crushable, small-cube, lightweight and slower-moving SKUs in the next pick position zone.

Other features of automatic pick machine SKU location are:

- SKUs are family grouped.
- SKUs are allocated as (a) one SKU per pick sleeve or lane and (b) for fast-moving SKUs, multiple pick sleeves or lanes determined by the computer system to handle a floating pick position philosophy, lane capacity, and SKU movement.
- On-time and accurate replenishment and pick transactions at a pick position are verified.

Automatic Pick Position Replenishment

An automatic pick machine has a pick sleeve, pick lane, or a pick belt that holds a small inventory quantity. The automatic pick position also has high pick machine rates. To minimize stock-outs and maintain a customer service standard, on-time and accurate pick machine pick position replenishment is a requirement for a small-item order-fulfillment operation with automatic pick machines. Locations for automatic pick position replenishment (see Figure 4.57) are:

- At a SKU pick sleeve, pick lane, or pick belt
- In a ready reserve area
- In a remote reserve area

A SKU pick sleeve, pick lane or pick belt is an automatic pick machine pick position. An automatic pick position holds on average one or two master cartons with a 20 to 24 master carton piece quantity. To assist a replenishment employee make on-time, accurate SKU transfers from a master carton to a pick position, a mark is placed on each automatic machine pick sleeve. The mark is paint or tape at an appropriate location on the sleeve that holds a master carton SKU quantity. During the pick activity, when a mark becomes visible, it is a signal to a replenishment employee to perform one master carton replenishment transaction from a ready reserve position to an automatic machine SKU pick position.

An option for automatic pick machine replenishment is to have a photo-eye on each pick sleeve. When a photo-eye light beam is not broken by a SKU in a pick sleeve, the photo-eye sends a message to a microcomputer to activate an alarm that signals that a pick machine pick position has a low SKU quantity. A replenishment employee then completes a master carton replenishment transaction to the pick position.

To ensure accurate SKU replenishment, a SKU discreet identification human/machine-readable code is on each replenishment pick machine sleeve or pick lane structural support member. With a hand-held scanner, an employee completes a master carton and pick position barcode scan transaction to verify and send a message to a microcomputer that a master carton SKU piece quantity was replenished to a pick position.

An automatic pick machine replenishment design factor is the master carton ready reserve inventory location and ready reserve concept. Having a master carton ready reserve position ensure that a master carton inventory quantity is available for automatic pick machine replenishment. During automatic pick machine mechanical/electrical down-time, the ready reserve position also provides SKU pick positions to complete COs.

In an automatic pick machine module there are many SKUs, each of which requires a master carton ready reserve position. The most common master carton ready reserve concepts are carton gravity flow rack bays, decked standard pallet

Figure 4.57 Automatic pick machine replenishment layouts.

rack, decked wide span rack, or standard shelf bays. Design options for a ready reserve area layout are to have the ready reserve pick side face:

- The automatic pick machine replenishment side. This layout provides a direct SKU master carton flow from a ready reserve position across the replenishment aisle to an automatic machine SKU pick position. The layout provides a low number of SKU faces in a ready reserve area.
- The pick aisle between two ready reserve racks. This concept has a pick aisle between two ready reserve rows and pick aisle intersection with an automatic pick machine replenishment aisle. This layout increases a picker walk distance to complete a replenishment transaction and provides the greatest number of faces in a ready reserve area.

A ready reserve area provides back-up SKU pick positions. Between an automatic pick machine replenishment side and the ready reserve positions, there is sufficient space for a temporary non-powered conveyor pick line or pick cart travel path. This layout assures a picker with a paper pick document access to all ready reserve pick faces.

In an automatic pick machine area design, the remote reserve locations are pallet racks or decked pallet racks with forklift truck aisles that flow from the remote reserve positions to ready reserve positions. Remote reserve pallet rack rows and aisles have a perpendicular or parallel flow to the automatic pick machine ready reserve positions.

Replenishment transactions can be employee-controlled activities or microcomputer-controlled. To ensure accurate and on-time SKU transfer from a remote reserve position to a ready reserve position and for replenishment to an automatic pick machine pick sleeve, pick lane, or pick belt, an automatic pick machine operation uses a microcomputer-controlled replenishment instruction. The replenishment instructions are paper document or RF device instructions.

The automatic pick machine uses a trash conveyor to dispose of empty cartons, filler material and paper trash. The preferred trash conveyor travel path is above the ready reserve positions. The ready reserve carton flow rack bay, standard pallet rack, or standard shelf structural members are designed with the capacity to support a trash conveyor travel path. In an automatic machine layout, the main trash conveyor travel path options are:

- Trash conveyor is on an automatic pick machine replenishment side. The trash conveyors waterfall onto an intersecting main trash conveyor travel path. The main trash conveyor travel path transports the trash to the trash disposal area.
- With two automatic pick machines, one automatic pick machine trash conveyor is the main trash conveyor travel path that has the other automatic pick

machine trash conveyor has a waterfall conveyor that transfers trash onto the main trash conveyor.

Customer Order Entry and Automatic Pick Machine

An automatic pick machine consideration is CO entry and automatic pick machine activation. An order-fulfillment operation automatic pick machine that handles a high-volume or large CO number, receives CO SKU pick or release instructions from a microcomputer. With these operational characteristics, a host computer sends all CO pick requirements to microcomputer. Options for CO entry to an automatic pick machine are:

- Direct from a host computer to an automatic pick machine. This is not preferred due to the fact that the host computer could be busy with other functions and not be capable of transferring CO pick requirements to an automatic pick machine microcomputer.
- The host computer downloads CO SKU pieces to an automatic pick machine microcomputer that interfaces with an automatic pick machine. This arrangement means that as an automatic pick machine receives a SKU piece quantity that a CO is available to an automatic pick machine.

An automatic pick machine is activated to pick SKUs from a pick position for a CO. Options for activation of the pick machine are:

- CO sequence that was downloaded from a host computer to an automatic pick machine computer. This download method has an automatic pick machine start a CO pick process. An automatic pick machine completes all CO picks directly into a carton or tote or loose onto an open space between two cleats on a belt conveyor travel path. After loose SKUs are sent to a pack station, they are placed into a CO tote or carton. A partial tote or carton is sent on the take-away conveyor from the automatic pick machine area to another pick area. A completed CO is sent from the automatic pick machine pack station to the ship area.
- A bar code or RF tag CO discreet identification on a CO tote/carton that is transported over a conveyor travel path. A conveyor moves CO totes or cartons from another pick area to an automatic pick machine area or from a CO identification station to an automatic pick machine area. Totes travel on a FIFO basis past a code reader onto an automatic pick machine. The code reader sends each CO discreet or coded tote to an automatic pick machine computer, which activates an automatic pick machine to release sequenced CO SKUs from pick positions into a CO tote or onto an open section between two cleats on a rough top belt conveyor travel path. With an automatic pick machine that releases SKUs onto a belt conveyor travel path, at the discharge

end of the conveyor SKUs fall into a tote or carton. This concept is used in a small-item order-fulfillment operation that has a manual CO pick concept send empty or partially full discreetly identified CO containers to an automatic pick machine where an employee enters the CO discreet identification into the automatic pick machine.

Types of Automatic Pick Machine

Automatic pick machine concepts (see Figures 4.58 and 4.59) include:

- Itematic
- Automatic item dispenser or Robo Pick
- A-frame

Itematic Automatic Pick Machine

An Itematic automatic pick machine is a stand-alone SKU pick machine. Each Itematic automatic pick machine is a microcomputer-controlled, single CO pick concept that handles a specific size, range, and number of SKUs. In an Itematic pick machine, an individual SKU is assigned to a specific pick lane. A replenishment employee places a SKU quantity into an assigned pick lane, which begins on the back end of an Itematic pick machine, horizontal to the floor and ends at the Itematic pick head. When the SKU pick lane is full, the Itematic pick machine is ready to release or pick SKUs for a CO. During a CO pick process, a microcomputer-controlled picking device that has a single CO SKU requirement travels along the Itematic pick side. As the pick device arrives at a SKU pick lane that is on a CO, the computer triggers an Itematic machine picking device mechanism (head) to release or pick one SKU from the pick lane onto a take-away conveyor travel path. Picked SKUs travels on a rough top belt conveyor travel path from the pick machine pick area. SKUs are transferred:

- Onto a smooth top belt conveyor travel path
- Into a CO shipping carton or tote for transport to a pack station, where they are transferred, collected, or packed into a CO shipping tote or carton

To optimize packer productivity, a divert blade on a smooth belt conveyor travel path is used to divert a CO (slug of SKUs) onto a specific section of a pack station slide.

The Itematic pick machine is located on a ground-level floor or on a mezzanine level. The other design requirements are sufficient area for replenishment and the master carton ready reserve positions flow directly to the Itematic replenishment side.

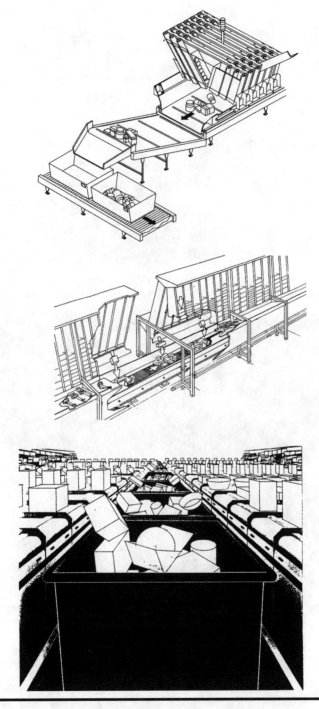

Figure 4.58 Automatic pick machine types: A-frame and Robo Pick.

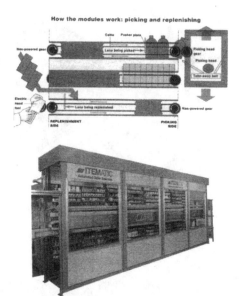

Figure 4.59 S.I. Itematic.

Automatic Item Dispenser or Robo Pick Concept

A Robo Pick or automatic item dispenser is another automatic small-item pick concept that is manually replenished and microcomputer-controlled to pick SKUs for a single CO. A Robo Pick concept has a central CO tote belt conveyor travel path with several banks, item dispensers, or cleated rough top pick belts on both sides of a central belt conveyor travel path. A cleated rough top belt conveyor travel path or dispenser ensures that proper space is maintained between SKUs and facilitates forward movement over the travel path. SKUs are picked and released either directly onto a belt conveyor or into a tote or carton.

As a CO tote or carton travels on a central belt conveyor between two pick or dispenser rows, a microcomputer activates the assigned pick belt or item dispenser to index or move forward. Forward movement permits one SKU to be released or picked into a CO tote or carton or onto a central rough top cleated belt conveyor travel path. As the SKU moves forward it passes a sensor that transmits a signal to a microcomputer that verifies the CO SKU pick transaction. All partial or full CO totes/cartons are transported to the next pick area or to a pack station. At a pack station, a CO is prepared for customer delivery.

Manual replenishment to a Robo Pick dispenser is carried out from an aisle by an employee. The employee places one SKU into each empty belt lane or pick position. The replenishment aisle is between the master carton ready reserve position concept and the pick position.

The Robo Pick concept handles loose or packaged SKUs that are medium- to small-cube, medium-weight to lightweight, and medium- to slow-moving. The concept provides a FIFO UOP rotation and medium number of faces per pick line. The Robo Pick concept handles a 54-order batch and 20,000 pieces per hour.

A-Frame or Automatic Pick Machine

An A-frame automatic picking machine has pick sleeves that are arranged to release or pick CO SKUs onto a central cleated gathering belt conveyor travel path. An A-frame automatic picking machine is a computer-controlled machine that handles SKUs for one CO. All COs are held in a microcomputer that sequences COs to satisfy a CO delivery schedule. As a discreetly identified CO carton or tote passes a barcode scanner or RF tag reader, the microcomputer sends an impulse along the A-frame to release SKUs that are associated with a CO discreet identification. The A-frame SKU sleeve dispenser releases SKUs at a rate of five to seven pieces per second onto a rough top cleated belt take-away conveyor travel path. At the bottom of each dispenser or sleeve is a small forward-rotating device that has a small upward projection. As this device rotates forward, the upward projection engages the rear side of a dispenser moving a SKU from the pick sleeve onto a gathering rough top belt conveyor travel path. The SKU slides over a short sheet metal or hardened plastic slide that is curved at the discharge end to ensure smooth SKU transfer onto the conveyor belt. To ensure that SKUs are separated on a gathering belt conveyor the travel path is equipped with cleats that project upward by 0.25 in. above the surface and are spaced every 12 to 18 in. over the belt surface. This creates an open space between two cleats that restricts and controls SKU movement over the conveyor belt.

A number of A-frame automatic pick machine models are available:

- The Ultrapick A-frame
    - Handles a SKU 1 to 2 in. in length, 0.5 to 9.5 in. wide, 0.38 to 5 in. high, and weighing up to 3 lb.
    - Has 10 to 52 pick sleeves or dispensers per 10 ln ft. which is one pick module and up to 25 pick modules per gathering conveyor belt travel path.
    - Each module handles 1000 COs per hour or 3000 plus lines per hour.
    - A gathering belt conveyor travel path with a travel speed of 3 ft. per second and travel path maximum width of 39 in.
    - Utility requirements that are 480 VAC, three phase and 120 volt one phase 60 Hz and no compressed air or vacuum requirement.
    - A pick machine that has a 250 ft. long by 8 ft. wide area with an overall height of 8 ft. This area does not include a CO carton/tote in-feed conveyor travel path, travel path to a pack station area, and ready reserve replenishment area.
- The Soft pick A-frame
    - Handles a SKU 7 to 16 in. long, 4 to 8.5 in. wide, 0.125 to 3 in. high, and up to 1.5 lb.
    - Has 10 to 20 pick sleeves or dispensers per 10 ln ft. for a pick module and up to 25 pick modules per gathering conveyor belt travel path.
    - Each module handles 1000 COs per hour or 3000 plus lines per hour.
    - The gathering belt conveyor has a travel speed of 3 ft. per second. The conveyor belt travel path maximum width is 30 in.
    - Utility requirements that are 480 VAC, three phase and 120 volt 1 phase 60 Hz and no compressed air but a vacuum requirement of 60 lw.
    - A pick machine that requires a 250 ft. long by 12 ft. wide area with an overall 4 ft. height. This area does not include a CO carton or tote in-feed conveyor travel path, travel path to pack area, and ready reserve replenishment area.
- The Flex pick A-frame
    - Handles a SKU 2.5 to 7.25 in. long, 1 to 3.75 in. wide, 0.5 to 3 in. high and weighs up to 3 lb.
    - Has 10 to 52 pick sleeves or dispensers per 10 ln ft. for a pick module and that can have up to 25 pick modules per gathering conveyor belt travel path.
    - Each module handles 1000 COs per hour or 3000 plus lines per hour.
    - A gathering belt conveyor travel path with a travel speed of 3 ft. per second. The conveyor belt travel path maximum width is 39 in.
    - Utility requirements that are 480 VAC, three phase and 120 volt 1 phase 60 Hz and 95 psi compressed air and no vacuum requirement.
    - A pick machine that requires a 250 ft. long by 8 ft. wide area with an overall height of 6 ft. This area does not include a CO carton or tote in-

feed conveyor travel path, travel path to a pack area, and ready reserve replenishment area.

The A-frame pick sleeve dispenser replenishment takes place during pick activity. A replenishment employee removes a SKU pick sleeve holder from the dispenser, transfers SKUs from a master carton into the sleeve, and replaces the sleeve holder.

Picked SKU Transfer

Options for transferring picked SKUs from a gathering conveyor belt to a pack station (see Figure 4.60) are:

- SKUs are discharged onto a slide with a divert blade that directs picked SKUs to a specific section on the slide.
- SKUs are discharged into a captive CO tote or CO shipping carton.

Characteristics of a captive CO tote include:

- The bottom surface on the inside is a vertically moveable surface that is supported by coiled springs that allow the bottom to be adjusted to fit the SKU quantity or weight.
- The interior bottom surface is padded with a cushion.
- The transfer location has a fabric or plastic shroud that extends downward. The sides of the tote come up around the shroud, which acts as a funnel to direct the SKU into the tote.
- The conveyor travel path has a mechanical tilt section. As a CO tote arrives at a transfer station, the tilt conveyor section angles the tote to minimize SKU impact on the bottom of the tote.

Order Carton/Tote Make-Up and Introduction to a Pick Line or Pick Aisle Station

In a small-item order-fulfillment operation, an important pick activity is CO tote or carton make-up and introduction to a pick line, pick vehicle, or pick aisle. It is realized in a batched CO fulfillment operation with a pick/sort and transport concept or pick/transport and sort concept that a CO shipping carton is introduced at a pack station. The operation should ensure that:

- Picked SKUs are retained
- Cartons or totes move along a pick line, pick aisle, or automatic pick machine, and past all pick positions

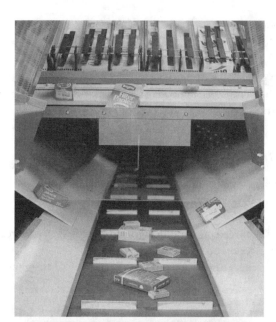

Figure 4.60 Automatic pick machine SKU transfer: onto a cleated belt and into a carton.

■ CO totes or cartons move from the last pick position to the next pick line, pick aisle activity, or to the pack area

If a CO pick container is a CO shipping carton, a carton moves from a manual or machine carton make-up station to a pick conveyor travel path, over a pick conveyor travel path, and from a pick conveyor travel path onto a completed CO take-away transport conveyor travel path. A completed CO take-away conveyor travel path transports a CO pick carton over a completed CO transport conveyor travel path from a pick area to the next activity station or pack station.

Factors to consider in determining how cartons are made up and introduced to the pick line include:

■ Carton material
■ Method for making up cartons
■ The location for carton make-up activity
■ Whether to use a one- or two-piece carton and (a) for a one-piece carton, to seal or not seal the bottom flaps; (b) for a two-piece carton, whether to include a cover on the bottom of the carton as part of the make-up activity or to have covers supplied at a pack station
■ On a pick line, with a one-piece carton, whether to have carton top flaps up or down
■ On some pick line applications, the method to be used for providing space on a carton surface for CO shipping label attachment
■ Carton type, that is whether to (a) use a new carton or (b) reuse a vendor carton
■ With some cartons, how to transfer and replenish cartons to the appropriate pick line workstation
■ Carton sizes or number
■ Corrugated material and type
■ Carton color
■ Company return address and CO delivery address label location or window printed on each carton exterior surface

Carton Type

Types of order-fulfillment CO carton are:

■ A one-piece carton with four side walls, four bottom flaps, and four top flaps. Prior to transfer of an empty carton to a pick line, an employee or machine seals the four bottom flaps with tape to provide a smooth surface for easy movement over a conveyor travel path or vehicle load-carrying surface. After all order-fulfillment activities, an employee or machine seals the four top flaps with tape to secure the carton and provide a smooth top surface.

■ A one-piece pop-out carton with four side walls, one integrated bottom surface, and top flaps. Prior to carton transfer to a pick line, an employee or machine pulls a pop-out carton bottom section, which becomes a secured bottom that is supported on the four sides. After all order-fulfillment activities, an employee or machine secures the top flaps with tape or a band.

■ A two-piece carton with four side walls, one integrated bottom surface, and one separate cover. Prior to carton transfer onto a pick line, a machine forms the carton bottom section.

How to Make Up Cartons

On a pick/pack line, the first activity station is a carton make-up station. If the operation picks loose, small-item flatwear into a captive tote, and uses a pick/sort and transport or a pick/transport and sort concept, carton make-up takes place at the pack station.

An employee or machine:

■ Removes a collapsed carton from a carton stack or from a machine in-feed sleeve
■ Forms the carton
■ Depending on the pick area procedure and carton type, seals the carton bottom exterior flaps with tape
■ Transfers a formed carton with the sealed or non-sealed bottom flaps onto a non-powered slide, powered conveyor travel path, or onto a carton stack

How a formed empty CO carton is introduced onto a pick line, pick aisle, pick vehicle load-carrying surface, automatic pick machine conveyor travel path, or onto a pack station table is determined by the order-fulfillment concept, whether single or batched COs; and by the carton sizes that are used in the order-fulfillment operation.

Manual Carton Make-Up Activity

In a manual carton make-up activity, an employee removes a collapsed carton from a stack and forms the carton. If it is a one-piece carton, the employee applies a tape strip or strips to the bottom flaps. With a pop-out one-piece carton, an employee pops out or pulls on a carton section to form the carton. A formed carton with a secured bottom and side walls are transferred to a carton stack, placed directly onto a pick line conveyor travel path, placed onto a pick vehicle load-carrying surface, or onto a pack table surface.

If a one-piece carton's bottom flaps are not sealed at make-up, the last pick line activity seals the bottom flaps. Unsealed flaps create a problem when the empty carton travels over a conveyor travel path or as it is transferred from a pick line to another activity station. In a high-volume order-fulfillment operation, the preferred carton seal concept is to have a start station employee or machine seal carton bottom flaps.

The disadvantages are low employee productivity, increase tape use or additional tape strips on a carton, handles a low volume, and with several tape strands on a carton, poor carton exterior appearance. The advantages are handles the widest carton variety, handles all tape types, no investment, and easy to relocate.

Mechanical Carton Make-Up Activity

A mechanical carton make-up activity has a mechanical device that removes a collapsed one-piece carton from a sleeve, forms the carton and applies a tape strip to the bottom exterior flaps. With a two-piece carton, the machine forms a carton with no tape by removing a corrugated sheet from a stack and forming it into a carton. Hot melt adhesive or special-designed slots and inserts hold the carton sections together. A formed carton is mechanically released from a form machine onto a powered conveyor travel path. As a carton moves forward, photo-eyes and a microcomputer send a "form a carton" command to a carton form machine. To maintain a continuous carton flow, an employee transfers collapsed carton bundles or corrugated sheets to a carton make-up machine.

To operate a mechanical carton make-up activity, an employee places collapsed cartons or corrugated sheets onto a make-up carton in-feed sleeve or section. During a pick activity, a carton make-up machine continually forms cartons and transfers them onto a CO take-away conveyor travel path. From a CO take-away conveyor travel path, a formed carton travels onto a pick line conveyor path, is placed onto a carton stack, is transferred to a pick vehicles carrying surface or transferred onto a pack station.

There are two types of mechanical carton make-up concept:

■ A fixed carton size make-up machine that forms one carton size. The machine has a lower cost, but requires an additional machine or modification to make up other carton sizes.
■ A variable carton size make-up machine that is adjustable to form several carton sizes. An adjustable carton size make-up machine has a higher investment but handles a wide variety of carton sizes. A carton make-up machine range is from 12 to 20 cartons per minute.

Prior to a carton make-up machine selection, you test your carton sizes on a carton form machine because there are extreme differences between carton sizes (largest size to the smallest size). The extreme differences can create problems and small-size cartons are difficult for a suction-based carton form machine to handle.

To optimize the carton make-up, pick line, and pack station activities, a pick line microcomputer sequences all COs for one carton size introduced to a pick line.

The disadvantages are some machines handle one carton size, investment, compressed air supply, and electrical power supply, considered permanent, and employee to in-feed cartons or corrugated sheets. The advantages are high volume, automatic carton transfer to a pick line, less tape waste, and easy to change carton sizes.

A carton make-up employee or machine transfers a made-up carton onto a powered conveyor travel path that feeds a pick line start station. As a made-up carton travels on a conveyor travel path, carton make-up activities are controlled by a photo-eye or by an employee who activates a stop and start button. For maximum carton make-up efficiency, a pick line pick activity has a carton slug for the same carton size.

A carton make-up machine transfers a made-up carton onto short powered conveyor travel path that feeds an employee carton stack station. As a made-up carton arrives at a conveyor travel path, an employee stacks made-up cartons. After a carton make-up machine or employee makes all cartons from a bundle, an employee transfers made-up cartons to a pick station or a pack station storage location. For maximum carton make-up efficiency, a pick line pick activity has a carton slug for the same size carton.

Carton Make-Up at a Pack Station

With a batched CO operation that uses a captive tote carton make-up takes place at the pack station. An employee transfers made-up cartons to the pack station or pack station storage location. For maximum carton make-up efficiency, a pick line pick activity has a carton slug of the same size.

The selection of the shipping carton size and application of CO discreet identification to a shipping carton occurs after completion of a CO pick or pick/sort activity. In a batched CO pick, sort, transport, and pack concept, there are multiple sortation and pack stations.

At a pack station that has a computer print the required carton size on a CO pack slip, display screen, or shipping label, the packer removes an appropriate collapsed carton from a carton stack to a pack table and makes up the carton or removes a carton from a stack that has been formed beforehand and places it onto a pack table. Alternatively, at a sort station CO pieces are discharged into a shipping carton that is then transferred to a pack station.

With a single CO pick/pack order-fulfillment operation, a CO shipping carton is introduced to a pick line or pick area prior to the first pick position. With this order-fulfillment concept, a CO shipping carton is transferred from an employee carton make-up station, carton make-up machine or queue conveyor travel path directly onto a pick line, pick vehicle, or placed into a captive pick tote.

One Carton Size for a Pick Line Activity

In a carton make-up activity that has one carton size on a pick line or one carton size diverted to a pack station, a host computer or pick line microcomputer arranges the CO introduction sequence CO SKU cube. This concept (see Figures 4.61 and 4.62) has other carton sizes used in the operation but COs for the same size carton are streamed onto a pick line or diverted to specific pack stations. After all COs are packed in the first carton size, a second or next CO carton size streams onto a pick line or are diverted to separate pack stations.

Advantages of this method are improved packer productivity and carton storage space at a pack station, by reducing different carton sizes, and time to refill and select a carton, minimize carton train transport problems over a pick line conveyor travel path, improve carton make-up productivity, and reduce change over time and control carton inventory.

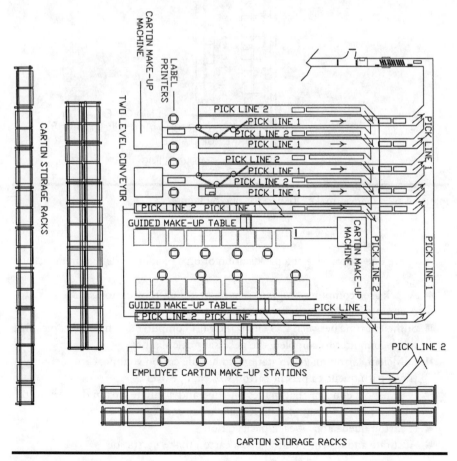

Figure 4.61   Carton make-up area for two pick lines.

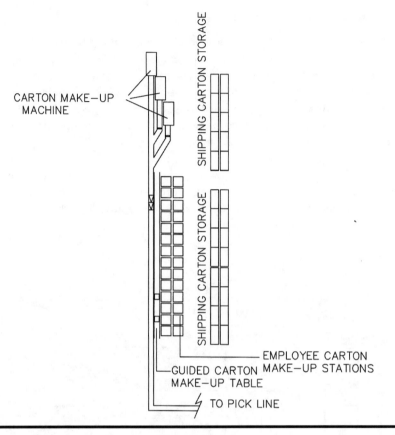

Figure 4.62 Carton make-up area for one pick line.

To have one carton size at a pack station or on a pick line, the requirements are:

- A pick area computer program that separates COs by carton size
- A computer program that cubes COs
- Sufficient time between CO entry and host computer download to a pick line microcomputer to complete the necessary calculations
- Sufficient carton make-up stations to handle carton volume, carton conveyor travel path queue to permit carton size change over
- A transport concept that makes on-time and accurate carton delivery to the pack stations and completed CO take-away concept
- A carton queue area at each pack station
- Accurate and on-time delivery to a carton make-up station

Mixed Carton Sizes for a Pick Line Activity

An order-fulfillment operation that has a mix of CO shipping carton sizes on the same pick line, conveyor travel path, or pick, vehicle load-carrying surface at one time is more complex. From instructions on a pick sheet, delivery label, or display screen, a pick line start or pack station employee collects the appropriate shipping carton size and transfers it onto a start station or pack table. COs and suggested CO shipping carton size is on a random basis.

The pick line design ensures:

- That all CO shipping carton sizes are available at a pick line start station
- That the correctly sized shipping carton is presented to a pick line start employee
  - The suggested carton size or identification is printed in a consistent location on each CO pack slip, shipping label, or a microcomputer-controlled display screen.
  - A computer downloads the CO sequence and associated carton size. As a CO enters pick line, a pick line computer has an empty shipping carton conveyor travel path move a suggested carton size forward from the various merge conveyor travel lanes onto a pick line start station. The merge conveyor travel lanes are queue conveyor sections for one shipping carton size and are fed by a microcomputer-controlled carton make-up machine or employee.

An employee introduces a suggested CO shipping carton size onto a pick line in the following way:

- The employee reads a suggested CO shipping carton size that is printed onto a CO pack slip or shipping label
- From a stack or a conveyor travel path with made-up cartons the employee removes a suggested CO shipping carton size
- The employee applies a CO shipping label to each carton exterior surface
- The employee ensures that a CO labeled shipping carton is returned in the correct sequence to a pick line conveyor travel

Options for selecting and transferring the correct carton are to transfer cartons from:

- A stack on the floor
- Several non-powered or powered conveyors
- A single overhead powered conveyor

If cartons are selected from a stack on the floor, an employee transfers stacks from a carton make-up area to a location that is adjacent to a pick line or pick vehicle start station. There is one carton stack for each shipping carton size. After a start station employee reads a CO pack slip or shipping label carton size, the start station employee removes the suggested carton from a stack. During this carton transfer activity, a start station employee places a CO identification label onto a shipping carton exterior surface. Per the order-fulfillment operation, an employee transfers a CO pack slip or sales literature inside a shipping carton and places a shipping carton onto a pick line or pick vehicle. When a carton is placed onto a pick line conveyor travel path or onto a pick vehicle load-carrying surface, a CO discreet identification faces the correct direction of travel to serve as a picker instruction.

The disadvantages are employee activity and floor area. The advantages are low cost, maximum carton size number, constant carton supply to a pick line or pick vehicle, and CO pack slip and sales literature are placed inside a customer shipping carton.

A multiple empty shipping carton conveyor travel path concept has an employee or machine make-up and in-feed one empty shipping carton size onto a zero-pressure queue conveyor travel path. To have an effective and on-time empty shipping carton delivery to a pick line start station, there are several conveyor travel paths. If possible, there is one conveyor travel path for each shipping carton size. The number of conveyors required can be minimized by sending one shipping carton size to a pick line. On a pick line, all COs that require the specific shipping carton size are picked and completed COs are sent to a pack station. Each made-up carton conveyor travel path transports one size CO shipping carton from a carton make-up station to a pick line or pick vehicle start station.

Carton conveyor travel path options are:

- One conveyor travel path that is set at the same elevation above the floor as a pick line conveyor travel path or order pick vehicle load-carrying surface. This empty CO shipping carton travel path handles your highest-volume carton size. Other empty shipping carton conveyor travel paths are straight conveyor travel path sections, incline and decline conveyor travel path sections, or curve conveyor travel path section to assure an empty shipping carton supply to a pick line or pick vehicle start station. The other conveyor travel paths handle medium- and slow-volume shipping carton sizes.
- Four conveyor travel paths that are floor level and set at the same elevation above the floor as a pick line conveyor or pick vehicle load-carrying surface. With this concept, as an empty shipping carton conveyor travel path ends at the start station, a start station employee has direct access to each empty shipping carton conveyor travel path and to an empty shipping carton. After picking an empty shipping carton from the empty shipping carton conveyor travel path, a start station employee transfers the empty shipping carton onto a pick line conveyor travel path or vehicle load-carrying surface. If more than four shipping cartons are required on a pick line, the low-volume shipping cartons have a pallet or carton flow rack locations that are within easy access to a pick line start employee.

- Many conveyor travel paths that are floor level and set at the same elevation above the floor as a pick line conveyor or pick vehicle load-carrying surface. With this concept each empty shipping carton conveyor travel path ends at a merge station for transfer onto a main empty carton conveyor travel path for transport to a pick line start station or vehicle load-carrying surface, a start station employee has direct access to each conveyor travel path and to an empty shipping carton. To ensure proper carton sequence on a conveyor travel path, a microcomputer sequences carton release onto the main conveyor travel path per the required CO carton size. If a wide variety of shipping carton sizes are required on a pick line start station, low-volume shipping cartons have carton flow rack locations.

The features are high investment with zero-pressure queue conveyor travel path and microcomputer program, floor area, fewer employees, few start station employee ergonomic problems, handles all carton sizes, and constant empty shipping carton size supply to a pick line start station.

A single line overhead powered conveyor or trolley travel path concept has an overhead powered chain and free trolley with a carrier conveyor travel path concept. Several carton tray carriers or hooks travel over a closed-loop travel path. A tray carrier travel path is past a carton make-up station and past a pick line or pick vehicle start station. Options for designing a carton tray carrier are a single hook or multiple hooks, clasp, or tray device that carries one or two cartons.

An empty shipping carton overhead conveyor travel path with a hook, clasp, or tray carrier concept has a closed-loop travel path that has empty carton carriers travel past a carton make-up station and a pick line start station or each pack station. At an empty shipping carton make-up station, a make-up employee or machine transfers an empty carton onto a vacant carrier hook or clasp, or empty tray. As the overhead conveyor travel path has the carrier hook or tray travel past a pick line start station or pack station, a pick line start station or pack station employee removes a customer order empty shipping carton from a carrier. With multiple pick line start or pack stations and a fixed carrier number, due to a first-come-first-arrive basis, there is a potential that:

- Overhead carrier empty hook or tray arrives at a pick line or pack station that has an employee to wait for a carrier hook or tray with a carton or suggest carton size
- With overhead carriers that have all carton sizes, a start station or pack station employee has to wait for a carrier hook or tray with a carton or suggested carton size

With a single concept, as an empty carton carrier or open space on a carton conveyor travel path arrives at a carton make-up station, an employee or machine transfers a made-up carton from the workstation onto an empty carrier or open space. Along a powered conveyor travel path, there are one or several carton make-up stations. The number of make-up stations is determined by the shipping carton volume, number of different carton sizes, carton make-up productivity, and packer productivity. A conveyor travel path that dead ends at a pick line start stationhas either a roller or drag chain conveyor surface. A closed-loop travel path has a roller, drag, or belt conveyor surface.

Features of a single line conveyor are:

- Empty shipping cartons arrive at a start station on a first-come basis. If an empty CO shipping carton delivery concept serves several pick line start or pack stations with a shipping carton size mix, this empty shipping carton delivery aspect has the potential problem to cause some pick line start or pack stations to wait for the suggested empty shipping carton size arrival.
- With one conveyor travel path or powered chain travel path, it is difficult to handle multiple shipping carton sizes.
- It is difficult to match the appropriate carton size on a travel path to a CO required shipping carton size.

Automatic Powered Conveyor Carton Transfer

An automatic empty shipping carton transfer concept has the host computer download a CO sequence and each associated CO shipping carton size to a pick area microcomputer. Each empty shipping carton travel path has:

- A carton make-up machine with a zero-pressure queue conveyor travel path from carton make-up to an automatic machine in-feed conveyor travel path
- Microcomputer-controlled belt conveyor travel path sections
- A merge table to release and singulate the appropriate shipping carton size from a carton make-up station onto a zero-pressure conveyor travel path

Prior to a label station, a conveyor travel path directs or skews an empty shipping carton to one conveyor travel path side and onto a belt conveyor travel path. On a belt conveyor travel path, a computer-controlled label machine is located along a pick line conveyor travel path. As directed by the computer, a label machine applies a discreet customer identification shipping label to a suggested CO order shipping carton. A label machine conveyor travel path transfers a labeled carton onto an automatic pick machine travel path.

With the automatic empty shipping carton transfer concept, each carton size make-up machine activity is controlled by a photo-eye on a carton conveyor travel path that feeds the merge table. As a carton travels past the merge table location, the photo-eye sends a message to the appropriate carton make-up machine to make up another carton. Per a CO cube or shipping carton size that was included in a host computer download to a pick area microcomputer, a powered belt conveyor travel path section prior to a merge table and the merge table surface forward movement controls the suggested CO carton size release, transfer, and skew onto an automatic machine in-feed conveyor travel path. A carton travels onto a carton label machine belt conveyor travel path. As a suggested CO carton size arrives at a label station, an employee or machine applies a CO shipping label to a shipping carton. After a CO discreet label is on a shipping carton, a CO labeled carton travels from a label station over a conveyor travel path onto an automatic pick machine start station.

The features are high investment, few employees, large floor area, and handles multiple carton sizes and a high volume.

Labeling a Customer Order/Shipping Carton

A discreet label or identification code on a CO pick/shipping carton distinguishes it from other CO cartons and is a pick line activity. During a manual pick activity on a pick-to-light line, a picker or a barcode scanner matches the discreet identification on a CO carton to a pick-to-light pick instruction. The match confirmed, a picker transfers a SKU from a pick position into a CO discreetly identified pick/shipping carton.

With an automatic pick machine, after a CO carton identification label is read by a barcode scanner or RF device the microcomputer matches a pick instruction to a CO identification label and a pick machine transfers a SKU from a pick position into a CO pick/shipping carton.

In an order-fulfillment operation area, a CO discreet identification code on a CO pick or shipping carton serves to:

- Direct a CO carton to another pick zone or location
- Instruct an employee or automatic pick machine to complete a pick transaction
- Activate a check weight machine
- Activate scanning by a manifest scanner

Labeling in Batched or Single-Picked Customer Orders

A captive tote with a permanent discreet identification is used in a pick-to-light or batched CO pick and sort into a tote operation. With these concepts, a WMS or computer system associates a label on a tote with a CO number. Each tote has four permanent labels, two on the exterior and two on the interior of the tote. These label locations ensure a picker or packer has good line of sight to a CO identification label. At a pack station, a packer scans a tote label to identify items for a CO.

Customer Order Label Considerations

Factors to consider in designing CO labels are label size and material and how a label is placed onto a carton.

The label size depends on the width and height of the CO delivery address barcode. A barcode label with adequate quiet zones provide good readability or fast scanning. Black ink and white paper enhance readability. The size of the label also depends on the available space on a CO carton. In a pick/pack application, a label is placed on a carton flap or on the side that faces a pick aisle. The application should not cover the bar code. Information on the label should be easily recognized by pickers or there should be good line of sight for a barcode scanner. With ink jet application, a human/machine-readable code applied to a corrugated surface should be easy to recognize and not blend into the corrugated material of the carton.

A CO label should have the following characteristics:

- It should be printed on paper with self-adhesive backing
- It should contain all company information
- It is not reusable
- It should be applied in the proper orientation to a carton

Some operations use a reusable, permanent license plate, transport unit (TU), or code on a captive tote. The orientation of the label on the tote should be in the direction of travel and degree of tilt and on both sides to ensure line of sight for a barcode scanner. After a labeled tote passes a barcode scanner and the WMS microcomputer attaches its discreet number to a CO number, the tote number becomes a CO identification as the tote travels over a pick/pass line. The relationship between a discreetly identified tote and a CO identification permits a permanently labeled tote to satisfy pick line requirements.

Label Placement

Factors that determine how a CO label is attached to a CO carton are:

- Container size and material
- Order-fulfillment concept
- Label size and type

In a small-item order-fulfillment operation, shipping carton size varies and some operations use a paper or plastic bag.

In a manual or automatic pick concept, a single CO carton moves on a conveyor travel path through a pick area. Before the carton arrives at the first pick station, it needs a CO identification, which directs manual or automatic pick activities. Actual placement of a label onto a pick carton is either a manual or machine activity and the label is placed onto a carton side or top flap to ensure proper employee or barcode scanner line/RF tag or sight.

When the label is applied manually this is done at a pick line start station. Labels that have been preprinted on sheets or rolls are used. The labels have CO identification that match a CO sequence identification. Empty cartons arrive at a pick line start station in the sequence that was directed by the WMS computer and as each carton arrives an employee removes the CO label and places it onto the carton. For a pick/pass pick line, using rolls of preprinted labels in a dispenser ensures proper CO sequence, minimizes label handling problems, and reduces trash from self-adhesive labels. In a batched CO pick and sort fulfillment operation, preprinted labels are placed in a stack at the pack station. To assure good final sortation activity, each CO label is printed onto a pack slip or sheet. Alternatively, CO identification labels can be printed on demand. With this concept, as a carton arrives at a pick line start station, a CO label is printed. A pick line start employee removes the label from the print machine, removes the self-adhesive backing and applies the label to a pick carton. In a batched CO pick and sort operation, as a CO captive tote or first SKU arrives at a pack station, a packer hand scans a bar code. A scanning device communicates the barcode information to a microcomputer and either one printer prints a combined CO label and pack slip, or a CO label is printed by one printer while a second printer prints the pack slip.

In an operation that uses a machine to apply a self-adhesive label, CO cartons en route to a pick line start station are skewed to one side of the conveyor travel path, where there is a label application station.

With existing machine label application technology, options for placement of the CO label are on the front or side of the carton, or on the top (see Figure 4.63).

If an order-fulfillment operation has a pick/pass line that transfers picked SKUs directly from a pick position into a CO shipping carton, the tag placement on the top flap of a carton is the preferred option. With the side barcode/RF tag label application machine, a barcode label/RF tag is applied to a location on a CO pick carton side that permits a picker or barcode scanner/RF tag reader device line of sight to easily to read a human-readable or machine-readable code. With machine labeling, the ability to interface with and apply a barcode label to the flaps of cartons with various heights is a major consideration. Options to address this include:

■ Setting the machine to handle the size carton that is used most, or use a standard carton size with a peel-off label that can be reapplied for shorter carton heights.
■ Setting the machine to handle the tall cartons and have a picker lift short cartons to match a CO identification or number to the pick-to-light display identification. With a machine label side apply concept a side barcode label/RF tag location does not interfere with carton tape and final barcode label location satisfy your delivery company scanning requirements.

In an order-fulfillment operation that has a CO completed at a pack station, label placement onto a one-piece or two-piece carton is completed at a pack station. In operation conveyor travel paths have side barcode scanners.

TOP LABELED
PICK CARTON

SIDE LABELED
PICK CARTON

FRONT LABELED
PICK CARTON

Figure 4.63 Pick line carton label locations: top flap, front, and side.

With a pick/pass order-fulfillment operation with a barcode CO identification label on a top flap as a pick instruction, for small cartons, the label options are:

- Peel-off label, pick into a standard-size carton, and complete the CO at a pack station
- Print a second label at the pack station
- Ink jet spray a CO identification number on the top flap of the carton, then at a pack station, place a CO delivery label onto the carton

With short cartons or cartons with a small top flap, a computer-controlled ink spray has the capability to spray a CO number onto a pick carton. As a carton approaches an ink jet spray station, all cartons are skewed and aligned to ensure controlled travel through an ink spray station. At an ink spray station, the microcomputer enters a CO identification number into an ink jet spray device. An ink jet spray pick device sprays black ink onto the carton exterior surface or flap. To allow sufficient time for the ink to dry, a carton travels on a zero-pressure conveyor travel path to a pick line start station. Prior to using an ink jet concept, ink jet spray tests are conducted on your carton exterior surface to assure that the required human/machine code clarity is per your standards.

The ink jet concept features are: does not require an employee, applies a code in a consistent location on a carton exterior surface or top flap. With this concept, it is difficult to handle a wide carton mix but on a pick line employees lift short cartons, there is no self-adhesive backing, and there is no need for trash handling.

Inserting a Customer Order Pack Slip

When an order-fulfillment operation uses a CO pack slip with a CO carton, a pick line or pack station employee places a pack slip into the CO shipping carton. Options are:

- The pack slip is inserted at a pick line start station or first pick location. This approach means that CO pack slips are preprinted or printed on demand and are sequenced to match the CO sequence on a pick line. During the pick activity, SKUs and filler material are placed on top of the pack slip top. At a CO delivery location, SKUs and filler material are removed from the carton before the pack slip. This approach may cause problems as a result of having to unpack the carton to obtain the pack slip.
- The pack slip is inserted after the last pick position and prior to a carton fill station. In this approach a CO pack slip is also preprinted or printed on demand in the same sequence as the CO sequence on a pick line. During the pick activity, SKUs are placed into a carton and at the last pick station a CO pack slip is inserted on top of the SKUs then filler material is added. At a CO delivery location, filler material is removed from a carton to obtain the CO pack slip. This approach minimizes the difficulties that may result from unpacking the SKUs to obtain the pack slip.
- The pack slip is inserted after the filler material. During the pick activity, SKUs are placed into a carton and filler material is placed on top. With this approach, a CO pack slip is preprinted or printed on demand in a sequence that matches the sequence of COs on a pick line. At a CO delivery location this concept is least likely to cause problems. With this method, the transport concept must be designed to ensure that the CO pack slip is retained in the carton.

In a pick/pass into a carton operation, at a start station a customer pack slip is attached to or placed inside a CO carton or captive tote. If there is a pick-to-light failure, the CO pack slip serves as a CO pick document. In a batched CO pick/sort, transport and pack or pick/transport, sort and pack operations, a packer places a CO pack slip into a CO carton.

With a high-volume, ready-to-ship vendor carton operation, an envelope label or slapper label concept is used. With this approach, a CO pack slip and sales literature are enclosed inside a paper envelope with a plastic window or inside a clear plastic envelope, which permits an employee or barcode scanner to read the CO delivery code. As the ready-to-ship carton passes an envelope apply station, a machine or employee applies an envelope to the side or top of the carton.

Pack Slip Information

Each CO pack slip has discreet identification for that CO and important information that matches the picked SKUs, such as customer name and discreet identification, company name and address, date, CO number, individual listing for each SKU on a CO along with description and quantity. The CO pack slip print and transfer sequence matches the CO carton sequence on a pick line or CO SKU sortation at a pack station.

In a pick operation, options for inserting the pack slip into a CO carton are:

- To manually insert the pack slip. An employee reads a CO discreet number on a pick carton, matches it to a CO pack slip number, and transfers the CO pack slip inside the carton or clips the pack slip to the side flap or wall of the carton.
- To have a machine insert the pack slip. After a barcode scanner reads a CO discreet number on a carton, a mechanical device transfers by air pressure or vacuum a CO pack slip into the carton.

Which process an operation selects is determined by:

- The type of pick operation, that is (a) pick/pass or (b) pick, sort, and pack
- The pick line or pick area layout factor
- The shipping carton volume
- Company policy, governmental codes, and delivery company requirements regarding use of a pack slip
- Available facility space
- Economics and available labor

Manual Customer Order Pack Slip Insert

On a pick/pack line, pack slips can be manually inserted:

- As a pick line start station activity
- From a pick position, by a picker
- At a pack station, by a packer

Pack slips that are inserted at a pick line start station are preprinted as directed by a WMS computer download to a pick line microcomputer, and the pack slips are stacked at a pick line start station; or they are printed on demand as a CO carton enters a pick line. An employee verifies that, as a carton enters the pick line a CO identification has been attached to it, and then matches a pack slip to the identification on the carton before placing the pack slip in the carton.

The advantages of inserting the pack slips at this stage are the start station designs permit the insertion of the pack slip to be an easy and simple task, with minimal impact on pick line start employee productivity; it does not require a pick position, and does not add a non-productive pick on a pick line.

When a CO pick slip transfer to a CO shipping carton occurs on a pick line as a pick activity at the first pick position, a picker has a pick instruction to transfer a CO pack slip from a pick position to a CO pick carton. The picker verifies that the CO identification matches the CO pack slip. The concept has CO pack slips preprinted in the CO download sequence or printed on demand after a CO identification on a carton is read by a barcode scanner/RF tag reader on the pick line.

With a pick/pass into a captive tote on a pick line or with a pick, sort, and pack order-fulfillment concept, pack slips are inserted into a CO shipping carton at a pack station. A pack station employee verifies the accuracy of a pick and sort activity by comparing the actual SKUs picked for a CO to a paper or visual display pack slip. With a CO pack slip printed on demand, a CO pack station packer transfers a CO pack slip into a CO shipping carton, either before or after the SKUs have been packed.

The disadvantages are ensuring that the CO carton label and associated CO pack slip sequence have the same numerical order, requiring an employee to read. The advantages are low cost, handles all carton sizes, handles all document types, and used on pick pass lines and at batched CO pack stations.

Mechanized Customer Order Pack Slip Insert

A mechanized or automatic CO pack slip insert concept has a mechanical device that releases or transfers a pack slip into a CO carton. A CO carton with discreet identification travels on a conveyor past a barcode scanner, which sends data identifying the CO to a pack slip print machine microcomputer. The carton travel path is on a first-in-first-out basis, and the microcomputer ensures that the CO identification matches the download sequence. The pack slip machine removes a printed CO pack slip from a stack and inserts it into a CO carton.

If the CO pack slip insert concept is to print and insert a pack slip on demand, a computer downloads all CO discreet identification bar codes in numerical sequence to a pick label printer microcomputer. After a CO discreet identification is attached to a CO carton and the carton travels past a barcode scanner, the CO identification is sent to a CO pack slip insert machine computer that controls a CO pack slip printer. This data transfer activates the printer to print the CO pack slip which is then inserted into the CO carton.

The print on demand concept handles an equal volume, but has a higher cost, and requires a queue conveyor. In comparison, the preprinted concept features lower investment and less need for a queue conveyor.

Advantages of the automatic versus manual concepts are that the automatic concept is most effective on a pick/pass line with a pick line start station and not effective in a pick, sort, and pack order-fulfillment operation.

SKU Pick Position Replenishment

The objective of a SKU pick position replenishment activity is to ensure there is adequate inventory in a pick position before SKUs are withdrawn for a CO. SKUs are transferred in the correct quantity from a storage position or from a ready reserve position to a pick position. In other terms, the objective of pick position replenishment activity is to minimize the occurrence of a no-stock or out-of-stock (stock out) condition at a SKU pick position on a pick line, in a pick aisle, or at an automatic pick station.

A *no-stock* condition occurs when SKU inventory is physically in a pick position but the WMS or on-hand book inventory file does not reflect it. When this situation occurs and there is a demand for that SKU for a CO, the WMS computer fails to print a pick transaction instruction for a picker or pick machine to pick a SKU from that position. In an order-fulfillment operation, this situation creates a WMS inventory control problem because SKUs are allocated to another pick position when there is inventory remaining.

A *stock out* or *out-of-stock* condition occurs when a WMS or book inventory file does reflect SKU inventory in a pick position but there is no inventory actually in the pick position. When this situation occurs and a CO requires a SKU, the WMS or computer does print a pick transaction instruction, and a picker or pick machine attempts to pick the SKU from a pick position with zero inventory. The out-of-stock condition creates a non-productive picker or pick machine activity because the picker or pick device traveled to a pick position and could not complete the pick transaction. In addition, the delay in filling the CO could result in a dissatisfied customer.

A replenishment activity is completed before a SKU inventory is depleted in a pick position. Designing the replenishment activity takes the following factors into consideration:

■ The SKU pick position philosophy or strategy that determines the pick activity
■ The storage location, which is selected to ensure quick and accurate transfer to a pick position
■ The pick position SKU piece capacity or pick position type, which determines the SKU piece replenishment quantity

- Timing of the replenishment activity
- The method used to transport SKU replenishment pieces from a storage position to a pick position
- SKU replenishment transaction verification and updating the WMS or inventory files

Pick Concept Impact on SKU Replenishment Activity

The pick area layout and equipment impacts the pick position replenishment activity. Basic pick concepts can be grouped as manual, mechanized, or automatic.

In a manual concept, a picker walks or rides to a pick area, where pick position concepts are shelves or decked racks arranged in rows with a pick and replenishment aisle between two rows or carton flow rack rows with pick and replenishment aisles. Shelves or decked racks have the following characteristics: there are nominally eight SKUs per shelf bay or decked rack bay; there are separate SKU storage aisles; no ready reserve positions in a pick aisle; the pick positions hold C and D or slow- to medium-moving SKUs, and a replenishment employee transports SKUs from a storage position to a pick position. For maximum employee replenishment productivity, SKU replenishment transactions are separated for each pick aisle. With this criterion, floor-level positions or positions that can be reached by a replenishment employee have SKUs for that pick aisle. This ensures SKUs can be easily withdrawn from a storage position and replenishment transactions in a storage aisle are for pick positions in that aisle. This is efficient and cost-effective, with few replenishment errors, few employees, few congested storage and pick aisles, and large SKU number and low pieces per SKU.

To complete a replenishment transaction for carton gravity flow rack pick positions, for high-volume SKUs, an employee transports SKUs from a storage position to a replenishment aisle, and pick position replenishment transactions are completed from the replenishment aisle. Medium- to slow-volume SKUs are transported to the replenishment aisle and placed directly into a pick position, and extra pieces are placed into temporary ready reserve positions.

For maximum employee productivity, SKUs are separated by pick zones or sequential carton flow rack bays, with 10, 20, to 30 SKUs per flow rack bay. For maximum employee productivity, SKUs that are allocated to one pick zone are allocated to the same storage aisle and SKU storage allocation mirrors the pick zone sequence.

The ready reserve positions can be (a) parallel to a pick area, (b) perpendicular to a pick area, (c) short aisle, and (d) long aisle.

S.I. Cartrac Concept

For use in a small-item order-fulfillment operation, the Cartrac is an electric motor-driven vehicle with shelves that has that travels over a closed-loop travel path and can bE-stopped and started by an employee. The travel path takes the Cartrac past pick and replenishment stations. Each shelf or carrier on a shelf has a SKU. As the Cartrac arrives at a pick station, a picker stops the carrier by stepping on a pedal and completes a single CO or batched CO pick transaction. The required SKU quantity is removed from the carrier and transferred onto an assigned CO location. After all SKU pick transactions, the picker releases the carrier for travel to the next pick station and the next carrier with a new SKU travels to the pick station. If a carrier has a SKU that is not required on a CO, the carrier travels past the pick station. The pick activity and carrier stop and start activity are repeated for all SKUs on a CO order or batched COs.

At a replenishment station, an employee or forklift truck removes an empty pallet and replenishes the carrier with another UOP. After an employee or forklift truck has replenished a SKU onto a carrier, the carrier is released for travel to the first pick station. The disadvantages are capital investment, fixed travel path, singlulated carrier over a fixed travel path can create queues on a travel path, and nonproductive time at a pick station and mini load or pallet carrier with shelves carries a small inventory quantity. The advantages are machine paced concept, minimal UOP, equipment and building damage, low maintenance, easy to operate, easy to expand, and handles a wide UOP mix.

Parallel to a Shelf, Decked Rack, or Pallet Rack Pick Area

In this arrangement (see Figure 4.64) the storage rack rows are parallel to a shelf, decked rack, or rack pick area. With a parallel arrangement, there are at least three aisles in the layout: two turning aisles at the end of each rack row end and a middle aisle. The middle aisle leads to the replenishment aisle or pick area. This arrangement increases the number of trips a replenishment employee makes and travel time between the storage area and replenishment aisle, or pick area. In a rectangular pick area that has SKU replenishment between a ready reserve area and pick area, there is an increase in trips and travel path distance. The feature requires additional replenishment employees and allows fewer replenishment transactions to be completed per hour.

Parallel to Carton Flow Rack Pick Area

This arrangement has storage rack rows parallel to a carton flow rack pick area. With a parallel arrangement to a carton flow rack pick line, the ready reserve positions that face the carton flow rack bays have the fast-moving SKUs. There are at least three aisles in this layout, with two turning aisles at the end of each rack row and a middle aisle that leads to a replenishment aisle or pick area. For fast-moving SKUs, this arrangement decreases the number of trips a replenishment employee makes and the travel time between the ready reserve position and the pick position.

READY RESERVE POSITIONS LOCATIONS

RESERVE RACKS PERPENDICULAR TO PICK POSITIONS

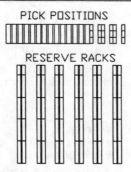

RESERVE POSITIONS PARALLEL TO PICK POSITIONS

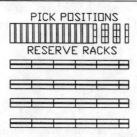

Figure 4.64 Pick position reserve racks: parallel to pick positions and perpendicular to pick positions.

For medium- to slow-moving SKUs, it increases the number of trips and travel time. With the 80/20 rule, this feature has fewer replenishment employees and improves replenishment transactions per hour.

Perpendicular to a Shelf, Decked Rack, or Pallet Rack Pick Area

A perpendicular arrangement has the shelf, decked rack, and ready reserve rack rows and aisles and product flow in a straight line from the storage area to a replenishment aisle, or shelf, decked rack, and pallet rack pick positions. With this layout, each aisle provides direct access from the storage area to a replenishment aisle or pick aisle. In a rectangular pick area, the perpendicular arrangement provides a great trip number per aisle and the replenishment employees have fewer turns at the aisle ends. If a ready reserve aisle has only SKUs that are in a corresponding pick aisle, UOP flow and employee trips are direct between two areas.

Perpendicular to a Carton Flow Rack Pick Area

With this layout, each ready reserve aisle has SKUs in carton flow rack bays that face a ready reserve aisle. This arrangement provides access to a replenishment aisle or pick aisle. If a ready reserve aisle has only SKUs that are in a corresponding pick aisle and the fast-moving SKUs nearest flow racks, the UOP flow and employee trips are direct between the two areas. In a rectangular pick area, the perpendicular rack row and aisle arrangement provides a great trip number per aisle and the replenishment employees have fewer turns at the aisle ends.

Short Aisle to Any Manual Pick Area

A short-aisle ready reserve area has the storage rack rows and aisles run the short dimension (width) of a rectangular area. This ready reserve rack row and aisle concept has turning aisles at each aisle end. On a per square foot basis, a short aisle ready reserve area concept provides fewer SKU faces per aisle, low replenishment employee productivity, and, if SKUs in a pick aisle exceed a ready reserve aisle, it has several ready reserve aisles.

Long Aisle to Any Manual Pick Area

This layout has the shelf, decked rack, or ready reserve rack rows and aisles run the long dimension of a rectangular area. There are cross or turning aisles in the middle of each rack row. On a per square foot basis, a long-aisle ready reserve area provides quick access to other aisles, more SKU faces per aisle, high replenishment employee productivity.

Replenishment to an Automatic Pick Concept

An automatic pick machine (A-frame, Robo pic, or Itematic) is used in an operation with medium- to fast-moving SKUs with small to medium cube. The SKU facings are very dense, with many SKUs per linear foot and each replenishment lane holds one or two vendor cartons with a nominal 12 pieces per carton. The ready reserve location is opposite the replenishment lane charge end. SKU replenishment is a computer-controlled activity. During machine downtime, the ready reserve replenishment positions serve as pick positions.

In an automatic pick operation, there are visual controls to signal the need for a pick lane replenishment. This is usually or photo-eye and a flashing light to signal the need for replenishment or a mark on a replenishment sleeve.

Automatic pick positions have ready reserve positions located directly behind an automatic pick machine replenishment side. A replenishment aisle is between an automatic pick machine and ready reserve positions. A replenishment aisle has sufficient width to permit an employee to perform all replenishment transactions and as required to set up a temporary pick transport conveyor and complete all pick transactions. An automatic pick machine ready reserve concept is a carton flow rack that is perpendicular or parallel to the replenishment aisle. A preferred flow rack arrangement is parallel replenishment rack design. This design minimizes a replenishment employee travel distance and replenishment SKU positions face the replenishment lanes.

When there arre more pick positions than replenishment positions, a carton flow rack design is perpendicular to the replenishment aisle. This design provides the greatest number of replenishment faces per aisle but results in a slight decrease in replenishment employee productivity because of an increased walk distance from the pick aisle to a replenishment aisle. The SKU replenishment sequence along the ready reserve faces has the same sequence as the pick machine SKU sequence.

SKU Allocation in a Pick Area Impacts the Replenishment Activity

SKU allocation to a pick position on a pick line or in a pick aisle has an impact on replenishment activity. A basic SKU storage/replenishment/pick location principle is to keep the travel distance between them as short as possible. Keeping the distance between two order-fulfillment locations as short as possible increases dual cycle transactions and improves hit density (number of SKUs per aisle per CO) and hit concentration (number of picks per SKU for a CO or batch pick wave). This feature reduces non-productive time and improves replenishment and picker productivity. Strategies for deciding on SKU allocation to pick positions includes random allocation, using the ABC principle or Pareto's law, family grouping, and putting fast-moving SKUs in one pick area (the golden highway). These are discussed elsewhere in this chapter.

Replenishment Aisle Design, Reserve Pallet Area, and SKU Allocation

A small-item order-fulfillment operation receives pallets, master cartons, and single items in master cartons and fulfills COs with single items. With an operation that handles SKUs of different types, the storage area and replenishment equipment uses a transport concept and replenishment methodology that handle a pallet, master carton, loose single SKUs, or items in totes or cartons. In a single-item operation, the facility has a 25-ft. clear ceiling height that is designed for an elevated floor level. Elevated floors are used for carton make-up, additional pick positions, value-added activities or other order-fulfillment activities, such as CO pick check, CO carton pack, or CO carton seal and manifest.

With a carton flow rack pick concept, a pick position replenishment aisle is a separate aisle. With a shelf or decked rack pick concept, a pick position aisle is a combined pick and replenishment aisle. With both pick concepts, a replenishment employee uses the replenishment aisle to transfer cartons or single items from a transport concept or vehicle load-carrying surface into a SKU pick position. Accurate and on-time SKU pick position replenishment minimizes stock-outs and results in high replenishment employee and picker productivity. Small-item or flatwear pick positions are:

■ Carton flow rack pick positions. A carton flow rack concept has a separate replenishment aisle. Because a carton flow rack is a pick position for fast moving, high-cube, and heavyweight SKUs, the dimension of the carton flow rack replenishment aisle has an impact on pick area layout.

■ Standard shelf or decked rack pick positions. A standard shelf, decked rack, or static pick position or push back rack concept handles medium- to slow-moving, small-cube, and lightweight SKUs. Pick positions have the open side of the shelf or decked rack bay face pick and replenishment aisles. This means that a pick aisle is the replenishment aisle and does not have a major impact on a pick line or pick aisle layout but can impact employee productivity.

Options for designing a carton flow rack replenishment aisle as a separate aisle are:

■ As a storage area turning aisle. With this concept, a replenishment vehicle and employee remain in an aisle to perform a replenishment transaction and residual SKU pieces are returned to the storage area or placed into a ready reserve position. This concept occupies a small area in the facility but results in frequent replenishment trips between the storage area and replenishment aisle for each pick position. As part of the storage area vehicle-turning aisle, it is a potential safety hazard.

■ As a replenishment aisle between the pick position replenishment side and a storage rack row. This storage rack row is parallel to the pick position replenishment end. In this very narrow guided aisle a high ride picker truck operator performs the necessary storage and replenishment transactions. This concept uses a small area, but has frequent replenishment transactions and it is difficult to have more than one vehicle in a replenishment aisle.

■ As a replenishment aisle between a pick position replenishment side and a low- or zero-pressure queue conveyor travel path. In the aisle, as a carton or tote is transported over a conveyor travel path, at the appropriate location a replenishment employee transfers a carton or tote to complete a replenishment transaction. This concept uses a small area, but requires an employee to move cartons or totes between the conveyor and the pick position. With this concept, it is difficult to sequence carton or tote arrival at a specific location on a conveyor to match a replenishment employee's location in a replenishment aisle. For best replenishment performance, it has a recirculation conveyor travel path and good carton or tote identification for a pick position.

■ As a replenishment aisle between pick positions and a storage rack row or floor-stacked pallet position row that is parallel to pick positions. In the pallet positions, a forklift truck deposits pallets, cartons, or totes as close as possible to a pick position. In the aisle between the pallet racks and carton flow racks, a replenishment employee performs necessary carton or tote transfers to complete a SKU pick position replenishment. This concept uses a large area but has few replenishment trips between the storage area and the pick area and permits multiple replenishment employees to work in a replenishment area. To ensure as many SKUs as possible in a ready reserve aisle, a small-item order-fulfillment operation has a floor stack pallet or single rack row directly behind the flow rack pick line and back-to-back rack rows between a single rack row and the storage area turning aisle.

A pallet storage area can be above the pallet or carton flow racks. The racks would require structural support designed to hold the additional weight and standard pallet rack load beams with upright frames that have space to contain each carton flow rack bay.

Another important replenishment consideration is the reserve pallet location and design. The purpose of the reserve pallet area is to provide storage positions for SKUs so there is timely and accurate SKU replenishment from the storage area to a pick position. To achieve this objective, the reserve pallet area rack rows and aisles are designed parallel to a replenishment pick line side or perpendicular to it. We refer the reader to the previous section in this chapter.

Design factors for a pick line reserve area are:

- A square or rectangular facility
- The number of pallet storage positions
- Forklift truck type
- Building height

With this concept, all second-floor or second-level pallet positions have SKUs in the same sequence as SKUs in the pick line bays, so a replenishment employee can quickly complete a replenishment transaction to a pick position. This feature minimizes replenishment employee travel time, travel distance, and search time. A good WMS system minimizes replenishment trips between the storage area and replenishment aisle. The pallet positions have an elevation that permits a replenishment employee to complete a transaction that is not a complete pallet.

SKU Reserve Locations

In a small-item flatwear order-fulfillment operation, SKU reserve locations can be:

- Off-site, in a different facility. These locations are difficult to inventory and incur delivery truck shuttle costs between the two facilities.

- Remote reserve positions, which are in the pick facility but in an aisle that is not adjacent to a pick aisle or replenishment aisle. Pallets or carts are transported between a remote reserve position and a pick line area. A remote reserve position incurs transport costs from a remote storage area to a pick area but provides improved inventory control.
- Ready reserve positions that are SKU storage positions (master carton or tote) adjacent to, across from, above, or below the SKU pick position. The ready reserve positions are in a static decked pallet rack or shelf. In a carousel, Cartrac, Robo pick, A-frame, Itematic, or other mechanized or automatic pick concept, the ready reserve positions are in carton flow rack bays, decked pallet racks, or shelves or in a carton AS/RS that is arranged parallel or perpendicular to a pick concept replenishment aisle.
- On-line reserve positions, that is, SKUs that are held in the pick position beyond the CO demand for a work day. In a carton gravity flow rack, slide rack, or push back rack concept, the on-line reserve SKUs are pieces that are located in cartons or totes that are queued behind the carton or tote at the pick front.
- Pick positions contain a SKU inventory and are the locations where a picker removes a SKU to complete a CO pick transaction. Pick position utilization is estimated at 50 percent.

Pick Position Types

Pick positions can be (a) fixed or permanent and (b) random, floating, and variable. For additional information, we refer the reader to the previous section in this chapter.

Pick Position Replenishment Capacity

Another factor in designing a fixed replenishment position is the SKU piece capacity of the pick position. The factors that determine pick position capacity are:

- The minimum number of pick positions needed to satisfy the number of COs
- SKU physical characteristics—length, width, height, weight, fragile nature, ability to stack, and UOP classification, such as flammable, edible, toxic, or hazardous
- Pick position physical space or cube, that is, the internal dimensions between a pick position structural members
- Existing SKU quantity in a pick position and the pick concept—manual, mechanized, or automatic

A pick position is designed to hold a SKU quantity that satisfies a specific number of COs that occur within a specific time period. A pick position is designed to accept the number of SKUs in one vendor carton plus an additional SKU quantity determined by management as a safety stock. Different types of pick position have the capacity to hold different-sized SKUs:

- The smallest SKUs are held in drawers, small bins, totes, or boxes that are placed on a shelf, carton AS/RS, or carousel position or peg board concept.
- Medium-sized SKUs are held on standard shelves, one-deep decked pallet racks, full horizontal carousel position, S. I. Cartracs, and automatic pick machines.
- Large SKUs are held on carton flow racks or push back flow racks, slides or chutes, two-deep decked racks, or wide span shelves, standard decked cantilever racks and pallets, tier racks, or stacking frames on the floor.

SKU Replenishment Quantity

SKU replenishment quantity is the number of pieces that are transferred from a storage (reserve) position to a pick position and is determined by the capacity of the pick position to hold a SKU replenishment quantity. To ensure excellent inventory control and minimal SKU damage, the SKU replenishment quantity is one vendor master carton. The pick position replenishment quantity can be determined by an employee or by the WMS. Having an employee determine replenishment quantity is not a preferred option. The employee estimates SKU quantity and transfers them from a storage location to a pick position. After a replenishment transaction, any residual SKU inventory is placed on the floor or on top of a pick position and the inventory replenishment form is sent to the office for the inventory files to be updated. SKU replenishment selection and transfer to a pick position does not match SKU withdrawal or CO requirement, and there is the potential to lose or damage inventory. The employee handles the SKU twice, and there is inaccurate replenishment, and delayed inventory file update. The WMS or computer projection method is preferred because the replenishment activity is directed by a computer or RF device. A WMS program provides instructions for SKU replenishment priority, SKU piece quantity, and SKU storage location that corresponds to an order-fulfillment activity, and SKU replenishment quantity matches pick position capacity.

The objective of determining an accurate SKU replenishment quantity is to ensure in a pick position that there is excellent space utilization, SKU rotation, and sufficient SKU piece quantity for CO SKU quantity. Factors in determining SKU replenishment quantity are:

- SKU piece quantity in a storage position
- Number of SKU pieces in a pick position at the start of the cycle
- Update for replenishment transactions or deductions for CO picked SKUs
- Predetermined SKU capacity for a pick position
- Predetermined safety stock or SKUs in a pick position
- Historical or expected CO demand
- Ready reserve positions

SKU replenishment quantities are based on (a) minimum quantity, (b) maximum quantity, (c) capacity quantity, and (d) quantity to match CO SKUs or picked clean quantity.

Minimum Quantity

A minimum SKU replenishment quantity is the SKU safety stock quantity in a pick position. When a WMS computer determines that SKU inventory in a pick position has reached a minimum SKU piece quantity and all CO picked transactions are completed for a workday, it creates a minimum replenishment demand or quantity. This minimum replenishment SKU quantity is one reserve master carton and is a low-priority replenishment transaction because there is sufficient inventory in a pick position for the next CO pick wave.

Maximum Quantity

A maximum SKU replenishment quantity is based on the number of pieces in a pick position that will ensure excellent pick position utilization and fewest trips for replenishment. The maximum pick position quantity holds the minimum inventory quantity plus one master carton. Note that this maximum quantity is greater than one master carton and depends on pick position capacity. The maximum replenishment quantity has a medium priority because the pick position has adequate inventory to satisfy most of the demand for the next CO waves.

Capacity Quantity

Capacity SKU replenishment quantity is the maximum number of pieces that will fit into a pick position. This situation is created when a pick position SKU quantity reaches zero or is out-of-stock. The SKU quantity that is required for the pick position equals the pick position capacity. Capacity replenishment to a pick position receives the highest priority because there are not enough pieces in the pick position to satisfy COs.

Picked Clean Quantity

Picked clean SKU replenishment is made to a pick position that serves as a pick position for the next pick wave, or one day's CO SKU quantity. A picked clean SKU replenishment quantity is an exact quantity that is based on COs for the next pick wave or pick day. After CO pick transactions have been completed, the SKU piece quantity is reduced by CO pick transactions. After picking the last wave or CO SKU piece quantity, a pick position SKU piece quantity is zero. With a picked clean SKU replenishment concept, SKU replenishment to a pick position can be a different SKU and is the quantity needed for another CO pick wave or the next day's COs. A picked clean replenishment activity allows an employee to profile an empty pick line for the next pick day, which increases hit concentration and hit density and improves replenishment and picker productivity.

Timing of Replenishment

SKU pick position replenishment activity is best performed the moment a pick position becomes depleted, and is best replenished with a SKU quantity that maximizes the pick position space and optimizes the number of trips between a storage area and a SKU pick position. To achieve this objective, for best results all SKU pick position replenishment transactions are controlled by a WMS computer that determines the SKU quantity based on CO SKU quantities.

This it is important to determine the appropriate time for a pick position replenishment transaction to occur from the storage location to a pick position. SKU replenishment transactions can be controlled manually or by computer.

- Manually controlled replenishment transactions rely upon an employee to determine the time that a SKU piece quantity is moved from a storage position to a pick position. This is a random concept that is based upon the employee's experience and is not sequenced with order-fulfillment or CO pick activities. SKU replenishment is made to a partially depleted pick position. The disadvantages are no control of employee activity, low employee productivity, poor pick position space utilization, and no coordination with a CO pick activity. Advantages are it is low cost, low investment, and needs no employee training.
- A WMS computer-controlled replenishment transaction uses a computer or WMS system by a paper document or RF terminal to direct an employee to perform replenishment transactions between a storage location and pick position.
    - Paper document. A replenishment paper document lists all the SKU replenishment transactions in the order that they are to occur on an employee's shift. SKUs are listed by the identification numbers in a sequential order that is based on the anticipated time that a pick position will become depleted. The first column states the SKU storage location and the second column is for a replenishment employee mark that indicates that a SKU withdrawal transaction was completed on schedule. The third column states a SKU pick position and a fourth column is for a replenishment employee mark to verify replenishment transaction to a pick position.
    - An RF device replenishment concept has an employee complete a replenishment transaction to each pick position with a hand-held, wrist, or finger bar code scanner. A bar code scanner/RF tag reader directs a laser light beam at a pick position bar code/RF tag and an RF device or ter-

minal display screen shows SKU replenishment quantity and SKU storage location. After each SKU replenishment transaction, the employee scans a pick position bar code label/RF tag and SKU identification bar code label/RF tag. This verifies completion of the SKU replenishment transaction to a pick position. The disadvantages are management control and discipline, capital cost, and employee training. The advantages are accurate transaction records, excellent employee productivity, reduces lost inventory, improves space utilization and on-line or delayed transaction transmission.

Replenishment Concepts

SKU replenishment activity ensures that the correct SKU quantity is transferred from an assigned storage location to a correct pick position to allow a picker to complete a pick transaction for a CO. Manual replenishment concepts are:

- Slug replenishment. This method has a replenishment employee complete all replenishment transactions within one pick aisle before moving to another aisle. The disadvantages are potential stock-outs and lower employee productivity because the employee has to hand stack SKUs to make space for new pieces. The advantage is decreased employee travel time and distance between replenishment positions.
- Sweep replenishment. This method separates a pick aisle into zones. A replenishment employee starts in one pick zone, and performs all SKU replenishment transactions in that pick zone at predetermined locations and in sequential order, before moving to the next pick zone. Compared to other replenishment concepts, this method has the additional advantage that the concept closely complements a CO pick activity.
- Random replenishment. An employee carries out replenishment transactions that are based on his or her thinking. Disadvantages are pick position replenishment transactions do not complement picker transactions, low management control, low employee productivity, and if a pick position is not completely depleted, extra pieces are not placed in the best location. The advantage is that usually fast-moving SKUs are handled first.
- High to low SKU volume. The high to low pick replenishment concept is used for an order-fulfillment SKU group with two unique features: a high-volume pick area and a low-volume pick area that follows the 80/20 rule and customer service standard to complete COs within two hours. A high to low replenishment concept has a WMS program sequence a pick line set-up or replenishment that is based on your CO SKU volume. Pick line set-up or replenishment transactions are sequenced with the highest CO SKU volume as the first SKU for the replenishment activity. This means that the stor-

age area completes all UOP withdrawal and transport transactions to move the SKU pieces from a storage area to the pick line replenishment area. The next storage area replenishment transaction is for the second-highest SKU. This sequence is followed for each CO SKU. The high to low pick replenishment features are a WMS program to sequence the CO SKU and some ready reserve positions in the replenishment or pick area. The disadvantages are the cost of using a WMS program, the need for two ready reserve positions, and between the storage and pick areas, some travel path queuing. The advantages are the largest number of SKU pieces replenished to the pick positions, fewer trips between storage and pick areas, and faster pick activity.

Ready Reserve Concepts and Design Considerations

A critical aspect of pick position replenishment activity is to have SKUs ready for transfer into a pick position. Characteristics of a replenishment activity are that it:

- Handles a large number of SKUs with (a) many different physical characteristics and (b) different replenishment requirements that range from one to many pieces
- Covers a large area with many aisles and pick positions
- Occurs at the same time as a picker is completing pick transactions and, depending on the order-fulfillment concept, it occurs in a pick aisle with pickers in the aisle or in a separate replenishment aisle
- Is a dynamic activity that is performed by several employees

To optimize pick position replenishment functions a small-item flatwear operation has a ready reserve area adjacent to or behind the pick area. The ready reserve rack rows and aisles are arranged in a parallel or perpendicular flow to the pick area and the ready reserve aisle is short or long (described above). To ensure accurate and on-time SKU replenishment, the area design allows sufficient aisle space to perform the transaction. If an order-fulfillment operation has extremely fast-moving SKUs (A-type SKUs), the ready reserve positions are designed in the pick area replenishment aisle. This feature ensures that a SKU is ready and available for replenishment.

Ready reserve concepts are:

- Shelves or decked racks with a pallet or fork lift truck
- Shelves or decked racks with an HROS vehicle
- Multiple carousels
- Carton AS/RS
- Pallet AS/RS with a container handling device
- Sort link

Ready Reserve Locations in a Replenishment Aisle

In a high-volume small-item order-fulfillment operation, having a ready reserve location in a replenishment aisle is an important design factor. A ready reserve rack provides a temporary holding location for a SKU inventory quantity that did not fit into a pick position. A rack location ensures easy and unobstructed placement of extra inventory into a ready reserve position, and permits inventory tracking and quick transfer of SKUs from a ready reserve position to a pick position.

Factors in designing ready reserve rack positions within a replenishment aisle are the type of reserve position, that is, a pallet or hand stack, SKU number and associated SKU inventory quantity, building or free-standing mezzanine column locations, and transport equipment type.

Options for ready reserve rack positions in a replenishment aisle (see Figure 4.65) are:

- No ready reserve location, in which case SKUs are transferred directly to a pick position
- Back-to-back rack rows in the middle of a replenishment aisle
- A single rack row in the middle of a replenishment aisle
- A single rack row set off in a replenishment aisle
- Wire meshed decks on top of a flow rack

No Ready Reserve Locations on the Aisle

If there are no ready reserve positions on an aisle, it has floor-stack positions adjacent to the carton flow rack. A replenishment employee transfers the maximum SKU quantity into a pick position. Any residual SKU inventory is transferred from a replenishment aisle to a ready reserve position. This concept has a large SKU quantity in ready reserve locations and a WMS tracks SKU inventory. The features are a narrow aisle between a pick position and pallet, extra trips between a storage area and pick positions, which increase the potential for employee injury, building, equipment, and product damage, additional replenishment transactions, and increased potential to misplace or lose inventory.

Back-to-Back Rows in the Aisle Middle

The back-to-back ready reserve rack rows in the middle of a replenishment aisle has one rack row face a carton flow rack replenishment side. Between the rack row and the carton flow rack replenishment side is a forklift truck aisle to permit a pallet replenishment transaction. The forklift truck aisles occupy a large floor area, there are minimal pallet transport trips between the storage area and pick positions, which decreases employee injury and building, equipment, and product damage, fewer replenishment transactions, and, with a WMS system, decreased potential to

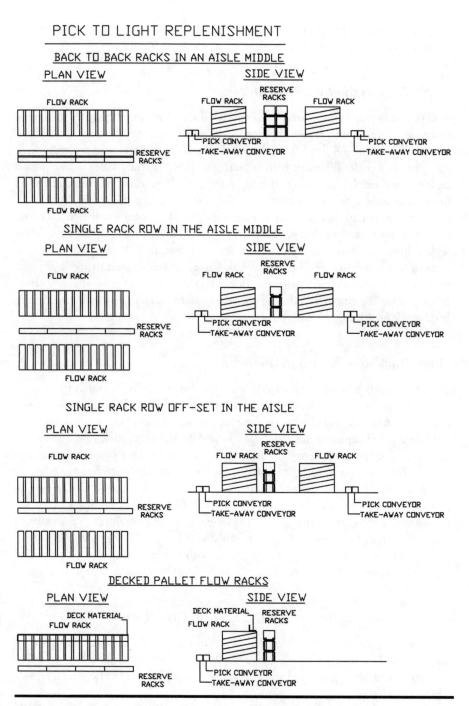

Figure 4.65 Pick line ready reserve racks: back-to-back in the middle of an aisle, single rack in the middle of an aisle, single rack off-set in the aisle, and on top of a deck flow rack.

misplace or lose inventory, permits on-time and quick replenishment to an empty pick position, and maximum pallet and hand-stack positions.

Single Rack in the Middle of an Aisle

A single ready reserve rack row in the middle of a replenishment aisle has rack bays that face both sides of a replenishment aisle. There is a barrier in the middle of elevated rack positions. A forklift truck aisle between the rack row and replenishment side permits a pallet replenishment transaction. Forklift truck aisles between each carton flow rack pick position and ready reserve rack row allow transport of pallets from the storage area. The forklift truck aisles occupy a large floor area, but reduce pallet transport trips between the storage area and pick positions, which decreases employee injury and building, equipment, and product damage. There are fewer replenishment transactions, and, with a WMS system, decreased potential to misplace or lose inventory, permits on-time and quick replenishment to an empty pick position, and requires fewer pallet positions. All elevated pallet positions have safety netting along the employee side and the barrier in the rack bay provides maximum hand-stack positions.

Single Rack Row Set off in the Aisle

This concpt has a single rack row in a replenishment aisle. Rack bays face each replenishment aisle and both replenishment carton flow rack sides. Between one pallet rack row side and the carton flow rack replenishment side is a forklift truck aisle for a pallet replenishment transaction and on the other rack row side or carton flow rack replenishment flow rack side is an employee aisle. The features of this concept are the forklift truck aisle takes up a small floor area, fewer pallet transport trips between the storage area and a pick position, which decreases an employee injury and building, equipment, and UOP damage, minimal replenishment transactions, and, with a WMS system, decreased potential to misplace or lose inventory, on-time and quick replenishment to an empty position, minimal pallet positions, and elevated pallet positions with safety netting along a personnel aisle side and with a barrier in a rack bay, maximum hand-stack positions.

Wire Meshed Decks on Top of a Carton Flow Rack

A single ready reserve rack row on top of a carton flow rack bay provides space for hand-stacked cartons on wire-meshed decks or decked pallet rack bays or pallet positions. This concept is used in conjunction with other carton flow rack replenishment options.

A very important consideration is to protect replenishment-side carton flow rack legs from damage by transport equipment. Carton flow rack pick position replenishment activity has a vehicle transport pallets in a replenishment aisle and drop a pallet in the replenishment aisle. A replenishment aisle is adjacent to the carton flow rack leg or parallel to a carton flow rack replenishment lane side. To minimize damage from pallets or a transport vehicle striking a carton flow rack leg or flow lane, an order-fulfillment operation installs and anchors a 6- to 12-in. high guard rail above the floor along a carton flow rack replenishment aisle side. The height of the guard rail protects against damage to carton flow rack legs and minimizes pallet hang-up.

Shelves or Decked Racks with a Pallet or Forklift Truck

This concept uses shelves or decked rack rows and an employee-operated powered pallet or forklift truck. The shelf layout has four to five position levels and a deck rack layout permits three to four position levels high with a 20-in. high opening for each SKU. An employee on a powered pallet truck, forklift truck, or tugger with carts, travels through the replenishment aisles. To improve replenishment employee productivity, a pallet truck has a shelf device or powered tugger pulls carts with multiple shelves. At a SKU storage location that has been assigned by a paper document or RF device, an employeE-stops the powered vehicle and transfers the SKU quantity from a ready reserve location to a shelf device or cart shelf. After completing all transfers, the employee travels from the ready reserve aisle to the replenishment aisle. In the replenishment aisle, the employee completes the replenishment transaction from the shelf device or cart shelf to a pick position.

The concept occupies a large area that is determined by SKU quantity, requires a number of employees, and has the potential for errors. It requires a medium investment, and handles a large number of fast-moving SKUs. It is difficult to ensure on-time replenishment of high-cube SKUs, but a WMS priority replenishment list minimizes this problem. If an operation handles a few to medium number of SKU and a few fast-moving/high-cube SKUs with a computer WMS replenishment program, it improves SKU replenishment activity because ready reserve positions can be sequenced to mirror pick positions, minimizing employee travel distances and travel time in the ready reserve and replenishment aisles between two transaction locations.

Shelves or Decked Racks with HROS

Shelves and decked rack rows with an employee-operated battery-powered high-rise order picker vehicle (HROS) use an entire building cube that is accessed by an employee on a HROS vehicle. A shelf or decked rack height is from 10 to 25 ft. The layout permits 5 to 12 position levels high, each of which is approximately 20 in. high, including load beam and deck material. This feature has a ready reserve area with a small footprint. As an employee rides on a guided HROS vehicle through a ready reserve aisle, he or she has the capability to move horizontally and vertically to each ready reserve position. The width of the cart or SKU-handling device permits it to be carried by the vehicle in the aisle between two shelves or decked rack rows. To ensure access to all SKU storage locations, a HROS vehicle transfers between replenishment aisles.

At an assigned ready reserve location, an employee transfers a SKU quantity onto the cart shelf or SKU-handling device. After transactions are completed or when a cart or handling device is full, the employee drives the HROS vehicle to an aisle end P/D station, where the cart or SKU-handling device is pushed from the HROS vehicle to a staging area or to a replenishment aisle. In the storage area, the HROS vehicle operator attaches an empty cart or device to complete another series of replenishment withdrawal transactions. In the replenishment area, a replenishment employee completes the pick position replenishment activity. With a high SKU number and pick volume, a HROS vehicle aisle directly behind the carton flow rack replenishment side is not preferred.

The HROS features are improved space utilization, smaller building area with a high ceiling, additional investment in a HROS vehicle and aisle guidance, and, for best performance, has a computer inventory control or WMS system.

Carousels

A single or multiple carousel ready reserve concept has stacked baskets that are attached to an overhead or floor-level powered, closed-loop chain that is propelled by an electric motor over a fixed travel path. In a carousel application, the movement of the baskets is controlled by a microcomputer. Each basket contains one SKU. Some carousel applications have a robotic injection and extraction device to move a specific basket between a carousel and replenishment transfer station. For maximum employee productivity, multiple carousels service one station and have conveyor travel paths that transport totes to the pick area.

After the SKU ready reserve position is entered into the carousel microcomputer, the carousel unit rotates the stacked baskets until the assigned basket arrives at the transfer station, the carousel stops, and an employee withdraws the required SKU quantity. With a multiple carousel concept, as the employee or mechanical device is completing a transaction from carousel A, the microcomputer activates carousels B and C to rotate. By the time withdrawal transactions are completed from carousel A, carousel B has completed its rotation and the appropriate basket is presented to the pick station. This feature permits the employee to complete the assigned withdrawal transaction from carousel B. Withdrawal transaction instructions are directed by a pick or sort to light, RF device, or paper document.

The SKUs are transferred from the carousel basket to a transport concept (conveyor or four-wheel cart) for travel from the ready reserve area to the replenishment area. In the replenishment area, an employee completes a replenishment transaction to the pick position.

The multiple carousel concept features are fewer employees, less employee injuries, with multiple carousels and computer directed rotation, high employee productivity, space utilization, fewer errors with paperless picking activity, and an investment.

Carton AS/RS

A carton AS/RS concept has a captive aisle stacker crane that is computer-controlled to travel between two shelves or pick positions rows. In each shelf or position is a tote, carton, or tray that contains one SKU. The shelf positions are the SKU ready reserve positions. Per a host computer CO download to the mini-load microcomputer, an AS/RS stacker crane is directed to complete a SKU withdrawal transaction. The stacker crane travels to an assigned aisle location and takes the appropriate carton on board. The stacker crane delivers the SKU either:

- Onto the outbound P/D station and, for a dual cycle, takes on board an inbound carton, or travels directly to the assigned storage location. From a P/D station the SKU is delivered by a conveyor or four-wheel cart to a replenishment aisle and into a pick position.
- Into an assigned pick position to complete a replenishment transaction. This works with a WMS system and stacker aisle between the carton flow rack replenishment side and the storage positions.

The features are accurate and fast replenishment, which is improved with dual carrier cranes, requires on-time replenishment cartons that are prepared for the pick position, good carton bottoms for travel over a conveyor and safety screen travel aisles.

The efficiency of a dynamic operation that has a pick-to-light CO pick concept with carton flow racks as the SKU pick position concept depends on accurate and on-time replenishment to a SKU pick position. The carton flow rack pick concept has:

- Carton flow rack bays with tote or carton flow lanes. The carton flow lanes are SKU pick positions that face a pick aisle. After a carton or tote is placed into a carton flow rack replenishment side, it moves over a conveyor travel path by gravity force to the flow rack pick side (see Figure 4.67 and Figure 4.68).
- Two aisles, one of which is a pick aisle and the other a replenishment aisle. In the pick aisle, per CO SKUs and pick instruction, a picker transfers the SKUs from a pick position into a CO carton. In the replenishment aisle, per a batched CO SKU requirement or demand the necessary cartons or totes are transferred from a storage position to an assigned SKU pick position.

A stacker crane replenishment concept has:

- An area for open vendor cartons. SKUs remain in the master carton or are transferred into a captive tote for transport to a pick position. If the vendor

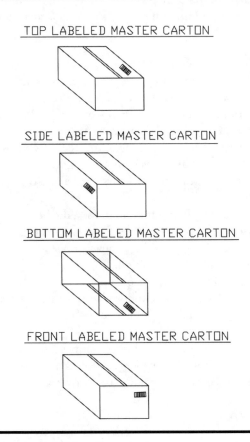

MASTER CARTON LABEL OPTIONS

TOP LABELED MASTER CARTON

SIDE LABELED MASTER CARTON

BOTTOM LABELED MASTER CARTON

FRONT LABELED MASTER CARTON

Figure 4.66   Master carton label locations.

carton is suitable, it is used to transport SKUs to the pick position. The open carton option is determined by the carton AS/RS crane carrier type, tote conveyor travel path concept, empty master carton, or trash conveyor travel path concept.
■ A crane P/D station.
■ One stacker crane aisle behind each carton flow rack row and storage positions or a stacker crane between two carton flow rack replenishment sides (see Figure 4.68).
■ An empty tote return conveyor in the pick aisle area, or a system to use the tote as a CO pick container.

After the host computer downloads CO SKUs to the pick area microcomputer, a storage area operation withdraws and transports the master cartons/totes from the

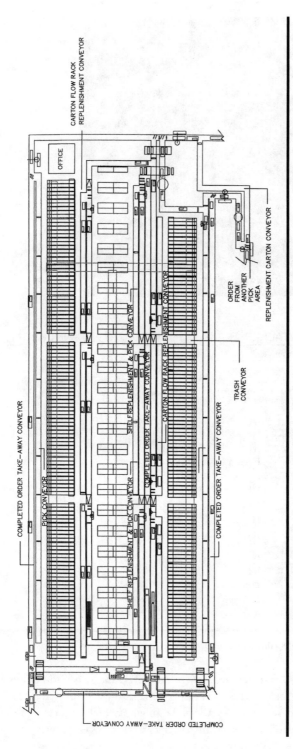

Figure 4.67 Carton flow rack replenishment with conveyor.

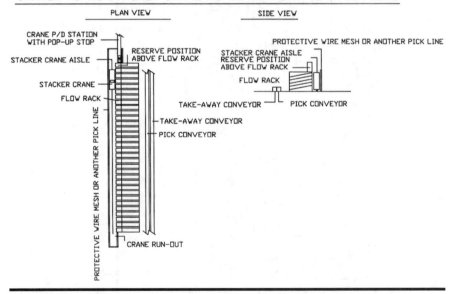

Figure 4.68 Crane replenishment aisle.

storage area to a SKU transfer area, where each SKU master carton quantity is sent to an open workstation. As the SKU bar code is read by a scanner, a SKU transfer quantity is indicated at the workstation. As an empty pick area captive tote arrives at the workstation, the employee transfers the assigned SKU quantity into the tote.

All poor-quality empty vendor master cartons are transferred to a trash transport concept. SKUs remaining in partially full master cartons or totes are returned to the storage area. At the workstation, the vendor or master carton or tote (license plate) is associated with a SKU. Full totes or cartons are queued on a conveyor travel path, per a replenishment cycle, totes or cartons are released to a transport conveyor. A transport conveyor moves totes or cartons past a scanner device from an open area over a travel path and are diverted to an assigned crane P/D station. Additional tote or master carton storage positions are above the carton flow racks.

After a tote or carton arrives at the appropriate P/D station, it travels past a scanner to a pick-up station. The scanner sends the carton or tote identifying information to the microcomputer that controls the stacker crane. At the P/D station, the stacker crane picks up a carton or tote and, per the microcomputer instruction, travels to an appropriate SKU replenishment position, where it transfers a tote or carton onto a SKU carton flow lane. In the carton flow lane, a tote flows over a conveyor travel path or lane from a replenishment side to a pick side.

During the pick activity, as a tote or carton becomes depleted, an order picker transfers it from a pick position to a conveyor that moves the empty tote to the open station. If the vendor or master carton is used, the empty carton is transferred from a pick position to the carton recycle or trash transport concept.

With a stacker crane replenishment concept, a carton flow lane SKU quantity is picked clean, which means at the end of the CO pick wave, there are no SKUs or totes in the flow rack. If there are SKUs or totes in the carton flow rack, they are sent to a storage area. A stacker crane replenishment concept features are few replenishment employees, small replenishment aisle, replenishment activity is controlled by a WMS computer, minimal hang-ups in carton flow lanes with plastic totes or high-quality cartons, master carton open area, and trash consolidation area. Note that the crane must be able to handle open cartons. The concept has a high cost, increased storage space utilization, improves accuracy and on-time replenishment, reduces employee injury and equipment/product damage, improved security, for optimum performance, each tote has a human/bar/RF tag symbology and with a properly designed conveyor travel path, operates 24 hours, 7 days a week.

AS/RS Concept with a Tote Handling Device

The AS/RS concept has a computer-controlled storage retrieval crane with a carrier that handles up to six totes, two storage position rows, a P/D SKU transfer station, and a tote conveyor travel path. An AS/RS crane is a guided vehicle that travels between two shelves or rack rows. Rack or shelf storage positions or stacks have a range from 40 to 100 ft. high and the roof and sides are supported by the racks. At the end of each aisle are P/D stations for tote in-feed and out-feed. A T-car transfers stacker cranes between aisles.

After the AS/RS computer receives the replenishment commands, it directs the AS/RS crane to travel horizontally and vertically down the aisle to an assigned storage position, where it takes a tote into a carrier. With multiple carriers per aisle transaction, the AS/RS crane travels to another location and takes another tote on board. With a full carrier or tote withdrawal transactions have been completed, the AS/RS crane travels to the P/D station, where it transfers all totes onto a transport conveyor and takes in-bound totes on board.

In a ready reserve application, a conveyor:

- Moves a tote from the P/D station onto a ready reserve pick station, where per the pick to light, RF, or paper pick instruction, the SKU quantity is transferred into a transport for transport from the reserve area to the replenishment area. With this concept, an AS/RS SKU tote is captive to the AS/RS concept.
- Moves an AS/RS SKU tote from the reserve area to the replenishment area. In a replenishment aisle, a replenishment employee follows the pick or sort to light, RF device, or paper pick instruction to transfer the SKU tote into a pick position. A replenishment employee transfers a SKU from a tote to a pick position to complete a SKU replenishment transaction.

An AS/RS concept features are few employees, capital investment, increased storage space utilization, standard tote, good-quality carton or tray, improves accuracy and on-time replenishment, reduces employee injury and equipment/ product damage, improved security, each carton or tote has a human/bar code/RF tag symbology, WMS or a computer program, good-quality master carton, a metal or hardened plastic tote or tray and properly designed conveyor travel path, operates 24 hours a day, 7 days a week.

Sort Link

A sort link concept has:

- A microcomputer-controlled carrier device that travels on a guided travel path.
- Vertical carrier lifts.
- Carrier storage lanes that are 30 or 60 ft. high and as many as 20 trays deep per lane.
- A roller carrier to carry one SKU. A roller carrier has the ability to connect together as a carrier train that is pulled between the various locations.

After cartons or totes are transferred onto a sort link tray, the tray is transferred to a P/D station where the microcomputer controls ensure that cartons or totes are secured on a tray. With this confirmation, the microcomputer directs an empty carrier to travel to the P/D station and take on board a loaded tray. With a tray on board the computer-directed carrier travels to a vertical lift device, which transports the carrier to the assigned storage level. At the assigned storage level, the carrier leaves the vertical lift and travels down the aisle to the computer-assigned storage lane. As the carrier arrives at the storage lane, a roller tray is pushed forward to lock with the existing roller tray, and the roller tray is pushed forward into the lane. After a deposit, the carrier travels to another computer-assigned location and completes another transaction. Per the host computer CO download, the micro-computer directs an empty carrier to a storage lane that has the required SKU. If the SKU is on the front roller tray, the carrier takes the roller tray on-board. If the SKU is on another roller tray, the microcomputer tracking system transfers the intervening roller trays to another storage lane. When the appropriate roller tray is at the storage lane exit location, the carrier takes the tray on board and travels down the aisle to a vertical lift, where it is moved to an assigned level and discharged onto a travel path for transfer to a P/D station.

In a ready reserve application, a conveyor:

- Moves cartons, totes, or trays from the P/D station onto a ready reserve pick station, where. per a pick or sort to light, RF device, or paper pick instruction, the SKU quantity is transferred from the sort link captive tray into a transport tote for transfer from the reserve area to the replenishment area. This concept has a sort link carrier with a tray captive to the sort link concept.
- Transports a sort link carrier with a tray from the reserve area to the replenishment area. With this concept, in the replenishment aisle a replenishment employee follows a pick or sort to light, RF device, or paper pick instruction to transfer the SKU into a pick position to complete a SKU replenishment transaction.

The sort link features are few employees, capital investment, increased storage space utilization, standard container, carton, or tray, improves accuracy and on-time replenishment, reduces employee injury and equipment/product damage, improved security, for optimum performance, each tray requires a human/bar code/RF tag symbology, requires a WMS or computer program, metal or hardened plastic tray, and with a properly designed conveyor travel path operates 24 hours a day, 7 days a week.

Replenishment Transaction Verification

A replenishment transaction verification concept ensures that the correct SKU piece quantity was transferred at the correct time from a storage position to an assigned pick position. SKU pick position replenishment transaction verifications can be carried out as a delayed update or on-line update, and with manual documentation or automatic transfer.

Delayed Update

A delayed replenishment transaction verification concept has an employee complete a SKU pick position replenishment transaction prior to the WMS or inventory control department update. With a delayed concept all pick position replenishment transactions and physical inventory relocations are completed and the WMS or inventory book entry is made at a later time. A delayed replenishment transaction verification concept is used in an order-fulfillment operation that has a problem with on-line data transfer from remote RF devices or the host computer has difficulty receiving data transfer. In an order-fulfillment operation that uses a delayed update concept, a replenishment transaction is recorded in a microcomputer that has a small memory or is marked on a document. The disadvantage is that it is possible for a second transaction to scheduled for an inventory quantity that has already been reduced by a previous transaction. The advantage is that it minimizes the host computer investment and on-line communication network cost.

On-Line Update

On-line SKU pick position replenishment transaction verifications have the following features:

- As a storage position SKU quantity is depleted, the withdrawal is entered in a WMS or inventory book file.
- As SKU pieces are transferred from the transport concept to a pick position the pick position SKU quantity is increased in the WMS or inventory book file. During SKU transfer, a SKU piece quantity on the transport concept is in a non-pickable location. An on-line inventory update concept has an efficient, on-time and accurate transport concept, SKU transfer to a pick position concept, and accurate and fast communication from an activity location by a hand-held RF device to a computer. This feature ensures on-time customer service and minimizes stock outs. Disadvantages are host computer has a large capacity and handles a large number of transactions.
- Communication network cost. Advantages are accurate and on-time transactions and fewer update errors.

Manual Document

A handwritten replenishment document concept has a replenishment employee use a manual or machine-printed form to record each replenishment transaction. After each transaction, the employee completes the appropriate line on a document. In addition to a place for the employee name and date, the form has:

- Three columns for each deposit transaction: (a) one for the SKU number, (b) one for the SKU piece quantity, and (c) another for the storage/pick position number
- Four columns for a withdrawal or replenishment transaction: (a) the SKU storage position number, (b) SKU number, (c) SKU piece quantity, and (d) pick position number

At predetermined times in a day, a partial or completed document is sent to the office and the appropriate inventory files are updated by a clerk or entered into a computer for inventory update.

Automatic Transfer

An automatic replenishment transaction verification concept uses a bar code scanning/RF tag reading and RF device communication device concept. Bar code scanning and RF tag reading device concept components are:

- A human/machine-readable label on each carton or tote and on each storage and pick position.
- A hand-held, wrist, or finger bar code scanning/RF tag reading device with a visual display terminal and a memory capacity in a scanner device or a RF communication network. The RF data transfer concept has a transmission terminal with an antenna that is attached to a vehicle or is attached to an employee and the storage/pick area has several receiving devices and a communication network to a microcomputer.

Design considerations are:

- Will the RF device be attached to a forklift truck or employee?
- What is the power source for the RF device and do you have spare batteries and RF devices?
- What is the RF wave in the geographic area, or is government approval for the band or wave required?
- How many and at what locations are RF receivers?
- Can a computer handle all operation transactions?
- What are the wire installation costs for your facility and what are ceiling and metal rack heights and do these affect the performance of the RF device?
- What are the SKU replenishment storage, pick position, manifest and other transactions per minute?

Small-Item Flatwear Picked SKU Transport Concepts

The objectives of a small-item flatwear transport activity are to move totes or cartons past the pick positions and to the next pick area activity or pack area, to move picked and labeled loose SKUs from the pick area to the pack area, and to do this in a constant flow that delivers COs to the assigned activity station.

Minimizing Travel Path Curves and Elevation Changes

One design objective in a small-item carton or tote transport concept for loose items is to minimize curves on the travel path, as well as elevation changes. Curves on a transport concept travel path are typically designed to avoid a building or equipment obstacle, provide a merge to another travel path, assure a proper divert, or due to building space limitations.

If a transport travel path concept has many curves, the layout design has high cost due to the need for additional drive motors or trains for each curve, increased facility electrical power or KVA, and, with a wide carton or tote mix, increased potential carton hang-ups or jams and additional space.

A vertical transport travel path is used to complete an elevation change on a conveyor travel path above or below a floor surface. An elevation change is made to avoid a building or equipment obstacle, service a workstation at a different elevation, or because of building space characteristics. An elevation change concept for loose small-items can be designed as:

- A powered belt conveyor travel path and a decline slide or chute
- A trolley path basket or pneumatic tube that has a travel path continuation
- A captive tote or CO shipping carton on a belt conveyor with a vertical lift and a decline slide, chute, or non-powered conveyor

A belt conveyor concept is a low-cost option that handles a high volume of loose SKUs, cartons, or totes, allows continuous piece incline or decline travel, and handles a wide SKU or carton mix. A slide or chute concept is also low cost, limited to a piece or carton decline concept. There is the potential for hang-ups in the chute and it handles a limited carton mix but wide loose SKU mix with consideration for a fragile/crushable and heavyweight/high-cube loose SKU mix. The vertical lift concept has a high cost, is set for maximum carton length and height, and requires short in-feed and out-feed belt conveyor sections. With this concept there is some difficulty in handling small and short cartons.

A trolley path with a basket handles loose SKUs over an incline or decline, has limited capacity, allows continuous travel, and requires horizontal run-outs. A pneumatic tube handles a small quantity of loose SKUs over an incline or decline travel path, allows continuous travel, and requires horizontal run-outs.

For additional in-house transport information, we refer the reader to the transport section in this chapter.

Transporting Totes or Loose SKUs from a Pick Area

All order-fulfillment operations have a requirement to move loose SKUs, totes, or cartons that have been picked for a CO from one pick area to the next or to a pack area. In a pick/pass order-fulfillment operation, the next activity is a quality and quantity check station for CO picked SKUs or a CO carton top seal station. In an order-fulfillment operation with a pick/sort and transport concept, the activity station is a quantity and quality check and pack station. With a pick/transport and sort operation, the activity is a shipping carton pack station.

A transport activity is the physical link between two locations in an order-fulfillment operation and it ensures that SKUs picked as loose items or in cartons or totes are delivered to the correct location, on time, and with minimal damage.

Design parameters for small-item transport are:

- Characteristics of the loose SKU or carton, such as length, width, height, weight, and whether the SKU is fragile or crushable
- Piece volume and carton number
- Order-fulfillment concept, methodology, and philosophy
- Induction and sort method
- Travel distance and travel path configuration
- Time to complete a CO order and delivery cycle or batched COs or wave
- CO number and pieces per CO or per batched CO group or wave

The transport function can be designed in one of the following ways:Employee carried

- Non-powered four-wheel cart
- Slide or chute
- Powered vehicles
- Powered belt conveyor travel path
- Trolley basket or bag on a manual overhead trolley travel path
- Pneumatic tube
- Tele-car

Employee Carry Concept

An employee small-item transport concept handles a low volume per trip and has the potential for employee injury. This is not considered an option for transport in a dynamic order-fulfillment operation. If transport is required to handle a small-cube, lightweight SKU or only a few SKUs, and an employee has a tote or apron and only needs to walk a short distance, an employee walk concept can be used in a pick area.

Four-Wheel Cart

A four-wheel cart is a transport vehicle for an order-fulfillment operation. SKUs are picked and sorted into an assigned CO shelf location or as discreetly labeled loose items in a cavity or tote. After all CO pick instructions are completed, the four-wheel cart is pushed over a variable path from a pick area to a pack station or to another transport travel path. An order-fulfillment operation that uses a four-wheel cart as a pick, sort, and transport vehicle, minimizes the SKU handlings. This method is preferred as a transport vehicle for the small-item order-fulfillment operation. To improve picker productivity, some four-wheel carts are equipped with a RF or pick-to-light concept.

The four-wheel cart was discussed in more detail earlier in this chapter.

Slide or Chute. A slide or chute is a fixed travel path concept for small items that has an elevation change between a charge end and a discharge end of a chute travel path. The entire travel path has a solid bottom surface and solid side walls. Loose small SKUs are moved by force of gravity from an upper elevation to a lower elevation. The travel path, slope, pitch or angle, surface, and side walls permit SKUs to flow over the length of the travel path. This flow is uncontrolled and there is potential for line pressure and damage with wide, heavy, fragile, or crushable SKUs mixed within a chute travel path. In a centralized manual or mechanized sortation concept, a chute is a transport section that allows picked SKUs to be transferred from a pick area to travel over a transport travel path to a sorter induction station and from a sortation travel path to a final sort or pack station.

A slide or chute has a travel path surface with a low coefficient of friction, travel path slope, structural support members for the underside and side, two side walls, an end stop, door, run-out or discharge end, and start/stop in-feed controls.

The bottom surface of the slide forms a travel path that extends from a charge end on one level to a discharge end on a lower level. The slope allows loose SKUs to flow over the travel path. To ensure a smooth flow, the travel path surface is even, smooth, and continuous for the entire length of the chute. In a long chute, the travel path may have sections, with the top section overlapping the next or lower travel path section. This overlay arrangement is similar to roof shingles.

Proper design of the charge and discharge ends ensures efficient SKU travel between two pick area travel path sections. In a chute transport application, the chute charge end is set at an elevation that is slightly below and under a feed conveyor travel path discharge end. Charge elevation and location to a feed conveyor travel path ensure:

- No SKU hang-ups at the intersection of the feed travel path and chute travel path
- From an in-feed conveyor, SKU forward movement and complete transfer of each SKU onto a chute travel path
- Sufficient force of gravity to ensure SKU transfer

The chute discharge end is designed to ensure that SKUs flow from a chute travel path onto the next workstation or travel path. In an order-fulfillment operation, the next pick area section is an induction or pack station. Design options are:

- A discharge end designed with an end stop or a hinged door. The end stop or hinged door is used with a chute that interfaces with a pack station that has the capability to receive SKUs from several chutes.
- When a chute discharge end interfaces with a pack station that only receives SKUs from one chute, the chute discharge end has a concave shape. The concave shape and run-out extend onto a pack station top. This design ensures that the SKUs are transferred in a constant flow and are properly located on a pack station top to minimize employee physical effort.

Factors that influence the design of a chute travel path surface are the mix of SKUs that will travel down the chute as well as their characteristics; the coefficient of friction between SKU exterior packaging and the travel path surface; building codes; elevation change between charge and discharge ends; the travel path length; available space; and costs. Chutes may be:

- Flat-bottomed
- Concave
- Flat-bottomed with a step in the travel path
- Spiral

Flat-Bottomed Chute or Slide

A flat-bottomed chute has a flat solid surface that extends between both side walls and over the entire travel path between the charge end and the discharge end. The chute has a straight, flat, even, and continuous travel path for its entire length. As loose small items flow over the bottom of the chute, there is the potential for damage to SKUs due to hang-ups on the surface as a result of jams, product queuing, and uncontrolled SKU flow. At an in-feed station, if a transport concept flows SKUs onto a chute with a wide bottom surface, a slight inverted V shape, and a high guardrail in front, the chute concept has the capability to provide UOP for two induction stations.

Concave Chute or Slide

A concave bottom surface is rounded or depressed in the middle and higher on the two sides. This concave shape directs small-item flow through the center of the travel path and minimizes UOP hang-ups. A concave chute travel path is a straight, flat, even, and continuous travel path for the entire length from the charge end to the discharge end.

Flat-bottomed and concave chute designs and SKU flow characteristics permit loose SKUs to flow over an entire travel path onto a mechanical sorter induction station. When loose SKUs queue on a chute travel path, they occupy approximately 30 to 40 percent of the total cubic area in the chute. The queue factors are:

- Loose small items are introduced into the chute at a constant rate at a higher elevation.
- Flow through a chute is at a fixed slope between two elevations.
- SKUs queue against a stop or guardrail on the chute bottom surface and at the discharge end.

Flat-Bottomed Chute or Slide with a Step in the Travel Path

A flat-bottomed chute can be designed with a step at the midpoint of the travel path. This type of chute has a flat-bottomed surface and side walls that are similar to a regular flat-bottomed chute except that there is a step down at the midpoint of the travel path. This separates the chute into two sections: a rear section above the step and a front section below the step.

The front section of the chute is 6 to 12 in. lower than the rear section. This feature permits queuing of additional loose, small SKUs on the chute travel path. The step chute bottom surface queues 45 to 50 percent of the total cubic area in a chute, whereas a regular chute queues at 30 to 40 percent.

Spiral Chute

A spiral chute is a specially designed flat or concave chute travel path. A spiral chute is a circular chute that declines from the charge end to the discharge end and is designed around a center post. A spiral chute design and product flow characteristic permit loose small items to flow over the center of the chute travel path and to queue against a stop or door. When loose, small items queue in a spiral chute, they occupy a very small 5 to 10 percent of the total cubic area.

Chute or Slide Degree of Slope

The degree of slope, pitch, or angle of the chute is determined by available floor space, the difference in elevation between the upper and lower levels, or between a charge end and discharge end, the coefficient of friction for small-item exterior packaging surface on the chute travel path, the shape, size, weight, and exterior packaging material, and start and stop controls for the transport concept that in-feeds the SKUs onto the chute.

Chute or Slide Materials

Materials that are used for the chute travel path include zinc or galvanized sheet metal, coated wood or pressed board, and fiberglass.

Sheet metal can be used for a chute with a long slide. The sheet metal is layered, with the front or top section overlying the lower or rear section. Some sheet metal surfaces have a spray or corn starch applied to it to prevent SKUs from sticking or becoming hung up on the travel path surface.

Sheet metal can be easily shaped to make turns, or formed with a concave bottom travel path surface or shaped in a spiral pattern. The chute has additional underside runners and cross-structural support members.

A coated wood or pressed board chute has a polyurethane or plastic coating on its travel path surface, similar to a kitchen countertop that is slippery. This coat-ing improves the flow of loose, small SKUs over the travel path. Coated wood or pressed board are used for a short travel path, which requires additional structural support runners and members because of the heavy weight. It is difficult to make turns or form this material into a concave chute.

A molded fiberglass chute creates a continuous, smooth, and even travel path and can be molded to form a straight, flat-bottomed, concave, or spiral travel path. If two fiberglass chute sections are required for a long travel path, the top section is overlaid onto the second or lower section. The features are very similar to a sheet metal chute; the exceptions are pieces move easily over the travel path, the increased strength requires fewer runners or support members. The fiberglass chute can be used for a longer length.

Structural Support

Structural support for the chute is provided by runners, which support the length, braces and beams that provide cross support. These are attached to the length and width of the chute underside and to the side walls. The structural support compo-nents include upright posts for a floor surface-supported chute or hangers for a ceil-ing-hung chute. Structural support is designed to accommodate the chute material, total weight of SKUs in the travel path, maximum weight for queued SKUs, and the seismic location.

Side Walls

The side walls are made from the same material as the travel path. They are attached to the bottom surface or are molded with the bottom surface part. Chute length and manufacturer standards determine the need for structural support and the number of braces. The chute sides direct the flow of small SKUs and when SKUs queue on a travel path the side walls ensure that loose, small items are contained within the travel path.

Charge and Discharge Ends

The design of the charge end accommodates either manual or mechanical transfer of SKUs into the chute.

When small items are transferred manually into the chute, the charge end is located adjacent to a pack station, at an elevation even with the pack station surface or slightly below it.

A mechanical transfer concept is used for SKU transfer from a powered belt conveyor onto a chute charge end. With this option, the conveyor travel path discharge end is extended over the chute charge end. The clearance between the chute charge end and the conveyor discharge end is determined by the belt conveyor frame depth and belt conveyor return travel path, and SKU characteristics, such as size and fragility.

The size of the SKU determines the clearance needed to allow the tallest or highest SKU to fall from the conveyor belt onto the chute without becoming hung up on the belt's return path. A fragile SKU restricts the distance that the SKU can fall from the conveyor belt onto the chute without becoming damaged from the impact on the travel path.

A mechanical transfer concept has a photo-eye control device located at the SKU chute charge end or transfer location. The photo-eye is angled across the SKU transfer location and when it is blocked by a SKU, the control system stops the in-feed conveyor belt. This feature minimizes potential product damage.

Control Devices

Photo-eyes are also located along the chute travel path to control SKU flow. The photo-eyes are located in at least three places along the chute, at the midpoint, the upper top section, and the transfer location.

The photo-eye control devices and microcomputer are designed to start and stop the conveyor depending on the length of the chute, UOP characteristic, and loose SKU line pressure. If the photo-eye in the midpoint of the chute travel path is blocked by loose SKUs, the photo-eye sends a message to the microcomputer to activate an alarm device, such as a flashing light or noise that notifies employees that the chute has SKUs queued on the travel path. If the photo-eye in the top section of the chute is blocked, it sends a signal to the microcomputer to stop the conveyor. If a photo-eye at the charge end is blocked by loose items, it signals the microcomputer to stop the transport conveyor. When SKUs are removed from the photo-eye line of sight, it signals the microcomputer to deactivate the alarm or to start the transport conveyor belt.

Powered Vehicles

A powered vehicle has a load-carrying surface but, for efficiency, loose, small-item SKUs are placed into a container. For a detailed review, we refer the reader to the discussion of powered vehicles and totes in this chapter.

Powered Belt Conveyor

A powered belt conveyor is an economical and efficient transport concept. A powered belt conveyor is used to transport a large quantity of loose SKUs or a wide SKU mix that meets specific characteristics. In a pick to conveyor belt order-fulfillment operation, loose SKUs are transferred from a pick position directly onto a moving conveyor belt travel path surface. A conveyor travel path has the following features:

- Sufficient horsepower to move a stopped and fully loaded conveyor belt forward.
- Sufficient width, with 1 to 2 in. between the edge of the travel path and the conveyor frame, to minimize SKU hang-ups on the side guards.
- A solid flat metal or plastic surface the full length and width under the travel path.
- Controlled SKU forward movement over the conveyor travel path
- If loose SKUs are diverted from a belt conveyor travel path, it has a smooth top surface.

A powered belt conveyor concept is designed to move loose SKUs across a horizontal or vertical travel path from the pick area to an induction area. Using the waterfall or belt conveyor curves, a belt conveyor travel path makes right or left turns or is divided into sections for a straight-run travel path over a long distance. The load-carrying surface is a belt (fabric with a rubberized covering) that moves between two pulleys. The belt is supported between the two pulleys by sheet metal or solid slider bed and wheel slider bed or combination sheet metal and roller or wheel slider bed. The belt is moved by an electric-powered motor through a drivetrain and sprockets pull the belt over an end pulley and tail pulley.

The belt conveyor frame is a preformed metal section that is either "toe-in" or "toe-out." A toe-in frame has the bottom frame lip face toward the conveyor belt, whereas a toe-out frame has the conveyor frame lip face out from the belt conveyor to the adjacent wall or aisle.

A metal conveyor frame steel gauge is determined by manufacturer standard, concentrated load weight on a travel path, conveyor component load weight, space between two conveyor stands or legs, and seismic location.

Steel conveyor frames have a coated exterior surface to reduce rust. The frame has predrilled holes at the load end, along both sides of the flange and at the tail section.

The predrilled holes are to attach splice plates, pulleys, motors and drives, guardrails, and other accessories. During installation additional holes are easily drilled.

To extend a slider bed belt conveyor length for more than one section, a splice plate is placed inside each conveyor section frame. Predrilled holes on the splice plate match the holes on the conveyor frame. Fastening the two conveyor sections together with nuts and bolts makes a longer conveyor travel path length.

Standard belt conveyor frame lengths range from 2 ft. 6 in. to 10 ft. or 12 ft. To ensure maximum conveyor travel path design flexibility and economics, conveyor design professionals attempt to use 5-, 10-, or 12-ft. frame sections. A belt conveyor frame width has a flat metal surface that spans between two fames. The width range for your belt conveyor travel path is determined by the SKU cubic foot volume and weight. Standard frame widths are 12, 18, 24, 30, 36, 48, and 54 in.

A belt conveyor frame height from the top flange to the bottom flange has a 5 to 6 in. depth. This depth allows the top belt and the conveyor belt return rollers to be housed within the space.

The SKU weight capacity and belt conveyor frame length determine the number of support legs that are required. A 10-ft. long section with a support leg at each frame end has a 240 lb./ft. load weight capacity. If the supports (legs) are every 5 ft. on the conveyor bed length, the load weight capacity is 960 lb./ft.

The load-carrying surface of the conveyor is designed as either a solid slider bed or a wheeled slider bed. A roller bed is used to transport totes or cartons, and for a discussion of roller beds we refer the reader to the tote or carton transport belt conveyor section in this chapter.

A solid slider bed has a box frame with a solid sheet metal surface that runs the full length of the travel path. The belt lies flat on the sheet metal. As the belt is moved by a drivetrain over pulleys and across the smooth surface created by the sheet metal, it provides a solid, smooth continuous surface for SKUs. If a conveyor travel path has two or more sheet metal sections along its length, to ensure a smooth travel path that does not interfere with the conveyor belt movement, the ends of the sheet metal are flanged or bent downward. A conveyor belt moving across a metal surface creates a drag caused by resistance to a belt that is loaded with SKUs as it moves across the sheet metal surface. This is known as the coefficient of friction and with a solid slider bed conveyor surface it is approximately 40 percent. A solid slider bed conveyor surface has the highest coefficient of friction but the lowest cost per linear foot.

A wheel slider bed conveyor surface forms a flat, smooth, and continuous belt similar to the solid slider bed belt conveyor surface but it has punched holes in the sheet metal. A skate-wheel protrudes through each hole. The skate-wheels are attached to an axle that is connected to the sides of the box frame. The skate-wheels are supported by middle support members. The skate-wheel axles are on 6-in. centers for the entire belt conveyor travel path length and each axle is attached to a conveyor frame at a specific height. The height allows a skate-wheel to extend upward approximately 0.25 in. above the solid sheet metal surface. The number of skatewheels per conveyor is determined by the width of the belt and conveyor frame. Standard skate-wheel spacings for wheel slider bed conveyor surface are: (a) a 12-in. wide slider bed has four skate-wheels; (b) an 18- or 24-in. wide slider bed has five skate-wheels; (c) a 30-in. wide slider bed has six skate wheels; (d) a 36-in. wide slider bed has six skate wheels; (e) a 42-in. wide slider bed has ten skate-wheels; and (f) a 48- or 54-in. wide slider bed has 12 skate-wheels.

As the motor, drive, and pulleys move the conveyor belt across the wheel slider bed, skate-wheels come in contact with the bottom of the conveyor belt. As the belt is pulled forward, the skate-wheels turn, lowering skate-wheel the coefficient of friction and reducing conveyor belt drag. A wheel slider bed has a coefficient of friction of approximately 10 percent. A wheel slider bed conveyor surface has the lowest coefficient of friction and a medium cost per linear foot.

Most belt conveyor manufacturers use a three- to four-ply conveyor belt. A three-ply conveyor belt has three woven fabric layers and a four-ply conveyor belt has four woven fabric layers. The one that is selected for an operation depends on SKU mix and characteristics, load weight, ply for existing belt conveyor in the facility, and drive and end pulley diameters.

A conveyor belt with thinner or lower ply has more flexibility and is used on a large-diameter pulley. A higher ply means that the conveyor belt is stiffer and is used with a small-diameter pulley.

A belt conveyor is made from a variety of woven fabrics, including cotton, cotton–nylon, rayon, or rayon–nylon.

Each layer is impregnated with rubber. The conveyor bottom surface rides across a solid sheet metal or skate-wheels on the slider, wheel slider conveyor bed surface, or rollers. A conveyor belt top or exterior surface is rubber, and forms the conveyor travel path surface for loose SKUs. The rubber absorbs the shock when loose SKUs are placed onto the conveyor belt and restricts movement of loose SKUs on the conveyor travel path.

If uncontrolled SKU movement on a conveyor belt travel path becomes a transport problem, cleats 0.25 to 0.5 in. on 12- or 18-in. centers are added to the belt conveyor surface. Cleats run the full width of the belt and create load-carrying spaces along the travel path surface. The space between two cleats creates a load-carrying surface for a specific SKU group or customer order. The cleats act as barriers that restrict uncontrolled SKU movement.

For best results and minimal SKU damage, the conveyor belt width is properly sized for the travel path width. Belt conveyor professionals agree that a belt conveyor width is 2 to 4 in. narrower than the width of the slider bed or frame. This means that a 30-in. wide slider bed conveyor travel path has a 26- to 28-in. wide conveyor belt.

When the conveyor belt is a straight path over a long distance, it may be necessary to have at least two belt conveyor sections. Belt conveyor manufacturers have their belt conveyor applications set at 100 to 150 ft. long. At the belt conveyor travel path distances, belt tracking over a travel path and pulleys is easier and accurate. A longer belt conveyor travel path creates belt tracking problems that create conveyor down time to correct. Two belt conveyor travel paths mean that each belt conveyor section has its own drive motor, drivetrain, belt conveyor travel path, pulleys, take-up device, and other components.

Belt conveyor factors that determine the need for multiple belt conveyor sections are:

- Belt width
- Belt type and ply
- Maximum transport weight and belt travel speed
- Bed or surface
- Manufacturer standards
- Horsepower and drive unit characteristics
- SKU length, width, and height
- Conveyor window

With multiple belt conveyor sections, SKUs are transferred using either waterfall belt conveyor sections or a non-powered roller or flat gap plate with chamfered cut edges between sections.

A belt conveyor waterfall concept has two belt conveyor sections on which SKUs travel in the same direction. With a waterfall belt conveyor, the first belt conveyor section overlaps the second section. With this concept, one belt conveyor section discharge end is at a higher elevation than the second belt conveyor charge end. This elevation change has sufficient clearance for the tallest SKU to fall from the belt-1 conveyor discharge end onto the belt-2 conveyor charge end. There are two ways to achieve an elevation difference:

- If the belt-1 section is at a high elevation, with the conveyor travel path surface level at the discharge end, the conveyor travel path is in several sections, to achieve the proper elevation at the final discharge location. The belt-1 conveyor travel path section is set at the highest elevation and each additional section is set at an elevation that is progressively lower until the belt conveyor travel path reaches its final discharge location. If the belt-1 travel path is at a low elevation, it is necessary to progressively increase the travel path elevation to the final waterfall location. If the travel path is long, to achieve the proper elevation at the final discharge location there are several belt conveyor waterfall sections.

With both options, the belt-1 conveyor section overlaps the belt-2 conveyor section at the discharge end. The number of waterfall sections are determined by the total conveyor travel path requirement and the previously mentioned design factors. A waterfall concept is used to transport a wide SKU mix (large SKUs and small SKUs) that are not crushable or fragile. To ensure that SKUs do not fall from the conveyor travel path, high side and front guardrails are used at the waterfall locations.

A non-powered roller or flat gap plate with chamfered cut edges that is placed between two powered conveyor travel path sections is used with a shorter travel path and low travel path window profile The two travel path sections and the top of the roller or flat gap plate are set at the same elevation. With this design, the belt-1 conveyor travel path section provides the power to propel SKUs across the non-powered roller or flat gap plate. The powered belt-2 conveyor section provides the power to pull SKUs from the non-powered roller or flat gap plate onto the belt-2 conveyor. This concept handles SKUs that will not fall through or jam in the small open spaces between the roller or flat gap plate and belt conveyor, and is capable of handling a wide SKU mix, including crushable and fragile SKUs. It is also easy to install.

Another component of a belt conveyor travel path is a curve in the travel path or a point where another travel path merges onto the travel path. A curve or merge is designed with a gap plate or waterfall and guardrails to ensure SKUs travel from the straight belt conveyor travel path to a curved belt or onto another powered conveyor.

Overhead Manual or Powered Trolley with a Basket or Bag

Another transport concept for picked SKUs is a manual or powered overhead trolley with a basket or bag. For additional details on this type of trolley and rail components and design, we refer the reader to the discussion of guided overhead vehicles in the section on "In House Transport Concepts" in Chapter 3.

During the design phase, the additional weight of the basket or bag and the weight of SKUs determine the design of structural support for the conveyor. In a new facility, these factors allow a manufacturer and architect to design the building ceiling and overhead trolley travel path structural support members to handle this new weight. A standard folding basket transport has a higher top of rail (TOR) above a floor surface.

SKU overhead trolley transport carriers (see Figure 4.69) can be metal folding baskets, circular metal baskets, or canvas folding bags that are either top-loading or front-loading.

Metal Folding Basket

A metal folding basket has two hooks at the top that extend upward, and coated wire-mesh sides, rear, and load-carrying surfaces. All the basket components are moveable. When the basket is not being used, the basket can be folded down to reduce less storage space. The load-carrying surfaces are made of electroplated steel and have a slight slope toward the rear, which helps to prevent SKUs falling from the basket. The shelves can be folded, which provides flexibility to the load-carrying surface cube. With the top shelf folded to the rear and secured against the rear, the metal basket creates one large opening to increase the cube.

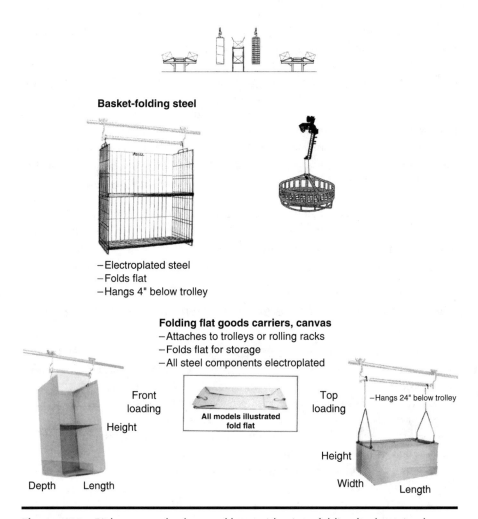

Figure 4.69 Pick concept baskets and bags: side view, folding basket, circular basket, front loading bag, and top loading bag.

To reduce the possibility of SKUs falling from a basket, the actual load-carrying capacity is approximately 50 to 60 percent of the theoretical load-carrying capacity or cube. Removable barriers across the opening increase the actual load-carrying capacity or cube. The hooks on the top of the basket permit an employee to hook the metal basket onto a trolley load bar. With a 36-in. long basket hook-to-hook dimension, an open or clear distance between two trolley necks is at least 36 to 42 in. wide. The trolley neck width permits the employee to attach and remove a basket from a trolley load bar. The folding rear surface is 36 in. long and runs the full height of the basket. The standard folding baskets are 18 in. deep or 24 in. deep.

Metal Circular Basket

A metal circular basket has one hook that extends upward from the basket center and attaches to a trolley load bar. A circular metal basket has two or three fixed coated wire-mesh levels, making the circular basket capable of carrying multiple SKUs or COs. An employee has access to the basket from any side. The basket is non-collapsible, which means that an unused basket occupies a larger storage space, or storage on a branch rail or spur, and medium load-carrying capacity.

Top-Loading Canvas Bag

A top-loading canvas bag has two hooks that are coated metal members and has four sides that collapse onto the bottom. Metal hooks are used to attach a canvas bag to a trolley bar. In a V-shape, the metal hooks extend upward from both sides. A canvas load-carrying surface has four solid sides and bottom. The four top sides have metal structural members that maintain the shape of the bag. The canvas bag is rectangular so that an employee transfers products from any side. A top-loading canvas bag length is a standard 36 in., with a clear span between two trolley necks that is 36 in. or more. Standard canvas bag dimensions are 18 in. deep and 26 in. deep and 18 in. high and 12 in. high.

The canvas bag has one load-carrying cavity and handles one loose SKU. When it is not being used, it is collapsed to occupy a small storage area, and it is light-weight, which permits one employee to transfer an empty canvas bag between trolley load bars.

Front-Loading Canvas Bag

A front-loading canvas bag has two metal hooks that extend upward from the bag and are attached to a trolley load bar. The solid sides and bottom create the SKU cavity. Metal hooks are coated and are on 18- and 30-in. center dimensions, which means a front-loading canvas bag's hooks fit between a standard trolley load bar length or the center-to-center distance between a trolley's two necks. Two hooks are positioned on a canvas bag to hang from a trolley load bar with a slight slope toward the rear. This feature reduces the risk of SKUs falling from the bag. A front-loading canvas bag has metal structural side members that are enclosed in the canvas material to maintain rigidity. A front-loading canvas bag is typically either 36 or 48 in. high, and 14 in. wide.

The two sides and rear determine the load-carrying capacity or cube. A front load-carrying canvas bag has an opening that is designed to have a shelf in the middle of the bag. This feature allows a front load-carrying bag to carry two SKUs or COs. A removable front barrier is used to keep SKUs in the load-carrying bag cavity. Like the top-loading canvas bags, the front-loading canvas bag is collapsible and easily handled by one employee. The disadvantages are that it is difficult to handle more than one SKU or CO, has a limited weight and cube-carrying capacity, and material can tear, and collect dirt and dust. The advantages are it requires less storage space and can be transferred between trolleys by one person.

Load Leveler

A basket-load leveler is a device that is attached to a trolley load bar and has two hooks to attach a canvas bag or metal basket hooks. A load leveler device is designed to act as a balancing device as a trolley load bar with a basket or bag travels over a decline or incline travel path. When a trolley with a basket or canvas bag travels over a vertical travel path, the load leveler keeps the basket or bag load-carrying surface in a level position. This feature means that SKUs are less likely to fall from the basket.

Pneumatic Tube Concept

A pneumatic tube moves small parts in a sealed carrier (container, capsule, or cylinder) that travels through a tube network and moves SKUs over a fixed enclosed overhead travel path. Lockable doors are located at the in-feed and out-feed stations to permit carrier charge and discharge. A power source to propel or pull the carrier is provided by an electric motor-driven system that creates a vacuum or air pressure in the tube network. A pneumatic tube network travel path is between two workstations and the travel path is a ceiling-hung, wall-supported or floor-supported concept. Pneumatic tube transport concepts are: (a) a push or propel concept in which air pressure forces a carrier travel through a tube network, and (b) a pull concept in which a vacuum force moves a carrier through a tube network between entry and exit stations.

A small-SKU powered pneumatic tube overhead transport concept has an air blower or vacuum pump, motor or air compressor, entry station and discharge station, tube or piping network and carrier, cylinder, capsule, or container.

An air blower (air compressor) or vacuum pump and motor supply air power to push or force a carrier through a tube network. A vacuum pump and motor create a vacuum in the tube network that pulls a carrier from an entry station to an exit station. An air blower or vacuum pump uses electric power to operate a three-phase motor with 10 hp. A pneumatic tube concept moves a total weight of 5 lb., including the weight of the carrier and SKU. A carrier moves through a pneumatic tube network at a rate of 15 to 20 ft./min.

The entry station in a pneumatic tube transport is located on the travel path where an employee places a cylinder or carrier into the pneumatic tube. To place a loaded carrier or capsule into a pneumatic tube network, an employee opens an access door and addresses a send location into a pneumatic tube dispatch device. The access door is a hinged door that is lockable and the passageway has sufficient length and width to permit the employee to transfer a carrier or capsule between the pneumatic tube and a workstation.

A pneumatic tube exit station is located on a network at a point that receives a carrier or capsule that was sent from an entry station. At the exit station, a carrier is discharged from the pneumatic tube travel path into a container at a workstation or remains in the pneumatic tube. Entry and exit stations are mounted on a wall or table. To reduce air compressor noise level, mufflers are located outside or within an enclosure or shroud. An audible or flashing light alarm is triggered to notify employees that a cylinder has arrived at an address workstation.

Each entry and exit station require electric power. If a pneumatic tube concept has a computer control system, the computer has a voltage (electric spike) protection or a dedicated electrical power line. At an entry station with a one-stop pneumatic tube concept, an employee opens a door and places a carrier into a pneumatic tube network. The employee closes the door and activates the pneumatic tube, sending the carrier to a workstation.

With a multiple-stop pneumatic tube concept, an employee enters a dispatch address into a carrier, by a dial, push button, or electronic logic method. After a carrier dispatch address is entered, the employee places the carrier into the pneumatic tube and sends the carrier to an assigned workstation.

A pneumatic tube carrier travel path has straight and curved sections. In a plan view, a pneumatic tube curve has a long radius curve section that has a 4-ft. bend to the turn and an off-set curve has a 3-ft. length. A plan view's pneumatic tube's lengths are confirmed by the pneumatic tube manufacturer and match the carrier length and width dimensions. A pneumatic tube enclosed travel path is airtight to ensure that the appropriate air pressure or vacuum is maintained through the tube. Air pressure or vacuum loss from a pneumatic tube network slows carrier travel speed or reduces air pressure or vacuum to a level low enough to make the pneumatic tube transport concept nonoperational. A pneumatic tube network is designed from preformed metal or plastic material with an interior that is a smooth, even, and continuous surface from the entry station to the exit station. The interior pneumatic tube diameter size is sufficient to accommodate a carrier traveling through the travel path.

The carrier, a capsule, cylinder, or other container, is the SKU- or load-carrying device. The carrier has a cone or round shape at the front and rear ends to ensure a smooth flow through the tube and permit a lockable access door at one or both ends. A cone-shaped front end is designed to ensure that the carrier travels through an enclosed travel path and reduces air resistance as it flows through the pneumatic tube. A tube transport moves a standard 5 lb. carrier over a 2000 ft. travel path. Carrier standard dimensions are:

- Overall inside diameter that ranges from 3 to 6 in., which is the UOP-carrying space
- Nine to 14 in. overall length
- Structural strength to carry a 5 lb. load

A pneumatic tube is designed as a one- or two-way transport concept. With a one-way transport concept, the air pressure (force) or vacuum moves the carrier in one travel direction only, from a carrier entry station to an exit station. A closed-loop carrier transport has separate air pressure or vacuum travel path systems. With this concept, one pneumatic travel path system is used to send carriers in one travel direction and a second system is used to send carriers through the same pneumatic tube in the reverse travel direction. Alternatively, the concept has two pneumatic tube travel paths, each of which has its own air pressure or vacuum system. With this concept, one pneumatic tube network is used for one-way carrier travel and the second pneumatic tube network is used for the reverse direction of travel.

Advantages of having a two-way pneumatic tube system in one travel path are lower piping costs and installation and it has a separate air system for both travel directions, which means lower costs to move carriers between two work stations.

The disadvantages of the pneumatic tube transport concept are that it moves a small SKU quantity, has a fixed travel path, limited travel distance, limited SKU mix, and requires an air compressor or vacuum device. The advantages are travel path has a small area, reusable carrier and code, minimal SKU and equipment damage and employee injury, easy to operate, and low maintenance.

Tele-Car

A tele-car is an electric-powered track conveyor and multi-wheel cart or tele-car that is an overhead transport concept. A tele-car is a self-propelled four-wheel cart that rides on an electric-powered track and moves SKUs in a sealed container across a fixed travel path from one location to another. This travel path is ceiling hung or floor supported.

A tele-car transport concept comprises a four-wheel or multi-wheel DC electric-powered vehicle or car, a container with a cover that is lockable and dispatch address location, modular aluminum track and structural support members, switching or divert sections, control center and power unit, and loading and unloading stations.

Four-Wheel Vehicle or Load-Carrying Cart

With a properly entered dispatch code, an electric-powered self-propelled four-wheel car travels over a fixed travel path from a dispatch station to an assigned station. A self-propelled car has four rubber wheels and a chassis or container platform. The four rubber wheels are spaced in a rectangular pattern on the platform underside and allow a car to move along the tracks. Each rubber wheel has an axle and bearings and its components support one quarter of the combined weight of the chassis, container, and SKU. The four rubber wheels are enclosed in a travel path or track cavity that ensures the car remains and is guided along a travel path. The guided travel ensures minimal equipment and building damage, high travel speeds, and incline and decline over a 90 degree angled travel path.

An electric-powered motor receives its electrical power (24 V DC) from two travel path buss rails. Four wheels enclosed in a travel path ensure that electric power (115 V) is transferred from the track to a tele-car. The electric power supply transfer occurs at predetermined intervals along a track. With a nominal car traffic volume on a horizontal track, the power supply locations are very 100 ft. on centers or the distance between two track locations. An electric-powered motor propels a car on four rubber wheels in a captive fixed travel path at 120 ft./min.

The chassis houses an electric motor, sprocket, axle support brackets, and container platform. A chassis load-carrying surface supports SKUs and container weight and ensures chassis attachment to a container platform.

Container- or SKU-Carrying Surface

The container- or SKU-carrying surface is securely attached to the chassis. A container has solid sides, bottom, and an attached lockable cover. During travel, the solid walls, top, and bottom ensure that SKUs are secure inside a container. As a car travels at high travel speeds, a solid enclosure reduces SKU damage or lost SKUs. The container cover has a device for an employee to enter a car destination address. Each magnet set has 0 to 9 as possible settings and a three-magnet set provides 999 possible different dispatch locations and is reusable. A container shape (length, width, and height) is determined by available travel path space, that is, width and height; SKU characteristics, that is, length, width, height, and weight; SKU volume per trip; SKU orientation inside a container; and available space for travel path run-outs at a loading and unloading station.

Design of the car travel path is influenced by the available clear space between the track and various building components (ceiling, columns, walls, and other items) or material handling equipment. SKU length, width, height, weight, volume, and orientation are design factors that determine a container shape. A car travel path clearance, SKU length, width and height characteristics, cube, and weight are affected by an electric-powered car standard load-carrying capacity. A car/container handles a SKU and container combined weight of 20 to 75 lb. Standard container configurations are lightweight and flat, large and heavy, and self-leveling.

Standard container sizes are: (a) 20 in. long, 13 in. wide, and 11 in. high; (b) 20 in. long, 13 in. wide, and 9.25 in. high to handle a mixed load; (c) 20 in. long, 16.5 in. wide, and 4.5 in. high to handle flat UOPs; (d) 18 in. long, 11 in. wide, and 4 in. high to handle low-profile UOPs, (e) 18 in. long, 11 in. wide, and 8 in. high to handle a mixed load; (f) 32 in. long, 15.25 in. wide, and 8 in. high to handle trays or a mixed load; (g) 32 in. long, 15.25 in. wide, and 12.5 in. high to handle large and heavy UOPs; and (h) 11.25 in. long, 12 in. wide, and 12 in. high with the self-leveling feature.

The container is a rectangle that matches the car chassis, which helps to ensure that the four wheels provide stability as the car travels over a travel path. The container height and width determine the travel path window. A standard travel path window is 13 in. high and 17 in. wide. When you design a car travel path layout through your facility, in addition to a container dimensions, you add a track and track structural support members. For a 13 in. high by 17 in. wide container, an electric-powered car travel path is approximately 21 in. high by 21 in. wide.

A self-propelled car has extruded aluminum modular tracks that provide a smooth, even, and continuous wheel travel surface with a low coefficient of friction and a cavity that holds the four wheels on the track, allowing a car with a full load to travel at 120 ft./min.

A manufacturer determines the car top direction on a travel path by locations for control rail, electric-powered rails, center insulator, and gear rack. Designs are:

■ Horizontal track sections with a car facing upward or toward the ceiling
■ Vertical car travel section
■ Horizontal car travel path with a car facing downward or toward the floor

A horizontal travel path can be designed as a single path or track, with window dimensions of 1 ft. 8 in. wide and 1 ft. 10 in. high; or as a double path. Window dimensions for the double path are 2 ft. 11 in. wide and 1 ft. 10 in. high. The path dimensions allow space for a container, chassis, track, and structural support members and 3 in. clear from building obstacles.

In a vertical car travel path design, window dimensions vary. Minimum travel path dimensions are 3 ft. 3 in. high and 3 ft. 3 in. wide. A vertical travel path may have a switch device that transfers a car from one travel path to another. In this case, window dimensions are 4 ft. high and 5 ft. 9 in. wide. Straight travel path sections are approximately 5 in. high by 8 in. wide and are available in standard straight 10-ft. long sections that can be attached to each other. If the travel path design has a travel path section that is shorter than 10 ft., the manufacturer adjusts the length in the field with square cut ends.

To optimize facility space utilization and to avoid building obstacles, straight sections and curves are used on a travel path. Horizontal car travel path curves have a radius of 2 ft. 6 in. to 3 ft. Curves on a travel path can be:

■ A 22.5-degree curve, with a short, straight travel path section at the lead and tail ends of the curve
■ A 45-degree curve, with a short straight travel path section at both ends
■ A 90-degree curve, to change a travel path direction from right to left
■ A 180-degree curve, which has two 90-degree curves that are connected with a short, straight travel path section

When the track design has a vertical travel path or elevation change to access an elevated floor workstation, a car in track concept is designed to service the workstations. Options for designing curves on a vertical track are:

- A 22.5-degree bend section; with an additional straight travel path section added between two vertical bends to create an elevation change.
- A 45-degree curve section.
- A 90-degree curve section, or, to make an elevation change within a short distance, two 90-degree vertical curves with a straight travel path section between them in either a Z- or a C-configuration
  - A Z-configuration is one in which the car makes an elevation change and continues its forward direction of travel. In this design, the car enters the travel path level, with the top facing up to the ceiling and exits the other level with the same orientation. A Z bend travel path has two 90-degree vertical bends. The interior radius of each curve faces a different direction. The elevation change with a Z-configuration travel path has a minimum distance of 4 ft. 6 in. To increase a vertical travel path between two different levels, length is added to the straight section between two vertical curves.
  - A C-configuration car travel path is one in which a car makes an elevation change and returns in the same direction of travel as the lower-level horizontal travel path. This design specifies that a car enters a C-configuration vertical travel path level with the top facing upward to the ceiling and exits the other level with the top facing downward or toward the floor. A C-configuration vertical travel path has two 90-degree vertical bends. In each curve, the exterior radius faces the same direction and the travel path is over the interior of both curves. To complete an elevation change, a C-configuration vertical travel path has a minimum distance of 2 ft. 3 in. To increase the vertical travel path between the two different levels, a straight section is added between two vertical bends.

In a standard design, the minimum distance between the bottom of the track support member and the ceiling ranges from 1 ft. 1 in. to 2 ft. 2 in. The distance variance is determined by container and workstation heights above the floor. With a transport concept where a car cover faces the ceiling, the standard distance from the floor to the top of the travel path ranges from 22.5 to 28 in. The dimension varies depending on container and workstation heights above the floor. Car in track transport design options are (1) a car that rides on top of a track, or (2) a car suspended below a track. With this travel path design, the container is below the travel path and has a different dispatch magnet setting location and lockable door location.

Car Switching or Divert Sections

In a multi-track, microcomputer-controlled car transport concept, a car switching or divert station section communicates to a microcomputer as to a delivery destination for each four-wheel car. Prior to switching track network and other four-wheel car traffic on a travel path, the microcomputer activates a switching device to transfer a four-wheel car from one travel path to another. The alternative travel path has the shortest travel distance from a dispatch station to an assigned station and avoids a four-wheel car queue. A car switching section has a track section that accommodates one four-wheel car and container length. After a four-wheel car is positioned and stopped on a switching section, a microcomputer activates a car switch track section to slide the car from one travel path section to an adjacent section. After the switching section is aligned with a new travel path, the microcomputer activates the car to travel forward onto a new car travel path. During the transfer activity, at a predetermined location before the switching station, the microcomputer ensures that other four-wheel cars are queued on the travel path.

Container-Loading and -Unloading Stations

A container-loading and -unloading station is located on a car travel path where an employee performs SKU transfer transactions between a workstation and car container. In a tele-car transport concept, four-wheel car container-loading and -unloading stations are located within an enclosed workstation that has fire walls and a car travel path equipped with fire doors. A container-loading and -unloading station can be designed in one of the following ways:

- A car-through station, in which a four-wheel car enters and exits over the same travel path. A single travel path section has a three- to five-container queue capacity. This design enables the transport concept to handle a high car volume, with the requirement that cars continually travel to the next dispatch station. Loading and unloading UOP transfer transactions take place on a car first-in, first-out (FIFO) basis. This car-through design requires the least area.
- A magazine station, which has two car-switching devices and a car container-loading or -unloading spur. As cars enter a magazine SKU transfer station, cars are switched from the main travel path onto a spur (short travel path). On the loading and unloading spur, an employee performs a SKU transfer transaction between a workstation and container. After all SKU transfer transactions, the car is moved forward to a second car-switching device, which moves the car from the loading and unloading spur onto the main travel path for travel to the next assigned workstation. This design allows continuous car travel on the main travel path while one car is at a workstation, gives a workstation the capability to handle high car volume. A magazine station requires a large floor area.
- A reentry station, which has a travel path 4 to 8 ft. in length that has the capability to queue three to five cars. Depending on the number of cars dispatched, cars are removed from the travel path or are reintroduced onto the travel path. Car transfer is on a first-in, first-out basis.

Control Module and Power Unit

The control center has a container-loading and -unloading station and a micro-computer and display screen to provide information on the status of the travel path or transport concept. The travel path status can be printed as a paper document or viewed on a graphic display. Information includes the location of each car on a travel path, container-loading and -unloading station status, car switch station status, container traffic and volume, car travel direction, and firedoor status.

An employee transfers SKUs into a four-wheel car container and ensures that the lid is closed and locked. A dispatch address is entered into one or all three address magnets. When the address magnets are set, the employee presses a send button on a control console. This action sends the car on the main travel path for travel to a receiving station. When it arrives at the receiving station, if the unloading and loading travel path section is full, the car activates an alarm at the receiving station. At the receiving station, an employee unloads SKUs from the container and completes a loading transaction or dispatches a car to another station.

The disadvantages are fixed travel path, small quantity, possible queues occur on a single travel path, and capital investment. The advantages are above-floor travel path, expandable, reusable container and address code, each loading and unloading station has a spur, car travel speed is fast and through a small window, and closed loop travel path returns all cars to a home base.

Moving Customer Orders Past Pick Positions

The second small-item, flatwear order-fulfillment transport activity is to move CO cartons past all pick positions. If an order-fulfillment operation is a pick/pass or batched order-fulfillment operation, a pick carton is transported by a picker, vehicle, or belt conveyor travel path through all pick aisles or past each pick position. If an order-fulfillment operation has an automatic pick machine, a conveyor travel path directs a CO pick carton past or under an automatic pick machine.

Customer Order Carton or Tote Transport Concepts

Transport concepts to move a CO pick carton past all pick positions are determined by:

- The type of pick tote or carton and its physical characteristics
- The pick concept, that is, single or batched COs
- The pick aisle, pick line, or automatic pick area layout and type of order-fulfillment operation, including picker or transport equipment to move a CO carton or tote

A pick and pack order-fulfillment operation transfers loose SKUs directly from a pick position into a CO captive tote or into a shipping carton. In this type of operation, a carton or tote must have the following characteristics:

■ A cavity to hold pre-cube SKUs picked for a CO.
■ A rectangular or square shape that fits onto a powered conveyor travel path, onto a vehicle load-carrying surface, or can be carried by a picker.
■ A solid, flat bottom exterior surface for small SKUs or, for large SKUs, a tote with a meshed, open bottom exterior surface or a CO shipping carton.
■ A conveyor travel path with sufficient length and width to have three rollers or skate-wheels under the carton bottom surface.
■ A powered conveyor, carton side wall height at lead and trail sides and weight to (a) activate photo eyes, (b) move over a travel path, (c) side wall shape and strength to permit queuing on a travel path without a carton/tote jumping from a conveyor travel path or becoming crushed, and (d) permit CO identification placement on a tote/carton flap or side. CO identification location has line of sight for a picker or barcode scanner.
■ A bottom surface that permits a completed CO to be transferred between two workstations.
■ A full tote or carton that can be easily handled by a packer or machine at a pack station.

Each pick concept requires a different pick line, pick aisle, or pick area layout. A dynamic order-fulfillment operation that handles a variety of types of SKU, including conveyable and non-conveyable, crushable, fragile, and high-value SKUs, has several order-fulfillment concepts. In a macro view, the layout of the order-fulfillment area depends on:

■ Number of SKUs
■ Number of SKUs allocated to each order-fulfillment conveyable and non-conveyable concept, which is influenced by physical characteristics, value, historical movement, and volume of each SKU
■ The compatibility of the order-fulfillment concept with available building space
■ The CO pick concept
■ The number of COs, pick lines per CO, pieces per line, and pieces per CO

In a batched or single CO fulfillment operation with shelves or decked racks, a picker walks or rides on a vehicle through a pick aisle and carries picked SKUs. After all pick transactions, with a single CO concept, a picker travels with the picked SKUs from a pick position to a CO carton or transport concept. With a pick/transport, sort/check, and pack batched order-fulfillment operation, picked SKUs are sorted into a CO tote or are transported loose on a belt conveyor to a sortation area. This type of order-fulfillment operation requires a large area for the following reasons:

- The pick aisles must be designed for two-way picker or vehicle traffic.
- The vehicle load-carrying surfaces are required to be wide.
- Pick positions are limited to a position that is elevated above the floor, that is easily reached by a picker or replenishment employee on an elevating vehicle.
- Each aisle has a limited number of pick positions.
- There need to be turning aisles at both ends of the pick aisle and/or a middle turn aisle.
- Pick position replenishment to occur as a picker completes a pick.

In an operation in which batched COs are picked into a carton or a single CO picked from carton gravity flow racks and packed into a CO tote or carton on a conveyor, the pick area has a medium requirement for floor area, including space for:

- Carton flow rack pick positions.
- Separate SKU replenishment aisles.
- A pick conveyor and a powered take-away conveyor for completed COs.
- With carton flow racks arranged perpendicular to a conveyor travel path, there will be a medium number of pick positions per aisle. With gravity carton flow racks parallel to a conveyor travel path, there will be a medium to low number of pick positions per aisle. With shelves that are arranged perpendicular to a conveyor travel path, there will be a large number of pick positions per aisle. With shelves parallel to a conveyor travel path, there will be fewer pick positions per aisle.
- Conveyor travel path turns.

In an automatic pick machine single CO concept, the pick area occupies a small square footage because the SKU take-away transport path is the automatic pick machine pick conveyor travel path. A conveyor travel path is a single belt conveyor that travels under or adjacent to an automatic machine picking device. For controlled SKU travel over a belt conveyor travel path, an automatic pick machine releases CO SKUs onto a belt with cleats or into a tote. The completed CO take-away transport conveyor services multiple pack stations.

The automatic pick concepts are:

- An A-frame pick machine, where the length and width of a SKU pick position are slightly larger than the SKU dimensions. The pick positions are arranged in vertical or angled sleeves that release SKUs onto both sides of a belt conveyor in the center. The sleeve discharge end is slightly above the surface of the conveyor. As triggered by a CO and microcomputer, an A-frame sleeve releases a SKU onto the conveyor without causing damage to the SKU.

■ A Robo Pic, which has several forward-moving short belt conveyor travel paths. Each travel path comprises a SKU pick position and a conveyor belt that is separated into sections with cleats that are the width of the conveyor belt and extend approximately 0.25 in. above the surface of the conveyor belt. Per a CO, a computer and CO tote or carton on a center conveyor travel path has the appropriate SKU pick conveyor belt move forward. When the tote or carton is directly under a pick position, the forward movement of the pick belt conveyor discharges a SKU into a CO tote or carton and activates a photo-eye to verify the pick transaction. A conveyor travel path transports the CO tote or carton from the automatic pick machine to an assigned pack station.

■ An Itematic pick concept has many SKU pick position sleeves that are slightly angled toward a discharge belt conveyor travel path. Per a CO and a micro-computer, an angled sleeve and pick device triggers the appropriate mechanical pick position sleeve to move forward and release one SKU onto a conveyor travel path. The conveyor travel path transports CO picked loose SKUs on a conveyor belt from the automatic pick machine to an assigned pack station.

Characteristics of the single-item automatic pick machine concept are that it handles a large number of SKUs, the pick area is small to medium, and there is a ready reserve SKU area that permits accurate and timely SKU replenishment to pick positions. In an automatic pick machine concept, ready reserve positions are gravity carton flow racks, shelves, or decked pallet racks that are arranged perpendicular or parallel to an automatic pick machine replenishment side.

Equipment for transporting cartons or totes includes:

■ Human carried
■ Non-powered three- or four-wheel vehicles
■ Powered three- or four-wheel vehicles
■ Non-powered conveyor travel path
■ Powered conveyor travel path

Each carton and tote transport concept has unique design and operational characteristics and a common design parameter. The common design parameter is that the physical characteristics of a carton or tote must match the transport concept. Objectives in designing a carton or tote are:

■ To allow a picker or automatic pick machine in a pick area to transfer CO picked SKUs into a carton or tote
■ To retain SKUs in a carton or tote on the transport concept
■ To allow a sortation employee to handle the tote or carton and transfer SKUs from the tote or carton onto an induction station or into a sort station
■ To be stackable or nestable in a sort or pick area, when not in use

- To be able to have a discreet human or human/machine identification label attached
- To be handled by an employee either full or empty

To achieve these design objectives, factors to be considered are:

- Material
- Solid walls and bottom surface with holes or open meshed walls and bottom surface
- Smooth side walls or ribbed side walls
- Perpendicular or angled wall to a travel path
- Side wall height
- Weight capacity, SKU length, width and height, and cube capacity
- Ability of an employee to lift handles
- A suitable location for discreet identification
- Nestable and stackable
- Exterior length and wall, direction of travel, and square or rectangle shape
- Ability to have a CO identification attached

Cartons for transporting SKUs are made of cardboard, chipboard, or corrugated or fiberboard. Totes are made of various types of plastic.

Cardboard or Chipboard Carton

A cardboard or chipboard box is the basic type of carton used for transporting SKUs. A cardboard carton is manufactured from pressed thin sheets of paper, fibrous material, and binding material, shaped according to your order-fulfillment requirements. A chipboard box is a two-piece carton with bottom and top sections formed by the manufacturer or machine formed at the order-fulfillment facility. In a dynamic order-fulfillment operation with a wide SKU mix, a chipboard box is not preferred for use as a pick and transport carton. Cardboard and chipboard boxes are used as pick and transport cartons for apparel, lightweight, and non-crushable SKUs. A cardboard carton is used as a CO pick carton and as a carton that is shipped from a facility to a CO delivery address. Each CO shipping carton is strapped or taped to ensure that it is sealed. A chipboard carton has limited use because it has a low weight capacity, can carry a small number of SKUs, and has low side wall and bottom strength.

Corrugated or Fiberboard Carton

A corrugated or fiberboard carton is a one- or two-piece carton. A one-piece carton is formed by the manufacturer and delivered as flat pieces in bundles to the order-fulfillment operation. The two-piece carton is delivered as pre-cut corrugated sheets that are formed and sealed at an order-fulfillment operation. When a carton cover and/or top and bottom flaps are sealed, a carton has four solid side walls, solid top, and bottom. In an order-fulfillment operation, a corrugated carton is used as a:

- Transport device for pick/pass CO picked SKUs.
- Carton for transporting picked and labeled loose SKUs or mixed SKUs for transport to a final sortation station.
- Completed picked, sorted, checked, and packed CO carton for transport from a pick facility to a CO delivery location. In a pick/pass order-fulfillment operation, an employee or machine seals the bottom flaps of a corrugated carton, which is used as a CO pick/pass container. In a sort and pack CO operation, a carton is formed at the pack station.

The interior of a corrugated carton is made from linerboard that is kraft brown corrugated material. The carton exterior material is also made from kraft brown material, bleached white or off-white linerboard. A corrugated carton side wall is secured to form one piece. Prior to SKU transfer into a corrugated carton, bottom flaps are sealed with tape or as a machine-formed bottom piece. After completion of the CO pick/pass or pack activity, the top flaps or cover is employee or machine sealed with tape or plastic bands. During transport to a CO delivery address, this feature ensures SKU security.

In selecting a corrugated carton for an order fulfillment operation, the following factors should be considered:

- SKU or CO characteristics, including length, width, height, and weight
- Packaging and sealing costs
- CO transport methods
- Internal and external dimensions
- SKU value
- Carton characteristics, including size, construction, weight, economics, and reusability
- CO delivery requirements
- CO identification attachment

The carton should provide the best protection for your SKUs at the lowest purchase and operating costs.

Corrugated linerboard or kraft board is made from:

- Pulpwood or fiberboard
- Four-drinier kraft linerboard made from virgin pulpwood
- Cylinder linerboard made from reclaimed wood fibers or a combination of reclaimed fibers and virgin pulpwood

- Corrugated linerboard made from a combination of straw, reclaimed fibers, and wood material

Types of corrugated board wall are:

- Single wall, which has a linerboard that is very thin and does not withstand line pressure on a powered conveyor travel path. Line pressure is the forward pressure that is exerted on a stopped carton by the cartons that queue and push against it on a powered conveyor travel path. With a zero- or low-pressure conveyor travel path, this carton line pressure is minimized.
- Single face, which is made by fluting or waving the corrugated medium. The corrugated medium is glued to the face. A single face carton increases SKU protection.
- Double wall, which has three linerboards, each separated by two corrugated fluted or waved members. Fluted corrugated members are glued to the linerboard. This multiple-wall container improves SKU protection and carton strength.
- Triple wall, which is designed with four linerboards and three flutes or waves. Each flute is glued to the linerboard. The multiple flutes and linerboard provide maximum SKU protection and side wall strength.

A-flute or wave corrugated member is attached to a linerboard or is glued between two linerboards. There are several types:

- An A-flute member has approximately 36 corrugations per ln ft., which is the highest flute or wave number between two linerboard connection points and provides the least number of flutes per foot. When a shock to a carton is transmitted in the thickness direction, this flute type has a greater capacity to absorb a shock. As a carton travels on a powered conveyor travel path, this ability to absorb shock provides high SKU protection.
- A B-flute has an estimated 51 waves per ln ft. and has the greatest number of waves per foot. This feature makes a B-flute more rigid but with less capacity to absorb a shock from an object. As a B-flute carton is traveling on a powered conveyor travel path, it has greater crush resistance, with maximum connection points for side wall and flute member.
- A C-flute has approximately 42 corrugations per ln ft. and has a medium number of flutes per foot. This feature gives the C-flute both the features of the A type and the qualities and advantages of the B-flute. If a zero- or low-pressure conveyor is used, a C-flute carton has a low probability for SKU damage.

Carton sections are joined together to form the four side walls and top and bottom surfaces. The sections are connected with tape, stitching, or glue.

- Tape. When carton sections are taped together, a 2- to 3-in. tape strand is used. Depending on your carton specifications, the tape is made from sisal, cloth, or reinforced fiber material.
- Stitching. These cartons are stitched together. This is the least popular method.
- Glue. One side wall is overlapped onto the other side wall section and moisture-resistant glue or adhesive is applied over the entire surface. This is a popular method to secure carton sections in the order-fulfillment industry.

Coated or Treated Carton

A coated or treated carton is designed to specifications and is coated with a resin substance that hardens and increases the rigidity and durability of the carton. A treated carton has four solid side walls, a solid bottom surface, and an open top. A coated carton is used as a captive carton in a pick/pass operation, in a batched CO selection operation to transport picked SKUs from the pick area to a sort/pack area, and as a pick position bin. The carton is designed with dividers inside to handle multiple SKUs or COs. When used on a powered conveyor travel path, a rough wood or pressed board bottom exterior surface increases carton durability and conveyability. In an order-fulfillment operation, a nestable coated or treated carton reduces the requirement storage area. Holes in the bottom surface permit dirt or dust to fall from a carton. The disadvantages are increased expense and the need for increased storage area. The advantages are increased durability, ability to handle multiple SKUs per carton, standard size, and to stack empty cartons.

Plastic Tote

A plastic tote is used as a captive tote on a pick/pass line, to transport loose picked SKUs in a batched CO selection operation or a pick position bin. A plastic tote has four solid or meshed side walls and bottom with an optional open top or an attachable lid. A plastic tote is manufactured to your specifications and can include the following features:

- Openings or handle grips on two or four sides to improve handling by employees
- Removable or lockable tops
- Slots in the sides so that dividers can be installed if the operation requires a tote that can be divided into sections
- A solid surface for discreet identification placement

The bottom of a tote can be open or closed, depending on the type of SKU and order-fulfillment operation. An open type has a basket weave pattern on the bottom surface that permits dirt or dust removal. A closed tote has a solid bottom surface and has several holes in the bottom to allow dirt and dust removal and to facilitate removing the tote from a stack. Unstacking holes reduce an employee's effort to remove a tote from a stack, by decreasing the vacuum effect that is created in the stack.

If a plastic tote is used on a conveyor travel path, the bottom exterior surface has a checker-square pattern to ensure positive traction. To reduce storage space at a pack station, stacked empty plastic totes are placed in a nested or criss-crossed pattern. The desired pattern is in your tote specification. If employees move totes across the warehouse floor, to minimize transferring dirt to SKUs, before placing one tote on top of another, an employee places a paper sheet on top of the SKU. To improve picker handling of plastic totes on a pick line, pickers wear gloves.

Plastic totes are manufactured from a variety of materials:

- Acrylonitrile–butadiene–styrene. This material has good impact value and compressive strength. A tote that is made from this material is stiff and brittle. If the tote side walls are likely to become damaged as the tote is moved, this material is not preferred for your small-item order-fulfillment operation.
- High-density polyethylene. This type of tote has excellent stiffness in various temperatures.
- High-impact polypropylene. This tote has excellent load strength and can be used in a wide temperature range. However, it has brittle side walls, corners and bottom surfaces that have a tendency to crack. If the tote side walls are likely to become damaged as the tote is moved, this material is not preferred for your small-item order-fulfillment operation.
- Fiberglass-reinforced polyester. A fiberglass-reinforced polyester or thermosetting plastic tote has good compressive load strength and resists wear. The disadvantages of this type of tote are higher cost and it weakens as the fiberglass becomes worn. Advantages are low damage probability and long life.

Some factors to be considered in designing a tote for an order fulfillment operation are:

- Smooth or straight side walls
- Ribbed side walls
- Lead and trail side wall shape and lips that permit stacking and unstacking
- Meshed side walls and bottom surface
- Solid side walls and bottom surface with or without holes
- Handles
- Open top
- Lid or cover
- Nestable
- Stackable
- Combination nestable and stackable
- Rigid tote

- Collapsible tote
- Checkered exterior bottom surface
- Three-side wall container
- Location for human/machine-readable discreet identification

A smooth side wall forms a continuous wall around the top and bottom and two ends of the tote, and provides maximum interior cube. Smooth-walled totes, with or without lids, permit a standard photo-eye design on a powered conveyor travel path, and during tote queuing on a powered conveyor helps to prevent a tote from falling off the conveyor.

A tote with ribbed side walls has interior and exterior walls with waves or indents. Waves are continuous from the tote bottom surface to the top lip and between two ends. A ribbed tote side wall provides maximum stacking strength to a tote but slightly lowers interior cube capacity. Prior to selecting a tote for use, you should verify that the proposed internal dimensions of the tote match your CO cube or SKU dimensions and that the external dimensions of the tote match your powered conveyor travel path design parameters.

The shape of the lead (front) and trail (rear) walls impacts the conveyor travel path. A corrugated carton or tote has straight walls that are perpendicular to the top and bottom surfaces and to a conveyor travel path. With a treated corrugated carton or plastic tote, lead and trail wall shapes can be straight or angled. A tote with straight and perpendicular lead and trail walls is less likely when queued on a conveyor to jump from the travel path. This type of tote is easily removed from a zero-pressure conveyor when queuing occurs, but harder to remove from a low-pressure conveyor. Straight lead and trail walls permit standard photo-eye installation. Lead and trail walls that are straight and angled from the bottom surface to the top are more likely to jump from the travel path when queuing occurs on a powered conveyor travel path. This design has an angle for photo-eye installation and permits stacking of totes.

Another optional design feature for a plastic or treated corrugated tote is an attached or removable lid that allows SKUs to remain inside. If a tote has a secured lid, your tote top lip and lid specifications include a lock device that is a plastic/metal numbered seal or commercial lock.

A nestable tote is a plastic or treated corrugated tote with four side walls that are designed with tapered or angled sides. Each side is wide at the top and narrow at the bottom. At a pack station, a tote with four tapered side walls permits an employee to nest empty totes when they are not being used, to reduce tote storage space.

If totes that contain SKUs are to be stacked, to reduce SKU damage and to ensure a secure stack, the nestable tote has a lid or paper inserted on the bottom of the tote and on the top of the SKU and an employee criss-crosses other totes on the stack. The tote side walls are designed with structural strength to support stacking. A nesting tote stacking ratio for a tapered tote has a range from 2 to 1 up to

5 to 1. A nesting tote ratio means that as one tote is nested inside another tote, an additional tote adds 1/5 tote height to a nested tote height.

When an individual nesting tote or tapered tote is moved on a powered roller or skate-wheel conveyor travel path that has line pressure, during tote queue on a powered conveyor travel path, there is a high potential for totes to jump a powered conveyor travel path. To minimize this problem, a low- or zero-pressure conveyor is used and photo-eyes are installed along the travel path to control tote travel. The conveyor travel path weight design specifications are stated to handle your nested tote stack weight.

A combination tote is a plastic tote that has both stacking and nesting features. There are two types:

■ A 90-degree stacking and nesting tote is designed with tapered side walls that allow empty totes to be nested together. The structural strength of the side walls permits a tote to be stacked at a 90-degree orientation on the top of another tote. To ensure a secure stack, the tote design has protrusions on the lid and grooves on each exterior bottom surface. When one tote is placed on another, the exterior bottom surface of the top tote engages the lip of the lower tote. To minimize the risk of the top tote sliding, the lip on the lower tote protrudes to fit in the grooves on the bottom of the top tote.
■ A 180-degree stacking and nesting tote has the top tote turned a half turn from its nesting position on top of the tote below it. The tote bottom ends and side walls have alternative siderails that are designed in one direction with one tote sitting on a lower tote top. When a top tote is turned 180 degrees onto a lower tote, rails are the stacking position.

When your order-fulfillment operation uses the stacking and nesting combination tote on a powered conveyor travel path, you ensure that the tote bottom surface is conveyable. On a pallet or cart, stack height and weight should not exceed the height that causes totes to fall from a cart or pallet.

Side walls that are corrugated, treated corrugated, or rigid plastic cannot be folded down the tote interior. To reduce the required storage space when not in use, a rigid side wall tote has a stacking or nesting feature.

In contrast, collapsible side walls on a tote can be folded down into the tote interior when not in use. When a collapsible tote is empty, an employee collapses the four side walls and reduces the required storage space.

A treated corrugated or plastic three-side wall tote is used to house small corrugated cartons on a powered conveyor travel path. At a pack station, the three-side wall feature permits an employee to easily transfer a carton from the tote. The advantages of a plastic tote are that it is reusable and has a long life, handles a wide UOP mix, is conveyable, nestable, and stackable, depending on the design, and has space for discreet identification attachment.

Tote or Carton Powered Conveyor Concepts

A tote or carton powered conveyor transport concept moves a carton or tote over a fixed travel path between two locations. The travel path can be horizontal or vertical (see Figures 4.70, 4.71, 4.72, 4.73, 4.74, 4.75, 476, and 4.77).

- Transport concepts that move a carton between two locations across the same floor level include:
  - Roller conveyor
  - Skate-wheel conveyor
  - Belt conveyor
  - Drag chain conveyor
- Vertical concepts move a carton between two locations on different floors. These include:
  - Belt conveyor
  - Vertical lift
  - Whiz lift

A dynamic order-fulfillment operation that handles small-item flatwear in a tote or carton very likely uses a powered conveyor as a transport concept. Fixed travel path length, travel speed, clearances, and curves are determined by your carton or tote characteristics and conveyor travel path design parameters and vendor UOP or CO volume.

A powered conveyor is the most efficient transport concept when a large number of cartons or totes is transported between two locations in a facility. If the operation dispatches cartons or totes to a workstation and there is a need to separate and queue totes before a pack or activity station, the low- or zero-powered conveyor travel path is the preferred transport concept. An area-powered conveyor is easy to operate, dependable, transports reusable totes/cartons, has low maintenance requirements, causes minimal damage to SKU, equipment, and building, causes little employee injury, and is low cost.

Powered conveyor transport concepts all comprise A-frame with floor stands, racks, wall brackets, or ceiling hangers, a drive motor and drivetrain that propels the conveyor travel path forward, end pulley and take-up devices, side guards, operational and emergency controls and devices, power and control panels, and accessories. Beyond that, there are many differences in powered conveyor concepts.

- A powered roller conveyor with a carton- or tote-conveying surface has an open space between two roller surfaces. There are four subgroups: (a) belt-driven live roller, (b) chain-driven live roller, (c) cable-driven live roller, and (d) line shaft and O-belt ring-driven live roller.
- A skate-wheel conveyor has open spaces between and adjacent to two skate-wheels and a powered belt in the middle of the conveyor travel path.

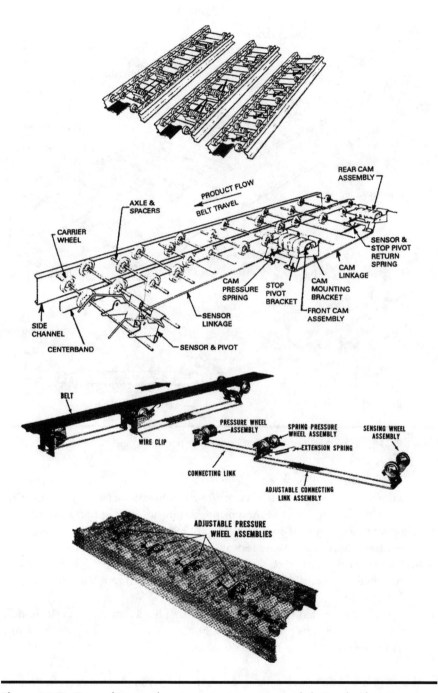

Figure 4.70   Type of tote and carton conveyor: powered skate-wheel conveyor.

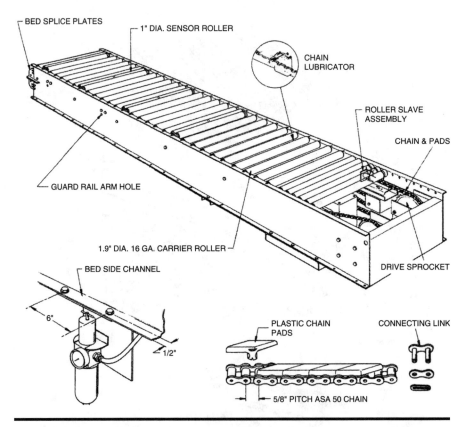

Figure 4.71 Type of tote and carton conveyor: padded chain.

■ Driven roller and powered belt and skate-wheel conveyor applications are defined to meet your specific carton/tote design parameters. The various levels for the forward force or movement pressure are (a) transport or maximum forward movement pressure conveyor, (b) low to medium forward movement pressure conveyor, or (c) zero forward movement pressure conveyor.

■ A powered belt conveyor that has one of the following: (a) slider bed, (b) roller bed or roller slider bed, (c) smooth top belt, (d) rough top belt, and (e) cleated top belt.

■ A powered drag chain or strand conveyor that has a drag chain in the middle of the conveyor travel path plus a minimum of two strands under a narrow carton/tote skate-wheel travel path surface.

The width of the travel path is determined by your carton/tote width and ranges from 12 to 61 in. The length of the travel path is determined by the distance between locations in the facility. To avoid building obstacles, other equipment or personnel walkways, a conveyor travel path design uses a curve, which varies from

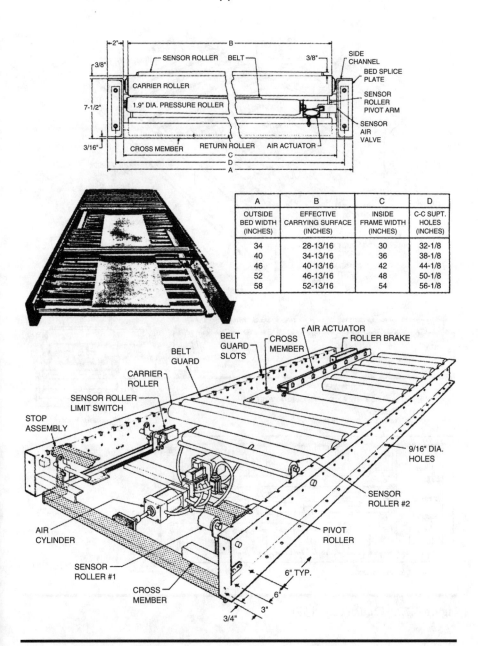

A	B	C	D
OUTSIDE BED WIDTH (INCHES)	EFFECTIVE CARRYING SURFACE (INCHES)	INSIDE FRAME WIDTH (INCHES)	C-C SUPT. HOLES (INCHES)
34	28-13/16	30	32-1/8
40	34-13/16	36	38-1/8
46	40-13/16	42	44-1/8
52	46-13/16	48	50-1/8
58	52-13/16	54	56-1/8

Figure 4.72   Air accumulation.

15 to 180 degrees. The height of the travel path height is determined by the maximum carton/tote height plus a conveyor travel path frame or motor and drive unit components and to interface with a pack station. Other powered conveyor travel path design factors are:

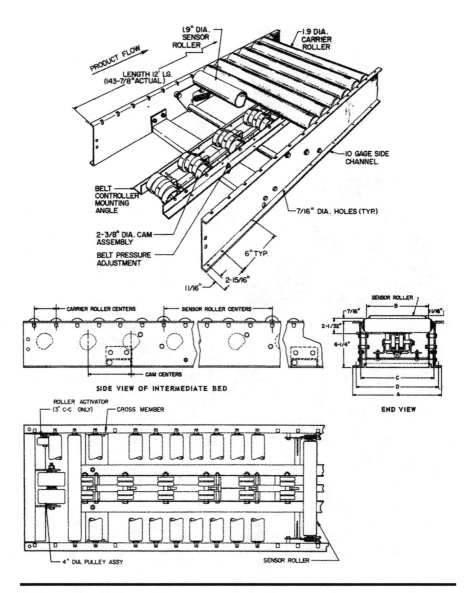

Figure 4.73   Cam accumulation.

- Firewall penetration and protection
- Ability to change travel direction
- How personnel cross a conveyor travel path
- To develop a conveyor travel path layout that shows how your proposed conveyor concept will look and operate to handle the volumes
- Develop a budget price for a conveyor travel path concept
- Prepare a bid

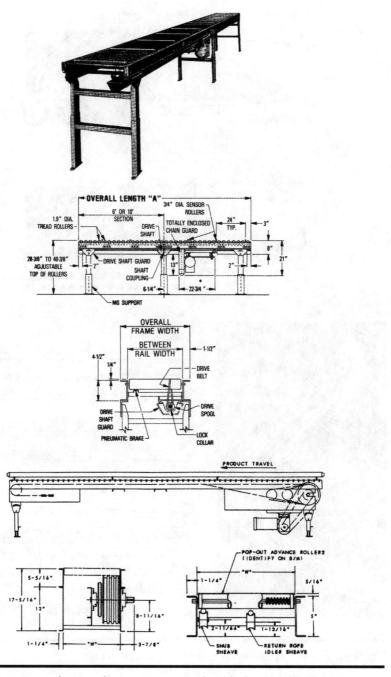

Figure 4.74 Type of tote and carton conveyor: line shaft and cable driven.

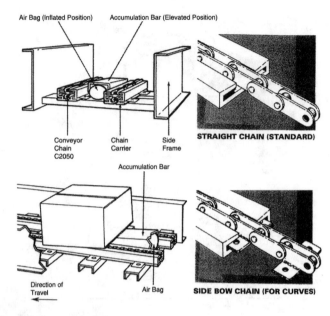

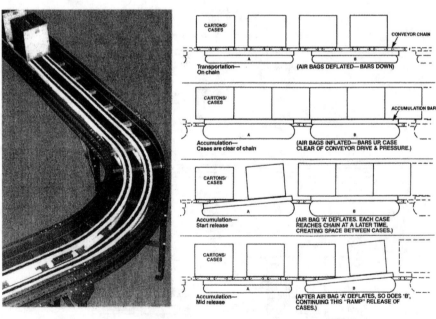

Figure 4.75 Type of tote and carton conveyor: drag chain.

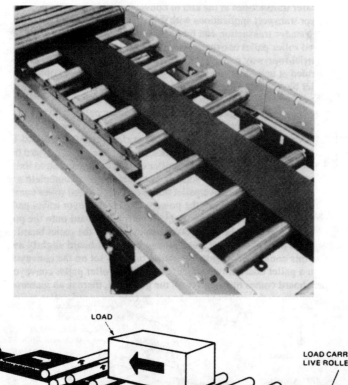

Figure 4.76 Type of tote and carton conveyor: belt-driven roller.

To design an efficient and effective powered roller, skate-wheel, or drag chain carton/tote conveyor travel path concept, guidelines are:

- Conveyor travel path layout with all charge and discharge stations.
- Conveyor travel path carrying surface.
- Skate-wheel or roller conveyor cover and spacing.
- Per your carton/tote width and height, determine a conveyor travel path width and frame depth and height.
- Angle, type, and width for curves.
- Travel path slope.
- Accessories on a conveyor travel path.

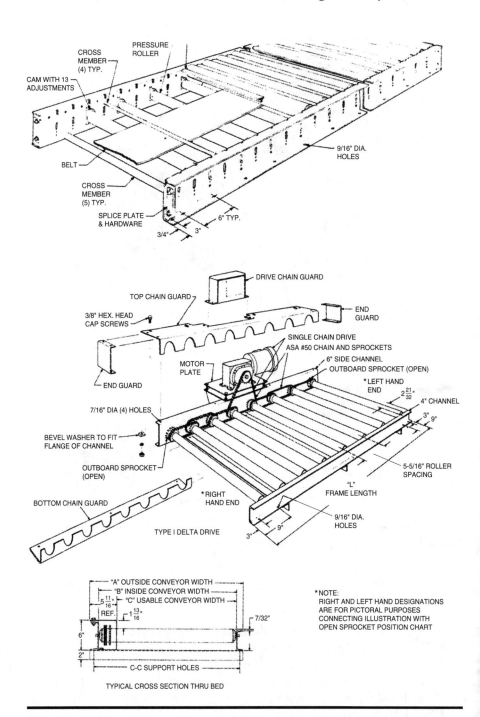

Figure 4.77 Type of tote and carton conveyor: chain-driven roller.

Live Roller Conveyor

A powered live roller conveyor is widely used to transport cartons or totes over a horizontal travel path between two locations. The smooth and continuous roller surface permits easy transfer of cartons between the travel path surface and a pack station or divert branch line spur. The rollers turn forward on an axle and propel the carton over the travel path. Types of powered live roller conveyor include (a) belt-driven low forward pressure live roller, (b) padded chain low- or zero-pressure live roller, (c) chain-driven live roller, and (d) cable-driven live roller.

Belt-driven live roller conveyor. The characteristic of a belt-driven live roller conveyor is that the bottom surface of a carton or tote comes in contact with and rides on the carrier rollers. Each carrier roller is turned on an axle and ball bearings. The two ends of the axle are placed into prepunched holes in the conveyor side frames. Depending on the transport concept, some rollers are driven or idle-type rollers. A driven carrier roller is a roller turned by a motor-driven belt that runs in the middle of the conveyor travel path, below the rollers. A driven carrier roller has adjustable pressure from cam number that pushes a motor-driven belt upward to come in contact with a driven carrier roller bottom side. A motor-driven forward belt movement forces a driven carrier roller to turn and propels a carton/tote over the travel path.

As a powered belt is moved forward in the direction of travel, live-driven carrier rollers are turned in the opposite direction of travel and move a carton/tote in the opposite direction of travel over a belt conveyor travel path. An idler roller is a carrier roller that does not have a cam under a motor-driven belt and a roller bottom does not come in contact with the motor-driven belt. With no motor-driven belt contact, an idler roller is not turned by the motor-driven belt, but an idler carrier roller does turn on axles as a carton/tote travels overit.

A closed-loop motor-driven belt receives its power as it travels through a drive motor and drivetrain assembly over the take-up device and end pulley or pulleys. This closed-loop motor-driven belt travel path is over pressure rollers, under a carton travel path to come in contact with a driven carrier roller bottom surfaces, and returns under pressure rollers to return and travel through a drivetrain.

A low forward pressure conveyor has driven carrier rollers on a conveyor travel path set with minimum belt pressure on driven carrier rollers bottom sides. There is minimum line or forward movement pressure between two cartons or totes that arE-stopped on a conveyor travel path. Low forward movement pressure means that there is very little belt pressure on driven carrier rollers.

Automated accumulation live roller conveyors. This group of conveyors has cam linkage actuated powered live roller queue conveyors and air-operated powered live roller queue conveyors that are (a) belt-driven and air-operated pressure rollers and (b) padded chain-driven and air-actuated or -operated pressure assembly.

For best queue conveyor performance, this type of conveyor is used to move cartons or totes with a rigid, flat, even bottom exterior surface, straight front, and rear side walls; all cartons or totes have bottom flaps in the direction of travel and there are three rollers under the shortest carton bottom surface.

Air-operated live roller mechanical or cam linkage accumulation conveyor. An air-operated live roller queue conveyor concept is separated into queue zones. Each queue zone has one sensor roller and driven carrier rollers. Driven carrier rollers have cam belt pressure assemblies. A driven carrier roller number is based on roller center spacings on the travel path. With a mechanical queue conveyor travel path, the queue zone (driven carrier rollers) is controlled by a sensor roller that is down line or ahead on a conveyor travel path. An automatic powered live roller mechanical or cam linkage queue conveyor concept is a queue conveyor concept that has a narrow carton/tote weight and size range. When your carton/tote length, width, and weight characteristics have a narrow mix, an automatic powered roller mechanical or cam linkage queue conveyor concept is an option.

An automatic powered live roller mechanical or cam linkage queue conveyor has line pressure or roller forward movement pressure on queued cartons or totes within a queue zone. A lead or front carton/tote has low or medium line pressure and other cartons/totes have low or minimum line pressure; this feature means that an employee has some difficulty to remove a carton/tote from a conveyor travel path.

Air-operated live roller queue conveyor. A live roller, air-operated queue conveyor travel path is separated into queue zones or conveyor frame sections. Each queue zone has two sensor and driven carrier rollers that are 3-in. center spacings along a conveyor frame length. Below driven carrier rollers are pressure belt rollers that are spaced on 6-in. centers. With an air-operated live roller queue conveyor concept, a conveyor frame length is a nominal 10 ft. long with one sensor roller at a discharge end and a second sensor roller approximately 5 ft. from the first sensor roller. Each sensor roller controls a queue zone for the trailing 5 ft.

An air-operated queue conveyor handles a wide range of carton or tote weights and sizes. When your carton/tote length, width, and weight characteristic variances are within a narrow range, an automatic powered live roller air-operated queue conveyor concept is an option.

A feature of the air-operated queue conveyor is that the line pressure (roller forward movement pressure) on queued cartons within a 4-ft. queue zone is low or zero pressure. This means that there is very low or zero (forward movement) pressure on the lead or front carton within the 4-ft. queue zone on a conveyor travel path, which allows an employee to easily remove a carton or tote from the conveyor travel path.

Air-operated padded chain queue conveyor. A live roller, air-operated padded chain queue conveyor concept is divided into carton/tote air-operated queue zones. Each queue zone has one sensor roller and driven carrier rollers. Driven carrier rollers have a corresponding air actuator assembly that regulates a plate under a padded chain travel path. Depending on the condition of the travel path, the padded chain is lowered or declined to come in contact with the carrier-driven roller's bottom surfaces. The number of rollers is based on the sensor roller and driven carrier roller center spacings. In a typical air-operated padded chain queue conveyor travel path application, a queue zone (driven carrier rollers) is controlled by a sensor roller that is ahead on a conveyor travel path. A live roller, air-operated padded chain queue conveyor handles the widest range of carton/tote weights and sizes.

A feature of the air-operated padded chain queue conveyor concept is the line pressure (roller forward movement pressure) on queued cartons/totes within a zone. As cartons/totes queue on a conveyor travel path, they have low or zero line pressure. This means that there is very low or zero (forward movement) pressure on the lead or front carton or tote within a queue zone on the conveyor and an employee can easily remove a carton/tote from the travel path.

Powered Belt and Skate-Wheel Conveyors

A powered belt and skate-wheel conveyor concept is a low-pressure, queue conveyor concept. A powered belt and skate-wheel conveyor has skate-wheels and a powered belt between the conveyor frames. The bottom of the tote or carton comes in contact with and rides on steel skate-wheels. Each skate-wheel turns on an axle with ball bearings and the axle is attached to the conveyor frames and has intermediate support members.

Skate-wheels are located below the powered belt travel path and are adjustable, so they are moved upward or downward against the bottom surface of a closed-loop powered belt. The belt is located in the middle of the conveyor travel path and rides on the adjustable skate-wheels. Having the skate-wheels in the raised position means that the forward-moving belt comes in contact with the bottom of the carton or tote, which is propelled forward over the travel path. the powered belt and skate-wheel conveyor is a low-pressure transport concept.

If a carton or totE-stops moving on the conveyor, this activates a sensor skate-wheel which is adjustable. The sensor skate-wheel and its associated mechanical linkage adjust the skate-wheels to a lower elevation. At lower elevation, the belt continues to move forward, but it is not in contact with the bottom of the carton or tote, so it is stationary on the conveyor travel path. This is why a powered skate-wheel conveyor is a queue conveyor.

Powered Belt Conveyor

A powered belt conveyor concept is preferred on a straight conveyor travel path that has a high volume of cartons or loose SKUs and controlled carton/tote movement over the conveyor travel path. When a conveyor travel path has a gap space pulled between two cartons/totes, two short belt conveyor sections accurately and consistently maintain the gap between the cartons/totes. Factors in designing a horizontal conveyor surface are belt conveyor surface type, belt ply, belt conveyor top, or surface characteristics, pulley characteristics, and carton travel path length.

Belt conveyor surfaces are: (a) belt on roller, (b) belt on slider bed, and (c) belt on slider and wheel bed.

Belt on roller. A belt on roller conveyor concept has the belt bottom surface travel over rollers that are evenly spaced along the conveyor bed or frame length. Each roller end is connected by axles that are attached to both sides of the conveyor frame. A belt load-carrying surface is turned by a drivetrain and is pulled over rollers that turn on axles. Turning rollers ensure a low coefficient of friction or drag on the travel path. The number of rollers on each conveyor frame section is determined by the shortest carton/tote length and it is a good guideline to have three rollers under the shortest carton/tote bottom surface length.

Belt on slider bed. A slider bed conveyor concept has the bottom surface of the conveyor belt travel over a smooth, continuous, and even sheet metal solid surface. If the travel path has several sheet metal sections along its length, they are overlapped to maintain the smoothest travel surface. As the conveyor belt is turned by a drivetrain, its bottom surface slides over the sheet metal surface. This method of conveyor surface support creates the maximum coefficient of friction or belt drag on a belt conveyor movement over the travel path.

Belt on slider and wheel bed. The belt on slider and wheel bed design is one where the bottom surface of the belt travels over a smooth, continuous, and even sheet metal solid surface, with a predetermined number of skate-wheels or rollers located on the travel path. The top surfaces of the skate-wheels or rollers protrude through the sheet metal surface to form the slider bed and wheel bed surface moving parts. A slider bed and wheel bed travel path provides a medium coefficient of friction or belt drag as a drivetrain pulls a conveyor belt over a conveyor travel path.

Line Shaft Roller Driven Conveyor

A line shaft roller driven conveyor concept is an economical carton/conveyor concept that is quiet, safe, and simple to operate and maintain. With a powered line shaft roller driven conveyor concept, the bottom of a carton or tote comes in contact and rides the driven carrier rollers. Driven carrier rollers turn on axles with ball bearings and axles attached to conveyor frames on center spacings that are based on the carton length. On one side of the roller is a groove in the exterior surface, over which a rubberized O-ring or band is routed (looped). With an O-ring drive belt looped over each roller top and under the locked line shaft spool, the O-ring drive belt forms an S-pattern between a roller and a line shaft spool.

As a drive motor and drivetrain and tooth sprocket turn the line shaft and each locked O-ring drive belt spool, each O-ring drive belt turns. An O-ring looped over a roller groove turns the roller forward. This forward movement propels a carton/tote over the conveyor travel path. Depending on the travel path layout, one drive motor has a line shaft section up to 100 ft. maximum. If a longer conveyor travel path length is needed another drive motor and drivetrain is added to the travel path.

If all rollers on the line shaft conveyor travel path are O-ring belt-driven carrier rollers and the line pressure exceeds the desired level during carton/tote queue, the number of O-ring belt-driven carrier rollers for each conveyor frame is adjusted to a lower number. To do this, an employee slips an O-ring belt from a driven carrier roller groove to an open space between a driven carrier roller and conveyor frame side.

Cable-Driven Roller Conveyor

A cable-driven roller conveyor concept does not permit carton/tote queuing on a conveyor travel path. To move a carton or tote over a conveyor travel path the driven carrier rollers are turned by an endless-loop chain that is threaded through a drive-train, through a cable take-up device, over a tail pulley, and across each roller snub sheave, and under a load-carrying surface over return sheaves to a drivetrain. Carrier and driven rollers are on (a) 0.5 in. centers, (b) 3 in. centers, and (c) 4.5 in. centers.

As a forward moving cable is pulled over snubs or sheaves, it comes in contact with a driven carrier bottom surface and turns a driven carrier roller forward, moving the carton/tote over the conveyor travel path. A cable-driven conveyor concept has application in a pick operation to transport lightweight cartons or totes.

Drag Chain or Strand Conveyor

A powered, drag chain or strand conveyor concept is a transport concept with some capacity for queuing. A drag chain conveyor has a carton or tote pulled by a motor-driven chain over a two- or three-strand roller or wear bar travel path. As the bottom of the carton or tote comes in contact with the top of the drag chain, the carton or tote is moved over the travel path. Two or three wear bars support the bottom of the carton or tote and provide a minimum coefficient of friction that allows the carton/tote to be moved over the travel path. A drag chain or strand-driven conveyor concept has application as a lightweight carton/tote transport concept.

Roller Conveyor Travel Path Curves

Roller curve types (see Figures 4.78 and 4.79) are:

■ A single lane conveyor curve, with one set of rollers between conveyor frames. A single roller curve can be either tapered or flat. A tapered roller conveyor curve has rollers that span the distance (length and width) between the conveyor frames. The roller diameter is smaller on the inside of the curve inside and larger on the outside. The tapered shape helps guide a carton or tote through the conveyor curve. The advantages are fewer jams as cartons or totes travel through a curve. The tapered curve requires a slightly more investment. A flat roller conveyor curve has standard rollers that span the distance (length

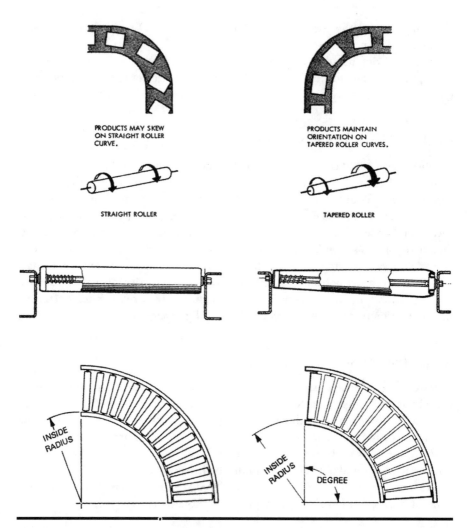

PRODUCTS MAY SKEW
ON STRAIGHT ROLLER
CURVE.

PRODUCTS MAINTAIN
ORIENTATION ON
TAPERED ROLLER CURVES.

STRAIGHT ROLLER

TAPERED ROLLER

INSIDE RADIUS

INSIDE RADIUS

DEGREE

Figure 4.78 Conveyor curves: straight roller and tapered.

and width) between conveyor frames. Flat rollers have the same diameter at the inside of the curve as at the outside. It is less expensive than a tapered curve but has potential for carton/tote jams as they travel through a curve.

■ A double lane roller curve has two sets of flat rollers and two beds with a middle support member. Each roller curve axle is connected to one conveyor frame (bed or channel) and to a middle support member. The middle support member for a roller axle has a roller set high. These features ensure carton/tote travel through a double lane roller curve, allows a conveyor surface at the same elevation between conveyor frames and allows a curved two roller set section to turn at different rates. This different roller turning rate (differential

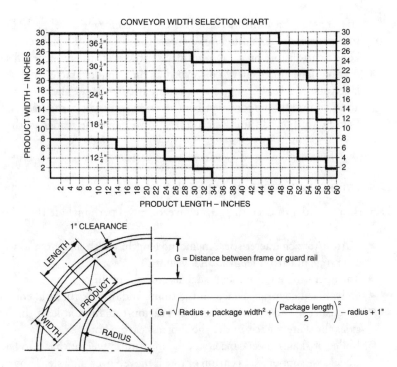

CONVEYOR WIDTH SELECTION CHART

PRODUCT WIDTH – INCHES

PRODUCT LENGTH – INCHES

1" CLEARANCE

G = Distance between frame or guard rail

$$G = \sqrt{\text{Radius} + \text{package width}^2 + \left(\frac{\text{Package length}}{2}\right)^2} - \text{radius} + 1"$$

FORMULA FOR DETERMINING DISTANCE BETWEEN FRAME OR GUARD RAILS

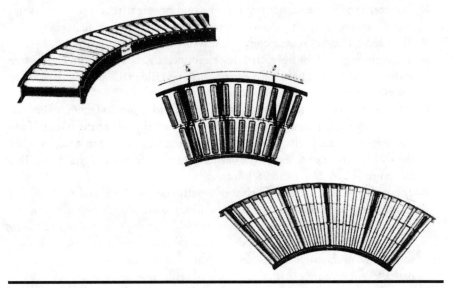

Figure 4.79 Curve chart: one, two, and three rollers wide.

effect) affects carton/tote through a curved conveyor travel path. During carton/tote travel over a differential curve, the differential concept effect tends to keep a carton/tote on a straight line as it travels through a curved section.

- A triple lane roller curve conveyor has the same characteristics as a double lane roller curve, but it has an additional curve lane. There are three rollers and two middle support members between conveyor curve frames.
- Belt curve. For additional information, we refer the reader to the belt conveyor section in this chapter.

Determining a Conveyor Travel Path

Factors to consider in deciding on a conveyor travel path include the following:

- Carton/tote characteristics, including length, width, height, and weight, and the range of carton lengths and weights.
- The operating environment, such as amount of dust.
- Galvanized coating surface on the frame and other structural components.
- To ensure a smooth and continuous carton/tote flow, at least three rollers or axles for skate-wheels under the shortest carton/tote.
- Roller or skate-wheel capacity is the maximum weight that a roller or skate-wheel axle supports as a carton or tote is moved over the axle. To avoid exceeding this capacity, the number of rollers or skate-wheel axles is increased.
- The conveyor frame capacity is the maximum weight that a roller or skate-wheel conveyor section supports between two stands or ceiling hangers.
- The need for curves on the conveyor travel path.
- The conveyor straight travel path is determined by the maximum carton/tote width, the clearance between a carton/tote and guide rails and the decision to use guide rails on the conveyor sides or frames.
- If there are no guide rails, the conveyor frame width should facilitate the widest carton or tote plus a reasonable distance from the side of the frame. If the conveyor has guard rails, the travel path should be designed to accommodate the widest carton or tote plus the desired clearance from each guard rail. This clearance is 1 to 2 in. from each frame.
- The incline or decline on the conveyor path and nose over and powered tail (transition points) determine the slope for an elevation change between the elevated end and lower end of the travel path. The slope range is from 5/8 to ¼ in. per foot and depends upon the carton/tote length and the conveyor surface. Manufacturer charts show a plan view length for a conveyor travel path elevation change, including powered tail and nose-over.
- Conveyor supports that are used to support the conveyor frames. The support concepts are (a) portable floor, (b) permanent floor, and (c) ceiling-hung supports.

- With a belt conveyor, crown and laggered pulleys are used to help ensure proper belt tracking. A crown pulley has a center higher than the pulley ends and laggered means that a pulley has rough grooves on its exterior surface.
- E-controls and stop/start device locations.

How to Cross a Conveyor Travel Path

For access and egress between a pack station and another facility location, the employee walk path and conveyor travel path intersect in the work area, and employees must be provided with a means to cross over or walk through the conveyor travel path (see Figure 4.80). Painted lines on the floor show employees where to cross a conveyor travel path. Concepts include: (a) stairs and a platform bridge, (b) a stile with a bridge, (c) a ship ladder with slats between rollers, (d) a lift gate, and (e) a bridge under the conveyor travel path.

The method that is selected is determined by the purpose of the bridge and cross-over, the conveying surface, the type of carton/tote, and the code and code height for a concept.

Incline and Decline Belt Conveyor Bridge

A conveyor travel path with an incline and decline belt conveyor combination creates a bridge between the floor and conveyor travel path support member bottom. The design has:

- From the floor level conveyor travel path a powered tail and incline belt conveyor top with a powered nose-over to create a bridge side
- A straight conveyor section for the bridge travel path
- From the elevated straight conveyor travel path, a powered nose-over and decline belt conveyor with a powered tail to the floor-level conveyor travel path

An incline and decline belt conveyor bridge concept moves cartons or totes between two locations and provides a path for employees to walk under an elevated conveyor.

Lift Gate

A lift gate section is a 3-ft. long section of a gravity conveyor travel path. In an open position, it gives an employee access to cross the conveyor travel path and in the closed position, it permits cartons/totes to travel over the conveyor travel path. Lift gates can be a (a) skate-wheel, (b) roller conveyor, (c) non-counterbalanced, and (d) counterbalanced type. For the best operation, a lift gate is spring-counterbalanced with handles on both sides. To ensure continued carton/tote flow across a lift gate section, a conveyor travel path design has:

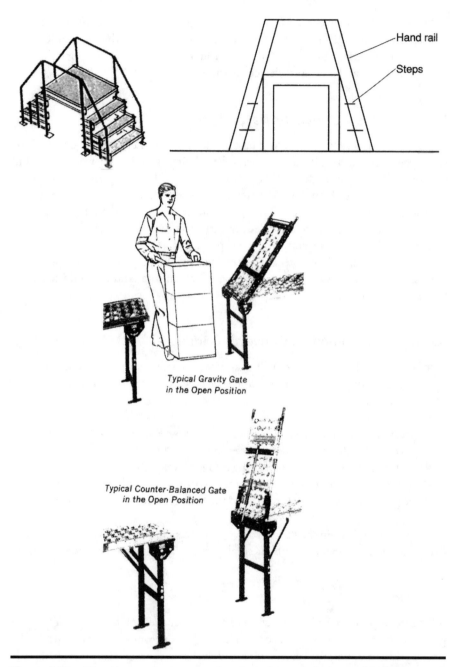

Figure 4.80 Crossing a conveyor travel path: stairs and platform, ship ladder, and spring-loaded lift gate.

- The lift gate lead end 1 in. higher than the discharge end.
- When a lift gate is used on a conveyor travel path, a 4-ft. long section ahead of the lift gate is designed with a switch that controls the electric motor that powers the belt conveyor or roller conveyor section. If the lift gate is in an open position, the switch turns off the powered conveyor section, preventing movement of cartons onto the lift gate. In a closed position, the lift gate permits electric power to the conveyor motor so that cartons/totes are propelled across the lift gate.
- When a lift gate is in the up position, a 36-in. minimum clear space is needed.
- A double conveyor travel path has double (side by side) lift gates.

For best operation, lift gate hinges are located at the conveyor charge end, have aluminum skate-wheels or rollers, anchored to the floor, a discharge end striker arm with striker cradles and cushion pads, and guard rails along both sides that do not become jammed when open. A lift gate is low cost and provides a path for employees to cross a conveyor travel path.

Stairs and Platform

In a stairs and platform concept, an employee walks up and over a conveyor travel path. Variations on this design include:

- Stairs and platform. Hardened metal stairs with handrails welded together leading up to or away from a platform. The metal surfaces are coated and the stairs and platform design meet local codes. The components provide a 3-ft. wide, or as required, bridge for an employee to walk over a conveyor travel path.
- A stile is similar to a stair and platform concept. Stile stairs to the platform are at a steeper angle.
- A ship ladder concept has slats between two rollers on a conveyor travel path and coated metal handrails. Rollers are installed between the slats, which are secured to each conveyor frame. As employees walk across a conveyor travel path, they place their feet on the slats. Before crossing the conveyor travel path, the employee makes sure that there is no carton or tote in or approaching the crossover path.

Horizontal Conveyor Travel Path Accessories

Powered live roller and some smooth top belt conveyor accessories are designed to improve carton/tote flow and control on a travel path, to minimize damage, and to reduce a conveyor travel path length. The accessories include merge and transfer devices, such as (a) a guide drum and guide wheels, (b) powered transfers, (c) powered turntables, (d) sweep plates, (e) a sweep rail or fixed diverter, (f) a traffic cop or controller, (g) stops, (h) lift gates, (i) divert devices, and (j) ball top transfers.

Merge and transfer devices are used where cartons or totes from one conveyor travel path intersect with and travel onto another conveyor travel path. Transfer devices include:

- Guide drums. A guide drum is used where a branch conveyor merges onto the main conveyor travel path. A guide drum is used to minimize the coefficient of friction or drag on a carton/tote side against guard rails at a location where two conveyor travel paths intersection. A guide drum is a 4-, 6-, or 8-in. diameter drum made from ⅜-in. thick tubing with two sets of ball bearings and a center tube and axle. The height of the guide drum is 10 ½ to 10 ⅞ in.
- Guide wheels. A guide wheel is used where a branch conveyor merges onto the main conveyor travel path. A guide wheel is used to minimize the coefficient of friction or drag on a carton/tote side against the guard rails at the intersection of the two travel paths. A guide wheel is a hardened plastic wheel 3, 4, or 5 in. in diameter, with roller bearings and a center tube and axle. The height of the drum is 1 ½ to 1 ¾ in.
- Powered 30- or 90-degree transfers. A 30- or 90-degree transfer device is used to transfer a carton or tote from the main conveyor travel path onto a branch conveyor or to move a carton or tote from a branch conveyor onto the main conveyor. travel path. The transfer can be designed as a skate-wheel, a moving belt, pop-up chain or a pusher or blade.

A powered turntable is designed to have a 180-degree travel path in a small area. The floor area is 4 ft. 6 in. by 4 ft. by 6 in. A powered turntable is a solid disk that has independent guard rails and drive motors.

A sweep plate is designed on a main conveyor travel path at merge or divert locations with branch conveyors. A sweep plate has skate-wheels and guard rails on both sides that direct a carton or tote from one conveyor travel path onto another.

A sweep rail or fixed diverter is a galvanized C-channel or skate-wheel component that is extended at an angle across powered roller or smooth top belt conveyor travel paths. A fixed diverter is secured to each conveyor frame and extends across the travel path. The carton/tote forward movement, travel speed, and diverter angle direct the carton/tote from one conveyor to another.

A powered, face deflector is used to transfer a carton or tote from a main travel path to continue to travel on the same elevation and from a branch conveyor travel path at a different elevation to merge onto the main conveyor travel path. Photoeyes and brake belt conveyor sections ahead of the deflector are used to control carton/tote travel.

A traffic cop or controller is a powered live roller or smooth top belt accessory. It has two hardened metal, spring-loaded, moveable arms that extend into and across each conveyor travel path. A traffic cop is designed to control cartons or totes traveling on separate conveyors as they merge onto one conveyor travel path. As a carton or tote on one travel path strikes the appropriate traffic cop arm, the arm is pushed to the side of the conveyor travel path the carton or tote to continue to travel on the conveyor travel path. The second traffic cop arm swings into the other conveyor travel path and prevents cartons or totes from moving onto the conveyor.

A stop on a conveyor travel path is designed to stop a carton or tote at a specific location as it travels on a conveyor. There are different types of stop: (a) a wood member between two rollers, (b) a roller end stop that is a small-diameter roller across a travel path, (c) an adjustable bar that is attached to one conveyor frame and has the ability to swing across a conveyor travel path and be locked onto the other conveyor frame, (d) a flat plate that is a preformed piece sheet on a conveyor travel path, (e) a flat plate with a crimp, (f) a flat plate with a crimp and an end plate, (g) a flat plate with a solid barrier or end, (h) a fixed bar, and (i) an air-operated pop-up stop.

Guard rails on a conveyor are used to guide a carton or tote as it travels over the conveyor travel path. The height of the guard rail is determined by the maximum carton/tote height that is to be transported on the conveyor travel path. Guard rails can be designed as:

- A solid, coated sheet metal guard rail, used on a loose SKU belt conveyor
- A wood slat, pipe, or angle iron attached to the conveyor frame
- One or two adjustable C-channels, with the closed section facing the conveyor travel path
- One or two single or meshed skate-wheel strands with adjustable levels

Other guard rail design options are:

- A flared lead end, in which the lead or front end guard rails are set at a slight angle to the outside of the conveyor travel path
- A non-flared lead end, which has the lead or front parallel to the conveyor travel path

A ball top transfer is used at a workstation or on a tight-radius, 180-degree curve. A ball top transfer is used at conveyor locations to facilitate easily movement of cartons or totes over the revolving ball top conveyor surfaces. A ball top transfer has many balls on the surface, each of which sets on ball bearings and is located in a housing on a bed. As a carton/tote is pushed or pulled by an employee, revolving balls permit it to move in any direction or over a guided travel path. A ball top transfer surface may be either a uniform ball pattern or a staggered pattern.

Vertical Conveyor Concepts

Vertical conveyors are designed to move cartons or totes between two floors, ensure the maximum number of pieces is delivered per trip, and move a vendor UOP or CO at the lowest possible cost.

Design parameters for vertical conveyors are clearances between the equipment, tote/carton, building structures, and conveyor travel path structural support members, vertical travel path slope, and interface with each horizontal conveyor travel path, required run-out, and total elevation change between each floor conveyor travel paths and safety features to include employee and fire wall or floor penetration and protection.

Vertical concepts include:

- A powered belt conveyor travel path that can be either straight or spiral
- A whiz lift
- A continuous vertical lift
- A decline slide or chute, with either a straight or spiral travel path
- A decline non-powered conveyor

Belt Conveyor

A powered belt conveyor has a low cost per ln ft., is easy to operate, easily interfaces with a horizontal transport concept, requires little maintenance, provides controlled carton/tote travel, and handles a constant, high volume. A vertical incline or decline belt conveyor concept (see Figure 4.81) can be (a) a permanent straight belt conveyor, (b) a portable straight belt conveyor, (c) a forward or reverse straight belt conveyor, or (d) a spiral belt conveyor.

A powered belt conveyor with a straight travel path has A-frame and structural support members, a powered tail that is a short horizontal belt conveyor travel path, or powered tail section on a lower elevation level, and a nose-over that is a convex section on a higher elevation level conveyor travel path, lock components, and underside guards, side guards, and controls.

The slope of the conveyor ensures controlled carton or tote movement and a constant flow between two floors. A belt conveyor travel path surface can be (a) a smooth top belt, (b) a rough top belt that increases the coefficient of friction between the belt conveyor surface and the carton or tote bottom, or (c) a cleated belt, which has upward extensions from the surface of the conveyor, which create spaces between cleats that act as a back stop for a carton or tote.

A spiral conveyor is centered around a post and gradually declines or inclines as the conveyor approaches the desired elevation or lower floor. The entire travel path length has exterior side guards. The conveyor surface is belt or hard plastic slats. A spiral conveyor travel path has less floor space, higher cost, and controlled carton/tote travel.

Whiz Lift

A whiz lift is an incline or decline vertical transport concept that requires a small lower- or upper-level surface area. The whiz lift is characterized by a powered slide

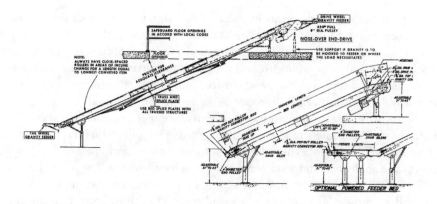

	Angles in degrees					
	5	10	15	18	20	25
Lift	Overall length, feet					
0 ft 6 in	5.7	10.3	9.4	9.0	8.9	8.6
1 ft 0 in	11.4	13.2	11.2	10.5	10.3	9.7
1 ft 6 in	17.2	16.0	13.1	12.0	11.6	10.7
2 ft 0 in	22.9	18.8	15.0	13.5	13.0	12.3
2 ft 6 in	28.6	22.3	16.8	15.0	14.6	12.9
3 ft 0 in	34.3	24.5	18.7	16.5	15.8	14.0
3 ft 6 in	40.0	27.5	20.5	18.0	17.1	15.0
4 ft 0 in	45.7	30.2	22.4	19.8	18.5	16.1
4 ft 6 in	51.4	33.0	24.3	21.3	19.9	17.2
5 ft 0 in	57.2	36.3	26.2	22.9	21.3	18.3
5 ft 6 in	62.8	38.7	28.0	24.5	22.5	19.3
6 ft 0 in	68.6	41.5	29.9	25.9	24.0	20.4
6 ft 6 in	74.3	44.5	31.8	27.5	25.4	21.4
7 ft 0 in	80.0	47.5	33.5	29.0	26.7	22.5
7 ft 6 in	85.7	50.0	35.4	30.5	28.0	23.6
8 ft 0 in	91.4	52.9	37.3	32.0	29.5	24.7
8 ft 6 in	97.2	55.7	39.2	33.5	30.9	25.8
9 ft 0 in	102.9	58.5	41.0	35.1	32.3	26.8
9 ft 6 in	108.6	61.4	42.9	36.5	33.5	27.9
10 ft 0 in	114.3	64.2	44.8	38.2	35.0	29.0
11 ft 6 in	125.7	69.9	48.5	41.3	37.8	31.1
11 ft 0 in	131.5	72.7	50.4	42.8	39.1	32.2
12 ft 0 in	137.2	75.5	52.3	44.3	40.5	33.3
12 ft 6 in	142.9	78.5	54.1	45.9	41.8	34.4
13 ft 0 in	148.6	81.2	56.0	47.5	43.3	35.5
13 ft 6 in	154.3	84.1	57.9	49.0	44.5	36.5
14 ft 0 in	160.0	86.9	59.7	50.5	46.0	37.6
14 ft 6 in	166.9	89.7	61.5	52.0	47.5	38.7
15 ft 0 in	171.5	92.5	63.5	53.5	48.8	39.8
15 ft 6 in	177.2	95.4	65.3	55.0	50.1	40.9
16 ft 0 in	182.9	98.2	67.2	56.5	51.5	42.0
16 ft 6 in	188.6	101.1	69.1	58.0	52.8	43.0
17 ft 0 in	194.3	103.9	70.9	59.5	54.3	44.1
17 ft 6 in	200.0	106.7	72.8	61.0	55.5	45.1
18 ft 0 in	205.7	109.6	74.6	62.5	57.0	46.2
18 ft 6 in	211.5	112.3	76.5	64.0	58.5	47.3
19 ft 0 in	217.2	115.2	78.4	65.5	59.8	48.4
19 ft 6 in	222.9	118.1	80.2	67.0	61.1	49.5
20 ft 0 in	234.3	123.7	84.0	70.5	63.9	51.5
21 ft 0 in	240.0	126.6	85.8	72.0	65.3	52.5
21 ft 6 in	245.7	129.9	87.7	73.5	66.5	53.5
22 ft 0 in	251.5	132.2	89.6	75.0	68.0	54.5
22 ft 6 in	257.2	135.1	91.4	76.5	69.5	55.5
23 ft 0 in	262.9	137.9	93.3	78.0	70.8	57.0
23 ft 6 in	268.6	140.8	95.2	79.5	72.1	58.0
24 ft 0 in	274.3	143.6	97.0	81.0	73.5	59.1

NOTE: Totals include 4-ft power tail and 30-in nose-over.

Figure 4.81   Incline conveyor: side view and chart.

or chute travel path and an elastic, metal-ribbed shroud that travels directly over the travel path and surrounds a carton or tote on the travel path.

Components of the whiz lift are structural support members and charge and discharge transition locations, in-feed slide or chute charge travel path and slope, an elastic, metal-ribbed shroud, and out-feed or discharge slide or chute end, drive motor and driver train, and controls.

Whiz lift structural support members provide support, rigidity, and stability to both the carton or tote and to the shrouded travel path. The structural component members are floor stands, upright posts, cross-members, and braces. The upright posts have angle iron and C-channel coated members with a foot plate. The foot or base plate has predrilled or punched holes for anchor bolt and nut attachment to the floor surface. The upright post member has pre-drilled holes to secure cross members and braces to upright posts with nuts and bolts. At the top floor level or discharge end, floor strands support the powered belt conveyor and the slide or chute travel path. On the ground-floor level or charge end, floor stands support the slide or chute travel path and horizontal powered carton conveyor travel path. The cross members and braces are added to upright posts as required to ensure rigidity and stability of the travel path.

The whiz lift travel path comprises a charge end, slide or chute, shroud, and discharge end. These components ensure that a carton or tote is transferred from a horizontal conveyor travel path onto the charge end of the whiz lift, travels over the whiz lift travel path, and is transferred from the discharge end onto a delivery floor horizontal conveyor travel path.

The charge end is designed to ensure that there are gaps between cartons or totes as they are transferred onto the whiz lift and that the cartons are positioned so that they are under the shroud. To achieve these two objectives, most conveyor manufacturers use an indexing belt conveyor concept or brake belt and meter belt conveyor concept. With a belt conveyor concept, a sufficient gap is pulled or created between two cartons or totes and the belt conveyor surface ensures maximum carton or tote forward movement onto a whiz lift slide or chute travel path.

The upper end of the whiz lift is a slide or chute with a curved angle piece of sheet metal or hardened plastic, which adds a low coefficient of friction to the bottom of a carton or tote. The curved angle ensures that forward movement of the conveyor exerts pressure on a carton or tote and pushes it onto the charge end of the slide. When a carton or tote is pushed forward, the carton or tote is surrounded by the shroud and is pulled onto the whiz lift travel path.

The upper end of the whiz lift interfaces with the horizontal conveyor travel path. If the elevated floor is the charge end, the shroud ensures the forward movement of a carton or tote onto the decline travel path. If the elevated floor is the discharge end, the shroud ensures forward movement from the whiz lift travel path onto the floor horizontal conveyor travel path.

An elastic, metal-ribbed shroud resembles envelopes that are back to back. This characteristic allows the shroud to handle a wide range of carton or tote sizes. The shroud is a closed-loop component, with two side sections that are pulled through a C-channel travel path or track. During vertical travel, the metal ribs fall in a downward direction due to gravity. As an empty shroud is pulled over a vertical travel path, the metal ribs fall flat against the slide or chute travel path. If a carton is within the shroud on the vertical travel path its top and sides are surrounded by the shroud. The forward movement of the shroud and the low coefficient of friction on the travel path surface pull a carton or tote over the vertical travel path. To minimize UOP damage, cartons have sealed top and totes have a secured lid.

The whiz lift drive motor and drivetrain ensure that the shroud is pulled over the travel path, pulling cartons or totes from the charge floor level conveyor travel path over a slide or chute travel path to the discharge floor. As a drive motor turns a tooth sprocket, the tooth sprocket pulls the open-link closed-loop chain forward that turns the elastic ribbed shroud pulley sprocket. As the tooth sprocket turns, the pulley interfaces with the two sides of the shroud and as the pulley turns, the shroud is pulled over the travel path. The drive motor and drivetrain pull a shroud with a carton or tote vertically over a 50-ft. travel path and can move 40 cartons/totes per minute.

A whiz lift has controls that stop and start the shroud and the horizontal conveyors on each floor. These controls are manual, E-stop push buttons or cords that are strategically located at each floor, and photo-eyes along each horizontal conveyor travel path.

The disadvantage of the whiz lift are that it does not move a high volume of cartons. Advantages are that it requires a small area, is easy to operate, and has few moving parts.

Vertical Reciprocating Lift or Conveyor

An incline or decline vertical reciprocating lift or conveyor concept (see Figure 4.82) is an electric, motor-powered fixed travel path concept. The fixed travel path has a load-carrying surface with sufficient space to support one carton or tote as it is moved between two levels in a facility.

The unique feature of a vertical reciprocating lift concept is that it is designed to move a carton or tote without an employee traveling on the load-carrying surface. This feature separates the vertical reciprocating lift concept from a standard freight elevator.

A vertical incline or decline reciprocating lift is made up of:

- The building area
- Structural support frame or members
- A number of masts that are either (a) straight, parallel to a wall or (b) at an angle to a wall
- Carriage, platform, or load-carrying surface and loading delivery locations
- Drive motor and drivetrain location
- Controls and safety features

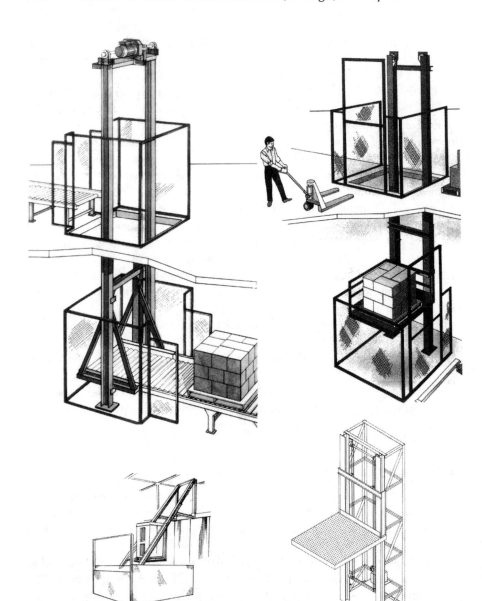

Figure 4.82 Vertical reciprocating lift.

An employee or horizontal powered belt or roller conveyor transports a carton onto the vertical lift reciprocating platform. As the carton travels on the horizontal in-feed conveyor travel path, a barcode scanner reads the carton barcode and sends the information to a microcomputer, which is programmed to move the vertical reciprocating lift to a transfer location. As the reciprocating vertical lift platform arrives at the transfer location, the carton moves forward over the conveyor travel path and is supported by the vertical reciprocating lift platform. After a sensing device verifies that a carton is properly positioned on the platform and the safety gate is closed, the platform is automatically moved from the charge location to the discharge location.

The reciprocating vertical lift platform travels over a vertical travel path to the discharge level, where the horizontal conveyor interlock or interface program ensuresthat the horizontal discharge conveyor travel path is ready to accept the carton and the protective enclosure gates are open. As the reciprocating vertical lift platform arrives at the location, its elevation matches that of the horizontal conveyor section or facility floor level sill. After the carton is transferred from the platform onto the horizontal conveyor, the reciprocating vertical lift is ready to be moved empty to another facility floor or remains at same location to move another carton.

The disadvantages are that the reciprocating lift has a fixed travel path, handles a uniform carton or tote load, is limited to interface with two or three floor levels, firewall or elevated floor penetration has fire protection and personnel safety features, and some models have an employee or powered conveyor unload and load cartons. The advantages are that it can be a manual or automatic concept, does not require an employee to operate the concept, handles a medium carton volume, flexible travel path configuration, two-way carton movement, and requires a small area.

Continuous Vertical Lift Concept

A powered continuous incline or decline vertical lift concept (see Figure 4.83) is designed specifically to move cartons or totes between two facility floor levels. One unique characteristic is that the continuous vertical lift is designed to move cartons or totes without an employee traveling with the carton. A continuous incline or decline vertical lift has flight bars or slats, which form carton- or tote-carrying surfaces or a platform that moves a carton or tote over a fixed travel path. As a continuous vertical lift travels between the charge and discharge or assigned facility floor level, the horizontal conveyor on the lower floor moves a carton forward onto one of the continuous vertical lift carton-carrying surfaces (series or bars or slats). An electric motor-driven shaft with two tooth sprockets continuously turns two closed-loop endless transport chain strands. Each transport chain strand is routed snugly over a drive motor tooth sprocket. The two transport chain strands and two secondary chain strands pull forward these carton-carrying surfaces from the lower floor level over the fixed vertical incline travel path to the upper floor level discharge location. The carton-carrying platforms travel over a vertical incline travel path and continuously arrive and momentarily stop at the appropriate elevated floor level to discharge location (one elevated floor level) or multiple elevated facility levels. After a carton on a continuous vertical lift carrying platform (series of slats) arrives at the appropriate discharge elevated facility floor level, a carton is discharged (transferred

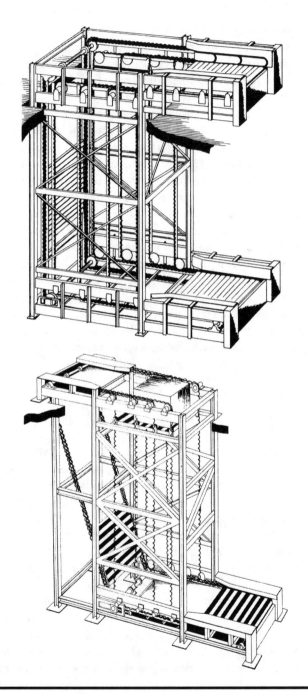

Figure 4.83 Continuous vertical lift: C-shaped and Z-shaped.

or moved forward) from the continuous vertical lift platform onto the elevated floor level powered horizontal carton roller conveyor travel path.

Depending on the elevated facility floor level design, carton flow pattern through the facility and operational requirements of each facility floor level, the continuous vertical lift configuration is a C- or Z-shaped travel path.

The components are:

- A fixed building vertical travel path location or shaft
- Structural support or frame members
- Two endless closed-loop open link transport chain strands with the two front links attached to a carton- or tote-carrying surface
- Two endless closed-loop open link secondary transport chain strands for supporting the carton-carrying surface over the vertical travel path
- A drive motor and drivetrain with a shaft that has four tooth sprockets, two for the main carton transport chain strands and two for the secondary carton transport chain strands
- Idler tooth sprockets at key chain travel path locations
- A series of flight bars or slats that are the carton- or tote-carrying surface
- Carton platform surface lift dogs or bars
- Lower- and upper-level horizontal conveyor travel path sections
- Wire-meshed or solid vertical travel path guards that include run-outs, and, in most continuous vertical lift applications, the lower floor level horizontal conveyor travel path has a dog house or fire protection device
- Optional vertical lift travel path configurations include:
  - A C-shaped configuration that services a single facility floor level or facility with multiple floor levels horizontal conveyor travel paths
  - A Z-shaped configuration services a single facility floor level horizontal conveyor travel path
- Travel path and safety controls

A Z-shaped travel path is one in which a carton on a facility floor leaves the horizontal conveyor travel path, enters a continuous vertical travel path facing one direction of travel, moves over the vertical lift travel path and is discharged onto a horizontal conveyor travel path facing the same direction of travel. In a Z-shaped continuous vertical travel path configuration, the design characteristics and travel path configuration are one carton charge level and one discharge level.

A C-shaped travel path is one in which a carton on a facility floor leaves the horizontal conveyor travel path, enters a continuous vertical travel path facing one direction of travel, moves over the vertical lift travel path, and is discharged onto a horizontal conveyor facing a different direction of travel. In the C-shaped continuous vertical travel path configuration, the design characteristics and travel path configuration are one carton charge level and multiple (up to five) discharge levels.

Factors that determine whether a Z- or C-shaped configuration is selected are:

- Facility elevated floor design and location
- Available space on each floor level
- One or multiple facility floor levels or delivery locations
- Operational requirements for carton one-way or two-way travel over a continuous vertical travel path between the various facility floor levels
- Operational requirement for a specific carton direction of travel on a facility floor level
- Carton size

As a carton travels on a horizontal conveyor, the continuous vertical lift concept interlock or interface program ensures that a vertical lift arrives at a transfer location, where the carton front and the front continuous vertical lift platform slat members start to interface. As the continuous vertical lift slat members move forward over the travel path, the carton moves forward and becomes supported by the continuous vertical lift slat members and the two carton-carrying platform lift dogs are supported by the carrying lift bar. Sensing devices on the lift verify that the carton is properly positioned on the platform, which is then automatically moved upward by the forward movement of the two transport chain strands and two secondary chain strands.

As the vertical lift arrives at the elevated floor or discharge level the interface program ensures that the horizontal conveyor travel path is ready to accept the carton. As the lift arrives at the transfer location, the carton front end and discharge horizontal powered conveyor travel path start to interface. As the carton moves forward it is supported by the horizontal conveyor travel path. After a carton is transferred from the platform slat members onto the horizontal conveyor travel path, the two load-carrying platform lift dogs are disengaged from the load-carrying platform secondary chain strands lift bar and the two secondary transport chain strands travel over the turn tooth sprocket. As the two main transport chain strands travel over the travel path, the load-carrying platform slats are flat and are traveling toward the drivetrain.

The disadvantages are high capital investment, fixed travel path, it handles a uniform carton or tote. The C-shaped configuration is limited to five discharge levels, and the Z-shaped has one discharge level. The lift penetrates the fire barrier in the facility, and there are personnel safety features. Two-way travel over the continuous vertical travel path requires additional controls and conveyor travel paths. The advantages are automatic concept, minimal employee activities, Z- or C-shaped travel path configuration design, and requires a small area.

Forklift Truck

A forklift truck vertical transport concept moves cartons or totes that are stacked onto a pallet. With the pallet on the forklift truck forks, the pallet is raised to a higher floor level, where the forks extends the pallet through a hand rail and kick plate opening or onto an open space in front of a safety gate, and deposits the pallet. On the elevated floor a pallet truck transports the pallet to the assigned location.

Design considerations are:

- The elevation of the forks above the lower-level floor that is required to complete a pallet transfer to the higher level
- Enough floor area to permit the forklift truck to complete a right-angle turn
- Forklift truck fork extensions have sufficient depth to deposit the pallet on the upper floor
- Proper personnel and fire protection on the upper floor
- Options for transferring the pallet onto the higher floor, including
  - A one-way bi-parting gate
  - A coral handrail with kick-plates
  - A double-acting safety gate

The forklift truck is not the best vertical transport concept for a high-volume, dynamic operation, but is considered for a low-volume operation with few moves.

Freight Elevator

A freight elevator is an electric-driven motor concept that moves cartons or totes that are stacked on a pallet. The freight elevator uses a hydraulic system, geared system, or gearless system with a rope to move a platform vertically through a fixed travel path between two or more facility floor levels. The freight elevator is designed so an operator or employee controls the vertical movement of the elevator platform and opens and closes the elevator door. To move an elevator platform, an operator activates a hydraulic oil pump, geared winding drum, or high-torque drive motor to turn a shaft with a sheave that turns and pulls a rope and moves a counterweight.

In most facilities, a freight elevator is used as a vertical transport concept for pallets, forklift trucks, or other equipment. If the operation has an automatic vertical transport concept, a freight elevator is a back-up transport concept, and is used to transport personnel.

Decline Slide or Chute

A decline chute and slide concept is a non-powered travel path for loose SKUs or cartons/totes. Movement of the SKUs or cartons/totes is uncontrolled between two locations. Components include:

- A carton/tote travel path that is flat, concave, and spiral
- Charge and discharge locations
- Side guards, an underside runner, and structural support members

- Photo-eyes to control in-feed
- Horizontal conveyors on each floor

The slope of a typical chute is between 10 and 30 degrees. The travel path surface can be sheet metal, galvanized or coated sheet metal, plastic or fiberglass, orplain or coated wood. Chutes are discussed in more detail in the section on transport concepts earlier in this chapter.

Decline Non-Powered Roller or Skate-Wheel Conveyor

A decline non-powered roller or skate-wheel conveyor concept is a non-powered travel path has along which the movement of cartons/totes between two locations is not controlled. Components include the carton/tote travel path, which is made up of non-powered rollers or skate-wheels, charge and discharge locations, side guards and structural support members, photo-eyes to control in-feed, horizontal conveyors on each floor, and a run-out on a lower elevation level. A non-powered roller or skate-wheel slopes between 3 and 6 in. for every 10 ln ft.

Carton/tote travel over the decline can be controlled in one of the following ways:

- A tilted carton/tote travel path on which the side of the carton/tote rides against the side guard rail, which increases the coefficient of friction
- Restricting the turning of the roller or skate-wheel, which increases the coefficient of friction on the bottom of the carton/tote
- Plastic strips that hang down into the conveyor travel path and rub over the tops of the cartons/totes to increase the coefficient of friction

Changing Direction of Travel

The direction a carton or tote faces on a conveyor can be changed in one of the following ways:

- Manually
- Using a mechanical turntable
- Using a roller conveyor curve with an inverted tapered roller
- Using a powered turntable
- Using two sets of skewed rollers and two extended short side guards

A tote or carton can be turned by an employee, who physically lifts and turns the tote or carton from a conveyor travel path and returnsit to the conveyor. This action rotates the direction of travel so that the carton label faces a different side as the carton continues to travel on the same conveyor travel path. Disadvantages are low volume, requires an employee, slow activity, and employee injuries. The advantages are no investment and no floor area requirement.

A mechanical turntable is an automatic section of a powered conveyor with sensing devices that receives a carton or tote, turns it 180 degrees, and powers it from the turntable back onto the conveyor section. As a carton or tote travels on the powered turntable, the label faces the near side of the conveyor. When it leaves the turntable, the label is facing a different direction. To ensure controlled flow onto a turntable, a brake belt conveyor section ahead of the turntablE-stops and starts forward movement of the totes or cartons. The disadvantages are investment, control system, and maintenance. The advantages are medium to high volume, few employees, and no employee lifting injuries.

An inverted tapered roller on a curve has a specially designed 90 degree curve between an in-feed location and the out-feed conveyor. There are inverted and standard tapered rollers on the curve. The width of the curve is designed to handle the diagonal length of a tote or carton as it turns on the curve. A carton/tote on a main conveyor travels across a 45 degree curve onto a straight conveyor travel path oriented in the long direction with the label on the near side. As the carton or tote travels across the curve, the inverted tapered rollers start to turn the lead end. When the carton or tote has traveled completely through the curve, the lead end has become the tail end and the label is now facing the far side.

The disadvantage of the inverted tapered roller is that there needs to be space for 20 ft. between the two sections. The equipment requires a significant investment. The advantages are that it handles all types of cartons/totes, and a medium to high volume. No employees are needed to man this concept, and lifting injuries are reduced.

Another concept for turning a carton or tote on a conveyor has a skewed, live (powered) roller conveyor that discharges cartons onto a wide straight conveyor travel path section with an extended conveyor side guard, a short, straight, powered roller conveyor section, followed by a second skewed conveyor section that discharges cartons onto a wide straight conveyor travel path section with an extended conveyor side guard. As a carton or tote travels through the first skewed section of the conveyor, the lead end is turned 90 degrees. On section 2, the carton or tote travels in the wide direction of travel to section 3, where the lead end is turned 90 degrees. This means that after traveling over the three conveyor sections, a carton or tote has been turned 180 degrees.

The skewed live rollers and extended guard rail concept that has a specific arrangement. The conveyor travel path on the far side has skewed rollers that direct a carton toward the far side of the conveyor travel path and against a guard rail. As a carton or tote travels along the smooth, continuous guard rail, the lead end comes in contact with the rounded and extended far-side guard rail section that protrudes into the conveyor travel path. This contact, with the forward movement of the roller, and the skewed roller direction turn the carton or tote 90 degrees.

A turn module has two powered roller conveyor sections, each with a set of skewed pop-up wheels that turns a carton or tote. A carton/tote turn module is designed with:

- An in-feed conveyor section that has pop-up wheels set between two rollers perpendicular to the direction of travel of a carton or tote. To ensure controlled carton travel through the pop-up wheel section, each pop-up wheel entrance has a pop-up stop to stop or start carton travel onto an active pop-wheel section. When the wheels are activated, they rise slightly above the conveyor surface and come in contact with the bottom of a carton or tote, moving it from the first conveyor travel path to the second. When a pop-up wheel is not activated, its conveying surface is slightly below the conveyor surface and does not come in contact with a carton or tote.

- A second conveyor travel path directly behind or adjacent to the first conveyor travel path. The second conveyor travel path also has pop-up wheels, at a 45-degree angle to the rollers. As a carton or tote moves across the pop-up wheels, the pop-up wheels turn, and this changes the orientation of the carton or tote on the second conveyor travel path. The 180-degree turn means that the lead end on the first conveyor travel path becomes the trail end on the second conveyor travel path. The features are additional moving components, wide conveyor travel path, carton/tote is turned in a short travel path distance, and completes a 180-degree turn.

Moving Cartons or Totes to the Side of the Travel Path

In an order-fulfillment operation, a carton or tote on a conveyor is required to be on a specific side of the conveyor to complete an order-fulfillment activity. There are various techniques to control a carton or tote as it travels on a conveyor. These include sleeve-wrapped or taped rollers, angled deflectors, skewed rollers with a spring-loaded guard rail, pop-up rollers or wheels, and a tilted conveyor travel path.

Sleeve-Wrapped or Taped Rollers

A sleeve-wrapped or taped roller can be used on any roller-driven conveyor travel path. When a tote or carton must be positioned on one side of a conveyor travel path, a number of rollers are wrapped spirally with grit-covered tape. The tape pattern on each roller is toward the desired side of the conveyor. This tape pattern moves the tote or carton toward a conveyor travel path side. With each progressive roller, as the bottom of the carton or tote comes in contact with the grit tape, the carton or tote is directed across the travel path toward the desired side of the conveyor.

Angled Deflector

An angled deflector has a skate-wheel guard conveyor or channel guard rail that reduces the coefficient of friction on a carton/tote side. The deflector is located on the far side of the conveyor and projects out toward the near side. As a tote or carton travels on the conveyor (roller-driven) and comes in contact with the deflector, the travel speed, movement, low coefficient of friction on the smooth metal roller surface, and the deflector angle move the tote or carton from the far side to the near side of the conveyor.

Spring-Loaded Guard Rail

Skewed rollers with a spring-loaded rail concept has the near-side conveyor travel path raised to tilt toward a side guard, spring-loaded skate-wheel guard rail, and fixed guard rail. The concept is located on a section of the conveyor ahead of an order-fulfillment activity station. As a carton or tote travels onto a conveyor travel path section, skewed rollers and a slightly tilted conveyor travel path direct it from the near side to the far side of the conveyor. As a carton or tote travels over a conveyor section, an angled spring-loaded skate-wheel guard rail ensures that it travels over metal rollers with a low coefficient of friction and is directed toward the guard rail on the far side of the conveyor. An angled, spring-loaded skate-wheel is set for the widest carton/tote dimension. As a carton or tote completes travel across this conveyor travel path, it rides along the far-side guard rail.

Pop-Up Rollers or Wheels

Pop-up wheels or rollers are set in a series in an open space between two powered rollers. A pop-up wheel or pop-up roller concept is a conveyor travel path section ahead of an order-fulfillment activity station. As a carton or tote travels onto a pop-up wheel or roller conveyor section, the pop-up wheels or rollers come in contact with the bottom of the carton or tote directing it to the assigned conveyor travel path side. With a pop-up roller concept, rollers elevate and move a carton or tote from the main travel path onto a divert travel path. When a pop-up wheel or roller is not activated, it lies slightly below the conveyor surface and is not in contact with the bottom of the carton or tote. The features are directs a carton/tote to a proper conveyor travel path location, investment, additional conveyor moving components, maintains a conveyor travel path window, and completes a carton/tote redirection on a conveyor travel path in a short distance.

A Tilted Conveyor Travel Path

A tilted conveyor section concept is used on a non-powered, padded chain or line shaft live roller conveyor travel path. With a tilted conveyor travel path, the far side of the conveyor has a slightly higher elevation than the near side. As a tote or carton travels across a tilted roller conveyor section, its forward movement, the conveyor travel path slope, and solid metal roller surface causes it to move from the far side to the near side.

Small-Item/Flatwear Batched Picked SKU Sort Activity

Sorting small-item SKUs that have been picked as batched COs is the heart of a batched CO pick concept. It ensures that SKUs that have been picked and labeled for a CO are separated from the mixed CO group. Sort activity in a small-item operation includes (a) picker concepts, (b) central manual concepts, and (c) central mechanized concepts.

Each sortation concept involves a method to determine a batched or grouped CO quantity, batched or grouped CO release, or picker control concept and picked SKU transport concept. The concepts differ in the method that is used for pick and sort instructions, picking CO SKUs, and completing the sort activity.

With a batched or grouped CO handling concept, a pick, transport, and sort concept is designed according to the piece quantity that is handled at each workstation. The workstation that handles the lowest piece volume establishes the productivity for the entire small-item order-fulfillment operation. For an order-fulfillment operation with a sortation activity, methods to determine the number of SKU pieces that make up a batch or group are picker driven or sortation station driven. For additional information on the batched pick concept, we refer the reader to the previous section in this chapter.

A transport concept moves batched CO picked SKUs either in a tote or loose from a pick area over a non-powered or powered travel path to a sort induction station. For additional transport concept information, we refer the reader to the transport section in this chapter.

Sort Instruction

Each small-item flatwear sort concept has a human- or human/machine-readable sort instruction. With manual pick sort concepts, a SKU sortation instruction is a human-readable code that is printed onto a paper document, label, or precoded SKU. A paper document or label indicates a CO discreet sort number.

With a manual, centralized sort concept, the small-item sort activity has:

- Human-readable label (or precoded SKU) on each CO picked SKU. Labeled SKUs are mixed in a tote or loose on a belt conveyor. Each SKU has a paper sort label that is attached to its exterior surface. If all SKUs have a SKU code, the code is the sort instruction.

Batched CO picked SKUs are separated by SKU code. Each unique SKU is picked into a specific tote or into a specific compartment in a separated tote. With the latter approach, a SKU sort instruction is either a human-readable printed paper document or a bar code label/RF tag, which is scanned/read and used to activate a sort location or sort-to-light display.

With a mechanized, centralized sort concept, a small-item sortation activity has either a human/machine-readable code on each label or a SKU code attached to the exterior of each SKU and in the sort is read by a bar code scanner/RF tag reader.

Paper Printed Document

A paper document is used in a sort in a pick aisle, or at a pick aisle end concept that has a machine-printed document for each CO batch. A paper document has columns and rows. The first column shows each SKU pick position. Each subsequent column across the top of the paper document is a CO discreet alpha or numeric identification. At a sort station, this CO discreet identification appears on a CO sort location. The first row along the side of the document is used to print a CO discreet alpha or numeric identification. Each subsequent row is used to show a pick position for a SKU on a CO in a batch. On a paper document, a computer prints only SKUs that are on a CO and in a specific pick area. Across each row under each column is the SKU piece quantity for each CO.

To determine a batched picked SKU sort location, a picker or sort employee reads the SKU pick position on a paper document and looks on a paper sort document to determine a CO SKU. Per a CO SKU quantity that appears under each column, a picker or sort employee locates a CO discreet number on a sort cart/area bin and transfers the correct SKU quantity from the pick tote into a CO sort location. This sort activity is repeated for each SKU on a printed sort document and for each tote or unique SKU.

Sort or Put to Light

With a sort or put to light instruction, the sort concept has each SKU in a separate tote or tote section and, for each SKU, a bar code/RF tag either on a paper document or on the SKU itself. A tote containing the SKUs is delivered from a pick area to a sort station, where each CO discreet identified carton or tote has been assigned (scanned) to a specific human/machine discreet identified sort location. Each CO sort location has a sort or put to light display and a paper printed pack document is attached to each CO shipping carton or sort location. After a sort employee removes a paper document with a bar code or a bar code labeled/RF tagged SKU from a pick tote, with a hand-held scanner/reader a sort employee hand scans the bar code/RF tag. A bar code scanner sends bar code/RF tag information to a computer. Per a CO SKU, the computer activates the appropriate sort or put to light display at each CO sort location. The activated sort or put to light display indicates a CO or sort location that has a SKU and CO SKU pieces. After a sort employee transfers the correct number of SKUs from a pick tote to a CO sort location, he or she presses the appropriate CO sort completion button on the sort or put to light display. The activity signals to a computer a CO sort activity completion for a particular SKU.

On a sort-to-light display screen, the sort employee is instructed to repeat the pick activity if the CO required two or more of those SKU pieces. If the display screen is blank or shows zero, the CO has a different SKU, or is completed, in which case the employee transfers the completed CO to a take-away conveyor travel path.

Machine-Printed, Self-Adhesive Sortation Label

A self-adhesive, machine-printed label is a human/machine-readable label that is created by a computer-controlled printer. For each piece on a CO batch, the printer prints a label with alpha characters and numeric digits plus a bar code/RF tag that discreetly identifies a SKU pick position and a CO sort location. Important information on a CO pick label face includes:

- SKU pick position and SKU description
- Customer name and address
- CO discreet alpha or numeric identification or sort location
- Bar code/RF tag

In a pick area, a picker reads the SKU pick position on the label face, travels to the pick position, removes the backing from the self-adhesive label, and applies it to a SKU piece. Per an order-fulfillment concept, a picked and labeled SKU piece is placed into a tote or directly onto a belt conveyor for transport from the pick area to the sort/pack area.

In a manual sort/pack concept, after a tote arrives in the sort area, an employee picks up a SKU, reads its discreet sort identification, and matches the SKU identification location to a sortation location discreet identification. With a match, a sort employee places a SKU into the appropriate CO sort location.

In a centralized, mechanized sort concept, SKUs arrive at an induction station either in totes or loose on a conveyor. At the induction station, depending on the mechanized sort concept, an induction employee either places:

- Each labeled SKU, with a label face up, directly onto a sort concept load-carrying surface or travel path section.
- Each SKU with label face up onto an induction conveyor belt for transfer onto an empty mechanized sort concept load-carrying surface or section.

Precoded SKUs

A precoded SKU is a SKU that has a human/machine-readable code attached to each piece. A vendor or company employee at the receiving dock ensures that each vendor-delivered UOP has a human/machine-readable code attached.

Customer Order/Picked Piece Sort Location

A small-item, flatwear apparel sort concept has an employee or mechanized sort device to separate an individual CO identified SKU from a mix of SKUs that have been picked for several COs. This sort activity collects SKUs that are picked and labeled for a CO into the CO discreetly identified sort location. This CO location is a temporary holding location until the sort process has been completed for a CO batch.

In an order-fulfillment operation that has a sortation activity, CO sort location options are:

- A sort side or window with a minimum of two or three openings, a solid bottom surface, two solid or meshed opening side walls and a front opening that has a removable barrier that is solid and runs either the full height or half the height of the window. Alternatively the barrier is solid, with a wire mesh opening or clear plastic window that permits SKUs to flow onto a pack table surface.
- A sort side or window and structural support members and surface to support a CO shipping carton. Cartons have an open top with flaps down, or an open top facing toward a sort window with flaps up.
- An alternative is an open plastic container facing up.

CO location design factors are SKU physical characteristics, including length, width, height, weight, crushable or fragile, peak CO volume, cube and pieces, sort method, to check a CO picked and sorted SKU accuracy, design parameters, and operational activities to place filler material in a CO shipping carton and to seal the carton and CO shipping carton make-up, replenishment, and identification activities.

Sort Location or SKU-Holding Device

In both manual and mechanized sort concepts, SKUs are sorted by CO into a queue location that is a CO sort location. The interior space of a sort location is designed to hold a SKU cube or piece quantity that is associated with a CO SKU piece quantity or cube.

SKUs Held in a Sort Chute/Slide

One sort location concept is to hold SKUs in a chute, with two solid or meshed side walls, a solid bottom surface, a front opening barrier, and an optional solid top. After a batched CO sort activity is completed, a sort/pack station employee performs a final sort. This final sort activity has a packer separate the SKUs for each CO according to a paper document (CO pack slip). After a final CO sort activity, the packer transfers the appropriate SKUs into a CO shipping carton.

In a centralized, mechanized sort concept, each sortation location has one sort window interface with a sort transfer device. A sort window is separated into three sort locations by fixed-position solid divider or a solid moveable flipper into three sort locations. With a chute and door at the chute discharge end, the sort location is designed to hold a SKU cube or piece volume one third to one half of the sort location cube. A step in the middle of the chute increases SKU or cube capacity. For additional information, we refer the reader to the section on chutes earlier in this chapter.

The sort location holding concept is made up of the following elements: (a) a human- or human/machine-readable code on each SKU; (b) a packer to complete final sort and transfer activity; (c) good picker productivity because of multiple locations within each sort location; (d) during SKU transfer from a sort location to a pack station, a packer performs a check activity to ensure the accuracy of the CO picked and sorted; (e) the packer determines shipping carton size and makes up a CO shipping carton, or the computer suggested carton size that is printed on a CO pack slip or shown on a display screen, and made-up cartons are delivered to a pack station; (f) there are few sort locations, but they hold a large SKU capacity or cube and have photo-eyes that detect if the chute is full or partially full; (g) a large sort and pack area; and (h) the sort activity has a low impact on SKUs.

SKUs Held in a Customer Order Shipping Carton or Captive Tote

With this sort location concept, an employee or mechanized sortation concept transfers assigned SKUs from a sort concept into an assigned CO shipping carton or captive tote. A CO sort location has structural support members to provide sufficient rigidity and stability to hold the weight and physical size of a CO carton or captive tote. Options for using a CO shipping carton or captive tote as a sort location are:

■ With the carton open side facing up, with flaps up, or a two-piece carton with no flaps. The carton is designed to hold the maximum computer-projected CO SKU cube, pieces, and weight. This can be used in a manual sort-to-light concept and in all mechanized sort concepts. The sort and hold location has a high window to handle a carton with flaps up and has a longer SKU drop from a discharge point into the carton.

■ With the carton open side facing up, with flaps down. This concept has the same design parameters and operational characteristics as a carton with flaps up, except that it has a shorter sort hold location window and the shorter SKU drop from a discharge point into a carton minimizes UOP damage.
■ With a captive tote. This concept has the same design parameters and operational characteristics as a carton with flaps down.

Features of the concept are (a) human- or human/machine-readable sort label or sort to light, used with a manual pick, label and sort concept; (b) an employee removes full cartons from a sort location and places empty cartons in the sort location; (c) it provides maximum sort locations per aisle; (d) has the greatest impact on a SKU; (e) requires a small area because it is possible to have sortation locations stacked four to five high; and (f) it improves packer productivity.

Manual-Batched Customer Order Pick and Sort Concept

Manual pick and sort concepts have a picker with a paper document or pick label pick and sort SKUs for a batched CO into a temporary CO sort location. In a pick area, this temporary CO sort location is a fixed or mobile shelf, tote, or bin. Manual pick and sort concepts have the following characteristics:

■ A paper printed document or paper label as a CO SKU pick and sort instruction.
■ The number of COs that can be batched is limited by the number of sort locations.
■ CO SKUs have a cube that can fit into a sort location.

Each CO sort location has a human/machine discreet identification that matches a CO discreet human/machine-readable identification on a pick instruction paper document or label machine-readable for a hand-held scanner. In a manual pick and sort application, at a sort location any overflow CO picked SKUs are placed into a CO human/ machine-readable identified overflow carton/tote in a separate location on a sort device.

Batched CO pick and sort concepts are:

■ A four-wheel moveable cart with shelves/bins on a pick aisle or at the end of a pick aisle
■ A multi-shelf location at the end of a pick aisle
■ A multi-shelf basket located at the end of a pick aisle that is attached to an overhead trolley travel path
■ A master sort tote on a conveyor travel path

Four-Wheel Cart with Shelves or Bins

SKUs for a CO are picked into a mobile four-wheel cart with multiple shelves or bins. A pick and sort employee:

- Receives the batched paper pick and sort instruction document
- Writes a batch number onto a cart placard
- Places each CO discreet identification on each cart shelf or bin sort location

Verifies that all written identifications match a CO discreet identification on a pick instruction paper document

The picker then pushes a four-wheel cart into a pick aisle and walks to the first pick position that appears on the paper document. The picker removes the number of that SKU piece indicated on the paper document from a pick position and per the paper document CO quantity, sorts or transfers the SKUs from the pick position into a four-wheel cart CO sortation location. After each SKU pick and sort activity, a CO picker/sorter places a mark in the appropriate location on a paper document or adjacent to a SKU pick position. This batch CO pick and sort activity is repeated for each SKU that appears on a pick and sort paper document.

When a pick and sort employee has completed all picks and sorts for batched COs, the paper document is placed onto a cart placard and the cart is pushed to the next pick zone or to the pack station area. A picker/sorter obtains the next batch wave paper document and prepares a new four-wheel cart for another CO pick and sort activity. If self-adhesive labels are used in a pick and sort concept, a picker/sorter places the appropriate label onto a SKU and transfers the labeled SKU into a CO sort location.

Shelves or Bins at the End of a Pick Aisle

To ensure good pick and sort employee productivity, there are two shelf sort locations on both ends of a pick aisle. After a picker receives a batched pick and sort instruction paper document, he or she writes each batch number onto a separate shelf placard, writes a CO discreet identification onto each shelf or bin sortation location, and verifies that all written identifications match a CO discreet identification on a pick/sort instruction paper document. After this activity, the picker walks into a pick aisle to the first pick position that appears on a paper pick document and removes the indicated number of SKUs from a pick position. The picker then walks to a pick aisle end and per a paper document CO ordered quantity, sorts or transfers the appropriate number of SKUs to a shelf or bin sortation location. After each SKU pick and sort activity is completed, a picker/sorter places a mark in the appropriate location on a paper document or adjacent to a SKU pick position. Pick and sort activity is repeated for each SKU on the pick and sort paper document.

When a picker/sorter has completed all picks and sorts for the batched COs, the paper document is placed onto a shelf placard. The picker/sorter then obtains the next batch wave paper document and prepares a new four-wheel cart for another CO pick/sort activity. While a pick and sort employee is completing a pick and sort transaction for the second batched CO group or wave, a master sortation employee with a four-wheel cart or other transport device transfers CO picked and sorted SKUs from a temporary pick aisle end sort location to a master sort transport device. A master sort employee transfers each batch paper pick/sort paper document to a master sort transport device. In a check/pack area, a paper pick/sort document is used to verify a picked SKU quantity and quality. If self-adhesive labels are used in the pick and sort concept, a picker/sorter places the appropriate label onto a SKU and transfers the SKU into a CO sort location.

The features of the mobile cart concept are few UOP handlings, few errors, few employees, and slightly wider pick aisle.

An Overhead Trolley with a Multi-Shelf Basket

An overhead trolley with a multi-shelf basket concept has a picker/sorter receive the batched paper pick and sort instruction document and write each CO batch number onto an overhead trolley multi-shelf placard. The pick and sort employee writes each CO discreet identification onto a shelf location and verifies that each sort document and shelf location discreet identification match. The picker/sorter then walks in a pick aisle to the first pick position that appears on the paper pick document, removes the correct number of SKUs and transfers them to various overhead trolley shelf basket cart CO sort locations. After each SKU pick and sort activity is completed, the picker/sorter places a mark in the appropriate location on the paper document. This batch CO pick and sort activity is repeated for each SKU on the pick and sort document.

When a picker/sorter has completed all picks and sorts for a batched CO, a paper document is placed onto a basket placard and an overhead trolley with a shelf basket is pushed to the next pick zone or to a CO pack station. A picker/sorter obtains the next batch paper document and prepares a new overhead trolley shelf basket for a new batch CO pick and sort activity.

Multi-Compartment Sort Tote on a Conveyor Travel Path

A multi-shelf or multi-compartment tote on a conveyor travel path concept has a picker receive a batched paper pick and sort instruction document. The picker writes a batch number onto a tote placard and writes each CO discreet identification onto a separate tote section. The picker verifies that each CO sort location matches the paper document CO discreet identification. After this check, the picker walks into a pick aisle to the first pick position that appears on the paper document. Per the paper document SKU pick quantity the picker removes the appropriate number of SKUs from a pick position, walks to a tote on a conveyor travel path and, per the paper document, sorts the appropriate SKU quantity into each CO container sortation location. After each SKU pick and sort activity completion, a picker/sorter places a mark in the appropriate location on the paper document. This batch CO pick and sort activity is repeated for each SKU that appears on a pick and sort paper document.

When a picker/sorter has completed all picks and sorts for the batched CO, the paper document is placed onto the tote placard and the tote is moved onto a take-away conveyor travel path for transport to the next pick zone or to a pack station. A picker/sorter obtains the next batch or wave paper document and prepares a new multi-shelf or compartment tote for new batched CO pick and sort activity.

Disadvantages of the manual paper pick and sort concept are the need for a picker to read, small piece volume or cube, with two to three pieces per CO, 16 to 20 COs per batch, four-wheel cart, trolley basket or conveyable tote, and paper concept. The advantages are low cost, handles slow-moving and small-cube SKUs, is easy to implement, and used with other manual pick concepts that have a different pick methodology.

Centralized Manual Sort Concepts

The centralized manual sortation concepts have batched CO picked SKUs delivered as loose SKUs in a tote or on a belt conveyor travel path to a centralized sort area. Delivery concepts for batched CO picked and labeled SKUs are:

- Loose, unlabeled SKUs unique to a single CO, in a four-wheel cart cavity, trolley basket section, or in a conveyable tote
- Loose, labeled SKUs for mixed COs, in a four-wheel cart cavity, trolley basket section, or in a conveyable tote
- Labeled, loose SKUs for mixed COs, on a belt conveyor travel path

At a centralized sort station, a sort employee handles each SKU by removing it from the transport concept, reading the SKU sort instruction, matching the SKU sortation code to a sort location code, and transfering the SKU to an assigned CO sort location. Depending on operational procedure, a CO sort location is a temporary holding location or a CO shipping carton or captive tote. When a centralized manual sort concept uses a four-wheel cart or overhead trolley basket as a transport concept to move picked SKUs from a pick area to a sort area and through a sort station, for an efficient and productive sort area the design factors are transport device pre-sort and a queue area for each sort location, complete travel path through a sort area, staging area for empty transport devices, and along a sort travel path an early exit travel spur for empty transport devices.

Centralized manual sort concepts include loose, picked SKUs in a four-wheel cart cavity, trolley basket section, or in a conveyable tote that are sorted by a paper sort instruction document or a sort-to-light instruction.

Centralized Manually Picked and Sorted SKUs with Paper

For an operation where SKUs are picked and sorted for individual COs, a centralized manual sort concept with a paper pick document has the following functions.

In the pick area, a picker picks the SKUs into one tote or a section within a tote. Other CO pick devices are a four-wheel cart or trolley basket with shelves or separated compartments. With this pick and sort concept, a picker places a batch number onto a tote, cart, or trolley basket section and walks or rides to a pick position. Per the pick instruction document, for each CO, a picker transfers the required number of SKUs into a separate tote, cart, or basket section, places the pick/sort document into the tote and transfers it onto the travel path. To control a batch, each CO batch has a separate color or coded tote.

In the sort area, a sortation employee ensures that (a) the CO sort location number has proper CO discreet identification and has a CO pack slip or (b) each CO sort location has a CO shipping carton or captive tote, proper CO discreet identification, and CO pack slip.

As the totes, carts, or baskets arrive in the sort area, a divert employee looks at the tote color and reads the batch number on the tote, cart, or basket. Per these two factors, a divert employee transfers a proper tote, cart, or basket from a main sort conveyor travel path onto an individual sortation station conveyor queue travel path. To assure proper sort, each sort station divert location has a batch number or color at a sort location. As totes, carts, or baskets arrive at a sort station, a sortation employee removes a pick/sort document from a tote, cart, or basket and verifies that a batch number on a tote, cart, or basket matches a sortation station batch number. With a sort code match, a sort employee removes a SKU from a tote and looks at a sort document for the first CO discreet identification that has a SKU. When a sort document CO discreet identification matches the sortation location CO discreet identification, the employee transfers a SKU quantity from a pick tote, cart, or basket into a sort location. A sort employee repeats the sort activity for all SKUs in a tote, cart, or basket and for each section. When a sort activity is completed for a CO batch, a sort employee moves to the next sort station or transfers totes or cartons for a a completed CO onto a CO take-away transport concept. After all first batch CO picked and sorted SKUs have cleared the sort station, the sort station is ready to accept a new CO order batch.

Components of a centralized manual sort concept are:

■ With a tote, cart, or basket, a main powered roller conveyor in front of the sortation area and an adjacent employee walkway.
■ A divert travel path at each sort station.

- At a sloped divert conveyor lane discharge end, a low-pressure or zero-pressure powered conveyor travel path that directs totes onto a sort station.
- A sort station travel path along the front of a CO sort location. For totes, this travel path is a non-powered, low-pressure, or zero-pressure conveyor and for carts or trolley baskets a designated travel path.
- CO sort locations.
- A CO shipping carton make-up and empty transport device staging area between sort stations.
- A low-pressure or zero-pressure CO take-away conveyor travel path along the back side of the sort location, for transport of completed COs to the next order-fulfillment operation activity. This CO take-away conveyor is over or under the main sort conveyor.

Centralized Manual Sort with a Sort-to-Light Instruction

A centralized manual sort of SKUs from a four-wheel cart, trolley basket, or tote with a sort-to-light instruction has the following features:

- Each SKU receives a bar code label/RF tag or is prelabeled, picked with a bar code label/RF tag, or picked by a paper document into a separate CO carton/tote.
- A sort employee has a hand-held scanner that reads the bar code and transmits the information to a computer.
- The computer activates all appropriate CO sort location sort-to-light displays, which indicate the SKU pick quantity for each illuminated location.
- Each CO sort-to-light location has a CO carton with a pack slip and sort or put to light display module. Each sort or put to light has a communication wire or fiber optic path to the microcomputer.
- A completed CO carton/tote take-away conveyor travel path.

Centralized Manual Sort with a Label Instruction

This concept involves a centralized manual sort of mixed SKUs each of which is labeled with a sort instruction. Each picked SKU in the four-wheel cart, trolley basket section, or tote has a pick/sort label placed on its exterior surface. The label has a CO discreet sort identification printed onto its face. During a sortation activity, this discreet identification is matched to a CO sort location discreet identification. A match is a signal for the sort employee to transfer a SKU from a carton/tote into a CO sort location. Pick/sort labels improve sort employee productivity and accuracy and make it easy to check employee productivity and accuracy.

Centralized Manual Sort from a Conveyor with a Label Instruction

The operational characteristics of a centralized manual sort from a conveyor concept are:

- The main sort conveyor travel path is a belt conveyor.
- From a pick area, specific COs are batched and placed onto a belt conveyor for transfer to a sort area. To maintain batch control in the sort area, at each pick zone picks are completed for one CO batch, or a picker in the last pick zone who picks the last SKU places a colored tote or signal item onto the conveyor travel path. Each pick zone has a specific colored tote or item. When all colored totes or items from the pick areas for a specific batch are received in a sort area, a sort employee indicates that the sort area is ready for the next CO batch of SKUs.
- SKUs are diverted from the main conveyor and waterfall onto a perpendicular belt conveyor travel path. This is a slow-moving conveyor that transports SKUs past all CO sortation locations so that the sort employee can complete sortation transactions. During a sort activity, an employee matches a sort label CO discreet identification to a CO sort location discreet identification. A match is a signal for a sort employee to transfer a SKU into a CO sort location.

Manual Small-Item Sort and Assembly Concepts

An objective of a manual sort and CO assembly is to pick individual SKUs for a CO group, to arrange them in sort and assembly locations, and per a CO to pick SKUs for each CO from a pick position into a CO captive tote or shipping carton.

Manual sort and assembly locations can be arranged in different patterns (see Figure 4.84): (a) straight line, (b) horseshoe-shaped, and (c) multiple aisles and rows. The sort and assembly locations are shelf positions, decked rack, carton flow rack, or a combination.

Transfer of SKUs picked for a CO to a sort and assembly location involves:

- Random SKU allocation to a sort and assembly position with no transfer transaction verification.
- SKU allocation to a sort/assembly position is verified with RF device transmission for paper transfer transaction.

SKU allocation to a sort and assembly position by a predetermined SKU identification number sequence, such as SKU inventory number last digit for the bay and next to last digit for the shelf or decked rack level

MANUAL SORT AND ASSEMBLY

MANUAL SORT AND ASSEMBLY STRAIGHT LINE LAYOUT

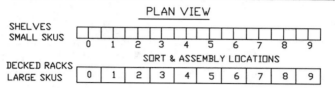

MANUAL SORT AND ASSEMBLY MULTIPLE AISLES AND ROWS

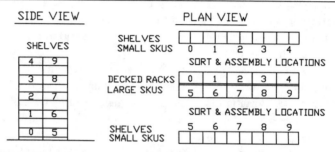

MANUAL SORT AND ASSEMBLY HORSESHOE OR U SHAPED LAYOUT

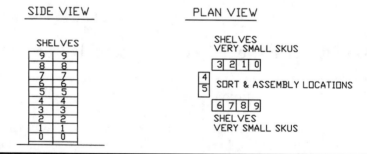

Figure 4.84 Manual sort and assembly: straight, multiple aisles and rows, and horseshoe or U-shaped shelves.

After all CO bulk picked SKUs are transferred from the pick carton/tote into a sort and assembly position, CO assembly employees are given CO pack slips. Pack slips have SKUs that were bulk picked and transferred from a pick carton/tote into a sort and assembly location. The assembly employee with a CO assembly instruction (RF or paper CO pack slip) reads a SKU identification number, determines the SKU sort and assembly position, walks to that position, removes a SKU quantity, and transfers it to a CO shipping carton or captive tote. The activity is repeated for each SKU on a CO pack slip or assembly instruction form and is repeated for each CO that was downloaded or sent from the host computer to a computer (printed on a CO pack slip) to a sort and assembly area.

Sort and Assembly Positions Arranged in a Straight Line

A straight-line sort and assembly area pattern has the sort and assembly shelves, decked racks, or carton flow racks arranged in one long row. This layout results in low employee productivity because they walk long distances between pick positions both to sort and assemble SKUs. It is difficult to handle a large number of SKUs, high cube or high volume, but handles a wide SKU mix.

Sort and Assembly Positions Arranged in a Horseshoe Pattern

A horseshoe-shaped sort and assembly area has shelves, decked racks, or carton flow racks arranged in a U-shape. This design features slightly higher sort employee productivity, because they walk shorter distances. SKUs are more easily transferred from one location to another, and it is possible to allocate odd-numbered positions to the left and even-numbered positions to the right, or in an arithmetic progression up one side and down the other. It is difficult to handle a large number of SKUs using this concept and it is preferred for very small SKUs.

Multiple Aisles and Rows Pattern

A sort and assembly area with multiple aisles and rows has shelves, decked racks, and flow racks arranged in short aisle and row segments. This design increases assembly employee productivity due to short distances between picks and return trips, easy SKU transfer position, potential short walk distance between two sort positions, creates higher sort employee productivity, It is difficult to handle a large number of SKUs or a high volume, but handles a wide SKU mix.

Random SKU Sort

Random SKU transfer to a manual sort and assembly position has an employee determine and transfer a SKU into a unique position in a sort and assembly area. After all CO bulk picked SKUs are transferred to discreet positions in a sort and assembly location, a CO assembly employee is given a CO or a CO group to assemble. A CO assembly employee with a CO assembly instruction format (CO pack slip or paper or RF device) obtains a SKU identification number or description and SKU sort and assembly discreet position, walks to the SKU position, removes the appropriate SKU quantity, and transfers them to a CO shipping carton or captive tote. This activity is repeated for each SKU on a CO assembly instruction (pack slip) and for each CO that was downloaded to the sort and assembly area.

The concept has low CO assembly employee productivity, because there is no routing sequence, employee confusion, double travel in aisles, and increased walk distance between two pick locations.

Random SKU Transfer with RF Device Verification

Random SKU transfer into a discreet sort and assembly position with RF device verification has an employee transfer a SKU to a unique position in a sort and assembly concept and verify each SKU transfer transaction with RF device verification. The RF device verification updates a SKU sort and assembly position in microcomputer files. Each shelf, decked rack, or carton flow rack position has a position identification that is easily recognized by an employee and is in the computer files.

Bulk picked SKUs are transferred to discreet positions in a sort and assembly area. Each SKU position is updated in computer inventory files, and CO assembly employees are given CO pick instructions or pack slips. The employee with a CO assembly instruction format (RF device or CO pack slip paper document) obtains the SKU identification number or description, scans the bar code label that appears on a CO pack slip and reads the RF device display that shows a SKU sort and assembly position, walks to a SKU position, removes and transfers a SKU quantity to a CO shipping carton or captive container. This activity is repeated for each SKU on a CO assembly instruction pack slip and for each CO that was downloaded to a sort and assembly area.

The features are higher CO assembly employee productivity, because there is a routing sequence that minimizes employee confusion, no employee double traveling, and can be used for a wide SKU mix.

SKU Transfer Based on the SKU Identification Number

Each SKU identification number is used to determine a SKU sort and assembly bay, and within a bay, each identification number identifies the shelf or decked rack level. The aisle identification is not required because the sort and assembly area has one aisle.

Numbering options are:

- The first digit of the SKU number to identify the bay and the last digit to identify the shelf or decked rack level. This concept is used when the first digits of a SKU identification are 0 through 9.
- The first digit to identify a bay and the second digit to identify a shelf or deck rack level. This concept is used when the first digits are 0 through 9.
- The last digit of the SKU number to identify a bay and the next to last digit to identify a shelf or deck rack level. This concept is used when the first digit number does not use all digits between 0 through 9.

If the first digit of the SKU number identifies the bay and the last digit identifies the shelf or rack level concept (see Figure 4.85), each aisle or bay in the sort area is associated with a number from 0 to 9. Thus, SKU number 10095 is located in bay 1

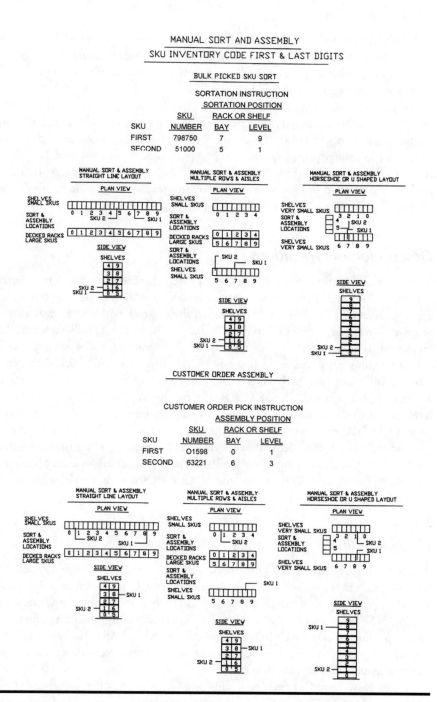

Figure 4.85   Manual sort and assembly by SKU code first and last digits.

and shelf level 5. This concept uses a large area, handles a greater SKU quantity, and handles a large cube. The same principle applies to using a combination of the first and second digits or the last and next to last digits to identify the bay and shelf or rack level (see Figure 4.86 and Figure 4.87).

In a one-aisle concept, when CO pack slips or pick slips are printed in a first digit SKU number arithmetic sequence for the bay it groups CO pack/pick slips by the first digit number from 0, 1, 2, 3, 4, 5, 6, 7, 8 and 9. Prior to the final pick process, a WMS printer print sequence has CO pack/pick slips distributed as a group to each bay that matches the first digit number. During the final pick activity, a CO pack/pick slip group in each appropriate bay improves picker productivity due to the fact that the first pick is at the bay with a CO pack/pick slip.

Description of Operations

The CO assembly instructions (CO pack slip) are the same as the sort instructions. With these concepts, a sort employee transfers SKUs by a SKU digit concept to an appropriate bay and shelf. Each bay and each shelf or rack within the bay is associated with a number from 0 to 9, as described above. After all CO bulk picked SKUs are sorted to an assigned shelf or decked rack bay position, CO assembly starts. A CO assembly employee walks to a bay and obtains a CO pack/pick slip from a stack. With a CO pack/pick paper or RF device, a CO assembly employee starts at the first pick position. The SKU digit directs the employee to the shelf or rack level and he or she transfers a SKU into a CO container. The employee reads the second CO SKU digit number and walks to a corresponding bay, identifies a corresponding SKU shelf or decked rack level and completes SKU transfer from that level into a CO carton or captive tote. The activity is repeated for each SKU on a CO instruction form (CO pack slip) and for each CO that was downloaded to the sort and assembly area.

Pick/Pack Slip Print Sequence

A WMS print sequence for CO pick/pack slips can be random, or by CO number, SKU identification number, or bay location number. CO pack slips or pick slips in a random print sequence means that a final assembly person or picker has random routing to SKUs in a CO. With a random CO pack slip arrangement, the first CO pack slip could require SKUs in bay 1 and the second CO SKUs in bay 9. From a central control desk, the random feature increases an employee walk distance from a central control desk to complete each CO.

Similarly, to have CO pack/pick slips in a CO identification sequence for the final assembly of SKUs for a CO is basically a random sequence. The features are the same as the random concept described above.

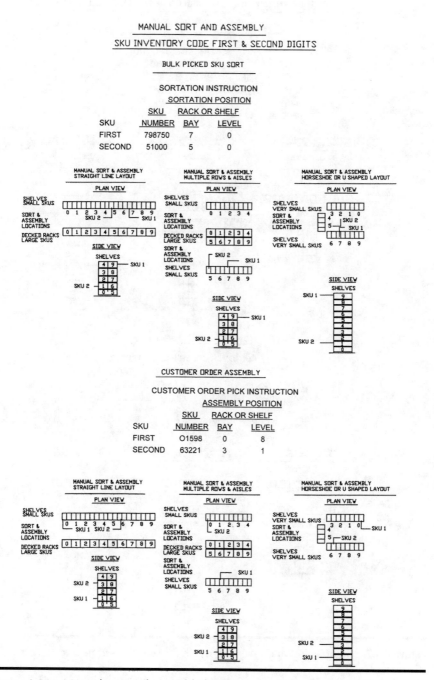

Figure 4.86 Manual sort and assembly by SKU code first and second digits.

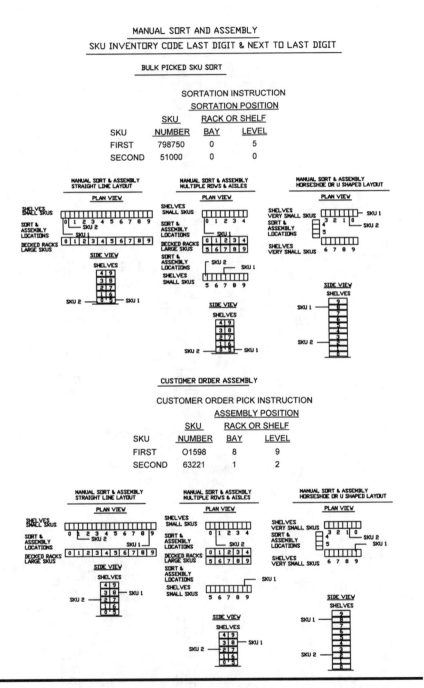

Figure 4.87 Manual sort and assembly by SKU code last and next to last digits.

To have CO pack slips or pick slips in a SKU identification number sequence is an arithmetic sequence in which all SKUs that have the number 1 as either the first digit or the last digit are printed in a group. With CO pack slips grouped into numbers and the first digit (or last digit) as a corresponding sort and assembly bay position, each group is placed into a bay. As pickers assemble COs, CO pack slips for the first SKU on each CO is in the same bay. This feature means increased productivity due to less walking for pickers to complete COs.

In a one-aisle concept, when CO pack slips are printed in the second-digit SKU number arithmetic sequence for the bay, it groups CO pack slips into the second or next to last SKU number from 0 to 9. Prior to the final pick process, the print sequence permits CO pack slips to be distributed as a group to each bay. This improves picker productivity due to the fact that the first pick is at the bay with a CO pack slip.

Very Small SKUs

A pick and manual sort concept for very small SKUs requires a low equipment investment but some cost in computer programs. For small SKUs such as jewelry or electronic parts, for a CO the computer prints a SKU bulk pick list. A bulk picker picks each SKU volume and transfers SKUs to a sort and assembly area. SKU sortation to sort positions is by SKU number last or first digit or RF device, reviewed in the previous section. After completion of all pick and sort transactions, for each CO a final CO picker picks SKUs from the sort positions. Sort and assembly design options include the following:

Straight line locations. A straight line location design has the sort and assembly shelf or decked rack positions arranged in a straight line from 0 to 9. This arrangement has the first sort position at the entrance to an aisle and each position number progressively increases to the required number of sort and assembly positions. With this arrangement, an employee starts at the first position and, for a CO, completes the withdrawal transactions from an appropriate sortation position. This straight line concept can increase the walk distance and time between two positions, minimizes the SKU hit density and hit concentration, and the aisle position number system has a progressive pattern.

Tunnel positions. A tunnel-shaped design has the sort and assembly shelf or rack positions as two rows with an aisle. Positions are numbered, for example, from 0 to 4 on one side and 5 to 9 on the other side of the aisle. The employee routing pattern has a horseshoe or U pattern, or even numbers on the right and odd numbers on the left, which is a straight line pattern. This tunnel arrangement has the first position at an aisle entrance and aisle position numbers progressively increase to complete the sort and assembly positions. The features are large potential to decrease the walk distance and time between two positions, increases the SKU hit density and hit concentration, number system has a progressive pattern, and easy to understand.

Combining Customer Order SKUs from Different Areas (Product Class)

If your pick operation has two separate facilities or SKU pick areas (SKUs from different product classes) and a CO has picks from both pick areas, prior to a CO pack activity, the multiple SKUs are:

- Sent to a sort area (reviewed above as manual or mechanized sort concepts).
- Combined into one pick tote/carton. To combine a CO with multiple SKUs (from different classes) into one pick tote/carton, pickers from two separate pick areas place each SKU into a CO carton/tote. The options are:
  - Pick directly into a CO pick carton/tote.
  - In each pick area to pick by wave to a staging location by CO number and transfer CO picked SKUs from the staging location into a CO carton/tote.

For a CO with SKUs from different pick zones, to pick directly into a CO carton/tote is a pick/pack or pick/pass activity. In a pick/pass activity, each CO carton/tote is discreetly identified and travels on a transport travel path past every pick zone. The transport travel path is a non-powered or powered conveyor or a four-wheel cart. As a CO carton/tote arrives at a pick zone the picker in the pick zone receives CO pick instructions (paper document with a carton or tote, RF device, or pick to light) and transfers the appropriate SKU quantity from a pick position into a CO carton/tote.

For additional pick/pass concept information, we refer the reader to the pick-to-light section in this chapter.

For COs with SKUs from different pick zones, SKUs picked and labeled for a CO are staged in each pick zone. SKUs are arranged in a CO identification arithmetic sequence, with the lowest number starting at the shelf left-hand position and increasing across the shelf to the right-hand position. To have additional staging shelf positions, the lower shelf is used with the same CO sequence. A CO staging shelf position is fixed shelves or four-wheel cart with shelves and staging position is adjacent to a CO travel path.

In the first pick zone, each CO carton/tote is identified with a CO numerical discreet identification that faces a pick area or staging area. As a CO carton/tote travels into a second pick zone, an employee reads a CO identification on a carton/tote and looks at a staging position. If a staging position has SKUs with the same CO identification label, he or she transfers SKUs from the staging position to a CO carton/tote. A CO carton/tote travel path is a zero-pressure conveyor with a stop device, four-wheel cart, or an overhead trolley with a basket/shelf with a stop/start control device. If a CO is not cube for a carton/tote, an overflow carton/tote is added to the conveyor travel path or with a shelf transport vehicle an overflow shelf is used to hold SKUs. If CO SKUs are very small (cube), a tote with separators is used on a transport concept to improve transfer productivity and minimize the number of cartons or totes on a travel path.

The design parameters are:

- Identification on each carton/tote and a pick label on each SKU.
- A computer to create pick waves and to pre-pick zones and staging area in each zone.
- Transfer employee activity.
- In each pick zone, transport concept has stop/start controls.
- Cube each CO or have transport device allow for overflow SKUs.

The features are a wide SKU mix, separates a pick area into zones, batch pick wave space is designed to handle a CO SKU volume and cube, employee to read, medium volume and employee number, good employee pick and sort productivity, flexible to handle volume fluctuations, and low computer cost, but requires a printed pick label.

An option to the wave pick, stage, and sort concept is a manual SKU sort concept that has a transfer employee use a paper-printed sort instruction to direct SKU transfer from a staging location into a CO tote, carton, cart, or trolley compartment. Components are each CO on a transport travel path, a computer-printed document, CO identification on each CO order tote, carton, cart, or trolley section, very high SKU volume that appears on most COs or a specific customer order group, SKU transfer location, and completed CO travel path.

The computer prints a paper document that lists each wave or batch of CO discreet identification in an arithmetic sequence. Before a CO shipping carton/tote is released, in each pick zone the required SKU quantity is transferred from pick positions to a transfer position adjacent to a CO carton/tote travel path, which permits an employee to easily transfer a SKU from a transfer location into a CO carton or tote. After a CO carton/tote is placed on the conveyor, it receives a discreet identification. Each discreet identification faces a pick aisle or the side that faces a transfer employee. The CO discreet identification corresponds to the CO identification on a computer-printed document.

CO cartons or totes travel at a slow speed, which permits a transfer employee sufficient time to read and match a CO carton/tote discreet identification to a paper document CO identification. With a proper match, a transfer employee moves the SKU quantity from the staging position into a CO carton/tote. SKU transfer activity is repeated for each discreet CO number that is on the paper document.

To increase CO volume, a double-wide CO carton/tote travel path transports two COs simultaneously. A double-wide conveyor concept has a near-side conveyor travel path that has CO cartons/totes that start or end with a specific digit or alpha character, such as 1 or A. The far-side conveyor travel path has CO cartons/totes that start or end with 2 or B alpha character. After a CO carton/tote travels through the pick zone, a transport concept moves it to the next pick/pass zone. SKU transfer into a CO carton/tote with a paper-printed document features are high volume, employee to read, excellent employee productivity, small building area, minimal computer investment, and best used for low value or promotional SKUs.

The disadvantages are handles a small to medium number of COs and only a few pieces per CO, SKU sort/pick instruction, management discipline and coordination between the pick zones and sort zones, potential sortation errors, human-readable code, and some SKU double handling. The advantages are low to medium investment, low impact on SKUs, can be installed in a low-ceiling building, permits batched CO pick activity, and handles a wide SKU mix.

Centralized Mechanized Small-Item Sort

There are a number of centralized mechanized small-item sort concepts. A centralized mechanized small-item sort concept requires queuing of SKUs ahead of each induction station and between a sort travel path and a pack station. A mechanized sort concept is laid out in arithmetic progression from an induction station with sort locations on both sides of a sort concept travel path. The sort locations have the even numbers on the right and odd numbers on the left. Functional areas are, for a pick area, a pick conveyor or other pick transport concept, conveyor or other transport concept for moving CO picked and labeled SKUs from a pick area to a sort area, a sort induction station, a sort travel path, no read divert station and return conveyor travel path, pack sort locations, clean-out sort location, relatch station and control panel, air compressor, mimic display, and electrical panel location.

Pick Conveyor

A pick conveyor or transport concept is designed to move CO picked and labeled SKUs, either as loose pieces or in totes on a powered conveyor travel path, cart transport concept, or belt conveyor travel path from a pick position to a transport conveyor.

Transport Conveyor or Concept

A transport conveyor ensures:

- That CO picked and labeled SKUs are moved from a pick area to a sortation area.
- That the conveyor travel path is designed to allow queuing or there is a staging area ahead of a sort induction station. Before each induction station, either:

- A tote transport concept with low-pressure or zero-pressure conveyor queue sections along which the totes are tilted to one side, at an angle that permits an induction employee easy access to SKUs.
- A chute that allows for queuing. Photo-eyes along the chute travel path control (start/stop) the in-feed powered belt conveyor.

Induction or In-Feed Station Conveyor

An induction or in-feed station conveyor is the location where each small-item discreet identification is entered into the sort concept computer. At an induction station, each CO picked and labeled SKU is transferred from the transport conveyor travel path or tote onto a sort travel path. Either an employee physically transfers each SKU onto a sort concept load-carrying platform or travel path, or there is a powered conveyor. An induction conveyor belt concept has at least four short belt conveyor sections that create a gap between SKUs and transfers each SKU to a sortation concept load-carrying surface or travel path.

The difference between the two SKU induction transfer concepts is that the conveyor has a higher induction rate. The induction conveyor concept ensures that as an empty load-carrying surface or open space on a travel path arrives at the induction station, a SKU is transferred forward onto the empty sort concept load-carrying surface. This feature increases the load-carrying surface utilization and sort capacity.

A belt conveyor does not require an empty tote or carton transport travel path, does not require an employee to transfer a transport device, and is less complex to design and operate. A tote, overhead basket, or cart transport concept is an empty transport device take-away concept, that requires an employee to lift a SKU and is and more complex to design and operate.

The SKU sort induction concepts are manual induction with a key pad, semiautomatic or hand-held scanner induction, and automatic or bar code scanning induction.

Manual or Key Pad Induction

In a manual concept, an induction employee reads the code on a SKU and enters it into a keypad, which transmits the SKU code to a microcomputer and tracking device. To achieve high induction employee productivity, a human-readable code has the largest type and minimal digits with a two or three digit maximum in a CO discreet identification. Alpha characters are not preferred on a manual induction label. The disadvantages are the need for employee training, the concept handles a low volume, potential induction errors, and greatest employee number and must rotate employees. The advantages are low investment and it serves as a back-up induction concept.

Semiautomatic or Hand-Held Scanning/Reading

In a semiautomatic or hand-held bar code scanning/RF tag reading concept, each SKU has a human/machine-readable code on its exterior surface. As an employee handles a SKU, he or she directs the light on the hand-held scanner at the bar code/ RF tag and then transfers the SKU onto a load-carrying surface or conveyor travel path. As a SKU passes the photo-eye on the sort conveyor travel path, the SKU is released to a sort conveyor system. As the SKU arrives at the discreet CO sort location, the computer, tracking device, and constant conveyor travel speed ensure that the sort device completes the sortation activity.

Disadvantages are additional investment and employee training. Advantages are reduced induction errors, higher volume capacity, and it serves as a back-up induction concept.

Automatic or Bar Code Scanning/RF Tag Reading Concept

In an automatic or bar code scanning/RF tag reading induction concept, each SKU has a bar code label/RF tag attached to its exterior surface. As the SKU travels onto the sort concept travel path, the bar code faces up or an employee ensures that the RF tag is attached. After an employee places the SKU directly onto a sort concept load-carrying surface or onto the induction conveyor belt, the SKU travels past the induction photo-eye to a bar code scanning/RF tag reading station, equipped with a fixed-position, moving or waving beam bar code scanner, which throws laser light beams onto the sort conveyor travel path. As a bar code labeled SKU is scanned (the light beam crosses the entire bar code), the scanning device or RF tag transmission is read by a reading device, sending each SKU discreet CO identification to a computer and tracking device. This information and the constant sort concept load-carrying surface travel speed ensure that the SKU is diverted at the assigned sort location from a sort load-carrying surface to a pack station.

A sort induction station can be designed as:

- A single induction station that is located ahead of the closed-loop sort concept travel path.
- A dual induction station, which is used on a rectangular or elliptical sort travel path. On each rectangle or elliptical travel path section is located an induction station that inducts pieces for the following straight travel path section. The dual-induction concept has double components, such as induction stations, scanners, photo-eyes, tracking devices, clean-out chute, and "did not read" stations. When compared to a single induction and sort concept, the dual-induction feature increases the induction and sort productivity by an estimated 80 percent. Disadvantages are additional cost and human/ machine-readable label. Advantages are few induction errors, few employees, and handles a high volume.

Sort Travel Path

A sort concept travel path has a load-carrying surface that carries a SKU from an induction station, past a bar code scanner/RF tag reader, to the CO sort station. Other sort travel path stations are "no read" station, "did-not-sort" station, clean-out station, and relatch station. A key component of the sort concept is the sort concept travel path. Types of sort concept include:

- Tilt tray
- Bombay drop
- Cross or moving belt
- Gull wing
- Nova sort
- Flap sorter
- Brush sorter
- Platform sorter
- Ring sorter

These are described below.

Transferring SKUs from the Load-Carrying Surface to the Sort Location

Techniques for transferring SKUs from the load-carrying surface into a sort location involve tipping, sliding, pushing, or dropping a SKU from the sort concept travel path. A sort concept station design has an arithmetic progression from an induction station to the last sort station. The sort concept stations are arranged with either the lowest number as the first divert station past the induction station, or locations with odd-numbered last digits are located on the left of the sort travel path and those with even-numbered last digits are located on the right. Sort techniques include:

- An active sorter, which has a powered induction station and sort conveyor travel path and a load-carrying surface from which a SKU is pushed or pulled into a sortation location. A cross/moving belt and flap sorter are active sort concepts.
- An active–passive sorter, which has a manual induction station, powered sort travel path, and a load-carrying surface from which a SKU is tipped into a sort location. The tilt tray, gull wing, Nova sorter, platform sorter, brush sorter, and ring sorter are active–passive sort concepts.
- A passive sorter, which has a manual or powered induction station, a powered sort travel path, and a load-carrying surface from which gravity force removes a SKU into a divert location. A Bombay drop concept is an active–passive sort concept.

A sort concept travel path is from an induction station to all sort concept divert locations. Sort concept travel path designs are categorized as straight-line or endless-loop.

Straight-Line Travel Path Design

A straight-line sort concept travel path runs in a straight line from an induction station to the last sort station. Past the last sort station, the travel path ends, a "did not sort" travel path declines to the floor, and the load-carrying surface returns empty to the induction station, under the sort travel path. Variations on the straight-line design (see Figure 4.88) are L-shaped and horseshoe or U-shaped layouts.

The straight-line travel path layout does not have a feature that automatically reintroduces "did not sort" SKUs to an induction station. In a straight-line design, all "did not sort" SKUs decline to floor and are manually returned to an induction station or transferred and scanned to an assigned pack station. The disadvantages are additional employees and low volume. The advantages are no dual induction and low investment.

Endless-Loop Travel Path Design

An endless-loop design has a travel path that starts at the lead end of the induction station and ends at the charge end. An endless-loop layout can be L-shaped, U-shaped, O or elliptical, or rectangular.

An endless-loop sort concept travel path has "did not sort" SKUs continue to travel past the last divert station and on to an induction station. As the SKU arrives at the induction station, its discreet identification is rescanned or reintroduced to a sort concept. To assure maximum sort concept load-carrying surface utilization and handling capacity, a sort concept computer has the ability to detect a repeated "did not sort" bar code. After the system reads the same SKU identification a specified number of times, the computer diverts the SKU onto a clean-out chute or onto a special divert station, from which it is manually transferred to a pack station.

The disadvantage is increased capital investment. The advantages are fewer employees, high volume, easy expansion, and permits dual-induction concept.

No Read or No Data Station

All sort concepts, and especially an automatic induction bar code scanning concept, has a "no read or no data" sort station and a powered conveyor travel path to reintroduce a "no read or no data" SKU (did not read a SKU bar code label) to an induction station. The first divert location past a bar code scanning/RF tag reading station is the "no read or no data" divert location, which is 12 to 15 ft. past the bar code scanning/RF tag reading station. This distance ensures time for communication between a scanner/reader device induction travel path photo-eye,

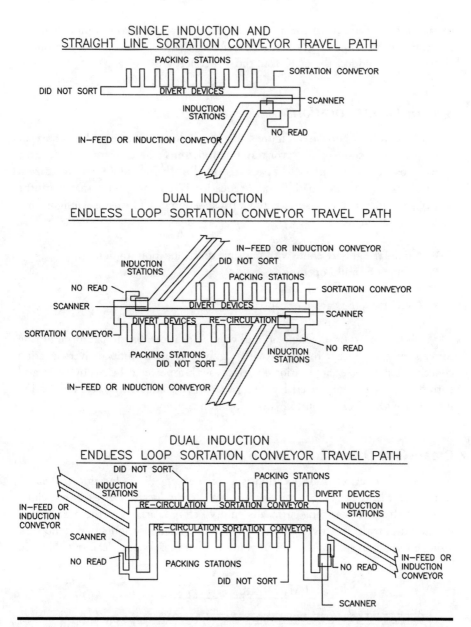

Figure 4.88 Mechanized sortation travel paths: straight line and single induction, endless-loop with single induction, and endless-loop with dual induction.

computer, and divert station. Controls are designed to have a "no read or no data" divert location that receives each SKU that does not have a bar code label or that has an unreadable RF tag. A computer instructs the sort concept travel path to divert the SKU onto the "no read or no data" divert location, where SKUs queue for an employee to return them to an induction station, or travel on a powered belt conveyor travel path back to an induction station. At the induction station, an employee reintroduces these SKUs to the sort concept.

Sort or Divert Locations

A sort or divert location (see Figure 4.89) has a chute or powered belt conveyor travel path. This chute or conveyor travel path permits a SKU to travel from a sort concept load-carrying surface to a pack station, directly into a CO shipping carton or onto a conveyor travel path. A divert location is designed to interface with the sort concept load-carrying surface and a pack station or CO carton. Designs for a sortation location are:

- A one-lane gravity chute with multiple divert locations optional
- A flipper or multi opening
- A belt conveyor travel path
- A CO shipping carton or captive tote

A tilt tray, cross or moving belt, gull wing, Nova sort, brush, platform, or ring sorter sortation concept interfaces with a gravity chute or flipper divert location concept. A flap sorter application interfaces with a powered belt conveyor travel path or CO carton/tote divert location concept. A Bombay drop sort application interfaces with a CO carton/tote divert location concept.

Clean-Out Divert Location

A clean-out or recirculation location is a sort concept divert location that has a sort concept load-carrying surface divert at a specific location. A clean-out chute ensures that a load-carrying surface is empty before it arrives at the next station on the travel path. It is usually located ahead of a relatch or induction station. In a single-induction station concept, this means that there is one clean-out chute; in

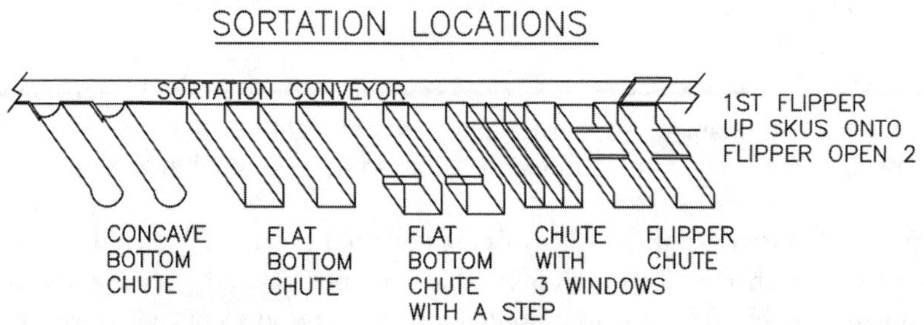

SORTATION LOCATIONS

SORTATION CONVEYOR

1ST FLIPPER
UP SKUS ONTO
FLIPPER OPEN 2

| CONCAVE BOTTOM CHUTE | FLAT BOTTOM CHUTE | FLAT BOTTOM CHUTE WITH A STEP | CHUTE WITH 3 WINDOWS | FLIPPER CHUTE |

Figure 4.89 Sortation locations.

a dual-induction station there are two clean-out chutes. After a SKU arrives at a clean-out chute discharge end, an employee physically transfers it to an assigned pack station or returns it to an induction station.

Relatch Station

A relatch station is a station on a tilt tray, Bombay drop, gull wing, or Nova sort concept. It is located past the clean-out chute and ahead of the induction station. After a load-carrying surface has been tilted at an assigned divert location or clean-out station, the relatch station ensures that the load-carrying surface has been righted and is ready to accept another SKU from the induction station.

Control Panel, Mimic Display, Air Compressor, and Electrical Panel

The control panel, air compressor, mimic display, and electrical panel are located in an induction station area and ensure that a sort concept is functioning properly and serve to indicate the status of the travel path to the manager. This feature permits the manager to start or stop the sortation travel path as necessary.

Various Centralized Mechanical Sort Concepts

Centralized mechanical sort concepts include:

- Tilt tray
- Flap sorter
- Bombay drop
- Nova sort
- Cross or moving belt
- Gull wing
- Ring sorter
- Brush sorter
- Side-tilting platform

Tilt Tray

A tilt tray sort concept is a fixed travel path powered by an electric motor or linear drive concept. The motor turns a shaft that moves a tooth sprocket drive and drive-train that propels a closed-loop chain. With a chain-driven tilt tray concept, each tilt tray is attached to a closed-loop chain that connects the trays together. A linear drive tilt tray is relatively new technology that has an electric current sent through a buss channel. Extending downward into the channel opening is a tilt tray carrying surface aluminum foot. The linear drive electric current traveling through the channel drives or pushes the aluminum foot forward through a channel travel path.

Each tilt tray is a load-carrying surface that carries a discreet customer labeled SKU. Long SKUs are carried by two trays. The tilt tray is so named because when it arrives at an assigned s sortation station it is tilted at an angle to transfer a SKU into a sort location. A tilt tray sort concept with at least two induction stations, standard tray size, and three locations per sort station performs approximately 8,400–10,000 sorts or tilts per hour.

A tilt tray travel path can be square or rectangular. A tray shape or chain travel path design allows each tilt tray to travel past the induction station and over a travel path to the appropriate sort station. At an assigned sortation station, a mechanical and electrical component tilts a tray.

Components of the tilt tray sort concept are the induction station, drive motor, drive unit, and sprocket or linear drive concept, closed-loop chain, take-ups, other tilt tray parts or linear drive concept, SKU tilt tray or load-carrying surface, sort location, structural support, side walls and guards, and electrical and computer control network.

Induction Station

An induction station is a sort concept location where an employee or powered conveyor section transfers a discreetly identified SKU onto a tilt tray carrier. SKUs are transported to an induction station either in totes or loose. The method is determined by physical characteristics of the SKU and other operational factors. At the induction station, SKUs are transferred onto a tilt tray carrier either manually or by powered conveyor belts. The latter provides greater productivity. SKU discreet identification is entered into a computer either by a manual keypad or hand held bar code scanner or an automatic bar code scanner/RF tag reader. To maximize return on investment and space utilization, an installation of a new tilt tray sortation concept should utilize a dual-induction concept. The advantages are higher piece volume, increasing capacity by 80 percent.

A single-induction tilt tray sortation concept (see Figure 4.90) means that all SKUs are inducted onto a tilt tray carrier at one induction station. All SKUs are transferred from a pick area via a transport concept to the same induction station. A tilt tray sortation travel path has one bar code scanner/RF tag reader or key pad SKU discreet identification entry station, one "did not read" station, one clean-out station, and one tray relatch station. Induction employees or a belt conveyor transfers each SKU onto an empty tilt tray carrier, which carries it to an assigned sort station on the left or right of the sort concept travel path.

A dual- or multiple-induction station concept has the following design features:

Figure 4.90 Induction options: manual and automatic induction.

- Two induction stations
- Two sort station areas between the two induction stations
- Two bar code scanners/RF tag readers or key pad entry stations
- One "did not read" station
- Two clean-out stations
- Two tray relatch stations

With the dual-induction station concept, one side of the tilt tray travel path is serviced by induction station one and the other side is serviced by induction station two. A dual-induction tilt tray sort concept maximizes the number of sort stations with a rectangular or elliptical travel path. Dual-induction stations can be designed to handle mixed SKUs or to have SKUs separated so each induction station handles one type of SKU.

Mixed SKUs are transported to either induction station and thus arrive randomly at either induction station. This means that SKUs are not separated for a specific sort station group beyond the induction station. This concept is less efficient than an induction station that handles separated SKUs, because a SKU intended for sortation section two may be inducted at induction station number one. The SKU then travels from induction station number one, past all induction station one sortation stations and past induction station number two before it arrives at the assigned sort location. This situation means that some tilt tray carriers travel from induction station number one past induction station number two unable to receive a SKU from induction two for sortation.

Separated SKUs that are picked and transported to a specific induction station result in an increase of volume by 80 percent. SKUs from induction station number one are sorted onto a sort station that is between induction station number one and induction station number two and SKUs from induction station number two are sorted onto a sortation station between induction station number two and induction station number one. The increase in volume results from the fact that most tilt tray carriers arrive empty at the next induction station, except when a "did not sort" occurs.

To facilitate SKU induction or transfer by an employee from the pick area transport conveyor travel path or tote onto a tilt tray carrier, the pick area transport conveyor travel path surface or the top of the tote is at the same elevation as the tilt tray travel path carrier or induction belt conveyor surface. After a SKU is placed or transferred onto the tilt tray carrier the control devices, tracking concept, and constant tilt tray travel speed ensure that the appropriate tray is tilted at the assigned sort station.

Factors to consider in designing the SKU transport concept include physical characteristics of the SKU, such as cube, weight, packaging or exterior surface, crushability and fragility. A tote transport concept is preferred for SKUs that have a high potential for damage. Empty totes are transported from the induction area back to the pick area. The tote transport concept can be a powered roller, skate-wheel, or strand conveyor travel path. Loose SKUs that are not fragile or crushable are transported on a powered belt conveyor with a slide. Either method requires stop and start controls on the travel path.

Communications between a tilt tray tracking device (a bar code scanner or RF tag reader) and microcomputer activate the assigned sort station tilt tray travel path lever so that it is elevated and comes in contact with the bottom of the tilt tray causing it to tilt at the assigned sort station. With the constant tilt tray travel speed, the tilt tray forward movement speed, smooth carrier surface, and gravity force, the SKU flows automatically from the tilt tray carrier onto the assigned sort station.

Electrically tilted trays have newer technology and generate less noise, have a short tilt window, and 45-degree tray tilt.

Tilt tray propelling concepts are motors, drives, and tooth sprockets with a closed-loop chain, or the newer linear drive, with an electric buss bar channel and a tilt tray with an aluminum foot that extends into a channel travel path.

Electric Motor, Drive, Sprocket, and Chain

An electric motor, drive, and tooth sprocket propels a closed-loop chain forward through a closed loop C-channel fixed travel path. Tilt tray design is determined by the following factors:

- Manufacturer standards
- Travel path length
- Number and angle of turns on the travel path
- Travel path slope and number of inclines and declines
- Maximum SKU load weight on each carrier tray assuming each tray on a travel path carries a maximum load
- Chain travel path surface

These factors create the coefficient of friction or drag on the closed-loop chain. If a tilt tray travel path has a high coefficient of friction, it requires a high-horsepower motor or multiple motors. A drive motor has a shaft that is connected to a drive unit. As the shaft turns, this drive motor and drive connection turn the drive unit. A drive unit has a sprocket with teeth or a screw (caterpillar) arrangement that extends outward and interfaces with the closed-loop chain, pulling the chain forward over the travel path.

The motor and drive unit can be bull wheel or caterpillar arrangements, or the newer linear drive.

- Bull wheel drive. A bull wheel motor drive unit design has a closed-loop chain that travels around a circular drive wheel and tooth sprocket side. The circular drive wheel has teeth that extend outward and interface with specific openings on the closed-loop chain as it travels past the motor drive unit. A bull wheel motor drive unit is first-generation technology and is hardly used on tilt tray concepts anymore.

■ Caterpillar drive. A caterpillar motor drive unit concept is second-generation technology. The motor drive unit design has a caterpillar drive located directly under the chain travel path. This design has a closed-loop chain that travels directly over the top of the caterpillar motor drive unit. A caterpillar screw turns and interfaces with the chain. As the caterpillar motor drive unit turns, an interface between a caterpillar motor drive unit and chain moves the tilt tray chain forward. A caterpillar drive unit location allows a tilt tray travel to have multiple caterpillar motor drive units. A caterpillar motor drive unit arrangement (a) handles 65 to 200 sorts per minute, (b) handles 20- to 40-lb SKUs. The SKU rides on a tilt tray with a 17- to 27-in. dimension on centers with a 2-in. gap between trays, and (c) travels approximately 180 to 220 ft./min. The advantages are less floor space, smooth chain start-up, less noise, fast chain travel speeds, long tilt tray travel path, several drive units per tilt tray travel path, and low maintenance.

■ Linear drive. A linear drive concept is new technology that propels a tilt tray carrier over a fixed travel path. A linear drive concept has an electric current that is sent through a buss bar and the electric current pushes an aluminum foot forward through the C-channel travel path. In a standard tilt tray concept, there are at least eight to nine buss bar sections to complete a tilt tray travel path, determined by the travel path linear feet, number of curves, SKU weight, number of declines and inclines, and manufacturer standards. Travel path sections are connected as one track, thus a tilt tray travel path is a closed loop. The advantages are smooth start-up, fewer moving components, faster travel speeds, lower noise level, less maintenance, and no chain slack or tense conditions.

Tilt Tray Chain and Chain Take-Ups

A closed-loop endless chain has hardened metal or plastic components with rollers or wheels that from the chain travel surface that comes in contact with a C-channel track. The chain components are hooked together to form an endless-loop chain with sections that are open to interface with the drive unit and to support a tilt tray carrier.

The C-channel track has a hardened metal or polyurethane wear bar on the interior surface and on the two side walls. The rollers or wheels ride on a wear bar to minimize friction or drag, and reduce wear, noise level, and vibration.

During installation, the chain is pulled through the entire track and connected together. The chain tension is adjusted to the proper level and the tilt tray carriers are secured to the chain on 17- or 27-in. centers, with a nominal 2-in. gap between trays. These features ensure that during induction there is sufficient space between trays, to allow accurate bar code scanning or RF tag reading and that there is proper space for SKU transfer at the sortation location.

A chain take-up device allows maintenance staff to maintain the proper chain length or tension as variations occur from wear and temperature changes. Take-up chain devices are located on the square or rectangular travel path turns, where there are no sortation or induction stations.

During warm summer months, in a warm geographical location, or after many hours of operation, the chains may stretch and become slack. This condition creates travel speeds that are off the designed speed and results in mis-sorts or maintenance problems. During cold months a chain contracts and becomes tighter. A tight chain creates tension and results in serious maintenance problems, such as broken chain sections.

The chain take-up devices are designed to move forward or backward from the vendor chain installation position or a specific start-up chain tension or length. A chain adjustment compensates for the condition of the chain and improves tilt tray operation.

"Did Not Read" Station

A "did not read" station and SKU re-entry travel path are the first sort station and are located approximately 15 ft. past an induction station. If a bar code/RF tag on a SKU was not read or a keypad entry was not complete, the tilt tray computer automatically tilts the tray that was not entered into the computer as it arrives at the "did not read" station, transferring the SKU from the tilt tray onto the sort station, from which it travels on a conveyor back to the induction station, where an employee re-enters the SKU onto a tilt tray travel path.

"Did Not Sort" Station

A clean-out or "did not sort" station is a location where all trays that failed to tilt at the sort station are moved into the clean-out sortation station where the tray tilts and deposits the SKU. At the discharge end an employee physically moves the SKU to an assigned sort station. A clean-out sort station assures that all tilt trays arrive empty at an induction station ready to receive another SKU. When most of the tilt trays arrive empty at an induction station, the tilt tray concept is able to handle a high SKU volume.

Relatch Station

A reset or relatch tilt tray station is located after a clean-out station and before an induction station. In this travel path location, a relatch tilt tray station ensures that all trays are ready or level to accept a SKU from the next induction station.

Tilt Tray or Carrier

Tilt tray or SKU carrier dimensions are 17 to 27 in. long with a width dictated by manufacturer standards. A tilt tray is manufactured from wood, plastic, or coated sheet metal designed to resemble the wings of a gliding bird. The sides are higher than the middle and the load-carrying surface is a rectangle or square. The wing-shaped design is intended to:

- Create a low-profile load-carrying surface
- Ensure that a SKU bar code/RF tag is properly presented to the bar code scanner or tag reader
- Create a slight cavity that holds the SKU on the carrier
- Provide a smooth surface for SKU transfer from the tilt tray onto a sortation location

The height of the side is determined by SKU features, chain travel speed and travel path slope, sort location, and manufacturer standards.

The base of the tilt tray is attached to a chain component or sits in a linear drive travel path. During tray travel, a tilt tray is in a flat position or level. At an assigned sort location or clean-out location, a computer activates a mechanical device (a lever) on the travel path that extends upward and comes in contact with the lead undercarriage. This contact and the forward movement of the tray tilt it toward the assigned side of the travel path, sliding the SKU onto a sortation location.

An alternative to the mechanical lever tilt mechanism is an electrical device that is computer-activated to elevate the side of the tilt tray before it arrives at an assigned sort location. As the tilt tray arrives at the assigned sort location, a tracking device, computer controls, and tilt tray carrier constant travel speed has an electrical component that triggers a hydraulic mechanical device that tips the tray at approximately a 45-degree angle. An electrical tilt tray features no knocking noise as the tray tilts, the carrier automatically resets the tilt tray, tilt is at a 45-degree angle, and tilt on a curve inside radius. There is no powered travel path section between a tilt tray travel path and sort travel path station and sorts to a narrower sortation window.

Sort or Divert Station

A tilt tray sort station is a location that is assigned for sorting COs. A sort location receives an assigned CO picked and labeled SKU from a tilt tray carrier. The tilt tray sort station surface has a smooth, continuous, and slippery exterior surface to allow a SKU to travel the full length of a chute. SKU exterior packaging material is designed to have a low coefficient of friction to the sort station surface. Sort station material options are (a) coated wood, (b) sheet metal, (c) hardened plastic, and (d) belt conveyor.

Sort stations are designed with sufficient length and width to hold the entire cubic volume of the SKU and are square or rectangular in shape.

Sort station characteristics are:

■ For faster travel path speeds, the sort window is made wider to handle SKU transfer from a tray.
■ To handle a wide SKU mix, the sort station has a wide window to accommodate variations in SKU weight and dimensions.
■ A solid bottom surface and solid sides with the bottom sloped downward, the full-length from the tilt tray travel path to the sort station front end. The sort station front end has a lip or front barrier to keep SKUs in the chute and allow SKUs to queue in a slide.

Queuing may occur when the entire sort station chute cube or space is not utilized or when there is uncontrolled line pressure on a front SKU that that has come to rest against the barrier. Sort station designs to improve space utilization and reduce SKU line pressure are to have a step in the middle of the sort station length, a sort station with multiple openings or flippers, or a belt conveyor.

A step in a sort station chute is located approximately half the distance from a sort station charge end to the front barrier. The step is designed to add to the difference in elevation between the lower and upper portions of the chute. When SKUs are queued against a barrier at the step, other SKUs in the sort chute ride over the top of the queued SKUs and continue to flow to the front barrier and queue there.

A chute that has multiple openings or flippers can also be used to increase space utilization in a sort concept. The chute is equipped with flippers that can be opened (to extend downward) or closed (level). The flippers are controlled by a wave mechanism or a computer. A flipper is placed in an open position to receive SKUs or in a closed position to allow SKUs to flow over it to an open flipper. Depending on the SKU bar code, the computer program opens or closes a flipper to ensure the SKU is deposited at the appropriate location.

The sort station design cube is determined by the cube quantity or the SKU volume that is sorted to the sort station. A 20 percent additional safety factor is allowed for the random mixed SKUs in a queue or the orientation of the queued SKUs in a sort station. If SKUs are sorted at a high speed for a wide mix (sizes and weights) and resulting in damage to lightweight or small-cube SKUs, the options are:

If possible to slow down the sort carrier travel speed to allow the sort concept to handle the SKU volume. With a wide variety (sizes and weights) in the SKU mix, setting the travel path at one speed that can handle both very lightweight SKUs and very heavy SKUs is difficult.

Install a plastic strip curtain on each sort travel path, at least one third the distance into the chute. Plastic curtain strips extend downward from a rod at the top of the sort station. The length of the plastic strips allows clearance for heavy and tall SKUs traveling along a sort station surface. When a lightweight SKU hits the curtain, the plastic strips slow it down, thus reducing potential damage and allowing the SKUs to queue in a sort station from the front.

A sort lane has photo-eyes that communicate the status of the lane to the tilt tray computer. The first photo-eye location is halfway from the front of the sort lane. When this photo-eye is blocked by queued SKUs, the computer receives a message and activates an alarm device so that a sort/pack employee responds and removes SKUs from the chute. SKU removal unblocks the photo-eye, which sends a message to the computer to turn off the alarm. A second photo-eye is located approximately two thirds of the distance from the front of the sort lane. When this photo-eye is blocked by queued SKUs, the computer does not activate the sort station tilt tray lever, thus reducing SKU or equipment damage from tilting SKUs into an overloaded sort station.

Structural Support, Side Walls, and Guards

Structural support, side walls, and guards are made of coated hardened metal and provide support for the drive motor, drive unit section, induction platform, tilt tray travel path, and sort stations. Floor base plates support and spread the load weight and there are anchor holes in the base plate for bolts and nuts to secure it to the floor. A tilt tray area should have a level floor to ensure a good installation. This reduces the need for shim plates to level the travel path upright floor support structures. A tilt tray travel path has a minimum 13-ft. clearance from the floor to the ceiling support.

The tilt tray side walls are solid, coated sheet metal or plastic structures that are attached to structural support elements on both sides of the travel path. The side wall members start from under the chain track and extend downward and upward to a location that is under a tilt tray carrier travel path. Side wall members are used to reduce the noise level from movement of the chain through the track and to provide employee safety.

The guards are solid, coated sheet metal, plastic, or wire-meshed. Per the manufacturer's standard, local code, or company specifications, guards are installed along both sides of a travel path to:

- Reduce the risk that a SKU will fall from a tilt tray carrier to the floor and cause damage or employee injury
- Reduce employee injury from a limb protruding into the tilt tray travel path
- Minimize the risk of a mechanical component falling from the travel path

If small SKUs accidentally slide from the tilt tray carrier, mesh netting is installed under the travel path. The netting extends a sufficient distance beyond the center line to the side and catches any item that falls from the tilt tray.

Electrical Components and Controls

Electrical and control network components include the bar code scanning device, E-stop devices, computer, photo-eyes, control and electrical panel, and air compressor. Collectively, these components ensure that SKU induction data, tilt tray carrier status, tray location on a travel path, and sort station status are communicated to the tilt tray computer, which controls the tilt tray moving parts. This communication allows the computer to start and stop the tilt tray and, in a flipper sort lane, to set a flipper to accept a discreetly identified SKU into the appropriate sort station. When activated by an employee, an E-stop device causes the electrical system to halt a tilt tray to minimize SKU, building, or equipment damage and employee injury. In a tilt tray concept, an overhead bar code scanner is a fixed-position moving-beam bar code scanner or RF tag reader. To ensure proper and accurate reads and communication, a bar code scanner has independent structural support to minimize vibration that may create "no reads" or misreads.

Flap Sorter

SKUs that are picked for COs are transported to an induction station, where a flap sorter concept moves them on a platform across a fixed travel path to an assigned sort station. A flap sorter travel path has many belt conveyor travel path sections that are each powered by an electric motor. As the SKU arrives at an assigned sort station, the conveyor section or flap angles downward, and with gravity, the SKU falls into the sort station. After a specific number of seconds, the conveyor section returns to a level position. In a level position, a flap sorter has completed a SKU sort activity and another SKU is transported to an assigned sort station. The travel path is from an induction station, over inactive flap sorter sections, and over the fixed belt conveyor sections to an assigned flap sorter belt conveyor section.

Components of a flap sorter are fixed, short powered belt conveyor sections, flap or flexible short powered belt conveyor sections, a SKU induction station, return conveyor section or "did not sort" station, drive motors, drives and pulleys, sort stations, structural support, and microcomputer and controls.

Fixed, powered short conveyor sections are located:

- At the induction location to ensure controlled SKU movement from a transport concept onto the flap sorter travel path
- On the flap sorter travel path to maintain controlled SKU movement from the sortation travel path onto a sort station
- On the flap sorter travel path after the last sort station to ensure "did not sort" SKUs are returned to an induction station

Fixed, short conveyor sections are 4 to 5 ft. in length.

Each flap conveyor section is designed like the fixed conveyor section, except that the lead end declines at an angle (downward) to create an open space between a declined travel path surface and the next fixed conveyor section lead end. This open space has sufficient clearance for the tallest SKU to pass through without striking the lead frame of the next conveyor section. This feature makes a flap sorter an active–passive sortation concept that sorts SKUs into a downward direction.

Forward movement of the flap sorter travel path and gravity transfer a SKU from the flap sorter into a sort station, located below the travel path. The sort station is another belt conveyor travel path or CO carton/tote.

The induction station has structural support, an induction area and employee work area, equipment, controls, the transport travel path discharge end, and the flap sorter travel path charge end. Induction station platform members are welded or connected together with nuts and bolts. The structural members are coated to minimize rust and dirt and have base plates anchored to the floor.

The "did not sort" return travel path is a powered belt conveyor travel path or a decline chute. The "did not sort" belt conveyor travel path has waterfalls or powered belt curve sections to return "did not sort" SKUs to an induction station.

The drive motors are electric powered, with drives and pulleys to ensure that the flap sorter powered belt conveyor travel path moves at a specific travel speed and in the correct travel direction to ensure the SKU sorts at the assigned location.

A flap sorter sortation station is a CO shipping carton, tote, or belt conveyor travel path. Sort locations are located directly below each flap sorter flexible powered belt conveyor decline discharge end. Sorted SKUs are queued in a CO shipping carton, tote, or transported on a belt conveyor travel path.

Travel path structural support members are welded or nut and bolt hardened metal members that are connected together. The members have a coated surface and each upright post base plate is anchored to floor. Structural members provide stability and rigidity to a flap sorter travel path.

Electrical and control components are photo-eyes, bar code scanner, keypad, computer, electrical and control panels, and required E-stops. Together the components ensure:

- That the SKU sort location or SKU identification number is entered into a sort computer
- That there is no travel path, product, or equipment damage or employee injury
- That an assigned SKU is sorted from a travel path at the assigned sort station or is returned to the induction station

After a SKU with a bar code/RF tag or human-readable code is delivered from the pick area to an induction station, an induction employee transfers it onto the induction conveyor section or directly onto the flap sorter travel path. When a SKU is transferred onto the flap sorter travel path, the bar code faces upward to permit the bar code scanner to read the bar code or RF tag reader to read the RF tag. If manual keypad entry is used at an induction station, an employee enters the SKU address. Both concepts, a bar code scanner/RF tag reader or key pad, send the SKU sortation address to the computer.

An induction option is to have the batched CO picked SKUs with only labels delivered to an induction station and an induction employee enters a SKU discreet identification or number into a computer. After an employee transfers a SKU to a flap sorter travel path, a computer ensures that the SKU is diverted or sorted to a first CO that has the SKU. Per the control and tracking system, a flapper sorter diverts the SKU at an assigned location.

As a SKU passes a bar code scanner/RF tag reader or key pad entry station, the conveyor divert controls, tracking device, and conveyor travel speed ensure that the assigned SKU arrives at the assigned sort station, where the assigned flap sorter flexible conveyor section angles downward to transfer the SKU from the flap sorter travel path onto a CO sort location. With this sort concept, there is some impact on a SKU because it is a waterfall concept.

The disadvantages are handles limited shaped, cubed, and weight SKUs, capital investment, maintenance, some impact on a SKU, and it is a single induction sort concept. Advantages are it handles a high volume, handles both a sort bar code/ RF tag, human-readable code or SKU discreet identification, handles batched CO picked SKUs, and sorts direct into a CO shipping carton, tote, or belt conveyor travel path.

Bombay Drop

A Bombay drop small-item sortation concept (see Figure 4.91) is used in a batched CO picked and labeled SKU order-fulfillment operation. A Bombay drop sort concept has platforms that are side mounted to an electric motor-powered, closed-loop chain. After a SKU is inducted onto a Bombay drop carrier, the computer, tracking device, and constant travel speed ensure that a Bombay drop sort activity occurs at an assigned sort location. It is possible to design a Bombay drop concept as a dual-induction concept.

Components of the Bombay drop sortation concept are SKU load carriers or trays, electric drive motor, pulleys and sprockets, structural support and frames, and computer and controls.

In a Bombay drop concept the load-carrying surface is a tray with two moveable flaps. Each tray is designed to support a SKU with the largest cube characteristics and heaviest load weight per a Bombay drop design parameters. Each tray handles one SKU. The flaps are attached at the far end to a metal rod. During transport from an induction station to a sort location the flaps are locked to form a flat, solid SKU carrier. For SKU sort at a sort location, the two flaps move downward to create an opening through which the SKU falls. After a sortation, the flaps return to the flat and locked position. Each tray flap has a trigger device. After a SKU is

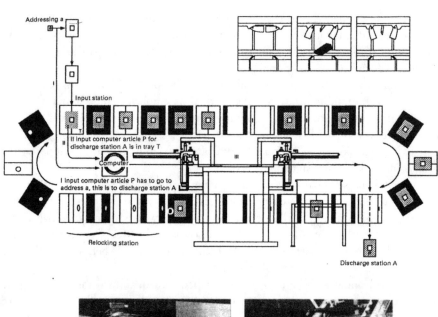

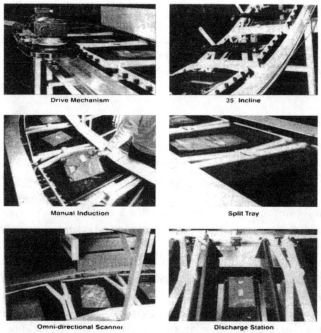

Figure 4.91 The Bombay drop.

discharged from the tray, the trigger device extends upward above the metal rod and comes in contact with an angled reset device that forces the tray trigger down and causes the flaps to fold upward to the center. Prior to transfer of a SKU onto a Bombay carrier, its discreet number is entered into a sortation concept computer. As the tray arrives at an assigned location, a computer-controlled device activates the trigger, and the trays unlock. The concept is so named because the tray flaps open in a manner similar to a military airplane's Bombay doors opening to drop a bomb. Gravity force allows the SKU to fall through the opening between the two flaps into a sort station below, which is a carton, tote, or belt conveyor travel path.

An electric-powered chain is designed as a closed-loop travel path that travels past an induction station, all sort stations, reset station, end pulley, and drive motor. Attached to the chain are 22 to 30 tray carriers with support members and travel wheels. When a powered chain is not traveling across a pulley, the chain is traveling in an enclosed track or C-channel. As the chain travels around an end pulley, the sprocket teeth ensure that it maintains its direction and speed. A drive motor is attached to a shaft that turns the tooth sprocket. The drive motor size or horsepower depends on the length, number of carriers, and maximum SKU weight on each tray.

Structural support is provided by angle iron and metal sections that are connected together by welds or nuts and bolts. All structural members are coated and have sufficient gauge to handle the static load weight and maximum dynamic load. To ensure stability and to spread the load weight, each frame post has a base plate that is anchored to the floor.

Other structural sections are the powered chain travel path, induction station, and sort stations. An induction station has a platform designed for the weight of an employee and the combined total weight of the SKU and travel path section.

A sort station is a flat surface that supports a carton, tote, or belt conveyor travel path. A sort station surface is designed to support the maximum combined weight of the carton and SKUs. The height above the floor allows clearance for the maximum height of the carton or tote so that it does not interfere with the tray flaps when they are in the open position as the tray travels past the sortation station.

Induction employees enter SKU codes into a computer system that identifies each CO SKU and sequences the SKUs in an arithmetic order that starts with sort station one and programs a Bombay drop sort concept for each CO SKU to sort a SKU to each CO sort station. With the tracking device and constant chain (tray) travel speed, the computer ensures that each SKU is sorted to its assigned station.

An entire batch of CO picked and labeled SKUs is queued on the transport concept for release as a slug to a Bombay drop induction station, where an employee is ready to induct a group of SKUs onto a Bombay drop sort concept. The employee enters the discreet number for each SKU into the induction concept computer. To complete a CO picked SKU entry, an individual tray carrier receives a SKU. As the tray passes the reading device, the scanner/RF tag reader verifies that the tray is occupied with a SKU. After verification, the computer addresses each assigned tray carrier to a specific CO sort location. As each tray leaves the induction station, it is tracked by the computer. When an assigned tray arrives at an assigned sort station, the tray carrier opens in the middle to drop the SKU into a CO carton, tote, or onto a belt conveyor travel path. As the tray continues to move forward on the travel path, the doors are relocked before it arrives at the induction station.

The disadvantages of the Bombay drop are that it handles a limited SKU mix and weight, has a high capital investment, and is a one-direction sort concept. The advantages are handles batched CO, uses SKU or CO discreet number, can be designed with dual induction and sorted by family group.

Nova Sort

A Nova sort concept is a modular tilt tray mechanism that has a closed-loop, fixed travel path. A Nova sort concept has similar design parameters and operational characteristics as the tilt tray sort concept, except that each Nova sort unit has its own drivetrain.

Components of the Nova sort concept are three to four tilt tray trains, with a 500-lb load-carrying capacity and the ability to tilt to the right or left of a travel path, a computer for each tray train, a fixed travel path that has an electric-powered buss bar, and on-board sensing devices to detect an employee in the travel path. A Nova sort vehicle is a ceiling-hung or floor-supported device, able to travel up slight grades and over 90- and 180-degree curves. A travel path directs a tilt tray train past an induction station and each delivery sort station. At the induction station, an employee transfers a SKU onto a vacant tray and keys a tray destination location into the computer. The computer assigns the tray to tilt and discharge the SKU into an assigned sort station.

After all trays are tilted at assigned sort stations and reset to the level position, the tray train travels on a travel path with each tilt tray train queuing prior to the induction station or continues traveling to the maintenance spur. On the maintenance spur a tilt tray train waits for the next assignment. The number of sort stations can be expanded by adding a track to the travel path.

The disadvantages are it handles a medium volume, has a fixed, closed-loop travel path that has potential train queues, and capital cost. Advantages are ceiling hung or floor-supported travel path, easy expansion, and unload/load at any station.

Cross or Moving Belt

A cross or moving belt sort concept (see Figure 4.92 and Figure 4.93) has individual belt conveyor carriers. Each carrier has a short, powered belt conveyor platform that is mounted onto four motor-driven wheels that are designed to travel over a closed-loop travel path past an induction station and past each sort station. At the induction

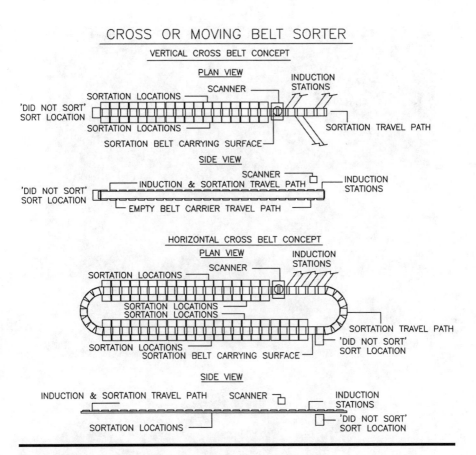

Figure 4.92 Vertical and horizontal cross belt sorters.

station a bar code labeled SKU is placed onto a vacant carrier. As the carrier travels under a bar code scanner device, the scanner reads and sends a bar code or SKU discreet identification (sortation location) to a computer. The computer, tracking device, and constant carrier travel speed ensure that the carrier short belt conveyor moves forward and arrives at the assigned sort station. The assigned sort station is on the left or right side of the travel path. Each sort location is designed with one, two, or three deep controllable chutes or flippers that cover a chute or CO sort location. Per the controls, the flipper chutes open to accept assigned diverted SKUs.

Design options for a moving belt sortation concept are:

- A vertical, endless, closed-loop travel path that resembles a track on a tank
- A horizontal, endless, closed-loop travel path
- An endless, closed-loop travel path that has an S configuration

Figure 4.93 A vertical cross belt sorter.

Moving belt sort concepts are similar in some respects to a tilt tray concept. Features are listed below:

- An electric, powered, motor-driven or extruded aluminum and electric-driven belt platform.
- A travel path and platform that varies in size from 5 to 8 in.
- A closed-loop, endless travel path that directs short powered belt platforms past an induction station and each sort location.
- SKUs are diverted to the left or right of the travel path.
- It accommodates a wide SKU mix.
- There is a single-induction station.
- A powered moving divert platform.
- SKUs are inducted onto a sort concept.

Vertical Cross or Moving Belt

A vertical cross or moving belt concept is a single-induction concept with an induction station, divert locations, and fixed travel path. The vertical moving belt is capable of handling "did not sort" and "did not read" SKUs. A single-induction station inducts SKUs onto a cross belt travel path at one location. A cross belt sort concept is designed with two employee or four automatic induction stations. Maximum manual induction capacity is 2000 trays per hour or with an automatic induction option 2500 belts per hour. The cross belt induction has the induction discharge end at the location where the vertical belt carrier is horizontal to the floor. As the cross belt carrier reaches this horizontal position, an induction station moves a SKU forward onto the carrier. A vertical cross belt concept has a maximum of 66 sortation stations per travel path side.

The travel path layout is a straight line from the induction station to a "did not read" location. Past this point, the cross belt travel path flows directly under a sort travel path back to an induction station. During travel from an induction station to a "did not read" location, the cross belt carrier faces the ceiling. After a "did not read" location, the carrier faces the floor. As the carrier makes a turn from a face-up position to a face-down position, any "did not read" or "did not sort" SKUs fall from the carrier into a tote. This travel path configuration resembles a tank or bulldozer track. A vertical moving or cross belt carrier sort rate is a nominal 8000 sorts per hour. A standard cross belt sortation carrier has a length of 40 in., a width of 28 to 40 in., 10 in. in height, and SKU maximum weight capacity is 16 lb.

Horizontal Cross Belt

A horizontal cross belt concept can be designed with single or dual induction. Dual induction increases the sortation capacity by a factor of 1.5 to 1.75. Other components are "did not read" or "did not sort" divert locations, four curves to complete the endless-loop travel path, and a medium or large floor area to complete the sortation travel path. With dual induction, this concept has the capacity to complete 16,000 sorts per hour.

As a load-carrying surface arrives at an assigned sort location, the computer activates the conveyor to turn in the proper direction that moves a SKU forward from the load-carrying surface onto a sort location travel path. Past the sort station, the load-carrying surface travels on to an induction station to receive another SKU. With a single-induction station, a moving belt sort concept handles 10,000 to 14,000 sorts per hour, with a SKU weight up to 44 lb. When floor space is limited and with a medium sort volume, a vertical cross belt sort concept is the preferred concept. With no constraints and a high sort volume, a horizontal cross belt sortation concept is the preferred concept.

The disadvantages and advantages are similar to those for a tilt tray sortation concept, except the additional disadvantages are that it is difficult to handle round objects, requires a high capital investment, and requires maintenance. Additional advantages are less floor space and easy expansion.

Gull Wing

A gull wing sort concept is an active–passive sortation concept that has many single-item carriers that ride on a closed-loop travel path. The carrier has a gull wing, or V-shaped, surface. The operational features, disadvantages and advantages are similar to those of a tilt tray sortation concept. After a gull wing carrier arrives at a sortation location, a tipper is activated to move upward to tip a carrier to the side. A carrier tipping action and gravity force allow a single small-item to slide from a carrier to a sort/pack station. A gull wing sort concept handles a wide variety of SKUs and performs 12,000 sorts per hour.

Ring Sorter

A ring sorter concept (see Figure 4.94) receives its name from the fact that the sort travel path is in a circle or ring shape. After a coded small-item flatwear SKU is inducted onto a ring sorter travel path, it travels on a movable belt conveyor surface. At an assigned divert location, a carrier powered belt moves forward and the SKU is transferred from the belt sort travel path to an assigned divert location. A ring sorter sort concept has an induction travel path and overhead bar code scanner, powered belt sort lanes and dividers, sort designs and stations, structural support, and sort platform and controls.

A ring sorter induction belt conveyor travel path and overhead bar code scanning device have several short, powered belt conveyor sections that ensure there is a gap between SKUs. With a proper gap, a SKU travels over a decline powered belt conveyor travel path that directs it under an overhead bar code scanning/RF tag

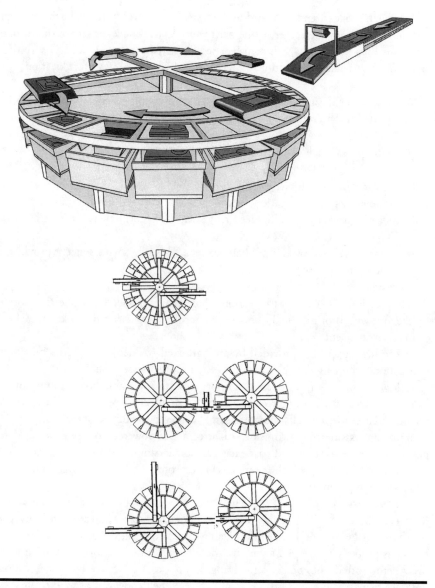

Figure 4.94 A ring sorter: one, two, and three sorters.

reading device and onto an open sort belt conveyor surface. The overhead bar code scanner/RF tag reader reads the SKU bar code and sends the data to a computer. The computer and the ring sorter constant travel speed ensure that an assigned SKU is sorted from the travel path onto a SKU sort location.

A ring sorter has four to eight powered belt conveyor sections, spokes or platforms. A ring sorter plan view looks like a wagon wheel with a center and four to eight spokes. Each spoke extends outward to the ring outside sort locations. Each conveyor section moves from the center to the perimeter. As the conveyor sections and solid platforms turn in a circle, each conveyor section arrives in turn at a SKU sortation location. At this assigned location, a SKU is transferred into a CO sort location.

- Ring sorter design depends on the dimensions of the SKUs that are to be handled by the concept.

Design options include the following:

- A basic design with one sort ring.
- Double-deep final sortation locations, with two-deep sortation locations for each ring sorter. A double-deep ring sorter concept increases the number of CO sort locations.
- A reversible in-feed powered belt concept. This ring sorter concept provides a large number of sort locations with a high sort rate.

Connection belt travel path sections that transfer SKUs from one ring sorter (spoke and wheel) or major sort concept to another ring sorter or spoke and wheel or final sort concept.

Structural support is provided by are hardened, welded, and coated members that support the ring sort platform and each CO sort location. To assure proper sortation alignment, the structural support legs are adjustable upward and downward. The sides of the ring sorter provide the structural support and rotating track for powered conveyor belts and platforms, center post, drive motor, and rollers. Components ensure that a ring sorter platform and powered conveyor belts rotate past an induction station and past each CO sort location. The solid side walls and platform surface minimize the noise level and protect against employee injury and UOP and equipment damage.

The CO sort location has hardened, welded, and coated metal members that form a cantilevered shelf. The shelf width matches a ring sorter powered belt conveyor width dimension. Length and elevation above the floor ensure that a CO sort carton or tote is properly positioned to receive the maximum number of pieces and that a SKU being sorted does not become hung up on a carton or tote lip. Sort location options are plastic tote, cardboard carton, powered belt conveyor, or solid chute.

Control devices include start/stop controls, E-stop devices, overhead bar code scanner/RF tag reader, and power and control panels.

Brush Sorter

A brush sorter concept is very similar to a tilt tray sort concept in the following ways:

- SKU induction options

- Trays or carriers that travel over a fixed, closed-loop travel path that directs a SKU under an overhead bar code scanner/RF tag reader and past each sort location
- Protective side guards along the travel path

Features unique to the brush sorter concept are:

- A flat carrier with (a) a diamond pattern to minimize uncontrolled SKU movement on the carrier, (b) a flat coated metal surface, or (c) a hardened wooden surface, and a lip on the trail or rear side of the tray relative to the direction of travel. The lip extends upward about 1 in. above the carrier surface and acts as a barrier to retain a SKU on the carrier.
- Each tray has a pusher brush at the far side. At an assigned CO sort location, a microcomputer activates the brush, which moves forward from the far side to the near side, acting as a broom to sweep a SKU from the tray carrier onto a CO sort location. The brush also acts as a back-stop. During induction of a lightweight SKU onto a tray, the brush retains the SKU on the tray.
- Along each travel path side is a protective side guard or skirt.

A brush sorter handles a SKU that ranges from 16 to 20 in. in length, 12 to 16 in. wide, with a maximum weight of 15 lb. It can sort 6000 single inductions and 11000 dual inductions per hour.

Tilting Platform

A tilting platform sorter concept has an in-feed station, computer, and a side-mounted tilt table carrying trays. The end of each tray is attached to a closed-loop chain that is driven by an electric motor that has a tooth sprocket and a second tooth sprocket that interfaces with the chain, tray leveler, and discharge locations.

After SKUs are singulated on the in-feed conveyor travel path, they are transported to the in-feed station, where each SKU is transferred onto an empty tilting platform. The tilting platform with a SKU travels over a fixed travel path and arrives at an assigned sort location, where the platform tilts forward, sliding the SKU onto a sortation location. A tilting platform sorter concept is used for a low- to medium-volume operation.

Customer Order Check Activity

The next step in a small-item flatwear order-fulfillment operation is a quantity and quality check. This check ensures that the number of SKUs and the SKU description matches a CO. At a check station, the CO SKU quantity, description, and weight are voice directed or appears on a digital display, label, or a paper document that permits a check employee or machine to verify the order-fulfillment accuracy. An employee-based method or RF device concept verifies both a CO picked SKU quantity and quality. A scale-based check concept verifies a CO picked SKU quantity.

When an order-fulfillment operation experiences customer complaints that customers have received shortages, overages, damaged or incorrect SKUs at a CO delivery address, an order-fulfillment operation implements a CO picked check activity and station.

A CO shortage is when a CO pack slip and the actual CO received at the delivery address do not match in terms of the number or type of SKU. Problems with a CO include SKU shortage, overage, damage, and incorrect SKUs. These result in dissatisfied customers because it costs them time and money.

CO check activities include:

- A manual CO check activity for SKU quality or SKU quantity, or both.
- A mechanized check activity that has a digital scale that indicates an actual CO order carton weight. An employee compares a computer-projected CO SKU and carton weight to the actual CO weight.
- An automatic or "check on the fly" activity that has a computer interface to a scale to compare actual CO picked SKUs and carton weight to the computer-projected CO SKU and carton weight.

A CO check activity is a post-pick activity that occurs in all manual, mechanized, or automated order-fulfillment operations. In a single CO fulfillment operation, the check activity occurs on a pick line or at a pack station. In a batched CO pick operation, the check activity occurs at a sort or pack station. In an automatic pick concept, the CO check occurs prior to arrival at a pack station or at the pack station. The procedure that is used for a CO check activity is determined by CO volume, SKU value and SKU mix, company policy, company order-fulfillment or CO delivery performance standard, and type order-fulfillment operation.

Carrying out a CO check activity requires use of facility space and adds time to a CO order delivery cycle time. With a small to medium CO volume, a manual or mechanized check concept is adequate to handle the volume. With a high CO volume, a mechanized or automatic concept is better able to handle the volume. For COs with high-value SKUs, 100 percent of the CO-picked SKUs are checked.

A CO that has small-cube, lightweight, powered, or liquid-filled SKUs mixed together increases the difficulty of checking COs cost effectively and efficiently using a manual, mechanized, or automatic check activity. With an employee CO check concept, it is difficult for an employee to locate all small-size SKUs in a CO carton that has this wide SKU mix, or with SKUs that have common shapes and colors. With a mechanized or automatic check concept, the difference between the projected and actual weight of a small-size SKU and a powder or liquid-filled SKU has the potential to exceed an acceptable or obtainable weight variance.

Company-established CO check and customer service standards determine the type and number of COs that are sent through a CO check station. A CO is sent through a check station based on a specific dollar value, customer history, and past delivery performance. With a manual or mechanized CO check concept, to select a CO for a check activity is more difficult. With an automatic CO check concept, a bar code scanner and computer identifies a CO for checking. If an order-fulfillment operation performance is acceptable in relation to delivery or service standards, a CO check station does not check 100 percent of COs, but performs a random check. If performance is not acceptable, the CO check may be on 100 percent of COs.

Quality and Quantity Check Concepts

Quality and quantity check concepts are based on a company's customer service standard and past CO delivery performance. Checks are carried out on a random sample, 100 percent of COs are checked, specific COs are checked, or there is no check of COs.

No Check Performed

The practice of not checking COs is based on an excellent record of past picker accuracy and CO delivery performance. To be effective and to ensure continued compliance of an order-fulfillment activity with company standards, there should be a periodic random check of COs.

Random Sample

A random sample check is carried out on 7 to 10 percent of the COs. If a random sample check satisfies service standards, the remaining COs (90 to 93 percent) are considered acceptable. A random sample check concept is used for any SKU pick operation that handles a small to medium SKU or CO volume in a medium-size floor space.

100 Percent Check

In this type of operation, every CO is checked. This concept is used in an order-fulfillment operation that has a high occurrence of overages, shortages, mis-picks, incorrect SKUs, and damage. A manual 100 percent check concept is used for any type of order-fulfillment operation that handles a small to medium SKU or CO volume in a large floor space. A "check on the fly" concept is a 100 percent quantity check activity.

Specific Customer Order Check

With a specific CO check concept, the company CO files indicate specific customers who have had a high frequency of problems with COs. To minimize future CO delivery problems, when one of these identified customers places a CO, the order-fulfillment operation has this specific CO sent through a manual quality and quantity check station. A specific CO check concept can be used in any order-fulfillment operation.

Order-Fulfillment Operation Type and Customer Order Check Concept

The type of order-fulfillment operation influences how COs are checked. There are three main types:

- A batched CO fulfillment concept in which SKUs for several COs are picked together and flow from a pick area, through a sort area to a pack station. At the pack station, per a CO pack slip, a packer completes a final sort of SKUs, checks the CO pack slip, and packs a CO.
- A batched CO fulfillment concept in which SKUs are picked for several COs at the same time and subsequently sorted into a CO assigned sort location. After batch pick activity, completed COs are transported in a four-wheel cart with shelves, large divided tote, or divided trolley basket from a pick/sort area to a pack area. In the pack area, a packer with a CO pack slip checks and packs SKUs for a CO.
- A single CO pick/pass concept in which a SKU that is picked for a CO is transferred from a pick position directly into a CO pick/ship carton or tote. During the pick activity, a CO pack slip is placed into the CO carton or tote and SKUs are checked for accuracy at a check station.

Batched Customer Order Check Concept

With any batched CO pick, sort, and pack concept, an order-fulfillment operation has all picked SKUs sorted by an employee or mechanized sort concept. A sort concept separates mixed SKUs into separate COs.

- CO locations and pack stations. At a pack station, an employee checks the pick and sort accuracy against a CO pack slip. The packer matches the actual description of each SKU to the description on a CO pack slip.
- A sort location contains a number of CO sort stations. With this sort concept, from a large quantity of batched SKUs, an employee or mechanized concept sorts picked SKUs for each CO to one sort location. This initial sort effort reduces the number of COs per sort location to a smaller predetermined CO

number and SKU number per CO. From this smaller batched CO picked and presorted group of SKUs, a final CO sort is a pick, sort, and check activity. A final check is completed by a packer who matches each SKU code to a CO pack slip code

Single Customer Order Check Considerations

A single CO pick/pack concept has a picker or automatic pick machine pick from a pick position per a CO. SKUs for a CO are transferred from pick positions directly into a CO pick/shipping carton, tote, or between two cleats on a belt conveyor travel path surface. CO picked SKUs are transported on a cleated belt conveyor or in a carton/tote to a manual, mechanized, or automatic CO check station.

Manual Check Concepts

There are two methods for manually checking COs: a detailed line item check and a SKU quantity check.

A manual, detailed quality and quantity check is performed by verifying that a SKU quality and quantity in a CO carton matches the CO pack slip. An employee takes the following steps:

- Physically removes the pack slip from a CO carton
- Handles each SKU
- Reads the description of each SKU on the CO pack slip
- Compares that description as well as the number of the SKU to the actual description and quantity of each SKU in the CO carton
- Verifies the CO check activity by placing a check mark adjacent to the SKU description and quantity on the CO pack slip
- Places the CO pack slip and SKUs into the CO carton

The CO pack slip should have a SKU description that is easily understood by a CO check employee and the quantity should be clearly indicated. In a paperless CO check activity, where a hand-held scanner is used at the CO check station, a host computer downloads to a microcomputer the CO SKUs and each SKU discreet code and quantity.

The manual check requires space on the building floor, and on a conveyor travel path or vehicle queue area. A manual activity has the potential for employee errors, and adds time to complete the CO.

With a batched CO pick, transport, and sort concept, the final CO check activity is completed at a pack station. A picker sorts picked SKUs into an assigned CO sort location and completes a pre-check activity. At the pack station, a packer with a CO pack slip transfers labeled SKUs from the sort location into a CO carton or tote. A second or final CO check is completed by a packer with a CO pack slip, who verifies that the actual CO SKU quantity and quality (description) match the CO pack slip.

A manual CO check is used in a batched CO fulfillment operation that has a pick label or SKU label and a sort concept. In a single CO or batched CO with a sort concept order-fulfillment operation, a manual SKU quantity and quality check is used to check a problem CO.

An effective and cost-efficient check activity has:

- A CO pack slip with clear and bold print
- Each SKU code, CO code, and pack slip SKU description and quantity clearly visible and understandable
- Sufficient workstation space or off-line conveyor travel path with adequate light fixtures
- Easy access to a problem CO station and to a pick area that permits resolution for problem COs
- Easy access to the main conveyor travel path for checked CO return to transport for travel to a pack station

A quantity check is an activity that involves a total piece count. A quantity concept has:

- An order fulfillment computer that cubes each CO pick/shipping carton size to determine SKU quantity per CO pick/shipping carton
- A check station or pack station where a packer physically counts each SKU in a CO carton
- A CO check or pack station employee who matches the actual SKU quantity in a CO carton or tote or on a pack station table to a CO pack slip SKU quantity

The total piece count method is used in all batched CO fulfillment applications that use a CO pick discreet or SKU discreet label with a sortation concept, in a manual and low-volume CO pick/pack order-fulfillment operation, or a high-volume operation that has a problem CO check activity.

The disadvantages are the need to have a computer cube program to determine SKU quantity per CO carton, and to ensure accurate data for the computer program. Projected SKU quantity per CO shipping carton needs to be clearly printed and visible on a CO pack slip. There is the potential for CO quality pick errors if an employee only counts the SKUs in a CO shipping carton. Advantages are few employees, high CO volume, easy to train with no reading, less building space, and less conveyor travel path queue.

Computer-Controlled Check Concepts

A computer-controlled CO check concept uses a scale to determine actual CO carton weight in relation to a computer-projected weight. Computer-controlled check concepts (see Figure 4.95) are:

- A stationary CO check activity in which an employee weighs a CO carton or tote and compares it to a computer-projected CO carton weight. If a CO has more than one carton, the cartons are queued and the total actual weight for a multiple-carton CO is compared to a computer-projected weight for that CO.
- A dynamic CO check activity in which a CO identification is scanned and sent to a computer and each CO carton is weighed with an in-line scale as it travels over the conveyor travel path. A single-line conveyor travel path has CO cartons with identification move forward from a scanner device to an in-line scale on a first-in-first out basis. The in-line scale has a communication link to a computer, which is linked to a divert device on the conveyor. The computer compares the actual CO carton weight to a computer-projected CO carton weight. This is referred to as an "on the fly" check.

A cost-effective and efficient computer-controlled CO check concept has the following design features:

- It ensures that each SKU weight and cube is accurately entered into the computer program. A computer program allocates the number of SKUs that fill one CO carton or matches the weight of SKUs that is acceptable to a CO delivery method. If a CO has two cartons, the computer allocates SKUs to a second carton.
- It ensures a CO pick or shipping carton internal dimensions, tare weight, cube, and fill utilization factor are accurate in the computer program.
- A SKU list is developed that shows actual and projected weight variance for each SKU and an acceptable projected to actual weight variance, plus or minus, is established.
- There is a computer interface between a host computer and an in-line check weight scale.
- On the conveyor travel path, ahead of the in-line scale, there is a gap pulled between two CO cartons/totes. Cartons or totes are signaled on a conveyor travel path, a scanner device reads the CO identification and transmits the data to a microcomputer, and each CO carton is momentarily stopped on the conveyor travel path ahead of the scale.
- It ensures that a CO carton actual weight is sent to a microcomputer or scale digital display. With a stationary CO container concept, an option is to have a computer-projected CO weight printed on a CO pack slip.
- It ensures that each CO carton exterior surface has a discreet bar code/RF tag.

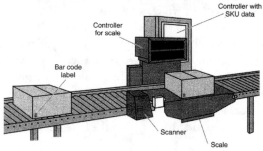

MANUAL CHECK WEIGH

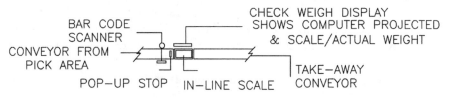

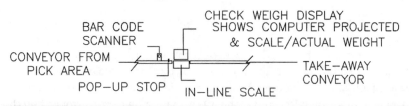

Figure 4.95 Customer order check concepts: check weigh and manual check weigh.

- It ensures that there is sufficient conveyor travel path to divert problem or "out of weight" CO cartons/totes, space to resolve a problem CO, and proper access for CO carton/tote reentry to a conveyor travel path.
- With multiple cartons or totes for a customer order, to be able to queue multiple carton weights and to ensure that a computer projects the total weight for a multiple-carton CO, that a scale or employee totals the weight for each CO and that there is sufficient queue conveyor travel path to handle the multiple CO carton/tote situation.

A stationary computer-controlled scale concept has a CO pick or shipping carton/tote travel over a conveyor travel path with an automatic or manually operated stop device and bar code scanner ahead of a scale station. When thE-stop device is activated, the other CO cartons/totes on the conveyor queue against thE-stopped carton/tote. This feature ensures that one carton/tote at a time travels past the check employee onto the scale section. With a stationary scale concept, a check employee verifies that the carton/tote actual scale weight is within variance to a computer-projected weight, and then pushes the carton/tote back onto the conveyor. If the carton/tote actual weight does not match a computer-projected weight, it is transferred to a problem station. As the employee moves a carton/tote from the scale, an employee or photo-eye deactivates thE-stop device to allow the next carton/tote in the queue to move forward past a bar code scanner onto the scale section. The bar code scanner is an employee hand-held scanner or fixed-position scanner. Since cartons or totes are spaced on the conveyor, with a fixed-position bar code scanner on the conveyor travel path, cartons pass the scanner in a FIFO sequence that allows the CO identification to be sent to the microcomputer and scale. The scale section is a non-powered roller conveyor or ball top transfer conveyor section. In a stationary scale concept, support structures are not attached to the conveyor travel path support members but are independently floor supported or hung from the ceiling.

The scale has a digital display screen that shows the actual weight of a CO carton or tote. A check weight employee compares this to the computer-projected weight printed on a pack slip or shipping label and on second digital display screen. If a CO shipping carton/tote, or the total for several CO cartons/totes, is within a check weight standard, the CO cartons/totes continue to travel on the conveyor to the next pick area activity station. If the weight is not within the standard, the CO cartons/totes are transferred from the conveyor to a problem CO station, where a check employee physically checks the SKU quality and quantity.

An "on the fly" concept is similar to a stationary check weight concept, except that it uses the latest scale and computer interface technologies. In this concept each CO has a discreet code that is associated with a computer-projected weight. The bar code/RF tag data is sent from a bar code scanner/RF tag reader to the computer. Prior to a pick activity from a host computer, a scale area computer receives and stores computer-projected weights for all CO identified cartons or totes. As a CO identified carton/tote moves forward over a conveyor travel path, a gap is pulled between cartons/totes. The lead carton/tote travels from a gap station over a conveyor travel path section that has photo-eyes and start/stop devices to ensure that only one CO carton/tote travels onto a scale. A bar code scanner/RF tag reader sends the CO identification to the scale computer. From its data files, the computer selects the computer-projected weight that is associated with that CO identification. The scale conveyor travel path surface is a roller or belt conveyor to ensure that a carton/tote is properly positioned on the scale to obtain an accurate actual CO identified carton/tote weight.

The actual weight of the CO carton or tote is compared to the computer-projected CO identification weight. If it is within an acceptable weight variance, the computer accepts the weight and the carton/tote moves from the scale conveyor section onto the conveyor travel path for transport from the scale station to a seal station. If the actual weight is outside an acceptable weight variance, the computer sends a divert command to a transport conveyor divert device and the carton/tote is transferred onto a problem CO station or line conveyor lane, where a CO check employee completes a manual check. After the check employee takes corrective action, the CO is physically transferred from a check station onto a conveyor travel path for transport to a seal station.

A problem CO or "out of weight" conveyor lane has a divert device located 15 ft. from the check weight station, and has sufficient space on the conveyor to handle queuing of a reasonable CO carton/tote volume. It is a low- or zero-pressure queue conveyor. The check employee has access to a pick area, and uses a CO pack slip that is on paper or a PC screen. There is a reentry station to transfer CO cartons/totes onto a take-away conveyor travel path.

A computer check concept is appropriate for a high-volume or dynamic operation, with a wide product mix.

Customer Order Pack Activity

A CO pack or fill activity occurs in any small-item flatwear order-fulfillment operation, whether it services a retail customer, catalog, e-mail, direct mail, or commercial business customer, and whether it is a manual, mechanized, or automated operation.

The CO pack activity ensures that merchandise is protected against damage or from being lost during delivery. How a CO is handled at a pack station depends on the particular industry. There are two major groups: the direct customer contact or catalog, e-mail, or direct marketing group and the retail store, plant, or commercial customer group.

If your business is in the direct market, e-mail, or catalog industry, the appearance of the package (container and label) is your customer's second contact with your company. The first contact was from your company catalog, e-mail message, or direct market visual presentation. Thus, the appearance of your delivery package is important. A CO package with a professional appearance improves your customer satisfaction.

Centralized and Decentralized Pack Activity

A typical CO pack activity includes the following procedures:

- Select and form a CO shipping carton/tote and apply a CO delivery address label to the exterior surface
- Transfer picked SKUs, company sales literature, and CO pack slip into the CO shipping carton/tote
- Perform a CO check activity
- Fill spaces in the CO carton/tote
- Seal the CO shipping carton/tote

The sequence of these activities varies depending on the type of pack station, whether the order-fulfillment operation is manual, mechanized, or automated, and the type of CO carton/tote. If the operation has a pick/pass into a CO shipping carton/tote concept, the CO pack activity is decentralized. This means the pack activity is separated into several activities at different locations along a pick line. After a computer-selected CO shipping carton/tote is formed and transferred to a pick line, at a start station and prior to the first pick position, a CO shipping label is placed onto a CO shipping carton/tote by an employee or label applicator machine. At the start station, an employee or insert machine transfers a CO pack slip into the carton/tote and picked SKUs are transferred from pick positions directly into the shipping carton/tote. Past the last pick position, a CO check activity is performed, and this is followed by a CO carton/tote pack or fill activity. At a pack or fill station an employee or machine transfers filler material into the CO shipping carton/tote to fill the spaces in the carton. The last pick/pass into a CO shipping carton/tote activity is to seal the carton/tote. As a tape or plastic strip is applied to the CO carton flaps, care is taken to ensure that it does not cover the delivery address label.

An order-fulfillment operation that uses a pick/pass into a CO captive tote or batched CO concept, has a centralized pack activity. In a centralized concept, CO pack activities are performed at one location. At this central location, a packer performs the following: (1) makes up a carton according to computer instructions on a CO pack slip; (2) completes a CO check activity and transfers SKUs and CO pack slip into the CO shipping carton, (3) adds filler material; and (4) applies a seal (tape or plastic band) to secure the carton top flaps and applies a CO shipping label to the carton exterior.

Pack Design Parameters and Operational Characteristics

An efficient and cost-effective pack activity is one in which the CO cube volume is matched to a CO shipping carton or tote that has the capacity to contain it and each pack activity is designed to handle a projected SKU and CO volume. Picker and packer productivity is improved by having a computer determine the total cube volume for a CO and match this cube volume to a CO shipping carton or tote. To achieve this, you obtain and maintain in your computer program:

- The exact cube, length, width, height, and weight for each SKU
- The exact internal dimensions for length, width, and height, cube for each shipping carton or tote SKU utilization factor (that is how much space the SKU actually occupies), and allowance for filler material for a shipping carton internal cube
- The total weight that is allowed by your CO delivery company
- A computer program that maximizes CO SKUs, computer-projected weight and cube

Knowing the SKU quantity for each CO carton determines the number of SKUs that are transferred from a pick position or pack station into a CO shipping carton/tote so that there is no overflowing damage to SKUs.

Actual SKU dimensions are very important. For example, an umbrella that is 30 in. long, 2 in. wide, and 2 in. high has a cube dimension of 120 in., but can such an umbrella fit into a shipping carton that is 10 in. long, by 3 in. wide, and 4 in. high? The answer is yes by a pure cube analysis, but no when a picker or packer tries to place the umbrella into the shipping carton. To obtain actual cubic dimensions (length, width, and height) and weight for each SKU, the options are (a) using vendor data, (b) a manual method, or (c) a mechanized method. This was discussed in detail earlier in this chapter.

Accurate shipping carton information required by a pack activity includes:

- Actual internal and external cube dimensions
- Weight or tare weight for an empty carton/tote with bottom flaps taped
- Company-determined occupancy rate or percent utilization for an internal shipping carton/tote cube

Internal cubic dimensions for a shipping carton or tote are a measure of the open space within a carton. Depending on the CO SKU size and shipping carton/tote internal size, a CO may require one or several shipping cartons/totes. For a multiple-carton CO, the cartons/totes may be different sizes.

External dimensions of a shipping carton are the exterior cubic area or the space it occupies on a transport concept or inside a delivery truck.

The tare weight of a shipping carton/tote is the weight of a cardboard, corrugated, or plastic carton/tote when it is empty. In a check weight concept, this shipping carton or tote tare is an important factor that is added to the computer-projected SKU weight.

A shipping carton/tote occupancy rate or utilization factor is determined by your order-fulfillment management team. The occupancy rate or utilization factor is a standard that is entered into your company computer cube or volumetric program, which uses the utilization factor to determine SKU pick sequence and SKU quantity per CO shipping carton/tote. Shipping carton/tote utilization factor is a standard minimum of 85 percent, with a maximum of 95 percent. The utilization factor ensures an excellent SKU quantity per CO shipping carton/tote, with minimal void spaces requiring filler material.

SKU weight is also entered into the cube or volumetric program. The program queues SKUs into a shipping carton or tote, and calculates a total SKU projected weight. The computer is thus able to allocate the SKUs for one shipping carton by obtaining the shipping carton/tote cubic utilization factor and the SKU weight that is allowed for allocation to one shipping carton/tote.

The pack station should be properly designed to handle the projected SKU or CO volume. The number of pack stations is based on projected productivity rates and associated volume. Each pack station, pack station equipment, and transport concept is designed to handle this projected CO volume.

An efficient and cost-effective pack activity must have an area in the pack station, on the conveyor travel path or cart concept, for queuing CO cartons, totes, or loose SKUs ahead of the pack table. If the CO SKUs are manually delivered to the pack station, here should be an adequate set-down area for in-feed carts that ensures a constant CO flow to a pack station and space for empty transport devices.

Equipment for queuing SKUs picked for a CO includes:

- A pallet that is either set on the floor or delivered by a powered roller conveyor
- A four-wheel cart
- An overhead basket or trolley
- A tote or carton on a conveyor travel path
- A slide or chute

A Pallet at the Pack Station

In an order-fulfillment operation with SKUs picked in bulk for COs, a pack station in-feed concept is designed to have a forklift truck or pallet truck deposit a pallet at a pack station. With this concept, as a pallet is depleted by the pack activity, a forklift truck or pallet truck removes the empty pallet and transfers a full pallet to the pack station. This exchange activity involves a transaction time during which a pack employee has some non-productive time.

If a pack activity has CO picked SKUs delivered on pallets to a pack station, a roller conveyor on the floor ensures a constant flow to a pack station. A powered or non-powered roller conveyor with an end stop is two to three pallets long, with one lane for inbound pallets and a second lane for empty pallets. To use minimum floor space with high and low side guards and bottom rollers, a pallet return lane has a pallet stand on a stringer side.

Overhead Trolley Delivery to a Pack Station

If a trolley basket is used to deliver COs to pack stations, the pack station has an overhead main trolley rail section that travels parallel to the pack station area. For each pack station, a pull-down switch to transfer trolley baskets from the main trolley line to an in-feed travel path to each pack station. A branch trolley rail is a spur or siding that has two trolley basket lengths for each pack station. Past the pack station, the branch trolley rail directs the trolley back onto the main trolley travel path. With this concept a series of 180-degree turns on the main travel path serve to transfer empty trolley baskets back to the pick area.

Cart Delivery to a Pack Station

If merchandise is delivered on four-wheel carts, the pack station has two cart flow lanes. Each cart lane length is two carts long, with one lane for inbound carts and the other lane for empty carts.

Totes or Cartons on a Conveyor

Cartons or totes on a roller or skate-wheel conveyor queue on a 10- to 12-ft. long conveyor section between a main area aisle and a pack table. The conveyor slopes toward the pack table and ensures a constant CO flow to the pack table. The COs are in totes or shipping cartons.

To improve packer productivity, the spaces above and below the conveyor sections are used for shipping carton storage. Special rack bays with full-depth separators are used to support and separate the carton types. This feature ensures a supply of cartons and if a CO is delivered in an incorrect shipping carton, a packer is able to make up the proper shipping carton with little time lost.

If a CO is delivered in a captive tote, an empty tote transport concept moves empty totes from the pack station to a pick area. Tote transport options are to stack totes in a pack station area and have one employee collect and transport them on a four-wheel dolly or two-wheel hand truck, or place them on a conveyor travel path within reach of the employee.

If your pack area is designed with a powered conveyor that delivers COs to pack stations, the travel path has pack stations perpendicular on both sides. As COs travel over the conveyor travel path, delivery options are:

- At each pack station, a photo-eye controls a CO carton delivery onto a powered conveyor travel path and diverts a CO at an assigned pack station.
- COs travel on a low- or zero-pressure conveyor travel path and a packer removes a CO from the conveyor onto a pack station where CO cartons are queued.

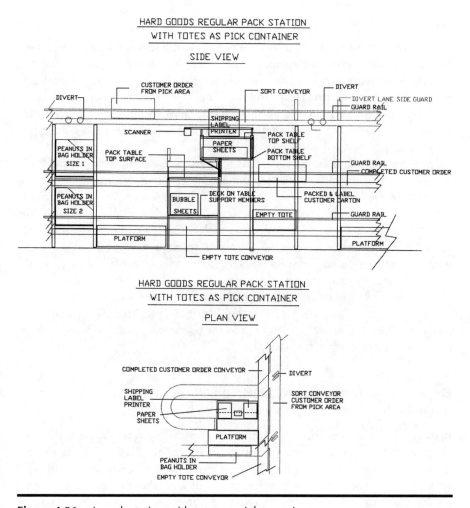

Figure 4.96 A pack station with totes as pick containers.

In a captive container or tote pick/pass or batched CO pick and sort concept, totes or containers are used to transport CO picked SKUs from a pick area to a pack or sort station. In a dynamic order-fulfillment operation, a large number of totes travel over the transport concept. The transport concept must be designed to queue empty totes in order to ensure a constant supply to a pick line start station and to provide a constant flow of completed COs in totes to a pack station (see Figure 4.96). To satisfy these objectives the operation has a procedure for in-feed and removal of empty totes.

This can be an employee who transfers empty totes between a transport conveyor travel path, totes diverted onto a declined non-powered roller or skate-wheel conveyor travel path, or totes discharged onto a powered conveyor with stacker and destacker machines along queue conveyor travel paths.

A Packer Transfers Empty Totes

An employee transfers empty totes onto or from a transport conveyor travel path. This is used in a captive tote operation or in a supply chain strategy in which totes are sent to the customer and the customer returns the tote to the operation. The concept has a tote transfer station and adequate empty queue conveyor travel path before the first pick station, and adequate queue conveyor travel path ahead of the pack or sort station. At the tote transfer station, there are tote stack/storage area and/or non-powered stacked tote storage lanes.

When there is an imbalance in empty tote flow between the pick area and pack/sortation area an employee at the empty tote transfer station can transfer totes onto or from the empty tote conveyor travel path. If there is a tote over-crowded condition on a conveyor travel path, at a tote transfer station an employee removes and places empty totes from a conveyor travel path. If customer-returned empty totes cannot be transferred onto a transport conveyor travel path, they are stacked on the non-powered tote storage conveyor lanes. If there is a shortage of empty totes on the conveyor travel path, an employee moves empty totes from the tote storage lane or area onto the powered conveyor travel path.

The concept features are less conveyor investment, preferred for a supply chain strategy that has customers return totes to the operation, employee-based activity, and management controls and discipline.

Empty Totes Diverted onto a Decline Non-Powered Roller/Skate-Wheel Conveyor Travel Path

A divert device on the main tote conveyor, located between the pack area and the pick area, diverts extra empty totes from the main travel path onto a queue conveyor lane, which is a declined non-powered skate-wheel or roller conveyor with photo-eyes and short belt conveyor sections. At the discharge end there is a mechanism that controls totes merging back into the main transport conveyor travel path.

Past the pack/sortation area, the main empty tote transport conveyor travel path has a high elevation, photo-eye, and a divert device. Empty totes accumulate on the main conveyor travel path and block the photo-eye, which activates the divert device to transfer empty totes from the main tote conveyor travel path onto the queue lane. The queue lane has stands to support the weight of the totes and ensure tote travel over the decline travel path. The empty totes queue against a pop-up end stop or belt conveyor section. If the line pressure of empty totes on the queue conveyor travel path is a problem, additional photo-eyes, pop-up stops, or short belt conveyor sections are added along the queue tote conveyor lane.

A photo-eye on the main empty tote conveyor controls the end stop or starts and stops the belt conveyor section. When this photo-eye is blocked by a tote, the pop-up stop extends up above the conveyor travel path or the belt conveyor section is stopped, causing the empty totes to queue. If the photo-eye is not blocked by a tote, the pop-up stop is lowered below the conveyor travel path or the empty tote queue lane discharges empty totes onto the main travel path. As empty totes are released from the conveyor travel path, the main travel path stop device restricts tote travel on the main travel path until all the totes are released from the queue conveyor lane.

The concept requires additional conveyor and controls investment, but does not utilize employees; totes are captive to the facility, and a more exact tote number.

Empty Tote Stacker and Destacker with Queue Conveyor Travel Path

The empty tote stacker and destacker concept has a queue conveyor travel path between the pack/sort area and the pick area. An empty tote conveyor connects the pack/sort area to a stacker machine or machines. Totes travel from the pack/sort area to the stacker area, where a tote stacker machine stacks or nests from two to ten totes as one unit. The number of totes in a stack is determined by the design team and depends on the physical characteristics of the tote, available space, and mechanical stacker and destacker equipment. The stacked totes are queued on a powered queue conveyor travel path, which moves totes to a destacker machine. The destacker machine transfers totes from the stack onto a take-away conveyor travel path. Single totes leave the destacker area and travel on the main tote conveyor to a pick area.

The concept features additional costs for the conveyors and tote stacker and destacker, no employees are needed in this area, totes are captive to the facility, and requires space.

Pack Station Conveyor Travel Path Concepts

In a pack area, captive totes or cartons are transferred from the main conveyor to a pack station. The transport concept is designed with divert devices (see Figure 4.97 and Figure 4.98). In a mechanized concept, divert travel path options are:

- From an elevated (above a pack table) conveyor travel path
- From a floor-level (below a pack table) conveyor travel path
- A U-shaped decline divert conveyor travel path
- An L-shaped divert conveyor travel path

From an Elevated Conveyor Travel Path

In this type of operation, an in-feed conveyor elevated about a pack station transports captive totes or CO shipping cartons to a pack table. Along the main conveyor

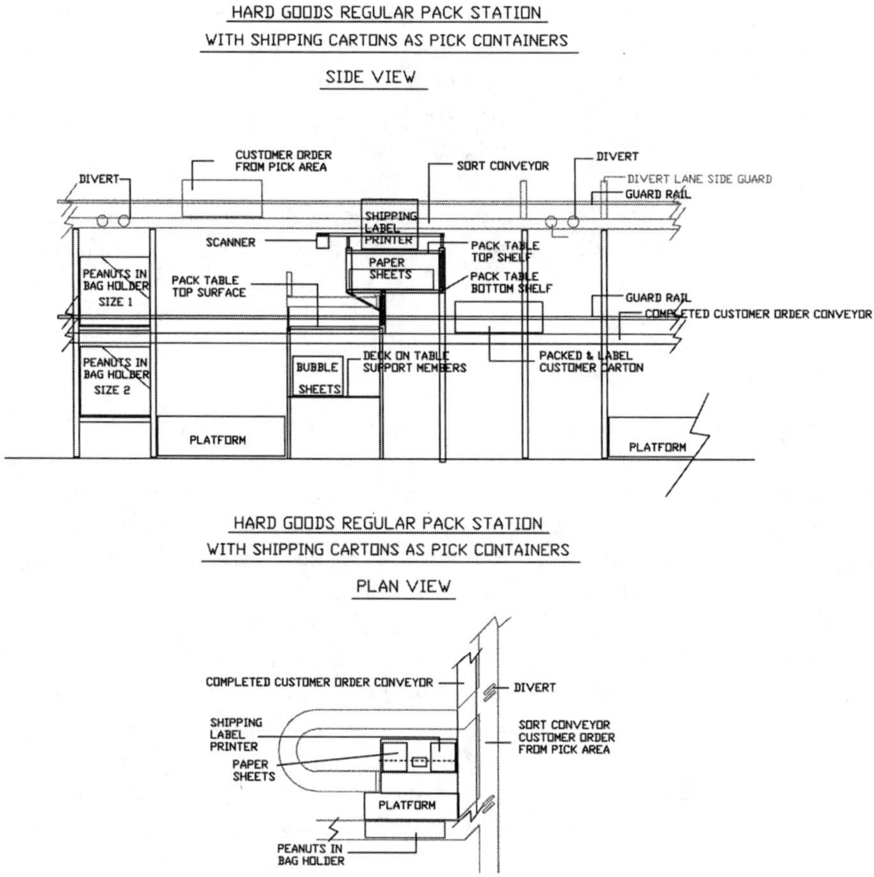

HARD GOODS REGULAR PACK STATION

WITH SHIPPING CARTONS AS PICK CONTAINERS

SIDE VIEW

HARD GOODS REGULAR PACK STATION

WITH SHIPPING CARTONS AS PICK CONTAINERS

PLAN VIEW

Figure 4.97   A pack station with shipping cartons as pick containers.

travel path, an assigned divert device diverts a carton or tote onto a decline conveyor travel path and onto the pack station. This decline conveyor travel path directs a tote or carton from the main conveyor travel path, to the front of a pack table.

From a Floor-Level Conveyor Travel Path

An incline in-feed conveyor travel path concept has a main conveyor travel path slightly elevated above the floor but below the pack station. A captive tote or CO shipping carton main conveyor travel path is slightly above the floor and adjacent to each pack station. Along the main conveyor travel path, an assigned divert device or transfer device moves a tote or carton up from the floor-level main conveyor travel path onto a conveyor travel path that is along a pack table front (non-employee

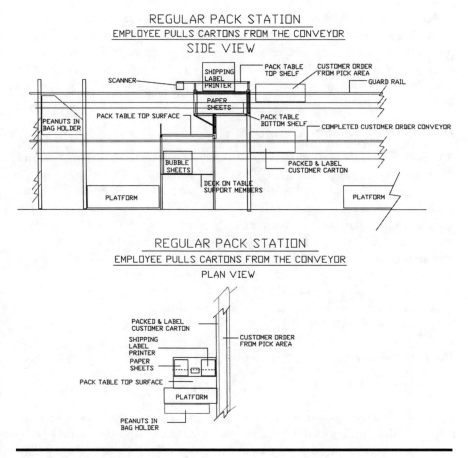

Figure 4.98 A pack station employee pulls cartons from the conveyor.

side). At the pack table, a right-angle transfer conveyor or curved conveyor travel path moves the carton or tote onto a pack table. The incline conveyor is a powered belt conveyor. If standard-height cartons are at a low elevation, there is space for additional conveyor travel paths and completed CO carton conveyor travel path is at a pack table or at a low elevation.

U- or L-Shaped Conveyor Travel Path

The captive tote or pick carton conveyor from the main conveyor travel path to a pack station is either U- or L-shaped. A U-shaped travel path is a powered or non-powered decline conveyor travel path that extends from a divert location on the main conveyor travel path. The divert conveyor travel path makes a curve, declining slightly as it turns, and ends at a pack table. The U-shaped conveyor travel path queues cartons or totes. The L-shaped conveyor travel path extends from the divert location on the main conveyor path, and makes a right-angled turn onto a pack table. The L-shaped conveyor has less capacity for tote or carton queue, and is slightly more expensive to install than a U-shaped conveyor. It occupies a smaller area, and with a wide tote or carton mix, possible hang-ups are less likely.

Loose Items in a Slide or Chute

A gravity slide or chute design has an elevation change between a charge end and discharge end. The elevation change and low friction on the travel path surface permits gravity force to move loose SKUs or cartons over the travel path. A chute cubic foot capacity is designed to handle picked SKUs for one CO wave. In the typical direct mail or catalog operation, a chute can handle 50 to 60 SKUs at one time. A chute does not get filled to more than half its capacity, that is, a 50 to 66 percent utilization factor. With a chute concept that queues loose SKUs in front of a pack station, there is no transport travel path for empty totes from the pack area back to a pick area.

Chute design was discussed in detail earlier in this chapter.

Pack Table Types

Pack tables may be either plain flat tables or specially designed.

■ A plain flat table top with a 3 × 6 ft. surface has fixed or adjustable legs and provides a packer with the necessary work surface to separate and pack COs. If an operation handles very small size SKUs, a 3 × 5 ft. table is an option. Shelves adjacent to or behind a pack table surface provide additional carton storage.

■ A specially designed table has drawers and shelves that are used as mail stuffer and carton storage locations. In addition, 3 × 4 ft. shelves are located behind a pack employee. Tables may be designed with the following features: adjustable legs, peg backboard, and paper roll hanger above, on, or under the pack table.

Shortages, Overages, and Damages at a Pack Station

A pack station design has an efficient and cost-effective concept to handle a CO that is found to have an over-pick, shortage, or damaged SKUs. In a catalog or direct-mail industry, when a problem with a CO occurs at a pack station, a packer transfers the SKUs and CO pack slip to a problem CO station, where they are reviewed and corrective action is taken by management to satisfy a CO. To ensure good packer productivity, in a pack station area there is a problem CO static shelf or four-wheel cart with shelves. The concept serves as a location that temporarily holds a problem CO. This permits management to collect problem COs and repick a CO, or obtain a customer's permission to send an order with a shortage that becomes a back order.

Customer Order Shipping Containers

A CO shipping container is a cardboard carton or a paper or plastic bag. Its function is to ensure that SKUs are protected during transport and provide a surface for a CO delivery label so the CO is not lost. The choice of CO shipping carton or bag depends on SKU type and number of pieces per CO, SKU value, SKU packaging, and delivery method. In an order-fulfillment operation that serves catalog or direct mail customers, the type of container that is used to ship a CO is a very important factor hat influences customer satisfaction. An order-fulfillment operation may use corrugated or plastic bags, or cardboard or chipboard boxes.

Corrugated or Plastic Bags

In some pick operations SKU size, physical characteristics, and value permit a plastic or corrugated bag to be used as a shipping container. Mechanized pack machines are used for a high-volume SKU. The concept has an employee feed SKUs onto a conveyor line and a machine forms a bag to enclose the SKU and applies an envelope label to the bag surface.

Corrugated Bag

A corrugated bag is referred to as a jiffy bag or envelope mailer with a bubble sheet or padded paper interior that is used to ship low-profile, non-fragile, or protected/encased SKUs. The bag has two sides that are sealed and an open front or mouth that is sealed with glue, tape, staples, or stitch. There are different kinds of corrugated bags:

- Standard heat and seal jiffy bag with bubble sides
- Self-sealed jiffy bag with bubble sides
- Tape- or staple-sealed paper padded jiffy bag
- Self-sealed paper padded jiffy bag
- Tape- or staple-sealed rigid mailer
- Self-sealed rigid mailer

The corrugated bags are available in a variety of sizes. The use of air bubble sheets or a paper-padded jiffy bag depends on the customer preference, cost, and codes.

A corrugated bag is sealed using one of the following methods: (a) self-adhesive or gummed tape, (b) staples, (c) stitching by a machine, (d) self-locking bag, and (e) self-sealing bag. The method selected for use in an operation is determined by economics, volume, shipping method, package appearance and customer preference, transport concept, and storage space.

In an order-fulfillment operation, a corrugated bag is handled by a separate sortation concept from a carton sortation concept. This is to prevent heavyweight cartons falling onto a package with a fragile SKU. In addition, delivery companies handle bags and cartons on different concepts.

Plastic Bag

A plastic or mailer bag is used to ship flatwear apparel and has plastic sheet layers that are pre-sealed along two sides and bottom with an open front or mouth. They are available in a variety of sizes.

A plastic bag option is a double self-seal bag that permits a customer to return a SKU in the bag. The plastic bag features are less weight, waterproof, minimal or no cushion or padding, difficult to handle on some powered conveyors, and mechanical sortation concepts and comparable cost.

Cardboard Wrap

With a cardboard wrap and seal concept, a flat preformed cardboard sheet is placed onto a powered belt conveyor in the path of the SKU. Photo-eyes detect the cardboard sheet as it arrives at the location where a pack transaction will be completed. An employee or robotic machine completes the following transactions:

- Based on a CO SKU cube. a preformed cardboard sheet is transferred onto the conveyor travel path. If a manual pack slip or SKU transfer concept is used, a painted square on the cardboard helps an employee to properly place a CO pack slip and SKUs on the cardboard sheet.
- A pack slip is placed on the sheet.
- The SKU is placed on top of the pack slip.
- An employee or robotic machine folds the cardboard sheet along the sides, front, and rear to form a four-walled shipping carton with a top and bottom surface. Adhesive or glue is used to secure the carton.
- A shipping label is applied to the carton. With singulated cartons on a conveyor travel path, a conveyor travel path has a first-in-first-out flow, customer labels are properly sequenced to match the CO picked and packed SKU flow.

The cardboard wrap and seal concept handles a wide SKU mix, including flat small-items, flatwear apparel, and folded GOH and increases employee productivity and accuracy.

Sandwich Bag

A sandwich bag concept has:

- One or two small SKUs with a customer identification label that faces upward
- A cleated belt conveyor travel path
- A bar code scanner/RF tag reader and communication network to a label print and apply machine
- In-feed devices to supply kraft single-flute corrugated or plastic rolls
- Hot glue adhesive and bag side press machine
- A machine to print and apply labels
- A take-away conveyor

A sandwich bag operation has SKUs for a CO picked and delivered to an in-feed station to a sandwich bag pack concept. At the in-feed station, an employee places a CO pack slip, sales literature, and individual SKU or SKUs with the label face up between two cleated sections on a belt conveyor travel path. If a SKU is flexible, a corrugated sheet is first placed on the in-feed conveyor travel path, and the CO pack slip, sales literature, and SKUs are placed on top of the corrugated sheet.

The SKU travels forward under a bar code scanner, which reads the SKU bar code or CO identification. On a first-in, first-out basis, this information is transferred to a microcomputer that sequences a printer to print a bar code shipping label for placement onto the exterior of the shipping bag. The CO is enclosed in two single-flute corrugated cardboard sheets supplied by paper rolls above and below the conveyor section. As the CO moves forward, the two corrugated paper sections form a sandwich with the CO in the middle. As the two corrugated kraft paper sections move forward, a hot glue adhesive is applied along the four edges to form a sandwich bag. After a predetermined length the corrugated paper is cut and forms a two-sided shipping bag. At this time the two sides are pressed together to form a sealed bag.

The sealed shipping bag moves forward to the shipping address label print and apply station. At a label print and apply station, a computer-controlled printer prints on a FIFO basis a customer shipping label and applies a CO delivery label onto a shipping bag. The packaged and labeled CO bag travels over a conveyor travel path past an inspection station or directly into a BMC.

A sandwich bag concept handles a variety of SKUs, increases employee productivity, and for customer returns the package requires tape. During operation there is down time to replenish paper rolls and check and adjust a bag seal.

Corrugated or Chipboard Cartons

Cartons used in a small-item order-fulfillment operation may be one piece with top and bottom flaps or a two-piece carton. Both types are corrugated cartons that have

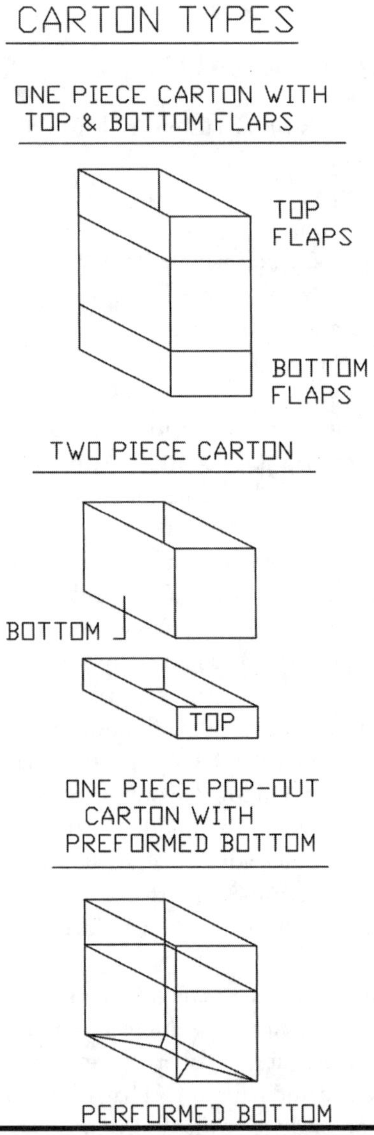

CARTON TYPES

ONE PIECE CARTON WITH
TOP & BOTTOM FLAPS

TOP
FLAPS

BOTTOM
FLAPS

TWO PIECE CARTON

BOTTOM

TOP

ONE PIECE POP-OUT
CARTON WITH
PREFORMED BOTTOM

PERFORMED BOTTOM

Figure 4.99 Shipping cartons: one piece with top and bottom flaps, two-piece, and pop-out one-piece.

a single or double flute or waves between two kraft paper layers. The various corrugated cartons (see Figure 4.99) are (a) chipboard box, (b) plain and straight side wall cartons, (c) creased or slotted cardboard cartons, and (d) pop-out cartons.

One-Piece Standard or Pop-Out Carton

One-piece carton has a corrugated sheet that is preformed by a manufacturer with four side walls and top and bottom flaps. At a pack station, a one piece carton has a packer or machine secure the four bottom flaps with tape. With the bottom flaps sealed with tape, on a pick line SKUs are picked/packed into a carton, or at a pack station, SKUs are transferred into a carton. After a check or packing activity, a packer or machine tapes the top flaps.

A pop-out carton has a packer push a non-formed carton from the sides. The carton is designed with bottom flaps that interlock and form a solid bottom surface when pressure is applied by the employee. For a small carton, a manually formed pop-out carton is handled more effectively than a standard carton formed by a machine.

The one-piece or pop-out carton is made up on a pick line or in a pack area, is a low cost option, and can be used with a wide SKU and CO mix. The disadvantages are the weight limit; on a pick line or on a conveyor travel path, top flaps up have a tall window; with flaps up, a pick or pack station employee has a higher reach to transfer SKUs into the carton; space requirements at a pack station.

Two-Piece Carton

A two-piece carton has a corrugated sheet that is preformed by a machine or manufacturer with four side walls and a solid bottom. At a pack station, an employee completes a pack activity and places a top or cover on the carton. The top is secured with tape or plastic straps. There is a form machine for the carton and cover, which are made up on a pick line or a transport concept moves cartons to the pack area. Two-piece cartons handle a wide SKU and CO mix, and hold many sizes with some weight limitation. On a pick line or on a conveyor travel path, the height of the carton side wall determines the travel path window, which is shorter than that for a one-piece carton with top flaps up, and a pick or pack station employee has shorter reach to transfer SKUs into the carton. On a pick line, a carton side wall receives a CO identification label or ink jet spray but for CO delivery, a CO delivery identification is placed on the cover.

Chipboard Box

A chipboard carton is used to ship low-profile, lightweight, non-crushable, and non-breakable small-items or flatwear apparel. A chipboard carton is a preformed, two-piece carton made from pressed corrugated fiber material. At a pack station and per company practice, a packer places a paper sheet in the bottom and a SKU is positioned on top of the paper. A pack slip and sales literature items are placed on the SKU, which is then covered with a sheet of paper and the box top section is placed over the bottom section and secured with tape or plastic bands. A packer places a CO delivery label onto the box exterior. The labeled box is transferred to a conveyor travel path.

An option to secure the top and bottom box sections is to have the chipboard carton moved to a plastic strapping station where one or two plastic bands are placed onto the box.

Use of a chipboard box is limited to specific SKUs, because it has low weight and cube capacity, and low side wall strength.

Plain Cardboard Carton

A plain cardboard carton is a one-piece carton preformed from a corrugated sheet that is secured together to form four sides. Each wall has flaps that extend upward and downward. On a pick line or at a pack station, before SKUs are placed into a CO shipping carton, an employee or machine closes the bottom flaps and secures them with tape. After the carton is packed with SKUs and filler material and sales literature and the CO pack slip have been added, the top flaps are taped to secure the carton and a CO delivery address label is placed on the carton exterior.

A small-item order-fulfillment operation with a wide SKU mix and varying number of pieces per CO uses several carton sizes to match a CO cube. In some companies, there are 12 to 34 different sizes. This type of carton has greater strength, and is more durable and withstands line pressure.

Creased or Slotted Carton

A creased or slotted carton is a specially designed corrugated carton with one or two creases on the four side walls, 1 to 2 in. below the top of the side wall. When a packer has a CO SKU cube that is less than the shipping carton cube, the creases make it possible for the packer to cut along each side wall and fold the side walls over. With four side walls folded, a CO shipping carton requires less filler material.

The creased or slotted shipping cartons are more expensive, but reduce the need to have different carton sizes, and require less filler material.

Handling Small-SKU Shipping Supplies

If an order-fulfillment operation packs small items into a CO shipping carton or bag, the options are:

- The small items are picked into a pick tote that is sent to a pack station and a shipping supply inventory is maintained at the manual pack station.
- Small items and shipping supplies are picked into a pick tote and sent to a pack station.

Shipping Supplies at a Pack Station

When shipping supplies are kept at a pack station, at each pack station there is a resident inventory for each shipping supply item. During the daily pack station operation an employee determines the shipping supply item quantity and performs necessary replenishment transactions. At a pack station, a pack station employee receives CO picked SKUs in a tote. To determine the shipping supply item, a packer reads a CO pack slip, shipping label, or display screen. Per the suggested CO shipping carton, a pack employee removes a carton from pack station. The features are each pack station has a shipping carton/supply holding space and for each shipping carton and supply item inventory quantity and minimal shipping carton and supply inventory control.

Shipping Carton Picked with SKUs

When an order-fulfillment operation has a shipping carton picked with a CO skus, the shipping carton and CO SKUs are sent together from a pick area to a pack station. In the pick area, each shipping carton is allocated to a pick position. Per a CO, a CO shipping carton has a pick instruction on a paper document, CO identification label, RF device display screen, or pick-to-light display.

Per a CO, an order-fulfillment employee picks a shipping carton from a pick position into a CO tote. After a CO SKUs are picked and transferred to a tote, the tote is transported to a shipping carton pick position, where a suggested shipping carton is transferred from a pick position into a CO tote. Together CO picked SKUs and the shipping carton are transferred from the pick area to a pack area. This procedure improves shipping carton inventory control, needs less employee time to replenish shipping cartons to each pack station, and results in a less crowded pack table.

Packing Small Items

If a small-item order-fulfillment operation places a small-item into a CO shipping carton or bag the options are:

- Conventional approach, or pack activity performed by a packer
- Mechanized approach, or pack activity performed as a mechanized activity
- Automated approach, or pack activity performed as an automatic activity

A conventional packer activity is one in which an order-fulfillment pick activity delivers individual or batched CO picked SKUs to a pack station, where a packer performs the entire small-item pack activity, including transferring SKUs, CO pack slip, and sales literature into a CO shipping carton or bag, placing the CO shipping label onto the shipping carton or bag exterior, and sealing the carton or bag. The conventional approach features low packer productivity, large number of pack stations, and does not maximize use of floor area relative to CO volume.

With a mechanical bagging or boxing operation, the pack operation has an employee and a programmable/mechanical bagging/labeling machine pack SKUs picked for a CO. With a mechanized bagging approach, picked SKUs and shipping supply items are delivered together to a pack station, where a host computer has transferred COs to a microcomputer that controls the bagging and labeling machine. After an employee transfers shipping supply items to a bagging machine sleeve or funnel, it is ready to perform bagging/labeling activity. As a mechanical bagging machine moves a bag forward, a label printer prints and applies a CO delivery address label to the bag exterior. A labeled bag is moved forward with bag mouth open. At this location, an employee transfers a SKU with a preprinted CO pack slip and sales literature folded around it into the bag. The bag is moved forward and a machine seals the bag mouth. A sealed and labeled bag is transferred onto a belt take-away conveyor travel path or is dropped into a BMC. With this first-in, first-out SKU and CO pack slip flow, a labeled bag and pack slip match. Features are small workstation and floor area, medium to high CO volume, pick accuracy check, CO delivery label placed onto a uniform location, minimal employee physical effort, medium investment, SKUs, literature, and pack slip match the equipment design parameters, and improves shipping supply item inventory control.

An automatic small-item bagging activity is one in which an employee transfer SKUs and shipping supply items to one location. SKUs, CO pack slips, sales literature, and shipping supply items are picked or printed and delivered to a pack station. Each SKU has a human/machine-readable code and the host computer has transferred CO data to the microcomputer that controls the automatic bagging machine. After an employee transfers a CO pack slip and sales literature onto a feeder and the shipping supply items onto a machine sleeve or funnel, he or she places a SKU onto the automatic bagging machine in-feed conveyor travel path, which is a cleated or finger belt conveyor surface that moves a SKU past a bar code scanner. The bar code scanner verifies the SKU as it moves forward to a transfer location. If a SKU bar code does not match, the conveyor travel path stops and an employee takes corrective action.

Along another conveyor travel path, sales literature and the CO pack slip are transported to the transfer location, where the CO pack slip and sales literature are combined with a SKU and transferred into a bag. The full bag is moved forward and receives a CO shipping label on its exterior surface. The machine seals the bag mouth and the bag is transferred onto a belt conveyor travel path surface or it is dropped into a BMC.

The bag movement along the conveyor travel path is a first-in, first-out flow pattern, and the SKUs and CO pack slips are sequenced to match. The features are small workstation and floor area, high volume, pick accuracy check, CO delivery labels are in a uniform location, minimal employee effort, investment, SKU and sales literature match, and shipping supply inventory control.

Customer Shipping Labels

A CO address label on a shipping container ensures that your CO delivery address is clearly visible and easily recognized on a carton or bag. We refer the reader to the more detailed discussion on labeling COs earlier in the chapter.

Carton Fill Activity

A CO carton fill activity has a packer or mechanical device fill empty spaces in a carton with filler material. This activity minimizes UOP damage during delivery truck transport from an order-fulfillment facility to a CO delivery address, and improves a CO carton seal and security.

Cartons may be bottom-filled before a SKU is placed in it and then top-filled, or spaces in the carton are filled after the SKU is put in (top-filled). Bottom- and top-filled cartons provide maximum SKU protection, but require additional storage space, pack labor and material cost, and add weight to a CO shipping carton.

The top-fill only concept provides some SKU protection, and requires less storage space, minimizes pack labor and material cost, minimum weight to a customer order shipping carton, and with CO returns, minimizes the returns handling labor and handling costs.

The fill station location can be either at the pack area or on a pick line. If the process is one in which COs are batch-picked, sorted, and packed or a single CO is picked into a captive tote, the fill activity is performed at a pack station. If the operation is a pick/pass into a CO shipping carton concept, the fill activity takes place on a pick line.

Fill at a Pack Station

Options for a fill activity located at a pack station include the following:

- In a batched CO fulfillment concept, SKUs are picked and delivered to a sort/pack/fill station. A sort/pack/fill approach is associated with a mechanized sortation concept.
- In a batched CO fulfillment concept, SKUs are picked and sorted before being delivered to a pack/fill station. This pack/fill approach is associated with a manual pick/sort concept.
- In a single CO pick or fulfillment concept, single CO picked SKUs are transported to a fill station or to a pack/fill as one CO separated from other COs or as loose SKUs (a) between two cleats on a belt conveyor travel path, (b) in a chute, or (c) in a captive tote.

The first pack/fill station concept is used for batched COs SKUs picked and delivered unsorted to a pack/sort/fill station. The pack station design features:

- An interface with a mechanized sortation concept
- An interface with a completed CO take-away concept that is a (a) conveyor travel path, (b) four-wheel cart, or (c) pallet
- A problem CO station
- Sufficient space (a) to perform a final SKU sort for a CO (b) to make up a CO shipping carton, (c) for filler material, working space and filler material reserve supply, and (d) for activities such as printing a CO pack slip and delivery label at a pack station

The second pack/fill station concept is used for batched COs with SKUs picked and sorted before being delivered to a pack/fill station. Pack station design elements are:

- An interface with a manual pick/sort and transport concept, either (a) a tote with separate compartments on a powered conveyor travel path or (b) a four-wheel cart with shelves and dividers
- An interface with a completed CO take-away concept: (a) a conveyor travel path, (b) four-wheel cart, (c) overhead trolley basket, or (d) pallet
- A problem CO station
- Sufficient space (a) to make up a CO·shipping carton, (b) for filler material, working space, and filler material reserve supply, and (c) for other activities such as printing a CO pack slip and delivery label at a pack station

The third pack/fill station concept is used for a single CO pick or fulfillment concept that has SKUs picked and transported to a pack/fill station or as one CO separated from other COs and as loose SKUs between two cleats on a belt conveyor travel path, in a slide, or in a captive tote. Pack station design parameters are:

- Sufficient queue space for both an in-feed and empty transport device: (a) tote non-powered or powered conveyor travel path, (b) overhead trolley basket, or (c) four-wheel cart
- An interface with a completed CO take-away concept: (a) a conveyor travel path, (b) four-wheel cart, or (c) pallet
- A problem order station
- Sufficient space (a) to make up a CO shipping carton, (b) for filler material, working space, and filler material reserve supply, and (c) for other activities such as printing a CO pack slip and delivery label at a pack station

Fill as a Pick Line Activity

When a CO shipping carton fill station is designed as a separate pick line activity, CO cartons are delivered on a conveyor travel path to a fill station. An efficient and cost-effective fill station has the following design factors:

- Filler material
- Working space that includes a dispenser that is easily accessed by a packer for filler material
- Sufficient space for a filler material reserve supply
- Prior to a fill station, sufficient low- or zero-pressure queue conveyor travel path and a fill station employee has the ability to start and stop a conveyor travel path
- Fill station design and equipment ensure a constant CO carton flow to satisfy the on-time CO ship schedule

Filler Materials

Materials used as filler material include the following (see Figure 4.100):

- Shredded or crushed/crunched paper sheets
- Peanuts made of Styrofoam™ or processed corn
- Styrofoam padding
- Air bubbles or air bubble sheets
- Foam
- Foam in a plastic bag
- Crushed cardboard or corrugated paper
- Cardboard carton with a plastic bed
- Preformed cardboard insert
- Meshed cardboard

The choice of filler material in an order-fulfillment operation depends on the following factors:

- Operational requirement to bottom/top fill or to top fill
- The economics of supplying the filler material and filling cartons
- Available building space for reserve supply
- Available workstation surface space
- Order-fulfillment pick/sort and pack concept
- SKU value
- CO delivery concept
- Ability to recycle or use filler material from CO returns or non-deliverable cartons

Paper Fill

Paper fill can be (a) a precut paper sheet, (b) a sheet from a paper roll, (c) shredded paper, or (d) machine-crushed paper. In an order-fulfillment operation, a paper filler material concept is preferred because of government code for catalog or direct

1 POLYSTYRENE OR PEANUTS
2 CORRUGATED
3 POLYURETHANE SHEETS
4 FORMED CORRUGATED
5 AIR BUBBLE SHEETS & BAGS
6 CELLUOSE WADDING
7 SHREDDED PAPER

Figure 4.100 Filler materials.

mail order-fulfillment companies to use a recyclable filler material. After a CO has been checked, a packer transfers paper sheets or crushed paper into a CO shipping carton to fill up the spaces. Per your order-fulfillment operations practice, a pack station activity has a packer place sheets or crushed paper into a CO shipping carton bottom and place the SKUs on the paper sheets or crushed paper.

With a pick/pack order-fulfillment concept, before the first pick position, an employee transfers crushed paper or paper sheets into a CO shipping carton. After the check station, a fill employee transfers crushed paper or paper sheets inside the carton to fill spaces.

If the operation uses paper sheets, they are precut and stacked on a pack station table or shelf above the table. The packer wears a rubber glove or uses a rubber tip on a finger to allow him or her to easily grasp the sheet.

When a paper roll is used at a pack station, the paper roll is located either above and to the front of the work surface or above and adjacent to the work surface. Machine-crushed paper is produced by a machine located above or adjacent to a pack station work surface or pick/pack line conveyor travel path. The machine is fed by a paper roll and has a paper length dial and paper discharge section on the front side. An employee sets the dial to determine the length of crushed paper. The packer estimates how much crushed paper will be need to fill spaces in a carton and as the machine discharges the paper the packer transfers it into the carton.

Shredded paper is also produced by feeding paper sheets into a machine that has many razor-sharp revolving wheels and rollers. The rollers move the paper forward to come in contact with the razor-sharp revolving wheels, which cut the paper into strips that are discharged to a packer. A disadvantage of shredded paper is that it produces dust that may become attached to the SKU.

Peanuts (Styrofoam or Processed Corn)

Peanuts are Styrofoam or processed corn manufactured in a peanut or a short tube shape. Styrofoam peanuts are considered by many governments as a non-recyclable material. Peanuts made of processed corn are considered recyclable, but can become stuck together, attract pests, and produce dust. At a pack station, peanuts are supplied in one of the following ways:

- From a vendor or on-site from a peanut processing machine that uses air compressors to move the peanuts into an elevated reserve peanut supply bag that has a charge location at the top and a discharge location along the bottom.
- For an order-fulfillment operation with a low CO carton volume, a packer transfers peanuts from a large tote into the shipping carton. In an operation with medium to high CO volume the peanuts are air blown through a pipe system and at each fill station, a packer has a hand-controlled stop and start discharge device. The peanut flow is a gravity forced funnel or air blower that directs peanuts into the carton. Alternatively, peanuts are packed and sealed in a plastic bag.

In a manual order-fulfillment operation with low to medium CO volume, loose peanuts are provided in a box, bin, or bag at each pack station. A packer uses a scoop or small empty box to transfer loose peanuts. During the pack activity a shipping supply replenishment employee takes peanuts to each pack station. This method is used in a small operation, it reduces packer productivity, there is some peanut waste, uses an employee for the additional replenishment activity, and spilled peanuts create a messy situation at a CO delivery location.

In an overhead peanut filler system, an air duct is used for peanut transfer through a duct or pipe system to a fill station for packer discharge into a CO shipping container. At many fill stations, a reclaim peanut system is used to capture peanut spillage and return peanuts to a reserve supply bag. The base of the pack or fill station is shaped as an inverted funnel or a sink, which catches spilled peanuts and directs them to an air blower that forces the peanuts through the tube system to a reserve bag. The peanut return system improves the appearance of a fill station, improves work-station safety, and improves peanut reuse. To ensure easy peanut flow tracking through an air duct system, plastic air duct sections are strategically located on straight sections and curves. An overhead peanut concept installation consideration is the peanut bag structural support. Prior to purchase and installation, the building architect reviews the additional load weight impact on the facility ceiling structural support.

If a peanut fill concept is used at a pick/pass conveyor travel path that has a bottom fill station and standard, known SKU cube occupancy in a CO shipping carton, the carton passes under a peanut fill station along the travel path. The carton is paused momentarily at the fill station. A timing device or sensing device controls peanut flow into a carton.

A packer controls peanut flow from a funnel into a shipping carton using either a manual or automatic method.

- ■ A manual peanut flow concept has an overhead peanut funnel and a nozzle with lips that are controlled by a lever operated by an employee. At a pack station, the CO shipping carton is located directly under the peanut funnel discharge end. The packer squeezes the flow control levers to open the lips of the funnel, allowing peanuts to flow from the funnel mouth into the shipping carton. The packer releases the levers to close the lips and stop the peanut flow.
- ■ An automatic concept has an overhead peanut funnel with a nozzle. The lips are controlled by a push button. At a pack station, the CO shipping carton is located directly under the peanut funnel discharge end. A packer pushes the peanut flow control button or a sensing device while a time control device controls peanut flow. When activated, an air blower pushes the peanuts into the carton. The packer releases the push button to stop the peanut flow. The automatic flow concept is an additional investment, but reduces packer hand effort and ensures a constant peanut flow.

In some direct mail and catalog order-fulfillment operations, 5 to 25 percent of the volume of returned CO cartons is occupied by peanuts. At the station where returned packages are opened, a peanut return system similar to that used to recover peanut spillage at a pack/fill station permits reuse of these peanuts. This feature reduces new peanut purchase.

Peanuts, Shredded Paper, or Shredded Cardboard in a Paper or Plastic Bag

The filler material, which can be peanuts, shredded paper, or shredded cardboard, is sealed inside a plastic or paper bag. The bags are supplied at the pack station in a bin. A packer transfers a bag from the holder into the carton. A supply replenishment employee refills depleted pack stations. The concept requires space at a pack station, but there is good packer productivity and minimal waste. When the package is opened at the delivery location, there is no mess.

Styrofoam Sheets or Padding

Thin Styrofoam sheets are placed inside a carton before the SKUs. Alternatively, the sheets, or thicker Styrofoam padding, are placed around each SKU piece before the SKUs are placed in a CO carton. The disadvantages are that the material is more expensive, difficult to reuse, some governments consider the material non-recyclable, low fill employee productivity, and additional space at a fill station. The advantages are improved SKU protection and less mess at a customer delivery location.

Air Bubbles

An air bubble concept has a machine that is located above or adjacent to a pack station or pick/pass line conveyor travel path. An air bubble machine is fed in back by a plastic sheet roll with preformed pockets. Air bubbles are discharged on the front of the machine. Air bubbles can be made before they are needed and stored at a pack station or made on demand.

The disadvantages of air bubbles is that it requires space at a pack station for the machine, as well as for storing preformed bubbles and reserve plastic rolls. The advantages are good fill employee productivity if the bubbles are made before they are needed, and good protection for fragile SKUs.

Air Bubble Sheets

An air bubble sheet is a thin plastic sheet with air bubbles along one side. A plastic sheet is sent through machine that uses heat and air forced into the plastic sheet to produce air bubbles on the sheet. At a pack station, air bubble sheets are queued in a trough or are produced on demand. With an air bubble sheet concept, a packer wraps an air bubble sheet around a SKU and places the SKU into a CO shipping carton. The air bubble sheet may be secured with tape. If there is any space in a CO shipping carton, additional air bubble sheets are placed in the carton bottom and on top of the SKU.

Foam in Place

A foam in place fill concept is manual or automatic. A manual foam in place concept has a packer spray foam into the bottom of a carton. A SKU is placed onto the foam, which forms around the SKU. As the foam hardens, additional foam is sprayed onto the SKU to surround it. An automatic foam in place application concept has an automatic fill station along a powered conveyor travel path. After a SKU is placed into a carton, the conveyor moves the carton to the fill station, where sensing devices control the flow of foam from a dispenser to fill the spaces inside the carton. When an appropriate quantity of foam has been sprayed into the carton, it is moved forward on the conveyor to a carton seal station.

If bottom fill is required in the CO shipping carton, the foam in place concept has the following process:

- A carton is moved to an automatic fill station and the foam is sprayed into the bottom.
- The carton is moved forward over a conveyor travel path to a SKU transfer station.
- An employee transfers a SKU onto the foam.
- At the last fill station foam is added on top of the SKUs.

Disadvantages are the fill material is expensive and difficult to reuse. An automatic concept adds to the conveyor cost. Some governments consider foam nonrecyclable. Advantages are SKU protection, minimal waste, and an automatic concept requires few employees.

Foam in a Plastic Bag

The foam-filled bag concept has a machine at each fill station that flows foam into a plastic bag, then seals and cuts the bag. The formed plastic bag is transferred from the machine into a shipping carton. After the bag is placed inside a carton, the foam expands in the plastic bag and forms around a SKU. The disadvantages and advantages are similar to those described above for the foam in place concept.

Crushed Cardboard or Corrugated Paper

This fill concept uses a machine that is located above or adjacent to a pack station or pick/pass line conveyor travel path to crush the cardboard or corrugated paper.

The machine is fed in back by a cardboard or corrugated paper roll and has a dial that controls the length of the crushed material discharged from the front of the machine. A packer sets the dial then starts the machine. The packer transfers the crushed sheet of cardboard or corrugated paper into a CO shipping carton or onto a trough for later use.

Another option is to have the filler material preformed into cone-shaped pieces, which are stored in a container or an overhead funnel at a fill station. An employee controls the funnel to release the filler into a CO carton or a fill station employee physically transfers the material into the carton.

The disadvantages are an on-demand concept has low fill employee productivity, it does not provide the maximum protection for fragile SKUs, at a pack station, it requires workstation surface space and reserve cardboard or corrugated paper rolls.

Cardboard Carton with a Plastic Bed

A cardboard carton with a plastic bed or liner has a plastic sheet attached to the four sides and hanging from the top. As SKUs are placed inside the carton, they rest on the plastic sheet, which supports the SKUs prevents them coming in contact with the bottom of the carton. After batched CO pick/sort and check activities or during single CO pick/pack activity, a packer ensures that SKUs are properly and securely arranged inside the carton on the plastic liner or bed. Top fill material is added to the carton.

Preformed Cardboard Insert

Three cardboard insert concepts are described.

- Notched (slotted) cardboard or corrugated paper side wall and insert. At the interior corner of the carton there is an additional cardboard piece with notches or slots and a specially designed cardboard insert with a lip on each corner. At each corner are two support members with slots or notches that are the full height and secured to the adjacent side walls. The cardboard insert is designed to fit within the carton interior with the lip on each corner inserted into a slot on the carton. This secure insert retains SKUs inside the shipping carton. A packer makes sure that SKUs are properly arranged inside a CO carton then a fill station employee transfers a cardboard insert into the carton, pushing the insert down until it rests against the tallest SKU. The lips on the corners of the insert are extended into the notches on the carton, securing the insert at a proper position and creating an open space between the insert and carton top. The insert protects the SKU from contact with the carton top after the flaps are sealed.
- Block insert concept. A standard shipping carton is used, with a cardboard insert that is a block or folded insert. One or two inserts are the full width

and length of the carton interior. A packer arranges SKUs inside a carton and a fill station employee transfers one or two cardboard block inserts into the carton. The employee pushes the insert with the flat surface downward until it rests against the tallest SKU, with the blocks or folds extended upward to the carton top. The feature secures an insert at the proper position and creates an open space between an insert and carton top so that SKUs do not come in contact with the top of the carton after it is sealed.

■ Punch hole. If an order-fulfillment operation has one or two small pieces per order, a cardboard insert with prepunched holes is placed inside an empty shipping carton. With the edges of the insert on the bottom of the interior, a packer pushes SKUs into the perforated holes on the insert. The insert reduces horizontal movement of the SKU in a carton and the carton top and bottom flaps secure the SKU against vertical movement.

The three concepts described above add to the expense of a shipping carton, but reduce the use of filler material and minimize shipping carton weight.

Meshed Cardboard or Cushion Pack

A cushion pack or meshed cardboard is a single-flue corrugated sheet that is sent through a mesh machine that forms the corrugated sheet into a meshed sheet. The machine makes many small cuts in the cardboard sheet. When the sheets are spread open the cuts give it a fish-net appearance. At a pack station, a packer transfers the meshed cushion into a carton or the meshed sheets are placed in a trough for later use. The material is flexible to fill spaces, is able to surrounds a SKU, and layers provide a cushion. It creates dust that can become attached to a SKU.

SKU Shrink Wrapped onto an Insert

A SKU is shrink wrapped onto a specially designed cardboard insert with extensions on its four sides. The shipping cartons have slots along the bottom of the four side walls. The width and height of the slots match the extensions on the inserts.

One of two methods can be used to pack the shipping carton.

■ An insert is pushed down into a carton interior until its four extensions are locked into the four slots on the carton. A packer places a SKU and CO pack slip into the carton and applies a CO delivery label in the correct location, then sends the carton to a shrink wrap tunnel. As the carton travels through the tunnel heat is applied into the carton so the film secures the SKU to the insert. After the carton exits the tunnel, its top flaps are sealed by a packer or by a machine.

■ An insert with SKUs is moved to a shrink wrap tunnel. As the insert travels through the tunnel heat is applied to the film wrap to secure the SKUs to the insert. After the insert exits the tunnel a packer transfers the insert to a carton and pushes it down into the carton until the four extensions are locked into the slots. A packer places a CO pack slip into the carton, applies a CO delivery label in a correct location and the top flaps are sealed by a packer or machine.

This concept is expensive, but reduces shipping carton size and inventory, and requires less filler material.

Customer Order Carton Seal Activity

An order-fulfillment operation services customers in any industry and has COs delivered by a delivery truck. A properly sealed carton minimizes SKU damage and pilferage, gives a good impression when it arrives at the customer, and helps to ensure customer satisfaction. The objective of a carton sealing activity is to ensure that a CO carton is sealed by your company and the protective seal is broken by the customer. During transport from a pack station to a CO delivery dock and through your supply chain, a sealed shipping carton provides a conveyable bottom surface and a short travel path window.

CO cartons may be sealed with tape or plastic bands, using a number of methods (see Figure 4.101).

Crisscross Tape Application Concepts

Tape strips are used to secure a the top and bottom flaps of a one-piece carton or the top of a two-piece carton onto the bottom. The tape application concepts include (a) a manually operated gummed-tape machine, (b) an electric gummed-tape machine, (c) manually applied self-adhesive tape, and (d) a semiautomatic or automatic self-adhesive tape machine.

Manually Operated Gummed-Tape Machine

A manually applied gummed tape concept uses tape that contains glue on one side. Moisture is applied to the glue to activate it. Water-based gummed tape can be activated using water at ambient temperature or heated water. At ambient temperature, if there is a delay or very long tape strips are used, the moistened glue may become dry and not secure a seal. With heated water the moistened glue retains its adhesive qualities longer. A pack employee applies the gummed tape that has been cut to the correct length and moistened to activate the glue to a carton flaps or cover. When the moisture evaporates, the glue seals the flaps or cover. An employee determines the length of tape needed or a machine is set to precut the tape to a specific length.

Figure 4.101 Carton seal concepts: self-adhesive tape, manual gummed-tape machine, and electric gummed-tape machine.

If a packer decides on the length of the tape, the concept uses a manually operated tape machine with a handle. Tape is fed through a moisture applicator and one pull on the machine handle provides a standard tape length.

The disadvantages are low packer productivity, tape waste, water supply, and space on a pack table. The advantages are low cost and flat surface.

Electric Gummed-Tape Machine

A packer sets a desired tape length on a tape machine selector dial/button and presses a hand or foot operator button. With preset tape lengths for each dial setting, the electric tape machine automatically dispenses a desired tape length. A packer applies the tape to a carton flaps or cover. Additional disadvantages are higher cost and electric outlet. Additional advantages are higher packer productivity and less tape waste.

Self-Adhesive Tape

Self-adhesive tape can be used to secure a carton. The tape is dispensed using one of the following methods: an employee uses a tape dispenser with an attached cutting section or holds a tape roll and cuts the tape with a knife; or tape is applied by a machine.

Manual Tape Application

A pack employee applies a self-adhesive tape strand to a carton's exterior flaps or cover. The packer uses a knife or a tape dispenser with an attached cutting edge to cut a tape strand to the desired length. The disadvantages are tape waste and employee determines the length. The advantages are handles all carton sizes, can be performed in any location and used in any operation.

Semiautomatic and Automatic Tape Concepts

With a semiautomatic tape application concept, a packer folds a carton flaps or places a cover over the bottom and transfers the carton onto a tape machine surface.

In an automatic tape application concept, a carton with a cover on or a carton with top flaps move along a conveyor and come in contact with two fixed angled arms that extend outward from a tape machine. At the entrance to the machine, a gap is pulled between two cartons. The carton flaps are folded down and the carton is moved forward under a tape applicator.

The semiautomatic and automatic concepts can be designed to handle a fixed carton size or a mix of sizes. A fixed carton size represents your highest-volume carton size. Most tape machines can then be manually adjusted to accommodate a new carton size. The disadvantage of using a fixed carton size machine is that and odd cartons need to be manually taped or taped on a separate line. The advantage is that the machine handles 12 to 20 cartons per minute.

A tape machine that handles a carton mix with different lengths, widths, and heights requires a higher capital investment and has a slower tape rate of 10 to 15 cartons per minute because the machine tape head adjusts to each carton height. The advantage is its ability to handle most carton sizes.

With either a fixed or random carton size tape machine concept, carton seal operations are similar and they have several components. The first component is a carton conveyor travel path concept that ensures a single carton is indexed forward to a tape machine. The carton travels on the conveyor across an automatic tape dispenser and the top is sealed. Side grippers or rough top belt conveyors provide power and controlled carton forward movement across the tape area.

An in-feed conveyor travel path ensures that cartons are lined up properly, queued, separated, and indexed forward onto the tape machine. An out-feed conveyor moves taped cartons onto a take-away conveyor travel path.

The disadvantages are difficult to handle a wide carton variety and sizes on a conveyor travel path, some downtime to replace a tape roll but downtime is minimized with tape cartridges and cost. The advantages are less tape waste, few employees, handles a high volume, and tape is applied in a standard location on all cartons.

Carton Strap Concepts

A plastic strap concept to seal a CO carton uses plastic bands or straps placed around the carton length or width. Variations are two straps on a carton length and width in a crisscross pattern, or two straps on a carton width in the same direction. The plastic strap concept is used on a CO shipping carton that may or may not be sealed with tape.

Plastic straps may be applied manually or using a mechanical or automatic concept.

- In a manual concept, an employee applies the straps to a CO carton. The disadvantages are low employee productivity and all straps are not evenly placed onto a carton. The advantages are wide carton mix, low carton volume, low cost, and can be performed at any location.
- In a mechanical strap concept, a carton travels over a conveyor travel path to a strap station or machine that bridges the conveyor travel path. The strap machine has one section that is elevated above the conveyor and a second section that is under the surface. This design allows the machine to place a strap around a carton. As a carton arrives at the strap station, an employee moves it from the conveyor across the middle of the strap machine and presses a foot or hand button that activates the strap machine to apply one strap to the carton. Two straps are applied on a carton in the same direction or in a crisscrossed strap pattern. To have the straps crisscross, after the first strap is applied an employee turns the carton and applies a second strap. For two straps in the same direction, after the first strap is applied an employee moves the carton forward and applies a second strap. The disadvantages are investment, electric outlet, and medium or high employee productivity. The advantages are medium to high volume, increases employee productivity, less strap material waste, and accurate strap pressure on a carton
- An automatic strap concept is similar to a mechanical concept in that it has a section that bridges a conveyor travel path. The automatic strap concept pulls a gap between two cartons on a conveyor travel path because the machine requires a few seconds to strap a carton. The preferred strap machine con-

veyor is a roller conveyor. To ensure a proper gap, a carton gap device or brake and meter belt conveyor sections are used before a strap machine. Cartons that are used in this concept are uniform in size. If straps in a crisscrossed pattern are required two automatic strapping machines are used on the conveyor travel path. After a carton receives one strap from machine one, the conveyor travel path turns the carton in the proper direction to receive a second strap from machine two. The various concepts that are used to turn a carton are reviewed in the transport section in this chapter. For two straps in the same direction on a carton, one strap machine is used to perform the function. At a strap station, after the first strap is applied to a carton, the machine indexes the carton forward and applies a second strap to the carton. Cartons are transferred to a take-away conveyor travel path. Disadvantages are investment in the machine and conveyor travel path. Advantages are straps are in a uniform location, handles a high volume, and one or two straps can be applied in any direction to a carton.

Inserting Sales Literature and Special Delivery Labels

Depending on company practice, the customers' wishes, and country and delivery company practice, sales literature, delivery documents, and labels are placed on a CO shipping carton exterior surface or inside a carton. Some examples are program or advertising guides, discount or special-price documents, hazardous product label, day/time to deliver instructions, and news releases about a company or new UOP or programs. In an order-fulfillment operation, these items are transferred into or onto a CO shipping carton or bag:

- In a pick/pass operation at a start station, SKU pick position, fill station, or label station
- In a pick/pack or a pick, sort, and pack operation, at a pick station or at a pack station
- In an automatic pack concept as an item that is automatically inserted into a CO delivery container or at a pack station

The options were reviewed earlier in this chapter.

Customer Order Pack Slip, Manifest, and Load Activities

The next small-item flatwear order-fulfillment activity is a CO pack slip, manifest, and load activity. In most operations, the activities are sequential activities that are interrelated. With a pick/pass order-fulfillment concept, a CO pack slip is inserted at a start station, at a pick position, or at a check/fill station. With other order-fulfillment concepts, a CO pack slip is inserted in a CO shipping carton or bag at a pack station. For additional information on this option, we refer the reader to the discussion of the pick/pass concept in this chapter.

The CO manifest activity has an employee or bar code scanner read and list each CO shipping carton or bag discreet identification. After the manifest activity, an employee transfers a manifested shipping carton or bag into a delivery company BMC or delivery truck. The manifest activity ensures and verifies that a CO discreet identified shipping carton or bag is listed and transferred from the order-fulfillment facility onto a CO delivery truck and to a delivery company.

Inserting the Customer Order Pack Slip

A packer or machine transfers a printed document into or onto a CO shipping carton or bag. The activity can be manual or mechanical. For review, we refer the reader to an earlier section in this chapter.

Customer Order Shipping Container Manifest

A CO shipping carton/bag discreet identification manifest activity has an employee, bar code scanner, or radio frequency reader to record a CO carton or bag discreet identification. A CO discreet identification is a human-readable, machine-readable, or human/machine-readable discreet code on each CO shipping carton or bag. Per your CO delivery company, a paper copy, diskette, on-line, or delayed communication network, a CO carton or bag discreet identification list is sent from an order-fulfillment facility to a delivery company office. Options include a handwritten manifest list or a computer-printed list that is obtained from a hand-held or fixed-position bar code scanning/RF tag reading device.

Manual or Handwritten Manifest

With a handwritten manifest option, an employee writes onto a manifest document each customer name, CO discreet identification, and package weight. At a shipping dock, a transport company vehicle driver verifies the package count and CO discreet identification and signs the manifest document. This document provides the order-fulfillment operation and delivery company with a shipment record. The record indicates that a CO package was shipped from an order-fulfillment operation and is a CO truck delivery company's responsibility. An option has preprinted CO discreet identifications on a list and an employee and the transport company vehicle driver place a check mark adjacent to each CO discreet number. The disadvantages are low CO carton or bag volume, possible errors, and handles limited transport delivery methods. The advantages are low cost and easy to implement.

Two-Part Label Manifest

With a two-part, preprinted label manifest concept, at a pack station a two-part label is attached to a CO shipping carton or bag exterior surface. Each label section has a CO delivery address, CO discreet identification, and package weight. Prior to loading a CO carton on to a delivery truck, a delivery truck driver signs and removes one label section that is retained by the order-fulfillment operation; the other label section remains on the shipping carton and is retained as a shipping record. The disadvantages are low CO container volume and handles limited transport delivery methods. The advantages are low cost, accurate record, and easy to implement.

Hand-Held Bar Code Scanning/RF Tag Reading Manifest

A hand-held bar code scan/RF tag read manifest concept has a CO discreet bar code label/RF tag on each CO carton or bag. A manifest employee directs a hand-held scanning device light beam onto a CO carton/bag bar code label/RF tag reader which receives a signal. The data is held in memory for later download to a micro-computer or is sent directly to a microcomputer. After a bar code label is read by a hand-held bar code scanning device or/RF tag reader receives a signal, the CO carton or bag is transferred to the assigned shipping conveyor or BMC for transfer into a CO delivery truck. For good employee efficiency, a bar code is easy to identify and the bar code is over squared. To reduce scanner/reader damage, an employee has a holder for the scanner/reader that is attached at his or her waist or uses a scanner/reader attached to a spring-loaded cord or mounted on a structural member.

If a hand-held bar code scanning/RF tag reading manifest concept is used on a shipping sortation and conveyor concept, a CO carton or bag bar code label/RF tag scan or read activates the shipping conveyor sortation concept to divert a CO identified carton to an assigned loading lane. The hand-held scanner concept handles a medium volume, is low cost, easy to implement, easy to relocate, and on peak days can be used at multiple scan locations.

"Scan on the Fly" Manifest

With a conveyor transport, sortation and direct load or "scan on the fly" manifest concept, each CO carton has a bar code label/RF tag. With proper carton/label orientation on a conveyor, the cartons are spaced and travel under or past a bar code scanning/RF tag reading device. The bar code/RF tag data is sent to a microcomputer. At an assigned divert location on a conveyor travel path, the microcomputer activates an assigned divert device to transfer a CO carton from a conveyor travel path onto a CO delivery truck loading lane.

A conveyor travel path and "scan on the fly" manifest concept is designed to register each CO carton discreet identification and divert each carton to an assigned CO location and does not obtain each CO shipping carton weight. The CO delivery truck company manifest requirement is each CO shipping carton discreet identification. If the "scan on the fly" manifest concept has an employee off-load concept, the cartons are labeled on the top surface. A bar code scanner/RF tag reader reads and communicates label information to a microcomputer. The data is stored for later manifest list transfer to a delivery truck company. The carton continues to travel on the conveyor to the off-load station, where an employee reads a human-readable label section and transfers the carton from the conveyor travel path onto a BMC.

The "scan on the fly" manifest and automatic divert concepts are high volume, high cost, and easy to implement.

Scan and Weigh "on the Fly" Manifest

The most sophisticated manifest concept uses a conveyor transport, sortation, and direct load or scan and weigh on the fly system. A scan and weigh conveyor and sortation concept is designed and equipped to obtain each CO carton discreet identification weight. Each CO carton is placed onto a conveyor travel path and cartons are properly spaced. A conveyor directs the bar code on the carton under or past a combined bar code scanning/RF tag reading device and scale station. The bar code/RF tag data and the associated weight are sent to a microcomputer. At an appropriate divert location on the conveyor travel path, the microcomputer activates an assigned divert device to transfer a CO carton from the conveyor onto a delivery truck loading lane. Bar code scanning/RF tag reader concepts are (a) fixed position and fixed beam, (b) fixed position and moving beam, (c) fixed position and waving beam, and (d) RF tag, general area.

The features of this concept are the customer delivery truck company has one manifest for each workday or for each delivery truck, direct communication to a host computer or to a microcomputer to hold the information for later download to a host computer. The delivery truck company receives the manifest as on-line, diskette, or paper communication.

Loading Customer Order Shipping Cartons

A CO shipping carton staging, loading, and shipping activity ensures that a CO shipping carton or bag is placed onto a CO delivery truck. The loading activity can be designed to:

- Individually sort CO shipping cartons by delivery address or region
- Bulk load CO shipping cartons onto one delivery truck
- Unitize CO shipping cartons onto a pallet or four-wheel cart
- Direct-load shipping cartons into a delivery truck

Loading by Delivery Address or Region

If an operation delivers large numbers of cartons to different regions, customer orders are sorted by delivery address or region. The company uses a conveyor travel path and sortation concept, and the freight company uses a spoke and wheel (multiple terminal) CO delivery concept. A spoke and wheel concept has a truck pick up CO cartons from your order-fulfillment operation as well as from other businesses and brings all CO shipping cartons to a sortation terminal. At each freight terminal, CO cartons are sorted for each region and transferred onto a freight shuttle truck for delivery to the appropriate region. In a freight region terminal, CO cartons are sorted for CO address delivery.

If an order-fulfillment operation loads CO identified cartons by delivery region, a powered conveyor and sortation concept ensures that each CO carton (discreet identification) is diverted from the main conveyor onto a CO assigned divert lane. When cartons are sorted this way, the operation has a lower carton delivery rate due to the fact that CO cartons are consolidated by region onto a delivery truck. A full delivery truck travels from an order-fulfillment facility to a freight terminal in the CO address region and your CO shipping cartons by-pass a local (order-fulfillment) freight company terminal.

Bulk Loading onto One Delivery Truck

An order-fulfillment operation that has a small CO carton quantity for many delivery regions bulk loads CO cartons onto one delivery truck for later sortation at a delivery company freight terminal. To obtain the best CO delivery rate, CO cartons for mixed regions are sent to a local freight company terminal, where they are sorted to the various delivery regions and are combined with COs from other companies. The disadvantages are higher CO delivery charge and slightly slower CO delivery. The advantages are lower investment, minimal buiilding space, and used with any carton-loading concept.

Unitizing onto a Pallet or Four-Wheel Cart

Depending on your CO shipping carton delivery concept, cartons are unitized onto a pallet or four-wheel cart (BMC). If a small-item order-fulfillment operation has one or several pallets or BMCs for a specific region and your CO shipping cartons flow through a delivery company sortation terminal, to obtain the best CO carton freight rate you unitize CO shipping cartons on a pallet or cart by region. Unitized pallets or BMCs reduce the freight company carton sortation handlings because the freight company terminal handles pallets or BMCs instead of individual cartons. With a smaller CO shipping carton quantity unitized on a pallet or BMC, a pallet or BMC by-passes the freight company sortation concept. The freight company ensures a constant empty BMC supply to the order-fulfillment operation. If an order-fulfillment operation has a quantity of non-conveyable cartons or bags, they are loaded onto a BMC. Non-conveyable items are jiffy bags, carton dimensions that exceed the conveyor/sortation capacity, or cartons or bags that require special handling.

The disadvantages are building space, handling empty and full BMCs, potential employee injury, and lower delivery truck utilization. The advantages are low cost, improved CO delivery time, lower delivery charge, and handles non-conveyable CO cartons or bags.

Loading Directly onto a BMC or Delivery Truck Floor

When CO cartons are delivered over a wide geographical area or are sorted at the freight company terminal, the operation uses a direct-load concept. CO cartons travel from a sortation conveyor over a divert conveyor lane directly into a freight company BMC or delivery truck. In the delivery vehicle, an employee transfers cartons onto the vehicle floor or onto another carton. BMC features include no double handling of CO cartons, few shipping docks, high CO volume, and reduces loading errors.

Chapter 5

Trash Removal and Small-Item and Flatware Apparel Vendor and Customer Returns and Rework Concepts

Introduction

In a small-item or flatwear order-fulfillment operation, in addition to receiving, shipping, storage, pick and pack activities, other important activities are trash removal, vendor rework, and customer returns. The objectives of the chapter are to:

- Identify and evaluate the guidelines for designing a new trash-handling concept for a pick line and customer returns process area, or remodeling an existing facility. Understanding these is key to making your small-item and flatware order-fulfillment operation more cost-effective and efficient in meeting your company's customer service standard.
- The operational review manual and mechanized customer return processes.
- Describe return-to-stock manual sortation and routing patterns.

- Describe sort area equipment and layout.
- Identify activities to ensure an efficient and on-time return flow.
- Review vendor-delivered UOP rework locations and flows.

Topics discussed in this chapter include:

- Types of trash
- Trash handling design parameters
- Trash removal
- Trash container types
- Mechanized trash transport concepts
- How to extend a belt conveyor travel path
- Fire wall or elevated floor penetration and risk
- Pick line trash conveyor travel path designs
- Powered chain carrier types
- Trash disposal devices
- Vendor rework and customer order returns

Trash Removal and Disposal

Vendor cartons with filler material and CO return cartons create trash in an order-fulfillment facility that need to be consistently removed from a CO pick line or CO returns process area to a trash disposal area. Trash is created in the following areas of an order-fulfillment operation:

- The pick prep area or pick area, vendor master cartons and filler material
- The replenishment area, as vendor master cartons are cut and filler material removed
- Pick positions as vendor master cartons become depleted
- The CO returns area, as returned SKUs and filler material are removed from cartons and bags

In a typical small-item, flatwear apparel or GOH order-fulfillment operation, most of the trash is generated from the pick area and CO returns process area. These areas generate a large amount of corrugated paper or cardboard trash. If the facility generates large quantities of this type of trash, a mechanized trash-handling concept is considered in pick line or CO process areas.

Small amounts of metal trash are generated in the pick area and this is handled by a manual trash container concept. The pick area also generates minor quantities of plastic wrap, bubbles, foam sheets, peanuts, and paper.

In a CO returns process area, the type and quantity of trash that is generated depends on the filler material that is used in your CO pack activity.

- Recycled plastic wrap, bubbles, or foam sheets, which are handled by a container recycle transport concept or trash concept
- Peanuts, which are handled by a recycle transport concept or trash concept
- Paper sheets, which are handled by a container or mechanized recycle transport concept or trash concept

From your existing order-fulfillment operation pick activity and CO returns process activity, your design team calculates the design year's trash volume and determines the volume of each type of trash from the pick activity area and CO returns process area. From these projections, your design team proposes a manual or mechanized trash concept.

Design Parameters

The trash-handling concept that is used in an order-fulfillment operation is based on the:

- Order-fulfillment activity area locations and trash disposal location
- Type of trash that is created from the various areas in an order-fulfillment operation
- Peak and average trash volume
- Available facility space for a trash collection and transport concept
- Capital investment
- Separate trash for a recycle program
- Trash hauler

Trash Removal Concepts

Manual trash concepts are used to handle trash from a low-volume order-fulfillment operation and to handle miscellaneous trash in a high-volume operation. Mechanized trash concepts use conveyors to handle a medium to high trash volume from a high-volume order-fulfillment operation.

Manual trash removal concepts are used in pick concepts in which an employee walks or rides to the pick positions, except a pick to paper or pick to light pick line concept and in CO returns processing activity that has six or fewer CO returns processing stations. Mechanized trash removal concepts are used in a pick to paper or pick to light pick line concept, stock to employee and automatic pick concepts, and CO returns processing activity that has seven or more processing stations. An effective and cost-efficient mechanized trash removal concept, employee activity stations in the pick area and the CO returns processing area are arranged in a line. With a pick line, to complete a pick a picker has a short walk distance and a completed CO take-away conveyor travel path. With a CO returns processing activity, CO return cartons are transported on a conveyor travel path to each process station and a CO returns processed or disposed small-item or flatwear apparel piece take-away conveyor travel path. On a take-away conveyor travel path, at a CO returns processing station, to ensure minimal CO returns of damaged, broken or leaking pieces are transferred into plastic containers.

Containers for a Manual Trash Concept

Containers that are used in a stationary or manual trash concept are plastic bags or empty vendor master cartons; plastic totes or containers; baskets; barrels; four-wheel wagons; self-dumping hoppers; and pallet cages.

Trash containers are strategically located in a pick area or CO returns area and permit an employee to easily transfer trash from a pick area, pick line, or CO returns processing station into a trash container. An employee with a transport vehicle brings an empty trash container and transfers a full container to the trash disposal area. When a plastic bag or empty vendor master carton is used as a trash container, an employee transfers the plastic bag or carton into a mobile container for transport from a pick line or CO returns processing station to the trash disposal area. A trash bag or carton is easy to handle, occupies a small space, can be hung from a pick position or CO returns workstation support member or in an unused space, and can be reused or disposed of with the trash.

A captive mobile trash container is a basket, barrel, four-wheel wagon, self-dumping hopper, or pallet cage that is pushed or pulled by an employee or transported by a powered vehicle from a pick line, pick area, or CO returns processing area, across the floor and to the trash disposal area. In the trash disposal area, trash is transferred from the container to the trash disposal concept. The trash containers are periodically washed or disposable plastic container liners are used to line the container. In a pick activity or CO returns processing area, plastic containers or totes with plastic liners are strategically located to collect and transport broken or leaking pieces to a trash disposal location.

The disadvantages are the need for more trash containers, limited trash capacity, an employee to handle the trash, labor expense to transport and clean trash containers, and it requires floor space. The advantages are low cost, flexibility to be located anywhere in a facility, handles all trash types, and minimizes employee injury from broken or leaking pieces.

Mechanized Trash-Handling Concepts

Mechanized trash-handling concepts mechanically move trash from a pick line, pick area, or CO returns station to the trash disposal location, where trash is either mechanically or manually transferred into the trash disposal system for removal from the facility. The mechanized trash-handling concept then travels back to the work area.

Mechanized trash-handling concepts are:

- A powered belt conveyor
- An overhead powered chain concept with a flat, hook, or clasp carrier
- A vacuum transport concept for shredded trash

Powered Belt Trash Conveyor Concept

A powered belt conveyor used for trash removal handles cardboard or corrugated trash. The conveyor has a rubberized, fabric surface that travels over a sheet metal surface, with an electric-powered and sprocket drive motor and drivetrain, end pulley, take-up device, side guards, stop/start controls, and E-stop devices.

The conveyor travel path is elevated over or runs below the floor of a pick line, pick area, or returns processing area. It has side guards along both sides of the conveyor travel path. At a transfer location and per the height of the tallest trash carton, the side guard farthest from the employee station is a 12-, 36-, or 48-in. minimum height above the belt. The side guard nearest an employee is approximately 6 in. above the belt and has a perpendicular or angle design. At a trash transfer or merge location where there is a high potential for employee injury, the side guard is higher to retain the trash on the conveyor or there is meshed netting under the transfer location.

A trash conveyor is designed as a horizontal travel path above or below the pick area. Depending on the pick line or returns processing area layout, the conveyor travel path can be designed to waterfall or make a curve around an obstacle. For long, powered conveyor runs, the travel path is separated into sections.

The load-carrying surface is supported between two end pulleys by three concepts:

- The conveyor travel path: (a) solid sheet metal or slider bed surface, (b) sheet metal with skatewheels or roller slider bed, and (c) rollers or roller slider bed.
- The conveyor sections: (a) waterfall concept or (b) gap plate or roller concept.
- The conveyor frame, bed, or channel to support the travel path with side guards or guard rails and structural support members.

- Other components of the conveyor are the electric drive motor, sprocket, and drivetrain; take-up pulley; end and tail pulleys; stop/start controls, E-stop devices, and power and control panels.

The conveyor frame or bed is a preformed, hardened and coated metal section. The metal frame sections are (a) toe-in shaped, with the bottom lip of the frame facing toward the conveyor belt and (b) toe-out shaped, with the lip facing out from the conveyor travel path.

The steel gauge used for the frame is the manufacturer's standard, determined by:

- The concentrated trash weight
- The conveyor component load weight
- The span between two conveyor floor stands or ceiling hung/hanger bracket support members
- Seismic location

A steel conveyor frame has a coated exterior surface to reduce rust and metal dirt. The frame has predrilled holes at the lead end, along both sides, and at the tail end. The holes are used to attach splice plates, end pulleys, drive motors, guard rails, structural support members, and other items. During installation, additional holes are drilled or punched into the conveyor frame as needed.

Standard sections for a conveyor frame range in length from 2.5 to 10 or 12 ft. long. To ensure maximum flexibility and economics, conveyor design professionals attempt to use 5-, 10-, or 12-ft. long frame sections. During installation, a conveyor frame section is field cut to fit the travel path layout. The conveyor frame width corresponds to the area covered by the flat metal surface or rollers. The width is determined by the volume of trash and the largest carton that is moved by the conveyor. Standard frame sizes or widths are 12, 18, 24, 30, 36, 48, and 54 in.

A standard trash conveyor has a 36- to 48-in. wide frame. The height is the distance between the top and bottom metal frame toes and is 5 to 6 in. deep. A conveyor frame depth allows top wheel roller or roller clearance and bottom conveyor belt return rollers housed in the space. The load-carrying weight of a conveyor section is determined by the number of legs for a floor-supported conveyor or hanger brackets for a ceiling-hung section. A 10-ft. conveyor section with a support or leg at the end of each frame has a 240-lb/ft. load weight capacity. If the floor supports/legs or hanger brackets are every 5 ft. on a conveyor frame length, there is a 960-lb/ft. load weight capacity. With a cardboard trash conveyor application, the load-carrying weight capacities permit standard floor support/leg or ceiling hanger and bracket design.

Structural support provides the conveyor and trash travel path window with sufficient clearance in a pick area or line and ensures that an employee can easily transfer trash onto a conveyor travel path. The conveyor belt is moved forward over the travel path and provides a solid surface for trash to be moved along the travel path. As an employee transfers trash cartons onto a trash belt conveyor, the belt is pulled through a drivetrain and forward over one or two end pulleys to move trash cartons from a charge area to a conveyor discharge end. Belt conveyor surfaces to transport trash are (see Figure 5.1):

- A solid sheet metal surface or slider bed
- A solid sheet metal surface with skatewheels extending through holes in the sheet metal or a wheel slider or combination bed
- A roller surface or slider roller

A conveyor frame with a solid sheet metal or slider bed provides a smooth, flat, continuous surface, on top of which the conveyor belt lies flat. The belt is moved forward by a drive motor and drive pulleys, over the end pulleys, and across the smooth sheet metal surface. A solid sheet metal or slider bed conveyor surface has approximately a 40 percent coefficient of friction. A slider bed conveyor surface with its wide, solid surface is preferred to transport trash.

A solid sheet metal surface with skatewheels or wheel slider bed has a flat smooth surface like the solid slider bed conveyor surface but has holes in the sheet metal. A skatewheel protrudes through each hole. The skatewheels are attached to an axle that is connected to both sides of the conveyor frame and are supported by middle

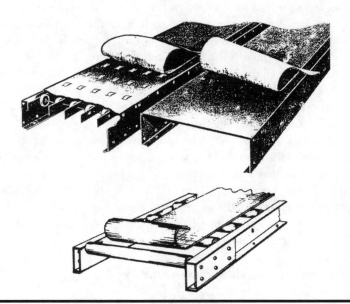

Figure 5.1 Belt conveyor types: slider roller bed, slider bed, and roller bed.

support members. Skatewheel axles are on 6-in. centers for the entire length of the conveyor travel path, and the axle is attached at a specific height to the conveyor frame. This height allows the skatewheel to extend upward approximately 0.25 in. above the sheet metal surface. The number of skatewheels on the conveyor travel path is determined by the width of the conveyor belt and frame. Standard skatewheel slider bed conveyor surface spacings are:

- 12-in. wide slider bed surface with four skatewheels
- 18- to 24-in. wide slider bed surface with five skatewheels
- 30-in. wide slider bed surface with six skatewheels
- 36-in. wide slider bed surface with six skatewheels
- 42-in. wide slider bed surface with eight skatewheels
- 48- to 54-in. wide slider bed surface with 12 skatewheels

As the drive motor, drivetrain, and end pulleys move the conveyor belt across the wheel slider bed conveyor surface, the skatewheels come in contact with the bottom of the conveyor belt. As the conveyor belt is pulled forward, the skatewheels turn. The turning skatewheels lower the coefficient of friction on the belt conveyor bottom surface and reduce conveyor belt drag. A wheel slider bed conveyor surface has a coefficient of friction of approximately 10 percent, has a lowest coefficient of friction, and a medium cost. A wheel slider bed concept is an option for moving trash.

A slider roller bed conveyor surface can be designed as:

- A sheet metal surface with holes that run the entire length of the solid metal frame surface. A short roller fits into each hole. The conveyor metal surface ends are angled downward so that the bottom surface of the conveyor belt moves across the slider bed conveyor travel path surface and the slider bed conveyor surface (sheet metal) does not interfere with the forward movement of the conveyor belt. The sheet metal and roller bed surface has the same design parameters and operational features as a skatewheel roller bed.
- Long rollers as an under-side support and axles are connected to both sides of the conveyor bed. A roller slider bed conveyor surface with one long roller that spans the width of the frame and a roller slider bed conveyor surface with staggered short rollers have rollers that have 1.9-in. diameters. Rollers are spaced along the full length of the slider roller bed travel path. Each roller extends approximately 7/16 in. above the solid slider conveyor bed surface. The one-wide roller slider bed has rollers on 36-in. centers and staggered short roller slider bed conveyors have rollers on 15-in. centers.

Standard one-wide roller slider conveyor surfaces are:

- A 12-in. wide conveyor bed with an 8-in. wide roller

- An 18-in. wide conveyor bed with an 8-in. wide roller
- A 20-in. wide conveyor bed with an 18-in. wide roller
- A 30-in. wide conveyor bed with an 18-in. wide roller

Standard staggered roller slider conveyor surfaces are:

- A 30-in. wide conveyor bed with an 8-in. wide roller
- A 42-in. wide conveyor bed with an 8-in. wide roller
- A 52-in. wide conveyor bed with an 18-in. wide roller

A single-roller slider bed conveyor surface is used on a narrower conveyor and a staggered roller bed conveyor surface is used on a wide belt conveyor travel path. A roller slider bed conveyor surface has the conveyor bottom frame rest on top of the rollers. As the drive motor, drivetrain, and pulley move the conveyor belt, the rollers turn. The turning action lowers the conveyor belt drag or coefficient of friction. A roller slider conveyor bed surface has a coefficient of friction of 25 percent, a low to medium coefficient of friction, and a low to medium cost. To extend a slider roller conveyor travel path length to match your conveyor travel path layout, the roller bed frame ends have predrilled holes that match the holes in the splice plate ends. When a splice plate is placed inside a conveyor frame, splice plate holes and two conveyor frame sections holes match. Fastening two conveyor frame sections together with nuts and bolts makes a longer conveyor travel path and is considered for moving carton trash.

If your trash is medium- to large-sized cartons and the conveor belt width overlaps the two conveyor frames, a roller belt conveyor concept is considered for moving carton trash.

The load-carrying surface of the conveyor is three- or four-ply, depending on trash mix and features, load weight, ply for existing belt conveyors that are in your facility, and conveyor drive and end pulley diameters. A conveyor belt with the thinner or lower ply has more flexibility and is used on a large-diameter pulley. A high conveyor belt ply means a stiffer belt, which is used with a small-diameter pulley. The conveyor belt can be made of cotton, cotton–nylon, rayon, or rayon–nylon. Each ply or layer of the woven fabric material is impregnated with rubber, and rubber forms the top surface. The conveyor belt bottom surface, made of the rubberized woven material, rides across the solid sheet metal, rollers, or skatewheels. The rubber surface absorbs the shock from trash placed onto the conveyor and restricts trash movement on the conveyor surface as the belt is pulled forward.

If uncontrolled trash movement on the conveyor belt surface is a transport problem or the trash conveyor travel path makes a decline or incline, to control trash transport the rubberized surface has ripples or bristles across the surface. The ripples or bristles are perpendicular to the direction of travel and serve to restrict uncontrolled trash movement on the conveyor. This is referred to as a rough top belt surface. Another option to control trash on the conveyor surface is to have cleats on the conveyor. Cleats are placed on 12- to 18-in. centers and span the belt width. With this design, the space between two cleats creates a load-carrying surface. The two cleats become barriers that restrict uncontrolled trash movement on a belt conveyor surface and permits a higher slope to a travel path.

To have minimal trash hang-ups, a conveyor belt width is properly sized for a conveyor travel path width. A belt conveyor width is a nominal 2 to 4 in. narrower than the slider bed or frame width. This means that a 30-in. wide slider bed conveyor travel path has a nominal 26- to 28-in. wide conveyor belt.

For long runs, the trash conveyor is designed with several belt conveyor sections. In this case, to transfer the trash between two belt conveyor travel path sections, the trash has to travel over the gap or space between sections (see Figure 5.2). Design options for this transfer are to have a waterfall or dump, gap plate, or gap roller.

A waterfall or dump cascades the trash from one conveyor travel path onto another. This waterfall effect is designed with one discharge section overlapping or at a higher elevation than the second conveyor section charge station. This elevation change allows sufficient clearance for the tallest trash carton to fall from the first trash belt conveyor travel path discharge end onto trash belt conveyor travel path 2 charge end. The elevation difference can be designed by starting the trash belt conveyor 1 section either at:

- A high elevation above the floor, with the conveyor surface level at the discharge end. If the conveyor has a long travel path, to achieve the proper trash belt conveyor elevation at the final discharge location, there are several trash belt conveyor sections. The trash belt 1 conveyor section is set at the highest elevation and each additional trash belt conveyor section is set at an elevation above the floor that is progressively lower until the trash conveyor travel path reaches the final discharge location.
- A low elevation above the floor and progressively increases trash belt conveyor elevation to the required final waterfall location. If the conveyor has a long travel path, to achieve the proper elevation above the floor at the final discharge location, there several waterfall sections. in order to minimize damage from jams or hang-ups there are photo-eyes at key transfer locations to stop or start the trash conveyor. The waterfall sections resemble shingles on a roof. The number of waterfall sections is determined by the total travel path distance and the previously mentioned design factors. The waterfall belt conveyor concept is used to transport a wide trash carton mix.

The gap plate concept has a metal plate that has both ends angle cut. With the angle cut facing down, the gap plate spans the open space between two trash conveyor travel path sections. A gap roller is a non-powered small-diameter roller. The roller spans the open space between two trash conveyor travel path sections.

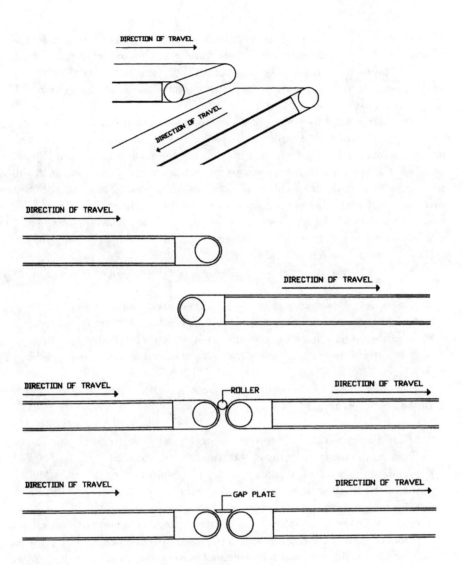

Figure 5.2 Design options for bridging the conveyor gap: waterfall, gap roller, and gap plate.

The gap plate and gap roller concepts also have photo-eyes at key locations to stop or start the trash conveyor.

A gap plate or gap roller is set at an elevation above the floor so that the top surface is at the same elevation as the belt conveyor surface. With this trash belt conveyor travel path design factor, trash belt conveyor 1 section provides power to propel a trash carton across a gap plate or gap roller. Powered trash belt conveyor 2 provides power to pull trash cartons from a non-powered gap roller or gap plate onto the second trash belt conveyor travel path. A gap plate and gap roller trash belt conveyor travel path transfer concept features are shorter overall trash belt conveyor travel path and a lower trash conveyor travel path window, preferred for trash cartons that do not fall through or get jammed (hung-up) in the very small space between a gap device and the conveyor belt, wide trash carton mix, easy to install, and photo-eyes at the transfer locations.

At the end of the conveyor, as a trash carton arrives at the trash disposal location, the travel path return pulley overlaps a chute. With the force of the trash forward movement and gravity, the chute guides the trash into a trash disposal container, thus permitting trash items to fall into a trash disposal system. A full trash line control in the trash chute has a control device to start and stop the trash conveyor travel path. The electric drive motor with a shaft, sprocket, and drive pulley provides power to turn a sprocket and the trash belt conveyor travel path. Standard conveyor motors with horse power ranges are (a) 1.5 hp, (b) 2 hp, (c) 3 hp, and (d) 5 hp that are C-faced 1750 rev/min.

Drive motors are available as:

- Fixed speed. Most fixed-speed motors are less noisy and lower in cost.
- Variable speed. Most variable-speed motors are noisy and most costly.
- Belt forward movement or travel direction, which is less costly.
- Forward and reverse belt movement or travel direction that is more costly.

A trash belt conveyor drivetrain has a tooth sprocket and drive chain. Drive unit tooth sprockets are:

- Small diameter, that is, 4 to 6 in. diameter, attached to an electric motor shaft. As the motor shaft turns, it turns the tooth sprocket.
- Standard 8- to 12-in. diameter drive pulley sprocket. As the drive pulley turns, the tooth sprocket turns, and the pulley moves the belt forward. A sprocket has teeth that extend outward and interface with a closed-loop drive chain, placed over two sprockets. The drive chain has openings that match the sprocket teeth size and center spacings. As the drive motor sprocket is turned by the shaft, the drive chain is pulled forward by the sprocket teeth over the drive motor sprocket. As the drive chain is pulled forward, the drive chain openings interface with a drive pulley sprocket to turn the drive pulley.

The location of the drive unit location on the trash conveyor travel path can be:

- Center drive, that is, in the middle of the travel path, on the underside of the conveyor frame and extending downward for about 24 in.
- End drive, that is, located at the end of the conveyor travel path, on the underside of the conveyor frame and extending downward for about 24 in. An end drive unit is preferred on a conveyor travel path to have a trash belt conveyor travel path surface pull trash forward.

Standard drive units have the ability to move a trash belt conveyor at a travel speed that ranges from:

- 40 ft./min to 80 ft./min at increments of 5 ft./min.
- 80 ft./min to 120 ft./min in increments of 10 ft./min. The faster travel speed is achieved with a special drive unit.

A trash belt conveyor travel path window allows for maximum trash carton height, clearances, and conveyor drive unit and frame.

A safety factor that is very important is that all drivetrain moving parts, especially the drive chain and two sprockets, have a protective cover. A powered belt conveyor drive motor with:

- A small horse power has a slow travel speed.
- A large horse power has a fast travel speed.
- A sprocket with a nominal drive pulley has a fast travel path speed. A large drive motor sprocket with a normal drive pulley sprocket has a slow travel path speed.

A take-up pulley, end idlers or tail pulley, snubber roller, and belt return rollers are devices to ensure that a belt conveyor surface has sufficient tension and accurate travel over a conveyor travel path. Standard 4- to 6-in. diameter pulleys are two common take-up pulley sizes. A trash belt conveyor take-up pulley is located on a conveyor travel path prior to the drive pulley, in the case of a center drive unit, or, in the case of an end drive unit, at the end of the travel path before the drive unit. During normal operation, the conveyor belt becomes stretched from its original installation length. A stretched belt reduces the ability of the drive pulley to turn the belt. To correct this, the belt has an adjustable take-up pulley that is moved forward (outward) from the original setting. This outward adjustment increases the length of the belt to compensate for the stretching. A manually adjusted take-up pulley moves (adds) 24 to 26 in. to lengthen the conveyor travel path. When a take-up pulley adjusts for belt stretch, there is no additional length to an actual trash conveyor travel path due to a take-up pulley is below an actual trash conveyor travel path.

A trash belt conveyor travel path has a tail pulley. A center drive unit has two end idlers. An end drive unit with a take-up pulley at one end has one end idler or tail pulley. An end idler ensures that a belt conveyor travel path surface moves from the top travel direction and under a travel path. With a center drive unit and end drive unit conveyor, one end idler ensures that the belt moves over the travel path. With a center drive unit conveyor, a second tail pulley and an end drive unit conveyor, a take-up pulley ensures that a belt is moving forward toward an undercarriage drive unit. A trash belt conveyor has a snubber roller. A standard snubber roller is 6 in. in diameter and is located on the underside of the conveyor frame. A snubber roller ensures that the belt travel direction is accurate and threaded over the drive pulley. As a trash belt conveyor surface pulls the trash forward over a travel path, the belt may creep to the left of right of the travel path. Options to straighten the belt if it moves to the right of the travel path are to move the right-side snubber roller forward or the left-side snubber roller backward. If the belt creeps to the left side, the options are to move the right-side snubber roller forward or the left-side snubber roller backward.

The trash conveyor belt return rollers are attached to the lower part of the conveyor frame. A non-powered roller supports the return belt conveyor. The belt conveyor pulleys factors are (a) laggarded or roughed surface or surfaces on the pulley face, (b) single or dual crowned pulleys, (c) made from hardened steel, and (d) greased and sealed bearings.

Safety devices, underside guards, accessories, and controls minimize trash hang up and employee injury, and increase trash handling. Safety devices are installed to minimize trash hang-ups and employee injury. The various safety devices are E (or emergency) stops, such as stop and start push buttons or pull cords, and wire mesh or solid covers for the underside of the conveyor frame. E-stops are placed at locations along the conveyor travel path where there is a high possibility for trash hang-up or employee injury. A red-colored push button or pull cord (lanyard) identify these as key safety devices. The control system has the capability to stop all conveyors within line of sight of a safety device or to stop the entire trash transport system. Conveyors are turned-off at the control panel and a flashing light is activated in the area where the employee pushed an E-stop device.

Criteria for stopping the conveyor are determined by your staff, conveyor manufacturer, and the complexity of the conveyor transport concept. To activate a conveyor concept, an employee corrects the problem situation, resets a safety device and restarts the conveyor concept at the control panel. An E-stop pull cord is more flexible than a push button and is more often used because it can be activated by an employee who is at some distance from a problem situation. An E-stop push button is a device for specific locations, such as at a control panel or on the ground floor below an elevated trash conveyor travel path.

Depending on the trash conveyor travel path layout and building features, the conveyor travel path may penetrate a fire wall or elevated floor surface. When a conveyor travel path penetrates a fire wall, most conveyor professionals use a dog house concept. A dog house concept has:

- Short belt conveyor sections, each of which has a specific travel path location and each belt has a control device that is controlled by a smoke or fire detector
- A fireproof shell
- A fire or smoke detector-controlled shutter door

When a conveyor travel path penetrates an elevated floor, most conveyor professionals use a dog house concept or a coral concept when the travel path does not penetrate a fire wall. A coral concept has kick plates and handrails and a conveyor nose-over and travel path.

Side guards run the full length of the conveyor travel path and ensure that trash or cartons remain on the conveyor. The height of the side guard is determined by the height of the tallest trash carton, one carton stacked on another carton height, and per your conveyor manufacturer standard.

Solid side guards are painted and made from zinc, galvanized sheet metal, or angle iron. The base or toe of the side guard is attached by tek screws or with brackets to the top of the conveyor frame. To ensure a smooth and continuous side guard surface along a conveyor travel path, the toe of the side guard faces outward (toward the aisle). Side guard sections can be installed with the first section overlying the second section similar to the shingles on a roof, or the two end sections are flared or angled out and away from the conveyor belt.

The angle of the side guard in relation to the conveyor frame can be perpendicular to the conveyor frame or at an angle to the conveyor frame. A flared or angled side guard is used on a conveyor travel path where trash is loaded onto the conveyor. As trash moves over the conveyor travel path, a flared or angled side guard deflects trash away from the loading location onto the conveyor travel path. An angled side guard is a solid sheet metal member that is approximately 6 in. long and intersects the conveyor frame at a 45-degree angle. This side guard height and angle permits an employee to easily transfer trash onto the conveyor. An angled side guard is used along a pick line or at a returns station.

A double-high side guard is a solid side guard with a height equal to the height of two trash cartons on the conveyor travel path. This situation occurs at a transition location between two trash conveyor sections, at a waterfall location, or at a transfer location. This extra height reduces the risk of trash falling from the conveyor travel path onto an employee.

In a pick line another side guard concept is to use wire-meshed or fabric netting along the far side of the side guard. As an employee throws a trash carton onto the conveyor, the netting or mesh restricts the movement of the trash so that it falls onto the conveyor.

On a pick line with many straight guard rail applications, a side guard has a saw-tooth design. This design has a guard rail opening in front of an employee returns station or on a conveyor near side. The far side of the conveyor travel path is a solid side guardm which ensures that trash remains on the conveyor. The travel path between two workstations has an angled or saw-tooth design that permits easy trash transfer through an opening by employees from both workstations and minimizes trash hitting an employee at a workstation.

The trash conveyor travel path design options are:

■ For a pick line concept (a) over a completed order container take-away or customer return carton conveyor travel path. The trash conveyor travel path is supported from the ceiling or floor and (b) supported by pick position structural support members.

■ For a batched customer order pick/sort concept, the trash conveyor is in a specific area.

Pick Line Trash Conveyor Design

In a pick and pack line operation, the preferred carton trash transport concept is a powered belt conveyor in a location that permits an employee to easily transfer an empty carton from a pick position onto the trash conveyor. In a pick line application, the trash conveyor is controlled by an E pull cord or push buttons and photo-eyes. Design options for a trash conveyor travel path (see Figure 5.3 and Figure 5.4) are:

■ Above a pick line and completed order take-away conveyor travel path
■ Above pick positions
■ Below a pick line and completed order take-away conveyor travel path

If the trash conveyor is above a pick line, completed order take-away, or customer returns carton conveyor travel path, the top of the belt is 5 ft. 6 in. to 6 ft. above the floor, or at an elevation above the floor that satisfies local code and employee ability to transfer cartons onto the trash conveyor. The conveyor has structural support from the floor or ceiling or is attached to pick position structural members. When a trash conveyor is supported from the floor, structural support upright members do not interfere with a completed order transfer from a pick line conveyor to a completed order or customer returns carton conveyor. With the ceiling-hung or structural member-supported trash conveyor, the support rod ends on the near side. Structural support elements that face a pick aisle are padded to minimize employee injury. Side guard considerations are:

■ For a single pick line, the far-side guard is at least 24 in. above the top of the belt. The pick aisle or near-side guard is 6 in. high and angled to the conveyor frame or is a straight high side guard with a saw-tooth design.

■ For a dual or mirrored pick line, both side guards have a pick aisle or near-side guard that is 6 in. long and angled to the conveyor frame or are straight high side guards with the saw-tooth design. From the pick position a pick line

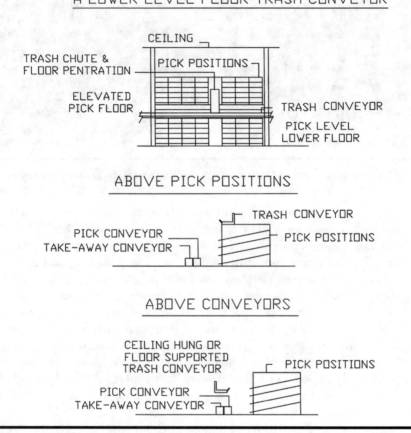

Figure 5.3 Trash conveyor locations: a chute from an elevated floor to a lower level, above pick positions, and above conveyors.

employee removes an empty carton, turns, and throws or transfers the carton onto the conveyor travel path.

A trash conveyor above the pick line, completed order take-away, or customer order carton conveyor travel paths provides easy access to the trash conveyor for employees to transfer a trash carton. The trash conveyor drive unit is typically located on the underside. The drive unit location does not interfere with employee activities. For a long travel path, a gap roller or gap plate is used to bridge the gap

Figure 5.4   Trash conveyor location above a pick position and take-away conveyor.

between two conveyor sections. This feature ensures continuous trash carton flow. All conveyor frames are level, and the entire conveyor travel path is level.

If the trash conveyor runs above the pick position or returns processing station, the conveyor travel path is supported by the static shelves, decked pallet racks, or flow racks or other structural members. The top of the conveyor belt is determined by the height of the structural support and employee reach. The elevation above the floor for the top of the belt is at least 60 in. above the floor, so that pick position structural support members do not interfere with trash transfer onto the conveyor or with CO returns carton transfer onto a take-away conveyor. At a pick position, after SKUs have been removed an employee throws the empty carton from a pick position upward and onto the trash conveyor. The far-side guards are high or have meshed netting and the near- or aisle-side guards are 6 in. high and angled. Gap rollers or gap plates are used for on long conveyor travel paths.

A trash conveyor above the pick positions or CO returns processing station is designed to provide easy access to the trash conveyor. The drive unit is typically located on the under side so that it does not interfere with employee activities and is not an obstruction to CO pick carton, tote, or CO return SKU movement. For a long belt conveyor travel path gap rollers or gap plates are used to connect conveyor sections. This ensures continuous trash carton flow. The trash conveyor frame and travel path are level; pick position structural support members are designed to support the additional dynamic and static loads. With a mirrored or dual pick/pack line application or double lines of CO returns processing stations, there are two trash conveyors and, depending on the pick area design, the height of the side guard allows trash to be received from an elevated pick line.

A trash conveyor that is below the pick line or CO returns processing workstations is used in a pick/pack application or CO returns processing station that has double-stacked pick lines or in a CO returns processing section. This means that there is one pick/pack for the A or B SKUs on the ground floor and a second pick area with static pick positions for C SKUs on an elevated floor, or the returns processing area is above or below another activity that creates trash. A ground floor pick line or processing workstation area has the same design factors as the trash conveyor above a pick line described above. Additional considerations are that the far-side guard rail should have sufficient height to direct trash carton flow from an elevated pick line onto a trash conveyor travel path. To minimize trash falling from a trash conveyor travel path, the ground-level front-side guard rail is straight and tall, designed with solid sides. The opening or transfer location on the elevated floor is located directly above the ground-level trash conveyor travel path. Conveyor travel path sections are controlled by a photo-eye. As trash is transferred through a trash chute opening on the elevated floor, the photo-eyes stop and start the ground-floor first or front conveyor section. This avoids trash falling between the side guards and landing on the second trash conveyor section that is moving forward, and minimizes trash jams or trash falling from a conveyor travel path.

Overhead Powered Chain Conveyor and Carrier Trash Concept

An overhead powered chain conveyor and carrier trash concept is designed with load-carrying devices or carriers to move empty vendor cardboard or corrugated cartons from a pick area or CO returns processing stations to the trash disposal area. Components of the concept are a closed-loop, powered chain with a drive motor, drivetrain, and take-up devices and downward extending pendants, a C-channel travel path, structural support members, a trash transfer station and controls, and E-stop devices. The carrier can be (a) a flat carrier, (b) a carrier with a single hook, (c) a carrier with multiple hooks, or (d) a clasp carrier.

An overhead powered chain conveyor and carrier trash concept is designed to have a closed-loop, powered chain travel path with carriers that travel through a pick area or CO returns processing stations and to a trash disposal area. The carriers are attached to pendants that extend downward and are spaced on 5- to 6-ft. centers. With ten to 12 carriers per minute, carriers are transported from a pick line or returns processing station area to a trash disposal. The downward-extending pendants, carrier design, and elevation of the powered chain travel path above the floor allows an employee to easily transfer trash from a CO returns process station or pick area to a carrier. The elevation above the floor ensures that there is a 7.5 ft. clear employee travel path between the floor and the bottom of the trash carrier.

An overhead powered chain trash carrier is designed to securely carry a trash carton from a CO returns processing station or pick line area to a trash disposal station. Design options (see Figure 5.5) are (a) a flat, shelf, or tray carrier, (b) a single- or multiple-hook carrier, and (c) a clasp carrier.

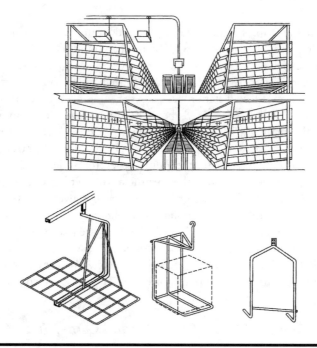

Figure 5.5 Overhead towveyor: double carton carrier, single carton carrier, and hook carrier.

Flat, Shelf, or Tray Carrier

A flat, shelf, or tray carrier has a coated metal neck and a wire-meshed load-carrying surface. The load-carrying surface or tray has a slight slope (5 degrees) to the rear of the carrier. The slope and carrier pendant design helps retain cartons on the carrier. the neck of the carrier is at back of the tray and extends upward to connect with an overhead powered chain conveyor pendant or dog.

The carrying surface has the surface area (length and width) to carry one or two empty trash cartons. Some metal carriers have two metal strands at the rear that run from the tray to a pendant and serves as a carton backstop. An option is a metal lip that extends approximately 1 in. above the load-carrying surface perimeter. The lip serves as a barrier to keep empty cartons on the carrier. As a flat carrier passes a returns processing station or pick line, an employee performs a carton transfer transaction by placing or removing an empty carton from a CO returns processing workstation or pick line onto a carrier.

A wire-meshed carrier concept is designed as an employee transfer activity or to automatically dump or tilt at a trash transfer station. An automatic carton dump or tilt concept design considerations are wire-meshed carrier has two underside runners, decline powered conveyor travel path, and wire-meshed carrier travel path has curved slides that interface with wire-mesh carrier underside runners.

As a carrier travels on an automatic dump carton powered conveyor travel path, the bottom runners come in contact with slides on the travel path. As the runners travel over the slides, the travel path rises up and declines. This elevation change causes the carrier to tilt forward and dump cartons into a trash container. Automatic dump features increase the cost of the wire-meshed carrier cost and reduces employee effort for trash removal activity.

Hook Carrier

A hook carton carrier has a single or double metal hook extending upward from an overhead trolley chain dog or pendant. The end of the hook has an upward pitch or bend to ensure that an empty carton remains on the hook. As a hook carrier travels past a pick or workstation, an employee removes an empty carton and transfers it onto a carrier hook. The disadvantages are single hook carries one carton and the double hook carries two cartons and has a high, clear ceiling for a travel path window. The advantage is a narrow travel path window.

Clasp Carrier

A clasp carrier is a single-carton carrier device that has a metal section connected to a dog or pendant that extends downward from a powered chain, a flat metal strip, and a saw-tooth clasp that faces the metal strip. The employee inserts a trash carton flap between two clasp components to secure the carton, which hangs downward. As a clasp device arrives at a trash carton disposal station, an employee performs a carton transaction to remove a trash carton from a carrier into a disposal container. The disadvantages are carries only cartons and high ceiling for a tall travel path window. The advantage is a narrow travel path window.

Shredded Trash Vacuum Transport Concept

This is discussed in the section below on shredded trash disposal.

Disposing of Carton Trash

The next order-fulfillment activity is to dispose of the carton trash. Trash can be disposed of as a bale, shredded trash, compacted trash, or in an open hopper. Before selecting a trash removal concept, each trash removal concept should be reviewed for your trash volume with your order-fulfillment operations trash handler to provide a cost estimate and to ensure that the proposed trash disposal concept is acceptable to your trash handler and community. After carton trash has been removed from a master carton open/pick/CO returns area to a trash disposal area, the next step is carton trash disposal or removal from a facility. Carton trash removal or disposal options are (a) manual bale, (b) hopper, (c) shredder, (d) compactor, and (e) vertical or horizontal baler with either a manual or automatic option.

Manual Bale Trash Concept

In a small-volume order-fulfillment operation that has a pick line or returns processing area that creates a low trash volume, a trash removal option is to manual bale carton trash. An employee breaks down or flattens the cartons and stacks them flat. The stacks are tied together with twine to make a bundle, which is placed onto a pallet or four-wheel cart. The disadvantages are floor space, labor, and low trash volume. The advantages are that it conserves storage space and improves future trash handling. There is no investment and no electric wiring is required. Trash is handled on a continuous basis.

Open Hopper Trash Concept

A hopper trash removal concept has a hopper provided by a trash company located in an open dock door. An employee opens the door and throws carton trash into the open hopper. When the hopper is full, the trash company picks up the full hopper for transport from the facility and delivers an empty hopper. With the hopper concept, an order-fulfillment operation ideally has two hoppers and two hopper locations. When a full hopper is transported from an order-fulfillment operation the second or empty hopper handles carton trash. In a facility with only one hopper, when a full hopper is removed from the facility, there is no trash container, which creates double-handling of trash and a messy condition.

The disadvantage is that the concept handles a low volume of trash.

Trash Shredder Concept

A trash shredder concept has a carton trash conveyor, revolving blades that shred the cardboard, a vacuum or air-blown conveyor for the shredded cardboard, and a trash compactor. As trash travels on the conveyor travel path, which has a sensing device to minimize employee entry to the interior, it passes through revolving blades that are at a lower elevation on the travel path. The blades shred the cartons into very small strips. The shredded pieces travel along a path that has many holes and is enclosed in a second solid-walled tube. The cardboard pieces are pushed or pulled by a vacuum or air-blown force. At the end of the tubular travel path, the shredded cardboard is discharged into a compactor. Depending on the elevation change from the trash conveyor to the compactor, there are a number of vacuum sections located at intervals along the conveyor travel path. When a compactor is full, photo-eyes direct the trash into a second compactor. A compactor company picks up the full trash hopper and delivers an empty hopper to the facility. Before deciding to use a shredder trash disposal concept, you should consider the capability of the trash hauler to handle shredded trash. In addition, in an operation that uses this concept, self-adhesive tape is not allowed on the trash cartons because they will clog the revolving blades. This concept has difficulty handling some filler materials.

The disadvantages are increased cost, increased maintenance and employee safety controls, and trash hauler acceptance. Advantages are improved compactor cube use and few employees to handle a high volume.

Trash Compactor Concept

A trash compactor concept has an enclosed hopper that is temporarily connected to a facility by a set of hooks and a moveable ram. An employee, trash belt conveyor, or enclosed vacuum/air-blown travel path transfers trash into the compactor cavity. When carton trash in the cavity reaches a predetermined level, an employee or timing device activates an electric/hydraulic powered ram to move forward. The trash is pushed toward the rear of the cavity and is compacted against the compactor rear wall. When a trash compactor is full, the compactor company delivers an empty trash compactor container and picks up the full trash container. Disadvantages are cost, with one compactor in a facility, potential trash queue problem, additional safety controls and maintenance, and potential employee injury. Advantages are improved trash cube utilization, few employees, handles all trash types, and improves security.

Mechanical Baled Trash Concept

A mechanical baled trash concept (see Figure 5.6) has a manual or powered belt conveyor travel path that transfers carton trash into a bale compartment, an electric hydraulic ram that compacts cardboard trash into a bale, a bale tie device that places a metal or wire strand around the bale, and a transport concept to remove the completed bale. A baler can be used to bale cardboard or corrugated paper and plastic.

To improve the efficiency of your baler operation, cardboard and plastic stretch wrap have separate balers. To tie a bale, a baler device has an employee insert bale wires or strands into the appropriate places on a baler. New baler technology has a bale machine automatically thread the wires around a bale. A bale machine ties a bale size that fits onto a pallet or is handled by a forklift truck.

Baler selection factors are bale size, cardboard volume, size or dimensions and thickness, baler door open direction or bale discharge direction, bale discharge queue, normal- or high-speed ram, wire strand number and wire strand type, abil-

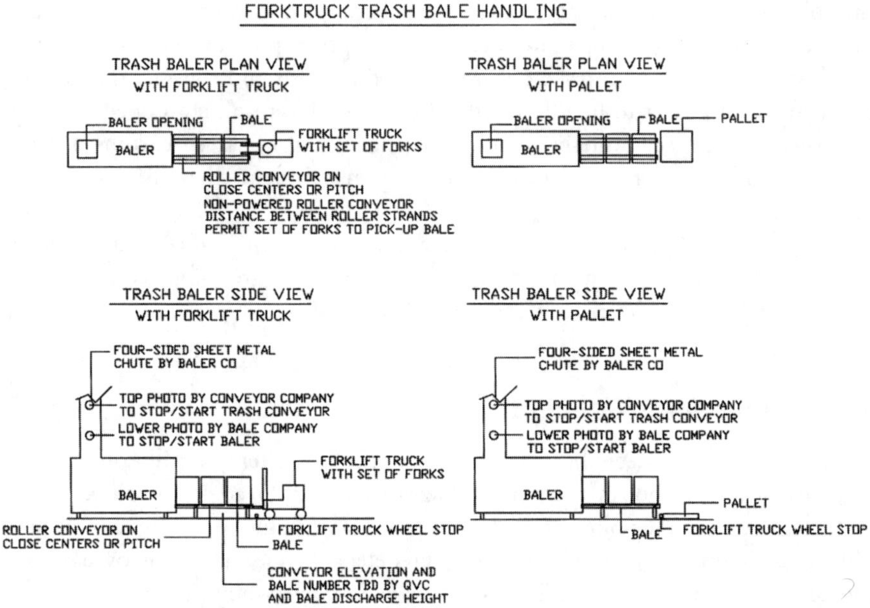

FORKTRUCK TRASH BALE HANDLING

<u>TRASH BALER PLAN VIEW</u>
WITH FORKLIFT TRUCK

<u>TRASH BALER PLAN VIEW</u>
WITH PALLET

BALER OPENING   BALE

FORKLIFT TRUCK
WITH SET OF FORKS

BALER

ROLLER CONVEYOR ON
CLOSE CENTERS OR PITCH
NON-POWERED ROLLER CONVEYOR
DISTANCE BETWEEN ROLLER STRANDS
PERMIT SET OF FORKS TO PICK-UP BALE

BALER OPENING   BALE   PALLET

BALER

<u>TRASH BALER SIDE VIEW</u>
WITH FORKLIFT TRUCK

FOUR-SIDED SHEET METAL
CHUTE BY BALER CO

TOP PHOTO BY CONVEYOR COMPANY
TO STOP/START TRASH CONVEYOR

LOWER PHOTO BY BALE COMPANY
TO STOP/START BALER

FORKLIFT TRUCK
WITH SET OF FORKS

BALER

ROLLER CONVEYOR ON
CLOSE CENTERS OR PITCH

FORKLIFT TRUCK WHEEL STOP
BALE

CONVEYOR ELEVATION AND
BALE NUMBER TBD BY QVC
AND BALE DISCHARGE HEIGHT

<u>TRASH BALER SIDE VIEW</u>
WITH PALLET

FOUR-SIDED SHEET METAL
CHUTE BY BALER CO

TOP PHOTO BY CONVEYOR COMPANY
TO STOP/START TRASH CONVEYOR

LOWER PHOTO BY BALE COMPANY
TO STOP/START BALER

BALER

PALLET

BALE   FORKLIFT TRUCK WHEEL STOP

Figure 5.6   Trash removal baler concept.

ity to store bales in a facility or to load bales onto a truck, floor ability to support a baler, available space and ceiling clearance, electric power supply availability, and cardboard trash in-feed concept and controls.

Baler types are (a) manual side feed and vertical ram, (b) manual side feed and horizontal ram, (c) manual top feed and horizontal ram, and (d) automatic top feed and horizontal ram.

With a manual feed bale concept, an employee transfers the cardboard into the baler cavity and operates the baler. A baler has a ram that compacts the trash, a baler control switch located away from the baler cavity opening, and a protective door that is closed shut to start the ram movement. This interlock feature protects employees against having limbs caught in the baler cavity. A manual feed baler is used in an order-fulfillment operation that has a small to medium trash volume.

An automatic top feed and horizontal ram baler is used in an operation that has a medium to high trash volume and has a belt conveyor travel path that moves and dumps carton trash into the baler cavity opening. Automatic baler components are:

■ A powered belt conveyor travel path that feeds trash onto a four-sided chute.
■ A chute that directs carton trash flow from a conveyor travel path to discharge into the baler opening cavity.

- A chute travel path has photo-eyes that detect partial and full status. When a photo-eye that detects the partially full condition is blocked by a trash item, a message is sent to the baler microcomputer to activate the ram. When a photo-eye that detects a full condition is blocked by a trash item, a message is sent to the conveyor microcomputer to stop the powered belt conveyor travel path.
- A bale machine that ties a bale which is transferred onto a non-powered or powered roller conveyor travel path. The bale conveyor travel path is long enough to queue several bales. During a baler design and layout, consideration is given to the trash carton in-feed location, baler discharge door location, and baler discharge door opening direction.

Compared to a compactor concept, a mechanical baler concept has twice the compaction ratio. A compactor has a 5-to-1 trash compaction ratio and a baler has a 10 to 1 trash compaction ratio. The higher trash compaction ratio means more trash is in each compacted bale.

Disadvantages are cost and additional safety controls. Advantages are less storage space, cleaner facility, bales are handled by a forklift truck, few employees, and minimizes trash hauling expense.

Customer Order SKU Rework Activities

SKU rework activities are created from CO returned SKUs and vendor-delivered SKUs that have been rejected by your quality control activity. These SKUs are not available for COs. Some SKUs can be reworked to meet your company's quality standard. This requires labor and material expense.

Rejected vendor-delivered SKUs are different from CO returns in the following ways:

- Vendor-delivered SKUs that require rework are not entered into your company inventory for sale
- Labor and material expenses for reworking vendor-delivered SKUs are reimbursed by the vendor
- SKUs from CO returns are not saleable inventory until they are transferred to a pick location

Customer Order Return and Rework Activity

In any order-fulfillment industry (catalog, direct market, industrial, or retail), CO return and rework are part of the operation's activities. CO return and rework volume varies from 5 to 38 percent of the volume that was shipped to customers. CO return and rework activities are very labor intensive and occupy a large area in the facility. They are designed handle surges, as well as a very wide SKU mix. and to have as short a delay as possible from the time a returned SKU is received to the time the reworked SKU is returned to a pickable position.

The difference between a CO outbound operation and a CO returns operation is that a CO return activity is a more complex and time-consuming activity and SKU queues occur more frequently.

A CO return is a CO ordered and shipped package that was returned by a CO or a delivery company (undeliverable) to an order-fulfillment operation. A CO return package has at least one piece per package and an industry average of two to four pieces per package. At an order-fulfillment operation, CO return SKUs fall into one of two major groups:

- Not opened—COs that were undeliverable packages or where a customer refused the packages
- Opened—COs that were opened and returned

An undeliverable or unopened CO return package is an important CO return SKU. An undeliverable CO package means that a customer did not break your company's CO seal on the package. Unless there is delivery or shipping damage, a SKU inside the package has good quality. In most cases, an undeliverable CO package does not require a returns processing employee to inspect the SKU. After the SKU is placed into a pick position, the update is made to an inventory file and the SKU is available for sale. If a SKU has been out of inventory for a short period of time, it is considered an out-of-stock situation. This means that an undeliverable SKU reduces out-of-stock SKU by one SKU. With a returns employee spending less time to process undeliverable SKUs, and thus has higher productivity, and there are additional pieces available for sale.

The second CO return type is a package that was opened, resealed, and returned by a customer to your order-fulfillment facility. A resealed package indicates that the customer has opened or broken your company's CO seal on the package. When this resealed package arrives at the order-fulfillment facility, it is assumed that the customer has removed a SKU from your CO delivery carton and that the customer has completed a SKU test, a fit, or an application.

With all packages that have been resealed and returned, a returns processing employee checks the SKU quality. The factors that determine the SKU quality are:

- If the SKU is in a manufacturer carton and whether the manufacturer's seal has been broken
- Whether, per a CO return statement, the customer has (a) used the SKU, (b) received a poor-quality SKU, (c) received a damaged SKU, (d) received the wrong color, size, or SKU

■ Whether the SKU was damaged during transport of the CO return to your facility
■ If a SKU is the type that comes in contact with human skin or is edible, it is disposed of using an approved disposal method

Common reasons for a CO SKU return are:

■ The wrong SKU was received, which is an order-fulfillment pick error or CO error.
■ The package was received late.
■ The SKU was received in a damaged condition.
■ The customer ordered several colors or sizes.
■ The SKU did not match the catalog or television picture or stated dimensions.
■ The SKU was out-of-date.
■ The SKU was recalled by the manufacturer.
■ At a company retail store, there was an over stock situation.
■ With the catalog and direct market order-fulfillment operation, (a) the customer was not home, (b) the package was not accepted by the customer, or (c) an incorrect CO delivery address.

The main objective of a CO return activity is to ensure that the CO return package is received at your facility. After opening a CO return package, the returns processing activity verifies:

■ That the returned SKU agrees with a CO pack slip or was ordered by the customer.
■ That the customer receives appropriate credit.
■ That the returned merchandise follows a prescribed route through the facility.

   ■ For quality SKUs, physical transfer from the CO returns area to a pickable position—each SKU is entered into the available and saleable inventory.
   ■ SKU is returned to a position that is not part of the available and saleable inventory location, because a typical CO return has one or two pieces; at a later time individual SKUs are transferred to a pickable position in a pick area. Per your pick area layout, this pickable position is either the present SKU pick position or another pick position that receives these quality returned SKUs.
   ■ A non-quality or non-saleable SKU is entered into non-saleable inventory (different inventory status) and is physically transferred to a temporary storage position.

From these temporary storage positions, non-saleable SKUs are assigned for:

- An outlet store or jobber
- An employee store
- A rework location
- Donation to charity
- Trash, because of damage, the SKU is out-of-date, or company policy regarding human consumable or cosmetic SKUs
- Return to the vendor
- Use as repair parts for another SKU piece

Vendor or Customer Order SKU Rework

A rework SKU requires modification or repair to turn it into a saleable SKU that matches company standards. SKUs for rework are vendor rework or CO return rework.

Vendor Rework

When a vendor-delivered SKU is received that does not match your company standards or is in poor-quality condition, this situation creates a vendor rework SKU. The receiving manager notifies the purchasing department, which directs your operation to either perform necessary rework activity or hold the SKU for vendor pick-up.

For inventory control, a vendor rework SKU is received and entered into a non-saleable inventory status. After the vendor rework (modification or repair) process is completed, the reworked sku is transferred from the rework area through your receiving and QA processes into a saleable inventory status, and to an assigned location.

Customer Order Return Rework

A CO return rework SKU is received and entered into a non-saleable inventory status and transferred to the rework area. After the required rework (modification or repair) process is completed, it is transferred from the rework area, into an available and saleable inventory status, and to an assigned storage/pick location. Vendor return SKUs tend to be larger quantities that are pallet or master carton quantities, whereas CO return rework SKUs are a wide SKU mix, in smaller quantities.

Customer Order Return Considerations

To improve your CO return processing activity, the important operational areas to consider are:

- Outbound or CO delivery factors
- Handling CO return package(s) at an order-fulfillment operation
- Customer order look-up table location

Outbound or CO delivery carton activities include:

- Accurate CO taking method.
- On-time and accurate order-fulfillment completion for a CO order/delivery cycle.
- Ensuring that a CO delivery package is properly addressed, that it is sealed, and the CO pack slip contains a CO return address label that has your company delivery address. To optimize a CO return process activity, your CO return label has a specific bar code and human-readable code for each SKU group, such as GOH, jewelry, houseware, or flatwear apparel. This feature is important for a single-line CO returns process and SKU flow.
- Ensuring that each SKU has your company discreet human/machine-readable identification label. A human-readable code is in alpha characters or numeric digits that are easily recognized and read by your CO returns processing employees.
- Ensuring that your order-fulfillment return facility address appears on your CO shipping label or delivery carton.
- A reliable CO delivery company and CO delivered package has proper filler material for a CO to repack a SKU if necessary.

Factors in deciding on the CO delivery return handling activities include:

- Whether to have a CO returns process flow that (a) separates undeliverable CO return packages from COs that have been opened, resealed, and returned or (b) mixes the undeliverable and opened, resealed, and returned packages
- Ensuring that the company return label in a CO package directs a CO return package to the preferred CO returns processing area
- Having a clear and understandable CO returns processing procedure
- Adequate facility space, equipment layout, and return SKU flow path and process to handle peak CO return package volume

Customer Order Return Process

A CO return process activity is the physical activity to move a CO return package and SKU through the CO returns processing area to a SKU final disposition. CO return process activities (see Figure 5.7) are:

- Whenever possible, your delivery company presorts your CO return packages by your major SKU classification.
- Delivery company vehicles with CO return packages are staged in your truck yard or unloaded onto a CO return conveyor concept or onto a CO returns dock staging area. If a CO return package label identifies a high-value SKU on a CO original order, the delivery company, in their CO returns handling process, separates these CO return labels/packages. As these CO returns arrive at your CO returns dock, high-value CO returned SKUs are unloaded to a secured area.
- Options for opening returned packages. (a) At a separate open station, a CO return package is removed from a CO return BMC or conveyor travel path and an employee opens the package, removes any filler material, and transfers the opened package back on to a conveyor travel path; or (b) at an unloading station, a CO return package is only opened and transported on a conveyor travel path to a CO returns processing station.
- Return filler material and empty CO returns carton-handling station options, including (a) at the opening station filler material is transferred to a recycle or trash transport concept. With a CO return package opened at a separate station, CO return SKUs are placed into a CO return carton and the carton is sent to a CO returns process station; or (b) at a processing station, filler material is transferred to a recycle or trash transport concept.
- At a processing station, CO and SKU return verification and CO return credit issue activity.
- Trash-handling options, including (a) separate open and filler material removal process station where, after proper CO verification an empty CO return carton is transferred to a trash transport concept; and (b) with a process station removing filler material to a transport concept, after proper CO verification an empty CO return carton is transferred to a trash transport concept.
- Depending on SKU quality, the SKU is (a) transferred or presorted to an assigned disposition container or (b) transferred as loose disposed SKUs onto a belt conveyor travel path for later sortation to an assigned container.
- Depending on the process for disposing of containers and SKUs, a process for transferring a SKU from a presort area to an assigned temporary available and non-saleable holding position and in inventory. The SKU is later transferred to the order-fulfillment available and saleable inventory in a storage/pick position.

RETURNS PROCESS CONVEYOR

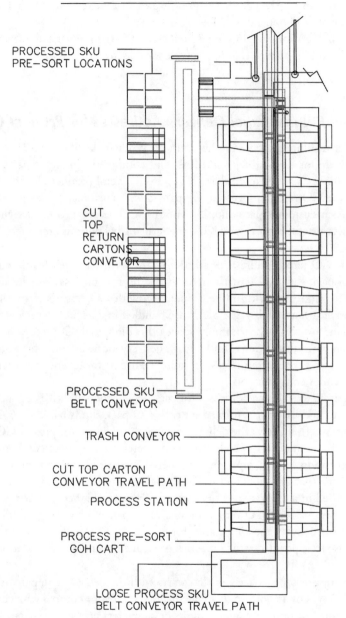

PROCESSED SKU
PRE—SORT LOCATIONS

CUT
TOP
RETURN
CARTONS
CONVEYOR

PROCESSED SKU
BELT CONVEYOR

TRASH CONVEYOR

CUT TOP CARTON
CONVEYOR TRAVEL PATH

PROCESS STATION

PROCESS PRE—SORT
GOH CART

LOOSE PROCESS SKU
BELT CONVEYOR TRAVEL PATH

Figure 5.7   Customer returns processing layout.

■ Depending on the method of disposal of containers and SKUs, activity for transferring a SKU from the presort area in a mixed tote to an appropriate available and non-saleable pick position and inventory.
■ The process for carrying out a final sort from the presort tote into a storage/pick position disposition to available for sale inventory status.

Separate Returns Label Delivery Addresses for Product Groups

If an order-fulfillment operation has multiple product classifications in one facility and COs are for single line, multi-line, and combined product classifications, it is likely that a CO return will have SKUs from several product classifications. To improve the CO return process, a company adds a customer return pre-addressed label to each customer pack slip. To improve CO return process activity and to provide security for high-value SKUs, the return label options are:

■ One CO return address for all SKUs. The return address is printed on each CO pack slip. CO returns arrive at one returns process dock location. In a high-volume order-fulfillment operation with CO returns from all product classifications, high-value SKUs are handled in a general return area, which means low security, maximum space utilization, high CO return unload and open productivity, fewer CO returns delivery locations, and, for a SKU product classification, an increase in-house transport for return to stock or return to vendor consolidation.
■ Separate CO return addresses for each SKU/product classification. The CO return address that is printed on each CO pack slip is based on SKU classification. This allows the delivery company to separate or presort CO returns by SKU classification and they are delivered to a separate company address (returns dock) or in separate containers. In a high-volume operation with CO returns that are separated by UOP classifications, high-value SKUs can be handled in a separate returns area, which means higher security, good space utilization, good CO return unload and open productivity, additional CO returns delivery locations, and, for a SKU UOP classification, minimal in-house transport for return to stock or return to vendor consolidation.

A CO return process activity that starts with the delivery company presorting return packages by major product classifications has your company machine-readable (bar code) or human-readable digits or characters printed onto your company preprinted CO return label. After the delivery company completes presorting of the CO return packages, the separated and unitized packages are delivered to the designated CO returns processing area.

The preprinted CO returns machine-readable and human-readable code identifies, for a single-item CO, the SKU class, size, or other criteria that were established by your order-fulfillment operation. With a multiple-SKU order (piece mix from different SKU classes), the SKU return code on the return label follows criteria that were established by your order-fulfillment operation. Examples of possible company criteria are:

- A multiple-SKU CO that contained jewelry and hard goods has a jewelry return address code printed on the CO return label
- A multiple-SKU CO that contained GOH and jewelry has a GOH return address code printed on the return label

Another delivery company CO returns package presort activity is to have them separate and unitize CO undeliverable return packages from other CO return packages. On a separate BMC or code, these undeliverable CO packages are delivered to your assigned CO returns dock.

Your company appropriate human/machine-readable sortation code or label characteristics are:

- For a single-line or -item CO, a CO return label has your desired SKU classification bar code as part of the human/machine-readable sortation return code.
- For combination COs (pieces or SKUs from different merchandise groups), a CO return label has the SKU classification bar code as part of the human/ machine-readable sortation return code. This code identifies the SKU with the highest value or per your company desired criteria.
- For high-value or -security SKUs, a separate human/machine- readable return code.
- The CO delivery label has a bar code as a human/machine-readable sortation return code.

If the delivery company does not presort CO returns packages, your CO return packages are randomly unitized and delivered to your CO returns dock. Randomly unitized and delivered CO returns has CO returns that have a wide SKU variety and package mix with both unopened and resealed return packages. With no presort activity, employee productivity at the CO returns processing station is low. In addition employees at the CO return presort and final sortation stations have lower productivity because they are handling different SKU classes and different-sized SKUs.

When the delivery company presorts at their terminal, CO return packages are unitized and delivered to your company returns dock. The separated and unitized CO returns on one BMC or conveyor travel path provides high processing employee productivity, returns presort and final sortation process high employee productivity, and faster SKU transfer to a pickable location.

Customer Order Returns Delivery Activity

Factors in determining the CO return delivery route to your order-fulfillment facility or to a separate CO return facility include the following:

- Your company practice regarding how CO returns arrive at your order-fulfillment CO returns dock. This CO returns activity is directed by your CO return address label and your delivery company procedure.
- Your delivery company ability to drop a CO returns truck in your truck yard staging area. Later in the work day, your yard truck or the delivery truck moves CO returns from your staging area to your CO returns dock position.
- SKU type and package characteristics, such as (a) large size, (b) small to medium size, and (c) high value.
- Return package device (a) BMC, (b) four-wheel cart, (c) cages with fork openings, (d) floor loaded on a delivery truck floor and CO returns cartons are transferred to a powered conveyor travel path, (e) palletized and secured on a pallet.
- Your outbound CO or package shipping method and your delivery company ability to drop CO delivery and CO returns trucks in your truck yard.

Customer Order Returns Package Delivery Truck Spot Concept

A CO returns package delivery truck spot concept controls the CO returns package delivery trucks in your order-fulfillment facility truck yard. The CO returns package delivery concept impacts your CO returns processing operation and temporary CO returns carton-holding area. Objectives of a CO returns delivery truck concept are:

- To ensure that your delivery company and your company CO returns operation handles your CO return package volume in a cost-effective and efficient manner
- To ensure a constant flow of CO returns package arriving at your CO returns operation and flowing through your CO returns process operation

A CO returns delivery truck concept options are to have your delivery company:

- Move CO returns trucks in your truck yard
- Drop a CO returns truck at an assigned dock location or temporary truck parking location

CO returns package delivery truck concept factors are:

- Your delivery company's ability to schedule and dispatch CO returns delivery trucks
- Your CO return package volume
- Your yard truck or a delivery company truck to move your delivery company trucks
- Ability to reuse an empty CO return package truck as a shipping truck

Your delivery company spots or drops a CO returns delivery truck at a CO returns unloading dock. When the CO returns delivery truck is empty, a delivery company truck transfers the empty CO returns delivery truck back to the delivery company terminal, moves an empty delivery truck to a facility shipping dock or to a truck yard holding area. The features are increased trips with full or empty delivery trucks between your facility and their terminal, increases transport cost, and CO return package queues are difficult to handle.

Your delivery company spots or drops a CO returns delivery truck in your truck yard or at a CO returns unloading dock. Per your CO returns operation requirement for a CO returns truck at a truck dock, your company yard truck moves a CO returns delivery truck from a yard parking location to a CO returns dock. After a CO returns delivery truck is empty, your company yard truck transfers an empty CO returns delivery truck back to a truck parking location or moves the empty delivery truck to your shipping dock. The features are the need for a company yard truck, increases a delivery company truck trips with full trucks between your facility and your delivery company terminal, decreases your transport cost, permits your operation to use an empty CO returns truck as a staging area and to improve scheduling your returns staff and improves a FIFO CO returns package processing operation.

Customer Order Return Package Handling Concept

The process for handling CO return packages impacts your delivery company and your CO returns processing operation and space. The objective of a CO return package delivery concept is to ensure that your delivery company and your company CO returns operation handles your CO return package volume in a cost-effective and efficient manner.

Factors in selecting a CO returns package delivery concept are:

- Your delivery company to utilize their facility, material-handling device, and delivery trucks
- Your CO returns operation to handle, stage, and control CO return packages flow on a material-handling device
- Ensuring that the equipment for handling CO return packages is managed by your CO returns process operation

CO returns package delivery methods are (a) stacked on a delivery truck floor, (b) four-wheel carts or BMCs, (c) wire-meshed cage with fork openings, or (d) stacked and wrapped onto a pallet.

Stacked on a Delivery Truck Floor

CO returns packages are stacked on a delivery truck floor and a full CO returns truck is moved from the freight terminal to your CO returns processing operation dock. There, a CO returns employee unloads the truck and transfers CO return packages onto a CO returns transport concept. A delivery truck or yard truck moves the empty CO returns truck to a yard staging location or to a shipping dock location.

The features of this delivery method are maximum CO return packages per delivery truck, an efficient and cost-effective CO returns package unloading activity, a delivery dock extendible conveyor and CO return carton transport concept, good yard management, additional truck parking locations and a yard truck to transfer empty and full CO return package delivery trucks between various yard locations, minimal facility area requirement, physical effort to unload a CO returns delivery truck, maximum time to unload a CO return delivery truck, and minimal investment.

Four-Wheel Carts or BMCs

A four-wheel cart or BMC CO return package concept is a square or rectangular container with casters/wheels under each corner. The container is made of fabric, plastic, wire-mesh, or sheet metal, with holes in the four side walls. One wall has a hinged and lockable door, manual push or pull handles, optional tow hitch and coupler, solid sheet metal or meshed hardened plastic or metal bottom member with holes, and non-nestable or nestable design.

On your delivery company terminal dock area, your CO return packages are transferred into a BMC. Full BMCs are transferred from the dock area into a delivery truck. When the truck is full, it is moved to your CO returns dock, where an employee unloads the BMCs from the delivery truck and transfers them to your CO returns dock staging area. A yard truck moves the empty CO returns package delivery truck from a CO returns dock position to a yard staging location or to a shipping dock position.

For long travel distances between a dock staging area and a CO returns process area, a BMC is designed to be towed by a powered vehicle or tugger. The BMCs and tow vehicles have a tow hitch and coupler, proper stability, proper caster and wheel arrangement, and properly designed aisle widths. These features permit a human or powered tow vehicle to tow a cart train between two facility locations. If you design your operation to tow a BMC train, intersecting, turning, and travel aisle widths are designed for the BMC train turning requirement.

After a BMS is unloaded in your CO returns operation, it is returned or recycled to your delivery company terminal or used in your shipping operation. To ensure maximum storage space for unused BMCs and maximum BMC number transport truck between your CO returns facility and your delivery company's terminal, the BMC is collapsible. If your CO returns operation has to queue load and empty BMCs in a large staging area, to utilize facility cube space storage, racks are placed in the area. Storage rack first elevated load beam levels are designed and used for BMCs and other storage levels are used for storage of other items. If your CO returns operation has limited staging area, an option is to use an empty CO return package delivery truck as a BMC staging area. In this case, the CO returns operation has three truck dock doors:

- One dock door for a delivery truck with full BMCs. During the CO returns process, this dock door is used by your unloading location.
- A middle dock door for empty BMCs staging area.
- One empty dock door for another delivery truck with full BMCs.

During the CO returns process, the dock door design permits your unloading location to have a constant flow of BMCs. As an employee moves a BMC from a full delivery truck, individual CO return packages are transferred to a CO return process open station. When a BMC is empty, it is transferred from an open station to the middle delivery truck. This dock and BMC concept features are BMC constant movement through your CO returns operation, limited facility staging space, and yard truck for yard full and empty delivery truck movement in a truck yard.

The BMC features are 10 to 20 carts or BMCs per delivery truck; for an efficient and cost-effective CO returns package unloading activity, your CO returns dock area has dock levelers or dock plates and one or two employees to move a full cart; if there is a long travel distance between the dock area and CO returns processing area, to have an efficient and cost-effective transport concept, your CO returns operation has a powered tugger, powered pallet, or forklift truck or employee to tow a BMC train; your facility area and storage rack floor level opening heights are designed to handle your BMC queues. The concept requires increased employee physical effort to move a loaded BMC, minimal time to unload a CO returns delivery truck, BMC cost, to reduce accidents, a delivery truck has BMC load-locking devices, reusable device, lockable for high-value items, and uncontrolled BMC movement can cause employee injury or damage.

Wire-Meshed Cage with Fork Openings

A wire-meshed cage with fork openings has rigid or collapsible side walls. One side wall is hinged to permit the full or half side wall to open, four corner posts that provide structural strength and cage stacking, matches a pallet dimensions, moved by an employee or powered pallet or forklift truck, and solid plastic sheet or cardboard as a liner.

At your delivery company terminal, your CO return cartons are transferred into a cage. Full cages are transferred into a CO returns delivery truck. When a CO returns delivery truck is full, it is moved from the freight terminal to your CO returns operation dock, where a CO returns employee with a non-powered or powered fork vehicle unloads cages from the delivery truck and transfers CO returns cages onto a dock staging area. After the cages have been unloaded from the delivery truck, the empty truck is moved to a yard staging spot or shipping dock location.

If your operation has a long in-house travel distance between two activities, the cage is placed onto a four-wheel tow cart that is towed by a powered vehicle or tugger. Each tow cart and tow vehicle has a tow hitch and coupler, proper stability, proper caster/wheel arrangement to permit a tow vehicle to tow a cart train. If you design your operation for a cart train, your intersecting, turning, and traffic aisle widths are designed for the cart train turning requirement.

After a cage is unloaded by your CO returns processing station, the empty cage is returned or recycled to your delivery company terminal. To ensure maximum storage space for unused cages and maximum unused cage transport between your CO returns facility and your delivery company terminal, cages are collapsible carts. If your CO returns operation has to queue cages in a large staging area, place storage racks in the area. The rack opening height is designed and used for the cages and for other order-fulfillment items. As previously mentioned in the BMC dock concept, a three delivery truck dock position concept is used to minimize in-facility cage staging space. The concept features are 10 to 20 cages per delivery truck, for an efficient and cost-effective unloading activity, it has a dock leveler or dock plate and an employee with a manual or powered pallet or forklift truck, with a long travel distance between a CO returns dock area and a CO returns processing area, to have an efficient and cost-effective transport concept, your CO returns operation tows a cart train with cages on each cart, facility area is designed to handle queues, minimal time to unload a CO returns delivery truck, cage cost and repair expense, reusable, stackable and collapsible, and with a hinge, reduces an employee reach into a cage.

Stacked and Wrapped on a Pallet

At your delivery company terminal, your CO returns are transferred onto a pallet. A full pallet is wrapped in plastic stretch film and the wrapped pallet is transferred onto a delivery truck.

When a CO returns truck is full, it is moved from the freight terminal to your CO returns operation facility dock, where an employee with a non-powered or powered pallet truck unloads pallets from the delivery truck and transfers them onto a CO returns dock staging area. The empty delivery truck is moved to a yard staging location or to a shipping dock location. If a returns operation has a long travel distance between two activities, a pallet is placed onto a four-wheel cart. The four-wheel cart's design and operational features are the same as the cage carts.

An empty pallet in your CO returns operation is transported to and used by your order-fulfillment operation. To ensure maximum storage space for unused pallets and maximum pallet transport in your facility, stacks are made from the empty pallets. As previously mentioned in the other concepts, storage racks are used in the area. The features are 10 to 20 pallets per delivery truck, for an efficient and cost-effective unloading activity, it has a dock leveler or dock plate and an employee with a manual or powered pallet or forklift truck, if there is a long travel distance between a CO returns dock area and CO returns processing area, to have an efficient and cost-effective transport concept, your returns operation tows a cart train with pallets by a powered tugger, your facility is designed to handle your queues, minimal cost, pallet is reusable, plastic wrap is thrown in the trash and requires labor and a trash container, additional expense for plastic wrap, and after plastic is removed from a pallet, potential CO return package damage from a CO return package falling to the floor.

Unloading Customer Order Returns

At a CO returns dock, packages are unloaded from the delivery company truck. The location of the CO returns unloading dock can be:

■ A returns building. If a CO returns processing activity has your delivery company truck transport CO return package trucks to a separate returns facility, the design factors are the operational support activities and costs associated with a separate facility, such as utilities, staff, and management, return to stock piece double transport and double handling to restock SKUs and increased time to complete a return stock transaction, security problem, additional transport and material-handling cost to combine CO returns with vendor returns, difficult to effectively use the facility space with CO returns peaks and valleys, and difficult to relocate employees to match CO return package volume.

■ A dock in the main building. If a CO returns activity has your delivery company truck transport a CO return package truck at an assigned facility dock, design factors are per your CO return package volume, improves facility space, management staff, some equipment, employee flexibility, and use in other order-fulfillment activities, minimum security problem, minimizes return to stock transport, material-handling activities and time, and reduces problems to combine vendor and CO return SKUs.

Customer Order Returns Package Staging Activity

Options for staging returns packages are to stage CO returns in your transport company delivery vehicle or to unload CO returns cartons or BMC onto your facility dock. When CO returns packages remain on the delivery truck, your CO returns operation is using the CO returns delivery truck as a queue area. Per your delivery company CO returns truck unloading criteria, your CO returns operation has two to four hours to unload a CO returns delivery truck. The concept allows an efficient and cost-effective activity and uses little facility space. It improves your ability to schedule a CO returns processing operation, FIFO CO returns operation, and interfaces with all CO returns material-handling devices or concepts.

Your CO returns operation should have adequate space in the dock area for unloading and staging CO return packages, to be able to handle your peak quantity of CO returns. After a CO returns delivery truck is spotted at your dock, an employee unloads CO returns packages or equipment from the delivery truck onto the CO returns dock area. To ensure tracking of the CO returns and a FIFO customer returns rotation, CO returns equipment is tagged with a color-coded and human-readable date label for each week day. With your CO return package business peaks and valleys, storage racks are placed in a CO returns storage area. The clear height between the floor surface and first load beam permits an employee to place CO returns equipment in a floor-level rack bay. Elevated pallet storage rack positions are used for processed CO returns, supplies, or other items. Staging CO returns packages on the dock is more efficient and cost- effective, but requires a larger facility area. The staging area permits scheduling of the CO returns processing operation, and color-coded tags permit a FIFO CO returns operation.

Returned SKU Sizes

In a catalog, direct market, or retail store operation, the CO returns process handles all CO return packages. In an efficient and cost-effective operation, to provide constant SKU flow and increased security for high-valued SKUs, the returns process operation is separated into:

- One location to handle very large-sized and large cube SKUs
- One location to handle medium- and small-sized and medium and small cube SKUs
- One location to handle high-value SKUs

An oversized SKU package can be processed in one of three ways (see Figure 5.8):

- At a dock. CO returns in large packages are separated and processed at a CO returns dock. An employee unloads all CO return packages from a delivery truck, separating large packages. The smaller packages are forwarded to a CO returns small SKU open station for CO returns processing. Large packages are placed onto separate material-handling Equipment, such as a pallet, four-wheel cart, or cage, as described above. These packages are sent to an open station and are opened, CO returned credit is processed, and the SKU is disposed of for return to vendor, return to stock, company store, or as damaged. The concept features are large packages handled by a separate concept, combines a CO return dock, and large package return activities, and has a low cost, CO return package transport concept is designed for small-items flatwear packages, easier to design a CO return large package flow and permits a CO return larger small package SKU volume handled on the remaining CO returns processing concept.
- At a separate opening station. Large packages are separated and processed at a separate opening station. An employee unloads all CO return packages from a delivery truck. From the dock, all CO return packages are transferred to a SKU opening station. As the packages are opened, an employee separates SKUs in large packages from the smaller packages. Small packages are forwarded to an open station and transported to a general CO return packages processing station. Large packages are stacked onto a material-handling device and sent to a separate CO returns processing station. The advantages are CO return large packages are handled by a separate concept. Prior to a CO return process station, the transport concept has a queue area.
- At a general processing station. All returned SKUs are transported to a general CO returns open station for processing. The CO returns processing station has a sortation location for small SKUs. Large SKUs are transferred onto a pallet, four-wheel cart, or pallet cage. This concept has lower employee productivity at the CO returns processing station because of the SKU mix.

LARGE SKU RETURNS PROCESS AREA

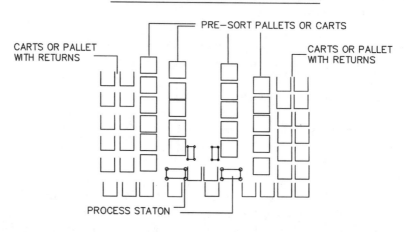

Figure 5.8 Customer returns processing station for large SKUs.

Opening Returned Packages

An employee at an open station:

- Slices or cuts the seal on a CO return package opens the carton flaps or bag end.
- Per your CO returns process operation, removes and transfers all filler material to the filler material recycle concept.
- Ensures that a CO return pack slip and SKU are returned to the package or placed into a return process captive tote.
- Transfers the package to a transport concept for travel to a CO returns processing station.
- Per your CO returns process, moves an empty BMC, cage, or pallet to a staging area and returns a full BMC, cage, or pallet to a CO returns opening station. This activity is not required on a conveyor travel path open concept.

A CO return package open activity has a CO return open employee with an industrial knife cut a carton or bag. To minimize injuries, the employee uses safety gloves.

Recycling Filler Material

Filler material is removed from a CO returns package and transferred onto a transport concept, which moves the filler material from a CO returns processing area to an outbound packing station or into a filler material reserve device. If filler material is not suitable for recycling, it is thrown in the trash.

The equipment that is used in recycling filler material is determined by the type of filler material and whether it can be reused. If the filler material is peanuts, air-filled bubbles, or crushed paper, recycle transport concepts are (a) an air-blown transport concept, (b) a belt conveyor, or (c) a large container.

For peanuts, each CO returns open station has a trough to capture peanuts and direct them into the enclosed peanut transport concept. Each workstation surface allows peanuts to pass through to the trough while retaining CO returned SKUs. Along the travel path at key straight locations and curved enclosed transport travel paths, clear travel path sections permit maintenance employees to view peanut flow or jams in the travel path.

In a small-volume operation, an employee dumps the CO returns package onto a wire-meshed surface that lies above a solid or plastic-lined container that retains peanuts. The container has four casters on the bottom, or fork openings for a forklift truck.

If your filler material is air bubbles, air bubble sheets, crushed paper, or crushed cardboard, a filler material recycle transport concept uses a solid or meshed container with four casters or fork openings or a powered belt conveyor travel path concept.

Transport Concepts

Options for designing a transport concept for moving CO returns packages from an open station to a processing station are:

- The return carton/bag. The CO returns package is used as the handling or transport package. The package is moved onto a conveyor or four-wheel cart for transport to a CO returns processing station.
- A captive container. A captive tote or tray is used to transport a CO return package from a CO returns open station to a CO returns processing station via the transport concept. After the returns are processed, the captive tote or tray is returned over a separate transport travel path to a CO open station or stacked and placed onto a material-handling device for later transport to a CO open station. A four-wheel cart. A four-wheel cart with shelves is moved by an employee. Each CO returns activity station has a cart in-feed lane and an out-feed lane. To have a cost-efficient CO returns processing operation, a CO returns process area layout has several cart lanes at each CO open station and CO returns processing station. At a CO returns open station, an employee transfers an opened CO return package from an open station to a four-wheel cart. At a CO returns processing station, an employee transfers a loose SKU from a process table onto an assigned cart.

■ A non-powered skatewheel or roller conveyor travel path. A non-powered roller conveyor has a charge end at a CO returns open station and CO returns processing stations are located along the conveyor travel path. At the discharge end there is a CO returns processing station on either side of the conveyor and an end stop. At the charge end, a CO returns station has a slight elevation to ensure that CO returns packages flow over the conveyor travel path. After a CO returns package is transferred to a roller conveyor travel path, an employee pushes the package forward and the travel path slope moves it over the non-powered conveyor travel path, pushing other packages already on the travel path forward. CO returned and opened packages are pushed forward until the line pressure restricts the ability of an employee to add a package at the charge end or a CO returns processing employee to remove a package. At a CO returns processing station, an employee pulls a package from the conveyor onto a CO returns processing station. Very small cartons or loose SKUs may cause problems on the conveyor, but these are minimized by placing them in a captive tote.

■ A powered roller conveyor travel path. A powered roller conveyor has a charge end at a CO returns open station and CO returns processing stations are located along the conveyor travel path. At the discharge end there is a CO returns processing station on either side of the conveyor. At a CO returns processing station, for easy package transfer, the preferred conveyor concept is a zero-pressure queue roller conveyor. A CO returns open employee transfers a CO returned and opened package from a CO returns open station onto the powered conveyor travel path. At a CO returns processing station, a CO returns employee pulls the package from the conveyor onto a CO returns processing station. In a dynamic returns powered conveyor concept, prior to a CO returns process station a conveyor travel path creates a gap (open space) between two CO returned and opened packages. The gap allows a mechanical divert device to transfer a package from the main conveyor travel path onto a CO returns processing station queue path. A photo-eye or other sensing device on each queue path when blocked by a diverted package deactivates the divert device.

■ A smooth top powered belt conveyor travel path. The belt conveyor also has the charge end at a CO return open station, with CO returns processing stations along the travel path and processing stations on both sides at the discharge end. The last two CO returns processing stations have a non-powered roller conveyor to provide a short queue distance and have a photo-eye, which when blocked by a package turns off the powered belt conveyor travel path.

Processing Returned Packages and SKUs

At a CO returns processing station, an employee removes a package from the transport concept onto a CO returns processing station (see Figure 5.9 and Figure 5.10). The employee follows the procedure outlined below:

- Scans a CO return label into a computer. A computer display screen shows the original customer-ordered SKU and quantity and verifies that the return is based on an associated CO.
- Removes a CO return SKU from its package or captive tote.
- Reviews the reason for the return and enters it into a computer.
- Determines the quality of the SKU and whether it will be reworked or disposed of.
- Applies a disposition label to each SKU. The labels are printed by a printer that interfaces with the computer.
- Determines if a SKU is in out of stock or back ordered status.
- Transfers the return package to a trash concept or places a captive tote on the assigned transport travel path or onto a material-handling device.
- Transfers a labeled SKU to a disposition or temporary holding concept.
- Issues credit to the customer.

The disposition label on a SKU serves as a pre-sortation or final sortation instruction. SKU disposition options are:

- Out of stock and back order SKUs are labeled BO.
- Return to stock, labeled RTS, or a label that is left blank. If the label is blank it is easily identified by a CO returns processing employee. A SKU with a blank label causes less concern when it is sent to another customer.
- Return to vendor, labeled RTV.
- Damaged, labeled D.
- Outlet store or company store, labeled OS or CS.
- Retail or jobber, labeled J.
- Charity, labeled C. Rework, labeled RW.
- Spare parts, labeled SP.

The most important SKU disposition classification to an order-fulfillment operation is the return to stock (RTS) or back order (BO) classification. A RTS or BO SKU is one that is in good or resale quality to complete a future CO. The operation had an out of stock, back order, or SKU temporarily not available situation and the returned SKU provides a SKU for a CO.

The SKU is transferred from the CO returns processing area to a pick position, entered into the WMS or inventory, and is available to satisfy an existing CO.

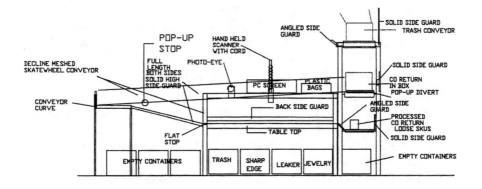

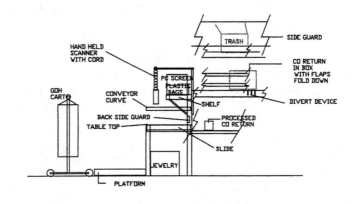

Figure 5.9 Customer returns processing station with divert and queue conveyor.

Customer Order Pack Slip Look-Up Table Location

To complete a CO, a host computer downloads details of the CO (CO identification, pack slip, and SKU piece quantity) to the WMS computer. The WMS computer uses the CO details to accurately pick, pack, and manifest the CO. If the CO is undeliverable, the delivery company returns the CO to the order-fulfillment operation. Undeliverable COs, as well as customer returns, require that the proper

CUSTOMER RETURNS PROCESS STATION
EMPLOYEE PULLS CARTON FROM CONVEYOR
FRONT VIEW

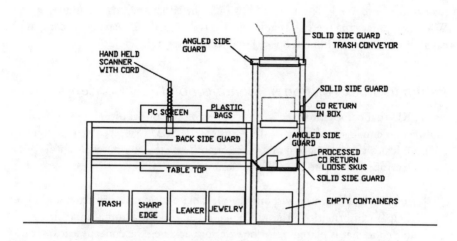

CUSTOMER RETURNS PROCESS STATION
EMPLOYEE PULLS CARTON FROM CONVEYOR
SIDE VIEW

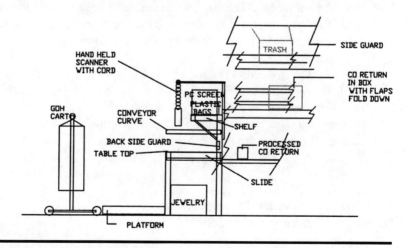

Figure 5.10 A customer returns processing station employee pulls cartons from the conveyor.

credit be issued to each customer for SKUs that were returned. To ensure proper credit and to minimize invalid customer refunds, each CO returns processing station matches each CO returned piece and CO number to the CO number and pack slip pieces in the company historical CO records. To complete a CO returns processing activity, an employee is required to verify a CO details. The CO detailed historical records or look-up table location is in either the host computer or the WMS computer. The method that is selected is designed for a particular company's capabilities, order-fulfillment operational features, and policies.

Return to Stock Position or Identification Label Concepts

Each SKU that comes into the returns processing station requires an identification to direct an employee who physically transfers the SKU to a pickable position and communicates the transaction and piece quantity to the computer inventory files. SKU identification or put-away instruction options are:

- An RTS label with or without the pickable position, return date, and other company information. The additional RTS label has a human/machine-readable code that is placed onto the exterior package. The concept features are:
  - With the second code, possible put-away employee confusion
  - Customer concern with a second label on a picked piece or after an employee places a labeled SKU into a pick position to have to remove the RTS label
  - Added computer time to look-up each SKU pick position and print time
  - Label print machine and label expense
- Use the existing SKU human/machine-readable code with a QA inspection label. The QA inspection label approach has preprinted QA labels at each process station; after SKU disposition a process employee places a QA inspection onto each RTS SKU. The QA inspection label has a human/machine-readable code QA inspection printed on the label face and is a different colored label. The concept features are minimal put-away employee confusion, minimal customer concern with a QA inspection label on a picked SKU, minimal computer time to look-up each SKU pick position and print time, and no label print machine and label expense.

Handling SKUs with Broken, Sharp Edges and Leaking Containers

At a CO returns processing station, an employee transfers all broken SKU pieces with sharp edges or containers with leaks (cracks in a container side) into a proper trash container (see Figure 5.10). This trash container is on the CO returns processing station floor. For these SKUs, the employee applies a returned SKU disposition label onto a paper document. At the end of the work day end or at intervals during the day, the documents with the labels are sent to the office for inventory update.

This serves to remove SKUs that could damage other SKUs or cause employee injury.

Customer Order Returned SKUs Not Presorted

In a returns processing activity in which there is no presorting of SKUs, all SKUs, regardless of disposition classification, are randomly transferred from the CO returns processing station into a chute, carton, tote, or onto a conveyor. At the conveyor travel path end, the mixed SKUs are transferred into a tote. Per your order-fulfillment operation inventory control concept, it is an option to have the SKUs allocated to a license plate carton or tote. After a tote is full, it is sent from the CO returns processing area over a transport concept to the storage/pick area. In the storage/pick area the tote is transferred to a storage/pick position, where individual SKUs are updated in the inventory and mixed SKUs in a tote are allocated to a storage/pick position. This means that mixed SKUs in one tote are ready to complete a CO.

Presort Concepts

At a CO returns processing station or at a conveyor travel path end an employee separates returned SKUs that have been labeled according to disposition into groups. Labeled SKUs are transferred into a preassigned tote. There is one tote for each disposition classification. The objective of the presort concept is to increase productivity when the SKU is returned to the pick area by minimizing time to sort and place SKUs in pick positions. Presort activities can take place at the returns processing station or at a remote location.

Presorted at the Returns Processing Station

At the CO returns processing station, an employee transfers a labeled SKU from into an assigned tote. SKUs are placed into the totes by predetermined criteria. Full totes are transferred from the sortation location onto a transport concept for movement from the presort area to the final sortation area or to the pick area.

Presorted at a Remote Location

A CO returns processing employee transfers SKUs labeled for disposition loose onto a conveyor. Damaged SKUs are transferred into a container at the CO returns processing station. The conveyor moves the SKUs from the CO returns processing station to a remote presortation location, where a manual or mechanized presort concept completes final sortation.

Presorted SKU Instruction Concepts

The disposition label on each SKU contains an instruction that is used to uniquely identify a particular SKU's disposition and each presortation concept opening, chute, carton, tote, or cart shelf. There are a number of ways to identify the disposition of a SKU:

- Return to stock (a) first digit or alpha character of the SKU inventory number, (b) last digit or alpha character of the SKU inventory number, or (c) SKU pick aisle
- Return to vendor (a) first digit or alpha character of the SKU inventory number, (b) last digit or alpha character of the SKU inventory number, or (c) vendor discreet number

Presorting Return to Stock SKUs by First Digit or Alpha Character of the Inventory Number

A SKU inventory number appears on each SKU and is a discreet number that is unique for each SKU. The SKU number has alpha characters, digits, or a combination. An inventory classification number has a unique first digit or alpha character to identify each SKU inventory classification group, thus a CO returns processing employee can use the SKU first digit or alpha character of the inventory classification number as a presortation instruction. Presortation locations with digits range from 0 to 9. Presortation locations with alpha characters are A to Z or 26 locations.

Thus, an RTS SKU with an inventory number 20001 has 2 as the sortation location. At the charge and discharge ends, each presortation chute, flow rack, shelf, cart shelf opening, or sortation location has an appropriate discreet number or alpha character. Each labeled SKU has a discreet number or alpha character printed on its face. This print is as large as possible and easy to read.

After your CO returns processing employee applies the RTS disposition label onto a SKU or the labeled SKU arrives at a remote presort area, an employee reads the first digit of the inventory classification number on the stock disposition label and transfers the SKU by disposition code to an appropriately numbered presortation location or tote. Full totes are transferred from the CO returns presortation area to a transport concept, which moves them to the storage/pick area. In the storage/pick area each piece is transferred to a storage/pick position and the inventory is updated for that SKU piece. Per your CO customer returns operational procedures, an individual SKU is finally sorted to a unique permanent or temporary position, or mixed SKUs are kept in one tote.

Presorting Return to Stock SKUs by Last Digit or Alpha Character of the Inventory Number

This is basically the same as using the first digit or alpha character of the inventory number, except the presort instruction is the last digit or alpha character. This method is used in a presort concept when the first digit or alpha character does not use all numbers between 0 and 9.

Presorting Return to Stock SKUs by Pick Aisle

SKUs that are presorted by the pick aisle number are labeled with the inventory pick position. A SKU pick aisle number is a discreet number that has alpha characters, digits, or a combination. Possible presort locations are 0, 1, 2, 3, 4, 5, 6, 7, 8, 9, and 10 and as many digits to identify each pick aisle. If the pick aisle identification uses an alpha character, an appropriate alpha character or characters (A to Z) is used as the presort instruction. If the pick area covers a large area with a large number of pick aisles, the pick aisles are separated into zones or one tote is used to collect for several sequential pick aisles. Each chute, flow rack, shelf, cart shelf opening, or presort location has the appropriate number.

An employee reads the RTS label pick aisle number and transfers the SKU into an appropriately numbered presort location and into a tote. Full totes are transferred from the presort area by a transport concept to the pick area, where SKUs are transferred from a tote to a pick position and the inventory file is updated. If a labeled SKU presortation concept presorts to a pick aisle group or to a zone, options are to send a full tote:

- Through an additional presortation area, where mixed SKUs for several pick aisles or a pick zone are separated into a tote for each pick aisle
- To a pick zone, where a final sort employee travels through the pick aisles and completes individual final SKU sort to a pick position

Presorting Return to Vendor SKUs by the First Alpha Character of the Vendor Name

A return to vendor first alpha character presort process uses the first alpha character of the vendor name as the presort employee instruction. If an order-fulfillment operation has a large number of vendors, the alpha characters are placed into groups, for example, (a) A–E, (b) F–J, (c) K–O, (d) P–T, and (e) W–Z. Each chute, flow rack, shelf, cart shelf opening, or each presort location charge and discharge end has the appropriate alpha character identification.

An employee reads the RTV vendor name first alpha character and transfers the SKU to an assigned presort tote or location. When a tote is full, it is transferred to an RTV storage area for final sort to each vendor's final sort position. To ensure good inventory control, as the SKU is sorted to a vendor location an on-line or delayed update is made to an inventory file.

Presorting Return to Vendor SKUs by First Digit or Alpha Character of the Inventory Number

This is basically the same as using the RTS SKU first digit or alpha character of the inventory number. The RTV SKU has it own sort locations. For additional information, we refer the reader to the RTS section above.

Presorting Return to Vendor SKUs by Last Digit or Alpha Character of the Inventory Number

In this concept, the presort instruction is the last digit or alpha character of the SKU inventory number. This presort concept is used when the first digit or alpha character does not use all numbers between 0 and 9.

Presorting by the First Digit of the Vendor Discreet Number

Using the first digit of the vendor discreet number as a presort instruction is basically the same as using the first digit or alpha character of the SKU inventory number. In your inventory each vendor is assigned a discreet number. The first digit of the vendor number is a discreet number that has been assigned to a specific vendor and the inventory computer system associates a SKU inventory classification number with a vendor number. If a vendor has several manufacturing locations, each vendor location has a discreet number first digit and presortation location. In an operation each vendor is assigned a specific identification number that means that each vendor SKU is associated with a vendor unique identification number. With a vendor identification number as the presortation position number, a vendor presortation location has the vendor first digit or number. Possible presortation location options are:

■ 1, 2, 3, 4, 5, 6, 7, 8, and 9

■ (a) 1–99, (b) 100–199, (c) 300–399, (d) 400–499, (e) 500–599, (f) 600–699, (g) 700–799, (h) 800–899, (i) 900–1,000, and (j) 0 and as many digits as necessary to identify each vendor number

An employee reads the RTV label (vendor specific identification) presortation position or number and transfers the SKU to an assigned presort location and into a container. When the container is full, it is transferred to an RTV storage area and the inventory is updated as the SKU is sorted into a vendor position.

Presorting by Last Digit of the Vendor Discreet Number

This is basically the same as using the last digit or alpha character of the SKU inventory number, except the presort instruction is last digit or character of the vendor discreet number.

Manual Sort Concepts

In a manual presort concept, to ensure high productivity and accurate presorting, each presort location has a human/machine-readable code on its front end and discharge end and is always in the same presort location (see Figure 5.11 and Figure 5.12). A manual presort location is (a) a chute, (b) a carton or tote in a flow rack, (c) a carton or tote on a fixed shelf or a shelf on a four-wheel cart, or (d) a pallet or wire-meshed cage.

An employee removes a SKU from the conveyor or tote. The employee reads the SKU disposition label, matches it to a presort location identification, and transfers the SKU to the presort position. In a dynamic and high-volume presort operation, pre-sorted SKUs are transferred into cartons or totes, which are transferred by a transport concept from the presort area to the manual, carton AS/RS, or carousel final storage/pick location area.

Mechanized Presort Concepts

A mechanized pre-sort concept has an employee remove a SKU that has been labeled for disposition from a CO returns processing tote, carton, four-wheel cart, pallet cage, or conveyor and inducts it onto a mechanized presort travel path. If the SKU is not manually inducted onto a sort travel path, a presortation travel path directs a SKU with its disposition label face up under a bar code scanner/RF tag reader, which sends the SKU information to the computer. The computer and a sort concept constant travel speed have a mechanized divert device transfer the SKU from the travel path into an assigned sort location. At the end of the sort location, sorted SKUs are queued or directed into a tote. When a chute becomes full, an

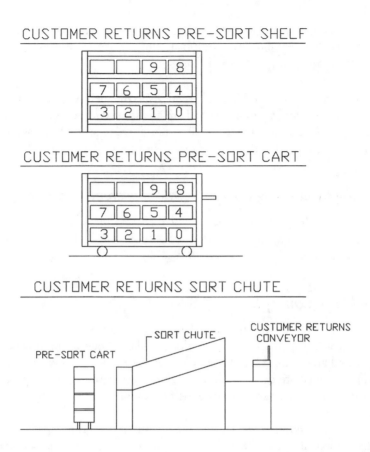

Figure 5.11 Customer returns presortation equipment: chute, multiple shelf cart, and multiple shelves.

employee transfers SKUs from the chute into a tote or replaces a full tote with an empty tote. In a mechanized sort application, a sortation location has mixed SKUs for an aisle. The totes are transferred by a transport concept from the presort area to a manual, carton AS/RS, or carousel final sort storage/pick location area.

The mechanized sort concepts that are used in the customer returns processing area are the same as those used to sort picked and labeled SKUs to a customer pack station. They include tilt trays, flap sorters, Bombay drops, ring sorters, and brush sorters. Mechanized sort concepts require additional space, utilities, high volume, human/machine-readable label, and higher capital investment. For additional information on these sort concepts, we refer the reader to Chapter 4.

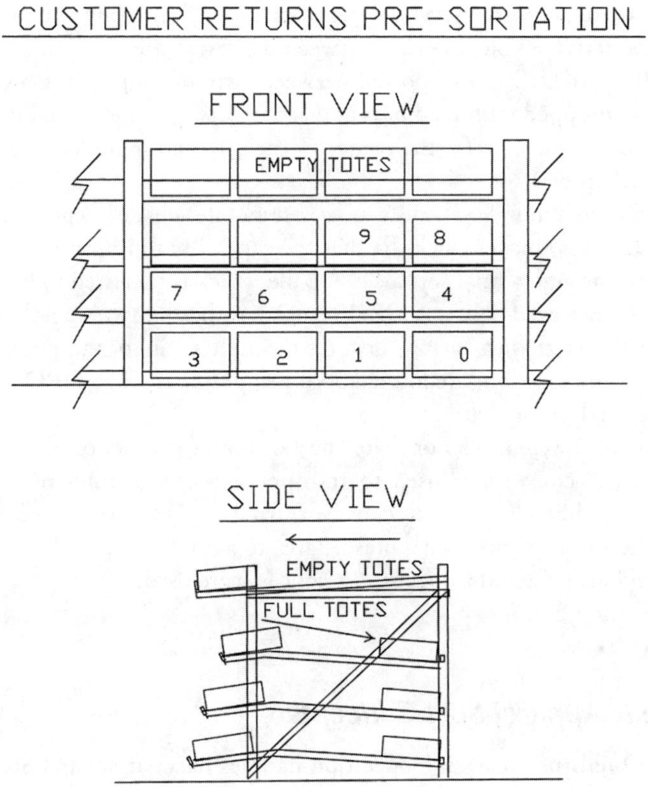

CUSTOMER RETURNS PRE-SORTATION

FRONT VIEW

	EMPTY	TOTES	
		9	8
7	6	5	4
3	2	1	0

SIDE VIEW

EMPTY TOTES

FULL TOTES

Figure 5.12   Customer returns presort flow rack.

Presorted SKU Transport Concepts

Transport concepts that are used to move presorted SKUs from a presort area to a temporary or final sortation location include:

■ A manual push cart with shelves or a GOH cart with load bars. A manual push cart has four wheels and one to five shelf levels. Each shelf has the capacity to hold one or two cartons or totes. Depending on the final sortation activity, a level shelf with the lip up permits SKUs to be handled as loose items on its surface. As a sort employee travels through a temporary or final sort aisle, he or she is able to see all the SKUs, which makes it possible to have multiple SKU transfers at one final sortation location. A manual push cart handles medium- and small-sized SKUs and the multiple shelf levels permit a cart to transport several cartons or totes, decreasing the number of non-productive employee trips between a presort area and final sort location. When SKUs are transported between two locations using a GOH vehicle, it is equipped with load bars and separators. For additional information on GOH carts, we refer the reader to the section on guided overhead vehicles in Chapter 4.

■ A powered mobile vehicle. A powered mobile vehicle is a powered pallet truck or (HROS) picker truck that has the capability to transport a pallet or pallet cage through a final sort aisle. A pallet truck is preferred to handle medium- and large-sized SKUs. A HROS truck has the ability to handle a pallet, pallet cage, or cart with cartons or totes through a wire or rail guide aisle. If final sortation locations are located above employee reach, a HROS truck handles the final sort activity.

■ A non-powered roller or skatewheel or powered roller or belt conveyor travel path. A conveyor is used to transport cartons or totes or to move loose, unsorted SKUs (on a belt conveyor) from a CO returns processing station to a presort area and from a presort area to a final sort area. In a final sort area, previously mentioned transport concepts are used.

Flatwear Apparel Sort Concepts

If the CO returns processing operation handles flatwear apparel SKUs, there is a very large number of SKUs to be handled. Each CO returns flatwear SKU receives a rework disposition label and flows through a rework or an additional process. At a CO returns presort station, flatware SKUs are presorted as one SKU group into one RTS carton or tote. When a carton or tote is full, it is sent from the presort area to a rework area for steaming, cleaning, and rebagging. In this rework area, flatware SKUs go through an additional inspection and a new disposition label is attached. After this preparation, flatwear SKUs are presorted (a) by SKU inventory classification number, (b) by pick aisle, or (c) to a pick zone. The concepts were described above.

GOH Customer Order Returns Sort Process

In an operation where SKUs are disposed of using a GOH device, design considerations are where to complete the sort into RTS, RTV, RW, or D categories. GOH sortation options are (a) mixed SKUs, (b) WMS position printed on a disposition label, or (c) last or first digit of the SKU inventory number.

Mixed Sort

A mixed SKU sort concept has all items sorted by its major classification (RTS, RTV, or RW) onto a load bar on the GOH. RTS, RTV, or RW SKUs are sorted into groups located between moveable load bar separators. When the GOH transport device is full, an employee moves it to a presort area for initial sort by product classification or directly to a final sort location.

WMS Position on the Returns Disposition Label

The disposed CO GOH returns SKU sortation by a WMS identified position on a returns disposition label concept has a returns process employee enter a GOH SKU inventory into a WMS computer and is not available for sale. The WMS computer identifies each SKU WMS position and prints a WMS position with existing SKUs onto a SKU disposition label. The labeled GOH piece is transferred as a mixed SKU with RTS, RTV, and RW SKUs onto a load bar or is presorted by disposition label first or last digit onto a load bar.

First or Last Digit of the SKU Inventory Number

This concept has each major classification (RTS, RTV, and RW) sorted by the first or last digit of each SKU inventory number On each load bar there are ten moveable separators with 0 to 9 digits on its face. After a sortation employee receives a CO disposal GOH return, a sort employee transfers the disposed GOH to an initial or final sort position. Each position is identified with a numbered separator. The hang bar separator number matches the GOH SKU inventory number first or last digit.

Initial Sort at a Processing Station

The initial GOH sortation takes place at a processing station and has a GOH hang bar on a static rail, trolley, or four-wheel cart. The hang bar is within easy reach of employees to complete the transfer. For best overall operational results, a sort concept that uses the last or first digit of the SKU inventory number is preferred at a processing station.

Initial Sort in the GOH Storage Area

If the initial sort activity takes place in the GOH storage area, a CO returns employee places all pieces randomly (RTS, RTV, and RW) onto a load bar on a static rail, trolley, or four-wheel carton. When the transport device is full or at intervals during the day, an employee transports the GOH pieces from a CO returns processing area to a GOH storage area. In the storage area, there is a GOH presort area and activity in which an employee sorts the GOH pieces by major groups, RTS, RTV, or RW, and by the last or first digit of each inventory number, or by GOH disposition label.

Final Sort at a Customer Order Returns Processing Station

Carrying out the final sort at a CO returns processing station requires a large area and is not recommended.

Final Sort in the GOH Storage Area

The final sort takes place in the GOH storage area concept (see Figure 5.13), where a returns processing employee completes the initial GOH piece sort activity. With the returns sorted into the three major UOP classifications and returns disposition or SKU inventory number, the final sort requires a small area and has high productivity. After GOH returns are final sorted to a trolley or cart load bar, a put-away employee moves to the trolley cart identified storage aisle or zone. Each trolley or cart contains GOH pieces that are in the particular storage aisle or zone. As a put-away employee enters a storage aisle or zone, he or she hand-scans a SKU. If the SKU exists in the aisle or zone, the hand-scanner display screen indicates the WMS identified position. If a SKU is new in an aisle or zone, the hand-scanner display screen indicates that an employee is to select a WMS identified position and physically transfer and scan a GOH piece to that position. At this aisle location, an employee looks at the load bar for additional SKU pieces that are the same. If there are additional SKUs, they are transferred to the storage position. The WMS scan transactions are sent to the WMS computer.

GOH Presort Concepts

A GOH concept is used for CO returns with very short SKUs. The GOH CO returns processing operations stars at a CO returns disposition station, where SKUs are removed from a CO return package, labeled with a rework disposition label, and hung onto a hang bar. At specified times or when a load bar is full, SKUs are transferred from the CO returns processing area to the rework area. Rework activities are steaming, cleaning, and rebagging. In this rework area, SKUs go through an additional inspection, attachment of a new disposition label, and return to stock activity. When this preparation is complete, SKUs are presorted by SKU inventory classification number, per pick aisle or to a pick zone. The concepts were described above.

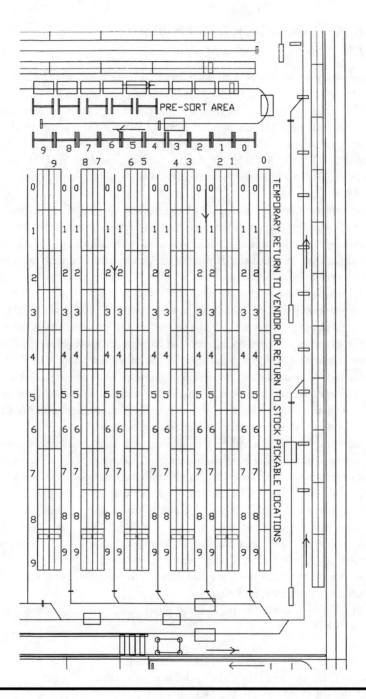

Figure 5.13 GOH returns presort and final sort.

A Non-Powered Trolley Travel Path with a Sort to Hang Bar Concept

A non-powered trolley travel path with sort to a hang bar concept design options are separate concepts for RTV and RTS SKUs or one concept for both RTV and RTS SKU. Trolleys are diverted from a main travel path onto a non-powered, U-shaped queue rail path. Depending on GOH RTS and RTV volume, available floor space and capital, the RTS and RTV trolleys are diverted to one queue rail or to separate rails RTV and RTS SKUs. As the trolleys complete the U-shaped curve, they queue against an end stop. Along each non-powered U-shaped queue rail there are carts with hang bars.

There are five carts on each side of the non-powered trolley rail travel path, to give a total of ten carts, one for each SKU inventory number last digit or first digit number or vendor alpha character. Each cart has a human-readable number, digit or alpha character and the carts are arranged in an arithmetic sequence. This scheme was reviewed above.

A sort employee picks a GOH piece from a RTS or RTV trolley, reads the sort instruction number or alpha character, walks from the trolley to a corresponding sort cart and transfers the piece to a cart or trolley load bar. An option to assist with the final sort activity, the trolley or cart load bar has ten moveable separators. Each separator is numbered 0–9 and arranged in a numbered sequence on the load bar. If your sortation concept uses another digit to complete a more detailed sort in the storage aisle, the second digit is used in the sort activity and allows the presort to get more similar SKUs in the same or a smaller load bar area. For additional information, we refer the reader to the CO small-item sort section.

At intervals or when the trolley or load bar is full, a sort employee moves a numbered cart or trolley with a hand-held scanner or paper put-away document from the presort area to the corresponding numbered aisle. In the aisle, the sort employee hand-scans the SKU label. The handheld scanner display shows an existing RTV or RTS location for the SKU. If the SKU is new, the employee walks to the proper location (second digit), hand-scans the RTV or RTS SKU code and GOH rail position code sends scan transactions to a computer to update the inventory file and places the GOH onto a rail. The activity is repeated for each RTV or RTS on a transport device; therefore the second or detailed sort activity reduces the put-away non-productive walk time and distance.

With a three-column and multi-lined paper document concept, the sortation employee writes:

- Under the first column each SKU code number
- Under the second column the GOH position number
- Under the third column the SKU quantity

Very Small Items or Jewelry Presort Concepts

To handle CO returns of very small SKUs, including jewelry, each SKU has an inventory number. Major groups (see Figure 5.14) are (a) watches, (b) earrings, (c) finger and foot rings, (d) chains, (e) bracelets, (f) pins and brooches, (g) packages or sets, (h) damage and return to vendor, and (i) rework.

In a jewelry CO returns processing operation, at a disposition station options are to:

- Mix SKUs in a tote and presorted at a remote location by pick aisle or pick zone for final sortation to a SKU pick position.
- Mix SKUs in a tote and at a remote location have SKUs pre-sorted by pick positions and transferred to the pick position.
- At a CO returns disposition station, to presort SKUs into separate totes by (a) SKU groups, (b) first inventory number, or (c) last inventory number. Each tote is for a specific pick aisle or pick zone.
- At a returns disposition station, to presort SKUs by groups into separate totes and have SKUs sorted into a temporary pickable location. Sorted SKUs are transferred from a temporary sortation location to a SKU pick position.

Mixed SKUs in a Tote Are Presorted by Pick Aisle at a Remote Location

A disposition station mixes SKUs in a tote that is sent to a remote location for presorting by pick aisle or pick zone. At a CO returns disposition stations, jewelry items are labeled for disposition, placed into a plastic bag and mixed into a tote. Full totes are transferred from a disposition station to a remote area for presorting into totes by pick aisle or pick zone. The presort area has a presortation location for each pick aisle or pick zone.

With a presort by pick aisle or pick zone concept, employees require a presort instruction. Options are:

- At a CO returns disposition station to have a CO returns disposition system interface with the inventory control program to have a pick aisle or pick zone printed onto a disposition label
- At the presort area, to have a RF device read a disposition label and on a RF display screen to indicate a SKU pick aisle or pick zone

A presort employee removes a SKU from a carton or tote, reads or RF scans the SKU disposition label, which instructs the employee to transfer the SKU into a specific tote for a disposition code pick aisle or pick zone. The activity is

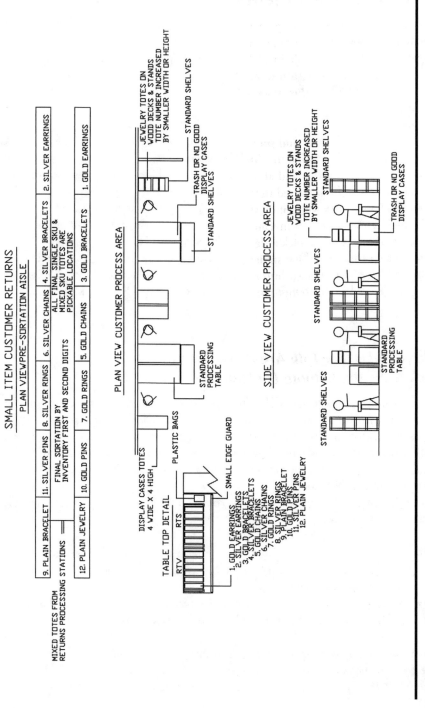

Figure 5.14 Jewelry or small parts presort.

repeated for each SKU that is in a disposition carton or tote. When a pick aisle or pick zone tote is full, it is transferred to an assigned pick aisle or pick zone for final sort to SKU pick position. This method increases SKU hit density and hit concentration per final sortation to a SKU pick location, minimizes employee walk distance and time between two pick locations, minimizes transport costs, and reduces confusion.

Mixed SKUs in a Tote Are Presorted by Pick Position in a Remote Area

At the CO return disposition station, SKUs are mixed in a tote and the tote is sent to a presort area for presorting to each SKU pick position. The area has a presortation location for each SKU pick position.

With a presort by SKU pick position concept, options for employee presort instructions are:

- At a CO returns disposition station to have a CO returns disposition system interface with the inventory control program to have a SKU pick position printed onto each disposition label
- At a presortation area, to have an RF device read a SKU disposition label and indicate a SKU pick position on an RF display screen

A presort employee removes a SKU from a tote, and reads or scans the disposition label. The disposition label instructs the employee to transfer a SKU into a specific tote for a disposition code or SKU pick position. The activity is repeated for each SKU in a disposition tote. When a pick position tote is full, it is transferred from the presortation area to an assigned pick aisle position. This tote transfer to a SKU pick position is a SKU final sort to a pick position.

During the presort activity, a SKU pre-sort position location is a discreet temporary location for a SKU. Per your company practice and inventory control program, a SKU presort position is a pickable location. This means that during presortation per COs and your inventory control program SKUs are available for COs.

Disposition Station to Presort Small SKUs by Pick Aisle or Pick Zone

At a CO returns disposition station where jewelry SKUs are presorted by pick aisle or pick zone, there is a tote for each pick aisle or pick zone. Partial or full totes are sent from a CO returns disposition station to a pick aisle or pick zone. The presort instruction options are:

■ At a CO returns disposition station to have a CO returns disposition system to interface with the inventory control program and have a pick aisle or pick zone printed on each disposition label
■ At a presortation area, to have an RF device read a SKU disposition label and indicate a SKU pick aisle or pick zone on an RF display screen

In a pick aisle or pick zone presort concept, a final sort employee removes a SKU from a tote, reads or scans the SKU disposition label that instructs him or her to transfer the SKU into a specific pick position. This activity is repeated for each SKU that is in a disposition tote. With this process an employee remains in an aisle to complete all final sort activities for a tote.

Disposition Station to Presort By Groups into Totes

At the CO returns disposition station SKUs items are presorted into jewelry groups into a tote. Full totes sent to a remote area, where SKUs are sent into a temporary pick position. A final sort transfers SKU into final sort totes, which are moved to a pick aisle for SKU (tote) transfer to pick positions.

A presort employee removes a SKU from a tote and reads or scans the disposition label. If a SKU is new or has been not been presorted to a temporary pick position, the RF device instructs an employee to transfer it into a vacant temporary pick position. Each SKU is in a disposition tote is transferred into a temporary pick position. When a temporary pick position tote is full and the presorted SKU quantity can fit into a permanent pick position, the tote is transferred from the temporary pick position to an assigned pick position in a pick aisle. If the presorted SKU quantity does not fit into a permanent pick position, the tote is transferred from the temporary pick position to another pick position in a pick aisle or zone. This tote transfer to a SKU pick position is the final sort to a pick position. During a presort activity to the SKU presort or temporary pick position, a pre-sort position is in your inventory control program as a discreet temporary pick position for a SKU, which means that the SKU is available for a CO.

Return to Stock Presort Locations

After a CO returns presentation process, the RTS sortation process options are to transfer the:

■ SKU final sortation to a SKU pick location
■ SKU to a temporary sortation location that has a pickable status
■ Tote of mixed SKUs to a pick position

Return to Stock Final Sort to the SKU Pick Position

Each tote, carton, cart, or pallet with presorted SKUs is moved from a presort area to a specific aisle in the pick area. In the pick aisle, RTS SKUs are transferred from the presort tote, carton, cart, or pallet to a pick position. SKU inventory quantity is updated in the inventory file and the SKU is ready for a CO.

The final SKU sort direct to a pick position concept (see Figure 5.15 and Figure 5.16) has:

- Sufficient mobile equipment and totes, carton, carts, or pallets to handle SKU dimensions and volume
- Sufficient SKU or carton, tote, cart, or pallet queue area in the customer returns processing area or pick area
- A transport concept to move SKUs and totes
- Return to stock instruction format that is either (a) a paper document or (b) RF device
- Handles mixed SKUs for one aisle in a tote, carton, cart, or pallet
- During a pick activity, an employee removes the disposition label

After a CO returns processing employee fills a tote with presorted or mixed SKUs, the tote is transferred from the presort area to the pick area. In the pick area, an employee places the tote onto a four-wheel cart shelf or removes SKUs from the tote. From a RTS label or after scanning a SKU bar code label, an employee is directed to a SKU assigned pick position. Arriving at the pick position, the employee transfers the SKU from the tote to the pick position and completes a SKU inventory update. The employee makes visual check to determine the SKU is the same as others in the tote or on the shelf and repeats the put-away activity until the tote is empty. If the next SKU is new, the SKU location is random and is in the same aisle. Disadvantages are additional time to complete a return to stock transaction, interfaces with paper document or paperless return to stock instruction, and advantages are requires less building area and equipment.

Return to Stock Presort to a Temporary Pickable Position

A temporary holding position is a pickable position within a pick area or specific aisles in a pick area. In a temporary pick aisle, RTS SKUs are transferred from a presort tote, carton, or cart to a SKU temporary pick position. The SKU inventory quantity is updated in the inventory file and the SKU is ready for a CO. After a time or when the temporary holding position is full, sorted SKUs or SKUs in a tote are transferred from the temporary holding location to a SKU pick position or a new pick aisle position. SKU sortation to a temporary pick position has:

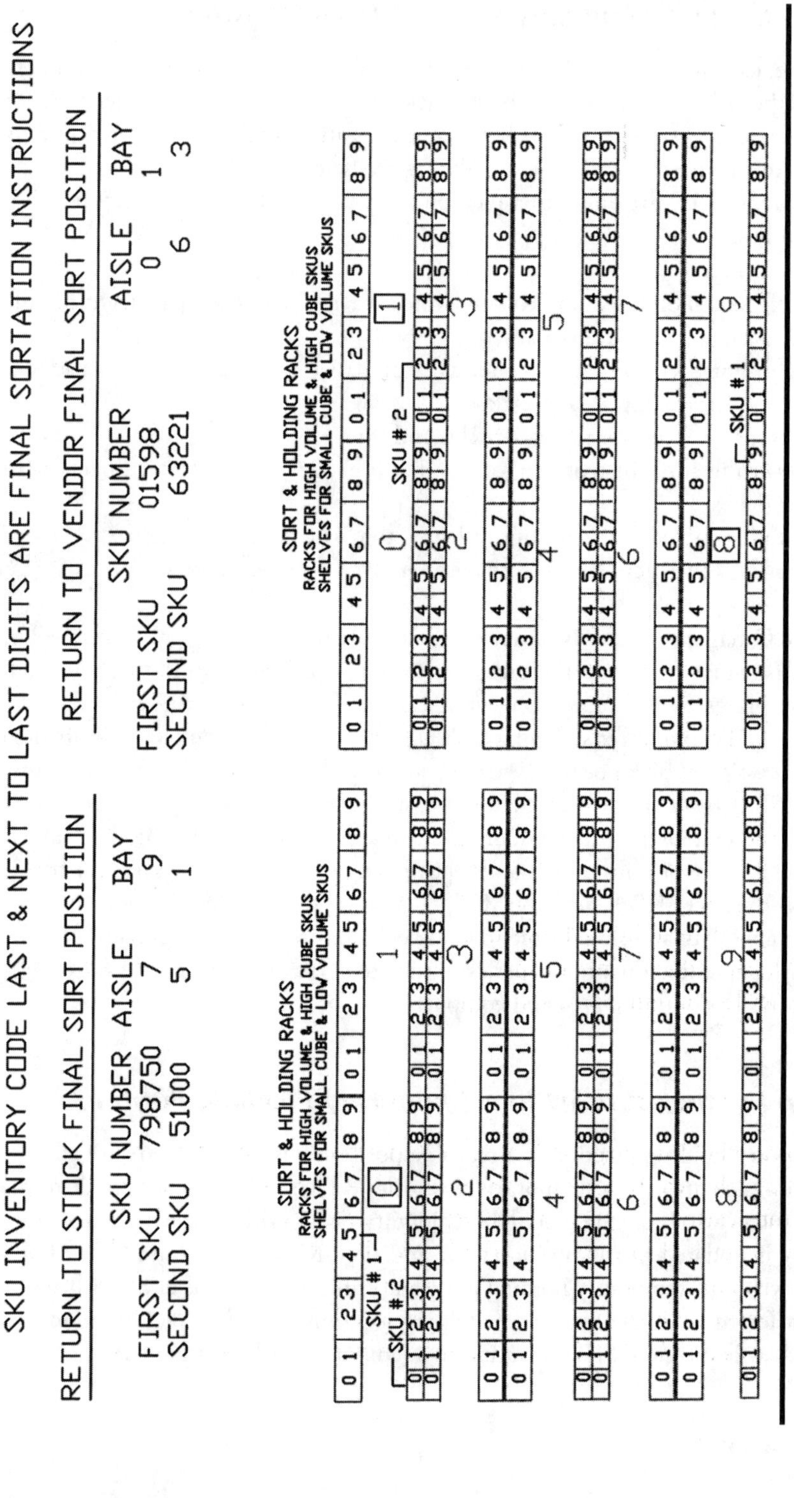

Figure 5.15 Final sortation by last and next to last digit of the SKU code.

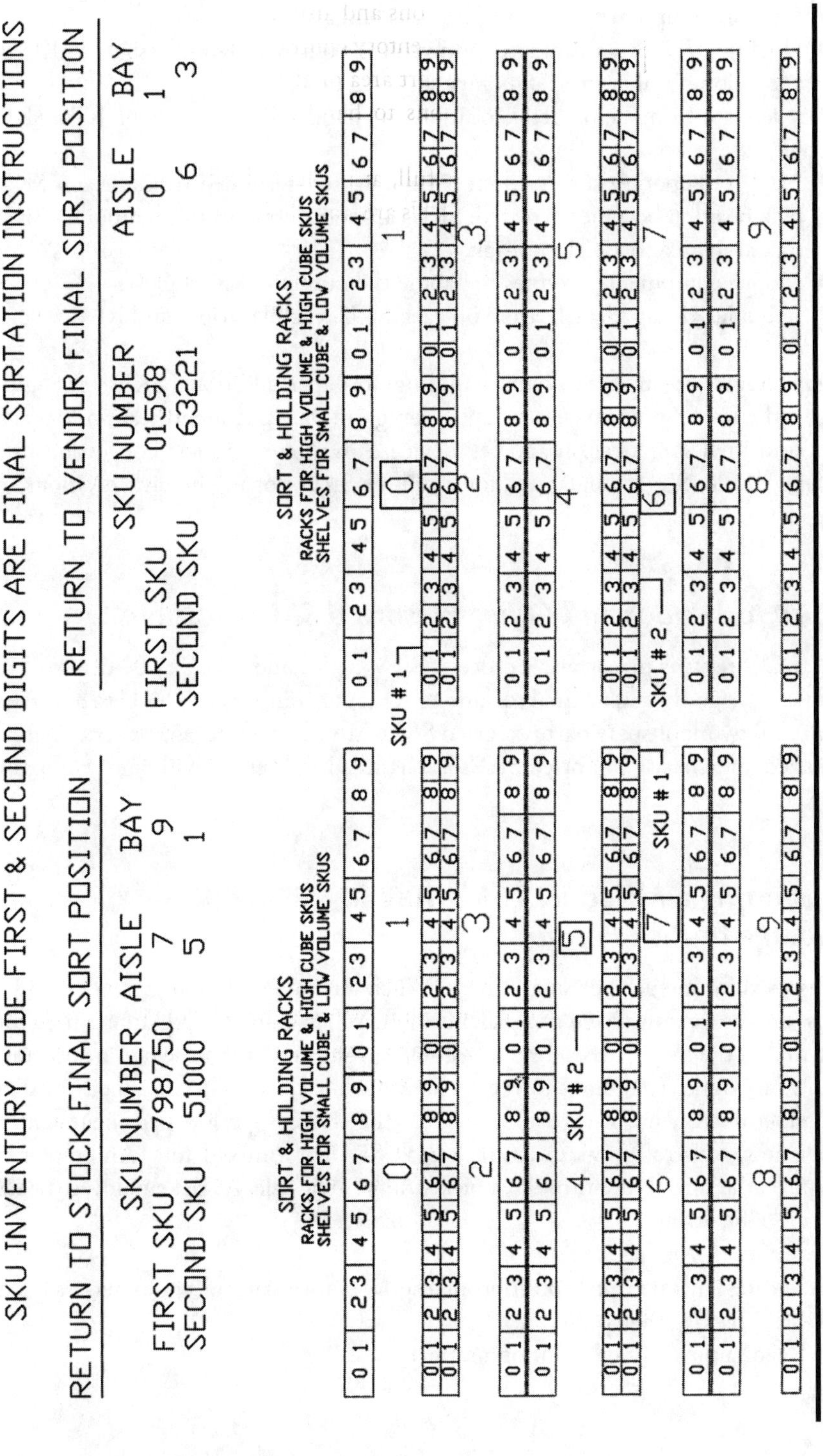

Figure 5.16 Final sortation by first and second digit of the SKU code.

- Separate temporary SKU sort locations and aisles.
- During SKU sort, a computer or inventory control program assigns a SKU to a location number in a temporary sort area or aisle.
- Adequate number of SKU locations to handle the number of RTS SKU pieces.
- After a temporary sort location is full, at predetermined times or as a SKU pick position becomes depleted, SKUs are transferred from a temporary SKU sortation area to a pick position.
- Mobile equipment and totes to handle the number of RTS pieces.
- Barcode/RF tag identification on each SKU, each location, and RF device.

Disadvantages are employee double handling, sort area, additional SKU sort equipment and computer, or inventory control program bar code identification for each SKU, and final sort location and RF device. Advantages are less congestion in a SKU pick aisle, SKU queue area, some double SKU slotting or pick positions in the pick area.

Return to Vendor and Other Presorted SKU Locations

After a CO returns presortation process, RTV SKUs and SKUs to be disposed of in other ways go through a final sortation process to transfer each SKU to a storage location. Rework or spare part presorted SKUs are sent to a storage or rework area. Damaged and charity presorted SKUs are handled in keeping with your company policy.

Transferring a Mixed RTS SKU Tote to a Storage or Pick Position

This is used for medium or small SKUs. When a mixed SKU tote is used for RTS SKUs, the tote is transferred to a pick position. When a mixed SKU tote is used for RTV SKUs, the tote is transferred to a storage position. During the presort process, as SKUs are transferred into a tote, each bar code label or RF tag is scanned or listed on a paper document for inventory update. The concept has a paper document or an inventory control program to track each SKU in a mixed tote. An employee transports the mixed tote through a pick/storage area aisle. As the employee travels in an aisle, he or she:

- Locates an open pick position or the inventory control computer-assigned SKU pick/storage position
- Reads a pick/storage position number

- Writes the pick/storage position number onto a paper document or with a RF device scans a pick/storage position identification label
- Writes a tote number onto a paper document or with a RF scans the tote label
- Transfers a tote into a pick/storage position

Per your inventory system, the inventory file update is on-line or delayed communication to the inventory control system. The concept features are quick return of SKUs to a pickable location, minimizes aisle congestion, high put-away employee productivity, minimal double handling, and in a mixed SKU tote to find a particular SKU, low-order picker productivity.

The RTV SKUs final sort concept has an employee transfer a SKU from a presort tote to a storage location. RTV final sort options are:

- Each vendor SKU returned to a fixed storage position
- Vendor SKUs mixed in a tote that is in a fixed position

Presorted Vendor SKUs to One Final Sort Position

Having RTV SKUs returned to a fixed SKU storage position (see Figure 5.17) means that for each SKU there is one position. RTV sort to a fixed SKU position has one tote for each vendor, which can be either presorted or mixed. In a vendor return aisle and as the employee removes a SKU from a tote, with a RF device scans a SKU bar code and the RF device shows a SKU location or a new location for a SKU. At a location, an RTV SKU is transferred to a fixed storage position and the SKU bar code and position bar code are scanned into the inventory files. A final sort employee travels through the aisle and completes the individual sort. The features are each SKU has a position, large final sort area, low put-away employee productivity, for vendor return consolidation is easier and quicker, and accurate inventory.

Vendor Mixed SKUs in a Tote Sorted to a Final Position

This concept has an employee transfer a tote with mixed RTV SKUs to a fixed storage position. A final sort employee travels with a tote through an aisle, and transfers the tote to a vendor storage position. Per an inventory file, a final sortation employee completes scan transactions, travels through an aisle, and completes tote transfer transactions. The features are minimal positions, smaller final sort area, improved put-away employee productivity, slower and more difficult SKU consolidation for a vendor shipment, and possible inventory control problems.

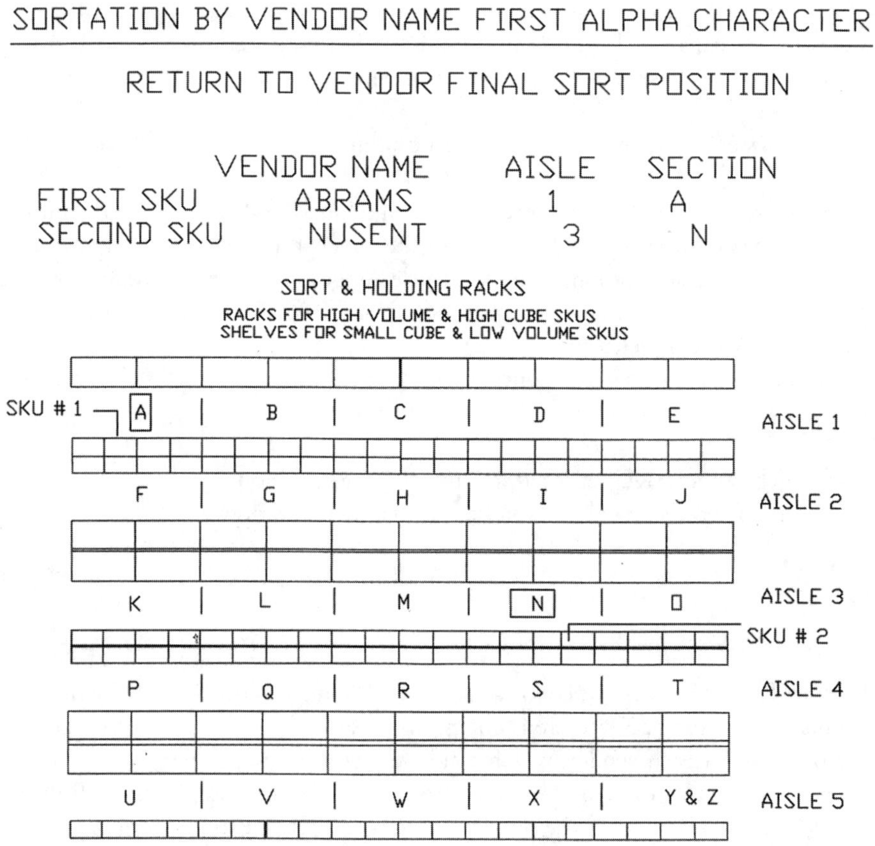
SORTATION BY VENDOR NAME FIRST ALPHA CHARACTER
RETURN TO VENDOR FINAL SORT POSITION
VENDOR NAME
AISLE
SECTION
FIRST SKU
ABRAMS
1
A
SECOND SKU
NUSENT
3
N
SORT & HOLDING RACKS
RACKS FOR HIGH VOLUME & HIGH CUBE SKUS
SHELVES FOR SMALL CUBE & LOW VOLUME SKUS
SKU # 1
A
B
C
D
E
AISLE 1
F
G
H
I
J
AISLE 2
K
L
M
N
O
AISLE 3
SKU # 2
P
Q
R
S
T
AISLE 4
U
V
W
X
Y & Z
AISLE 5

Figure 5.17 Final sortation by vendor name.

Manual Final Sort Concept to a SKU Fixed Position

An employee transports a mixed SKU tote through pick aisles. As the employee travels in a pick aisle, he or she picks up a SKU, reads or scans its RTS label, travels to a pick position, transfers the SKU into the pick position, and writes or scans the pick position label. This updates an inventory quantity in the inventory files. Manual final sort concept is used for the RTS activity in a manual order pick operation.

Mechanized Final Sort Concept and SKUs in a Fixed Position

A mechanized carousel or carton AS/RS concept has a presorted SKU placed into a fixed position. The RTS transfer to a pick position is completed at a mechanized pick concept front end. At this location a pick employee has a mixed SKU tote at an in-feed/out-feed station. As the pick employee removes an RTS SKU from a tote, he or she scans or enters the SKU number into a mechanized pick concept control panel. A mechanized pick concept computer directs the mechanized pick concept to withdraw an assigned tote from a storage position to an in-feed/out-feed station. After a tote arrives at an in-feed/out-feed station, an employee transfers a SKU into the mechanized pick concept tote. A microcomputer directs a mechanized concept to return the tote to an assigned storage location. Per your inventory system, file update is an on-line or delayed communication to your inventory system.

Manual Pick Concept and Mixed SKU Tote in a Pick Position

In a manual pick concept with a mixed SKU tote transferred to a pick position, an employee transports a mixed SKU tote through pick aisles. As the employee travels in a pick aisle, he or she:

- Locates an open tote position or the computer-assigned position.
- Reads or scans a pick position number.
- Writes a pick position number onto a paper document or scans a pick position identification label.
- Writes a tote number onto a paper document or scans a tote label.
- Transfers the mixed SKU tote into a pick position. Per your inventory system, the inventory file update is on-line or delayed communication to the system.

Mechanized Pick Concept Mixed SKU Tote in a Pick Position

In a carton AS/RS or carousel (mechanized) pick concept with mixed SKU tote transfer to a pick position, a pick employee at an in-feed location transfers a mixed SKU tote to an empty mechanized pick position. During a return to stock activity, an employee scans a tote bar code. A mechanized pick concept computer directs the mechanized concept to transport an empty tote to an in-feed/out-feed position, where an employee transfers the mixed SKU tote into a mechanized concept tote. The mechanized pick concept returns the mixed SKU tote to the computer-assigned storage location. Per your inventory system, the inventory file update is on-line or delayed communication to the system.

A Manual Concept with an RF Device for SKU Final Sort

A manual concept that uses an RF device in a final sort concept has a final sortation employee transfer a number of return totes for a particular pick aisle to cart shelves. Each tote carries an aisle number or pick area in human-readable format. A tote with the lowest number pick aisle is dumped onto the top shelf of a cart. A tote has mixed presorted SKUs that have pick positions in the same aisle. This tote dumping activity spreads mixed presorted SKUs over a shelf so that SKUs are easily seen by a return employee. After a tote is empty, an employee returns it to a cart and scans the tote bar code label/RF tag. An employee picks up a SKU, scans the bar code label/RF tag. An RF device shows the SKU pick position in a pick aisle. An employee pushes a cart to an assigned pick location, scans the pick position and RTV location bar code label/RF tag, and transfers or final sorts the SKU into a pick position or return to vendor location. The final sort activity is repeated for each presorted SKU on a cart shelf. Per your inventory system, an inventory file update is on-line or delayed communication to an inventory system.

A Mechanized Concept That Uses an RF Device for Final Sort

In a mechanized concept that uses an RF device transfer concept, a CO returns employee transfers a tote with presorted SKUs onto the mechanized pick concept in-feed/out-feed location, where SKUs are dumped onto a shelf. This spreads the SKUs over the shelf so that each SKU is easily seen by a sortation employee. An employee scans a SKU bar code label and the microcomputer activates a carton AS/RS, or carousel concept to move an assigned mechanized storage tote from a storage location to an in-feed/out-feed location. As the mechanized concept tote arrives at the transfer station, a barcode scanner or RF tag reader device reads a tote bar code and sends data to the microcomputer. A computer activates a transfer station display screen, which shows a SKU number. This directs a CO returns employee to ensure that the number matches a SKU number. With a proper match, the employee final sorts a SKU to a tote by scanning its bar code label/RF tag and scanning/reading a mechanized concept tote bar code label. The display screen shows the accurate SKU transfer is completed and the bar code/RF tag scanner transfers the data delayed or on-line to the microcomputer. If a SKU is not the correct SKU, the display screen shows the problem and an employee picks the appropriate SKU. As required by your inventory system, an inventory file update is on-line or delayed communication to an inventory control system.

Age RTS and RTV Temporary Hold Positions

Age RTS and RTV temporary hold positions are designed to ensure that active SKUs have positions in the RTS or RTV temporary holding area. It is very common in an order-fulfillment operation that the majority of CO return packages is received within 90 to 120 days after a CO delivery. A company historical SKU sale or CO delivery date and CO return SKU date determines a more exact number of days for a company or for each major product group to have the majority of returns sent back to the operation. The age RTS and RTV temporary hold position options are:

- A manual approach, in which an employee in a temporary sortation location writes the sort date onto a paper document on a tote or carton side
- The RTS or RTV label on each piece is removed and placed onto a tote or carton side
- With a scan put-away and computer capability to have the computer print the last date for a put-away transaction to each temporary hold position

The age concept ensures only active SKUs are retained in the temporary sort and hold area, improves space utilization and improves high put-away concentration and density that improves employee put-away pick consolidation.

Temporary Hold Position Full RTS or RTV Tote or Carton Handling

The objective is to have a process for moving full RTS or RTV totes or cartons from the temporary sort/hold position to a storage position. An employee determines when an RTS or RTV tote or carton is full and identifies the piece quantity per tote or carton. The employee hand-scans or writes the tote or carton identification and piece quantity per tote or carton before it is transferred to a storage position. The tote or carton identification, piece quantity, and new position are updated in the computer files. The concept improves space utilization, creates a small temporary hold area, minimizes consolidation pick activity, and improves put-away employee productivity.

Return to Vendor Consolidation Activity

Return to vendor SKUs are consolidated for packing and shipment from your distribution facility to a vendor facility. After a vendor has given permission to return a SKU, RTV employees consolidate SKUs in less than full master cartons into one carton onto a material-handling device. The employee lists the SKU quantity in each carton and seals the carton. Sealed cartons are ready for shipment to the vendor. The return to vendor consolidation activity is determined by how the SKUs are inventoried in the storage position and whether the SKU position is a manual or mechanized concept.

The return to vendor consolidation concepts are with a manual storage concept and a human-readable consolidation instruction format, to have an employee travel through the aisles and consolidate individual SKUs into a tote or carton, to have a mixed vendor SKU tote delivered to a sortation station. At the sortation station the RTV SKUs are separated into cartons and a consolidation employee travels through the RTV aisles directed by an RF device and totals the SKUs that are transferred into a carton.

Consolidation with a Paper Instruction

If a consolidation activity has a paper human-readable instruction format, an employee reads a vendor name from the paper document. With this information, the employee moves to an assigned storage position, transfers a SKU from the pick or storage position into a carton. Per your inventory system, an inventory file update is on-line or delayed communication to the inventory control system. The first return to vendor consolidation concept is a manual pick/storage concept with a paper consolidation instruction format. This concept has an employee travel through the aisles and consolidate (reverse pick) individual SKUs from a storage position in one carton and all full/sealed cartons are transferred onto a material-handling device. The features are low employee productivity, due to the increased walking distances between two locations, the employee has to read a computer-printed document, and handles a medium volume.

Mechanized Consolidation with a Paper Return Instruction

A mechanized concept with a paper consolidation instruction has an employee direct a mechanized concept to transfer a required tote from a storage position to an in-feed and out-feed station. At the in-feed and out-feed station, from a pick/storage tote, and per a printed instruction document, an employee transfers an assigned number of SKUs from a pick/storage tote into a carton. With this concept, if a SKU is not required, it remains in the tote and the tote is returned to a mechanized concept computer-assigned pick/storage position. The features are low employee productivity due to the walking distance, employee has to read, possible sort errors, and handles a low volume.

Manual Consolidation with an RF Instruction Format

The RF device consolidation instruction format has an employee travel through the aisles and consolidate SKUs into a carton. With this concept, an employee travels through the aisles and stops at an assigned pick/storage position. From the pick/storage position, per the RF device instruction an employee transfers the mixed SKU tote from the assigned pick/storage position onto the vehicle load-carrying surface. Arriving at the sort station, the employee picks a SKU from the tote and scans and transfers each appropriate SKU into a carton. The features are medium employee productivity, does not require an employee to read, minimal sort errors, handles a medium volume, and requires an investment.

Mechanized Consolidation with an RF Instruction Format

A mechanized pick/storage concept with a consolidation instruction format has a microcomputer that directs a carton AS/RS crane or carousel to bring a SKU tote to an in-feed/out-feed station. From a carton AS/RS or carousel pick/storage position, per the RF instruction, an employee transfers a mixed SKU tote into a return to vendor carton. The features are good employee productivity due to less non-productive time, requires an employee to read, minimizes sort errors, handles a medium volume, and requires an investment.

Consolidation from a Fixed Position

Consolidation or reverse SKU picking from a fixed pick/storage position has an employee or machine travel to each pick or storage position that has a SKU assigned to it. Per the inventory control system, one SKU is assigned to a pick/storage position. At each appropriate pick/storage position, the employee transfers the tote or SKU quantity from the pick position to a vehicle or on-board carrying device. Per the inventory control system, a SKU inventory transfer transaction is an on-line or delayed transaction. The features are during a return to stock process an employee has increased walking distance and time, minimizes pick position space or cube utilization, improves inventory control and slow-moving SKU allocation in a pick area, and with a return to vendor or final disposition activity, it requires minimal employee walking distance and time because SKUs are in a fixed pick/storage position.

Consolidation from a Mixed SKU Tote

This involves consolidation or reverse SKU picking from mixed SKU totes in randomly located pick/storage positions. An employee or machine travels to every pick/storage position that has a SKU assigned to it. Per the inventory control system, a SKU in a specific quantity is assigned to each position. At each appropriate pick or storage position, the employee transfers the tote or SKU quantity from the pick position to his or her vehicle or on-board carrying device. After an assigned SKU is transferred from a mixed SKU tote, the tote is returned to a pick/storage position. Per the inventory control system, the SKU inventory transfer transaction is an on-line or delayed transaction. The features are during a return to stock process an employee has a short walking distance and time because mixed SKU totes are placed in a pick/storage position, potential to maximize the pick position space or cube utilization, difficult for inventory control, and slow-moving SKU allocation in the pick area, potential pick errors, and return to vendor or final disposition activity, it requires the maximum employee walking distance and time due to SKUs are in a greater pick or storage position number.

Rework SKUs

The next small-item, flatwear, GOH and jewelry order-fulfillment activity is SKU rework. SKU rework activity modifies a rejected vendor-delivered SKU or a CO return SKU with off-standard quality to match your company's standard quality. The rework activity is completed in the facility or off-site by a third party or vendor.

Vendor Rework Activity Locations

The vendor SKU rework locations are:

- On the receiving dock. This is used to handle a small SKU quantity for vendor rework. The vendor compensates your company for the labor and material rework expenses. If SKUs have already been transferred to a storage position, they are removed from the storage area and transferred to an assigned receiving dock location. After a rework employee receives rework instructions, employees perform SKU rework activity in the dock area. The rework employee uses the vendor cartons as a rework table or a mobile table to support the rework SKUs and tools. The features are low employee productivity, due to a poorly designed work area and poor UOP flow over a rework area, occupies receiving dock area space, with moving pallets by powered mobile equipment in a receiving area, there is a potential for employee injury, possibility to have SKUs located in different locations, low employee productivity, no cost, and difficult to combine CO rework SKUs in a rework activity.
- In a specific area located in the receiving dock area or near a CO return/storage area. The vendor rework SKUs are transferred from the receiving dock area, storage area, or CO returns area to the specific rework area. The rework area is designed with two non-powered conveyor travel paths or lanes. The first lane flows the rework product from a pallet to the rework activity table. The in-feed conveyor travel path permits carton queue prior to the rework table. A rework table has sufficient space for an employee to perform the rework activity, for rework tools, and for an employee to replace the SKUs into a vendor carton or tote. The second non-powered conveyor travel path flows completed rework cartons from the rework table to a finished goods palletize station. At the out-feed conveyor end, an employee transfers the rework cartons onto a

pallet. As required by the number of rework stations and volume, employees are assigned to open rework cartons, transfer cartons onto a conveyor travel path, to perform the SKU rework and return the SKUs to a vendor carton, to reseal a completed rework carton, and stack the sealed carton onto a pallet. A rework area design option has a rework pallet that is placed adjacent to the rework table side and adjacent to another rework table side is an empty pallet. With this concept, a rework employee transfers a carton onto the rework table, opens a carton, removes SKUs, completes rework activity, returns a rework SKU to a carton, reseals a full carton and transfers a completed rework carton onto the other pallet. The features are higher employee productivity because of work area design and improved product flow over the work area, occupies a specific area, less potential for employee injury, with product queue and assigned employee activities, improved employee productivity, low cost, and combined CO rework SKU in one activity location.

■ Sent off-site to a vendor or third-party location. If your company sends the rework SKUs from your order-fulfillment operation to a vendor facility or to a vendor-assigned third-party facility, as a rework product is transferred from your facility, your SKU inventory is transferred from your file to the SKU vendor and is eliminated from your inventory. The vendor or third party completes the rework activity to the SKUs. When the rework product is returned to your order-fulfillment facility, reworked SKUs are received under a new purchase order and your receiving/QA personnel repeat their activities. The features are no cost, does not require facility space, and does not require labor.

Customer Order Returns SKU Rework Activity

In a CO returns processing operation, the CO returned SKUs that require rework are flatwear, GOH, and jewelry. CO return rework activity has your rework employee perform an activity or activities to ensure that the SKU quality meets company standards. After a SKU is reworked to meet company standards, it is transferred to a pickable position. A reworked SKU in a pickable position is available to complete a CO. Different SKU groups require different rework activities.

- Flatwear and GOH SKU rework activities are (a) remove protective packaging, (b) retain SKU disposition label, (c) inspect the SKU quality, (d) steam clean, remove lint, and press the SKU, (e) reseal the SKU in a protective package, and (f) place the old or new disposition label onto the SKU package exterior
- Jewelry SKU rework activities are (a) remove protective packaging, (b) retain the disposition label, (c) inspect the SKU quality, (d) buff and/or wash or

soak the SKU, (e) reseal the SKU in a protective package, and (f) place the old or new disposition label onto the SKU package exterior

These rework groups are sent from the return processing area to the rework area and, prior to being transferred from the rework area to a pickable position, they require a return to stock presort activity.

Index